Milicent
Schedules

CANADA AND THE WORLD

SECOND EDITION

GEOFFREY J. MATTHEWS
Chief Cartographer
Department of Geography
University of Toronto

ROBERT MORROW, JR.
Curriculum Co-ordinator
Wentworth County Board of Education

PRENTICE HALL CANADA INC.,
SCARBOROUGH, ONTARIO

To Helen and Lynne

Canadian Cataloguing in Publication Data

Matthews, Geoffrey J., 1932–
Canada and the world : an atlas resource

2nd ed.
Includes index.
ISBN 0-13-370073-9 (school ed.). — ISBN 0-13-305343-1 (trade ed.).

1. Atlases, Canadian. 2. Canada – Maps.
I. Morrow, Robert, 1942– II. Title.

G1115.M38 1994 912 C93-093543-8

Prentice-Hall, Inc., Englewood Cliffs, New Jersey
Prentice-Hall International, Inc., London
Prentice-Hall of Australia, Pty., Ltd., Sydney
Prentice-Hall of India, Pvt., Ltd., New Delhi
Prentice-Hall of Japan, Inc., Tokyo
Prentice-Hall of Southeast Asia (Pte.) Ltd., Singapore
Editora Prentice-Hall do Brasil Ltda., Rio de Janeiro
Prentice-Hall Hispanoamericana. S.A., Mexico

ISBN 0-13-370073-9 (school ed.) ISBN 0-13-305343-1 (trade ed.)

Publisher: Anita Borovilos
Managing Editor: Carol Stokes
Editors: Aileen Larkin, Rebecca Vogan, Denyse O'Leary, Pam Young
Production Manager: Theresa Thomas
Cover Design: Alex Li
Cartography: Department of Geography, University of Toronto

Printed and bound in Canada by DW Friesen

2 3 4 5 6 7 8 9 DWF 02 01 00 99 98 97 96 95 94

The publisher of this book has made every reasonable effort to trace the ownership of data and visuals and to make full acknowledgement for their use. If any errors or omissions have occurred, they will be corrected in future editions, providing written notification has been received by the publisher.

Official Québec Place Names

Name in the Atlas	Page No.	Official Québec Place Name
Appalachians	16	Les Appalaches
Chaleur Bay	85, 88	Baie des Chaleurs
La Grande Basin	30	La Grande Rivière Basin
Gulf of St. Lawrence	16, 44, 66, 85, 88	Golfe du Saint-Laurent
Hudson Bay	16, 60, 62	Baie d'Hudson
Hudson Strait	16, 60, 62	Détroit d'Hudson
James Bay	16, 60, 62	Baie James
Laurentians	16	Les Laurentides
Lièvre (River)	80	Rivière du Lièvre
Ottawa (River)	16, 66, 80	Rivière des Outaouais
Rouge (River)	80	Rivière Rouge
St-Jean	80	Saint-Jean-sur-Richelieu
St. Lawrence (River)	16, 44, 66, 80, 85, 121	Fleuve Saint-Laurent
Ungava Bay	16, 60, 62	Baie d'Ungava

CREDITS

Photography courtesy of:ACDI/CIDA: Anthony Scullion 136 t,l; (c) Advanced Satellite Productions Inc. 1994: 69 b,l. David Barbour 139 b,l; Crombie McNeill 140 t,r; Dilip Mehta 159 t,r. AP/Wideworld Photos Inc.: 151 b,l. Gary Birchall: 17, c,b & b,l; 18 c,r; 67 c,l; 74 b,r; 84 b,r. CN: 88 b,l. Canadian Pulp & Paper Association: 42 b,r; 67 t,r. (c) Canadian Sports Fishing Magazine: 67 b,c. Comstock: G. Gerster 129 b,l &139 t,r; H. Armstong 135 b,r; A. Tanner 155 b,r; S. Vidler 155 t,r & 175 b,l; G. Hunter 161 t,r & 170 c; G. Marché 162 b,r; H.A. Roberts 167-68 c; W. Gordon 169 c. Corporation de promotion industrielle, commerciale, et touristique de Sept-Îles inc.: 86 b,r. Government of Northwest Territories: 62 t,r; 61 b,l; 59 b,l. E.O.S.A.T.: 128 t,r; 134 b,r; 143 b,l; 158 t,r; 166 c. Gordon J. Fisher, Fisher Photographic: 71 c,r. K. Forbes: 18 t,c. (c) Robin White, Fotolex Associates: 27; 82 t,r. Ronald Fulton: 17 c,l; 64 t,r; 67 c,l. Geographical Visual Aids, V. Last: 17 b,r; 18 t,l; 37 b,c; 38 t,l; 61 c,r; 66 c,l; 73 c; 77 b,l & b,r; 82 b,l; 83 t,l; 130 b,l; 152 t,r;173 t,r; 174 c,r. Honda of Canada Mfg., Alliston, Ontario: 176 t,r. Hydro-Québec: 63 b,l. Inco Limited: 66 t,r. Landsat imagery courtesy of the Canada Centre for Remote Sensing, Energy, Mines and Resources Canada: 60 t,r; 75 b,l; 80 b,r; 85 t,l. Masterfile: Al Harvey 173 b,l; Lloyd Sutton 125 b,l. NASA: 28 t,r. Nova Scotia Tourism, Culture & Recreation: 86 t,r. Ontario Ministry of Agriculture and Food: 81 b,l. Ontario Ministry of Natural Resources: 38 b,r. Petawawa Foresty Institute/Forestry Canada: 37 c,r. George Philip Ltd. (c) 1991: 61 t,l. Ports Canada: 71 t,l. Province of British Columbia: 37 b,l; 72 b,r. Rainforest Action Network: 135 t,r. Reuters/Bettmann: 156 c,r. Société du Port du Québec/Port of Quebec Corporation: 79 t,l. Stadium Corporation of Ontario Limited: 80 b,c. Take Stock Inc.: (c) Angus McNee 77 t,r. The Stock Market Inc.: James Blank 126 t,r; D. Trask 129 c,t; Ed Young 126 b,r. Tony Stone Images: Paul Chesley 168 b,r. West Asian Department, Royal Ontario Museum: 160 t,r. Western Wilderness Committee/Gary Fiegehen: 41 b,l; WorldSat International Inc.: 171 b,l.

Acknowledgements

This totally redesigned edition of *Canada and the World: An Atlas Resource* is the result of an intensive research and production campaign to update, revitalize and expand topics of current interest. Again, the Department of Geography at the University of Toronto has produced the maps, working with a combination of manually created base material from the first edition and a new section of computer-generated world and world regional spreads.

We would like to thank the cartographers for adding their professional skills to the maps — Hedy Later, Diane Ferguson, Brigid McQuaid and Jane Davie. The creation of the world thematic section of the Atlas is a reflection of the cartographic and computer skills of Byron Moldofsky, Ada Cheung, Mariange Beaudry, Paul Degrace and Gerald Romme. Byron's expertise in directing the computer team is much appreciated. With their artistic and technical skills, the cartographers have produced the unique blend of maps, graphs and illustrations displayed throughout the volume.

Many others have assisted, before, during and after the work of the cartographers. To our teachers and mentors — James Forrester, Kenneth A. Stanley, the late Gordon E. Carswell, Dr. Lloyd G. Reeds and Don Revell — much gratitude is owed. Stuart Nicholson, Lloyd Payne and Lynne Morrow provided much needed data at the research stage. Mia London provided smooth efficiency as production co-ordinator. The editorial work of Aileen Larkin, with the assistance of Rebecca Vogan, Denyse O'Leary and Pam Young, changed rough work into a polished product. Clear direction and positive feedback were provided by Carol Stokes, Elynor Kagan, Anita Borovilos and Rob Greenaway.

This second edition builds on the strengths of the very successful first edition and maintains the clarity, utility and beauty that students and teachers have come to expect.

Geoffrey J. Matthews
Robert Morrow, Jr.

Authors' Note

Political boundaries, systems of government and the spelling of names are in a constant state of flux. Every effort has been made to use the best sources available; however there is a reliance on official sources using data and spelling provided by government agencies. The names of Canada's First Nations in legal and official data, and in historical sources, are often not the same names that First Nations themselves would use today. Changes are ongoing in many parts of the world, including Europe, the former Soviet Union, the Middle East and parts of Africa.

Teachers and students are encouraged to collect information from newspapers and journals and to contact embassies, government offices, non-governmental agencies and local First Nations organizations for current information.

In an effort to keep the users of *Canada and the World: An Atlas Resource* as up-to-date as possible, Prentice Hall will be updating this type of information and statistics related to both Canada and other countries on a regular basis.

The Cover

The projections used on the cover of this atlas are the Bartholomew Times (top and bottom maps) and the Hammer Aitoff (middle map). The top map is centred on the Pacific Ocean; the bottom map on the Atlantic. The middle map is an equal area projection which shows the South in a more correct proportion to the North.

Contents

CANADA — THEMATIC

CANADA — REGIONAL

WORLD — THEMATIC

WORLD — REGIONAL

North and Central America

South America/Africa

Eurasia

Australia and New Zealand

CANADIAN STATISTICS

GDP
HDI- human development index

WORLD STATISTICS

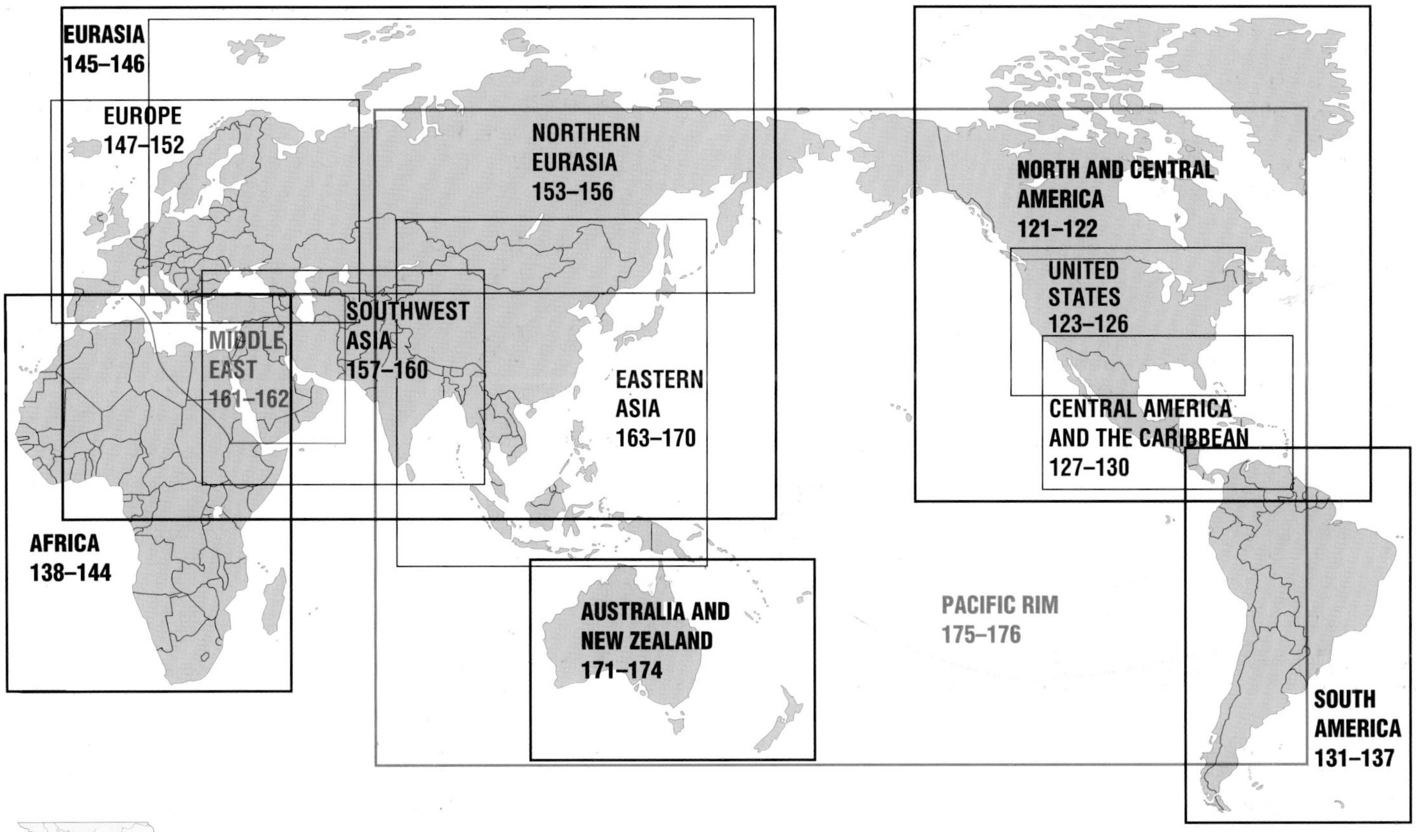
EURASIA
145–146
EUROPE
147–152
NORTHERN
EURASIA
153–156
NORTH AND CENTRAL
AMERICA
121–122
UNITED
STATES
123–126
SOUTHWEST
ASIA
157–160
MIDDLE
EAST
161–162
EASTERN
ASIA
163–170
CENTRAL AMERICA
AND THE CARIBBEAN
127–130
AFRICA
138–144
AUSTRALIA AND
NEW ZEALAND
171–174
PACIFIC RIM
175–176
SOUTH
AMERICA
131–137

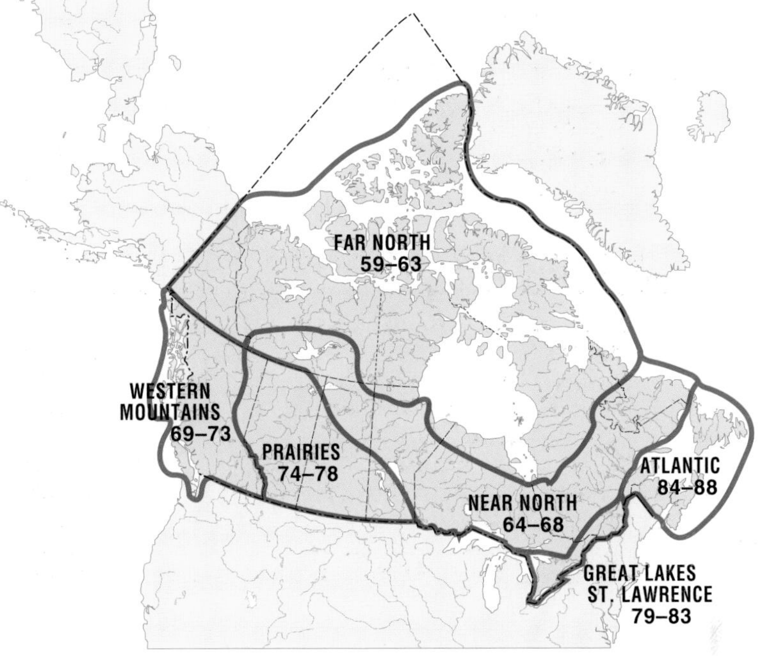
FAR NORTH
59–63
WESTERN
MOUNTAINS
69–73
PRAIRIES
74–78
ATLANTIC
84–88
NEAR NORTH
64–68
GREAT LAKES
ST. LAWRENCE
79–83

CANADA-THEMATIC

Map 1: Before 1600

All First Nations have creation stories that connect them to this land from the beginning of time. There is archeological evidence to suggest that migration from Siberia could have occurred during the last ice age (between 36 000 and 21 000 BC) when lower sea levels created a land bridge between Siberia and Alaska. The many distinct indigenous peoples in Canada developed diverse customs, languages and ways of life.

Note: As First Nations people assert control over their identity and representation, names change and are in constant flux. The names on this map are based on linguistic groups, contemporary political organizations and the historical record. They are not necessarily what people called themselves at contact. Due to the difficulty of finding valid sources, there may be misspellings, omissions and other inaccuracies.

Note: Two shades of red are indicated as "British." The darker shade identifies areas that were recognized as British colonies. The lighter shade identifies areas claimed by Britain, but not officially recognized as colonies.

Map 2: 1600–1690

- British
- French
- Disputed claim
- Spanish

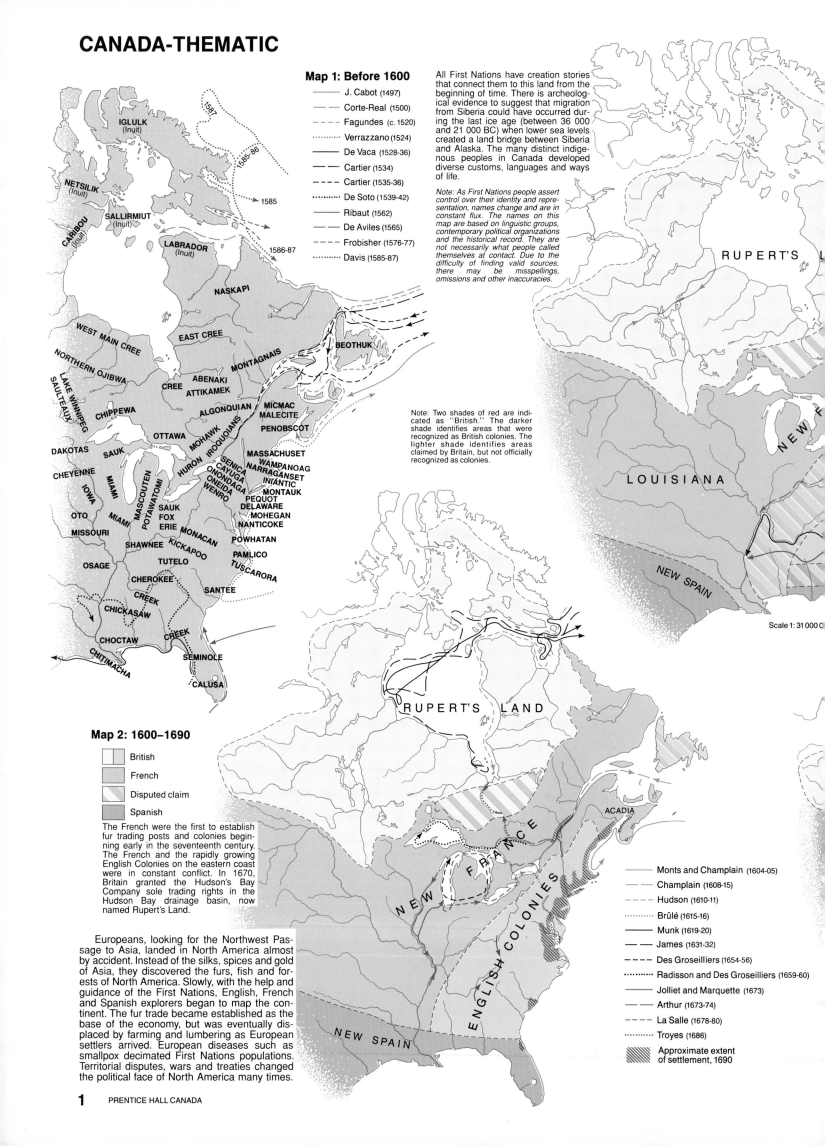

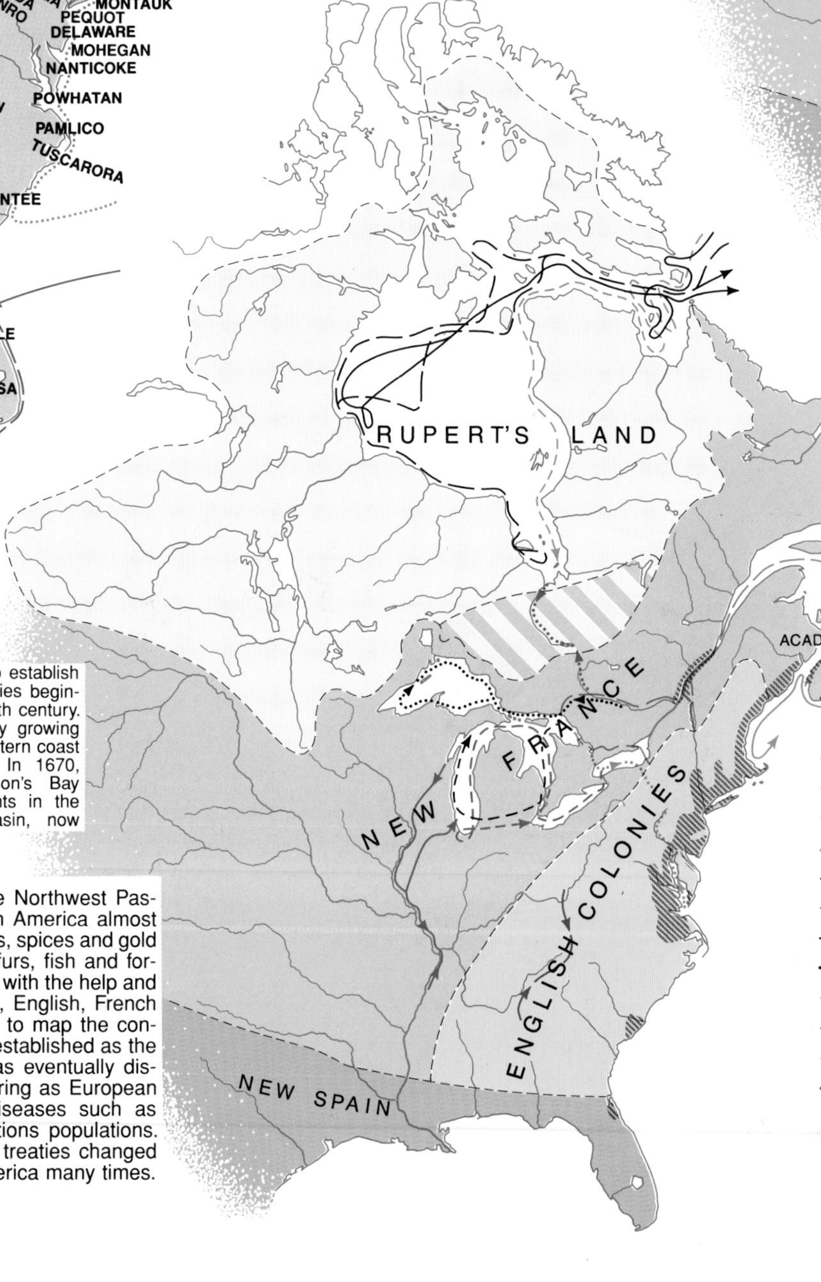

The French were the first to establish fur trading posts and colonies beginning early in the seventeenth century. The French and the rapidly growing English Colonies on the eastern coast were in constant conflict. In 1670, Britain granted the Hudson's Bay Company sole trading rights in the Hudson Bay drainage basin, now named Rupert's Land.

- Monts and Champlain (1604-05)
- Champlain (1608-15)
- Hudson (1610-11)
- Brûlé (1615-16)
- Munk (1619-20)
- James (1631-32)
- Des Groseilliers (1654-56)
- Radisson and Des Groseilliers (1659-60)
- Jolliet and Marquette (1673)
- Arthur (1673-74)
- La Salle (1678-80)
- Troyes (1686)
- Approximate extent of settlement, 1690

Europeans, looking for the Northwest Passage to Asia, landed in North America almost by accident. Instead of the silks, spices and gold of Asia, they discovered the furs, fish and forests of North America. Slowly, with the help and guidance of the First Nations, English, French and Spanish explorers began to map the continent. The fur trade became established as the base of the economy, but was eventually displaced by farming and lumbering as European settlers arrived. European diseases such as smallpox decimated First Nations populations. Territorial disputes, wars and treaties changed the political face of North America many times.

EUROPEAN EXPLORATION AND SETTLEMENT 1

Map 3: 1690–1713

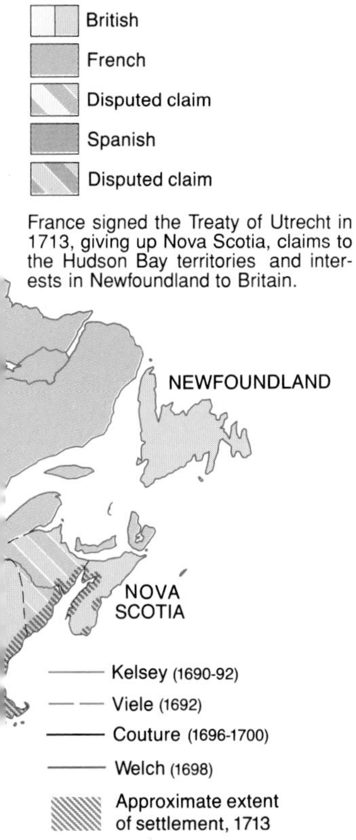

France signed the Treaty of Utrecht in 1713, giving up Nova Scotia, claims to the Hudson Bay territories and interests in Newfoundland to Britain.

Map 5: 1763–1784

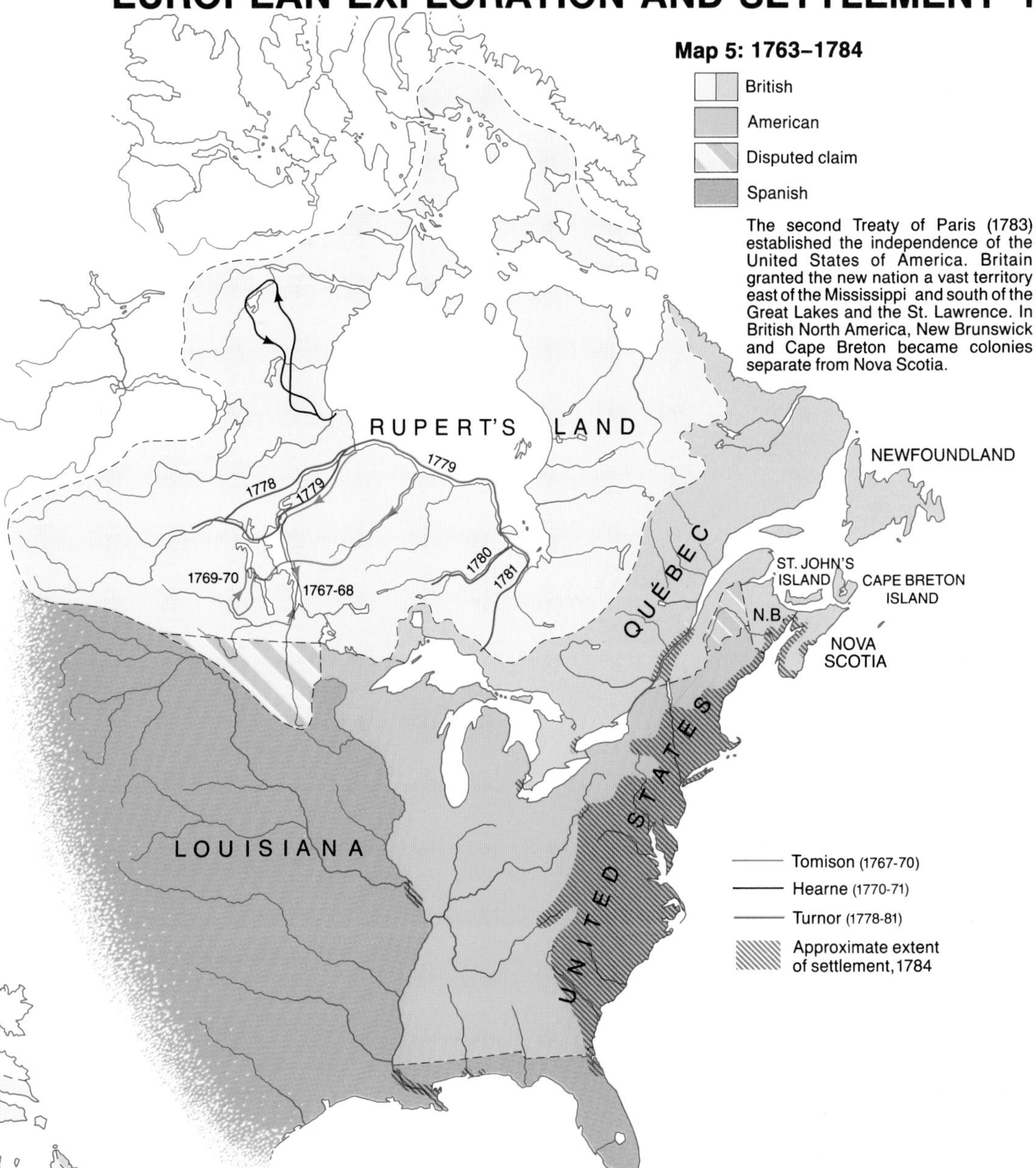

The second Treaty of Paris (1783) established the independence of the United States of America. Britain granted the new nation a vast territory east of the Mississippi and south of the Great Lakes and the St. Lawrence. In British North America, New Brunswick and Cape Breton became colonies separate from Nova Scotia.

Map 4: 1713–1763

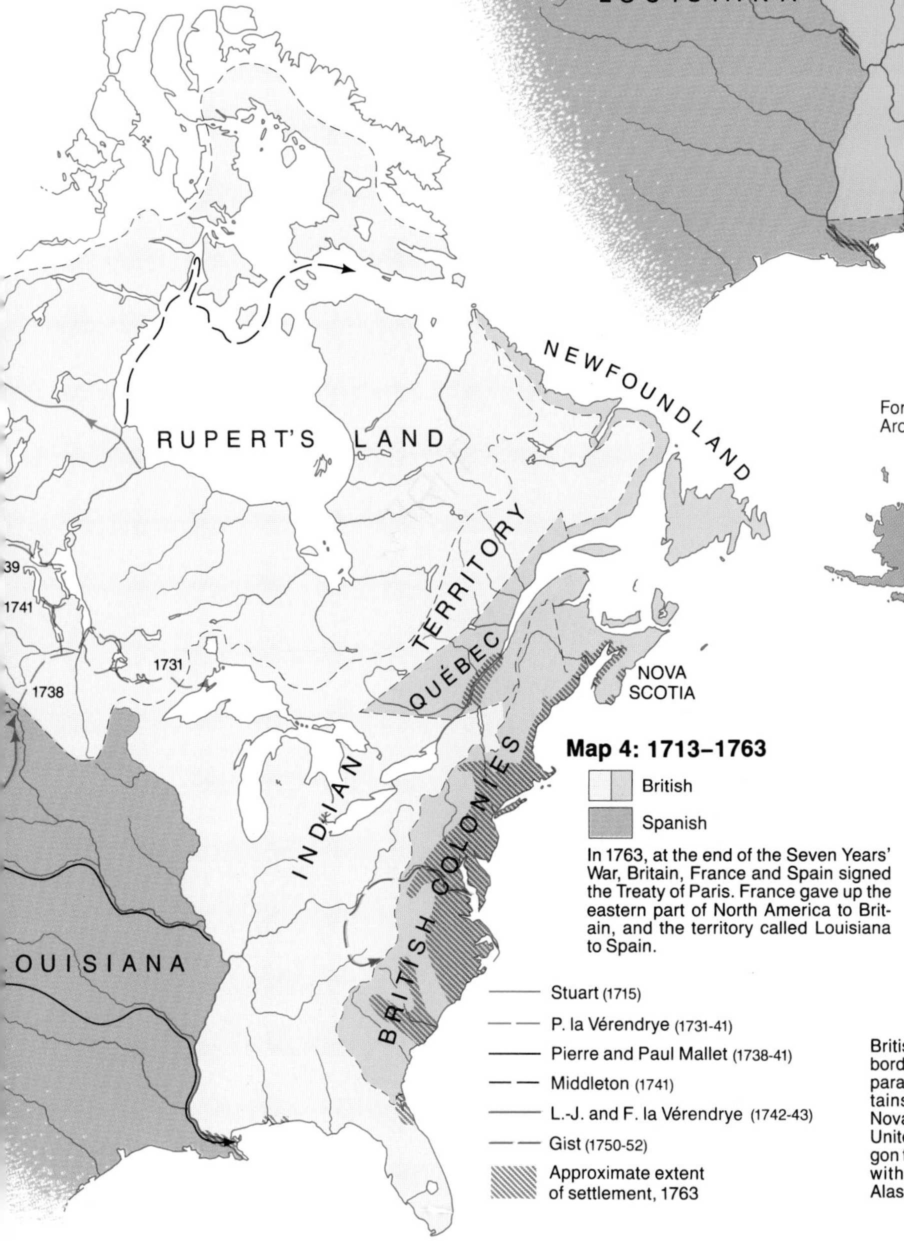

In 1763, at the end of the Seven Years' War, Britain, France and Spain signed the Treaty of Paris. France gave up the eastern part of North America to Britain, and the territory called Louisiana to Spain.

Map 6: 1784–1825

For more information on western and Arctic exploration, see pages 3 and 4.

British North America's international border was extended along the 49th parallel of latitude to the Rocky Mountains in 1818. Cape Breton rejoined Nova Scotia in 1820. Britain and the United States both occupied the Oregon territory, and in 1824 Britain agreed with Russia on the position of the Alaska boundary.

For explanation of terms, see glossary.

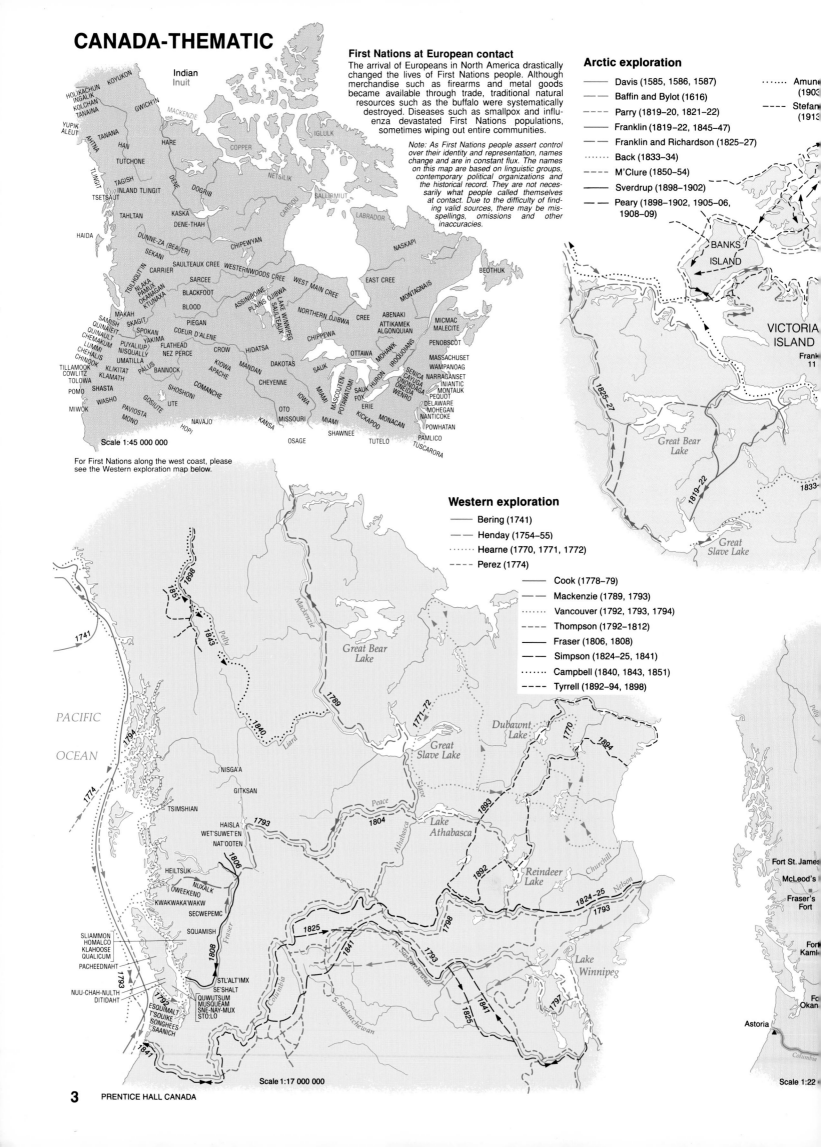

CANADA-THEMATIC

For First Nations along the west coast, please see the Western exploration map below.

First Nations at European contact

The arrival of Europeans in North America drastically changed the lives of First Nations people. Although merchandise such as firearms and metal goods became available through trade, traditional natural resources such as the buffalo were systematically destroyed. Diseases such as smallpox and influenza devastated First Nations populations, sometimes wiping out entire communities.

Note: As First Nations people assert control over their identity and representation, names change and are in constant flux. The names on this map are based on linguistic groups, contemporary political organizations and the historical record. They are not necessarily what people called themselves at contact. Due to the difficulty of finding valid sources, there may be misspellings, omissions and other inaccuracies.

Arctic exploration

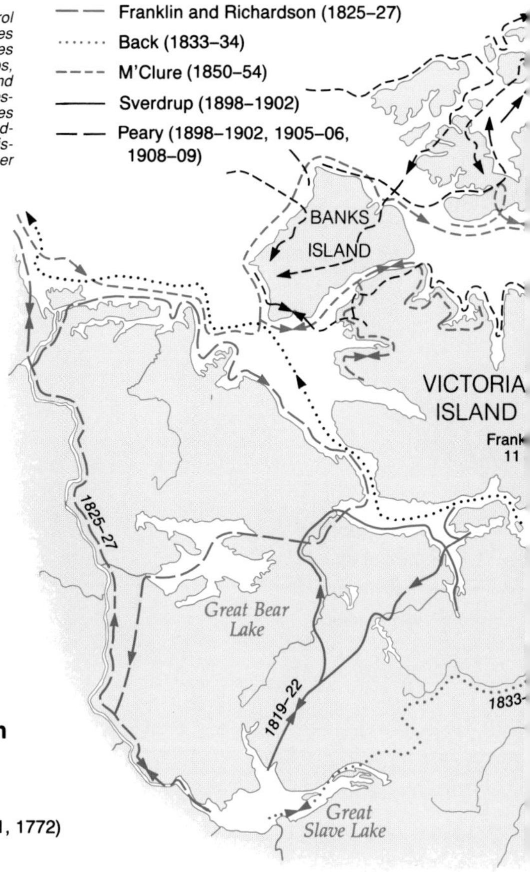

Western exploration

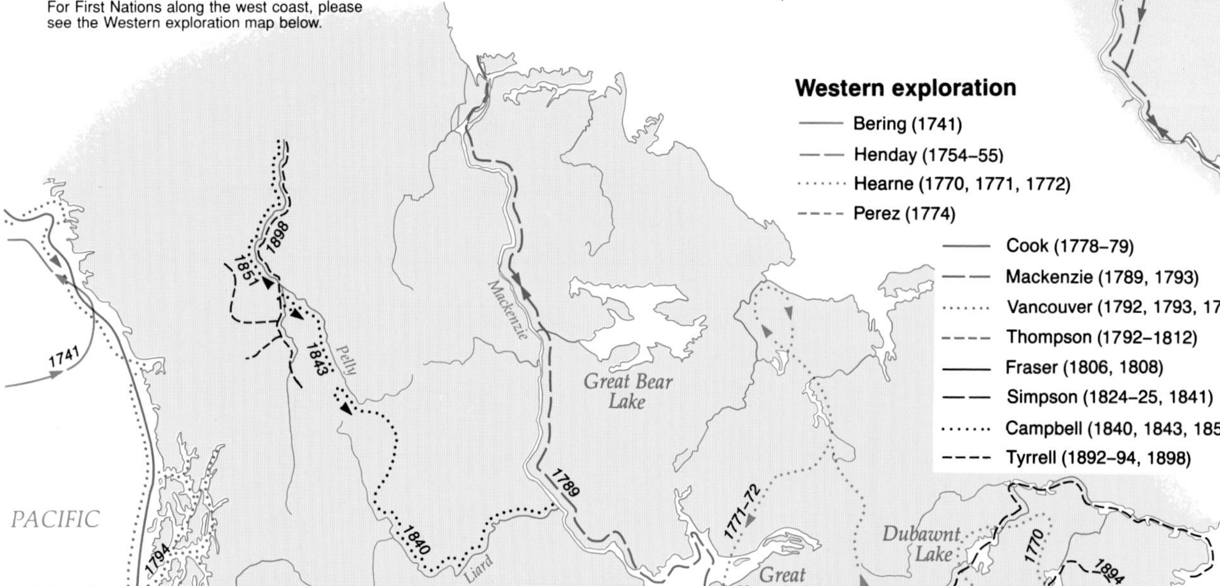

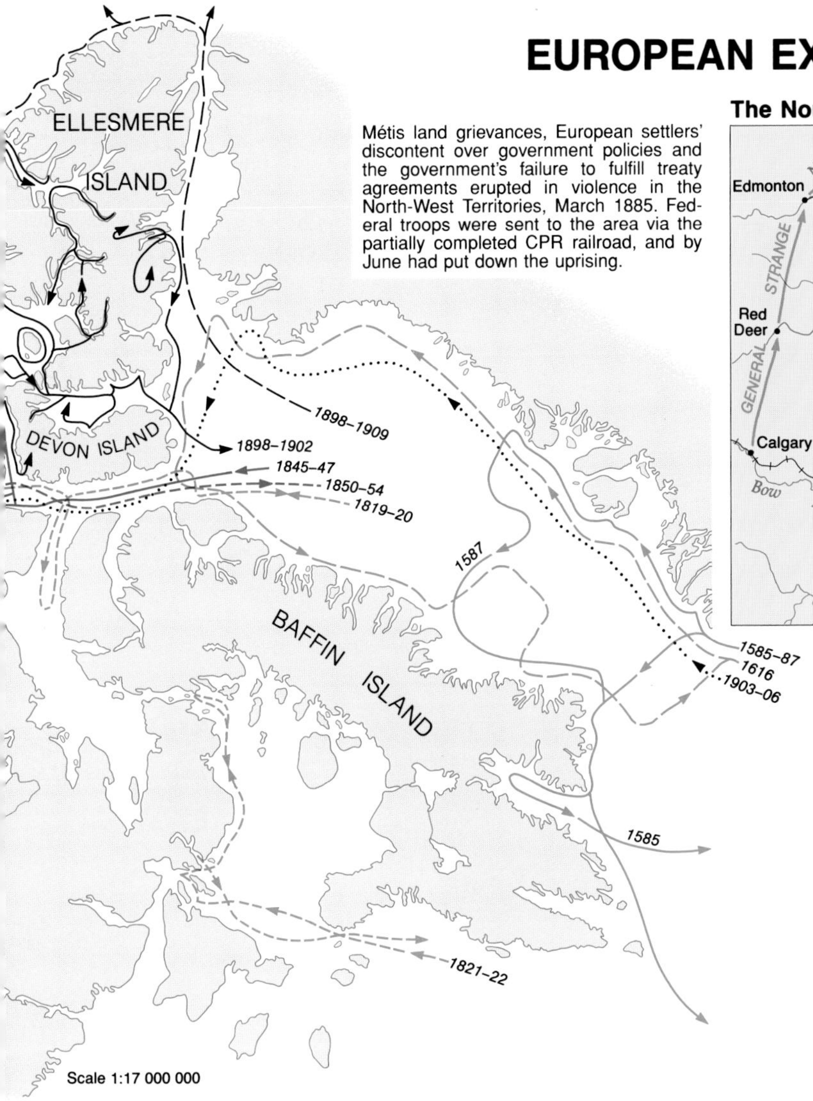

Métis land grievances, European settlers' discontent over government policies and the government's failure to fulfill treaty agreements erupted in violence in the North-West Territories, March 1885. Federal troops were sent to the area via the partially completed CPR railroad, and by June had put down the uprising.

The North-West Rebellion, 1885

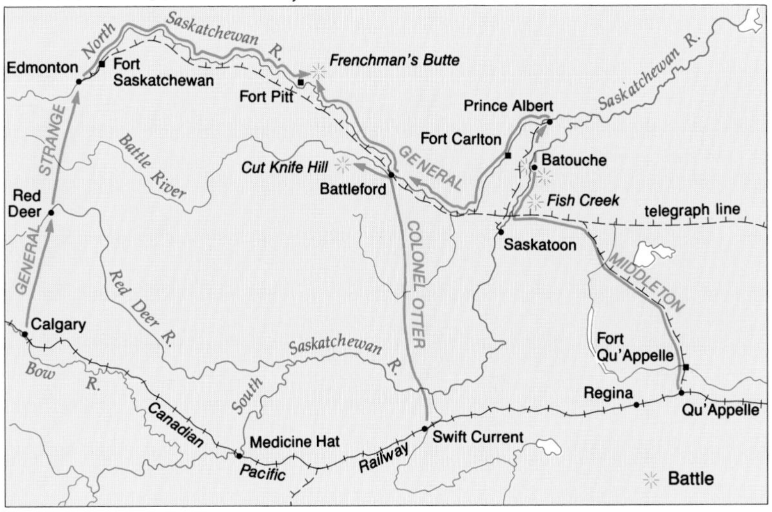

Railways in 1891

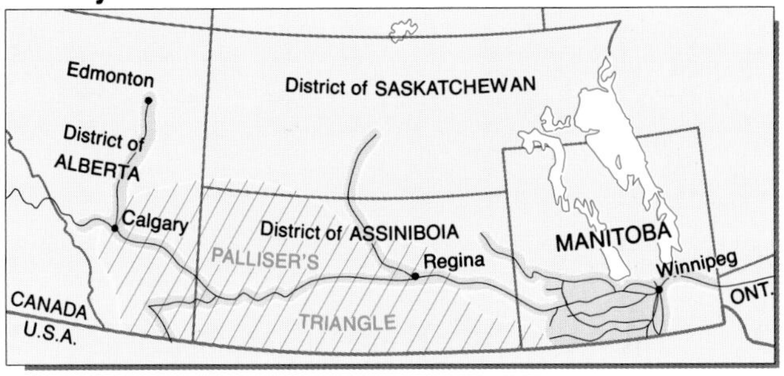

Railways in 1911

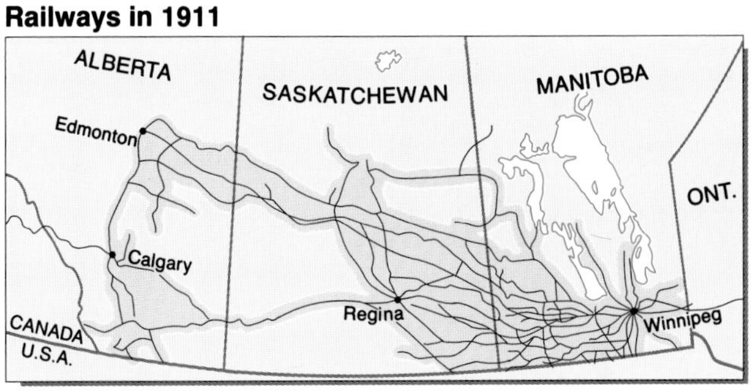

Railways in 1931

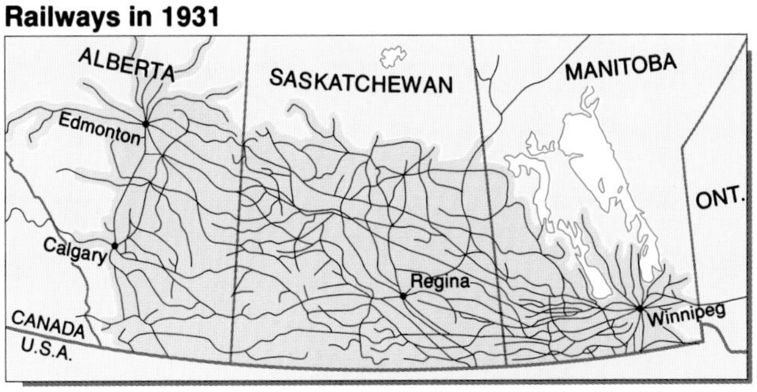

Early in the 1600s, European traders began journeying inland by canoe, exchanging goods for furs brought by First Nations. By 1760, the fur trade was the backbone of the Canadian economy. Competition for furs encouraged traders to push west all the way to the Rocky Mountains. As the interior was explored and trading posts established, the area became increasingly attractive to settlers.

In 1870, Rupert's Land was transferred from the Hudson's Bay Company to Canada and opened for settlement, signalling the beginning of the end of the historic fur trade. Completion of the transcontinental Canadian Pacific Railway (CPR) in 1885 helped to greatly increase the western population. As agriculture spread, side branches were built to serve the new farming communities.

Fur trade in the Canadian West

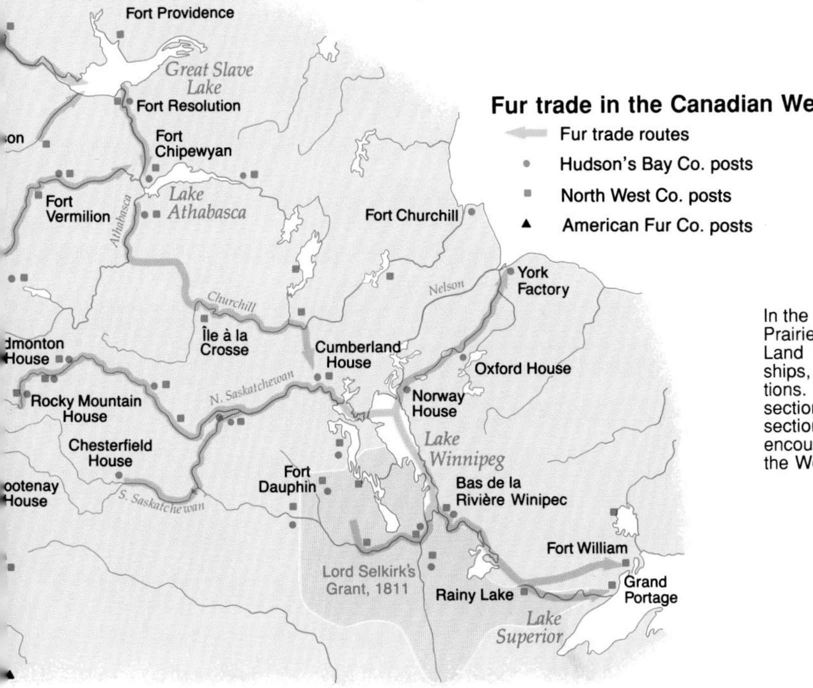

The Prairie township system

In the late 1800s, the grasslands of the Prairies were opened for settlement. Land was divided into square townships, each 10 km square with 36 sections. Settlers were given quarter sections (64 hectares) of land. Many sections were given to the railways to encourage completion of the railway to the West.

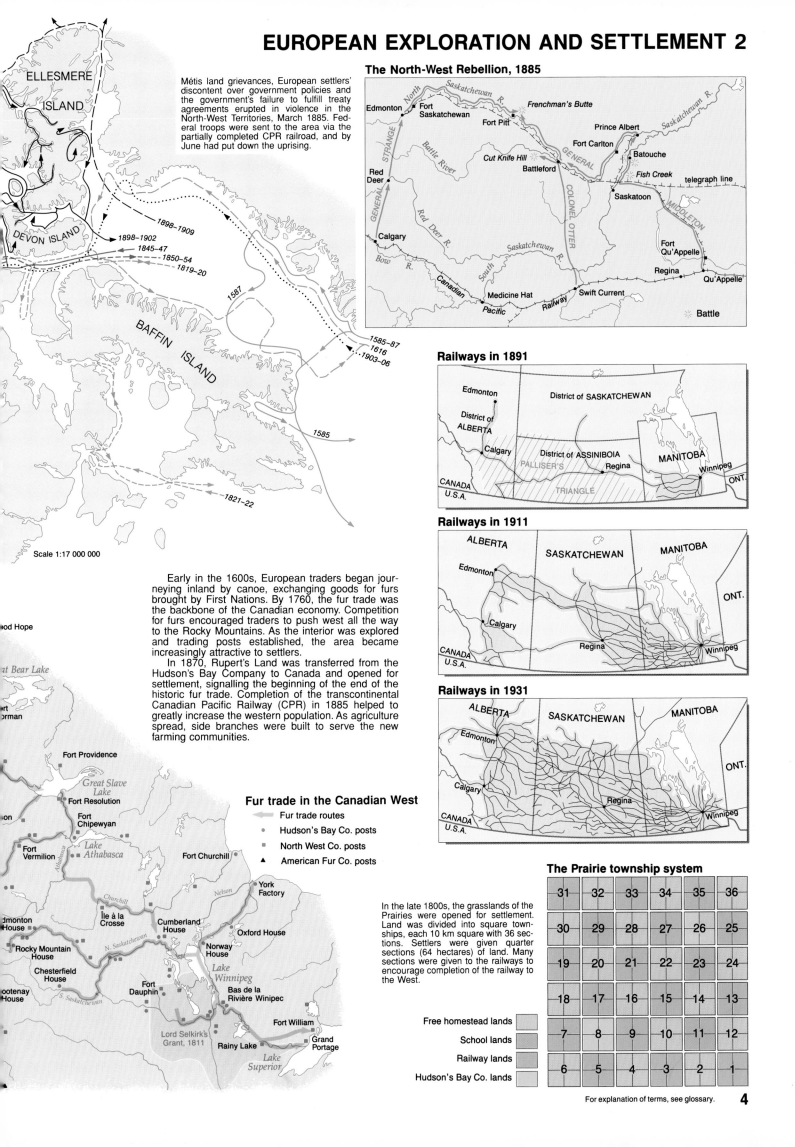

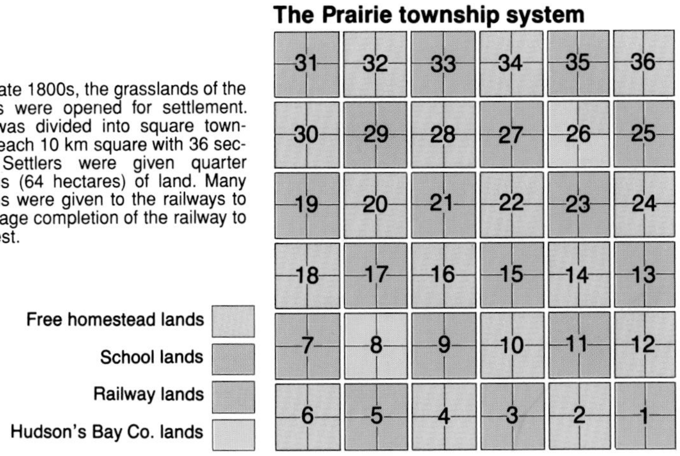

For explanation of terms, see glossary.

CANADA-THEMATIC

The sequence of maps from 1867 onwards adds a colour to the basic grey for each province or territory, as it achieves independence from Great Britain.

Map 1: Before Confederation, 1866

British Territory/ Colony | Danish | American | Russian

British Columbia reached its present boundaries in 1866.

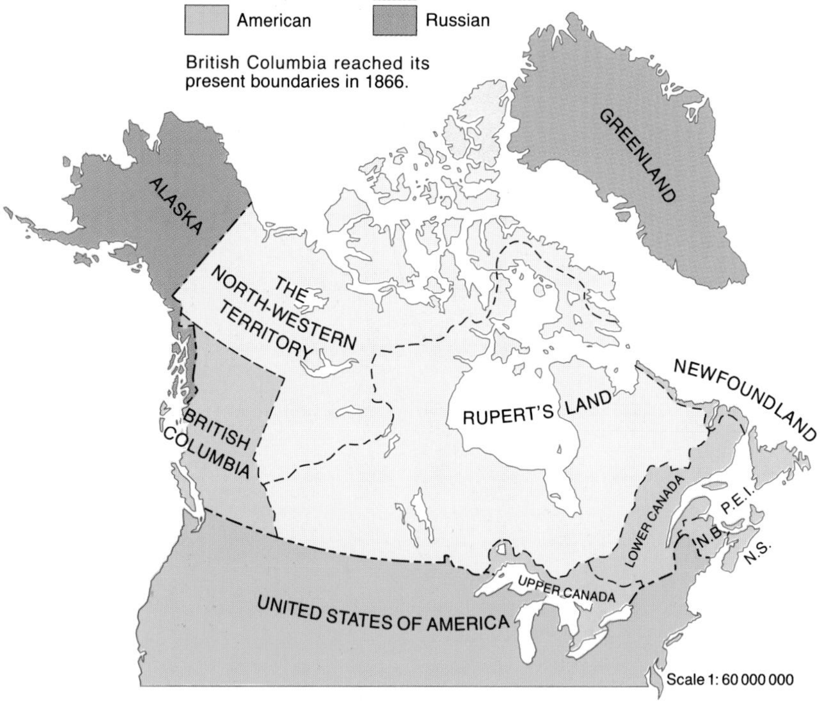

Map 2: At Confederation,

The Dominion of Canada formed by uniting Nova Sc New Brunswick and Can Two provinces were creat Ontario (formerly Upper ada) and Québec (form Lower Canada). Just ten before Confederation (Ju the United States purcha Alaska from Russia.

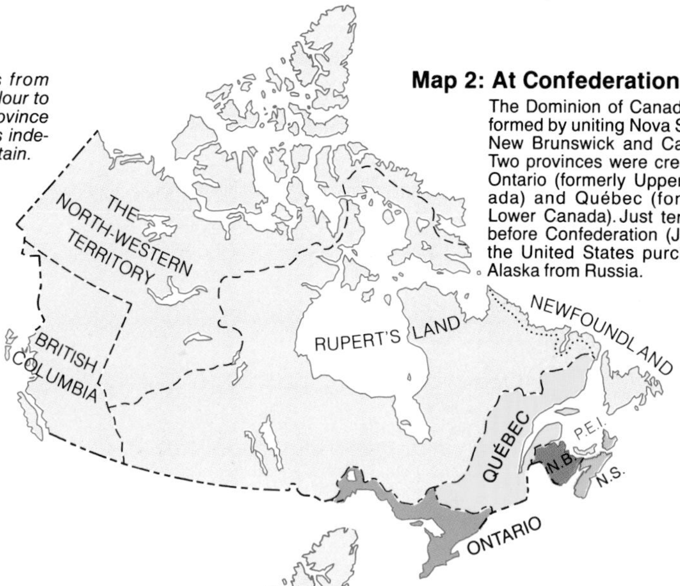

Map 5: 1876

The boundaries of On were changed in 1874 an District of Keewatin was ated in 1876, separate fro North-West Territories.

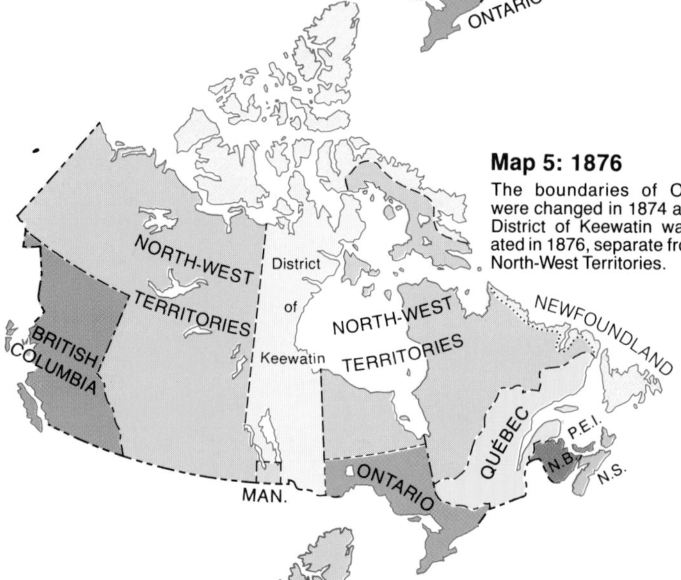

The Growth of Canada: Area and Population

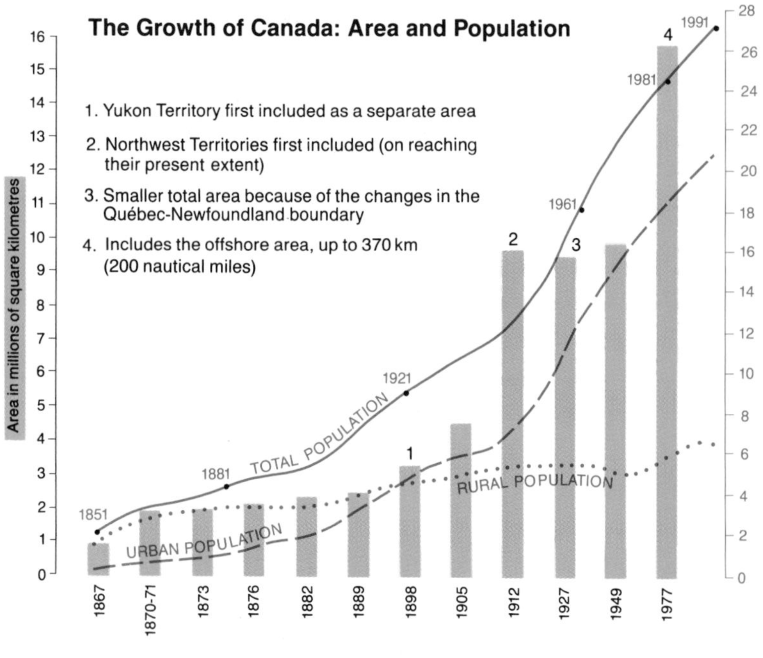

Map 8: 1898

The North-West Territo were divided further into tricts, and the Yukon beca separate territory. The n ern boundary of Québec extended.

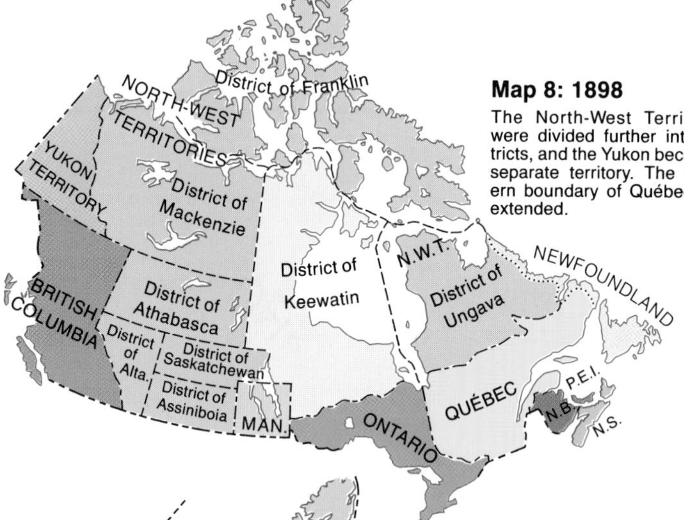

Map 10: 1912

The present boundaries of Ontario, Manitoba, and Québec were established, with the exception of the boundary between Québec and Labrador.

Map 11: 1927

Canada's boundaries w extended northward, and Imperial Privy Council dec on the Québec-Labra boundary. Québec, howe does not recognize this sion.

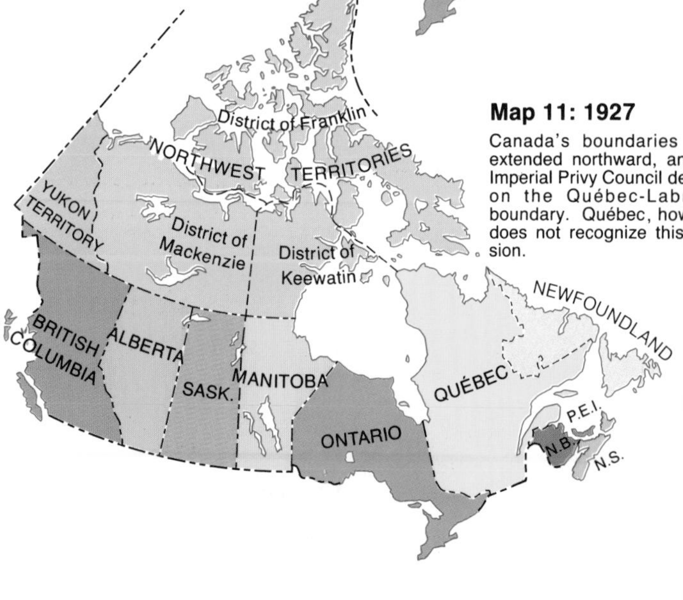

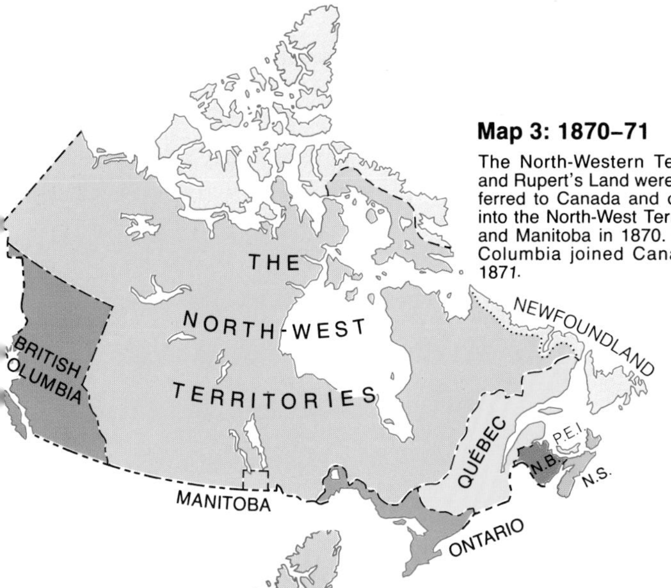

Map 3: 1870–71

The North-Western Territory and Rupert's Land were transferred to Canada and divided into the North-West Territories and Manitoba in 1870. British Columbia joined Canada in 1871.

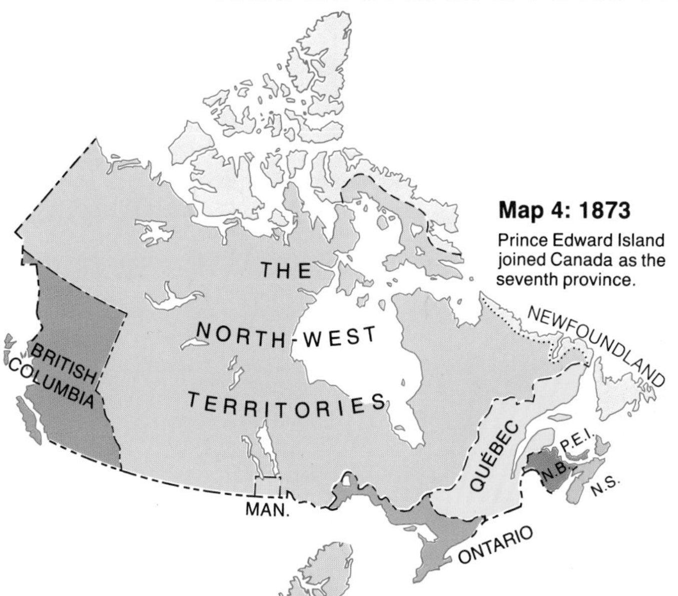

Map 4: 1873

Prince Edward Island joined Canada as the seventh province.

Map 6: 1882

Britain transferred the administration of the Arctic Islands to Canada in 1880. Manitoba was enlarged, but this was protested by Ontario. The North-West Territories were divided into several districts.

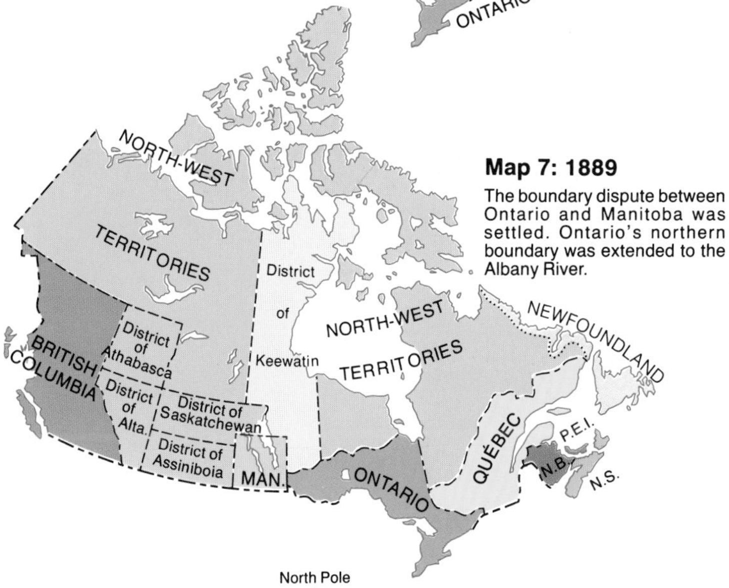

Map 7: 1889

The boundary dispute between Ontario and Manitoba was settled. Ontario's northern boundary was extended to the Albany River.

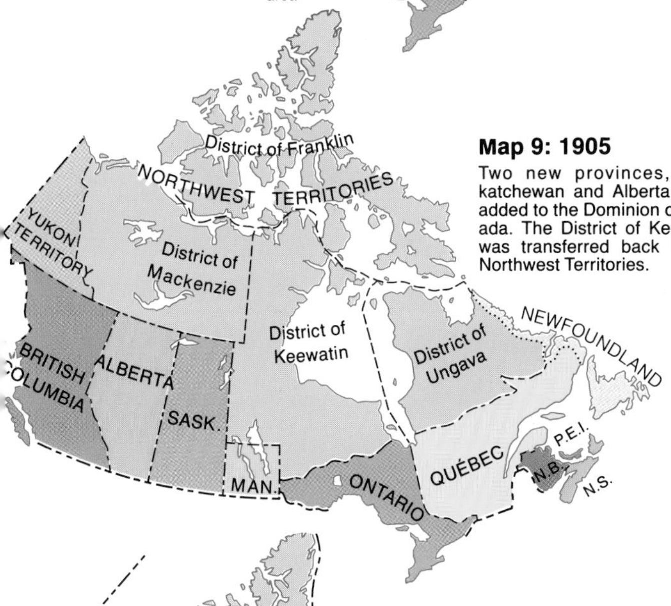

Map 9: 1905

Two new provinces, Saskatchewan and Alberta, were added to the Dominion of Canada. The District of Keewatin was transferred back to the Northwest Territories.

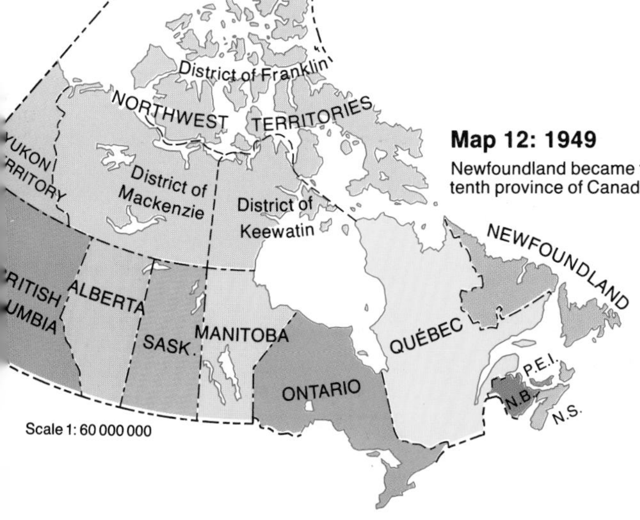

Map 12: 1949

Newfoundland became the tenth province of Canada.

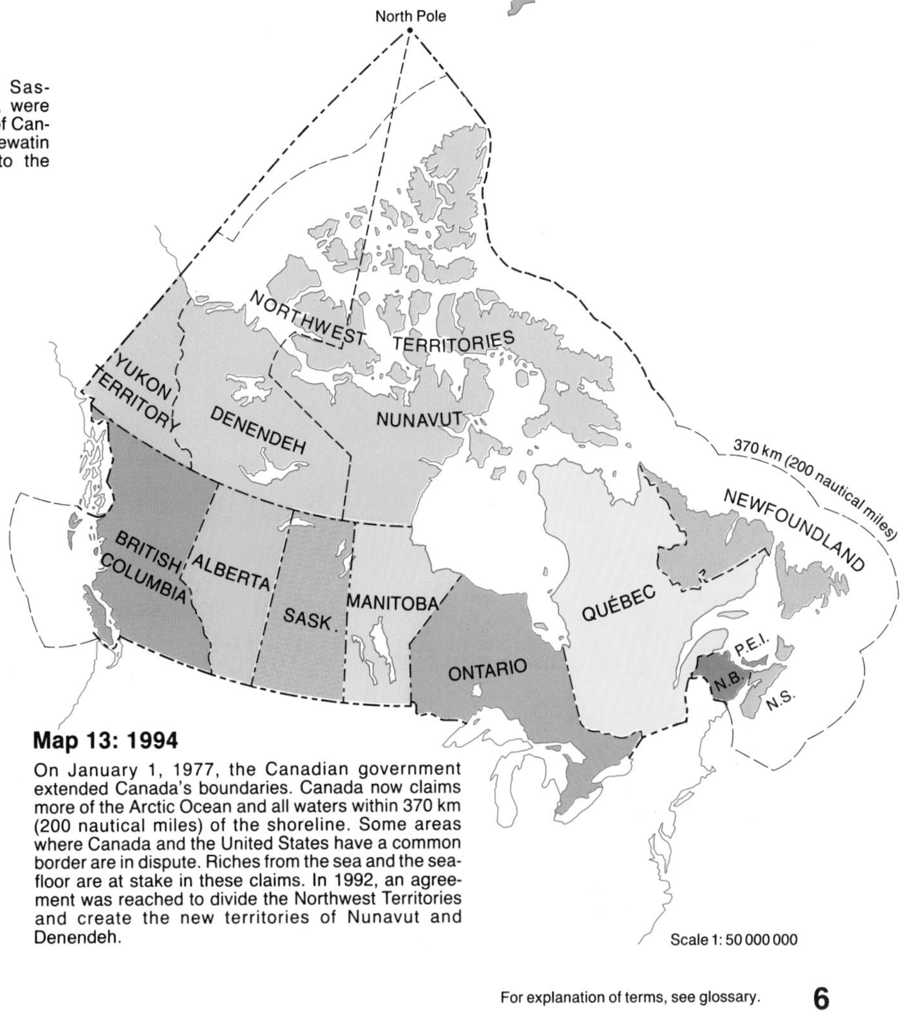

Map 13: 1994

On January 1, 1977, the Canadian government extended Canada's boundaries. Canada now claims more of the Arctic Ocean and all waters within 370 km (200 nautical miles) of the shoreline. Some areas where Canada and the United States have a common border are in dispute. Riches from the sea and the seafloor are at stake in these claims. In 1992, an agreement was reached to divide the Northwest Territories and create the new territories of Nunavut and Denendeh.

CANADA-THEMATIC

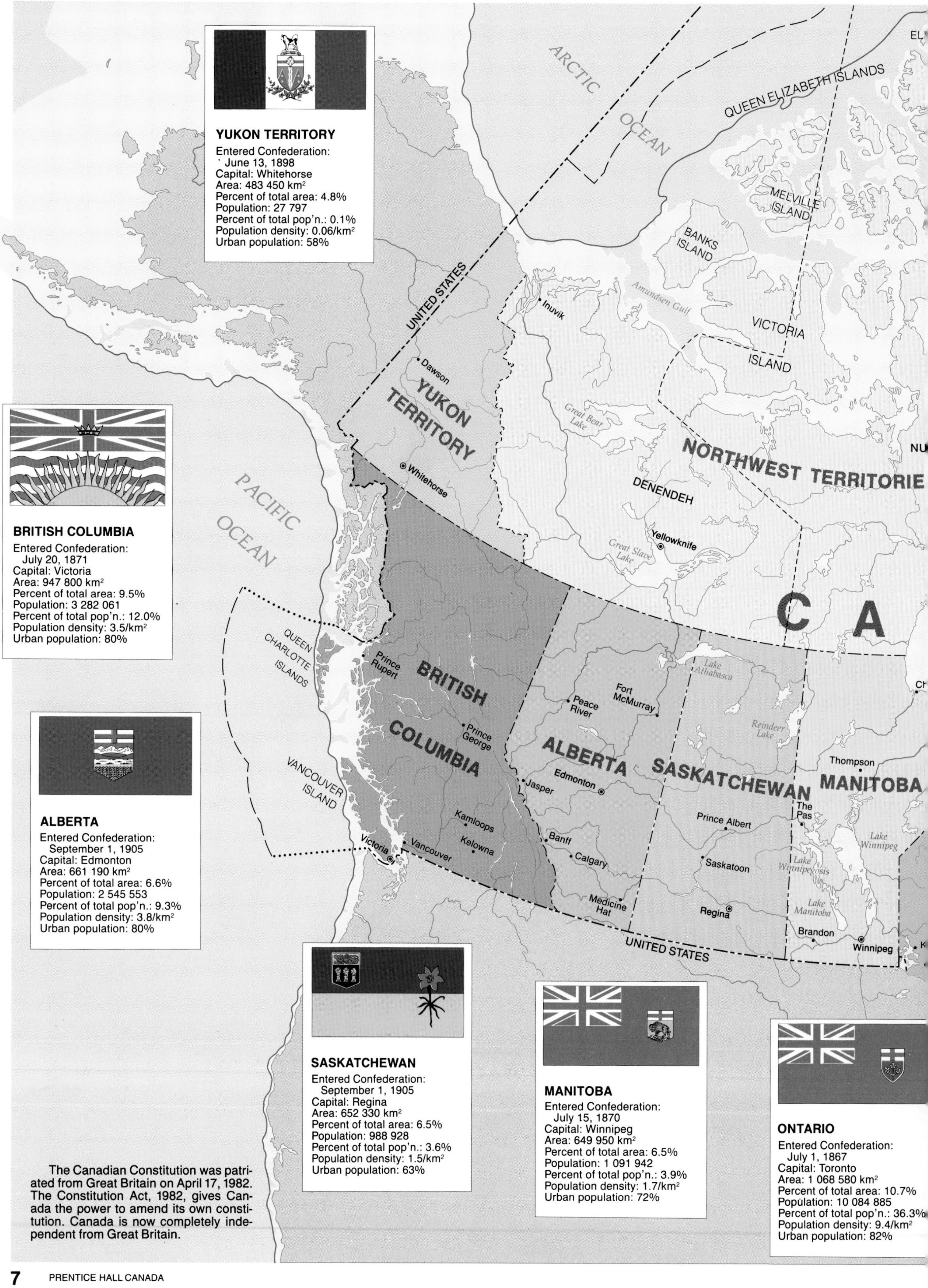

YUKON TERRITORY
Entered Confederation:
June 13, 1898
Capital: Whitehorse
Area: 483 450 km²
Percent of total area: 4.8%
Population: 27 797
Percent of total pop'n.: 0.1%
Population density: 0.06/km²
Urban population: 58%

BRITISH COLUMBIA
Entered Confederation:
July 20, 1871
Capital: Victoria
Area: 947 800 km²
Percent of total area: 9.5%
Population: 3 282 061
Percent of total pop'n.: 12.0%
Population density: 3.5/km²
Urban population: 80%

ALBERTA
Entered Confederation:
September 1, 1905
Capital: Edmonton
Area: 661 190 km²
Percent of total area: 6.6%
Population: 2 545 553
Percent of total pop'n.: 9.3%
Population density: 3.8/km²
Urban population: 80%

SASKATCHEWAN
Entered Confederation:
September 1, 1905
Capital: Regina
Area: 652 330 km²
Percent of total area: 6.5%
Population: 988 928
Percent of total pop'n.: 3.6%
Population density: 1.5/km²
Urban population: 63%

MANITOBA
Entered Confederation:
July 15, 1870
Capital: Winnipeg
Area: 649 950 km²
Percent of total area: 6.5%
Population: 1 091 942
Percent of total pop'n.: 3.9%
Population density: 1.7/km²
Urban population: 72%

ONTARIO
Entered Confederation:
July 1, 1867
Capital: Toronto
Area: 1 068 580 km²
Percent of total area: 10.7%
Population: 10 084 885
Percent of total pop'n.: 36.3%
Population density: 9.4/km²
Urban population: 82%

The Canadian Constitution was patriated from Great Britain on April 17, 1982. The Constitution Act, 1982, gives Canada the power to amend its own constitution. Canada is now completely independent from Great Britain.

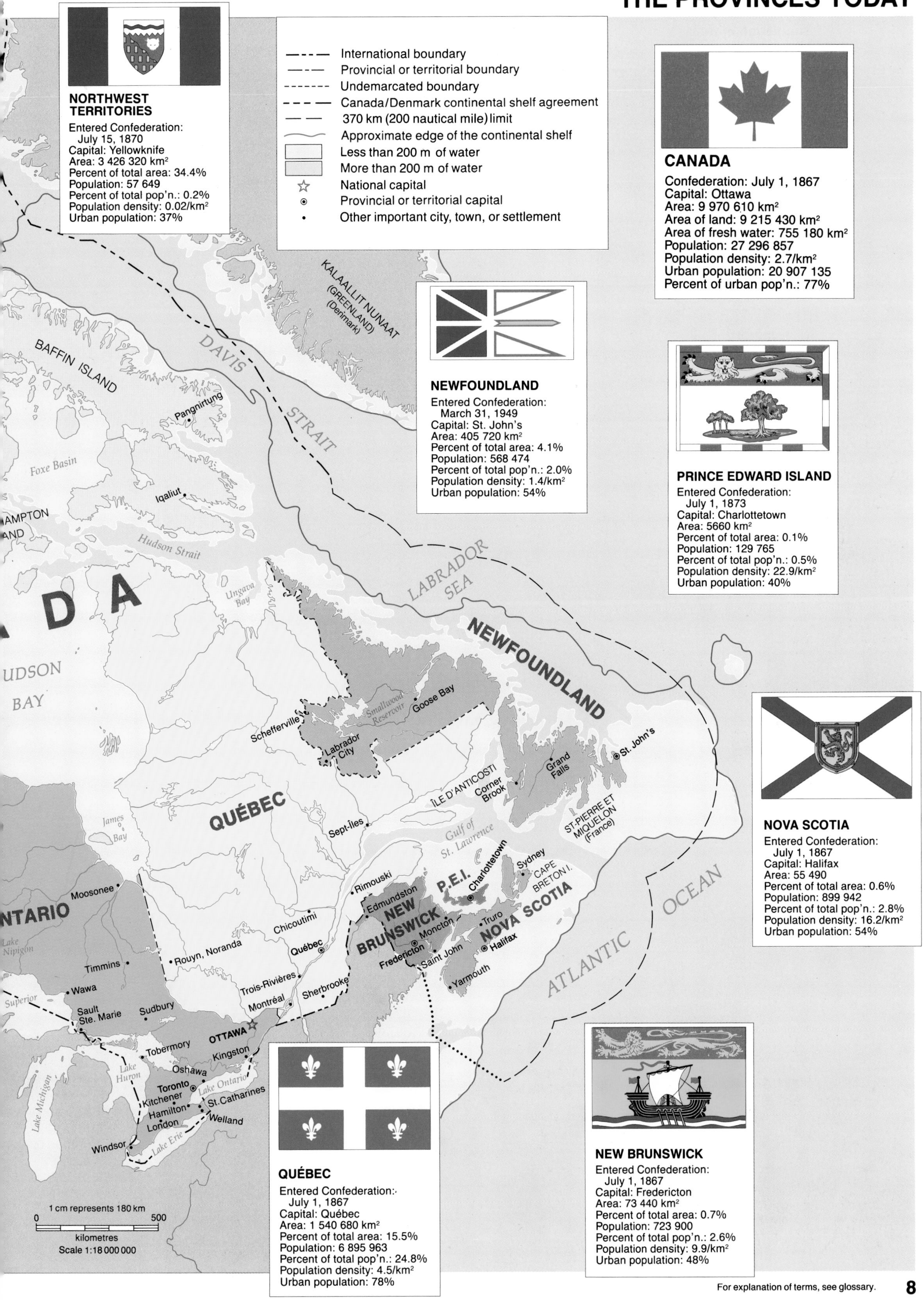

NORTHWEST TERRITORIES

Entered Confederation: July 15, 1870
Capital: Yellowknife
Area: 3 426 320 km²
Percent of total area: 34.4%
Population: 57 649
Percent of total pop'n.: 0.2%
Population density: 0.02/km²
Urban population: 37%

CANADA

Confederation: July 1, 1867
Capital: Ottawa
Area: 9 970 610 km²
Area of land: 9 215 430 km²
Area of fresh water: 755 180 km²
Population: 27 296 857
Population density: 2.7/km²
Urban population: 20 907 135
Percent of urban pop'n.: 77%

NEWFOUNDLAND

Entered Confederation: March 31, 1949
Capital: St. John's
Area: 405 720 km²
Percent of total area: 4.1%
Population: 568 474
Percent of total pop'n.: 2.0%
Population density: 1.4/km²
Urban population: 54%

PRINCE EDWARD ISLAND

Entered Confederation: July 1, 1873
Capital: Charlottetown
Area: 5660 km²
Percent of total area: 0.1%
Population: 129 765
Percent of total pop'n.: 0.5%
Population density: 22.9/km²
Urban population: 40%

NOVA SCOTIA

Entered Confederation: July 1, 1867
Capital: Halifax
Area: 55 490
Percent of total area: 0.6%
Population: 899 942
Percent of total pop'n.: 2.8%
Population density: 16.2/km²
Urban population: 54%

QUÉBEC

Entered Confederation: July 1, 1867
Capital: Québec
Area: 1 540 680 km²
Percent of total area: 15.5%
Population: 6 895 963
Percent of total pop'n.: 24.8%
Population density: 4.5/km²
Urban population: 78%

NEW BRUNSWICK

Entered Confederation: July 1, 1867
Capital: Fredericton
Area: 73 440 km²
Percent of total area: 0.7%
Population: 723 900
Percent of total pop'n.: 2.6%
Population density: 9.9/km²
Urban population: 48%

For explanation of terms, see glossary.

CANADA-THEMATIC

Population profiles

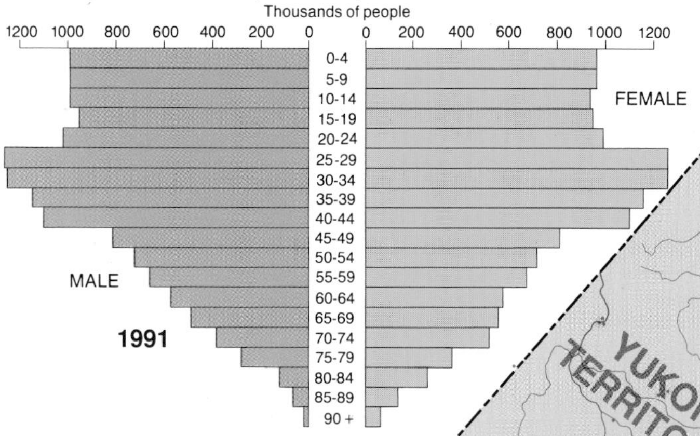

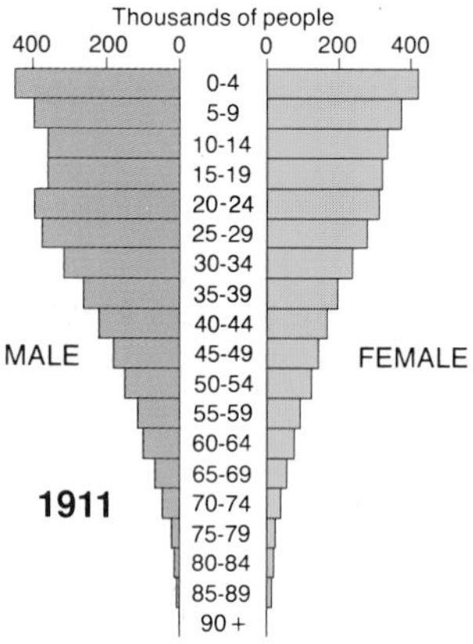

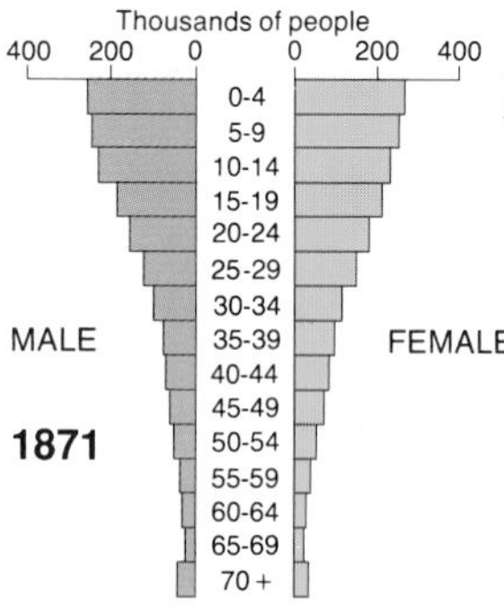

Population profiles show the structure of the population according to age and sex. Many factors affect the number of births and deaths, including war, economic depression, health care, birth control, and the role of the family.

Population distribution, 1871

One dot represents 200 people

Scale 1: 50 000 000

Percent population, 1871

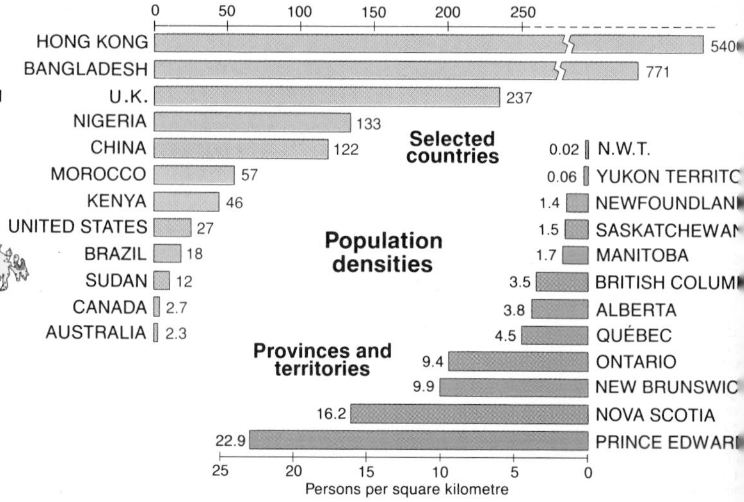

Population densities

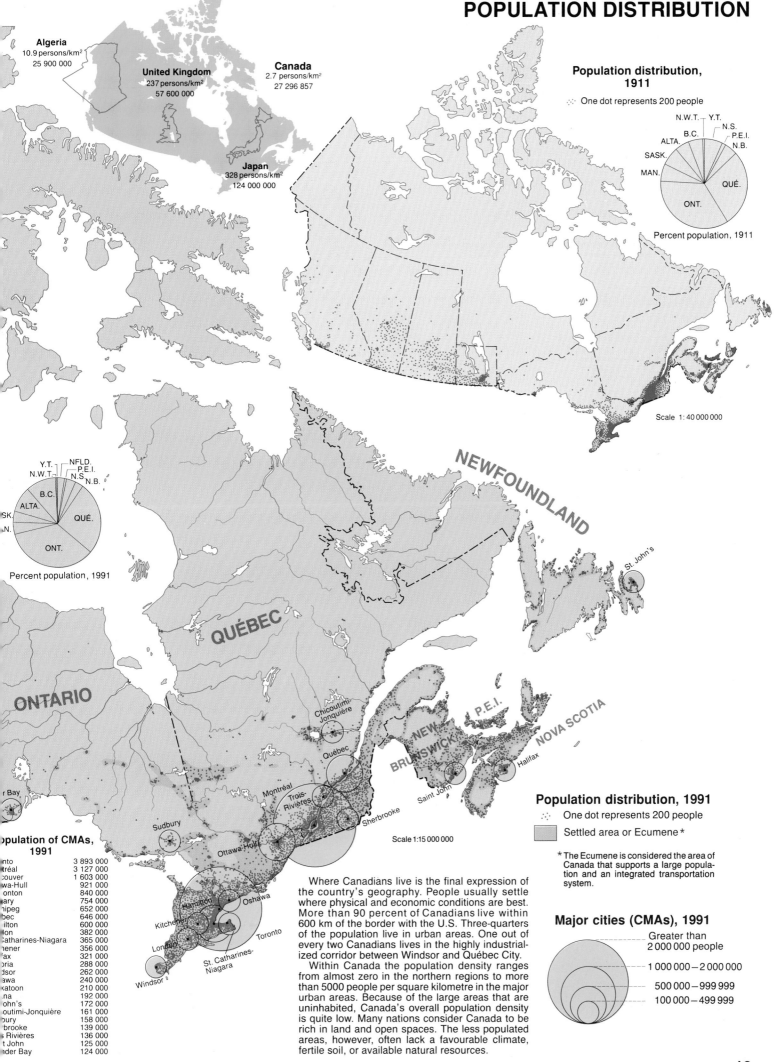

POPULATION DISTRIBUTION

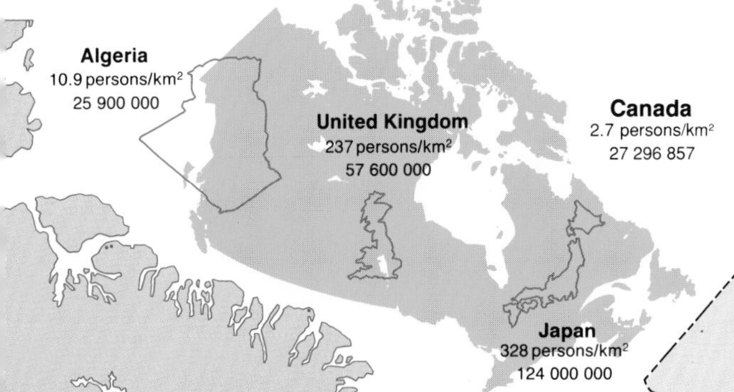

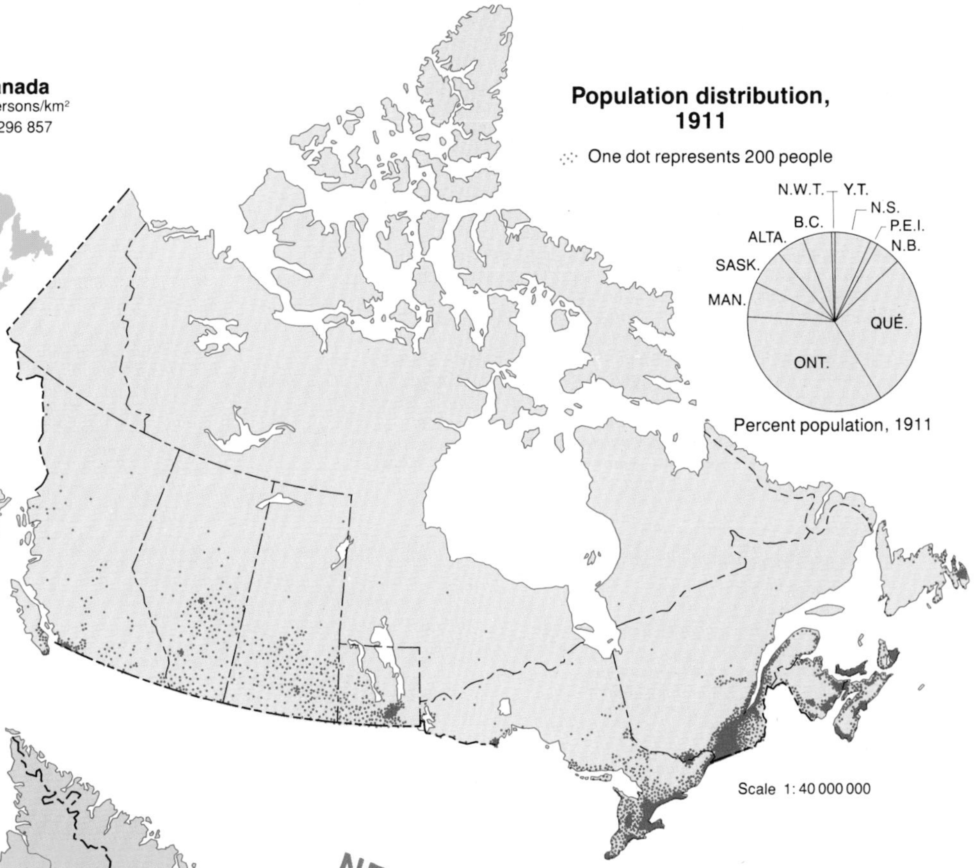

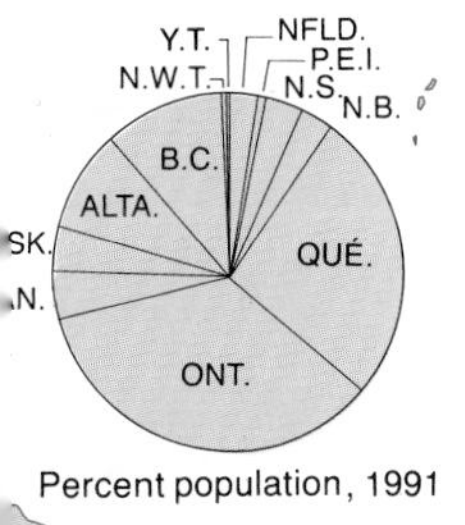

Population distribution, 1991

One dot represents 200 people

Settled area or Ecumene *

* The Ecumene is considered the area of Canada that supports a large population and an integrated transportation system.

Major cities (CMAs), 1991

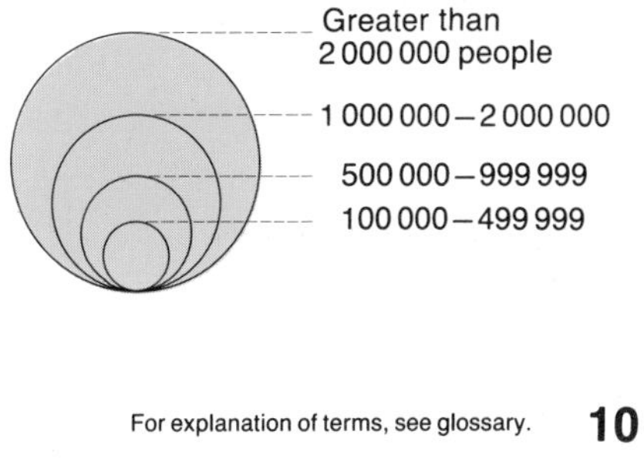

pulation of CMAs, 1991

nto	3 893 000
tréal	3 127 000
couver	1 603 000
wa-Hull	921 000
onton	840 000
ary	754 000
nipeg	652 000
bec	646 000
ilton	600 000
lon	382 000
atharines-Niagara	365 000
hener	356 000
fax	321 000
oria	288 000
dsor	262 000
awa	240 000
katoon	210 000
na	192 000
ohn's	172 000
outimi-Jonquière	161 000
bury	158 000
brooke	139 000
Rivières	136 000
t John	125 000
nder Bay	124 000

Where Canadians live is the final expression of the country's geography. People usually settle where physical and economic conditions are best. More than 90 percent of Canadians live within 600 km of the border with the U.S. Three-quarters of the population live in urban areas. One out of every two Canadians lives in the highly industrialized corridor between Windsor and Québec City.

Within Canada the population density ranges from almost zero in the northern regions to more than 5000 people per square kilometre in the major urban areas. Because of the large areas that are uninhabited, Canada's overall population density is quite low. Many nations consider Canada to be rich in land and open spaces. The less populated areas, however, often lack a favourable climate, fertile soil, or available natural resources.

CANADA-THEMATIC

The First Nations of Canada have diverse cultures, languages and lifestyles. The Inuit traditionally hunted and fished on the tundra. First Nations in Ontario and Québec hunted, fished and raised crops such as beans, maize and squash. In forested areas and on the Prairies, nomadic groups hunted and gathered. First Nations living on both coasts were also hunters and fished for their food.

Contact with non-indigenous peoples has resulted in dramatic social and cultural change for First Nations peoples. For example, finding new and suitable sources of income is a situation faced by most reserves, especially isolated ones. Today, land claims are a step toward recognized self-government and a means of realizing the rights to land and resources that belong to First Nations.

First Nations people and modern non-Native Canadian society have many shared ideas and materials. After a long-standing and active relationship, First Nations art, sports, politics, literature and customs have become part of the entire Canadian experience. There are also many Canadians of mixed ancestry. The Métis, for example, are descendants of First Nations people and Europeans (mostly French). They have established a separate identity as people of mixed heritage.

Population with First Nations origins

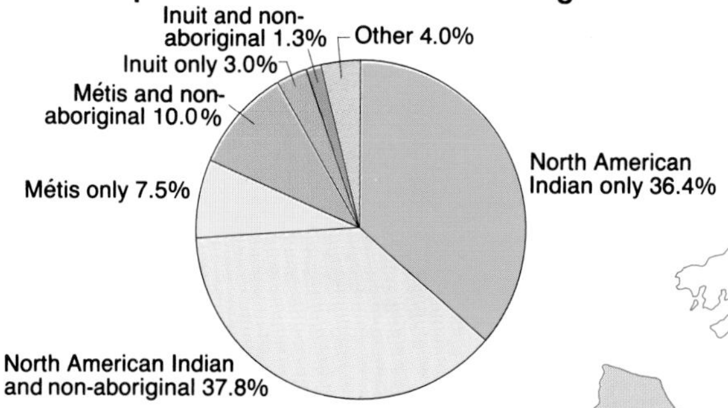

Total: 1 002 675

Comprehensive claims (See note below)

1. Nisga'a Tribal Council
2. Kitwancool Band
3. Gitksan-Wet'suwet'en Tribal Council
4. Haisla Nation
5. Association of United Tahltans
6. Nuu-Chah-Nulth Tribal Council
7. Council of Haida Nation
8. Heiltsuk Nation
9. Nuxalk Nation
10. Nazko-Kluskus Bands
11. Kaska-Dena Council
12. Carrier-Sekani Tribal Council
13. Alkali Lake Band
14. Taku Tlingit
15. Kootenay Indian Area Council
16. Allied Tsimshian Tribes
17. Council of Tsimshian Nation
18. Nlaka'pamux Nation
19. Kwakiutl First Nations
20. Sechelt Band
21. Musqueam Band
22. Homalco Band
23. Council for Yukon Indians
24. Inuvialuit Settlement Region
25. Gwich'in Settlement Region
26. Sahtu Claim Area
27. Deh Cho Claim Area
28. North Dene-Thah Claim Area
29. South Dene-Thah Claim Area
30. Tungavik Federation of Nunavut
31. Labrador Inuit Association
32. Innu Nation Claim Area
33. Conseil des Atikamekw et des Montagnais
34. Québec Inuit Claim Area
35. James Bay Territory*

*James Bay and Northern Québec Agreement and Northeastern Québec Agreement (Grand Council of the Crees of Québec and the Northern Québec Inuit Association; Naskapis of Schefferville)

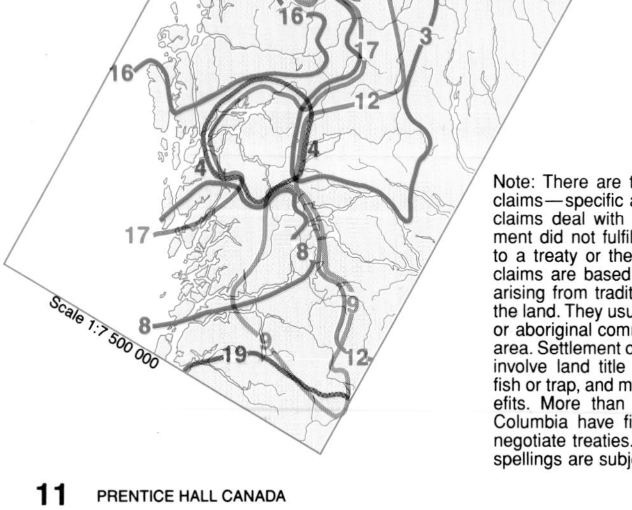

Note: There are two types of aboriginal land claims—specific and comprehensive. Specific claims deal with allegations that the government did not fulfill specific obligations related to a treaty or the Indian Act. Comprehensive claims are based on claims to aboriginal title arising from traditional use and occupancy of the land. They usually involve a group of bands or aboriginal communities within a geographic area. Settlement of comprehensive claims may involve land title (ownership), rights to hunt, fish or trap, and money or other rights and benefits. More than 42 First Nations in British Columbia have filed Statements of Intent to negotiate treaties. Land claim boundaries and spellings are subject to change.

First Nations populations in major cities

CMA	Total population	Métis population	First Nations population
Victoria	288 000	345	4130
Vancouver	1 603 000	4070	21 845
Edmonton	840 000	13 515	15 910
Calgary	754 000	4285	9870
Saskatoon	210 000	5585	6380
Regina	192 000	3720	7300
Winnipeg	652 000	14 990	20 255
Toronto	3 893 000	1430	12 920
Ottawa-Hull	921 000	1425	5195
Montréal	3 127 000	1675	5400
Halifax	321 000	—	1135

Population with First Nations origins, by province

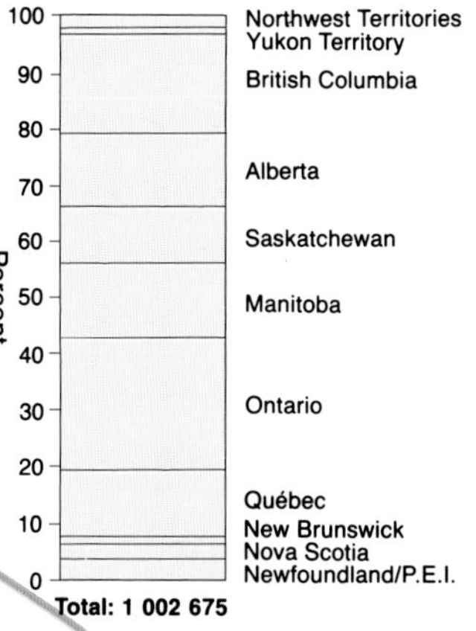

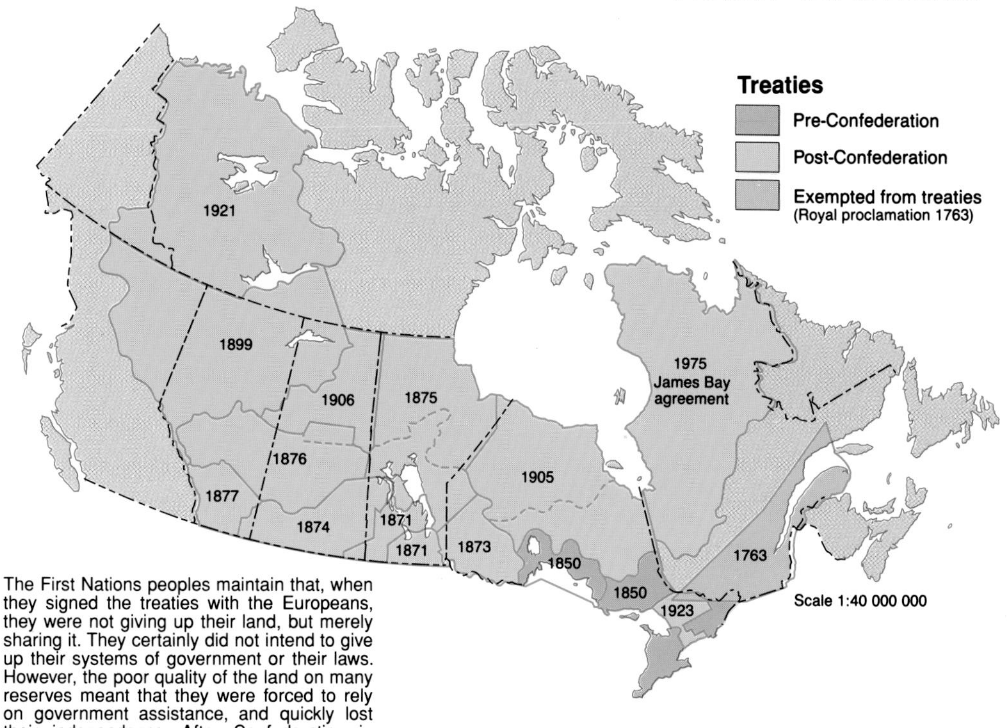

The First Nations peoples maintain that, when they signed the treaties with the Europeans, they were not giving up their land, but merely sharing it. They certainly did not intend to give up their systems of government or their laws. However, the poor quality of the land on many reserves meant that they were forced to rely on government assistance, and quickly lost their independence. After Confederation in 1867, the federal government assumed control over First Nations health care, education and reserve administration.

First Nations populations* by province

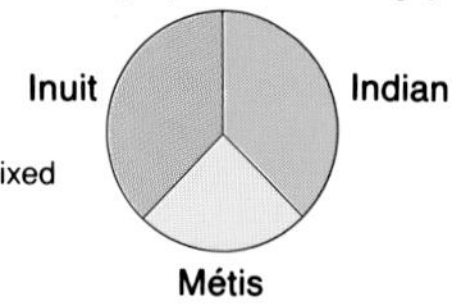

*Includes population of mixed origin as well as single First Nations origin

One dot represents 100 Indians
One dot represents 100 Inuit

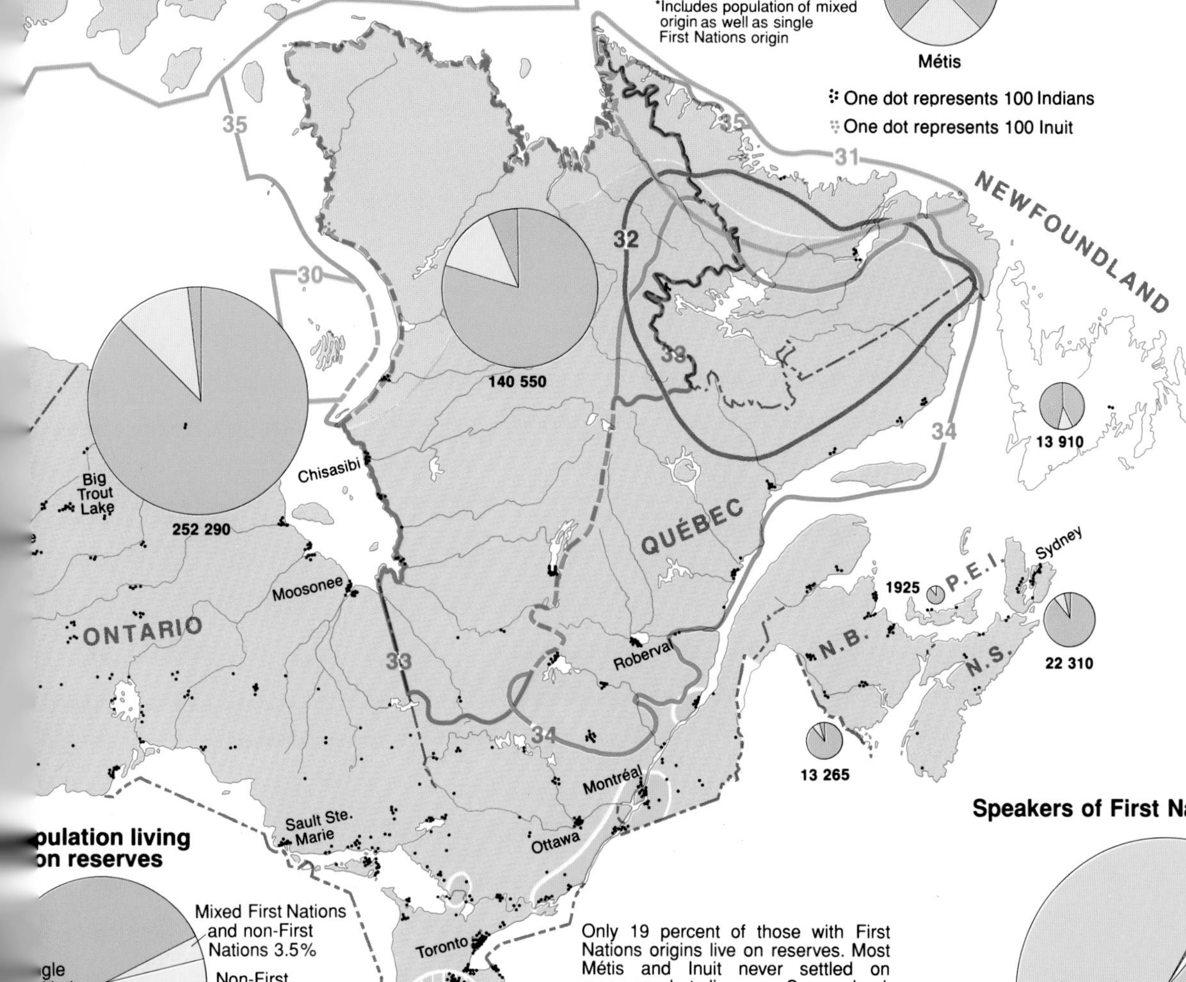

Language groups
(spellings may vary)

- **Algonquian** — Ojibwa, Blackfoot, Cree, Delaware, Abenaki, Malecite, Micmac, Montagnais-Naskapi, Potawatomi
- **Athapaskan** — Beaver, Carrier, Chilcotin, Chipewyan, Dogrib, Han. Hare, Kaska, Kutchin, Sarcee, Sekani, Dene-Thah, Tagish, Tahltan, Tutchone
- **Haidan** — Haida
- **Iroquoian** — Mohawk, Oneida, Onandaga, Cayuga, Seneča, Tuscarora
- **Kutenaian** — Kutenai
- **Salishan** — Bella Coola, Comox, Halkomelem, Lil'wat, Okanagan, Sechelt, Shuswap, Squamish, Straits, Thompson
- **Siouan** — Dakota
- **Tlingit** — Inland Tlingit
- **Tsimshian** — Coast Tsimshian, Nass-Gitksan, Southern Tsimshian
- **Wakashan** — Haisla, Heiltsuk, Kwakwala Ninitat, Nootka
- **Inuktitut** — Inuktitut

Population living on reserves

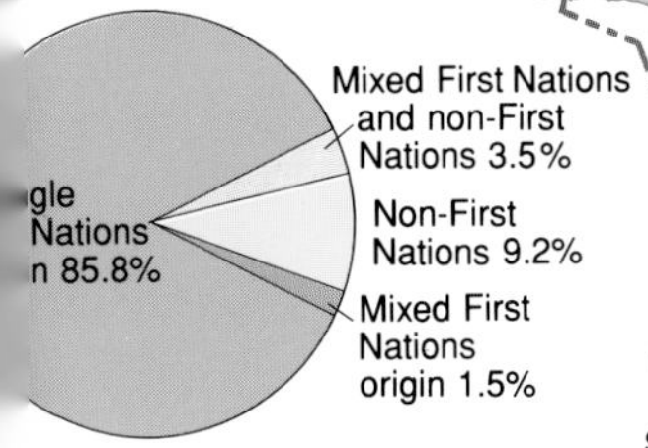

Only 19 percent of those with First Nations origins live on reserves. Most Métis and Inuit never settled on reserves, but live on Crown land. Today, approximately 50 percent of First Nations peoples live in cities and towns. Some reserves are becoming increasingly crowded due to high birth rates. Many young people are moving to cities where there are more employment and educational opportunities.

Speakers of First Nations languages

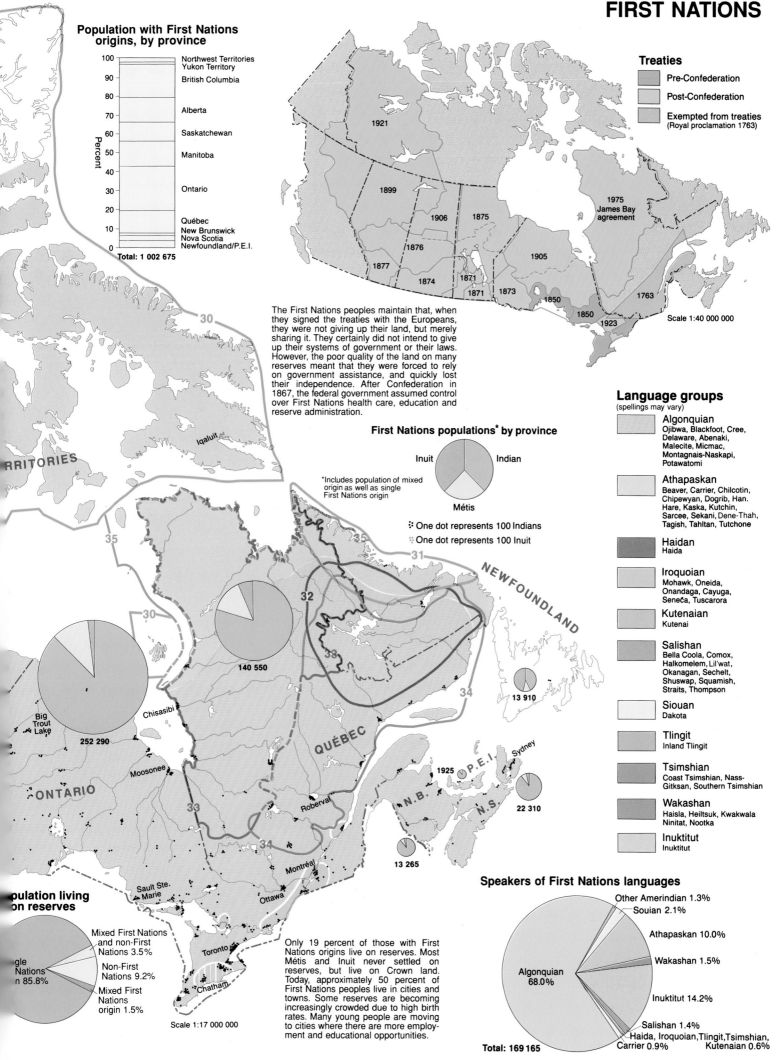

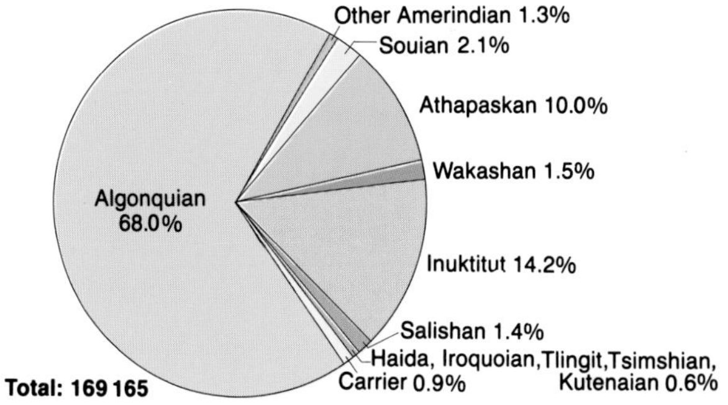

For explanation of terms, see glossary.

CANADA-THEMATIC

Canada is called a "cultural mosaic." It has many cultures and has two official languages. Customs and traditions from many countries enrich Canada's cultural environment.

According to the latest research, the ancestors of the North American Indians and Inuit may have arrived approximately 20 000 years ago. Some authorities suggest even earlier dates. Europeans began to settle in Canada in the early sixteenth century. Different ethnic groups often lived in clusters. They have given certain regions of Canada distinct cultural characteristics.

Today, the historical pattern of cultural groups is changing. People move within provinces and to different parts of the country, often because of job availability. However, Canadians have preserved many of their cultural differences.

Annual immigration

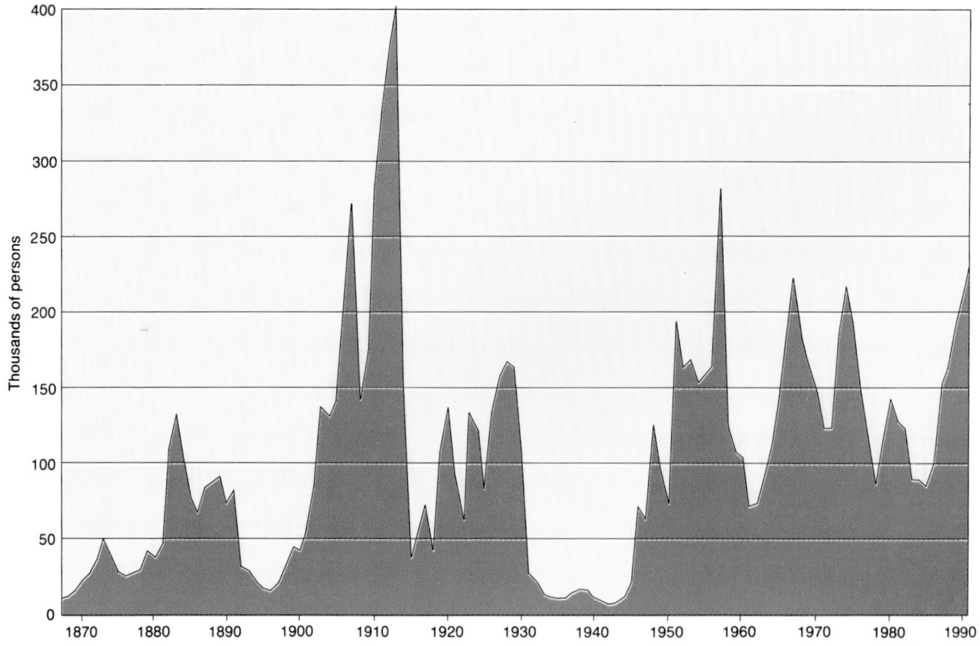

Source of immigrants, 1871

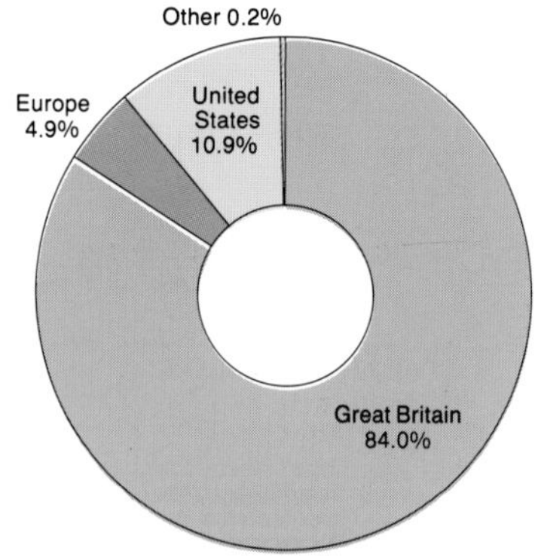

CANADA 594 027

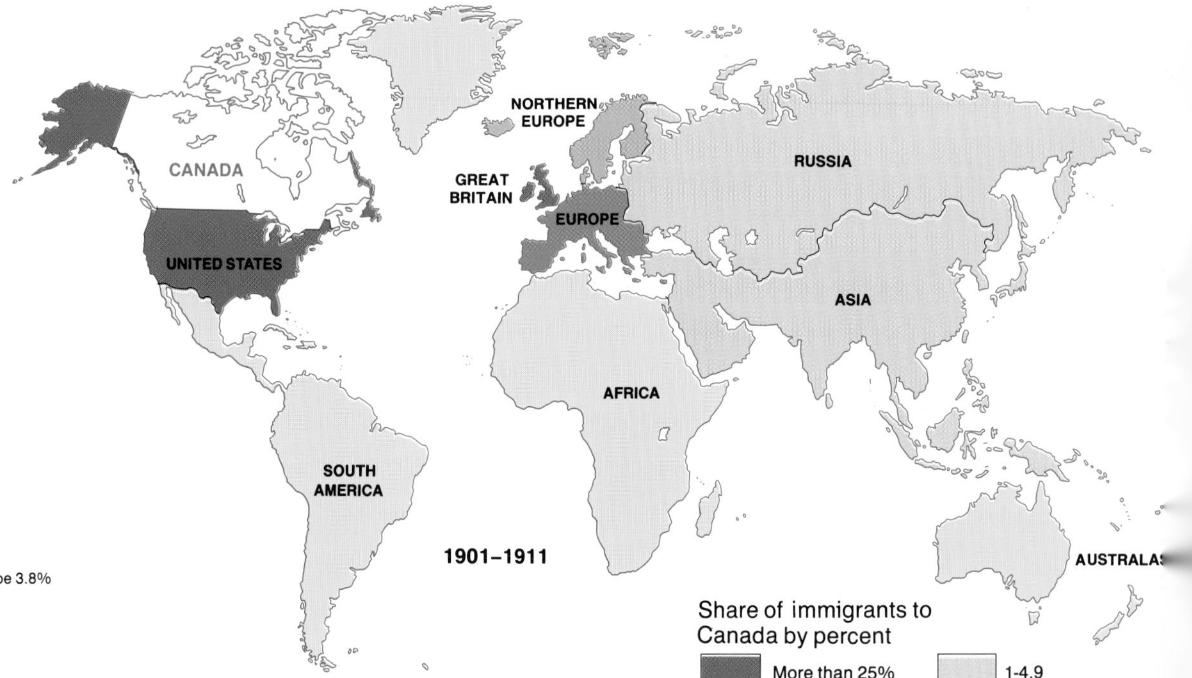

Source of immigrants, 1901–1911

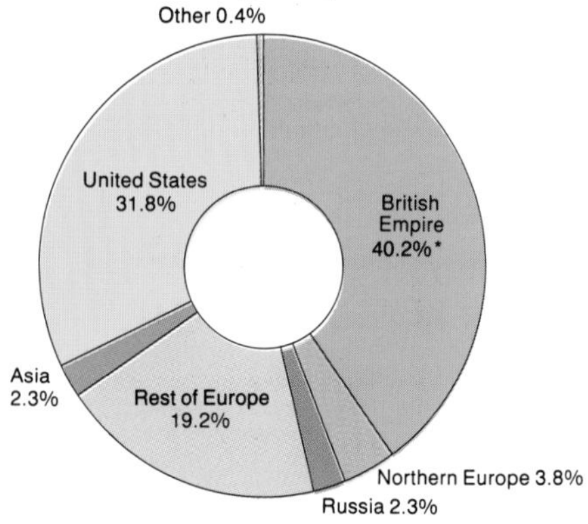

CANADA 1 681 041

*The majority came from the United Kingdom.

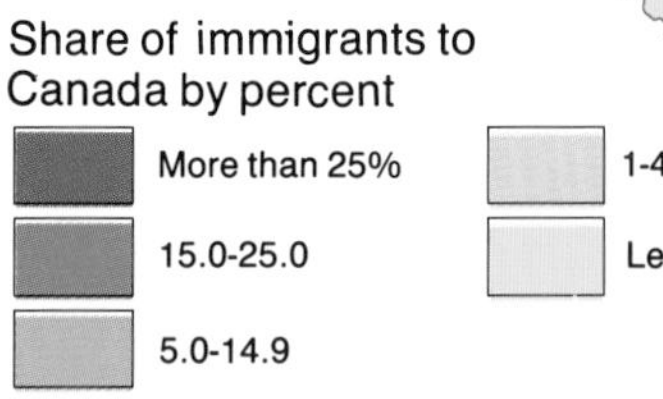

Immigration to Canada by region

Source of immigrants, 1981–1991

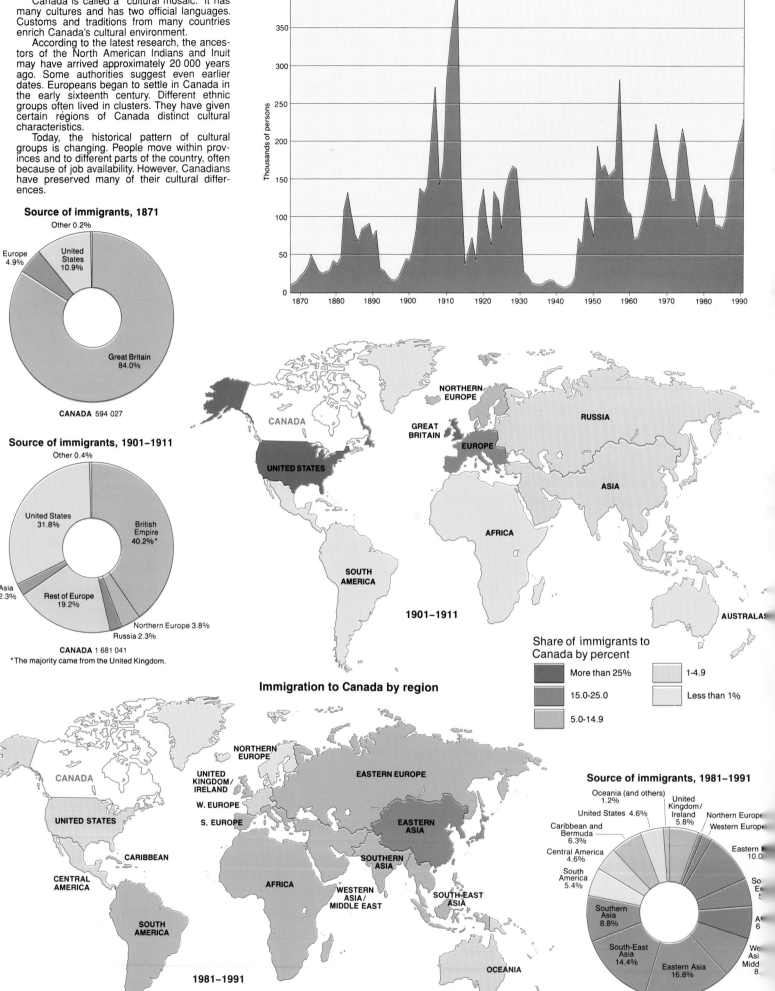

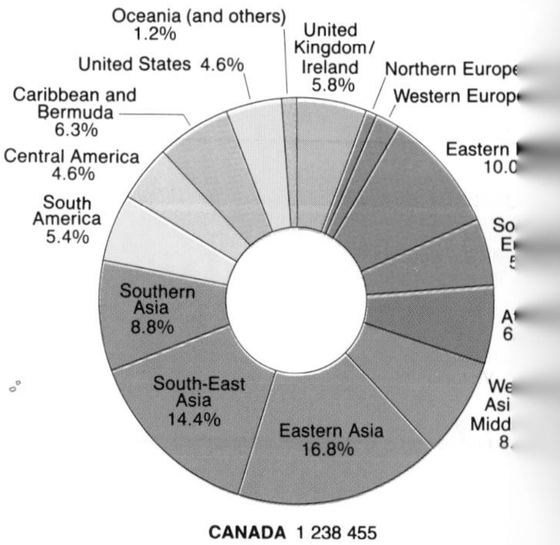

CANADA 1 238 455

Migration within Canada, 1992
5342
5008
3125
2051
3441
1867
1730
2070
10 949
30 328
11 528
BRITISH COLUMBIA
ALBERTA
17 741
12 978
1334
4816
QUÉBEC
4344
ATLANTIC PROVINCES
18 945
SASK. AND MAN.
27 596
6477
8882
17 177
13 284
ONTARIO
17 432
7927
19 706
13 292
1253
24 280
2761
5597
Ethnic composition of Canada
Scandinavian 0.9%
Portuguese 1.1%
East Indian 1.2%
Jewish 1.4%
Dutch 1.9%
Chinese 2.0%
Italian 3.9%
Native 2.1%
Others 11.7%
British 35.0%
French 33.8%
German 5.0%
First languages in Canada
YUKON TERRITORY
NORTHWEST TERRITORIES
ENGLISH 54.2%
INUKTITUT 28.7%
FRENCH 2.4%
BRITISH COLUMBIA
ALBERTA
SASKATCHEWAN
MANITOBA
ONTARIO
QUÉBEC
NEWFOUNDLAND
P.E.I.
N.B.
N.S.
FRENCH 81.2%
ENGLISH 8.7%
ITALIAN 1.7%
ENGLISH 98.4%
FRENCH 0.4%
ENGLISH 93.8%
FRENCH 4.2%
ENGLISH 93.1%
FRENCH 3.4%
ENGLISH 64.0%
FRENCH 32.7%
ENGLISH 83.0%
GERMAN 3.7%
UKRAINIAN 2.4%
CREE 2.0%
FRENCH 1.9%
ENGLISH 73.0%
GERMAN 5.8%
FRENCH 4.2%
UKRAINIAN 2.9%
CREE 1.9%
ENGLISH 75.1%
FRENCH 4.6%
ITALIAN 2.8%
CHINESE 2.0%
Scale 1:35 000 000
Official languages
More than 90% English
70-90% English
50-69% English
30-49% French
More than 80% French
Destinations of immigrants by province
More than 15%
10–15
5–9.9
1–4.9
Edmonton
Vancouver
Regina
Victoria
Calgary
Winnipeg
Thunder Bay
Sudbury
Montréal
Québec
Ottawa
Toronto
Saint John
Halifax
St. John's
Immigrants as a proportion of population in major cities
Origin of immigrants
All other countries
Caribbean
Asia
United States
Other Europe
U.K. and Ireland
Percent
100
80
60
40
20
0
1971
1973
1975
1977
1979
1981
1983
1985
1987
1989
1991

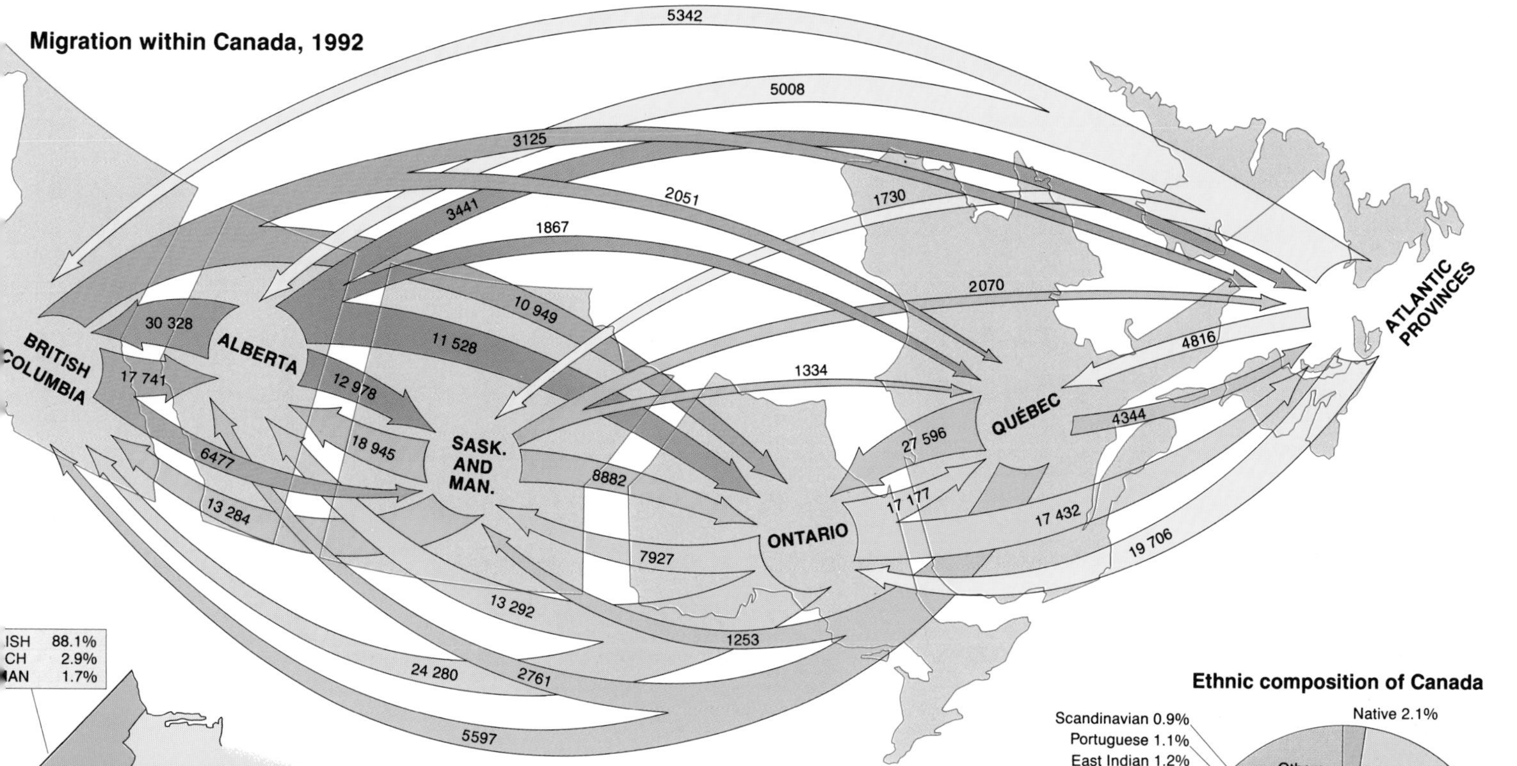

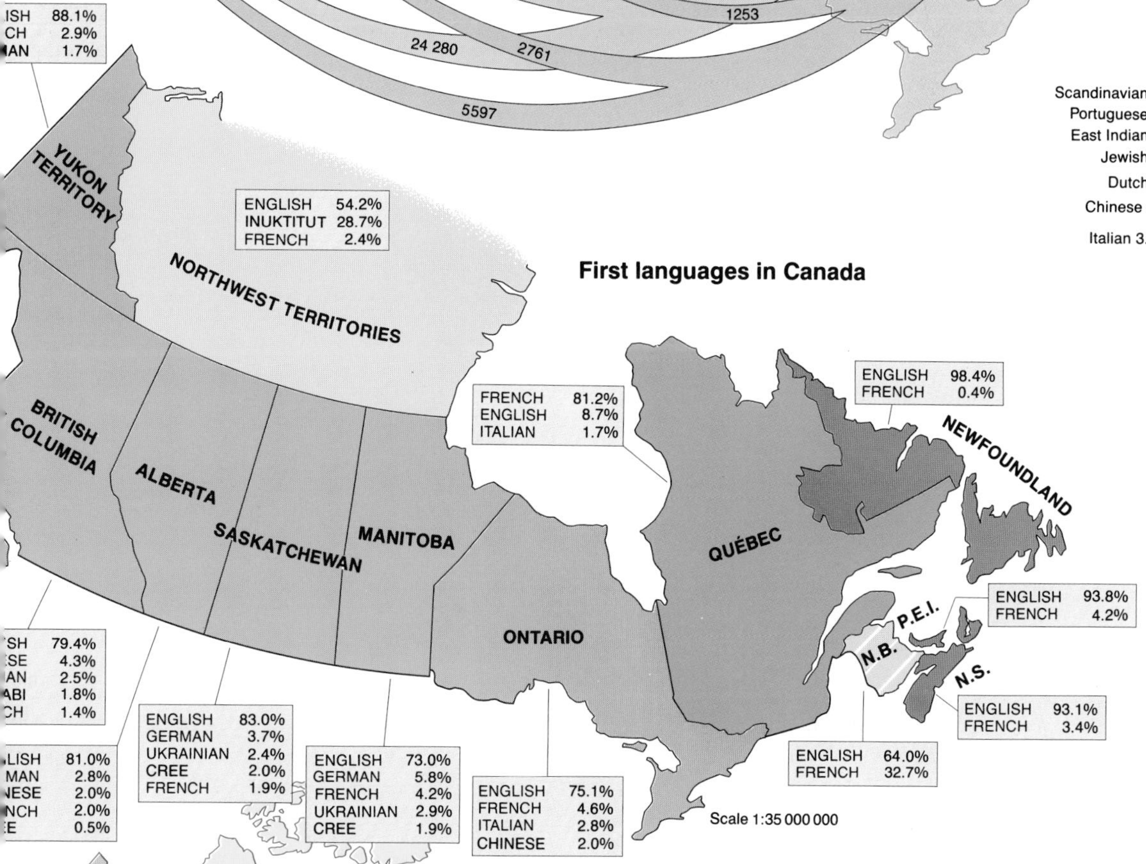

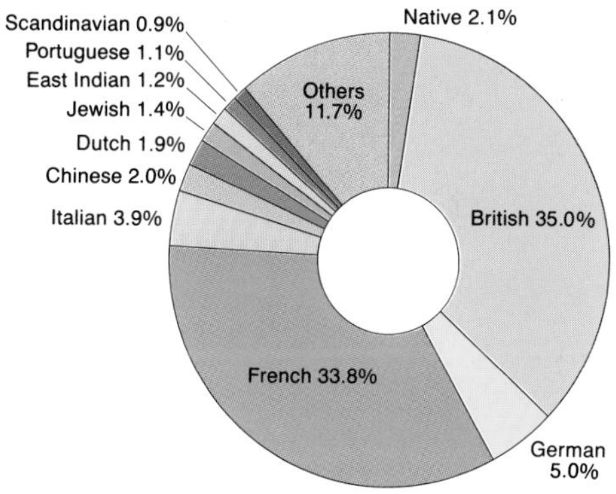

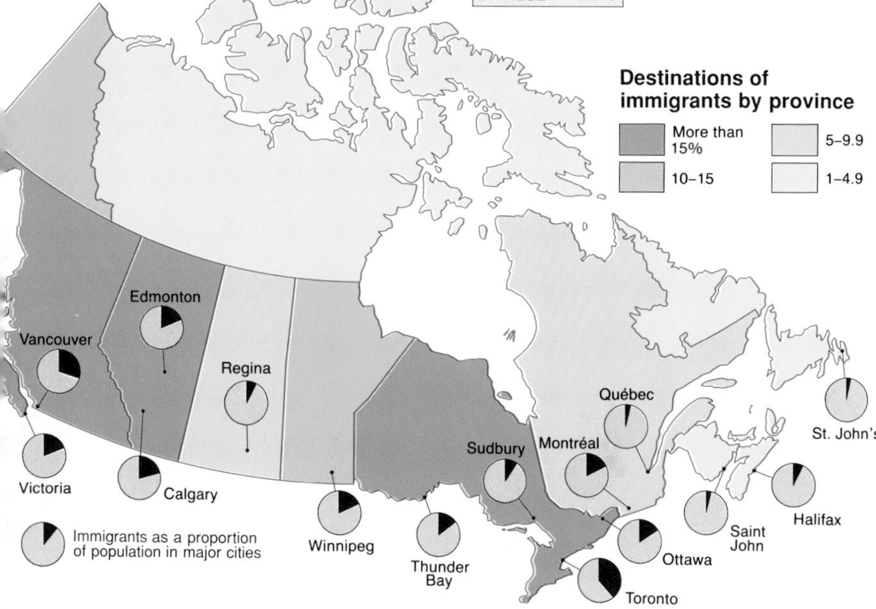

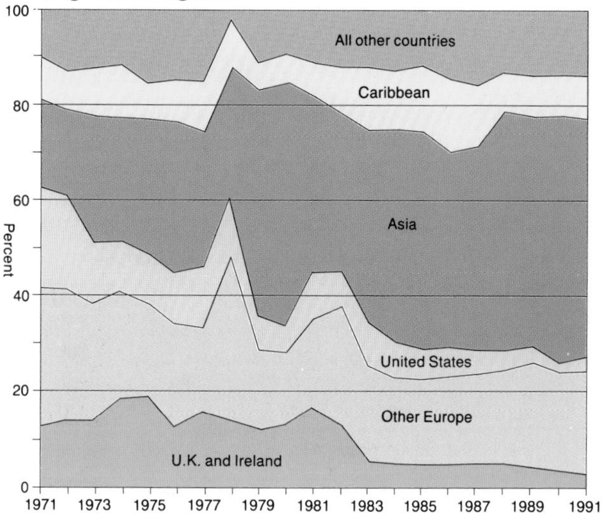

CANADA-THEMATIC

Cross section from Prince Rupert to Mittimatalik

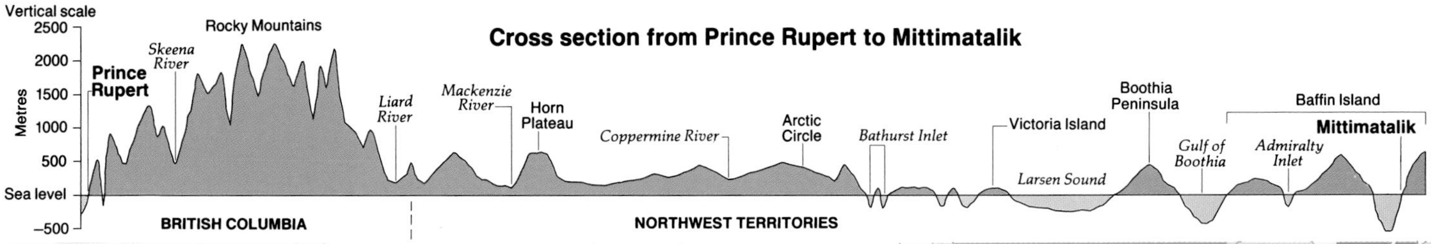

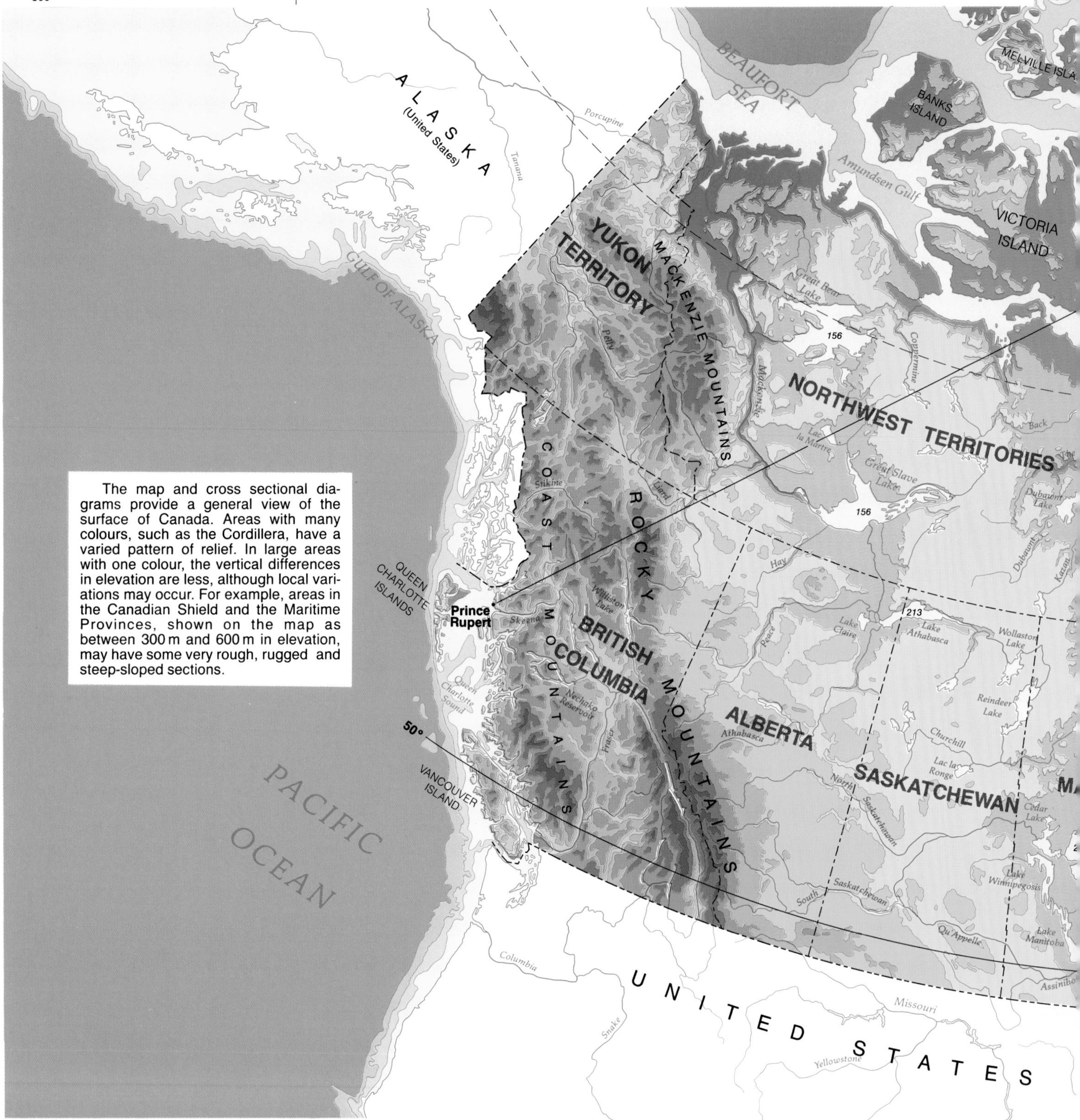

The map and cross sectional diagrams provide a general view of the surface of Canada. Areas with many colours, such as the Cordillera, have a varied pattern of relief. In large areas with one colour, the vertical differences in elevation are less, although local variations may occur. For example, areas in the Canadian Shield and the Maritime Provinces, shown on the map as between 300 m and 600 m in elevation, may have some very rough, rugged and steep-sloped sections.

The distance across Canada from east to west is more than 5000 km. The difference between the highest and lowest points in Canada is approximately 6000 metres. The vertical scale has been exaggerated greatly to emphasize these small vertical differences.

Cross section along 50° North Lati

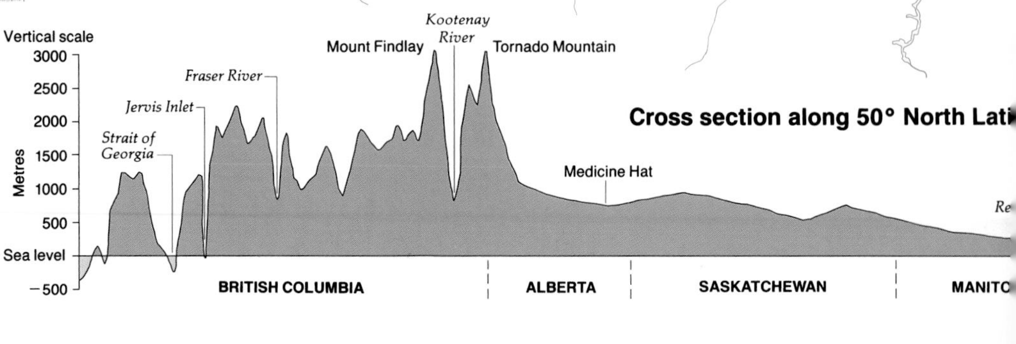

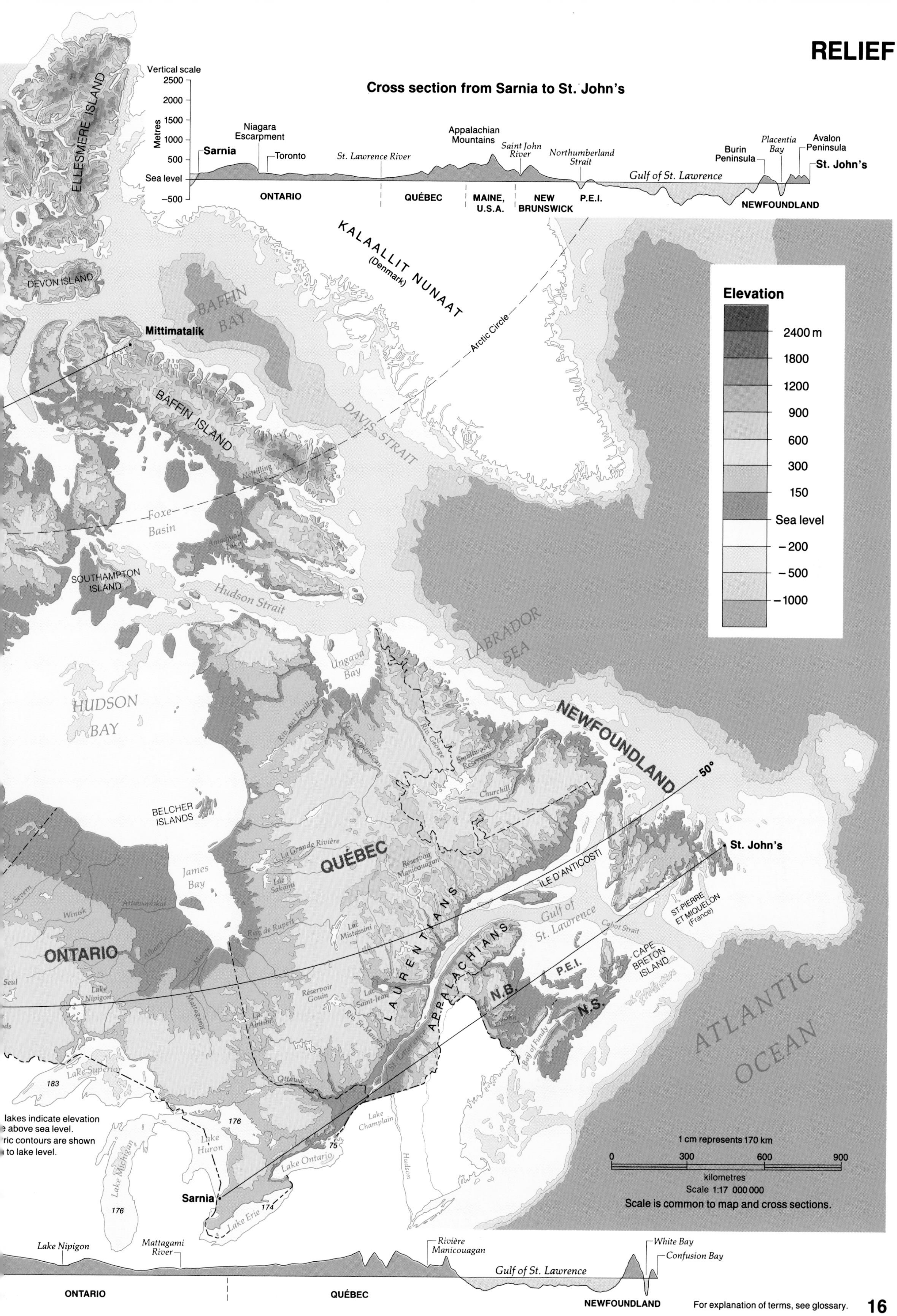
Cross section from Sarnia to St. John's
Vertical scale
2500
2000
1500
1000
500
Sea level
−500
Metres
Sarnia
Niagara Escarpment
Toronto
St. Lawrence River
Appalachian Mountains
Saint John River
Northumberland Strait
Gulf of St. Lawrence
Burin Peninsula
Placentia Bay
Avalon Peninsula
St. John's
ONTARIO
QUÉBEC
MAINE, U.S.A.
NEW BRUNSWICK
P.E.I.
NEWFOUNDLAND
Elevation
2400 m
1800
1200
900
600
300
150
Sea level
−200
−500
−1000
ELLESMERE ISLAND
DEVON ISLAND
KALAALLIT NUNAAT
(Denmark)
BAFFIN BAY
Mittimatalik
Arctic Circle
BAFFIN ISLAND
DAVIS STRAIT
Nettilling Lake
Foxe Basin
Amadjuak Lake
SOUTHAMPTON ISLAND
Hudson Strait
LABRADOR SEA
Ungava Bay
HUDSON BAY
NEWFOUNDLAND
50°
Rivière aux Feuilles
Rivière Caniapiscau
Rivière George
Smallwood Reservoir
Churchill
BELCHER ISLANDS
La Grande Rivière
QUÉBEC
Réservoir Manicouagan
St. John's
ÎLE D'ANTICOSTI
James Bay
Lac Sakami
Severn
Winisk
Attawapiskat
Rivière de Rupert
Lac Mistassini
LAURENTIANS
Gulf of St. Lawrence
Cabot Strait
ST-PIERRE ET MIQUELON (France)
ONTARIO
Albany
Moose
APPALACHIANS
P.E.I.
CAPE BRETON ISLAND
Seul
Lake Nipigon
Réservoir Gouin
Lac Saint-Jean
N.B.
N.S.
Mattagami
Lac Abitibi
Rivière St-Maurice
St. Lawrence
Saint John
Bay of Fundy
ATLANTIC OCEAN
Lake Superior
183
Ottawa
lakes indicate elevation
above sea level.
ric contours are shown
to lake level.
176
Lake Huron
Lake Champlain
75
Lake Ontario
Lake Michigan
Hudson
176
Sarnia
174
Lake Erie
1 cm represents 170 km
0
300
600
900
kilometres
Scale 1:17 000 000
Scale is common to map and cross sections.
Lake Nipigon
Mattagami River
Rivière Manicouagan
Gulf of St. Lawrence
White Bay
Confusion Bay
ONTARIO
QUÉBEC
NEWFOUNDLAND

CANADA-THEMATIC

Retreat of the Wisconsin ice sheet

Extent of the ice sheet, in years before present

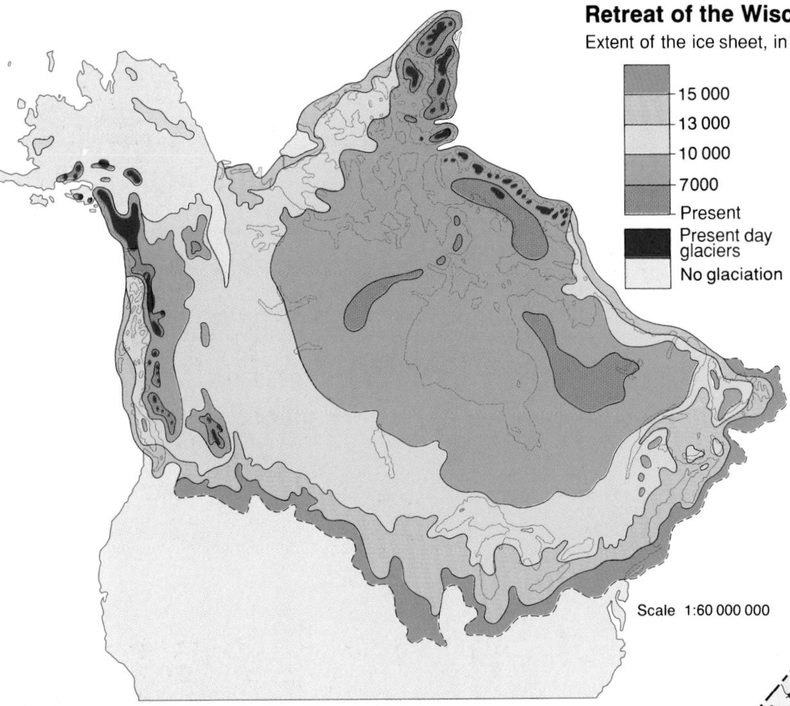

Over the past two million years, four distinct continental ice sheets covered much of Canada. They carried and later dumped unsorted glacial materials. The latest ice sheet, the Wisconsin, began to retreat about 15 000 years ago. Glacial meltwaters flooded low-lying areas in Ontario and southern Manitoba, reworking the glacial materials. Rocks were rounded and sorted by size and weight; glacial materials filled the valleys, smoothing the landscape. When the glaciers receded and the glacial meltwaters dropped to their present levels, these materials formed the basis for excellent soils.

INNUITIAN REGI

This region contains a variety of phy features, including glacier-clad moun and broad, flat plains. The arctic loc unites these different land forms. Fea such as permafrost, pack ice, glaciers rock deserts cover much of the A Islands. In future, this region may yiel gas and other minerals.

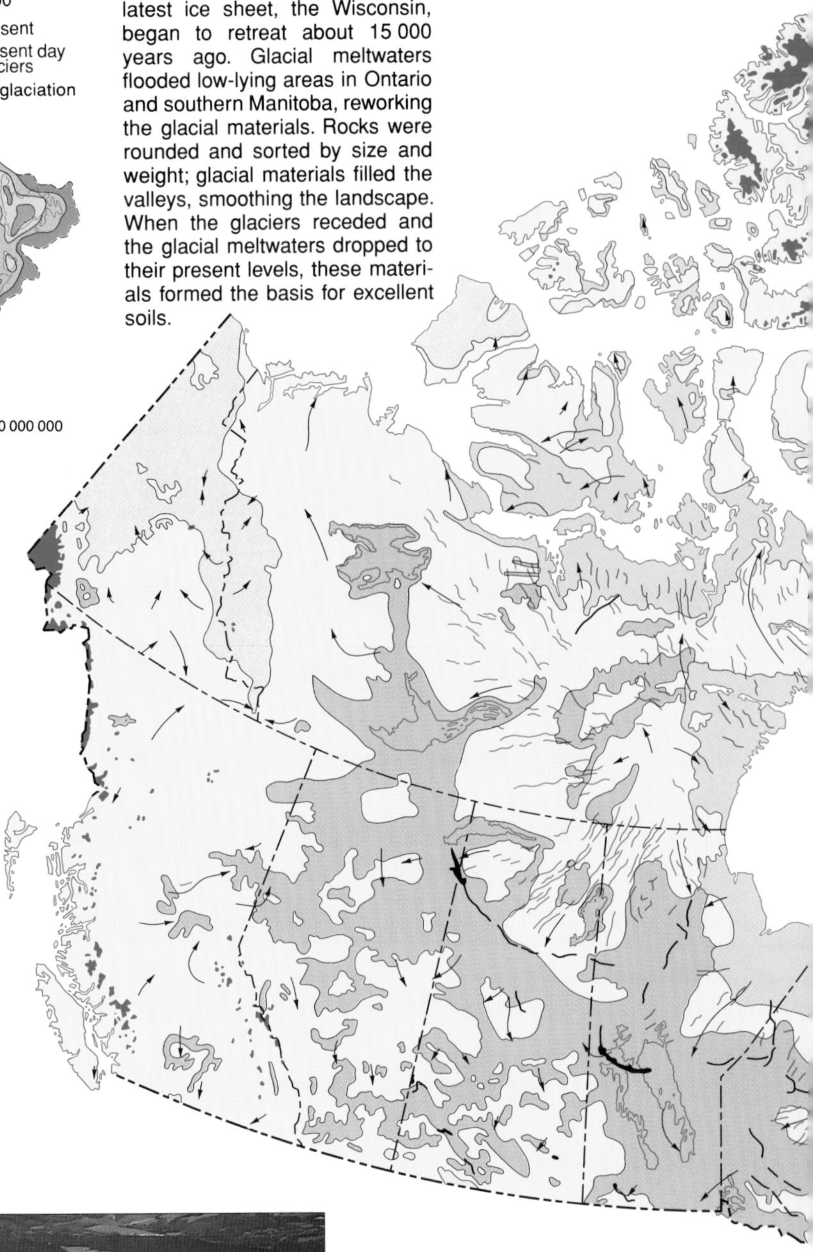

CORDILLERAN REGION

Mountain ranges which run parallel to the coast dominate this region. Between the Coast Mountains and the Rocky Mountains lies a series of plateaus, dissected by deep river valleys. Because of the rugged topography, people tend to live in the valley areas, particularly near the mouth of the Fraser River.

INTERIOR PLAINS

This region increases in elevation from east to west. Three prairie levels are separated by escarpments. The Manitoba lowland, once covered by a glacial lake, is very flat. The glaciated Saskatchewan plain is more rolling. The third prairie level, in southern Alberta, has flat plains, dry badlands and gentle hills. This region is one of Canada's most important agricultural areas.

THE CANADIAN SHIELD

Rugged, rocky, mountainous, picturesque – all of these terms describe the physical features of a region that covers almost half of Canada's land area. This region is rich in resources: forests for the pulp and paper industry; nickel, iron and copper for the mining industry; and water for hydro-electric power.

ST. LAWRENCE LOWLA

Sedimentary rock underlies this h populated region. Glaciers and thei waters have made the surface of th flat to rolling; transportation is eas farming is good. The Niagara Escar in southern Ontario and the Monte Hills in southern Québec are disti features of the landscape.

Physiography is the study of the earth's surface. It describes the major features of the earth that are influenced by bedrock and the landforms produced by erosion.

Flat-lying sedimentary rocks form plains, and ranges of folded mountains make up the Western Cordillera. Glaciers have greatly affected the surface of many areas. Glacial erosion is most significant in the Canadian Shield where the land was scraped bare and areas were gouged out, later forming lakes. Farther south, moraines, drumlins and till-covered plains were created by glacial deposition. Beaches along the edges of post-glacial lakes can also be found.

People can alter the physiography of an area. For example, the Holland Marsh in southern Ontario has been drained for agricultural uses. Hills have been levelled for urban development and strip mining purposes. Even some of the rivers, such as the Nechako in British Columbia, have been diverted for specific purposes.

ARCTIC LOWLANDS

This is a land of rock deserts, permafrost, mosses and lichens. The region hosts many migrating birds and large concentrations of wildlife: seals, whales, caribou, polar bears and musk oxen. Few people live in this area today; the potential for intensive use of the land is limited because of the fragile environment.

APPALACHIAN REGION

While these uplands are mostly forest covered, the broad Saint John River valley, the Annapolis Valley and the flat-lying sandstone region of Prince Edward Island have agricultural potential. The varied landscape, the ocean setting and the distinctive traditions of the people attract tourists to this region.

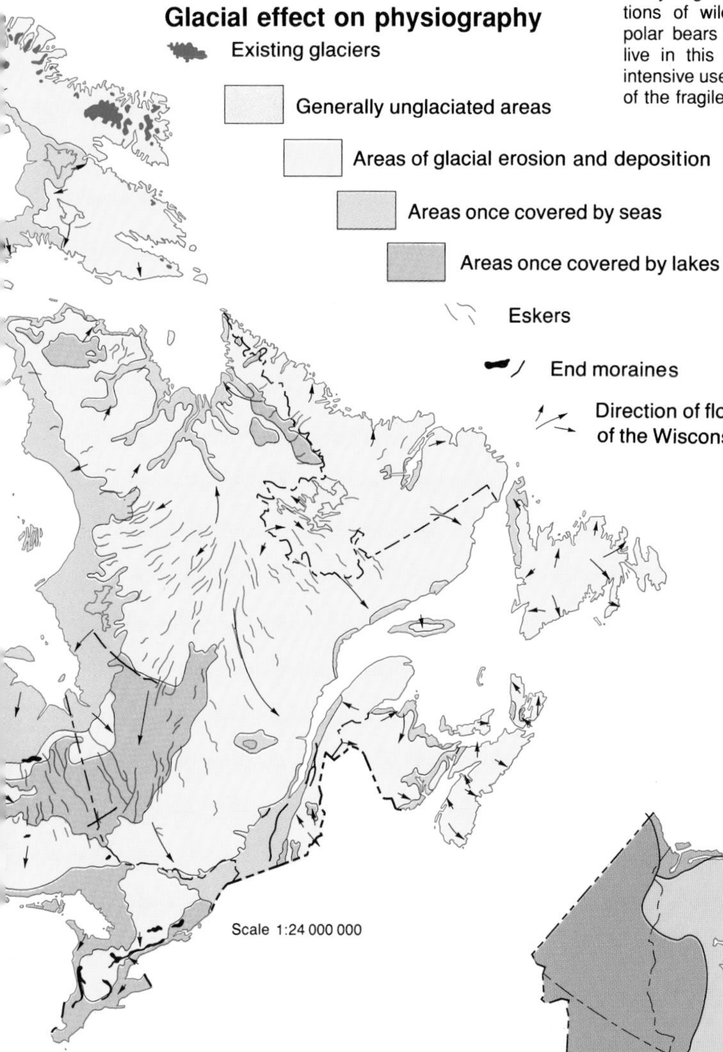

Physiographic regions

Canada can be divided into eight physiographic regions. These regions are based on the age and type of rock, the forces that have molded the rock and the surface features. The Canadian Shield is the oldest region. Shield rocks underlie the lowlands surrounding the Shield, but sedimentary rock and other materials have been deposited on top. On the outer edge of the lowlands are mountain ranges. The Appalachians, 400 million years old, are much older than the Western Cordillera and Innuitian ranges, which are less than 100 million years old.

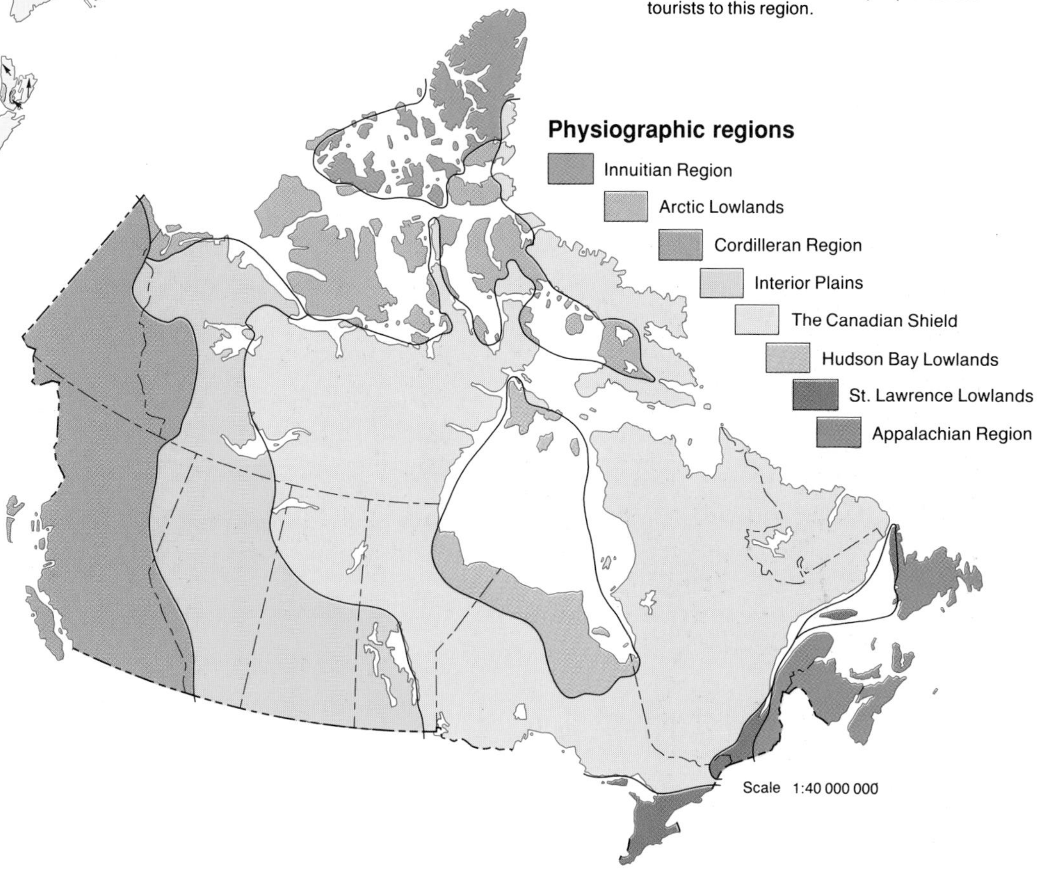

For explanation of terms, see glossary.

CANADA-THEMATIC

Geological time scale

The span of earth's history surrounds the map. Forms of life have changed greatly from the first single-celled organisms at the end of the Precambrian Era to the varied and complex living organisms of the last 4 000 000 000 years. Each break in the time scale represents 100 000 000 years. The scale emphasizes the most recent eras.

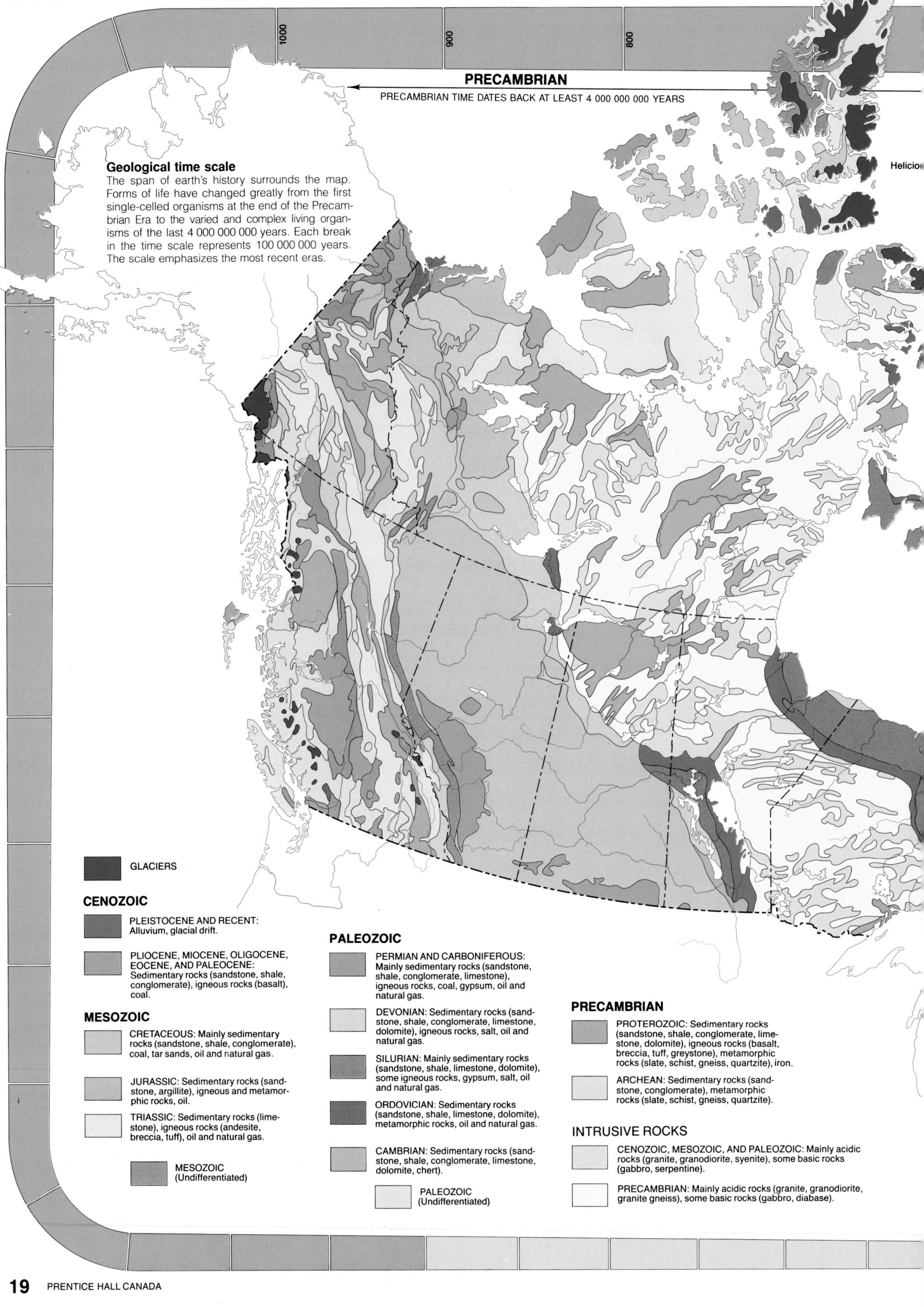

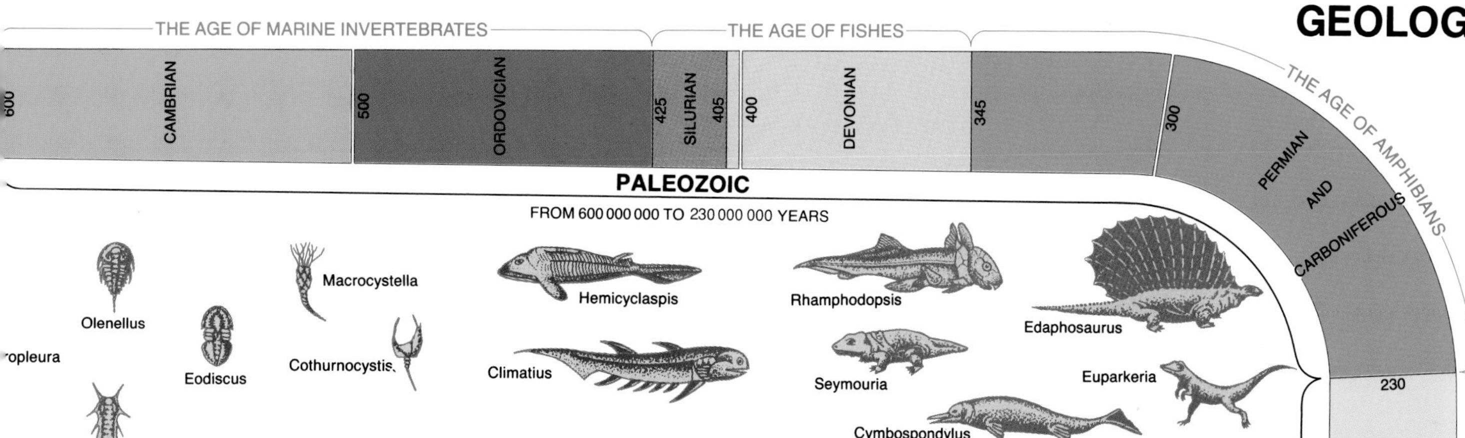

Geology is the science that deals with the physical nature and history of the earth. Geologists study the structure and development of the earth's crust, individual rock types and fossils.

Today, it is thought that the earth was created between 4 000 000 000 and 5 000 000 000 years ago. Most of the continental crusts were created from cooled magma at least 2 500 000 000 years ago. Erosion has worn down these mainly igneous rocks but there has also been some uplift as well. Sediments produced by this erosion were solidified by pressure and heat into sedimentary rocks. With more heat and pressure, some of these in turn have been changed (metamorphosed) into metamorphic rock.

Many rocks contain the remains, or the impression of the remains, of plants and animals. These fossils help scientists to estimate the age of rocks. The simplest forms of life — worms, jellyfish, and corals — date back over 600 000 000 years to the Precambrian Era. Great changes took place in the Paleozoic Era. Over a period of nearly 400 000 000 years, land animals, insects, land plants and forests appeared.

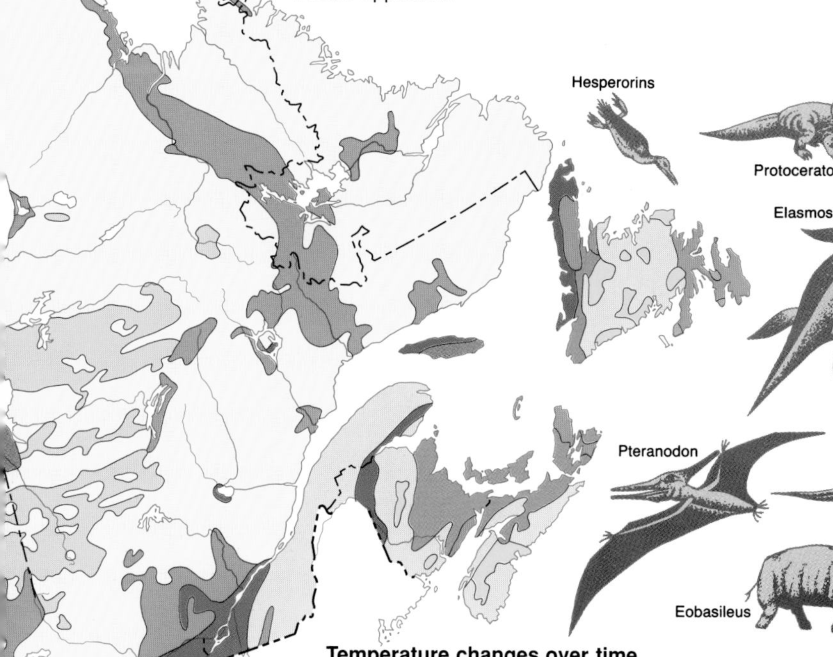

Scale 1:19 000 000

Temperature changes over time

Weather changes on a daily basis. Climatic changes are slower and less evident. However, small fluctuations in temperature can produce severe environmental changes for life on earth. If the atmosphere cooled by five Celsius degrees, the ice ages could return. If the temperature rose by a similar amount, many low-lying areas would be flooded as the polar ice caps melted.

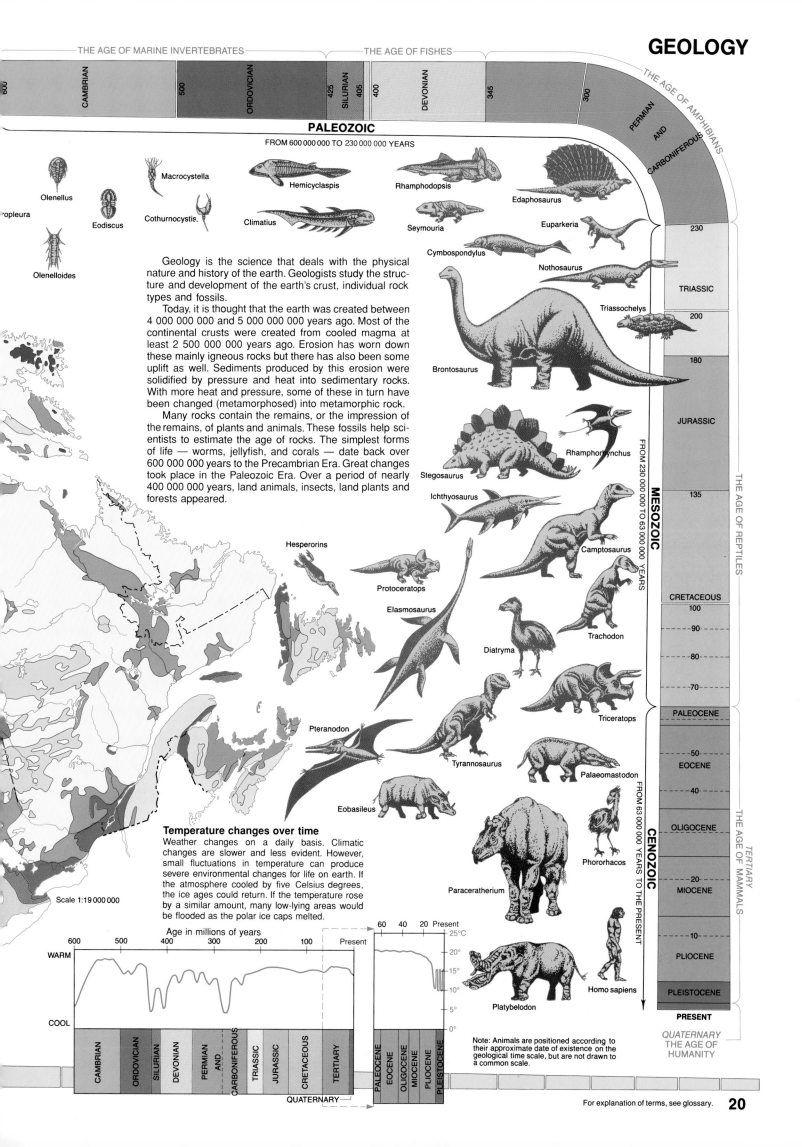

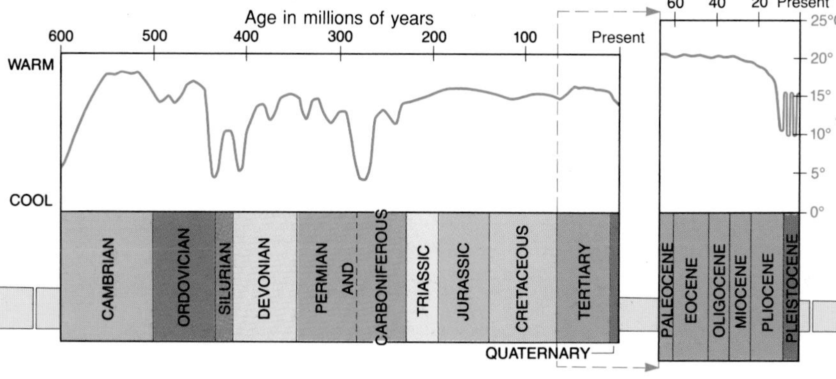

Note: Animals are positioned according to their approximate date of existence on the geological time scale, but are not drawn to a common scale.

For explanation of terms, see glossary.

CANADA-THEMATIC

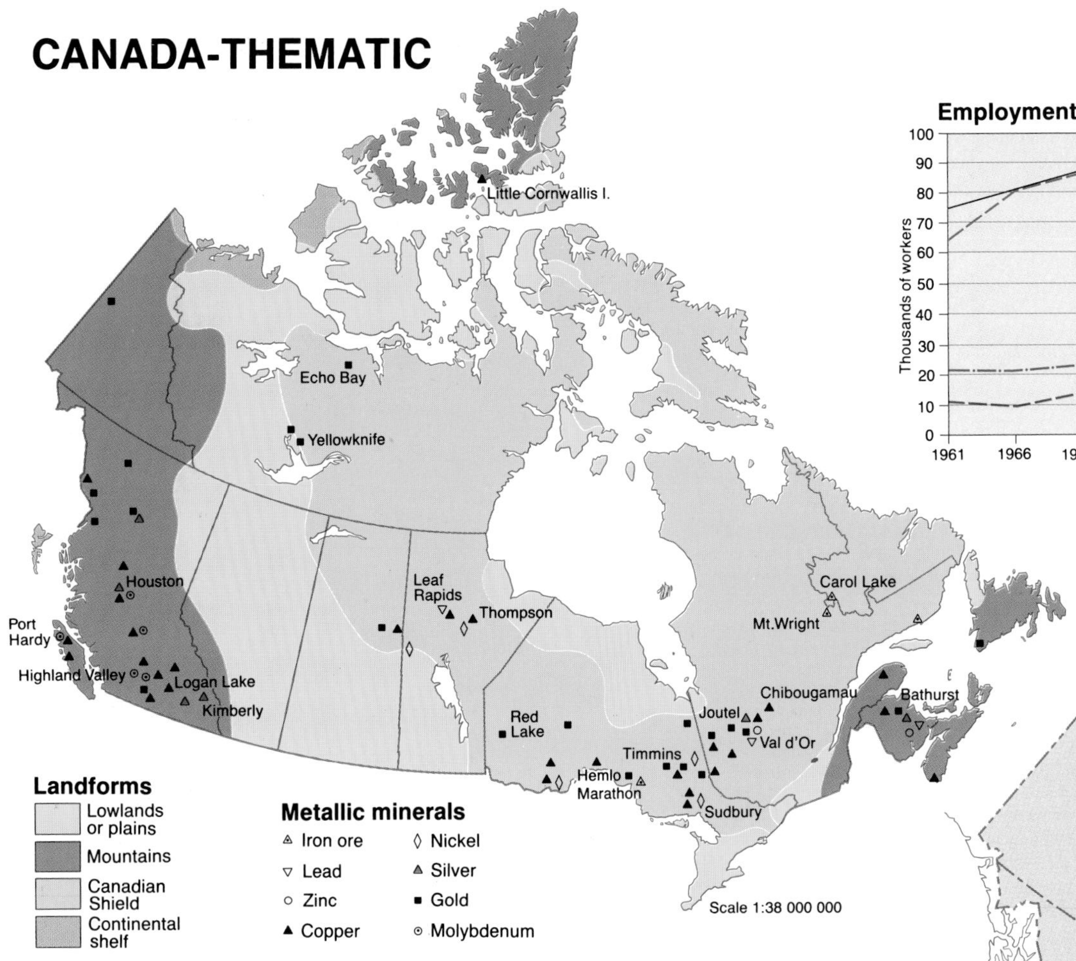

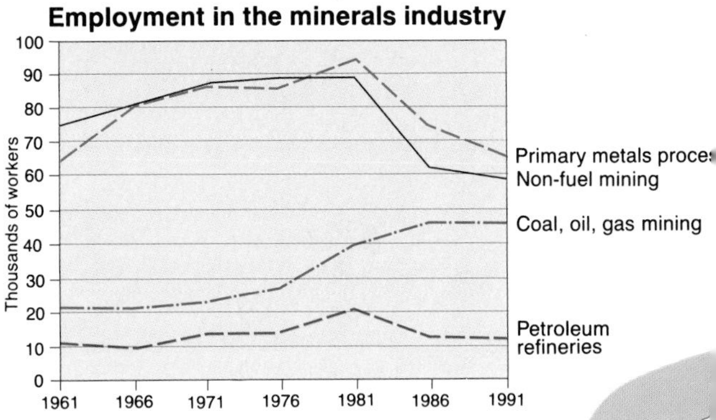

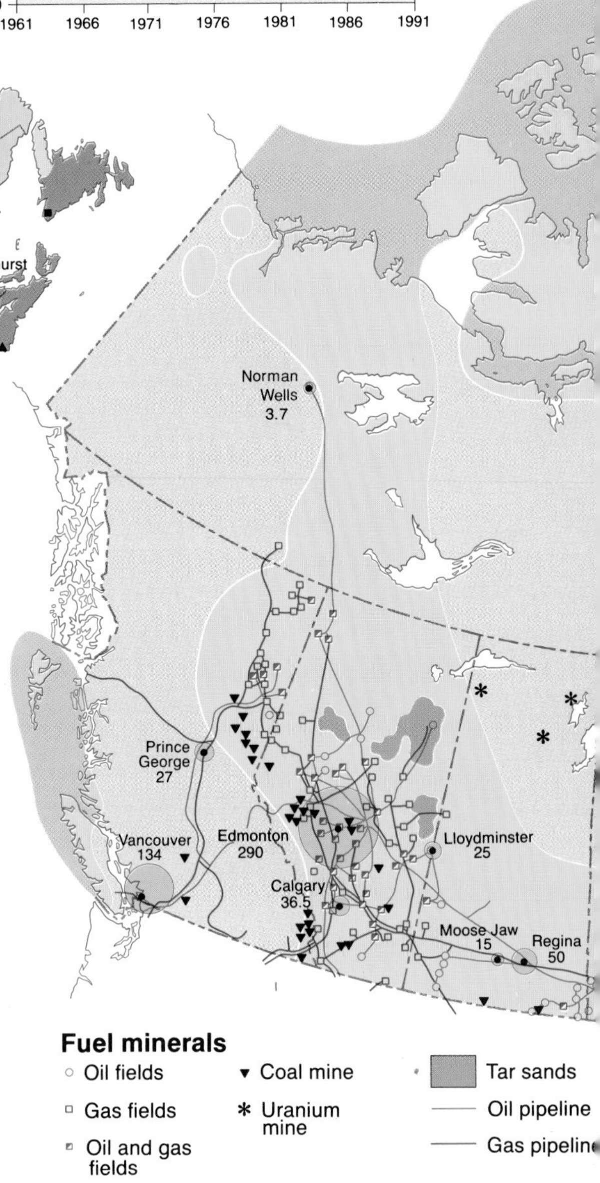

For economic purposes, minerals can be classified into four groups: *metallic* minerals, which include copper, zinc, nickel, silver and gold; *industrial* or *non-metallic* minerals, such as potash, gypsum and salt; *structural* minerals, including sand and gravel; and *fuels*, such as natural gas, oil and coal.

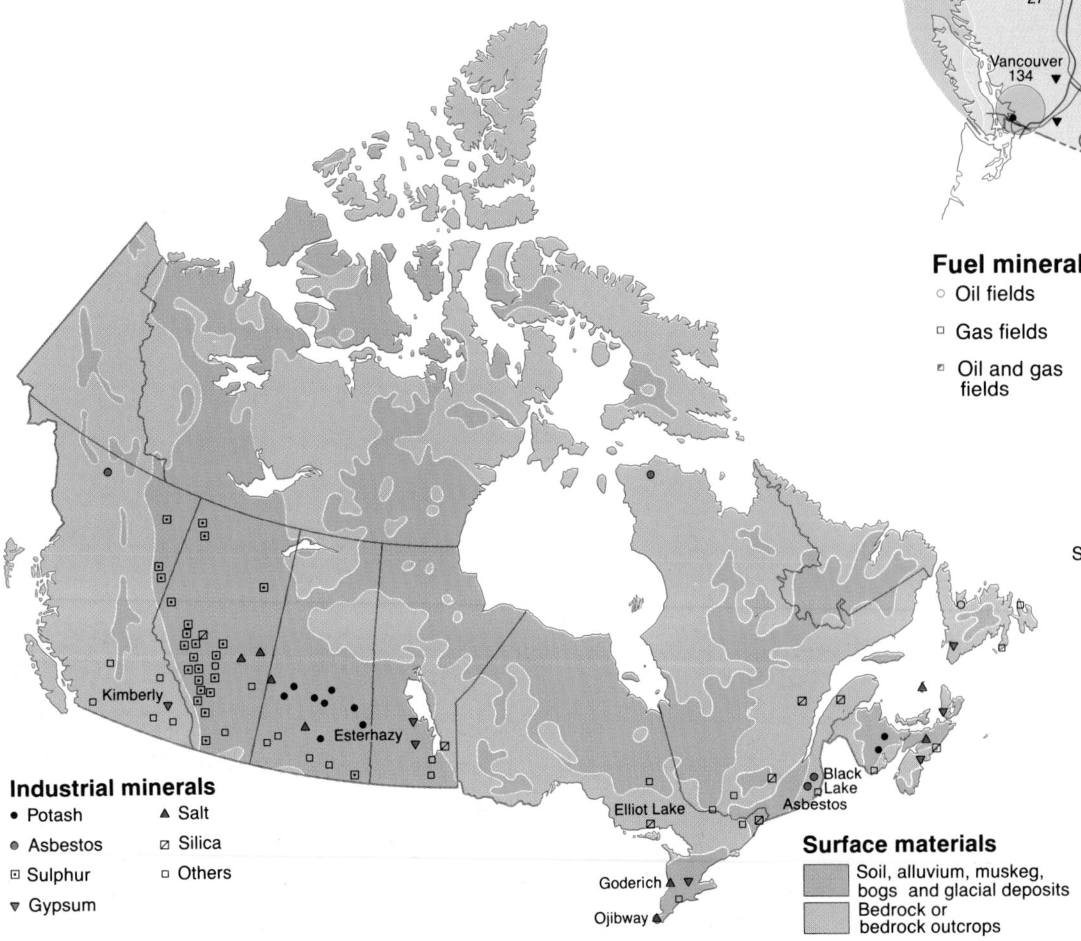

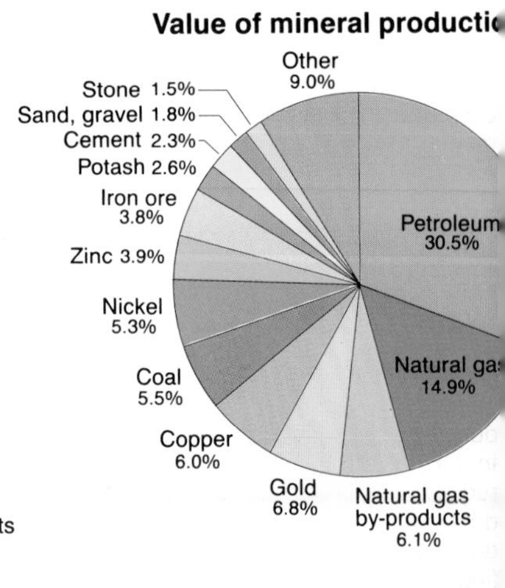

MINERAL RESOURCES

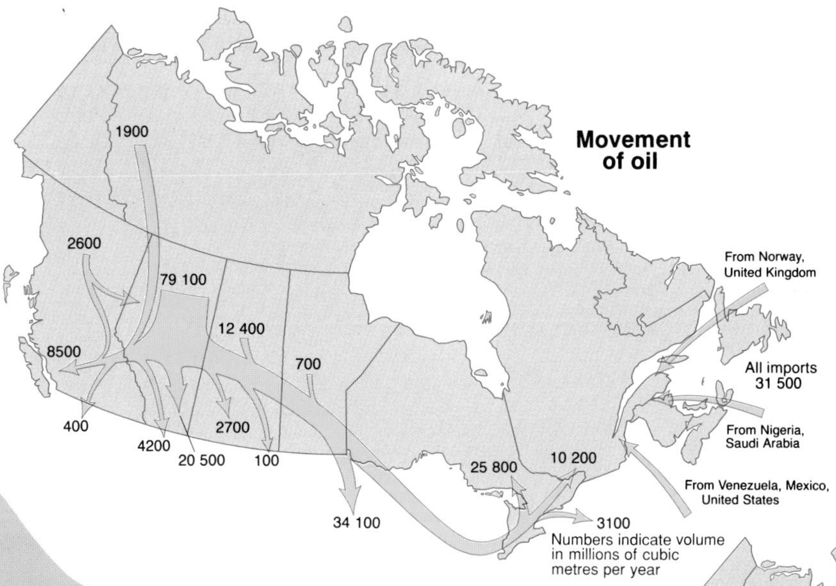

Value of exports*

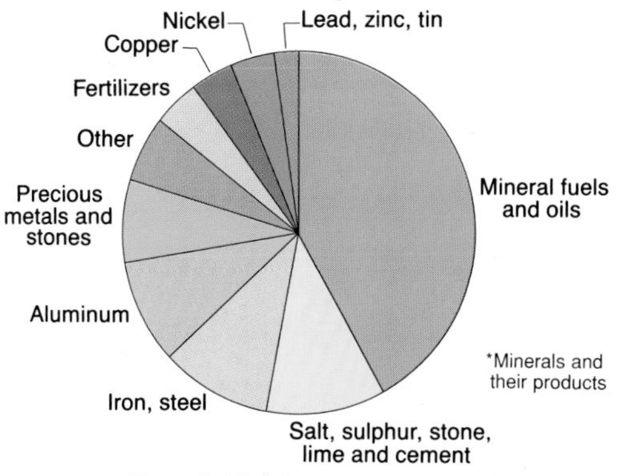

*Minerals and their products

Total: (1991) $37 093 300 000

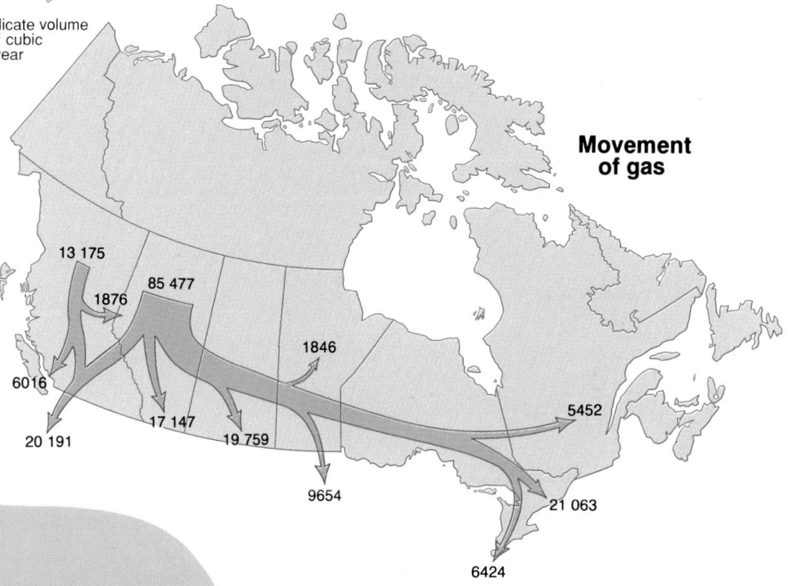

Canada exports about 80 percent of its mineral production. This makes the mining industry sensitive to both the economic health of its trading partners and fluctuations in international prices. Mine closings, with high job losses and the creation of "ghost towns," can often be traced to a fall in prices or reduced demand on the world market.

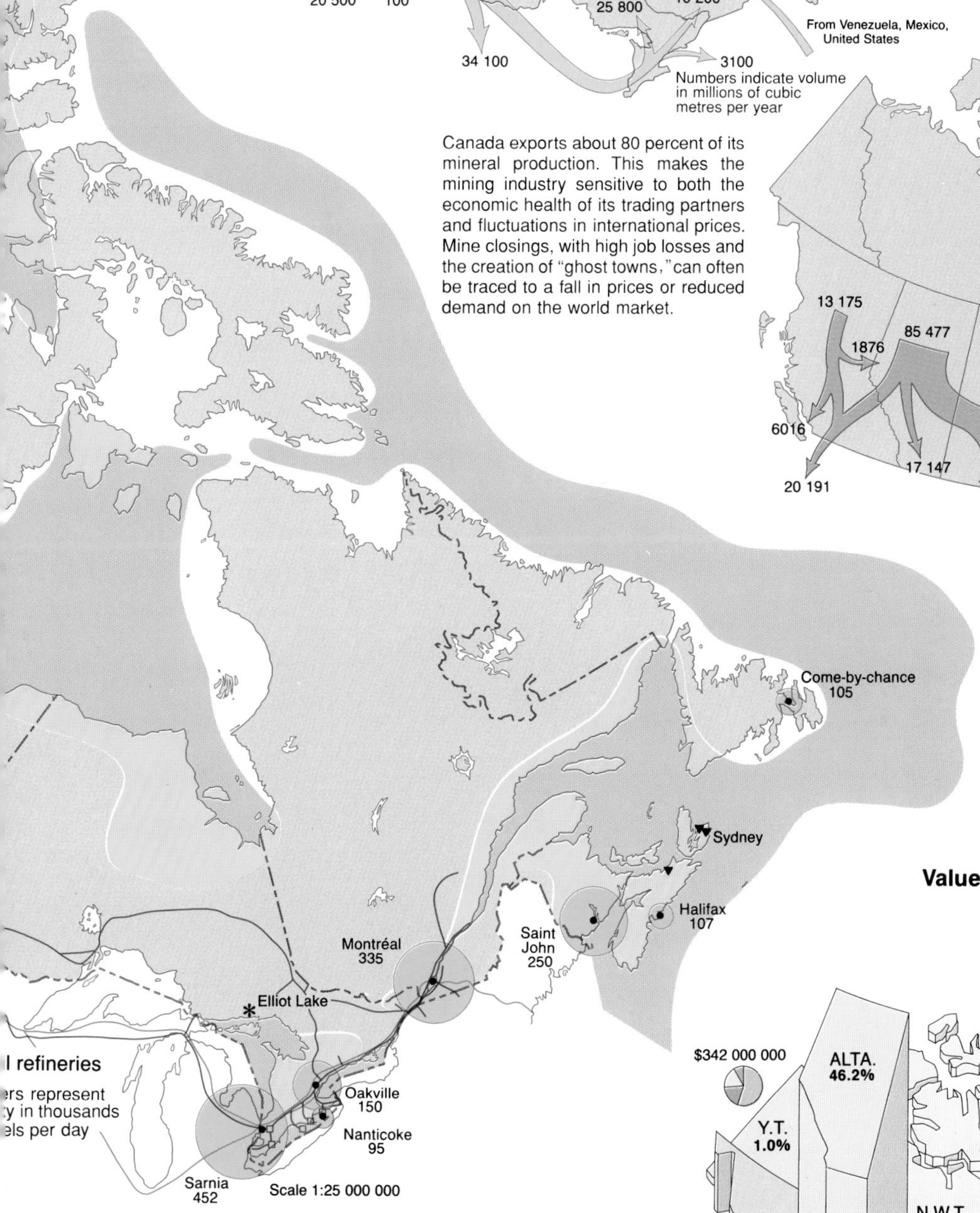

The mineral industry has been a major force in Canada's economic development. The discovery of minerals opened up northern areas and served as the economic base for many communities. The need to transport minerals to markets encouraged the development of major transportation routes such as the St. Lawrence Seaway and the Canadian Pacific Railroad. Exploration continues in many remote areas, especially in the Arctic.

Today, Canada mines over 60 types of minerals, more than any other nation. Canada is the largest producer of uranium and zinc and the second largest producer of nickel, cobalt, asbestos, potash and gypsum. In fact, Canada produces most of its own mineral requirements and has substantial reserves that await development. The growth of the mining industry depends on demand for minerals, world prices and finding enough capital to open new mines.

Value and provincial share of mineral production

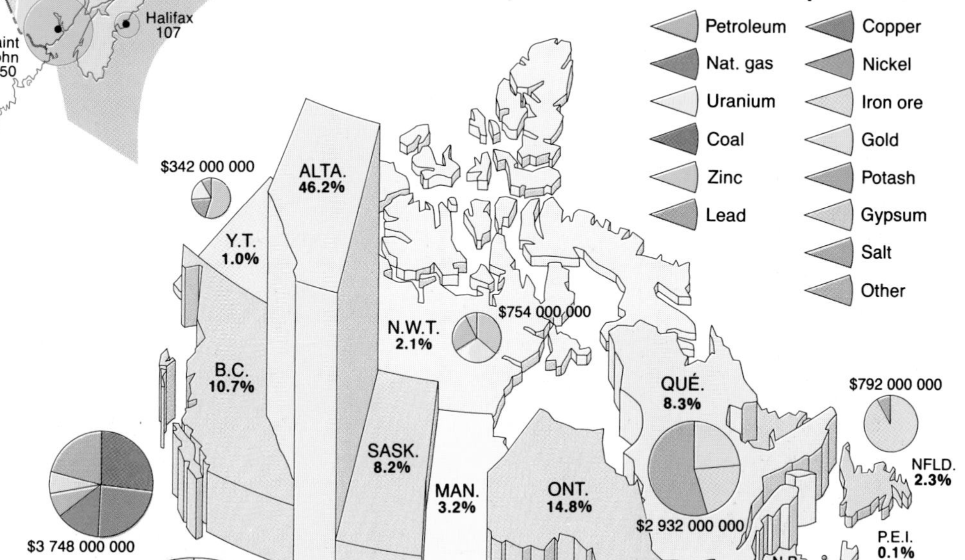

Total: (1991)$34 912 400 000

For explanation of terms, see glossary.

CANADA-THEMATIC

In Canada, the major sources of electrical energy are fossil fuels, moving water and uranium. New power projects are being planned or developed to produce electrical energy in the right place at the right price. These projects include large-scale coal-powered stations, nuclear power plants and hydroelectric mega-projects such as the James Bay power development. Canada, in spite of its small population, produces more hydroelectric power than any other country in the world.

Electrical consumption* per capita (kWh)

Norway	24 700
Canada	18 650
Iceland	18 000
Sweden	17 000
Luxembourg	14 000
Qatar	13 000
Finland	12 600
United States	12 100
Australia	8900
New Zealand	8400
Germany	7250
France	6500
Japan	6500
Russia	5870
United Kingdom	5700
South Africa	4000
Brazil	1708
Mexico	1360
Egypt	720
China	560
India	313
Nigeria	94

Top 10 countries

*Average annual

Total electricity demand

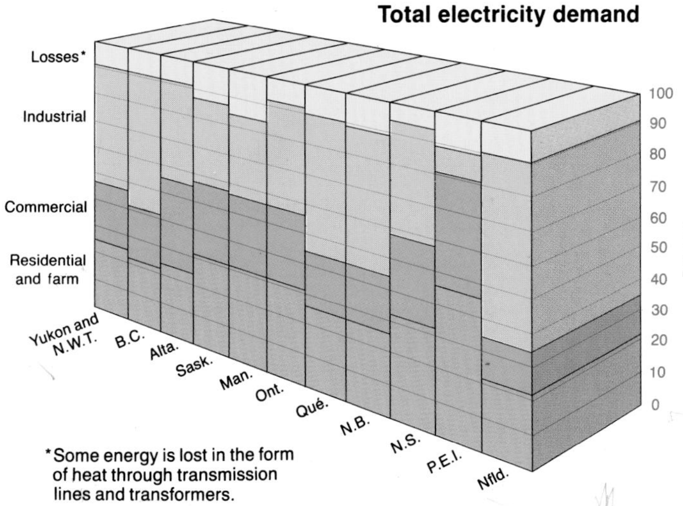

*Some energy is lost in the form of heat through transmission lines and transformers.

Primary energy sources

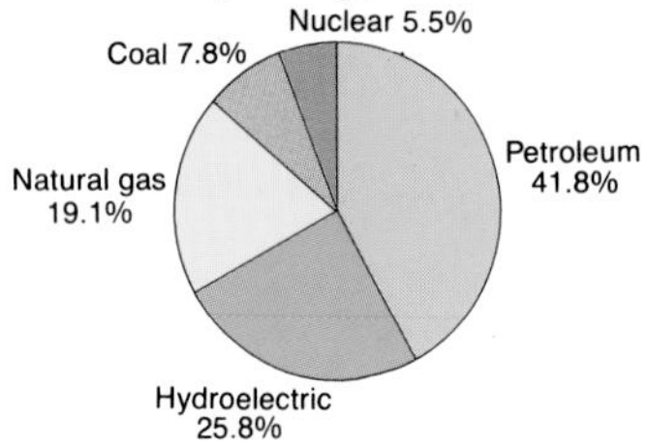

Household heating by region

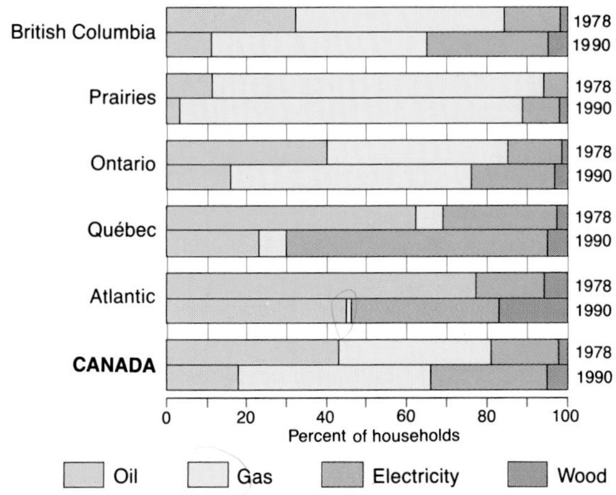

Annual emissions from electricity generation, 1989

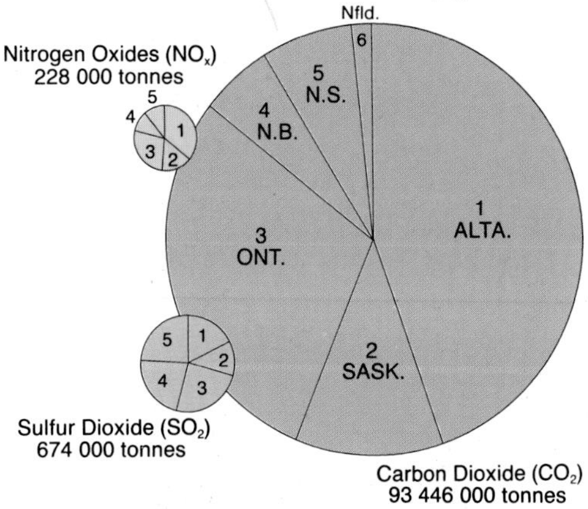

Other provinces or territories have zero or insignificant emissions.

Electrical Utilities' share of Canada's total emissions: Sulphur Dioxide 63%, Nitrogen Oxides 14%, Carbon Dioxide 22%.

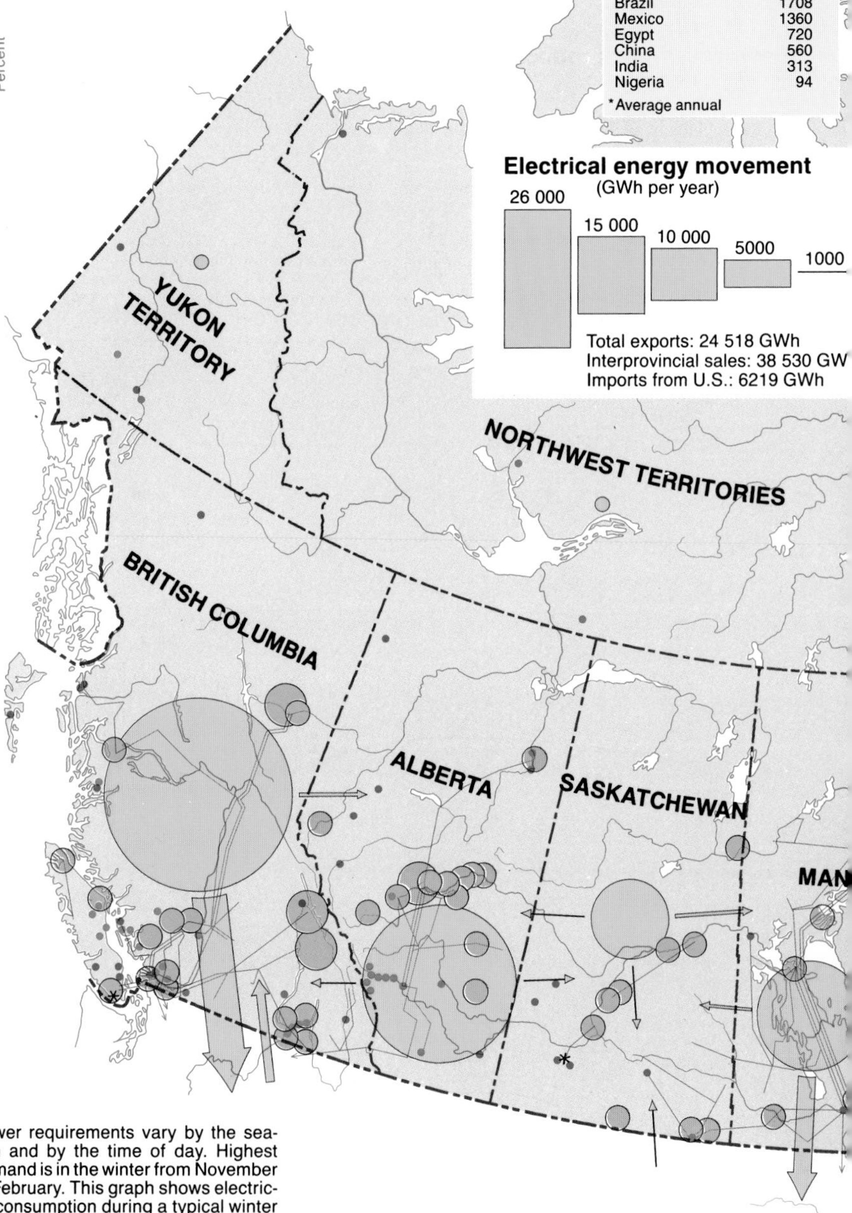

Electrical energy movement
(GWh per year)

26 000
15 000
10 000
5000
1000

Total exports: 24 518 GWh
Interprovincial sales: 38 530 GW
Imports from U.S.: 6219 GWh

Power requirements vary by the season and by the time of day. Highest demand is in the winter from November to February. This graph shows electricity consumption during a typical winter day in Ontario.

Daily pattern of electrical demand

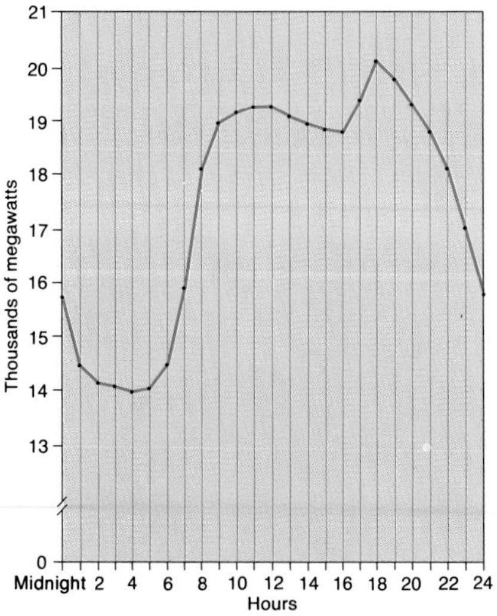

Cost of fuel for electricity generation

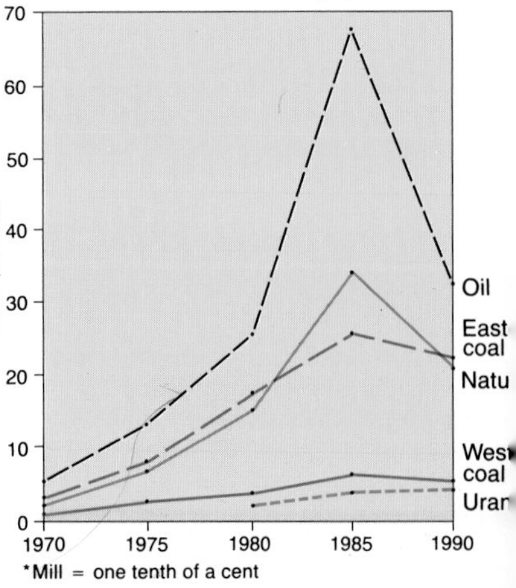

*Mill = one tenth of a cent

ENERGY CHOICES

nthly residential ctricity costs*

owknife	$135
rlottetown	115
nto	94
fax	90
ohn's	86
ehorse	83
ina	76
katoon	76
wa	73
couver	73
ericton	72
ary	69
onton	63
tréal	63
nipeg	61

*age cost per residential unit

Over the last decade, there have been increasing environmental concerns over the effects of burning fossil fuels and the long term costs of the disposal of nuclear waste. Large-scale hydroelectric projects have also caused controversy because they flood huge areas of land, destroy habitat and can cause high mercury levels in the environment. As a result, there is now an increasing focus on energy conservation and alternative energy sources such as the sun, wind and tides. Although these alternative sources have limited use at the moment, they are likely to become more important and less expensive in the future.

Electrical energy production
(GWh per year)

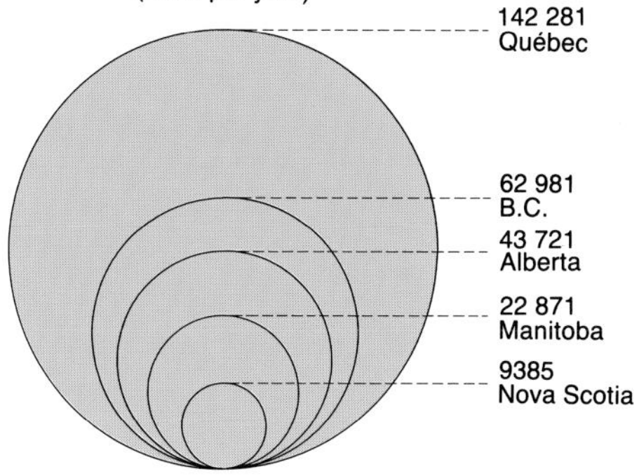

1 gigawatt = 1 000 000 kilowatts

CANADA 489 227

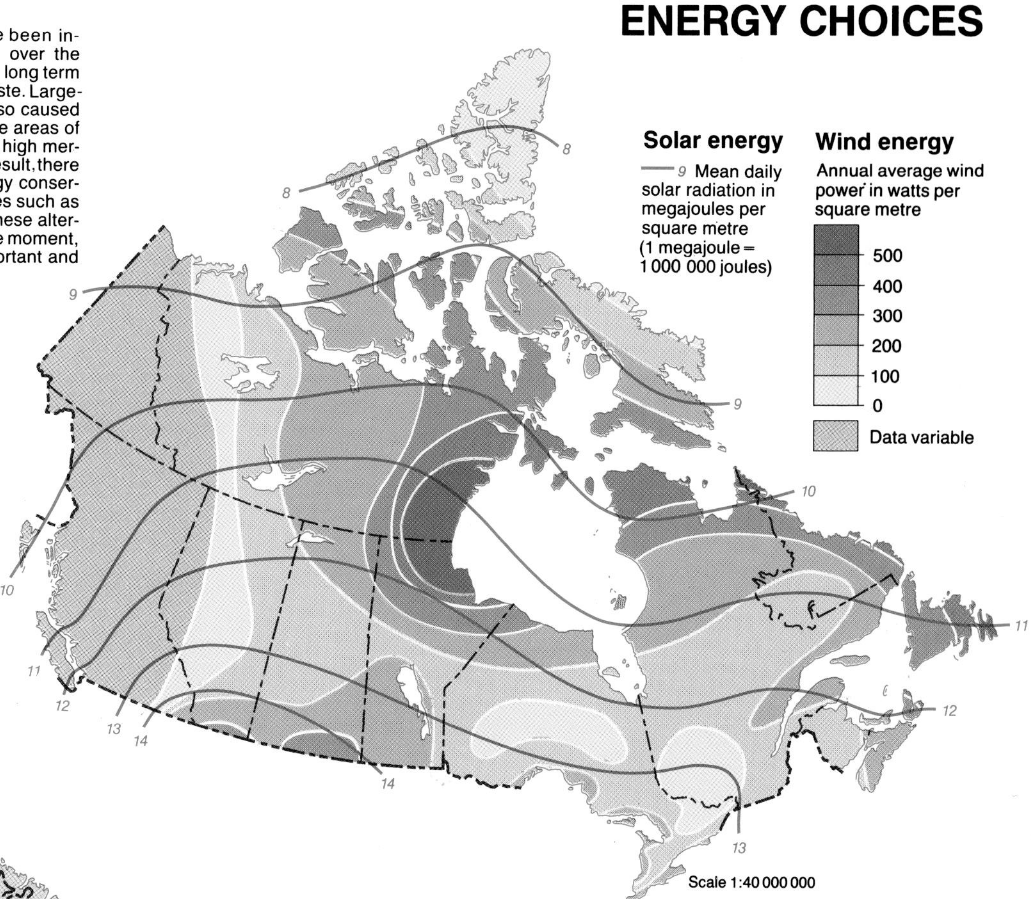

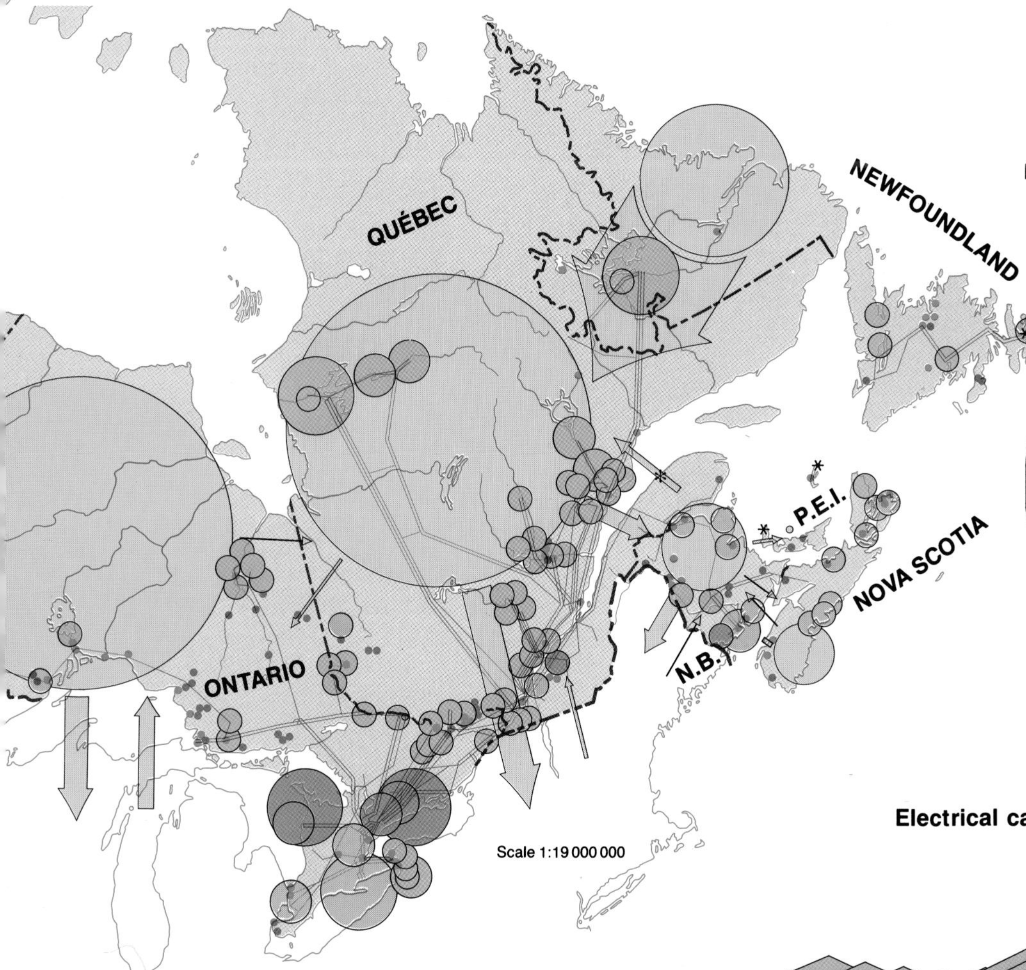

Electricity production by region

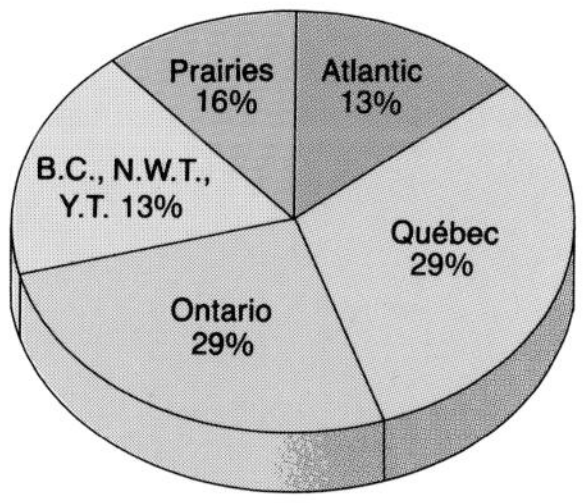

Electricity generation by fuel type

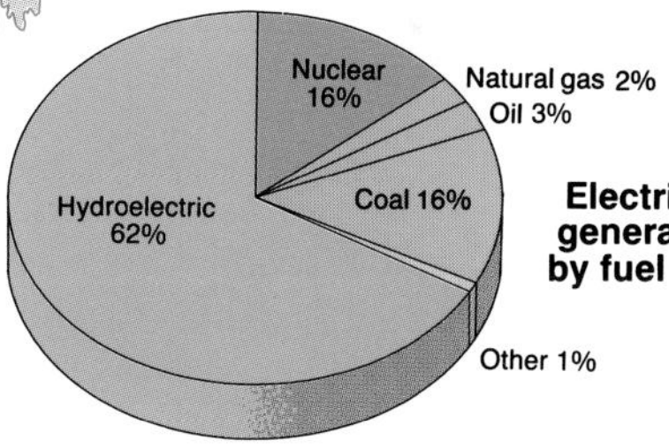

Electrical capacity

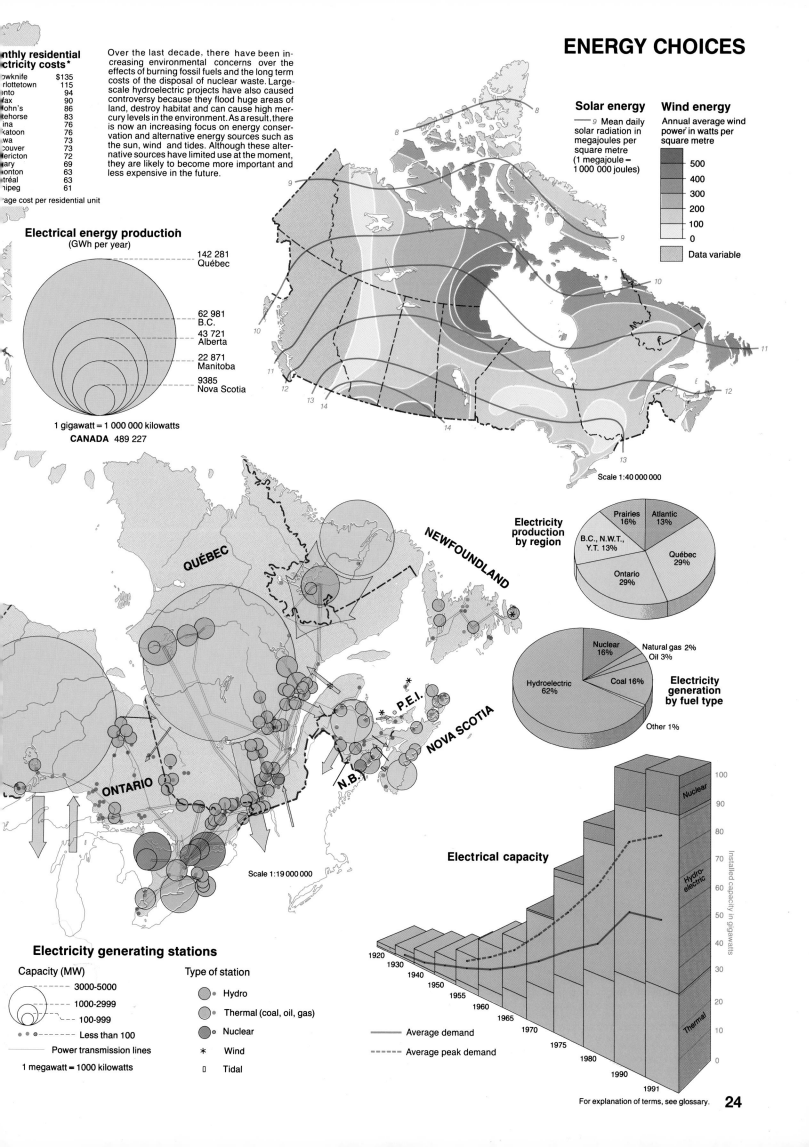

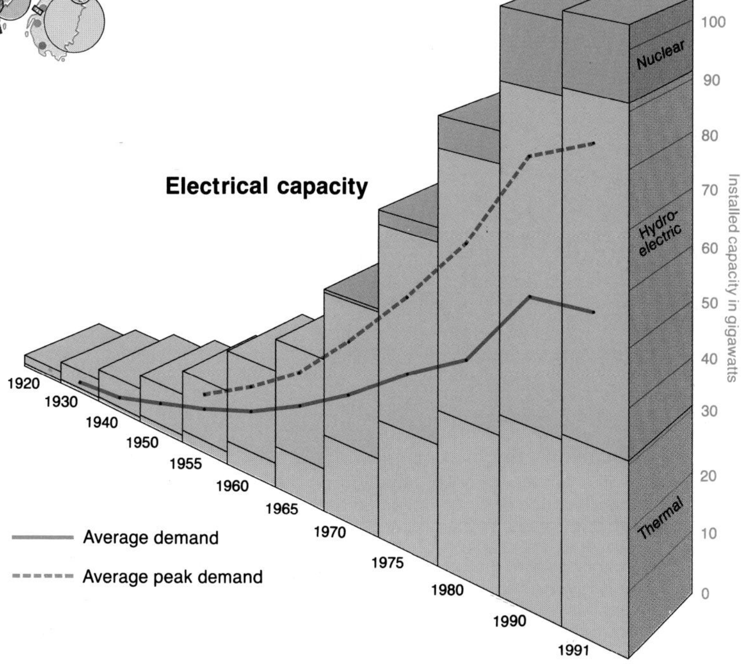

Electricity generating stations

Capacity (MW)

- 3000-5000
- 1000-2999
- 100-999
- Less than 100
- Power transmission lines

1 megawatt = 1000 kilowatts

Type of station

- Hydro
- Thermal (coal, oil, gas)
- Nuclear
- * Wind
- ▯ Tidal

For explanation of terms, see glossary.

CANADA-THEMATIC

Temperature

The average January and July temperatures are affected by several factors. The most important of these are latitude and continentality. The isotherms (lines that join points of equal temperature) reflect the influence of both factors. A continental climate occurs when there is little effect on temperature by a large body of water. A comparison between the two maps shows the wide temperature range that interior parts of Canada experience. This wide range is a characteristic of a continental climate.

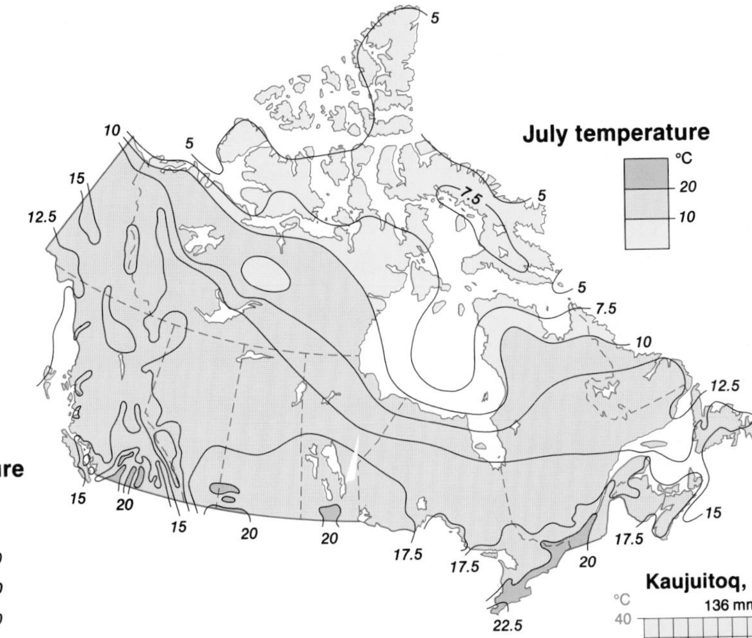

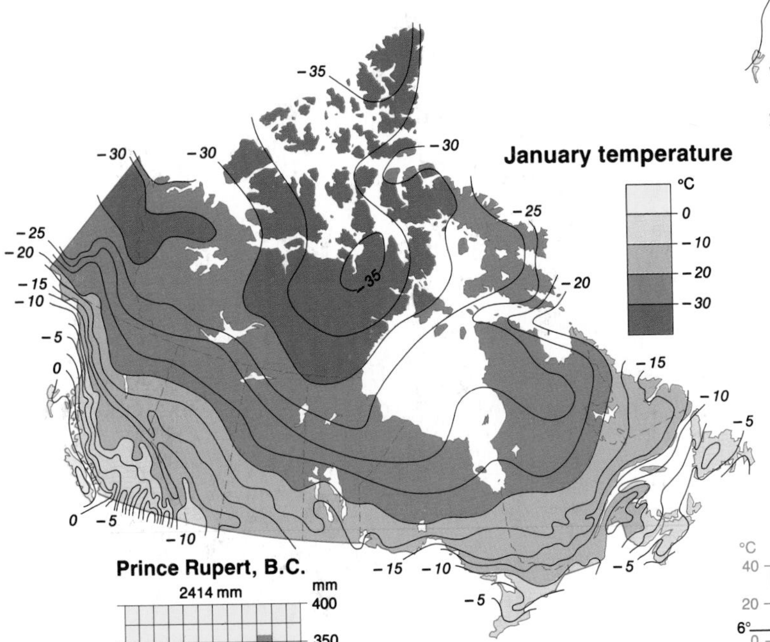

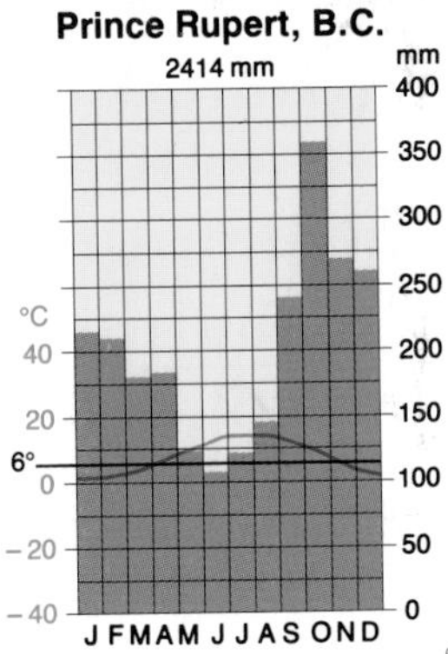

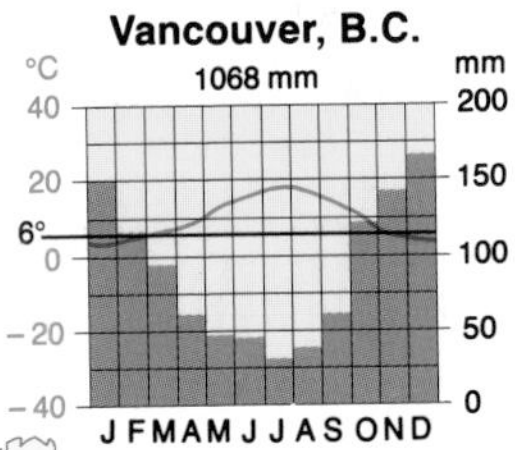

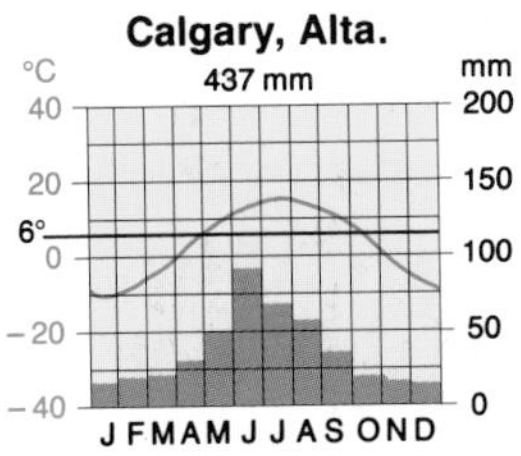

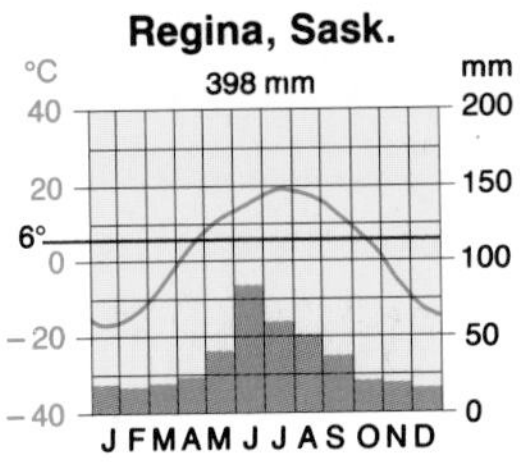

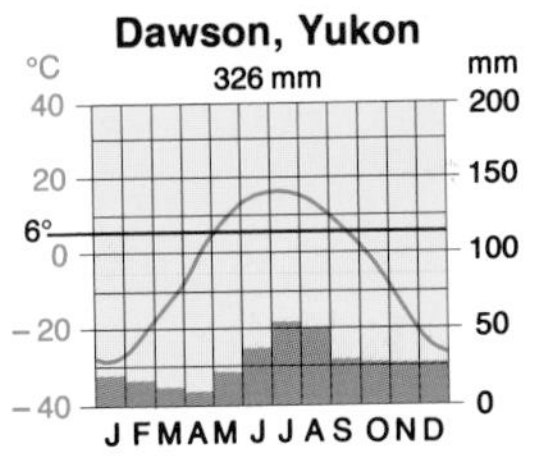

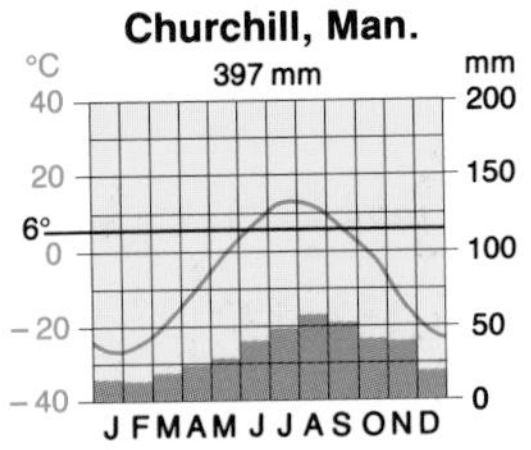

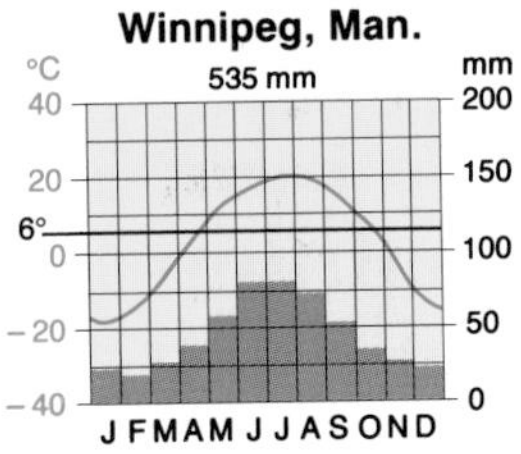

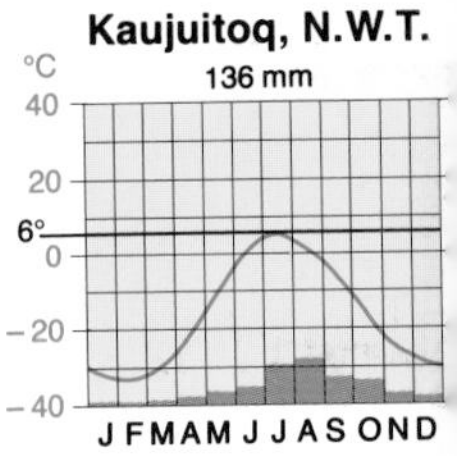

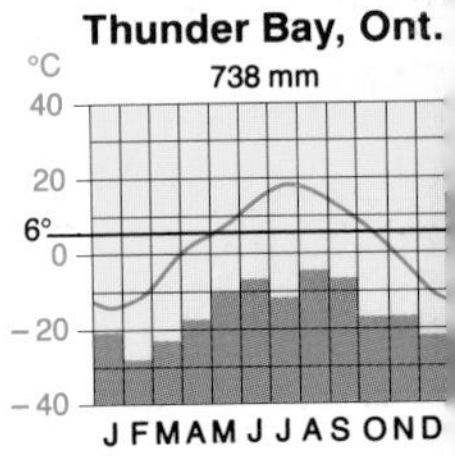

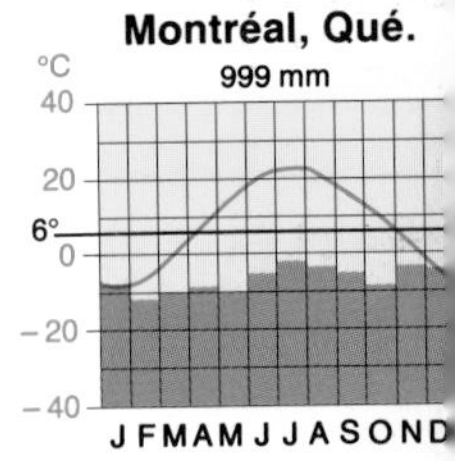

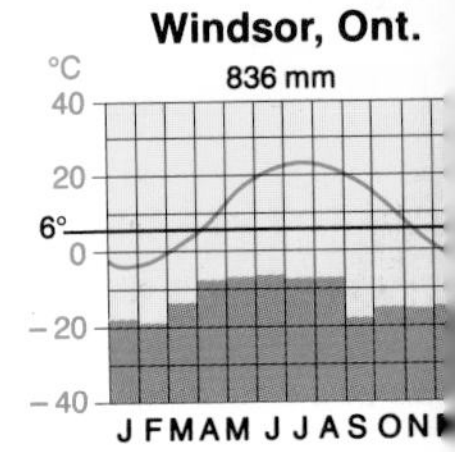

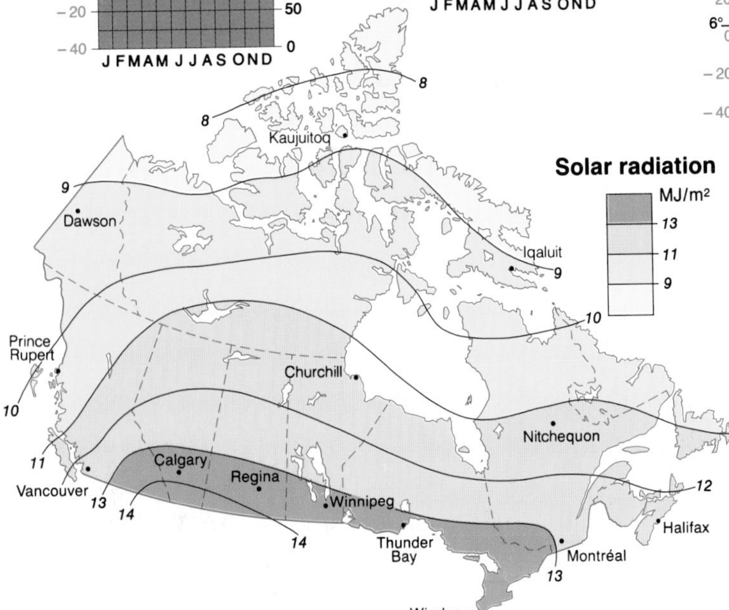

Radiation / Bright sunshine

The pattern of solar radiation on the map reflects the influence of cloud cover, latitude and the angle of the sun's rays as they enter the atmosphere and penetrate to the surface of the earth. A combination of factors affects the number of hours of bright sunshine in a given area each year. These factors include air masses, cyclonic storms and the number of daylight hours.

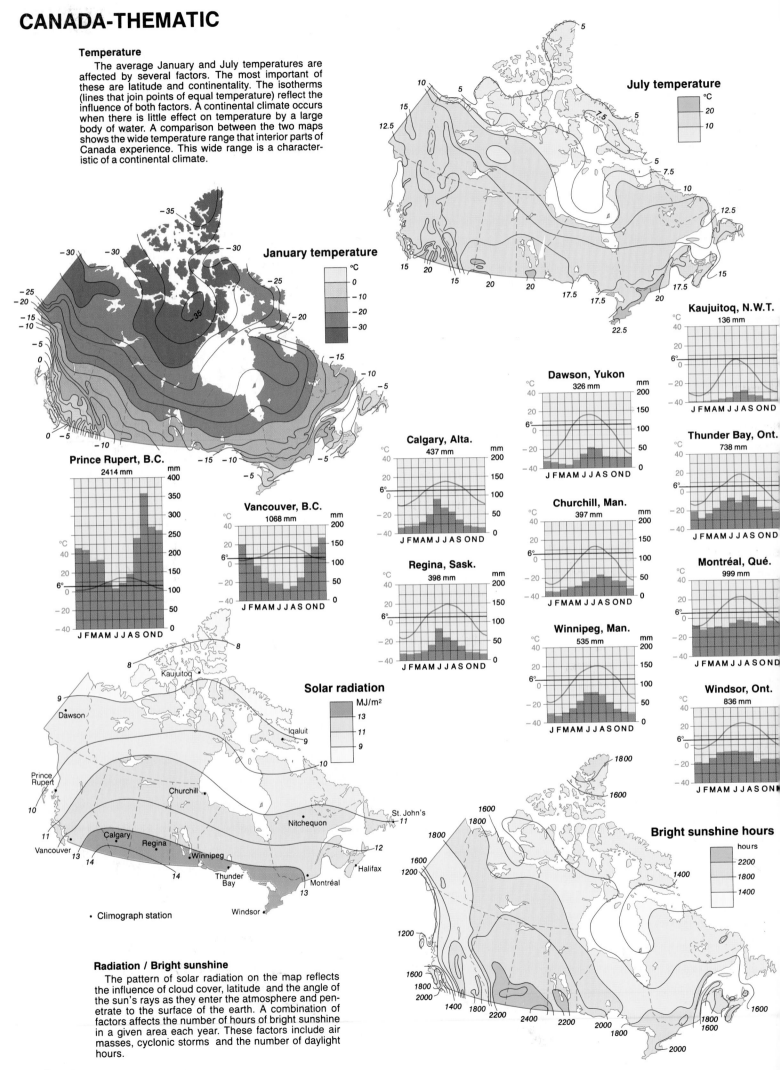

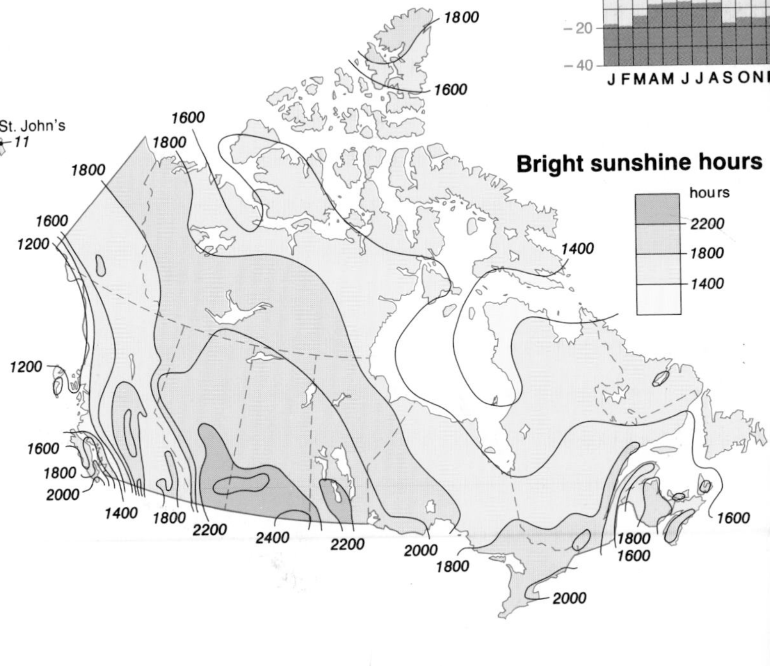

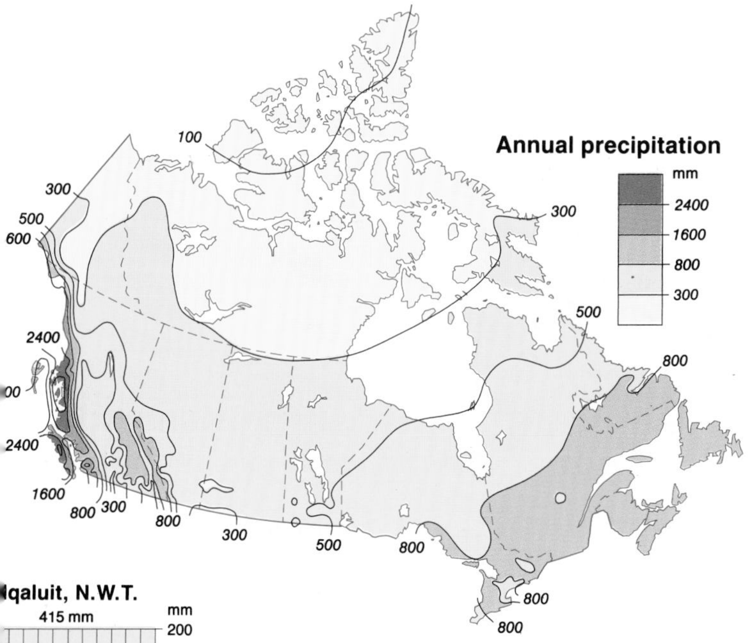

Precipitation / Snowfall

Precipitation includes all forms of moisture. Snowfall is included in precipitation, but the amounts have been converted to their moisture content using a 10:1 snow-to-moisture ratio. Several factors influence the pattern of precipitation in Canada–bodies of water, prevailing winds, air masses and relief. The Far North is almost like a desert. The low temperatures and lack of a source of moisture result in little evaporation and precipitation. The snowfall map illustrates how little snow actually falls in the Far North; most of the precipitation in this region falls as rain during the summer months.

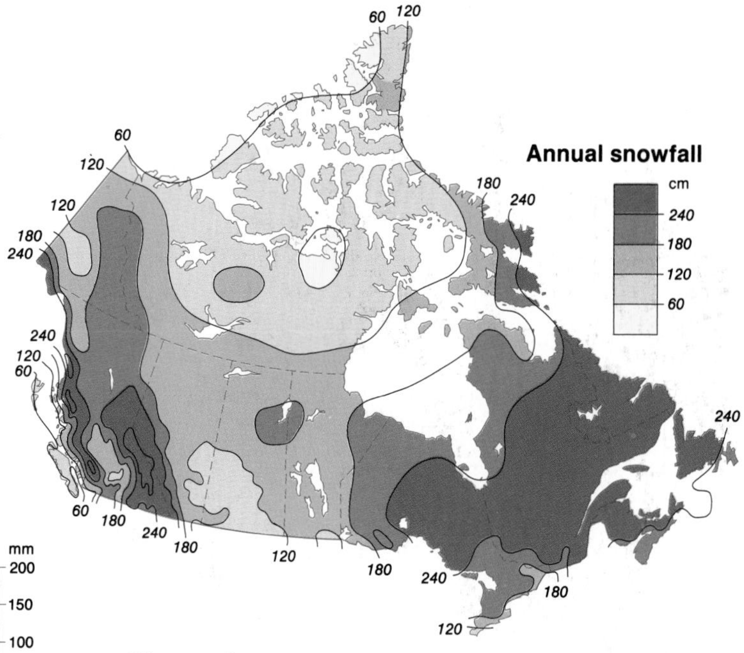

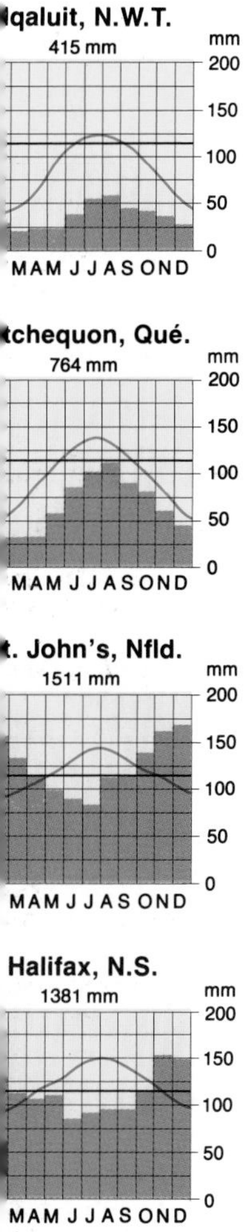

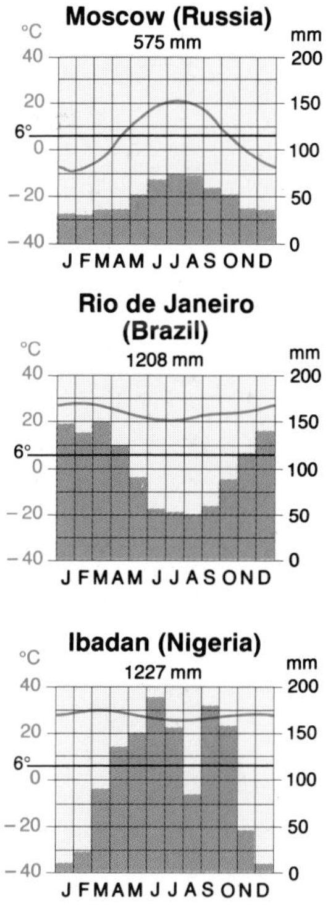

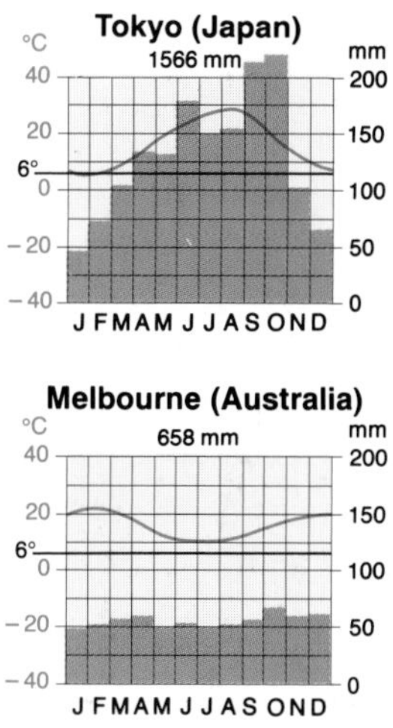

Climographs

Climographs show the average monthly precipitation with a bar graph, and the average monthly temperature with a line graph. (The growing season is indicated when the temperature line rises above 6°C.) The climographs are arranged in order of their geographical location (see the map on bottom left of page 25). Non-Canadian climographs are included for comparison.

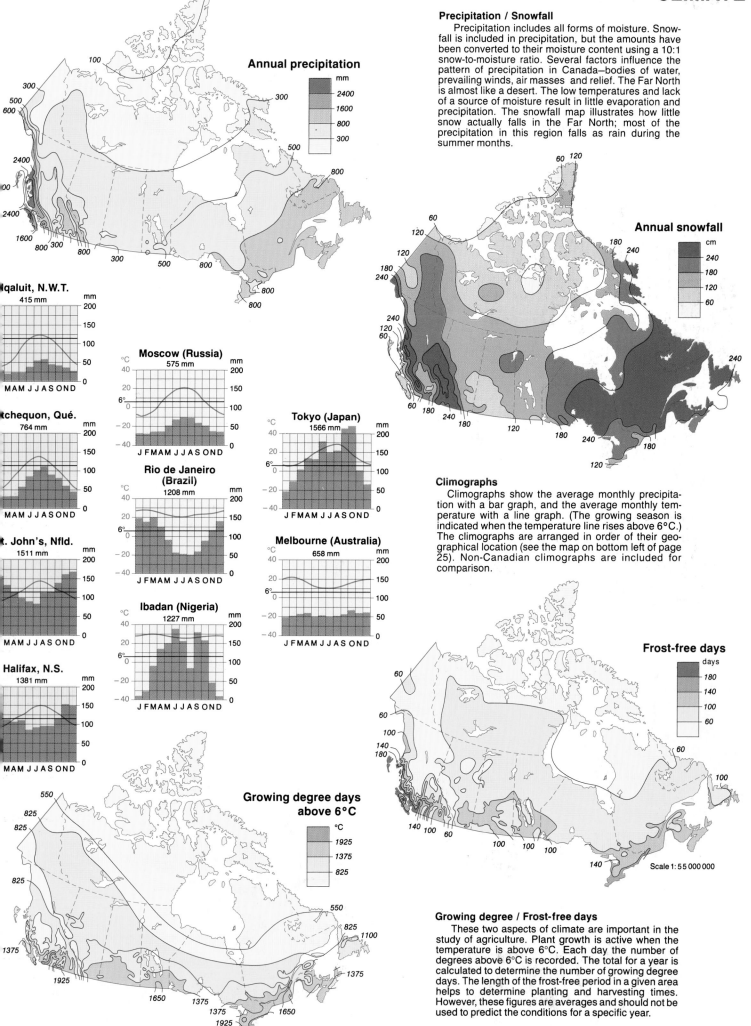

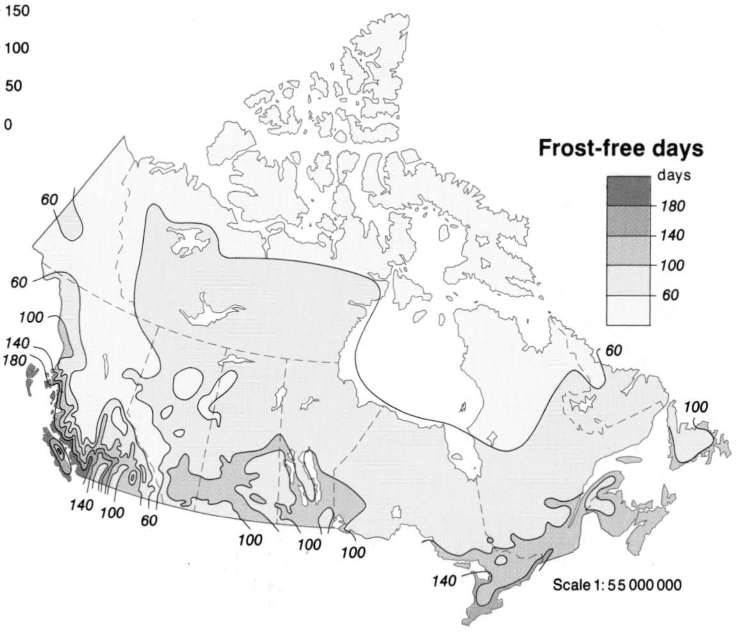

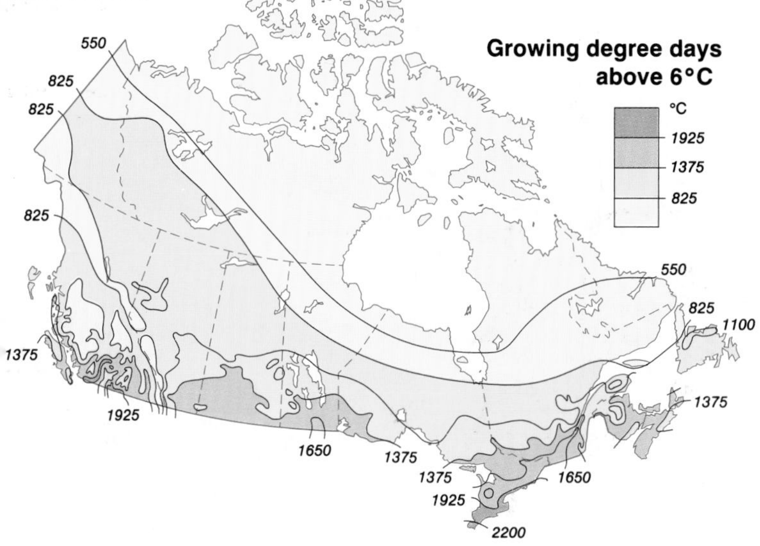

Growing degree / Frost-free days

These two aspects of climate are important in the study of agriculture. Plant growth is active when the temperature is above 6°C. Each day the number of degrees above 6°C is recorded. The total for a year is calculated to determine the number of growing degree days. The length of the frost-free period in a given area helps to determine planting and harvesting times. However, these figures are averages and should not be used to predict the conditions for a specific year.

CANADA-THEMATIC

Air masses

Air masses are large bodies of air in which temperature and moisture conditions are relatively similar throughout. They take on the characteristics of the surface over which they form, and carry these characteristics to "invaded areas" when they move.

Air masses can develop over land or ocean. The continental air masses in North America tend to be cold and dry in the winter, and warm and sometimes humid in the summer. Maritime air masses are affected by the temperature of the ocean below them. Warm ocean currents warm the air, increasing its ability to hold moisture. This effect, for example, helps to create the wet, warm climate of Canada's west coast. Where warm humid air passes over cold currents, such as the Labrador Current, moisture in the air condenses and can form heavy fog.

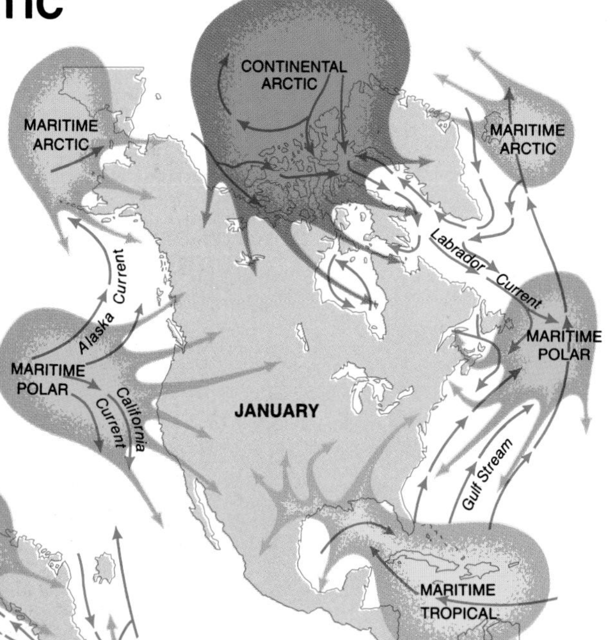

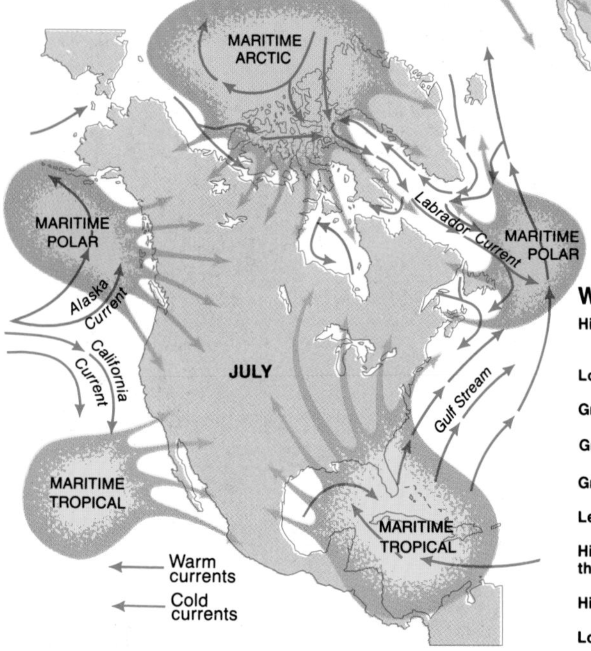

A weather satellite's view

Although we cannot control the weather, we can understand it better with the use of modern technology. Weather forecasts used to be based on human observations made twice a day. Today, weather satellites and sophisticated computers give meteorologists continuous information about weather changes. With more accurate data, forecasting has become more precise, and people can prepare for weather extremes.

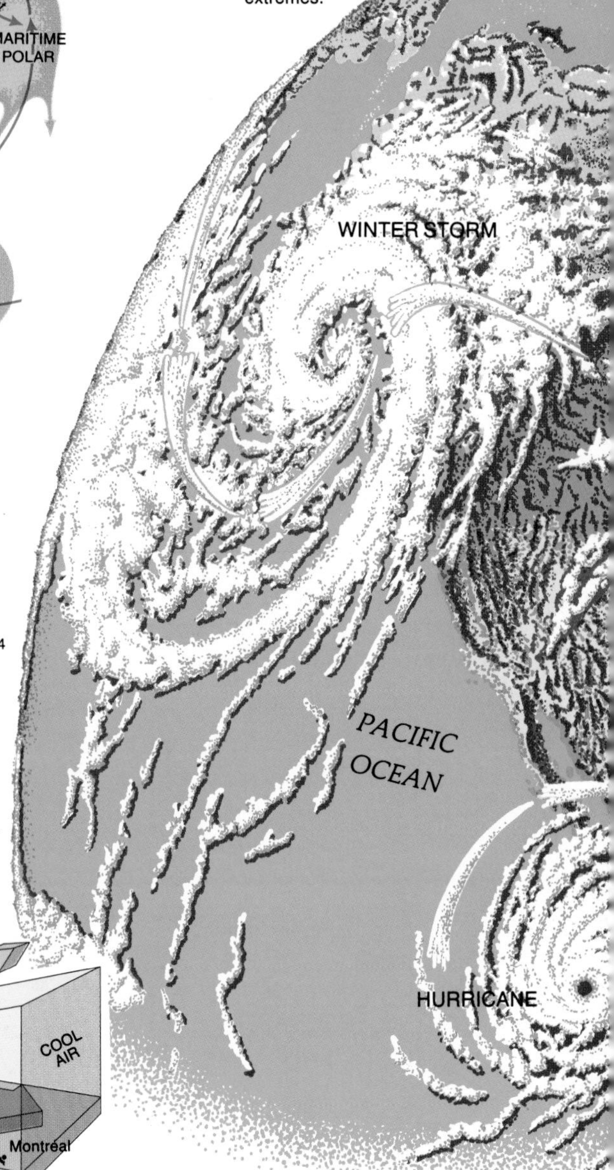

Weather extremes

Highest temperature:
45.0°C at Midale and Yellow Grass, Sask., July 5, 1937
Lowest temperature:
−63.0°C at Snag, Y.T., Feb. 3, 1947
Greatest precipitation in 24 hours:
489.2 mm at Ucluelet, B.C., Oct. 6, 1967
Greatest snowfall in 24 hours:
118.1 cm at Lakelse Lake, B.C., Jan. 17, 1974
Greatest precipitation in one year:
8122.6 mm at Henderson Lake, B.C., 1931
Least precipitation in one year:
12.7 mm at Arctic Bay, N.W.T., 1949
Highest average number of days with thunderstorms in one year:
34 in London, Ontario
Highest air pressure:
107.96 kPa at Dawson, Y.T., Feb. 2, 1989
Lowest air pressure:
94.02 kPa at St. Anthony, Nfld., Jan. 20, 1977
Highest wind speed sustained for one hour:
201.1 km/h at Cape Hopes Advance, Qué., Nov. 18, 1931

Weather fronts

Air masses move in order to equalize varying atmospheric pressures, from high pressure zones to low pressure zones. The surface or line where masses of air with different characteristics meet is called a front. Fronts can be classified as cold and warm fronts; a cold front occurs when a cold air mass moves into an area of warm air, and a warm front when the warm air mass moves into an area of cooler air.

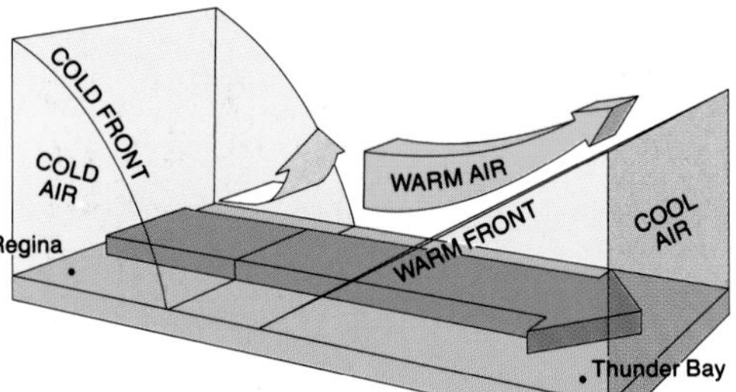

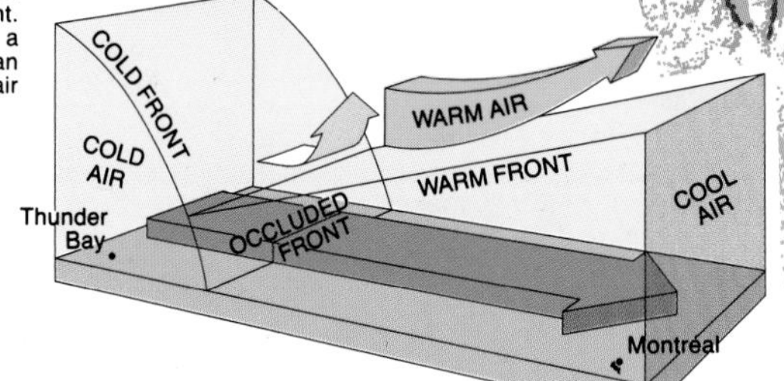

A cold front forces the warm air to rise quickly, causing heavy precipitation, often in the form of thunderstorms. Along a warm front, the warm air rises slowly over the cold air, creating widespread areas of gentle precipitation. Cold fronts move faster than warm fronts. An occluded front occurs when a cold front catches up to a warm front, forcing all the warm air aloft.

Storm front in southern Alberta

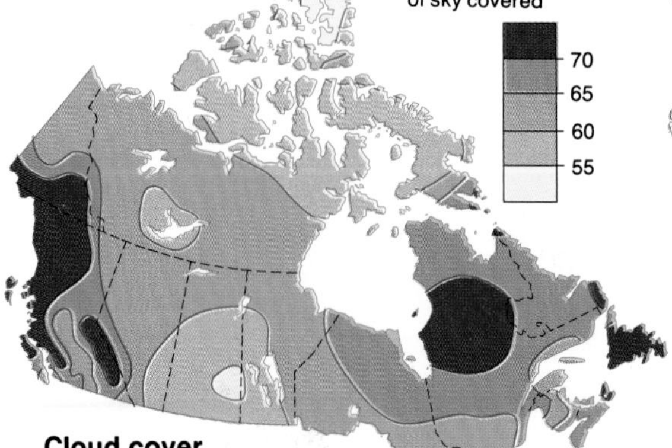

Cloud cover

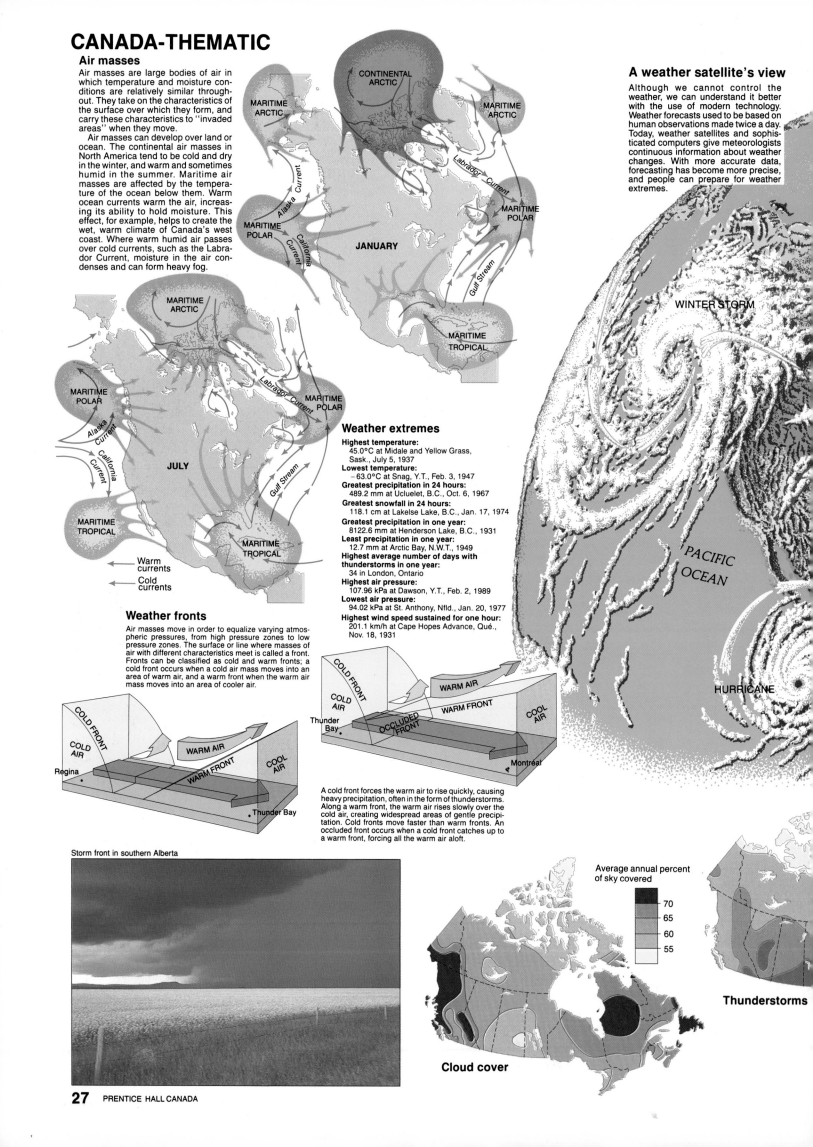

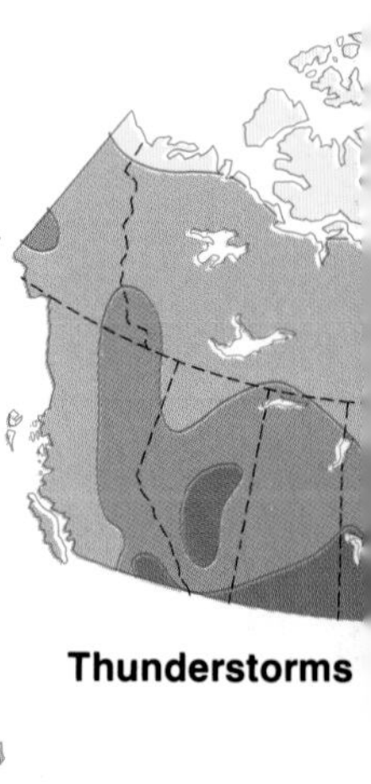

Thunderstorms

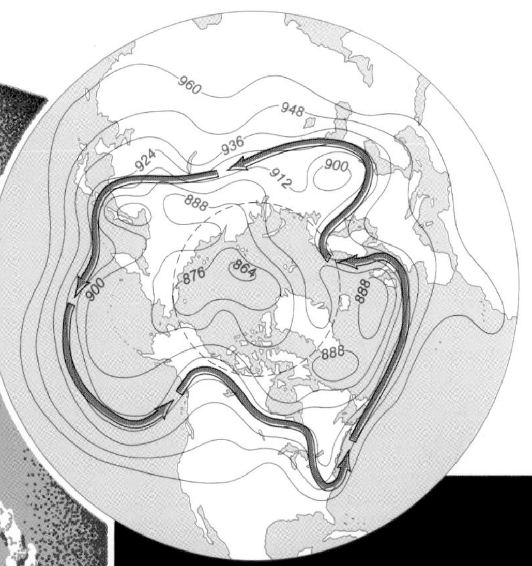

Jet streams

Jet streams are tube-like bands of high velocity wind found in the upper atmosphere. The two jet streams over North America blow from west to east at speeds of up to 450 km/h. The Sub-tropical Jet Stream separates tropical air from middle latitude air masses near 30°N. The Polar Front Jet Stream varies in location, as warm and cold air masses move north or south with the seasons. People who plan aircraft flight paths take the location of the jet streams into consideration.

The diagram to the left shows one possible location of the Polar Front Jet Stream. Isobars on the map show atmospheric pressure in millibars.

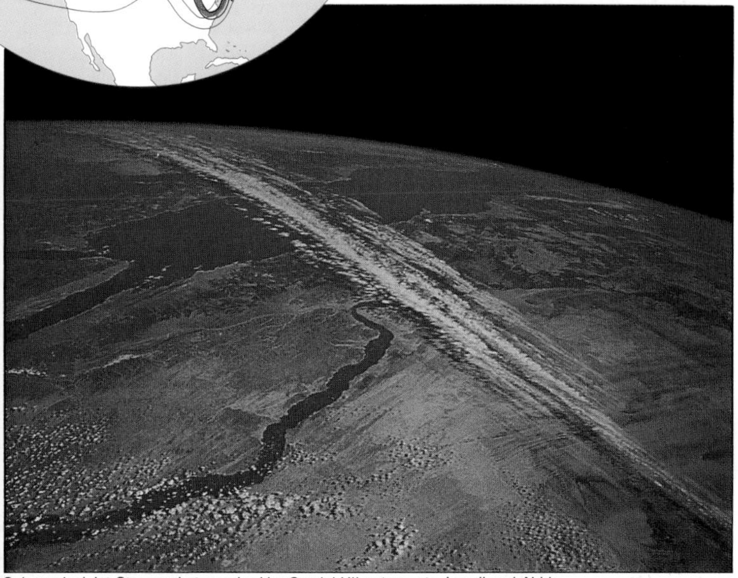

Sub-tropical Jet Stream photographed by Gemini XII astronauts, Lovell and Aldrin

Weather patterns form near the surface of the earth where the atmosphere is dense and heavy. Weather involves the elements of heat, moisture, pressure and wind, but the key element is heat. All weather changes can be traced to the differences in heating of parts of the earth's surface and the atmosphere above the surface. These differences are caused by latitude, relief and the nature of the earth's surface.

Since weather is an invisible layer of air that is always in motion, it is difficult to depict. Television weather forecasts use satellite images, computerized mapping and motion graphics to illustrate weather patterns. Meteorologists use these methods to predict atmospheric changes over short periods of time. Long-range forecasting is less accurate. However, thirty-day forecasts are still used to show general trends.

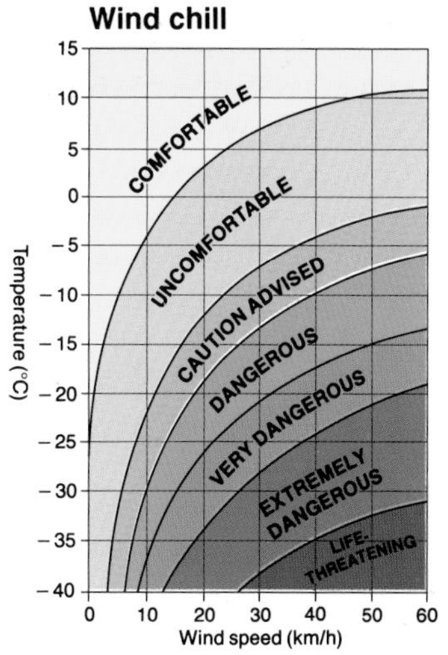

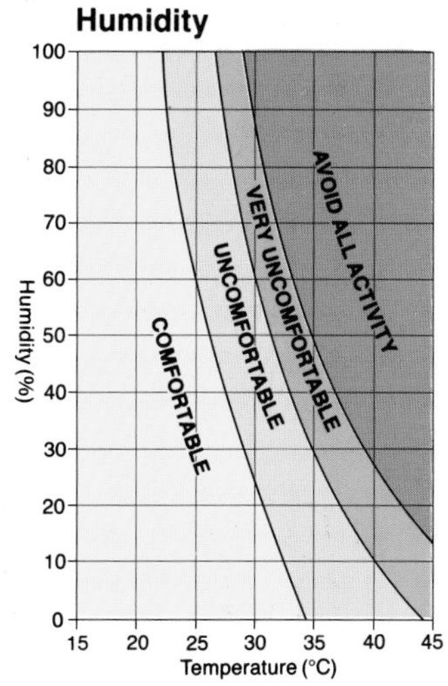

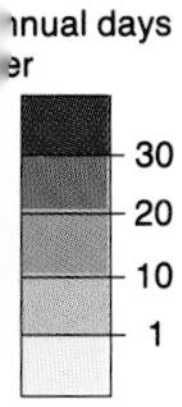

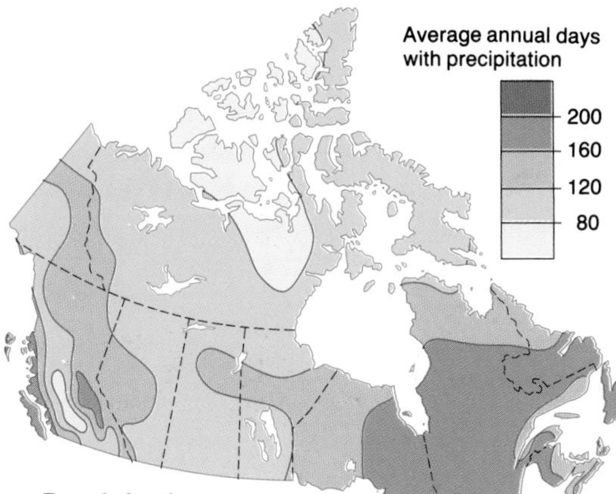

Precipitation

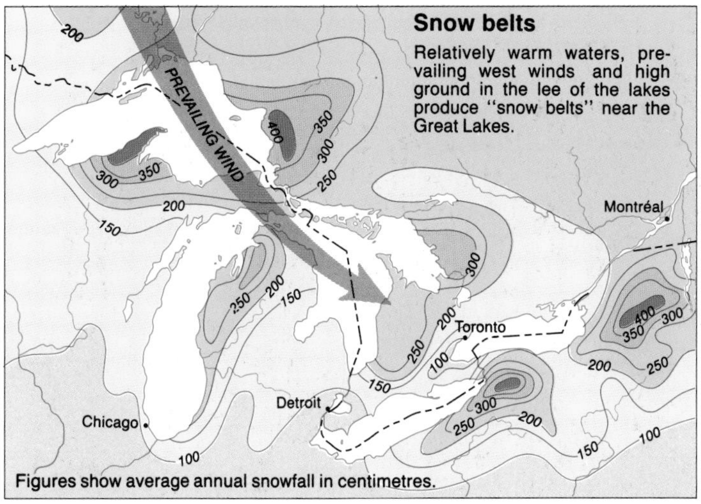

Snow belts

Relatively warm waters, prevailing west winds and high ground in the lee of the lakes produce "snow belts" near the Great Lakes.

Figures show average annual snowfall in centimetres.

For explanation of terms, see glossary.

CANADA-THEMATIC

World distribution of water

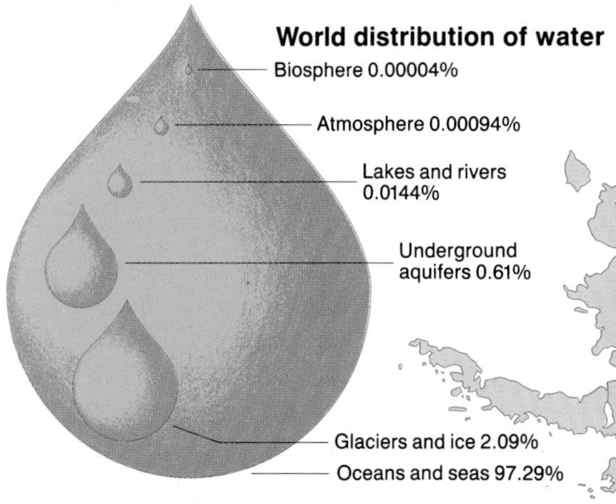

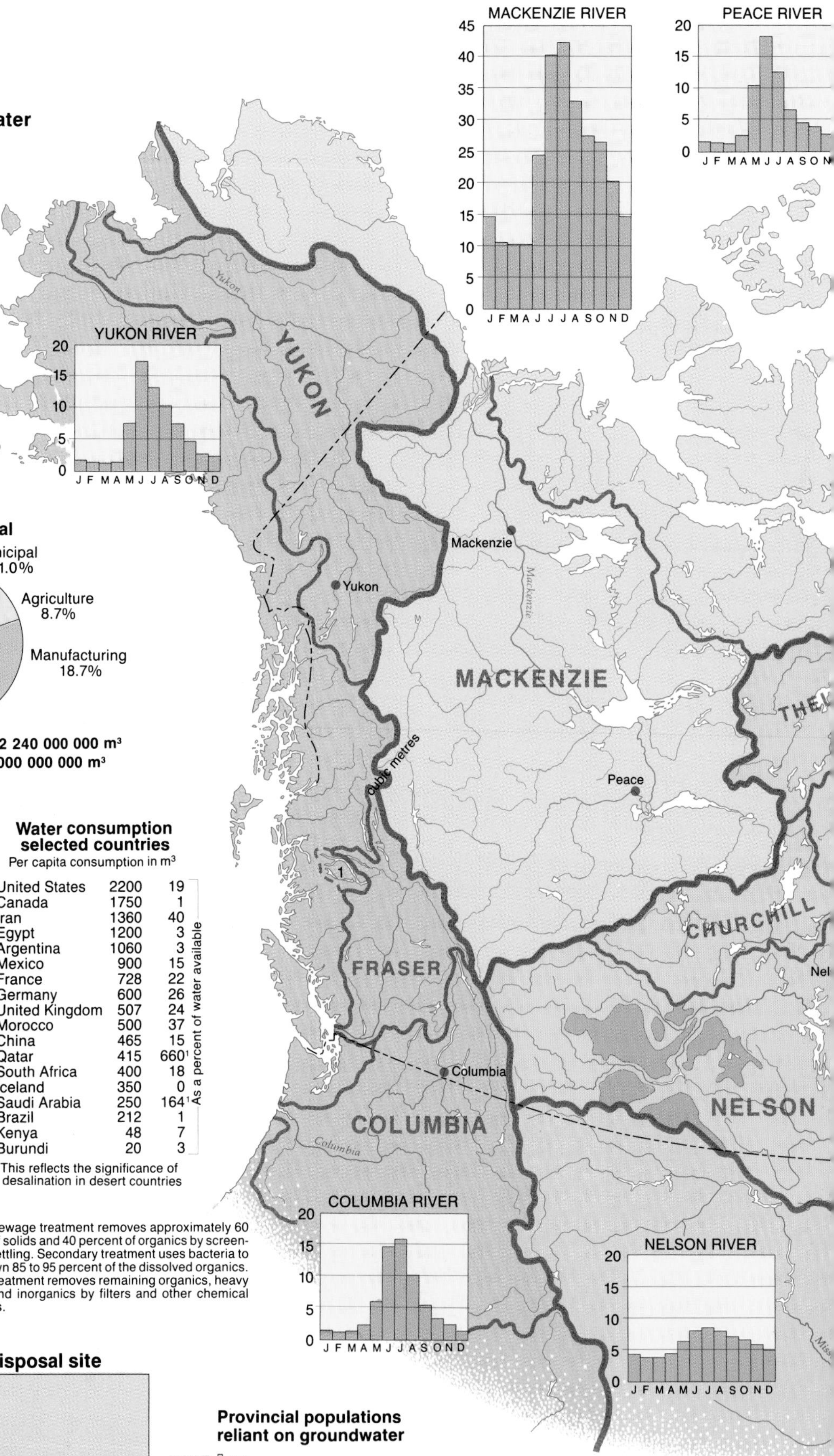

Water consumption

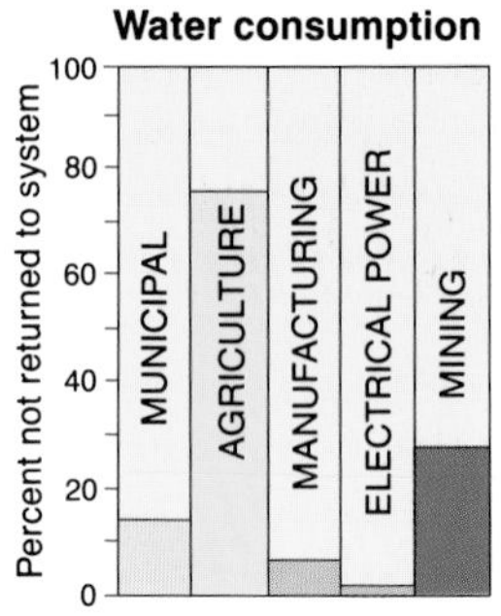

Water withdrawal

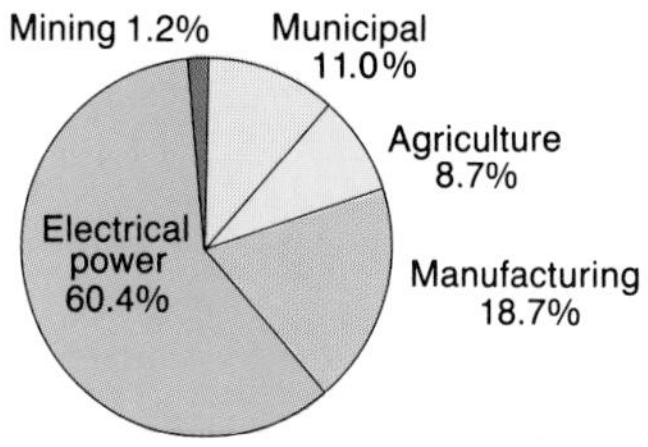

Water withdrawal: 42 240 000 000 m³
Water returned: 38 000 000 000 m³

Cost of water, selected countries

Country	Price
CANADA	$ 0.36
UNITED STATES	0.42
IRELAND	0.55
UNITED KINGDOM	0.66
SWEDEN	0.78
FRANCE	0.86
ITALY	0.93
GERMANY	1.33
AUSTRALIA	1.47

Prices in Canadian dollars per 1000 litres

Water consumption selected countries

Per capita consumption in m³

Country	Per capita consumption in m³	As a percent of water available
United States	2200	19
Canada	1750	1
Iran	1360	40
Egypt	1200	3
Argentina	1060	3
Mexico	900	15
France	728	22
Germany	600	26
United Kingdom	507	24
Morocco	500	37
China	465	15
Qatar	415	660[1]
South Africa	400	18
Iceland	350	0
Saudi Arabia	250	164[1]
Brazil	212	1
Kenya	48	7
Burundi	20	3

[1]This reflects the significance of desalination in desert countries

Municipal population served by sewage treatment

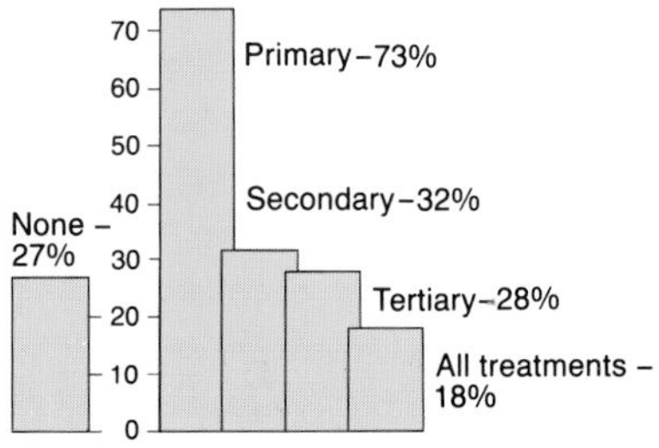

Primary sewage treatment removes approximately 60 percent of solids and 40 percent of organics by screening and settling. Secondary treatment uses bacteria to break down 85 to 95 percent of the dissolved organics. Tertiary treatment removes remaining organics, heavy metals, and inorganics by filters and other chemical processes.

Groundwater contamination from a waste disposal site

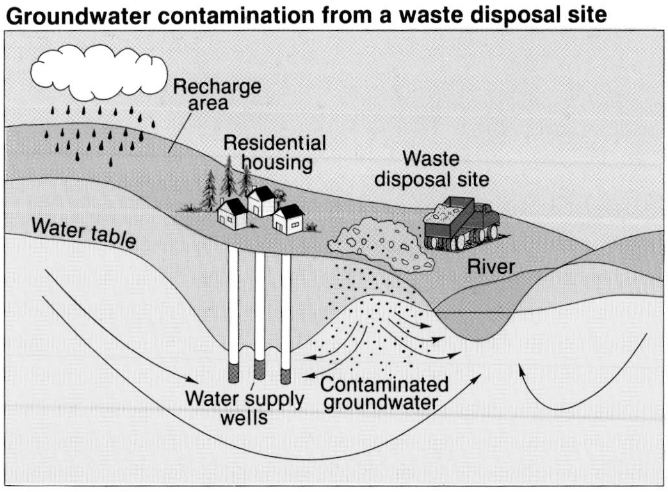

Provincial populations reliant on groundwater

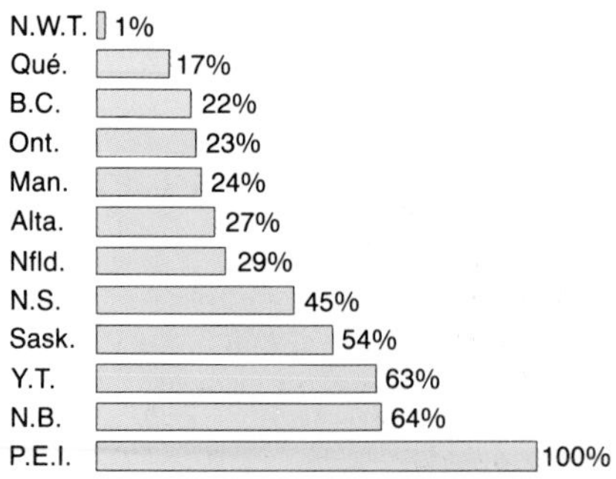

Drainage basins

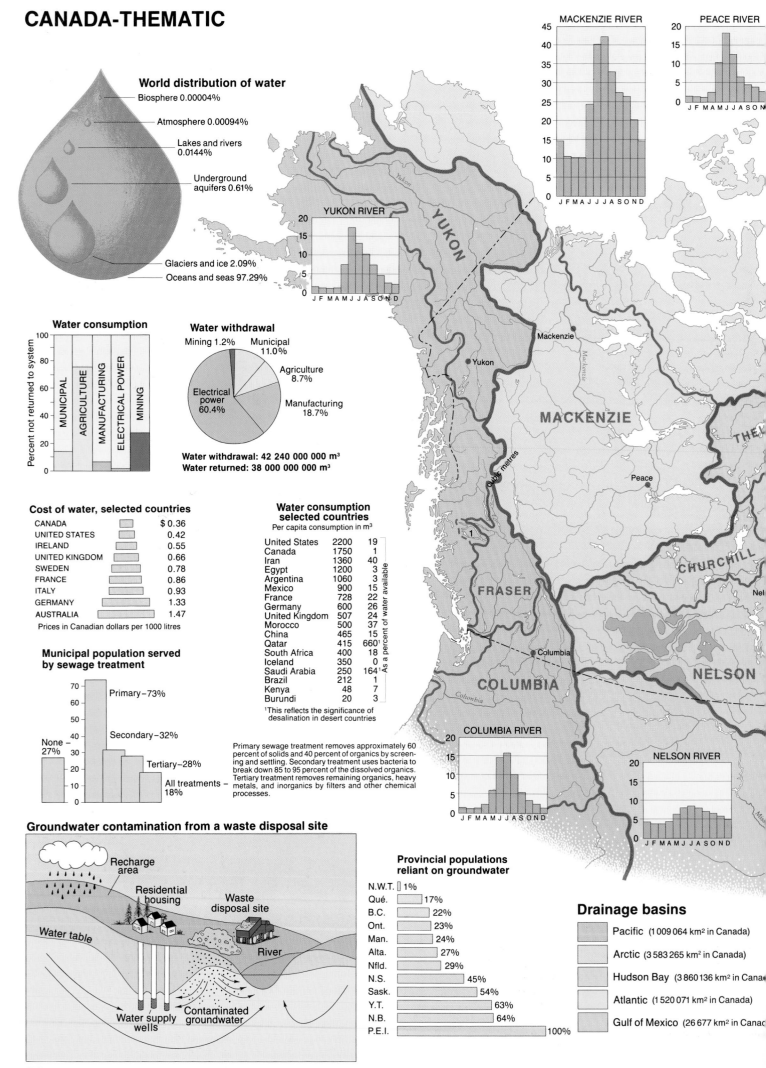

WATER RESOURCES

Water is essential for life. In the past, Canadians have taken water for granted as a result of its abundance in most regions. For many years Canadians have practised flood control, built hydroelectric dams, and dumped wastes of all kinds in waterways. Concerns about the environment and the increasing demand for water for irrigation and other purposes, have made Canadians realize the importance of water management. Controls on polluters and increased treatment of sewage have reduced the extent of the pollution. However, urban storm sewers and agricultural runoff that contributes nitrates, phosphorus, and pesticides continue to cause problems. Canada may have enough water for future needs; however, to protect water quality and ensure the health of the environment, Canadians will have to manage this resource wisely.

Precipitation can evaporate, sink below ground level, or run down the surface of the land. Moisture that sinks into the ground becomes ground water; the amount of ground water varies with slope, vegetation, temperature, and the porous nature of the soil or rock. Water that runs down the surface of the land and flows into streams and rivers is called runoff.

Average annual runoff

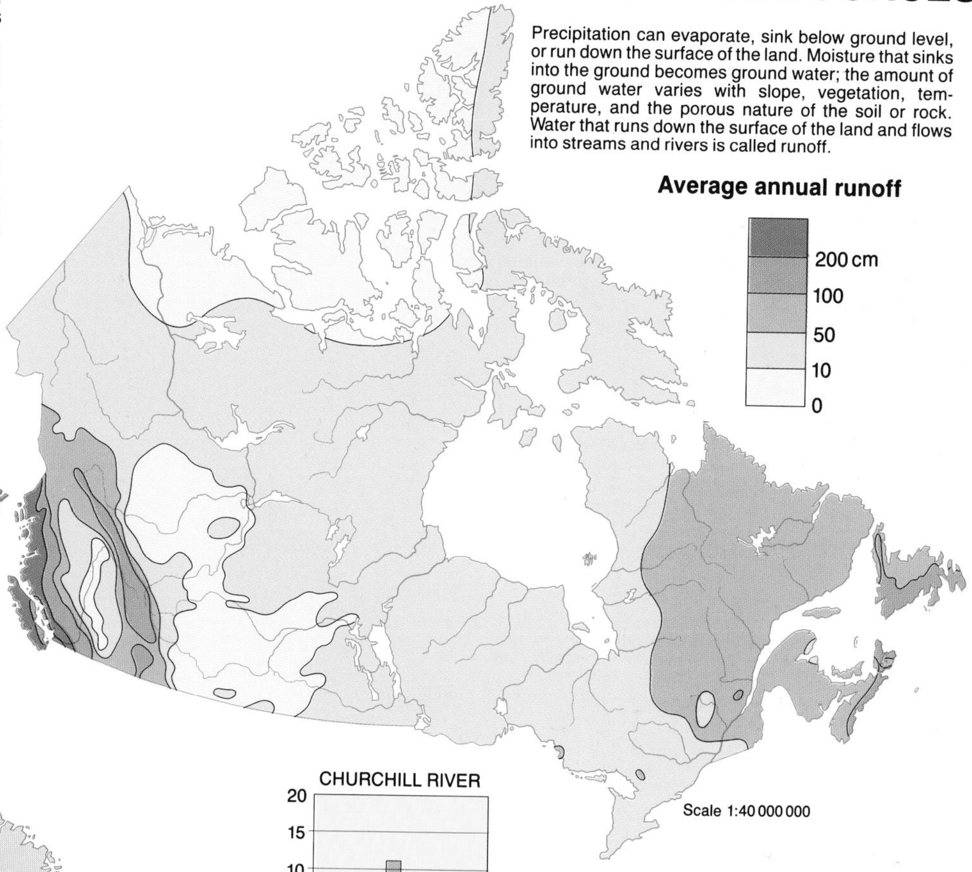

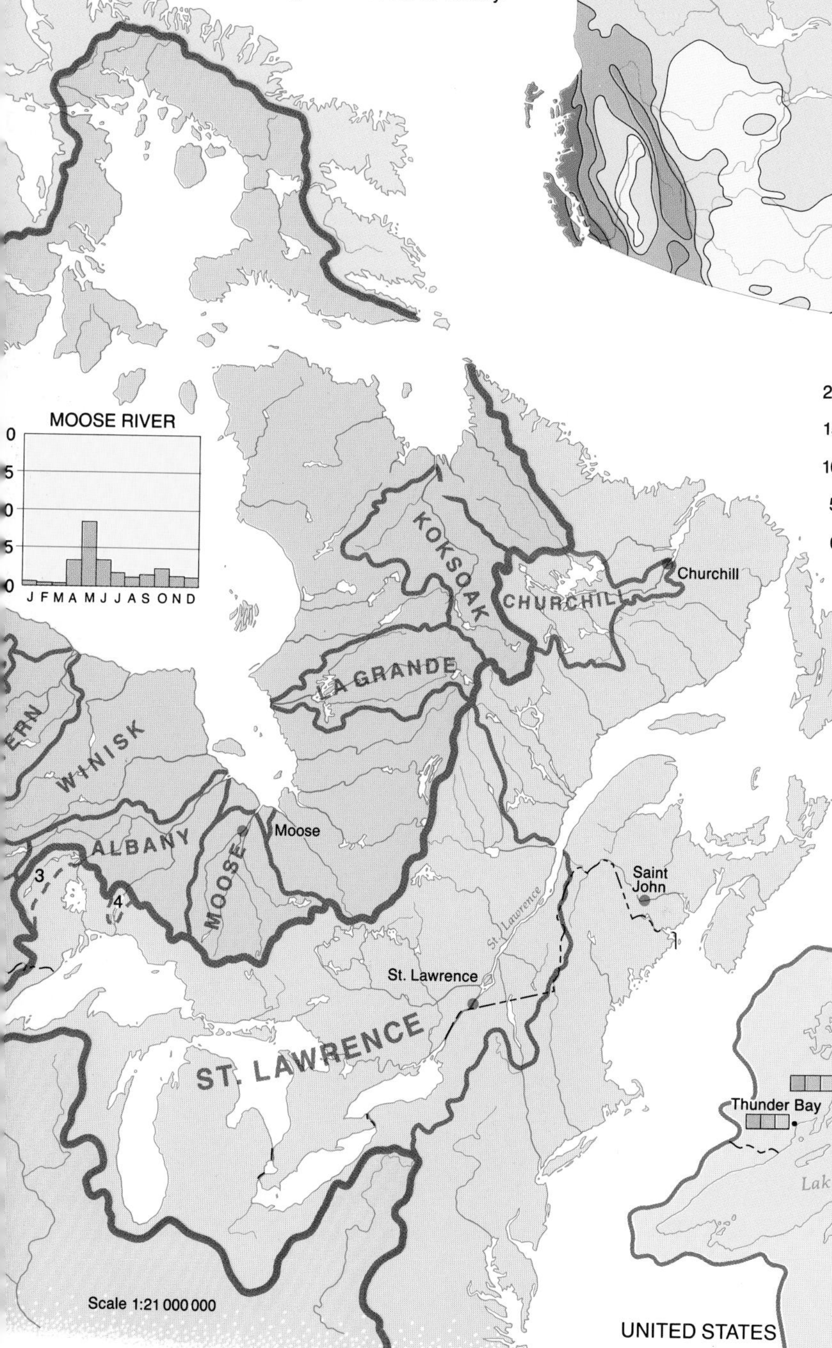

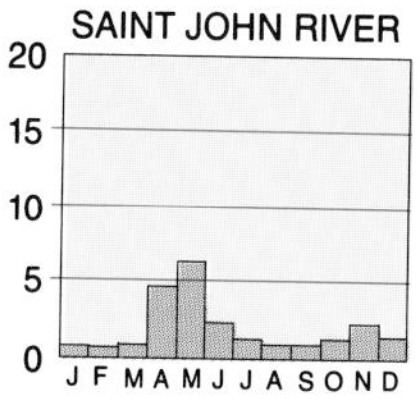

The five Great Lakes represent 18 percent of all the surface fresh water on earth. Over 40 million people live in the Great Lakes watershed. Destruction of wetlands, intensive agriculture, and heavy industrial activity have placed great pressure on this ecosystem. Over 360 different chemicals have been identified in the lakes, many of which are persistent and toxic. These chemicals bioaccumulate, or magnify, as they are passed among species in the food chain. More effective pollution-control legislation has reduced the input of many contaminants, but contaminated sediments and groundwater polluted by waste disposal sites will continue to be concerns in the future. Money and political will on the part of all governments bordering the Great Lakes will be essential to ensuring the health of the Great Lakes ecosystem.

Toxic substances in the Great Lakes

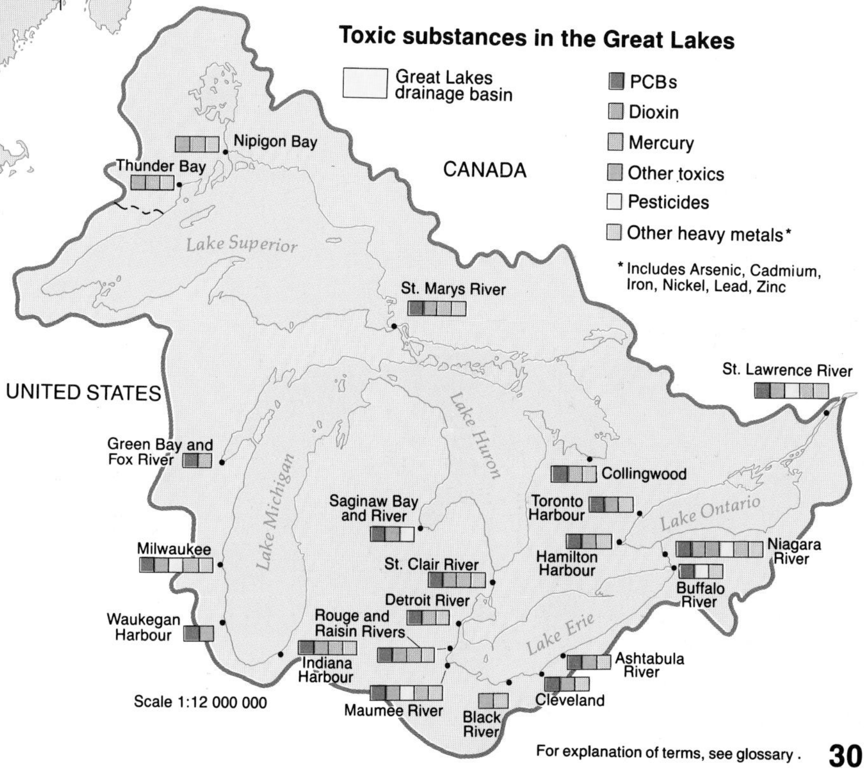

- Major drainage basin
- Minor drainage basin
- Artificially diverted drainage area
  1. Nechako
  2. Lake St. Joseph
  3. Ogoki
  4. Long Lake
- Internal drainage area
- Gauging station

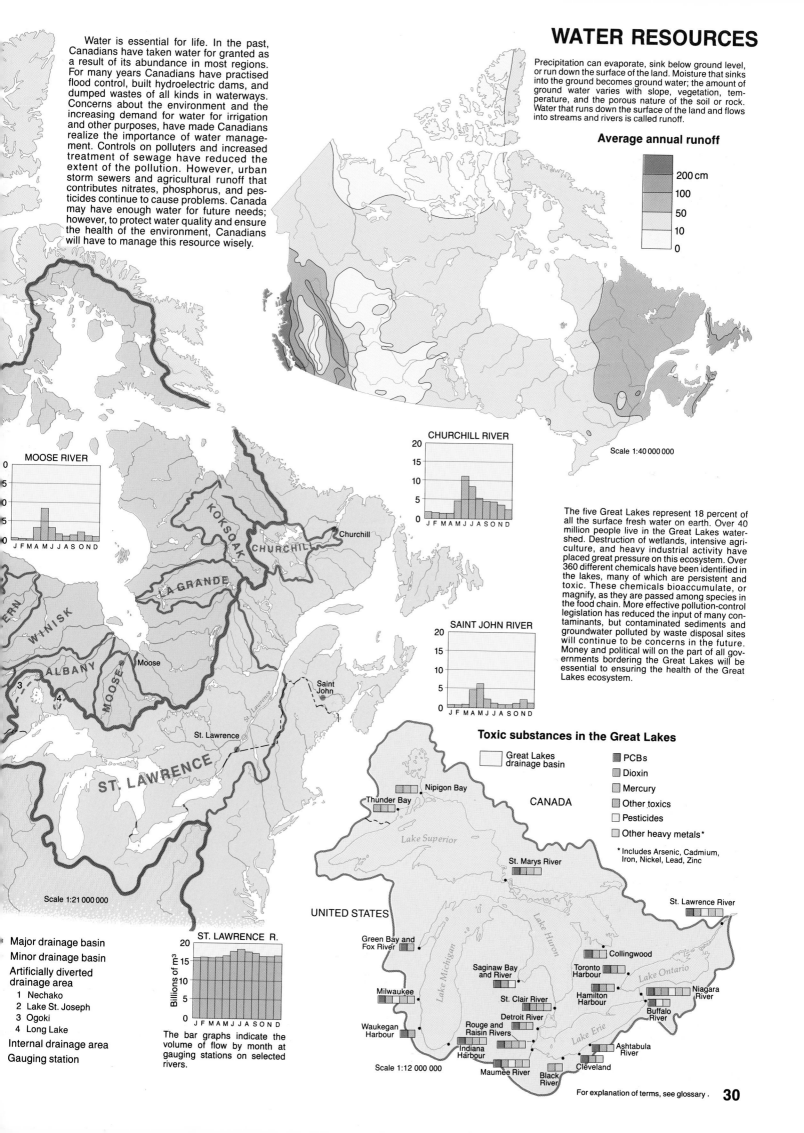

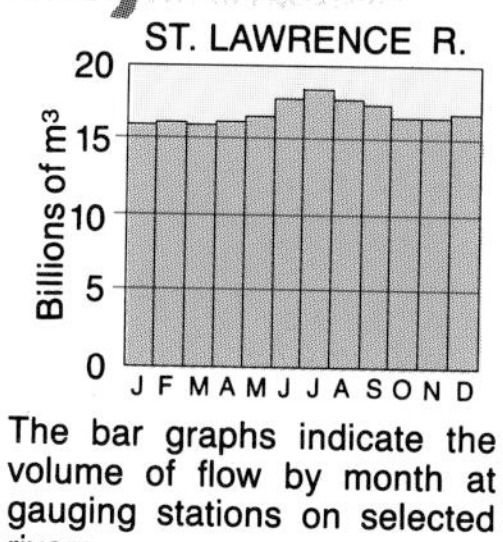

The bar graphs indicate the volume of flow by month at gauging stations on selected rivers.

For explanation of terms, see glossary.

CANADA-THEMATIC

Soils are grouped into seven classes, depending upon their potential for agriculture. The first three classes of soils — one, two and three — are considered capable of sustained production of commonly cultivated crops. Class one soils have no limitations for agriculture, while class two and class three soils have moderate and moderately severe limitations. These limitations can include climatic conditions, susceptibility to erosion, fertility, drainage and stoniness. The fourth to seventh soil classes include soils that are marginally arable to soils unsuitable even for pasture.

Soil capability classes one, two and three

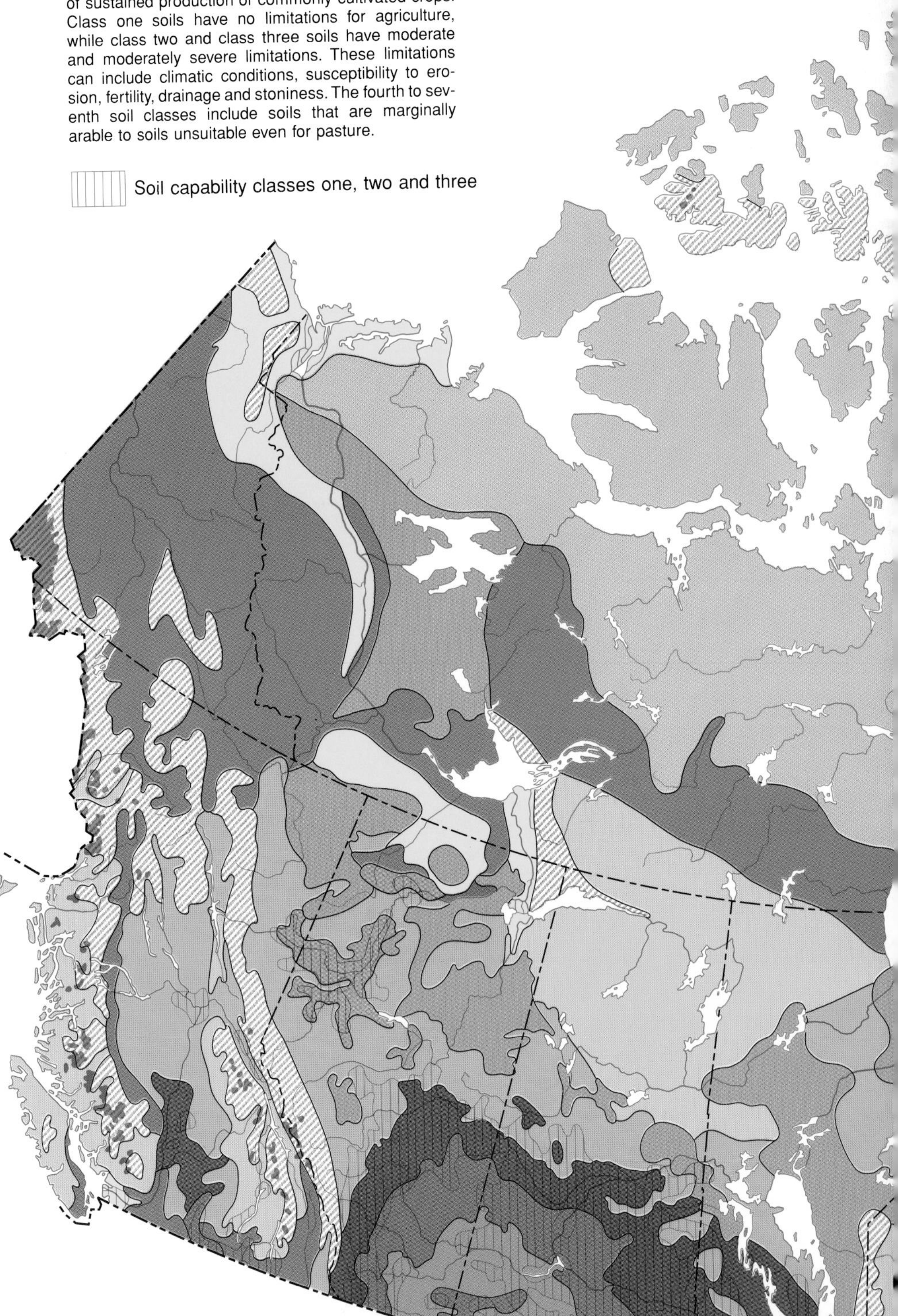

A typical soil profile

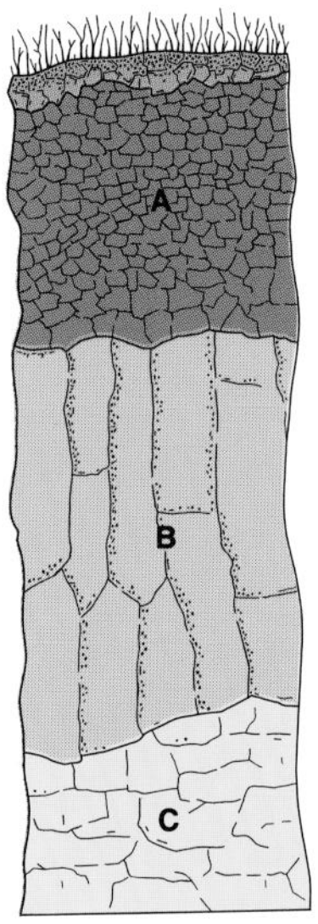

Organic layer

A horizon (topsoil)
The A horizon is mostly made up of partially decayed organic materials. At the bottom is a modified zone in which water, percolating down, has removed soluble minerals and fine particles.

B horizon (subsoil)
The B horizon is a combination of mineral and organic material. It contains the soluble minerals and fine particles from the A horizon. It is often possible to identify different layers within this horizon.

C horizon
The C horizon consists mainly of weathered "parent material," the mineral material from which the soil is made. This can be either bedrock or glacial deposits.

The profile above shows the general structure of soil. Individual profiles of some soil types shown on the map are given below.

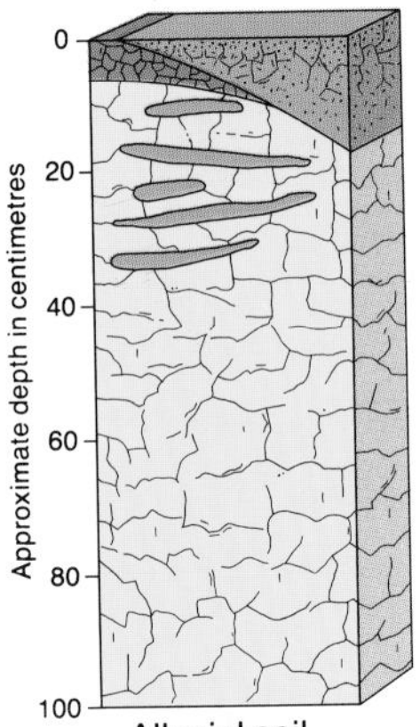

Alluvial soil

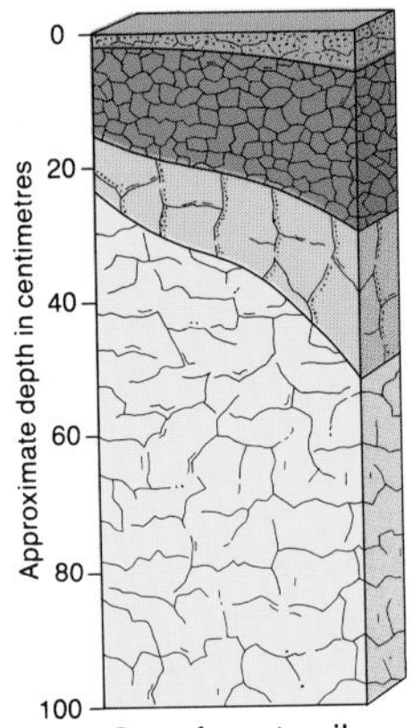

Grey forest soil

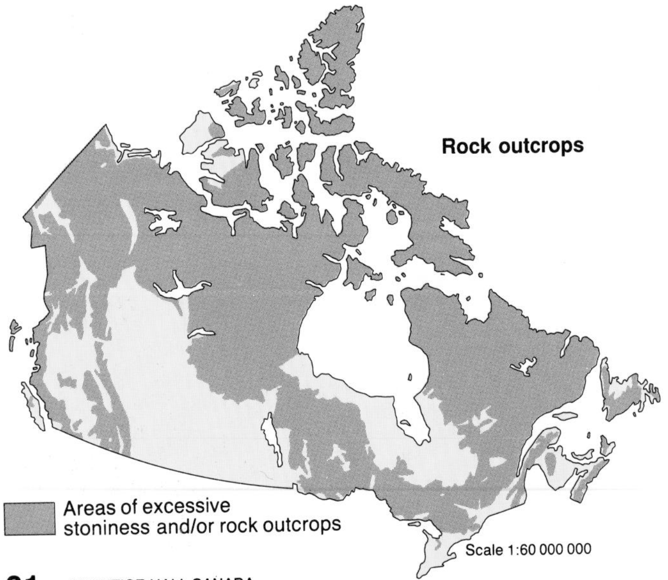

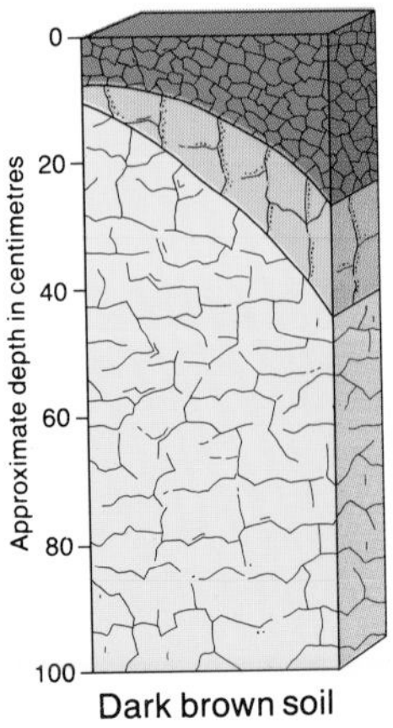

Dark brown soil

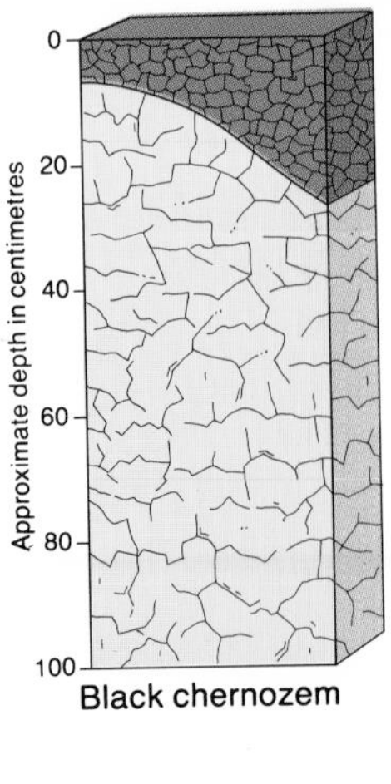

Black chernozem

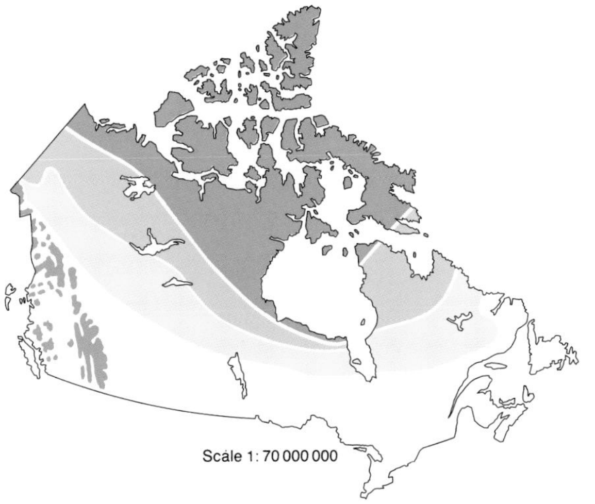

Permafrost

Permafrost is the condition of soil and rock when their temperature remains below 0°C for more than one year.

- Continuous permafrost
- Widespread permafrost
- Scattered permafrost

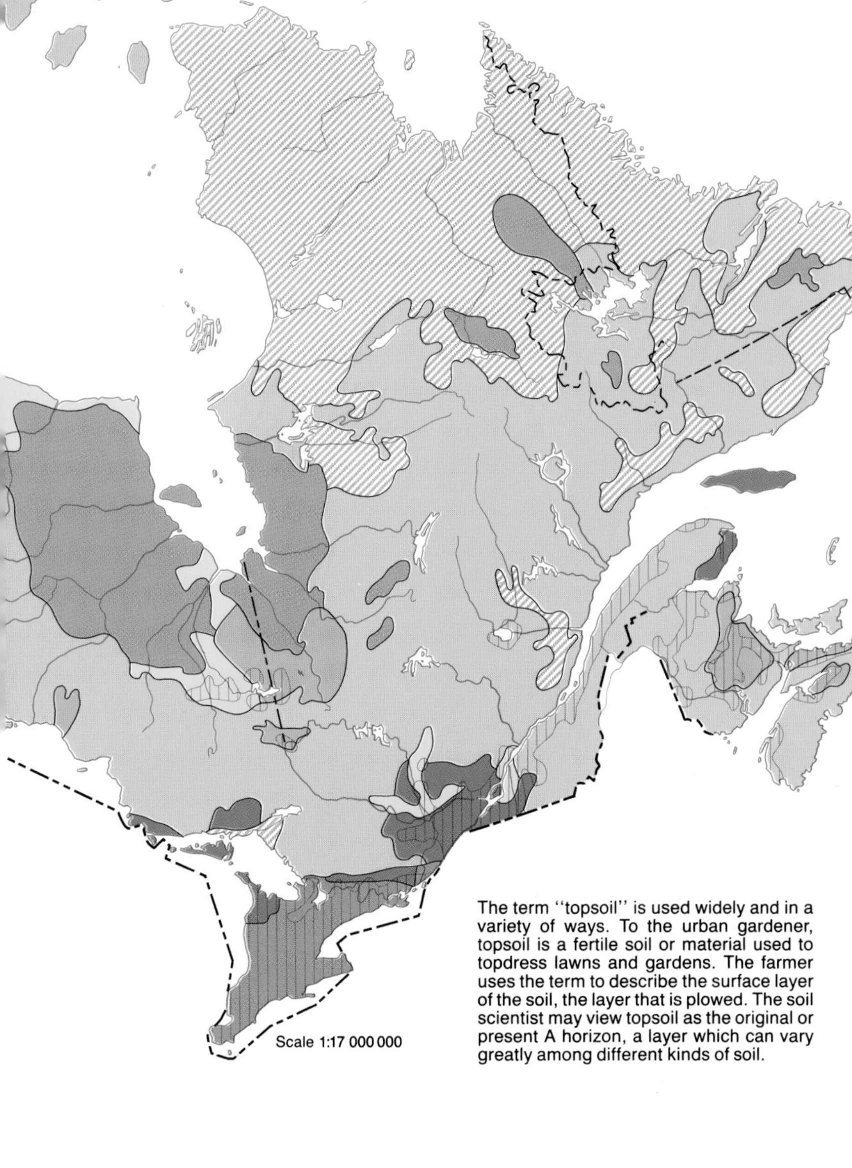

Arctic soils

- Tundra

Forest soils

- Podzol
- Grey-brown podzol
- Brown podzol
- Grey forest soil
- Black forest soil

Grassland soils

- Brown soil
- Dark brown soil
- Black chernozem

Azonal soils

- Dominantly organic soil
- Alluvial soil
- Dominantly rockland
- Glaciers

Soil is an important resource. The soils in different areas reflect climate, natural vegetation and parent material. To a lesser extent, slope and drainage also help to determine the type and fertility of the soil. Soils can be classified according to their characteristics. Those that have soil profiles reflecting the influence of climate and vegetation are called zonal soils. Azonal soils do not have soil profiles.

Leaching and calcification are two climate-related processes that are important in determining soil characteristics. Leaching occurs when heavy precipitation carries soluble minerals from the topsoil into the subsoil. This removes the nutrients that plants need, thereby making the soil less fertile.

Calcification occurs in dry climates. Water is drawn up to the surface by capillary action and evaporates from the surface. The minerals originally dissolved in the water are deposited in the topsoil. Although some minerals help plant growth, too high a concentration can poison plants.

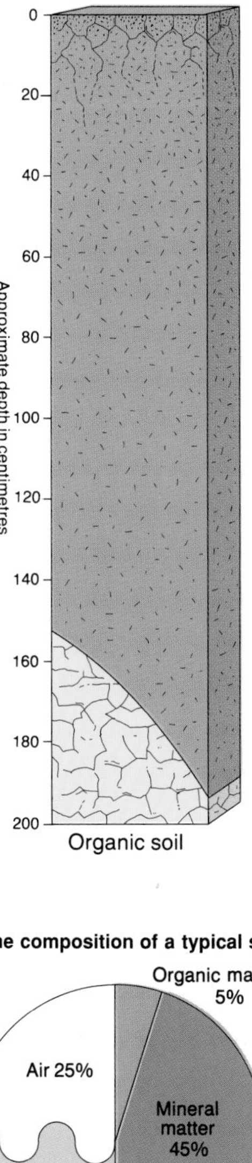

Organic soil

The term "topsoil" is used widely and in a variety of ways. To the urban gardener, topsoil is a fertile soil or material used to topdress lawns and gardens. The farmer uses the term to describe the surface layer of the soil, the layer that is plowed. The soil scientist may view topsoil as the original or present A horizon, a layer which can vary greatly among different kinds of soil.

Grey-brown podzol

Volume composition of a typical soil

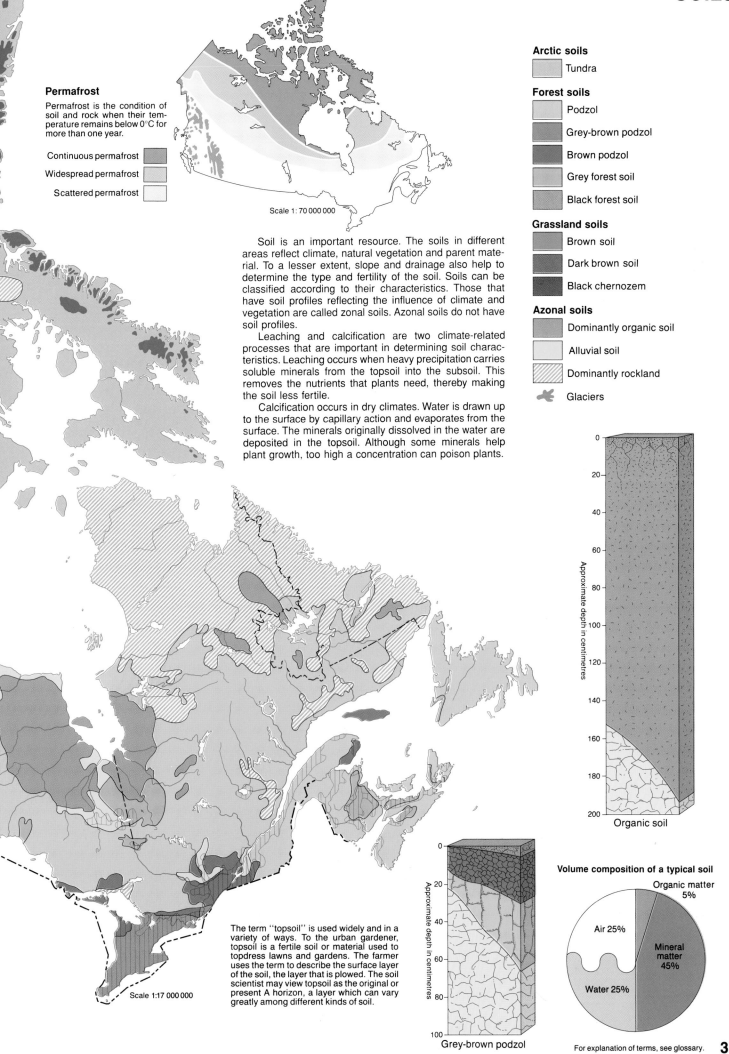

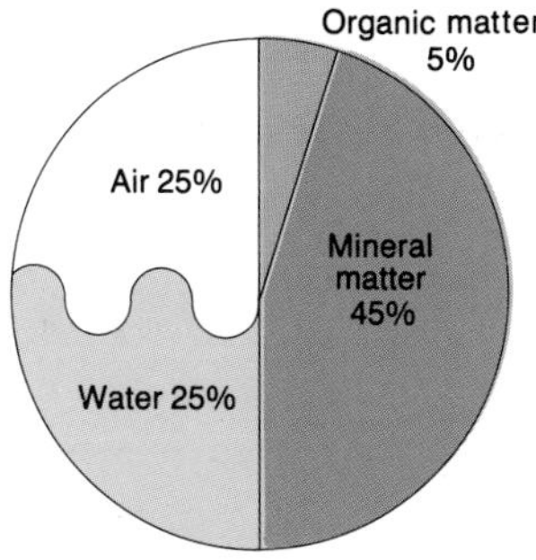

For explanation of terms, see glossary.

CANADA-THEMATIC

The changing farm

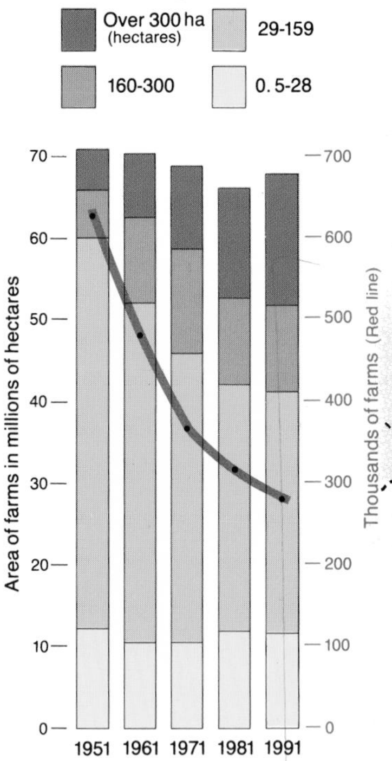

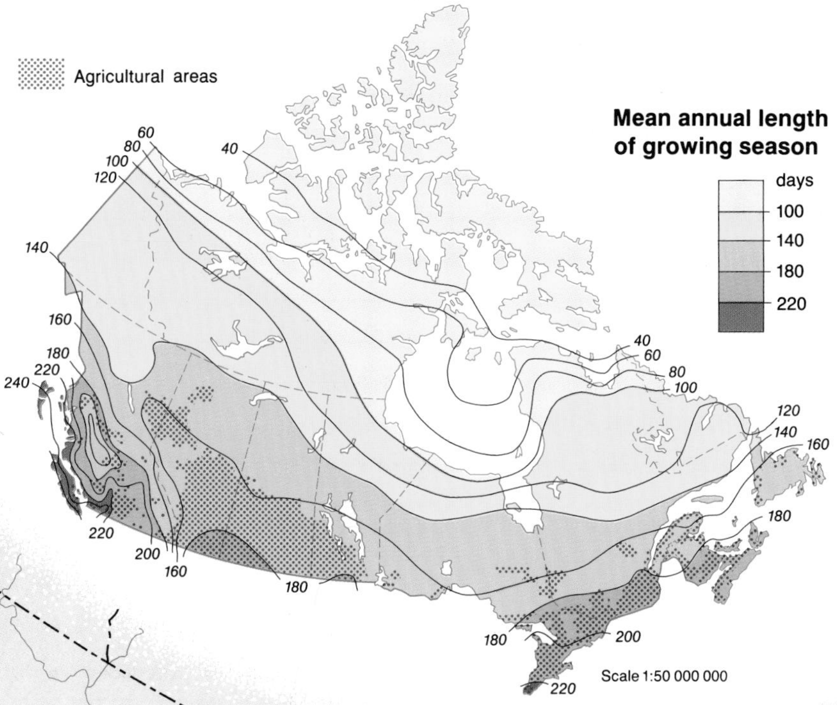

Total land in farms
67 754 891 hectares

Cropland
33 507 780 hectares

Summer fallow
7 920 948 hectares

Improved pasture
4 141 221 hectares

Other land
22 183 751 hectares

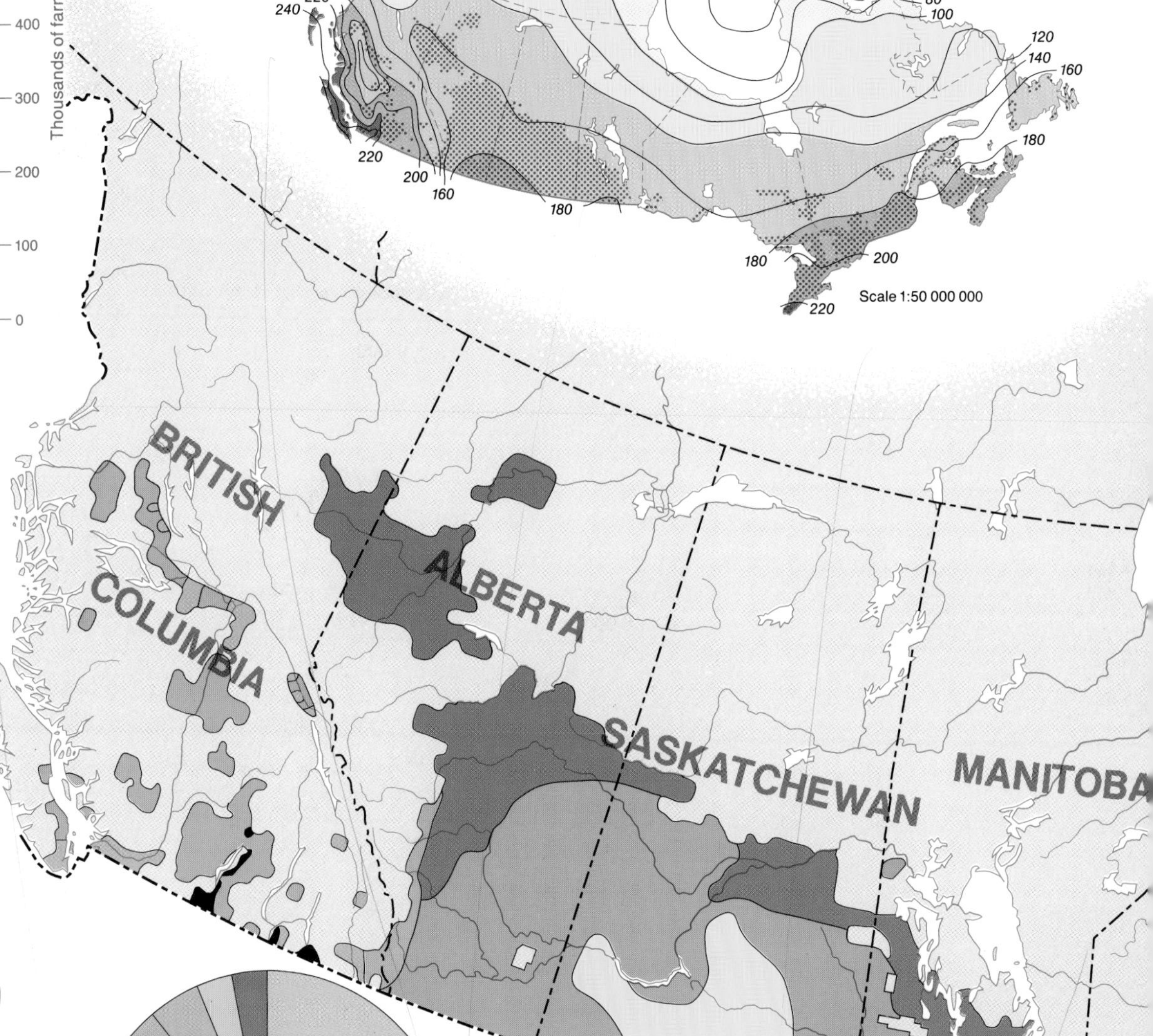

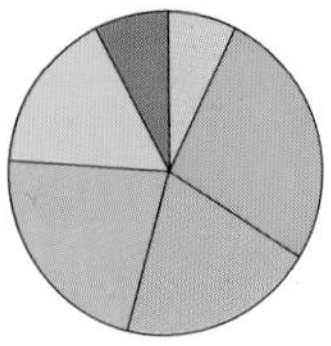

BRITISH COLUMBIA
$1 230 961 000

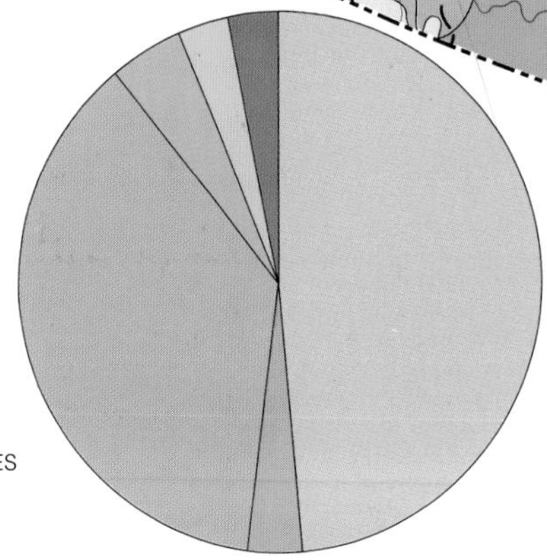

ALBERTA
$4 203 509 000

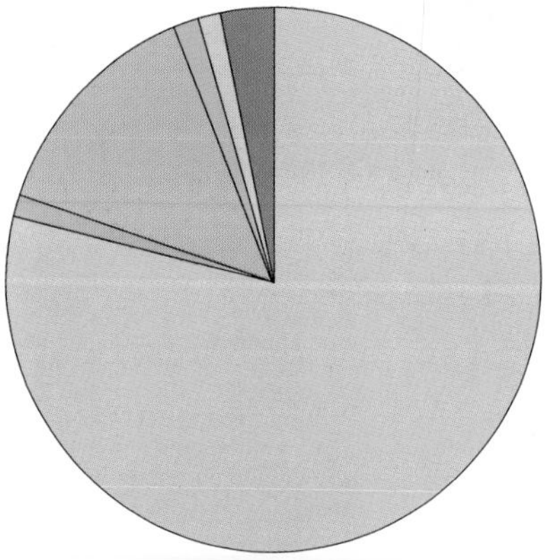

SASKATCHEWAN
$4 058 385 000

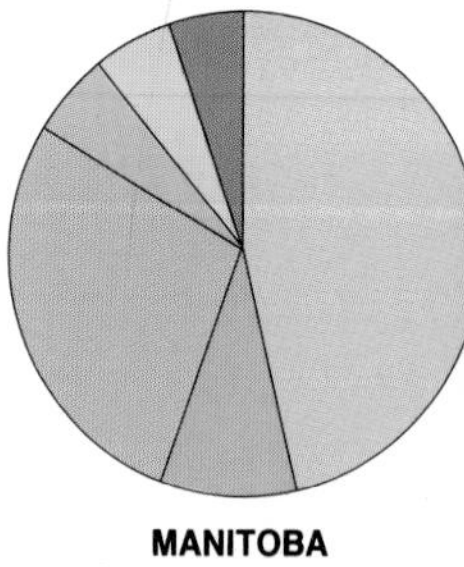

MANITOBA
$1 949 840 000

Average farm size

Canadian average: 231 hectares

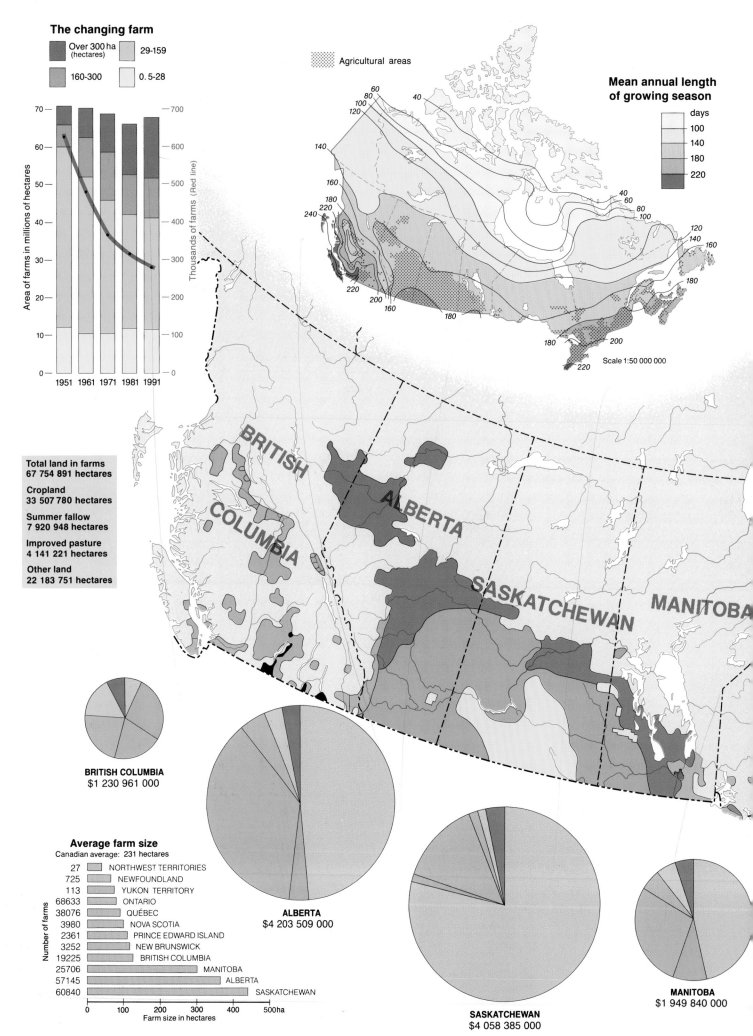

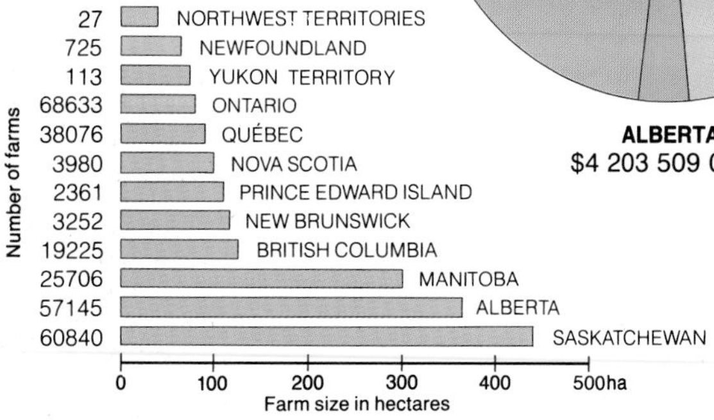

Most productive agriculture in Canada (90 percent by value) is found in the southern part of the Prairie provinces and in southern Ontario and Québec. Some products are associated with certain regions, for example, wheat with Saskatchewan. However, many areas of Canada practice "mixed" farming, producing combinations of crops and livestock.

Canadian agriculture is highly influenced by the global marketplace. Farmers need to produce excellent quality food at low prices in order to compete. Farms are generally large or highly mechanized, or both. However, there is increasing recognition that there are problems associated with intensive farming. Soil erosion, salinization and loss of soil fertility are difficulties facing many farmers, especially in areas where one crop dominates agriculture. These problems are compounded by growing public concern about the residues of pesticides in food and the environment, as well as the buildup of nitrates from fertilizers in water supplies.

Use of fertilizer and pesticides

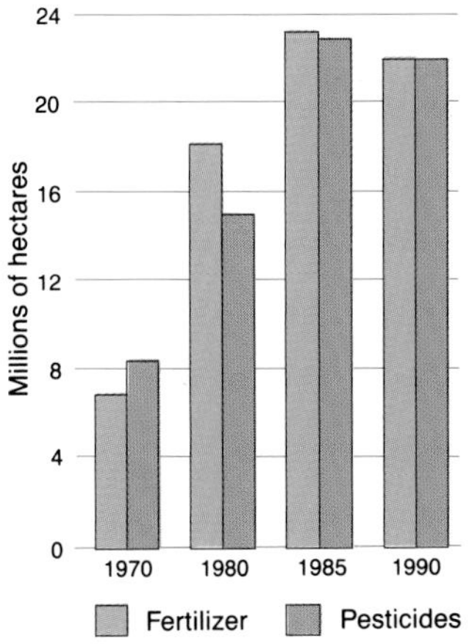

Value of provincial agricultural production

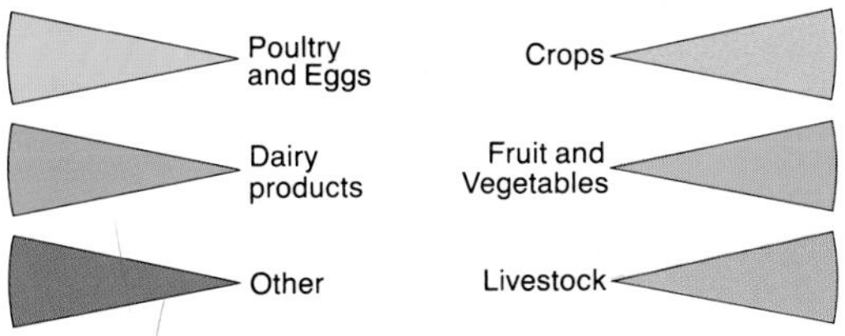

Types of farming

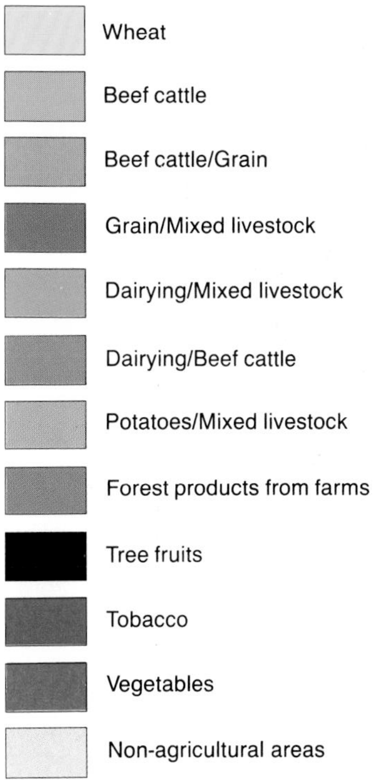

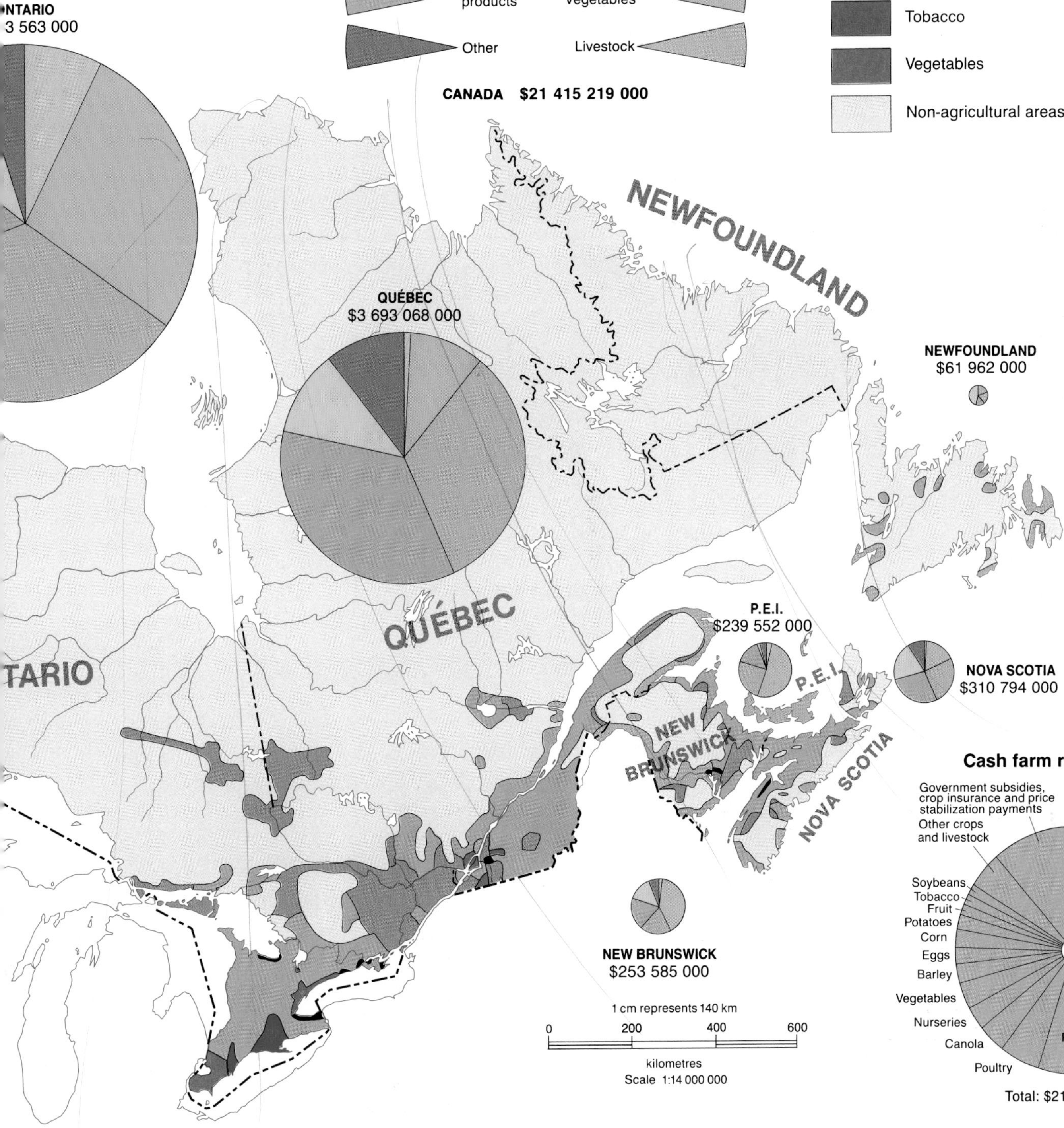

Cash farm receipts, 1991

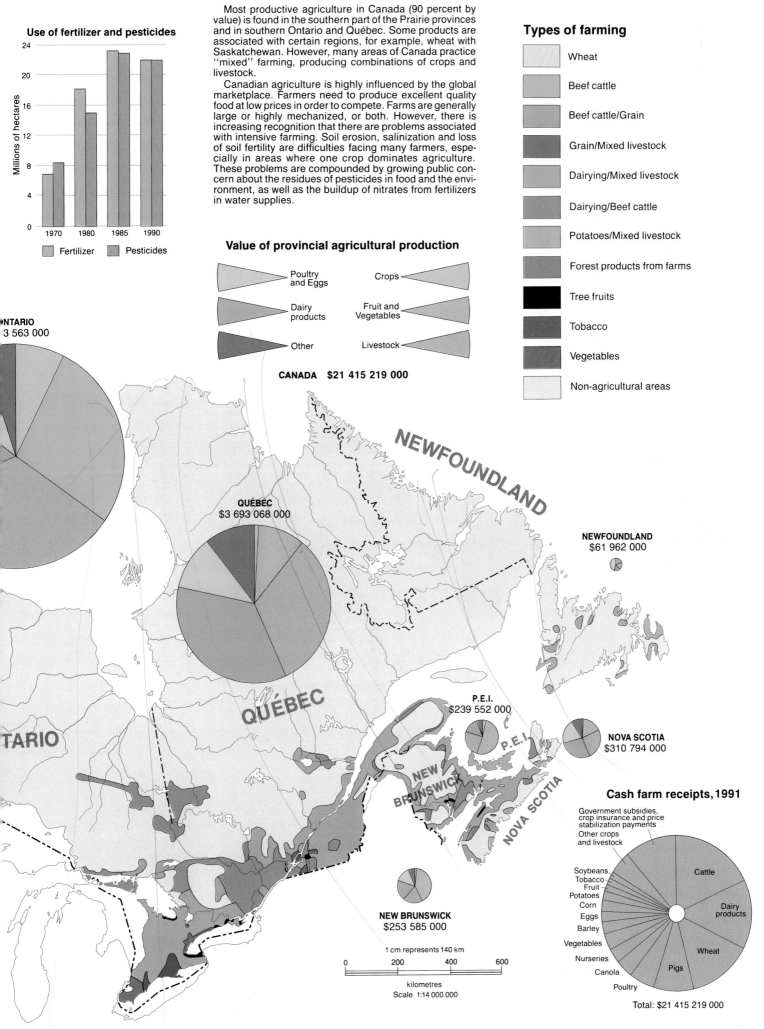

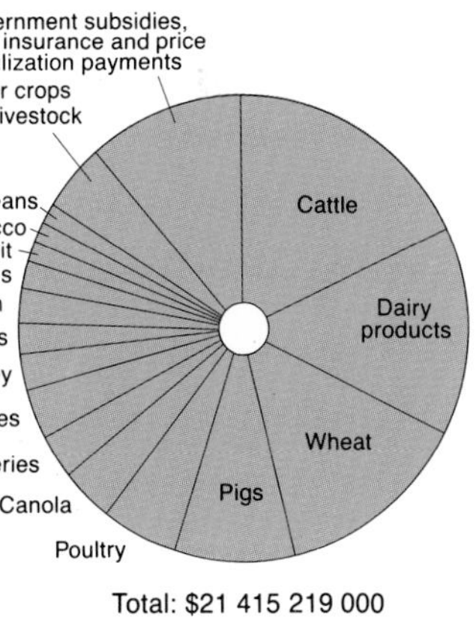

Total: $21 415 219 000

CANADA-THEMATIC

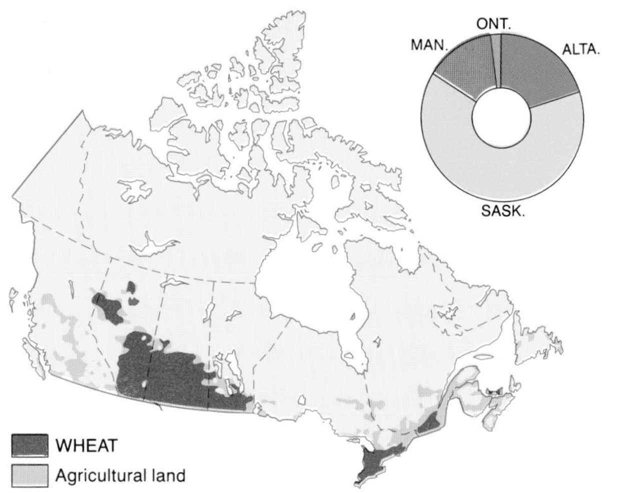

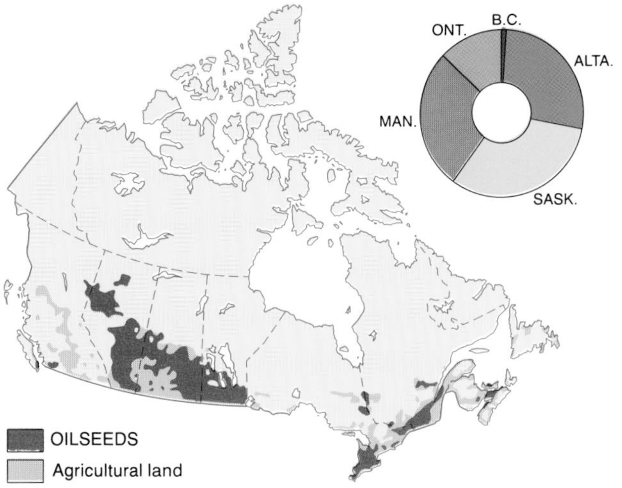

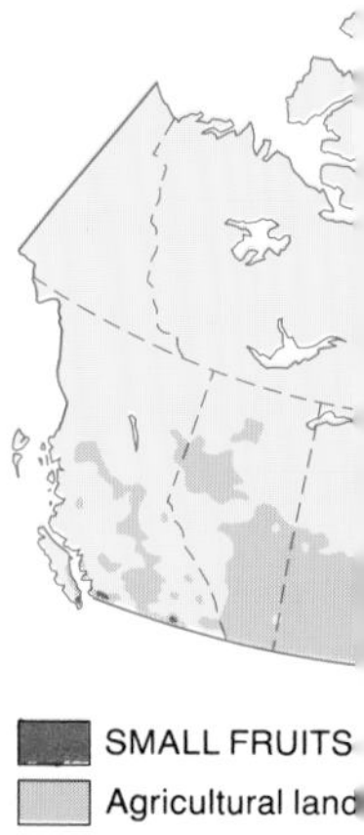

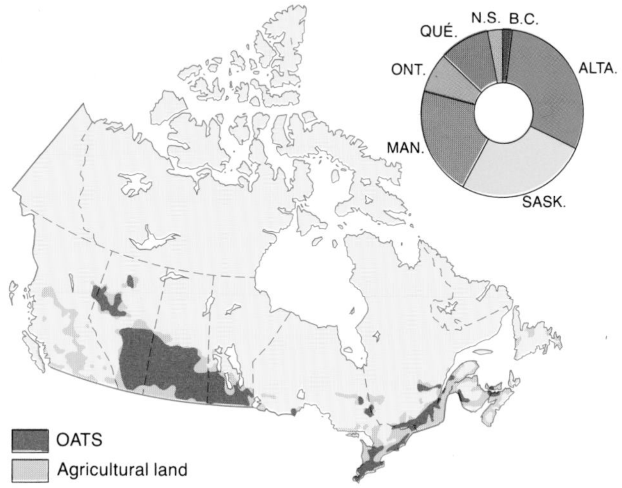

Value of agricultural imports

Total: $5 710 000 000

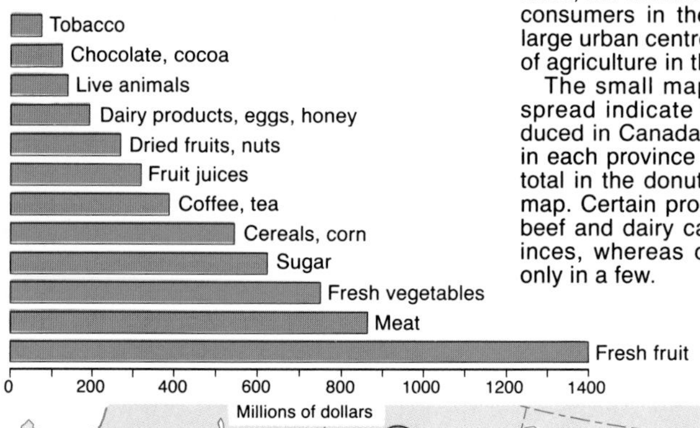

Agriculture is practiced in every Ca province. The land used for farming a diversity of agriculture reflect the physi tures, the climate and the market dema consumers in the area. The varied ne large urban centres often increase the d of agriculture in the surrounding areas.

The small maps around the edges spread indicate where each product duced in Canada. The value of the pro in each province is shown as a percen total in the donut graphs accompanyin map. Certain products, such as potato beef and dairy cattle, are found in mos inces, whereas corn and tobacco are only in a few.

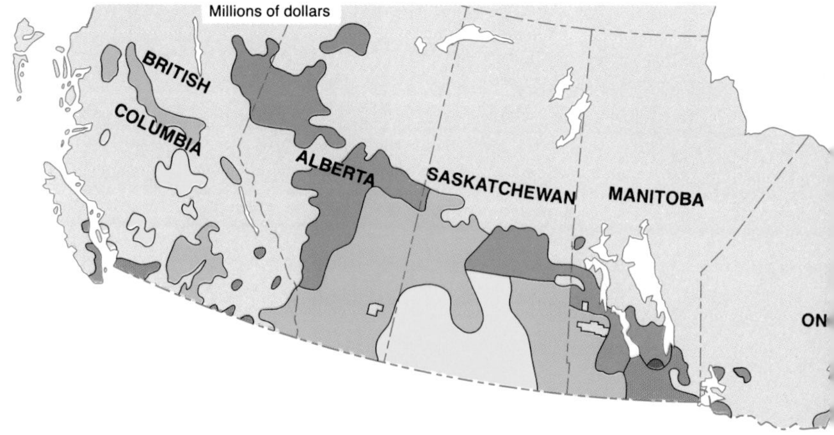

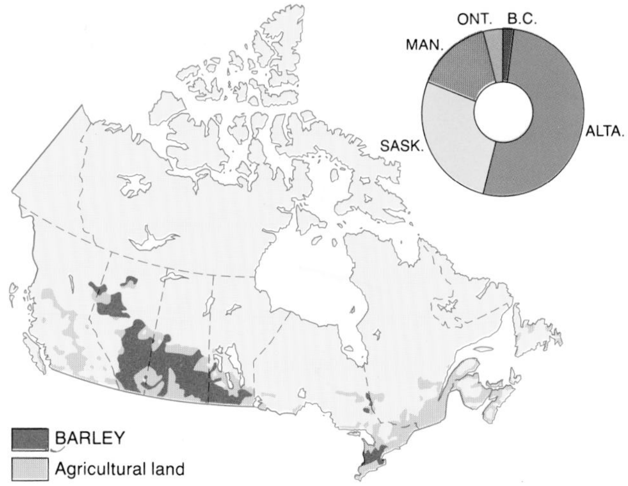

The diversity of agriculture

The number of products that are significant to the value of sales from the farm

9 or 10 | 7 or 8 | 4, 5, 6 | 2 or 3 | 1

The above map emphasizes t ber of products that are importa agricultural economy in any ar does not suggest, for examp southern Saskatchewan gro wheat. However, the volume a of other products do not contri nificantly to the area's econom

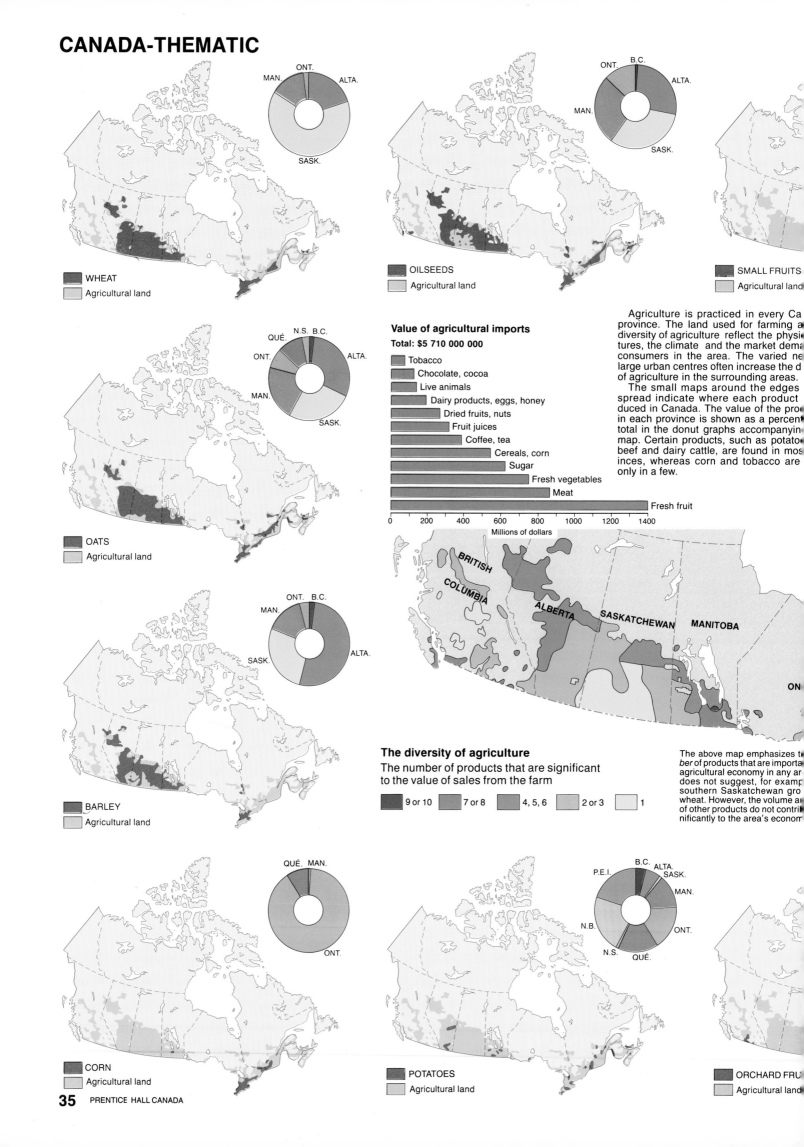

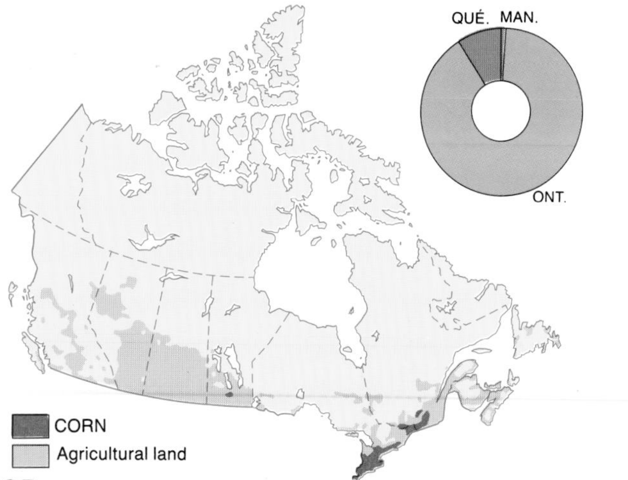

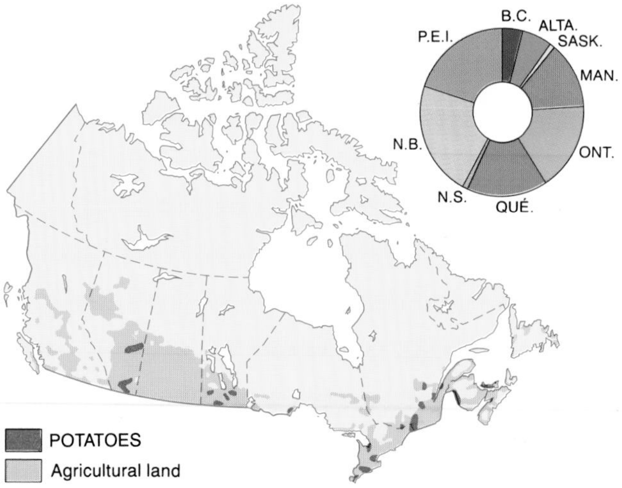

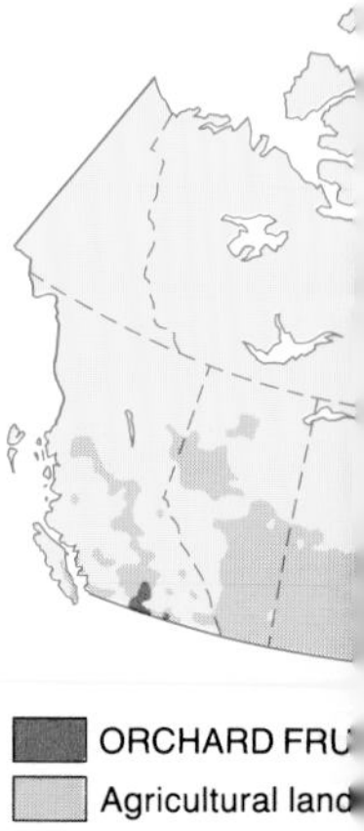

AGRICULTURAL DIVERSITY
N.B. P.E.I. B.C. MAN. N.S. QUÉ. ONT.
QUÉ. P.E.I. ONT.
TOBACCO
Agricultural land
N.S. N.B. P.E.I. B.C. QUÉ. ALTA. ONT. MAN. SASK.
BEEF CATTLE
Agricultural land
Millions of dollars
3800 3600 3400 1600 1400 1200 1000 800 600 400 200 0
Wheat
Meat
Live animals
Canola
Vegetables
Barley
Other cereals
Sugar
Tobacco
Dairy products, eggs, honey
Fruits
Other
Value of agricultural exports
Total: $8 993 000 000
lian agriculture is heavily influ- by the global marketplace. ts of agricultural products make se to one half of Canadian farm receipts. In order to protect its rs and keep them in business, la has developed a system of lies and management boards. ver, this can work against Cana- armers when other countries use strategies that keep Canadian cts out of their markets.
N.B. P.E.I. N.S. B.C. ALTA. SASK. MAN. QUÉ. ONT.
DAIRY CATTLE
Agricultural land
NEWFOUNDLAND
QUÉBEC
P.E.I.
N.B.
NOVA SCOTIA
Scale 1:25 000 000
Total labour force: 13 513 000
Agricultural labour force: 441 000
Labour force in agriculture
Newfoundland
British Columbia
Nova Scotia
New Brunswick
Québec
Ontario
Manitoba
Alberta
Prince Edward Is.
Saskatchewan
20 15 10 5 0
Percent
N.B. P.E.I. N.S. B.C. ALTA. QUÉ. SASK. MAN. ONT.
HOGS
Agricultural land
N.S. N.B. B.C. QUÉ. ONT.
N.B. P.E.I. N.S. B.C. QUÉ. ALTA. ONT. MAN. SASK.
SHEEP
Agricultural land
Provincial share of each commodity
N.S. N.B. B.C. ALTA. QUÉ. SASK. MAN. ONT.
POULTRY AND EGGS
Agricultural land
Scale 1:70 000 000
For explanation of terms, see glossary.

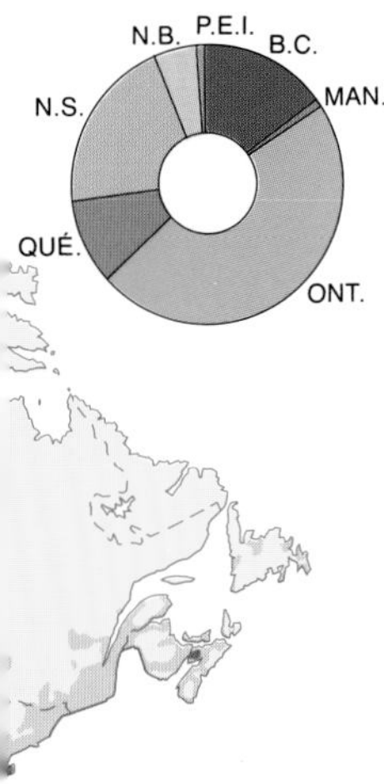

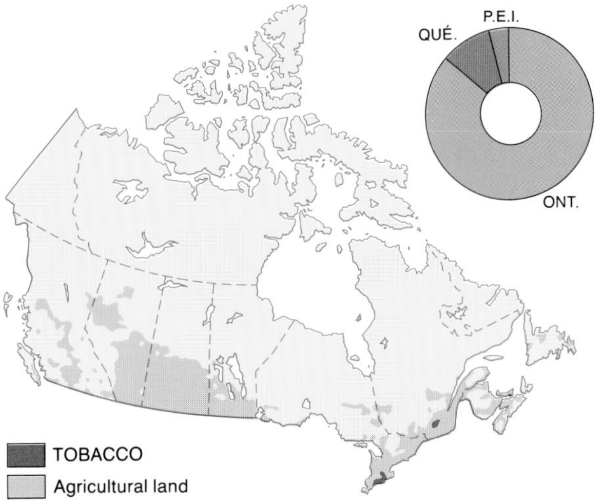

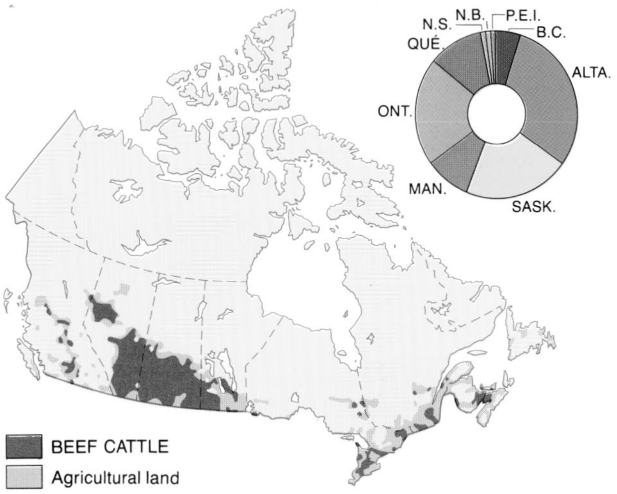

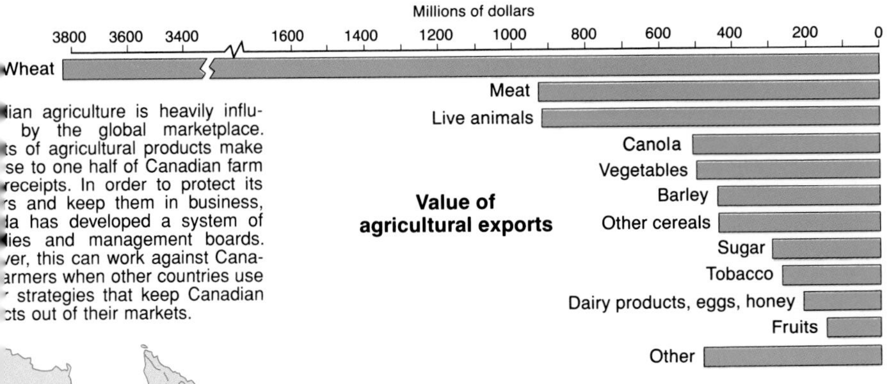

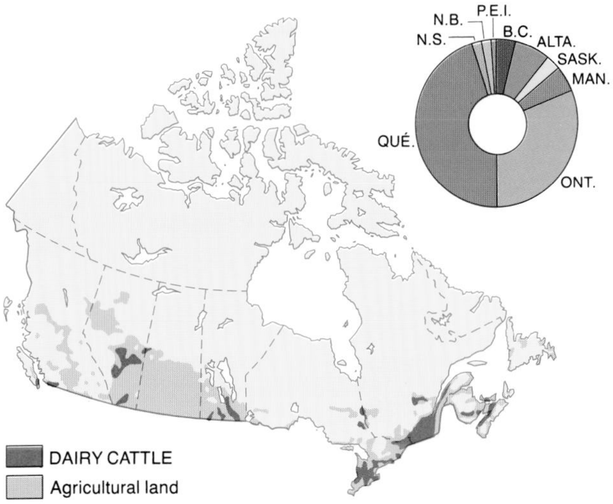

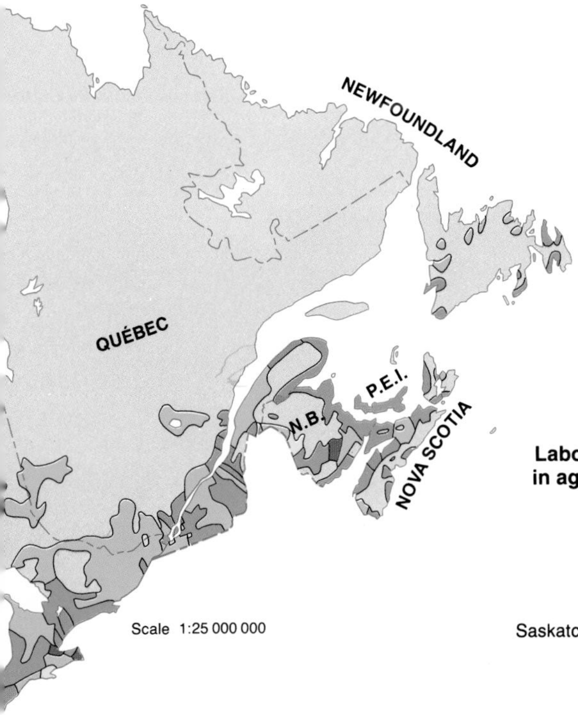

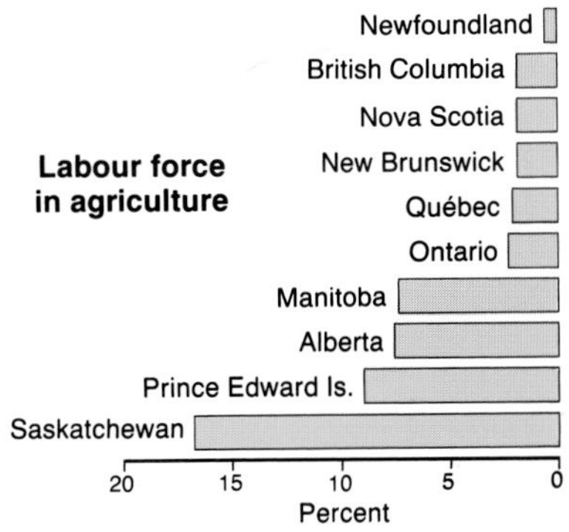

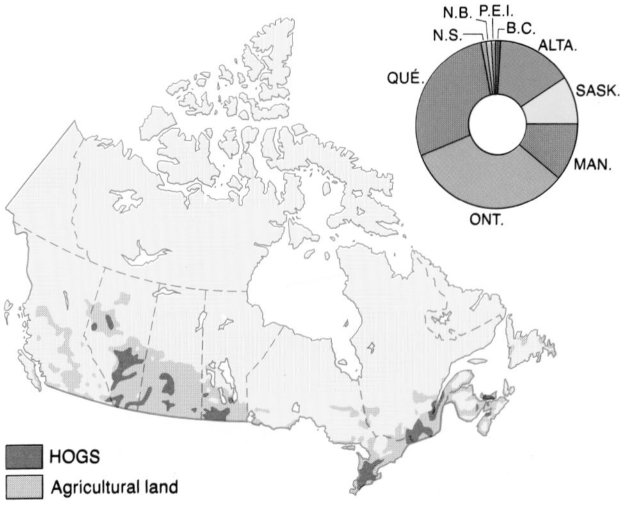

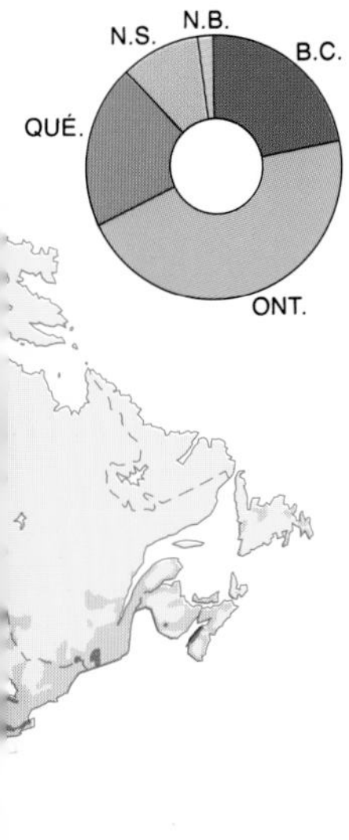

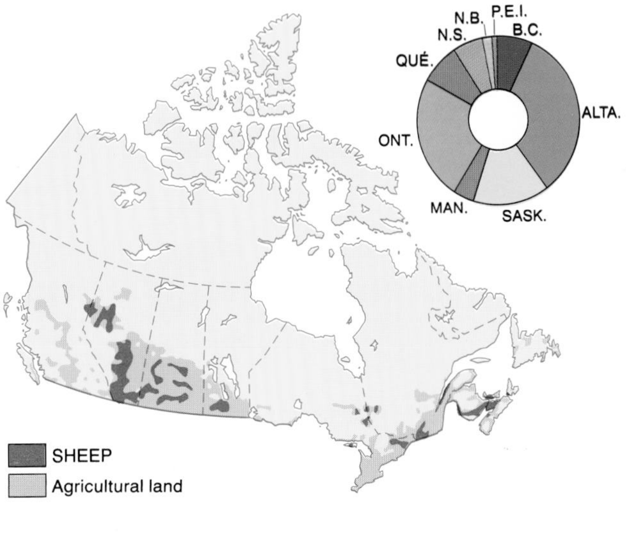

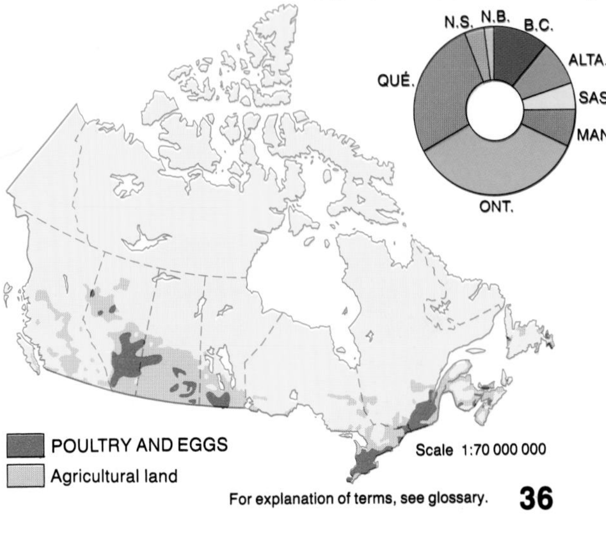

CANADA-THEMATIC

Most of Canada is covered by forest, tundra or grassland. These broad areas can be further divided into the vegetation regions shown on the map. The different regions reflect the influence of climate, particularly temperature and precipitation. Within each region, differences in the kind and extent of vegetative cover are caused by factors such as rock type, soil type, slope and drainage conditions. People have changed the state of the vegetation in many areas. For example, in grassland, parkland and broadleaf forest regions, agricultural and urban land uses have been introduced and the original natural vegetation has disappeared.

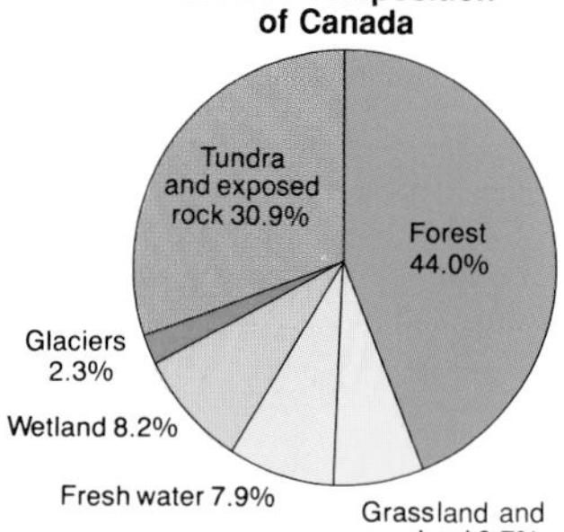

Ancient trees of the coastal forest, Vancouver Island

Grasslands of southern Saskatchewan

Boreal forest

NATURAL VEGETATION

Tundra of northern Baffin Island

Vegetation and characteristic flora

Tundra	Lichen/heath
	Dwarf shrubs/sedges/ lichen/heath
	Alpine sedges/ grasses and shrubs
Open woodland	Scattered needleleaf trees/ broadleaf shrubs/heath/grass
Bogs	Moss/sedges/ strings of needleleaf trees
Boreal forest	Needleleaf trees
	Mostly needleleaf trees with some broadleaf trees
Coastal forest	Large needleleaf trees
Sub-alpine forest	Needleleaf trees
High plateau/Pine forest	Mostly needleleaf trees with some broadleaf trees and grassland
Southeastern mixed forest	Mixture of broadleaf and needleleaf trees
Southern broadleaf forest	Broadleaf trees
Parkland	Broadleaf or needleleaf trees with patches of grassland
Grassland	Low, medium, and tall grasses
	Glaciers and permanent snow (no vegetation)

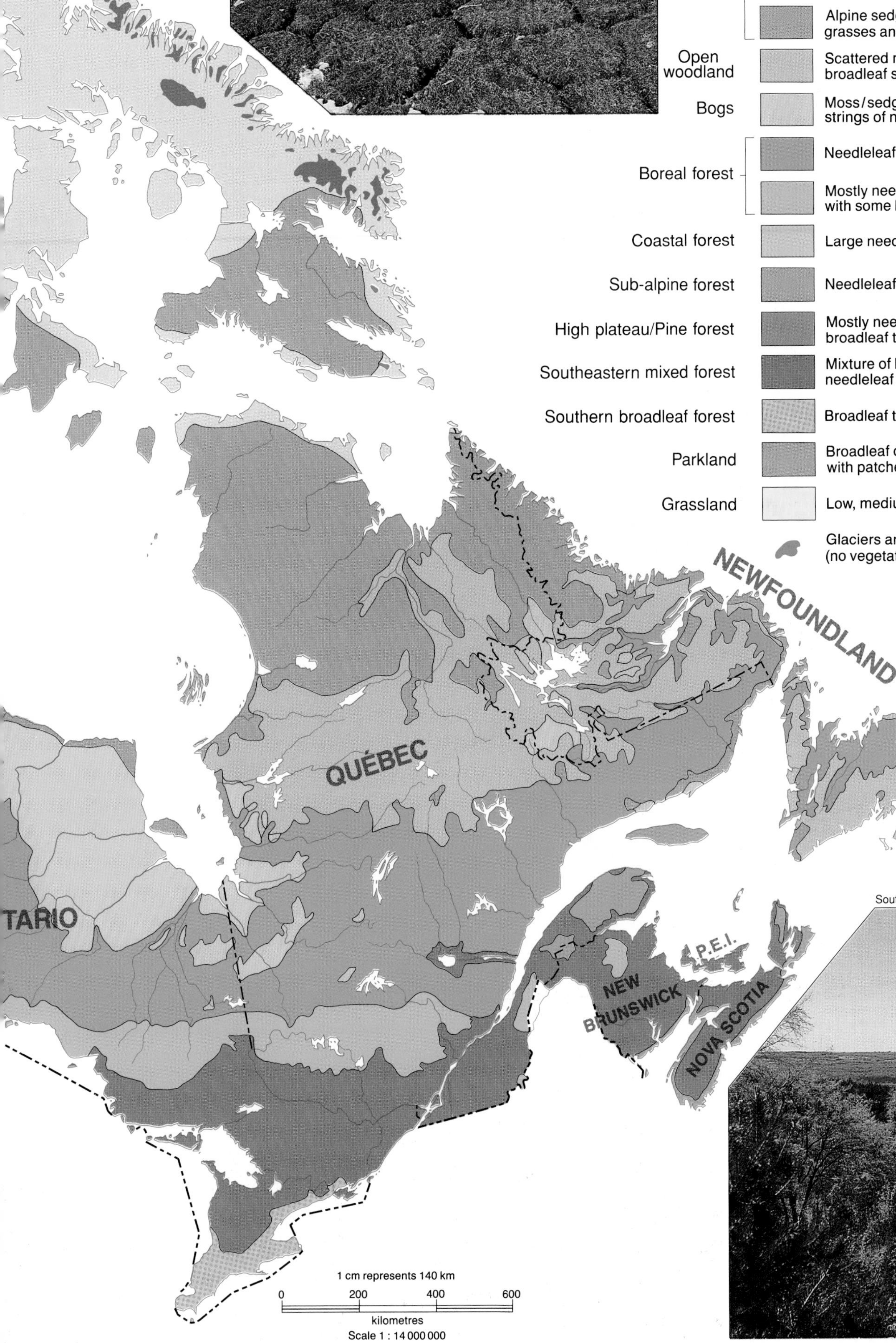

Southeastern mixed forest

For explanation of terms, see glossary.

CANADA-THEMATIC

Western Red Cedar

Douglas Fir

Lodgepole Pine

Western Hemlock

Trembling Aspen

Note: Trees are not drawn to a common scale. The names of hardwood trees are shown in red, softwood trees in black.

Black Spruce

Engelmann Spruce

Sugar Maple

Balsam Fir

Most coniferous trees (spruce, pine, fir) are called softwoods; many deciduous trees (maple, oak, cherry, hickory) are called hardwoods. Many areas of deciduous forest have been largely cleared of trees. Because of this scarcity of hardwoods, the furniture industry uses veneers (thin hardwood layers over softwood structures) more frequently.

Red Oak

Jack Pine

White Spruce

Eastern White Pine

White Birch

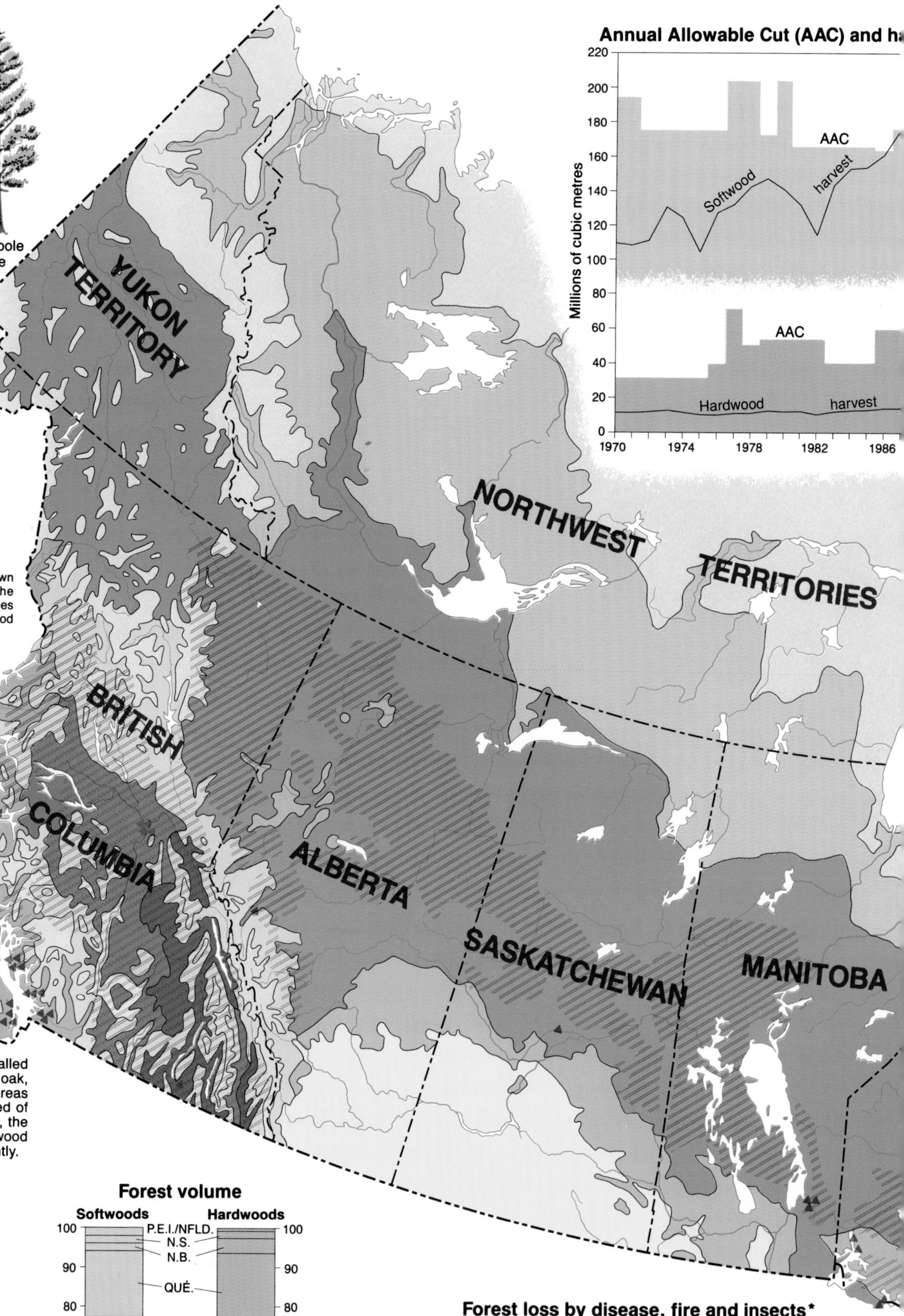

Forest volume

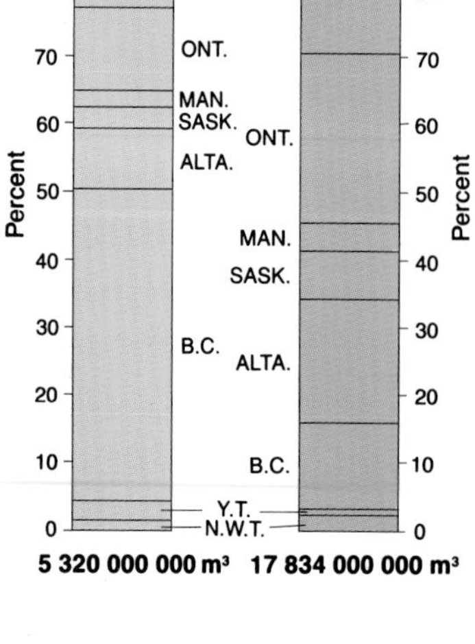

Forest loss by disease, fire and insects*
1989

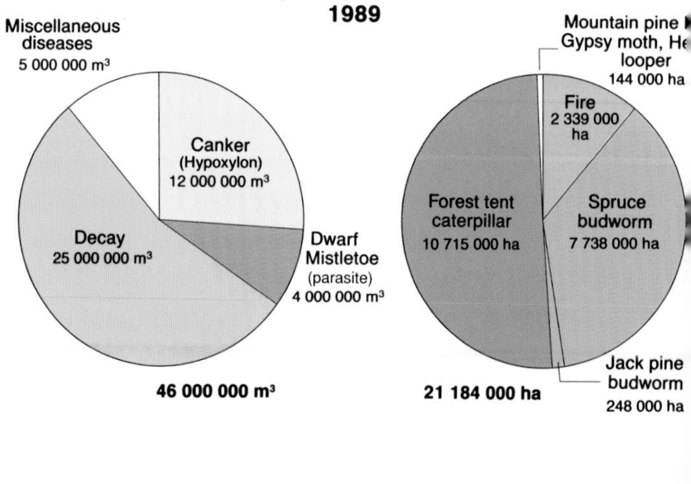

Canada has more forests than any other nation in the world. Forests cover close to half of Canada's land area, with each region having typical tree species and animal life. Standing forests provide wildlife habitat, prevent erosion, and offer a variety of recreational activities.

Logging and related forest industries directly employ 300 000 Canadians. About 900 communities across Canada rely at least partially on the forest industry for their livelihood. However, the logging of old-growth forests has become controversial.

Forests supply raw materials for many products, such as lumber, paper, plywood, cellophane, cartons and furniture. Canada ranks first in the world for production of newsprint, second for wood pulp and third for softwood lumber. Almost 50 percent of Canada's wood products are exported, mainly to the United States, the European Union and Pacific Rim countries.

World exports of forest products

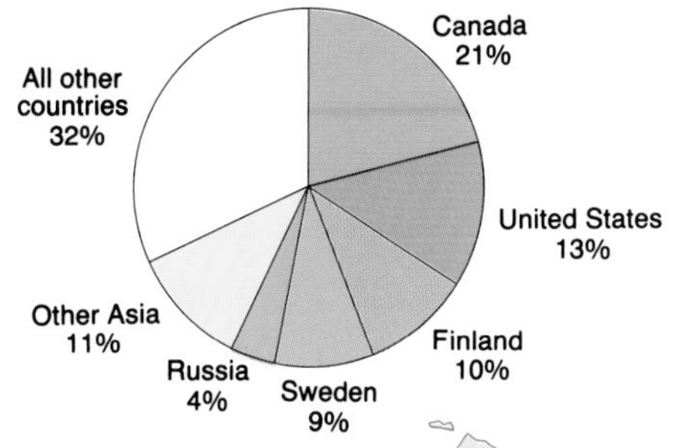

Forest regions and principal tree species

	Forest region	Principal tree species
Non-productive forest	Boreal–forest and barren ground	Black Spruce, White Spruce, Tamarack
	Boreal–forest and grassland	Trembling Aspen, Willow, Bur Oak
Productive forest	Boreal–predominantly forest	Black Spruce, White Spruce, Balsam Fir, Jack Pine, White Birch, Trembling Aspen
	Subalpine	Alpine Fir, Engelmann Spruce, Lodgepole Pine
	Montane	Douglas Fir, Lodgepole Pine, Ponderosa Pine, Trembling Aspen
	Coast	Western Red Cedar, Western Hemlock, Douglas Fir, Sitka Spruce
	Columbia	Western Red Cedar, Western Hemlock, Douglas Fir
	Deciduous	Beech, Maple, Black Walnut, Hickory, Oak
	Great Lakes-St. Lawrence	Eastern White Pine, Eastern Hemlock, Red Pine, Yellow Birch, Maple, Oak
	Acadian	Red Spruce, Balsam Fir, Maple, Yellow Birch
	Non-forested land	No major tree species

▲ Each symbol represents one pulp and paper mill with a production capacity of 300 t (tonnes) or more per day

Forest tenures granted to private companies

The forest regions map does not show the true extent of forest in Canada because some land has been converted to farmland or urban areas. Forest regions can be classified as productive or unproductive forest. Productive lands can produce commercially valuable crops of timber reasonably quickly. In unproductive lands, tree growth is slow, usually because of low temperatures, low precipitation or other extreme conditions.

orests in Canada

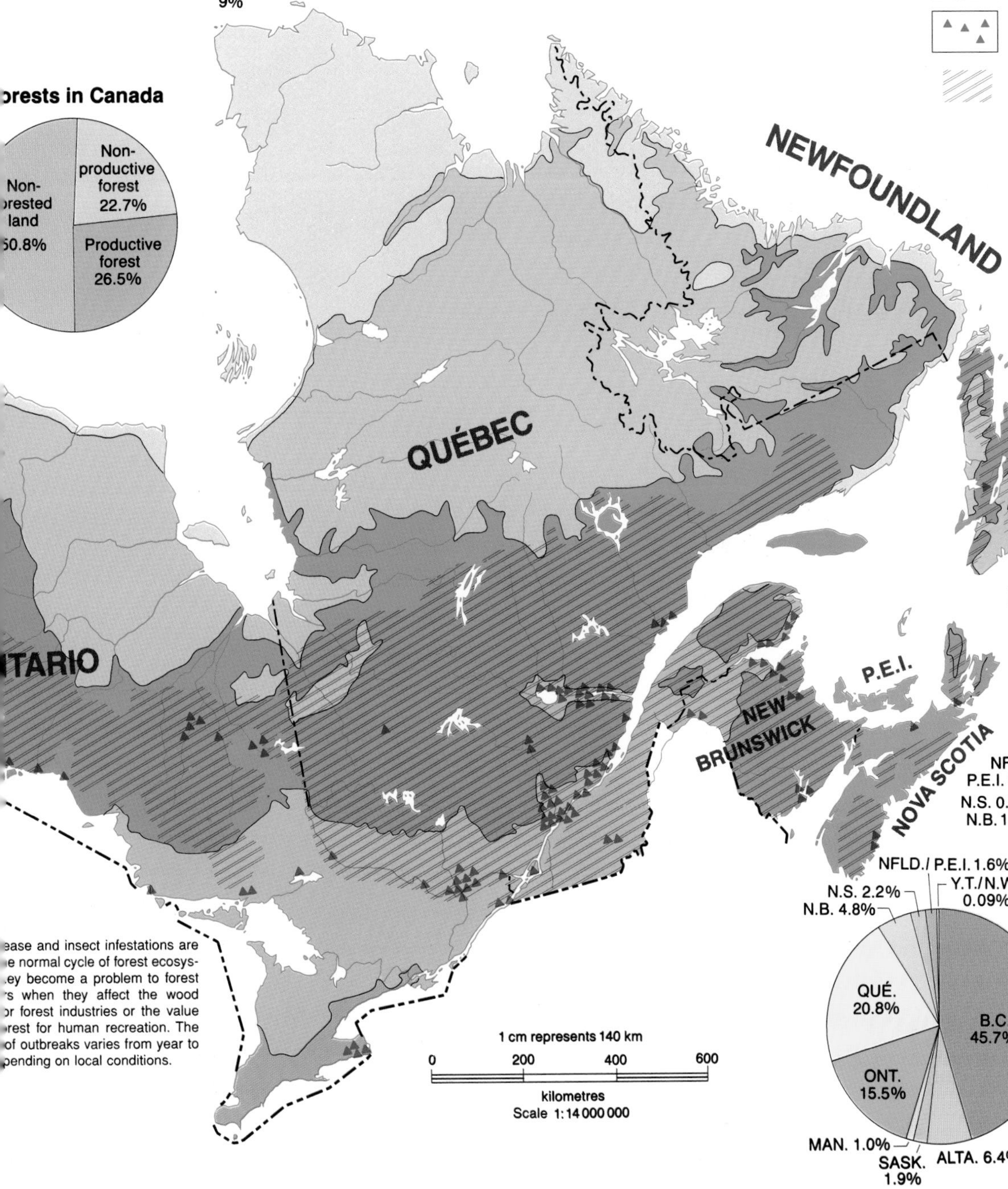

ease and insect infestations are e normal cycle of forest ecosys- ey become a problem to forest 's when they affect the wood or forest industries or the value rest for human recreation. The of outbreaks varies from year to pending on local conditions.

Provincial share of forested land

453 500 000 hectares

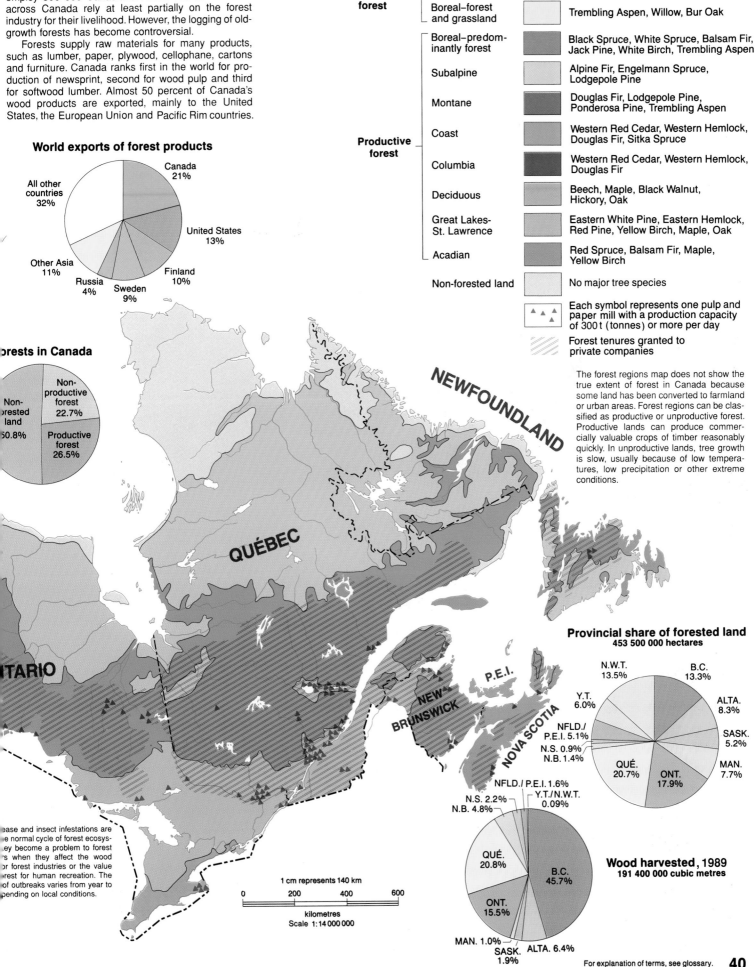

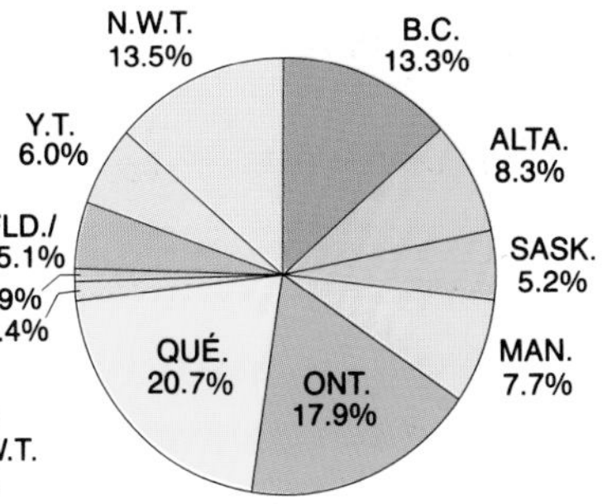

Wood harvested, 1989

191 400 000 cubic metres

For explanation of terms, see glossary.

Biogeoclimatic zones—southwestern B.C.

- Alpine zone
- Mountain Hemlock zone
- Wetter coastal — Western Hemlock subzones
- Drier coastal — Western Hemlock subzones
- Wetter coastal — Douglas Fir subzones
- Drier coastal — Douglas Fir subzones
- Wetter hypermaritime — Coastal Western Hemlock variant
- Urban areas
- Area of old growth temperate rainforest on Vancouver Island, 1990
- Parks *

Foresters classify forests into biogeoclimatic subzones and zones, defined according to vegetation, soils and climate. British Columbia has many different biogeoclimatic zones.

The forest of the Coastal Western Hemlock Zone is also known as temperate rainforest. These trees commonly live for 300 to 800 years, and grow to be as much as 95 metres high. In 1990, Vancouver Island contained 829 000 hectares of ancient, or old growth, temperate rainforest. This is approximately half of what existed on the Island in 1954. At historical rates of harvesting, all unprotected ancient temperate rainforest will be gone by 2022.

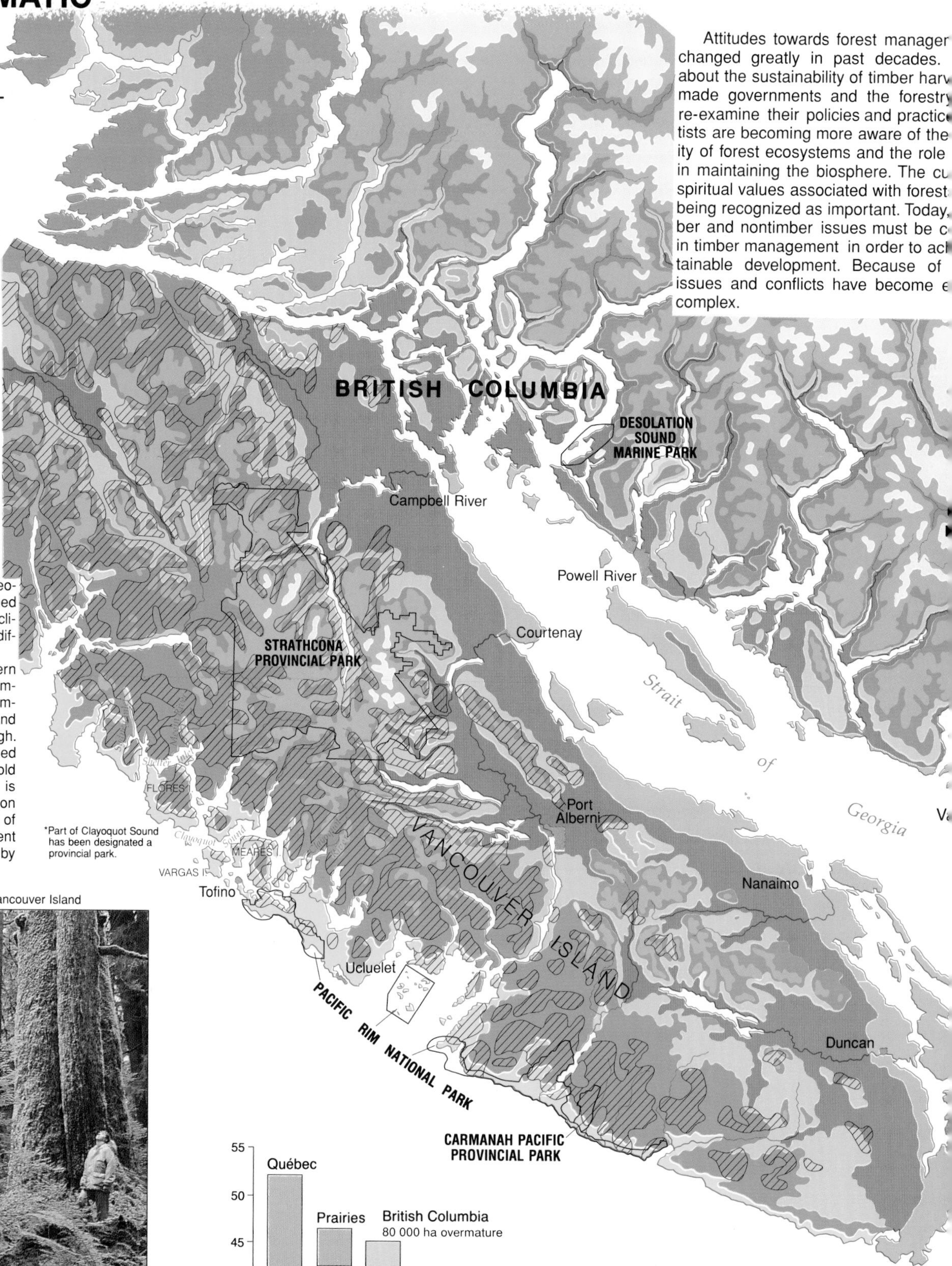

*Part of Clayoquot Sound has been designated a provincial park.

Attitudes towards forest manager changed greatly in past decades. about the sustainability of timber harv made governments and the forestry re-examine their policies and practic tists are becoming more aware of the ity of forest ecosystems and the role in maintaining the biosphere. The cu spiritual values associated with forest being recognized as important. Today, ber and nontimber issues must be c in timber management in order to ach tainable development. Because of issues and conflicts have become e complex.

Old growth forests in Carmanah Valley, Vancouver Island

Old growth forest is a term used to describe forest with old, and usually large, trees. Normally there are some dead trees, either standing or fallen, and an accumulation of organic matter on the forest floor. Old growth forests are complex ecosystems; they act as reservoirs of gene variation for the future, and allow scientists to study the way forest ecosystems function.

Old growth forests represent huge amounts of wood. In addition, the wood in the old growth forests may make up a large percentage of the wood suitable for harvest in an area for several decades. Taking old growth forests out of production may have severe social and economic consequences for local communities.

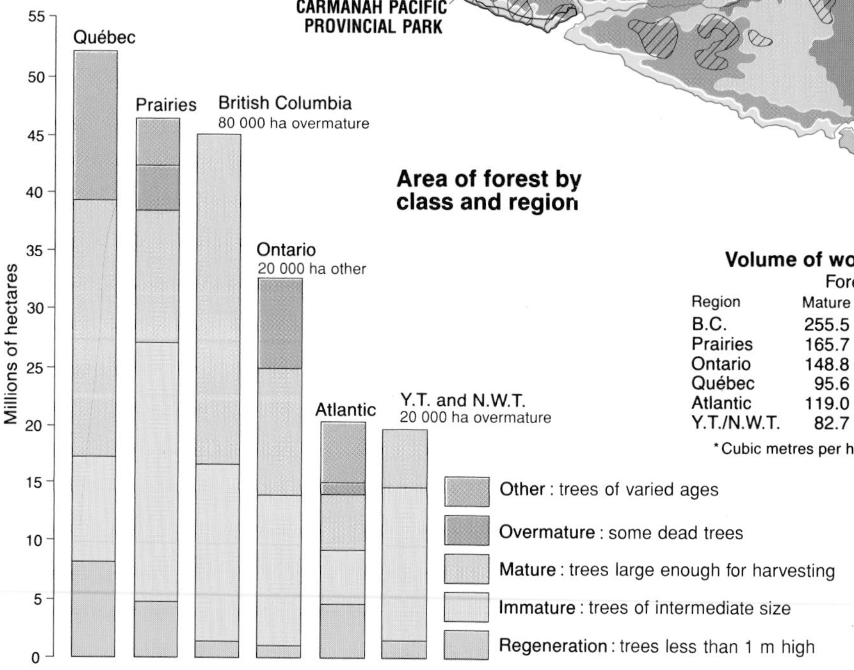

Area of forest by class and region

Volume of wood*

Region	Forest type: Mature	Forest type: Overmatu
B.C.	255.5	350.0
Prairies	165.7	191.5
Ontario	148.8	126.7
Québec	95.6	—
Atlantic	119.0	119.3
Y.T./N.W.T.	82.7	50.0

*Cubic metres per hectare

Harvesting methods

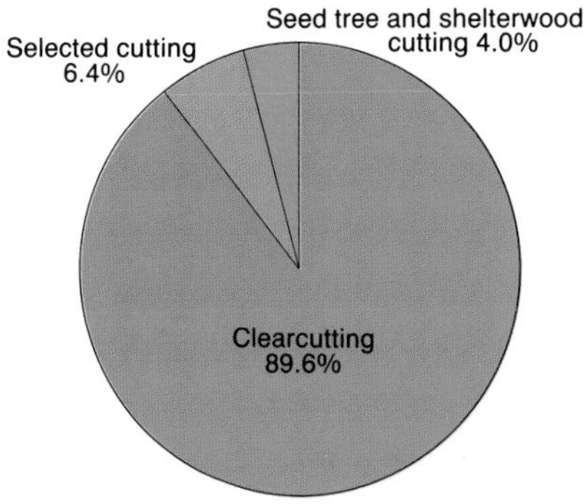

Single tree selection

Shelterwood

Seed tree

Clearcutting

Area of harvest and regeneration

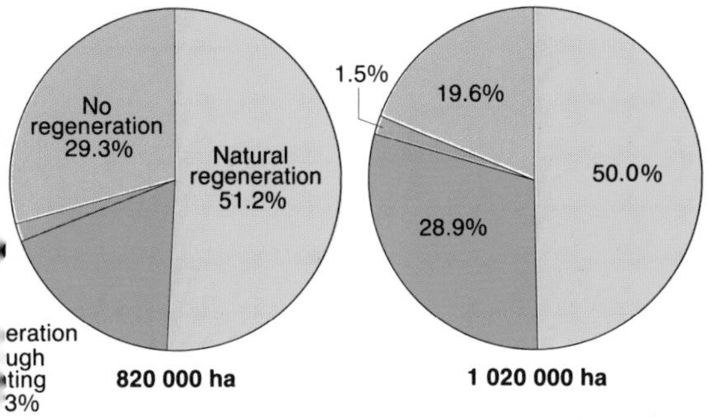

A clearcut area, southern Vancouver Island. The scale is 1: 120 000 (1 cm represents 1.2 km).

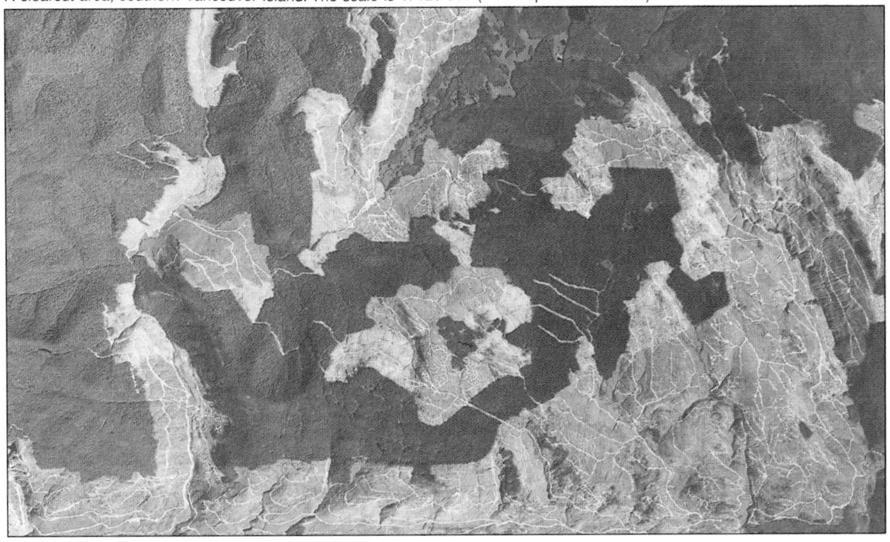

Tree species planted

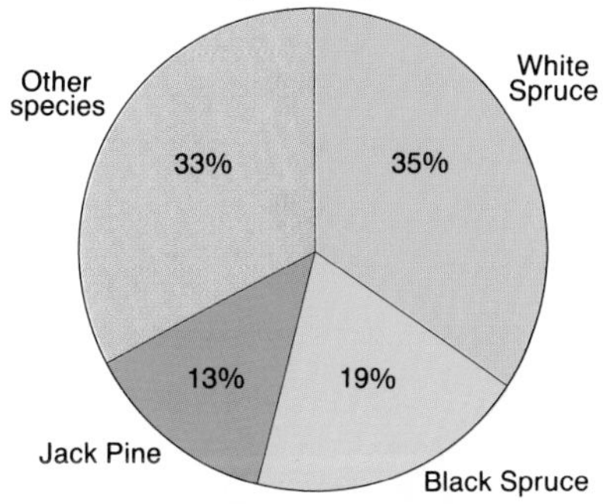

Total area planted (1988): 413 000 ha
Total seedlings (1988): 730 000 000

Stand tending activities

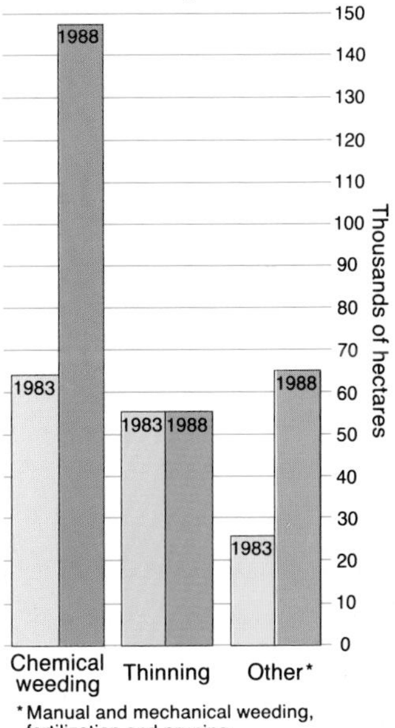

*Manual and mechanical weeding, fertilization and pruning

Expenditures on forest management

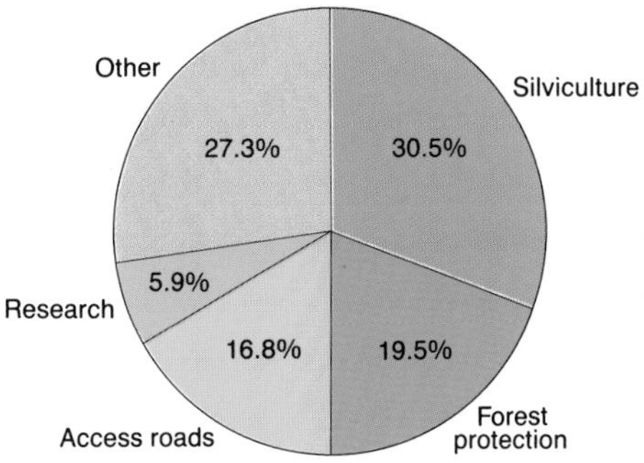

Total (1989): $2 200 000 000

A stand of pines being thinned

With increasing pressures on available forests, forestry companies are attempting to maximize yields by using intensive forest management. The most common form of harvesting is clearcutting, the removal of all trees. After harvesting, the area is prepared for natural regeneration, planting or seeding. Often tree species that have been genetically improved to grow faster are planted. The new trees are sprayed to reduce weed growth, insect infestations and disease, and are thinned to provide optimal light for growth. The fast growth of these trees means that they can be harvested after about 40 to 80 years.

There are concerns about the practice of intensive tree management, because it can result in reduced diversity of trees and animals and the loss of soil fertility. The chemicals used can have toxic effects. This method can also produce wood of poor quality and reduced resistance to disease, insects and climate change.

CANADA-THEMATIC

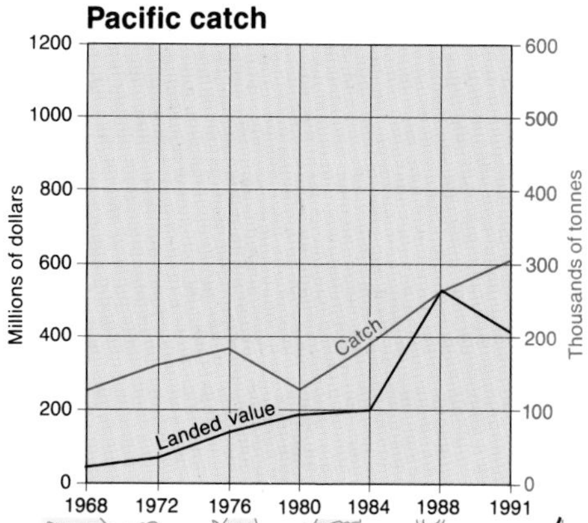

The fish in Canada's waters represent a resource that could be harvested indefinitely, if well managed. However, the many factors involved have made the management of fish stocks extremely difficult. For example, there are human-caused stresses such as pollution, overharvesting and destruction of habitat. Pollution has mainly affected fish species in inland waters or close to shore. Overharvesting by both Canada and other countries is believed to have caused serious population declines of northern cod. The destruction of spawning beds and the construction of dams have affected habitat. In addition, weather and ocean conditions can affect fish reproduction rates. There is also an element of uncertainty in the management process because fish stocks are difficult to measure accurately.

Major fishing countries of the world

Tonnes of fish caught, 1990 (live weight)

1	China	12 095 000
2	Former U.S.S.R.	10 389 000
3	Japan	10 354 000
4	Peru	6 875 000
5	United States	5 856 000
6	Chile	5 195 000
7	India	3 791 000
8	Indonesia	3 080 000
9	South Korea	2 750 000
10	Thailand	2 650 000
11	Philippines	2 207 000
12	North Korea	1 750 000
13	Norway	1 747 000
14	Canada (1991)	1 503 000
15	Denmark	1 513 000
16	Iceland	1 508 000
17	Spain	1 458 000
18	Mexico	1 401 000
19	France	897 000
20	Vietnam	850 000
21	Bangladesh	848 000
22	Brazil	800 000
23	Myanmar	744 000
24	Malaysia	603 000
25	United Kingdom	581 000
26	Morocco	566 000
27	New Zealand	565 000
28	Argentina	556 000
29	South Africa	536 000
30	Italy	525 000
	Total world catch:	**97 246 000**

Although Canada is not among the top ten nations in terms of catch, Canadian fish exports are generally very large. On average, only about one third of the Canadian catch is consumed in Canada. The major markets for Canada's seafood are the United States, Japan and Western Europe.

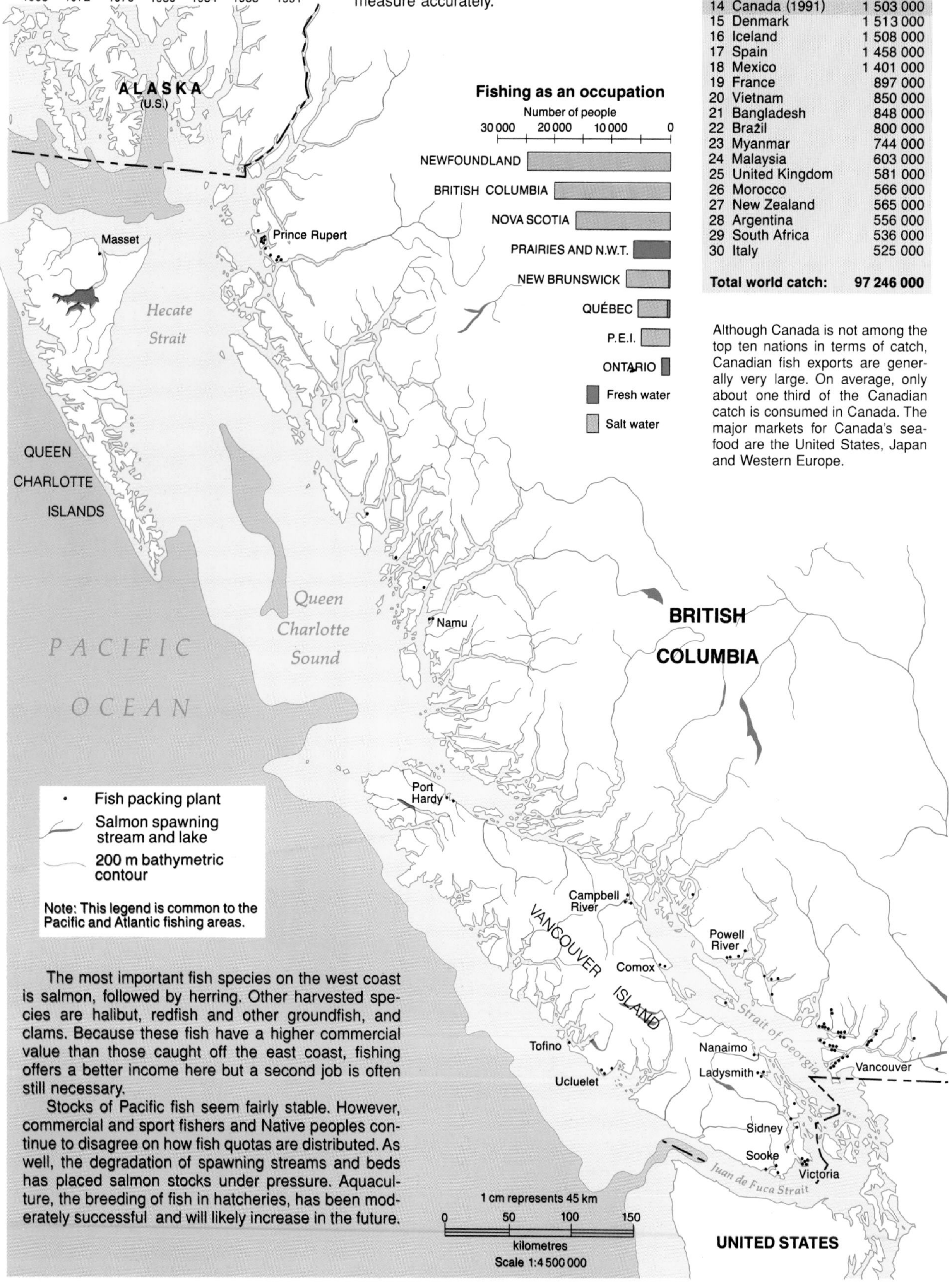

The most important fish species on the west coast is salmon, followed by herring. Other harvested species are halibut, redfish and other groundfish, and clams. Because these fish have a higher commercial value than those caught off the east coast, fishing offers a better income here but a second job is often still necessary.

Stocks of Pacific fish seem fairly stable. However, commercial and sport fishers and Native peoples continue to disagree on how fish quotas are distributed. As well, the degradation of spawning streams and beds has placed salmon stocks under pressure. Aquaculture, the breeding of fish in hatcheries, has been moderately successful and will likely increase in the future.

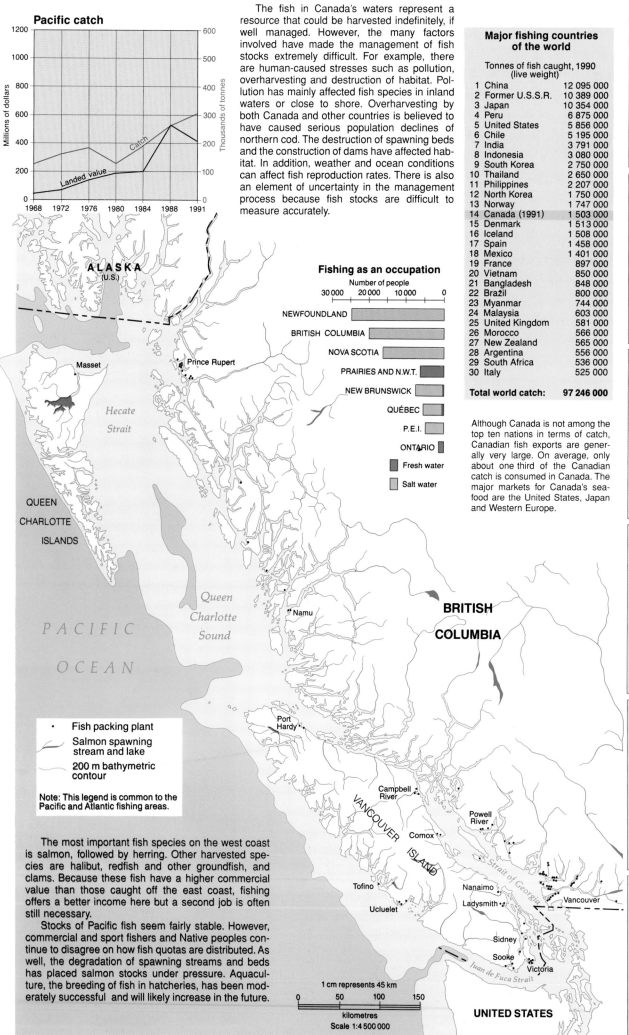

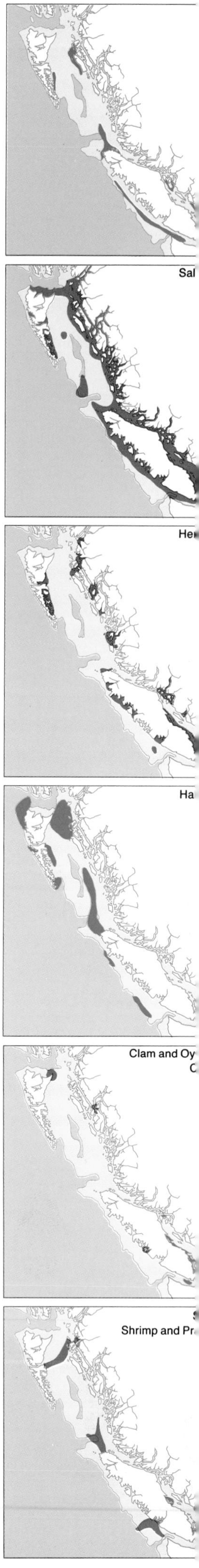

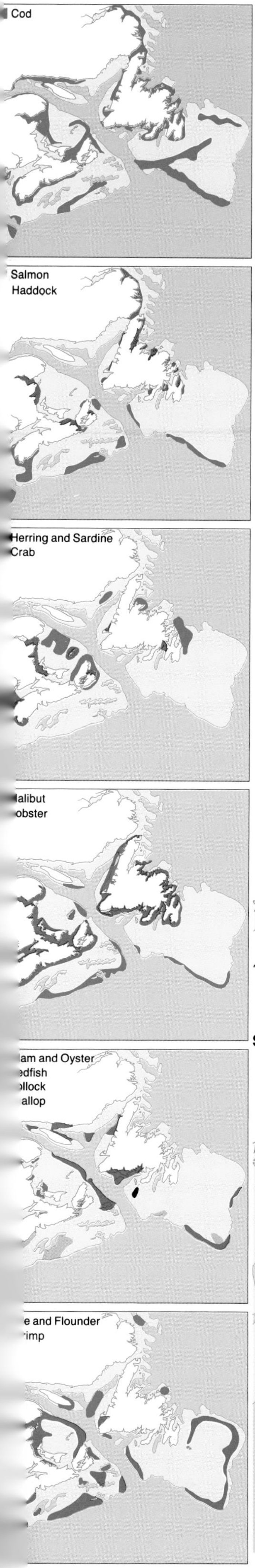

Traditionally, the most important species in the Atlantic fisheries have been cod and other groundfish. Harvests of other species such as herring, and shellfish such as lobster, scallops, shrimp, mussels and crab, have been increasing in recent years. In 1992, after a severe decline in stocks, a two-year moratorium on fishing for northern cod was declared. Other groundfish stocks were found to be decreasing rapidly as well, so allowable catches were cut back. The 30 000 vessels of the inshore fishery and the larger trawler ships of Canada and other nations have been seriously affected. Many processing plants sit idle, hurting the economy of the region even further. It remains to be seen whether and how quickly the cod and other groundfish stocks will recover.

Freshwater catch

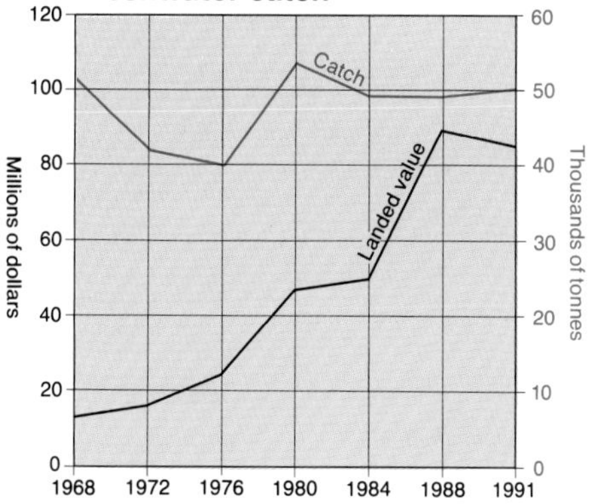

Atlantic catch

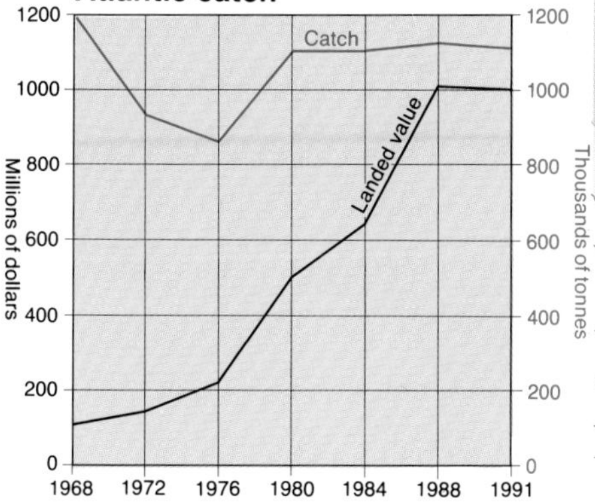

The marine food chain

The marine food chain shows how energy flows in the feeding relationships between plant life, animal life and humans. The sun injects energy into the system by causing plankton to grow. This energy is passed along the chain from level to level.

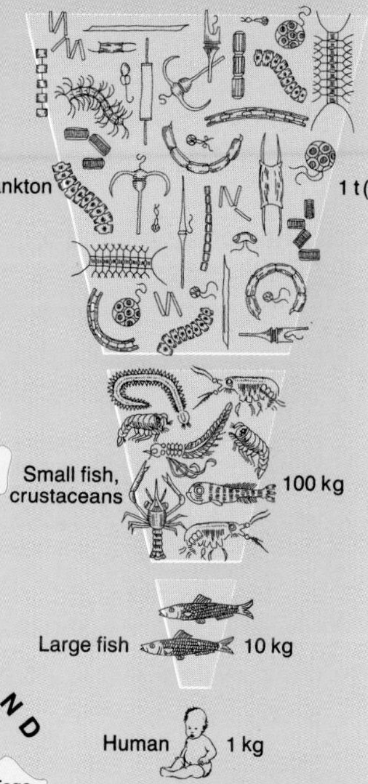

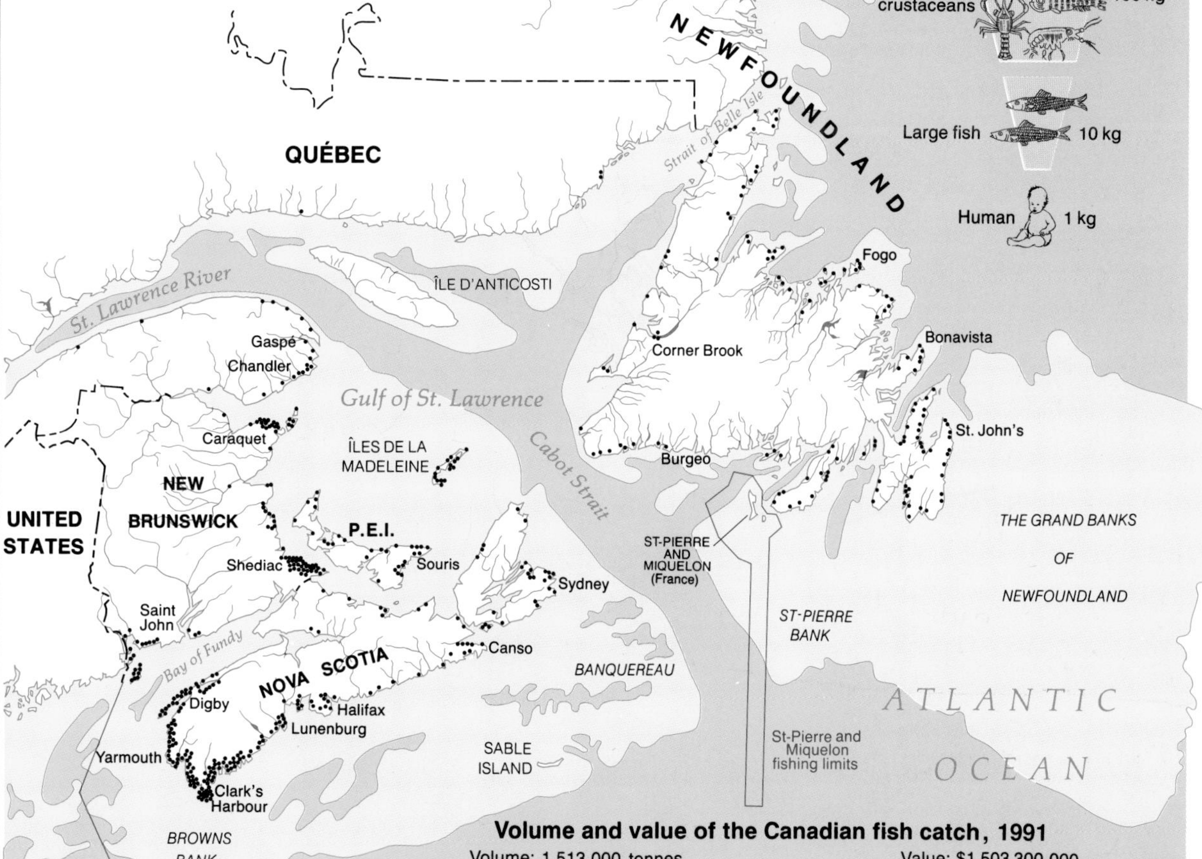

Volume and value of the Canadian fish catch, 1991

Volume: 1 513 000 tonnes

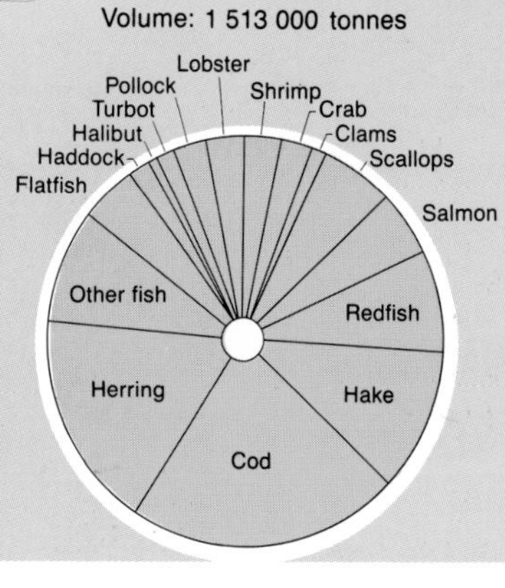

Value: $1 503 300 000

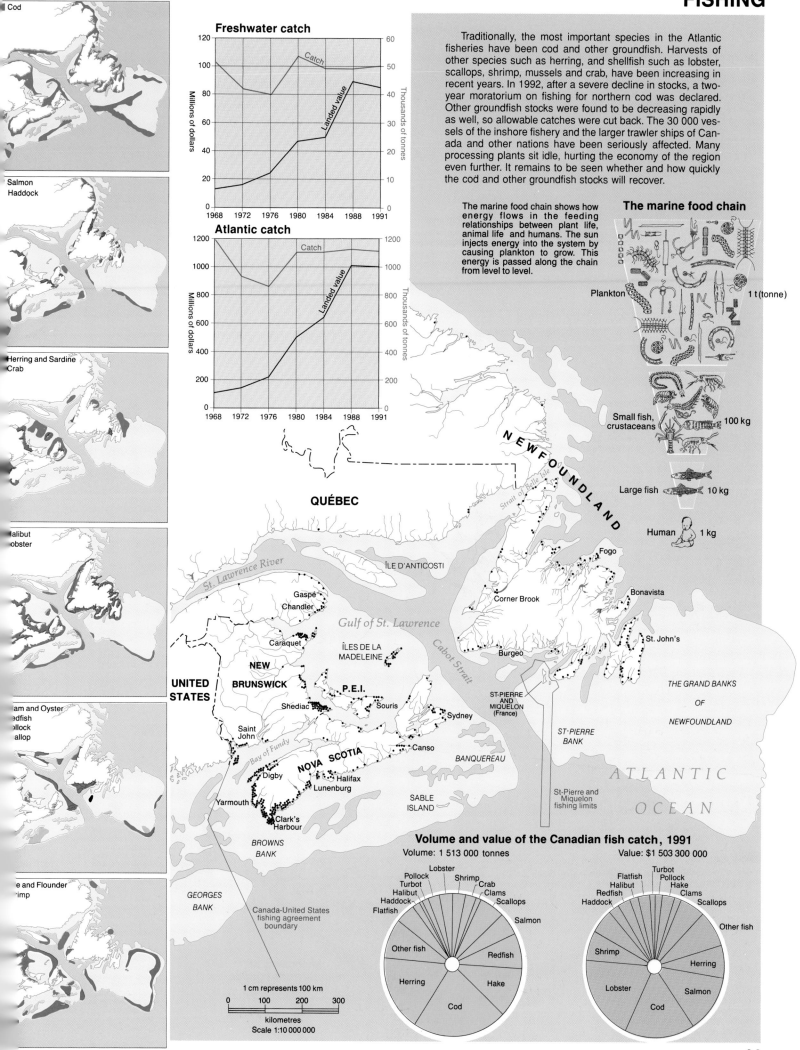

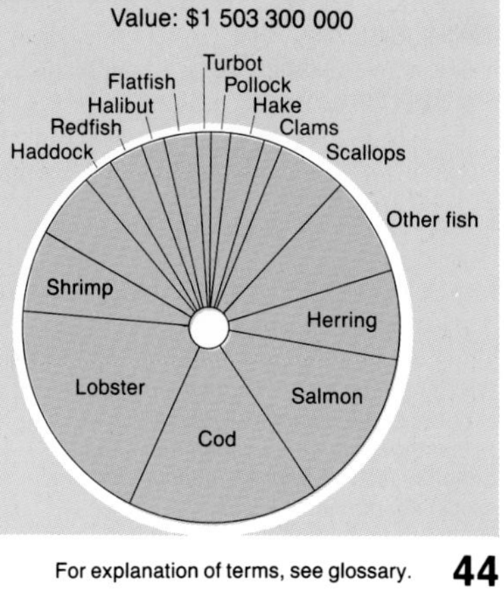

For explanation of terms, see glossary.

Value of manufacturing by province

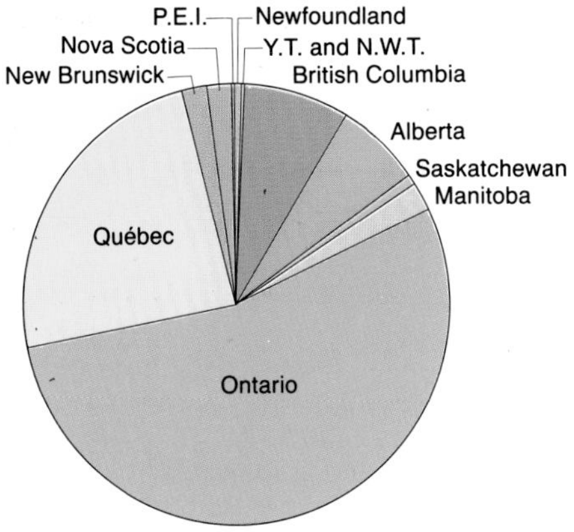

Value of manufacturing by CMAs

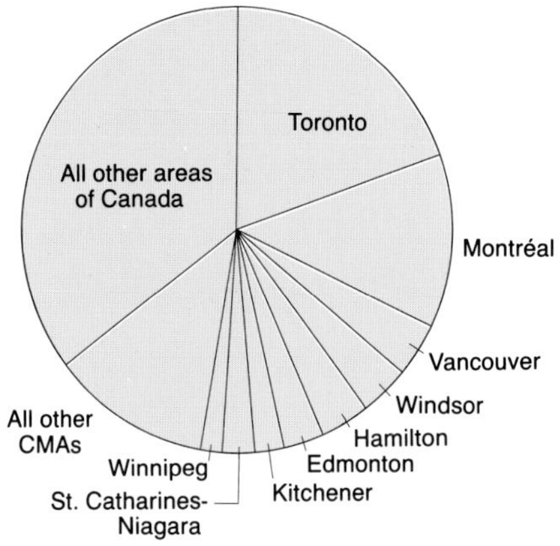

Manufacturing

Type of manufacturing* by selected CMA

- Food, beverage and tobacco products
- Leather, textiles, knitting mills and clothing
- Wood, furniture, paper, printing and publishing
- Machinery, transportation equipment and electrical products
- Primary metal and metal fabricating
- Rubber, plastic, petroleum, coal and chemical products
- Non-metallic mineral products and misc. manufacturing

*Based on the value of goods manufactured, the value added and the number of employees

Number of manufactu establishments by economic region

- More than 10 0(
- 2000–10 000
- 300–1999
- 150–299
- 75–149
- Less than 75

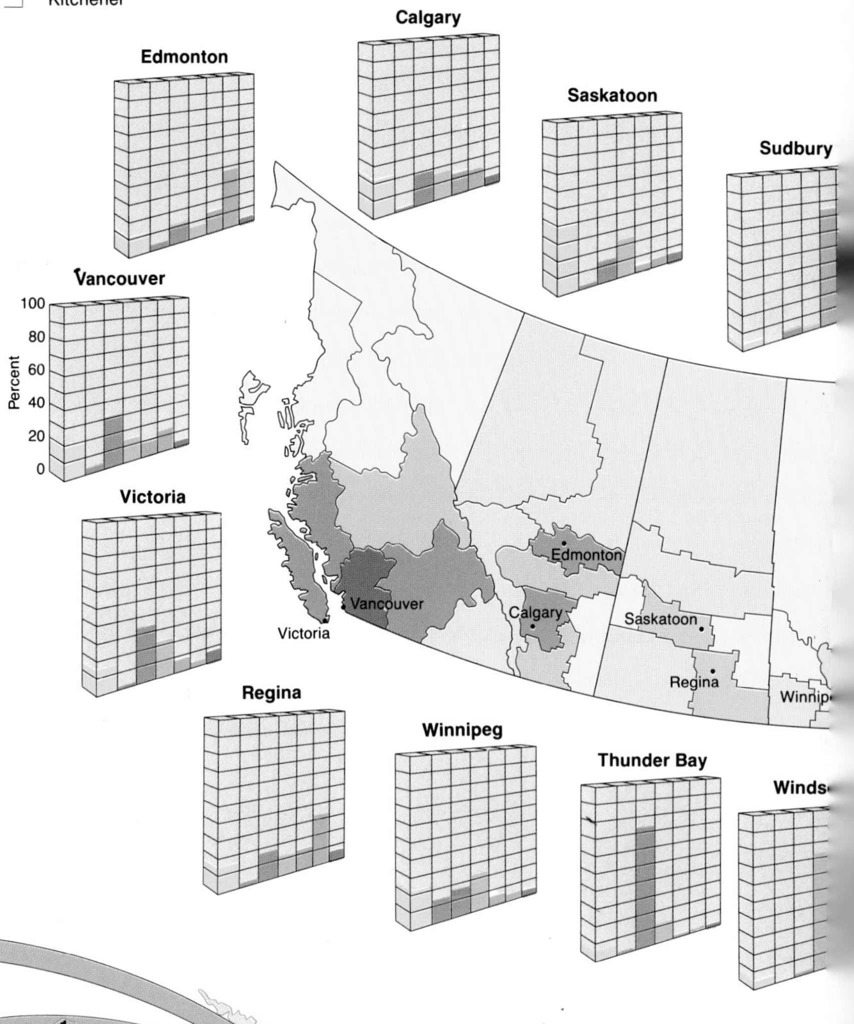

Regional distribution of manufacturing

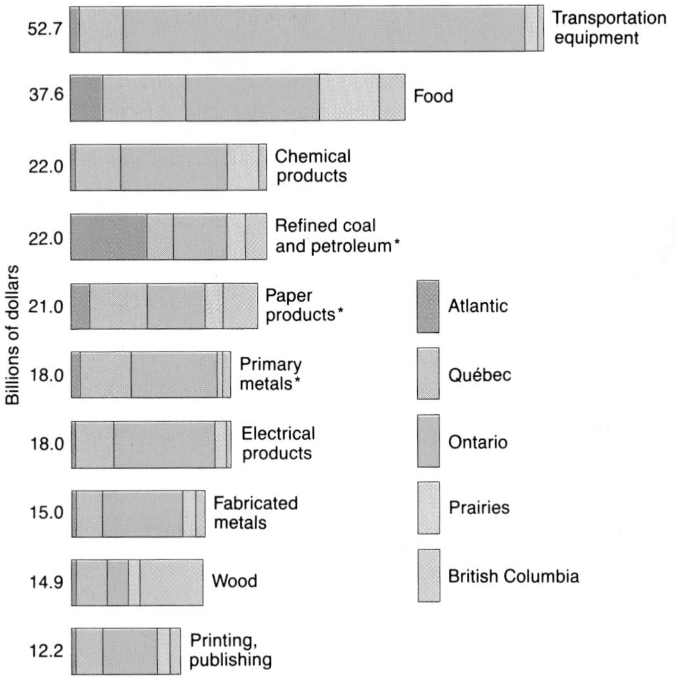

*Value estimated for Atlantic and Prairie regions

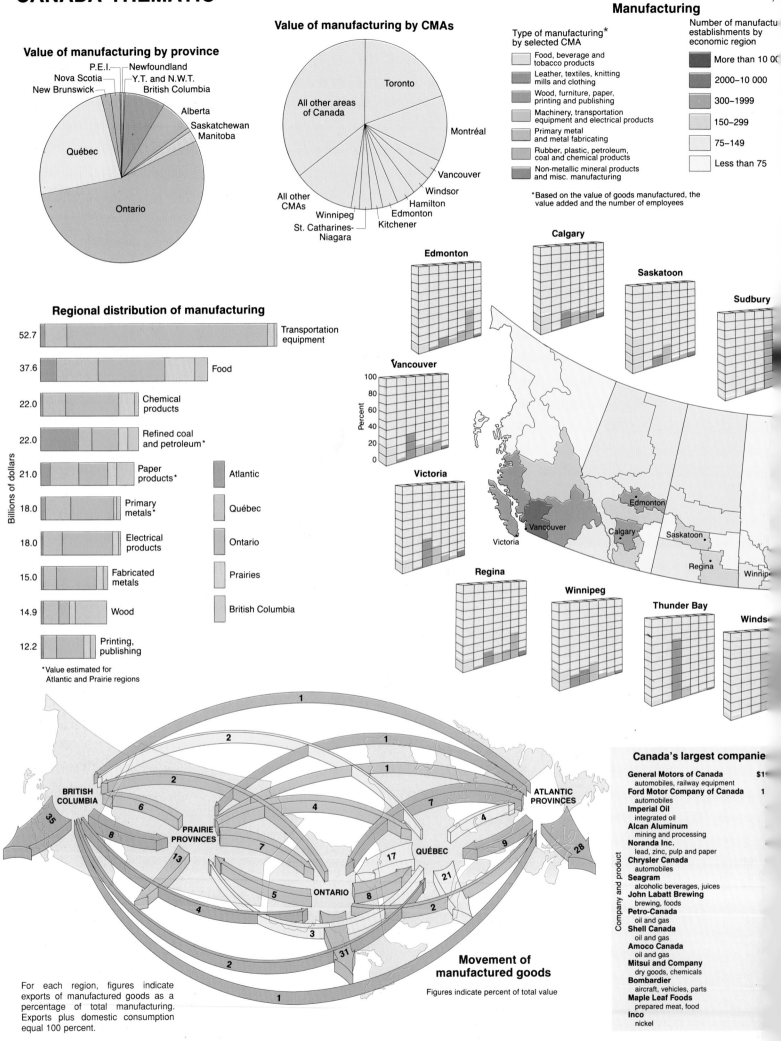

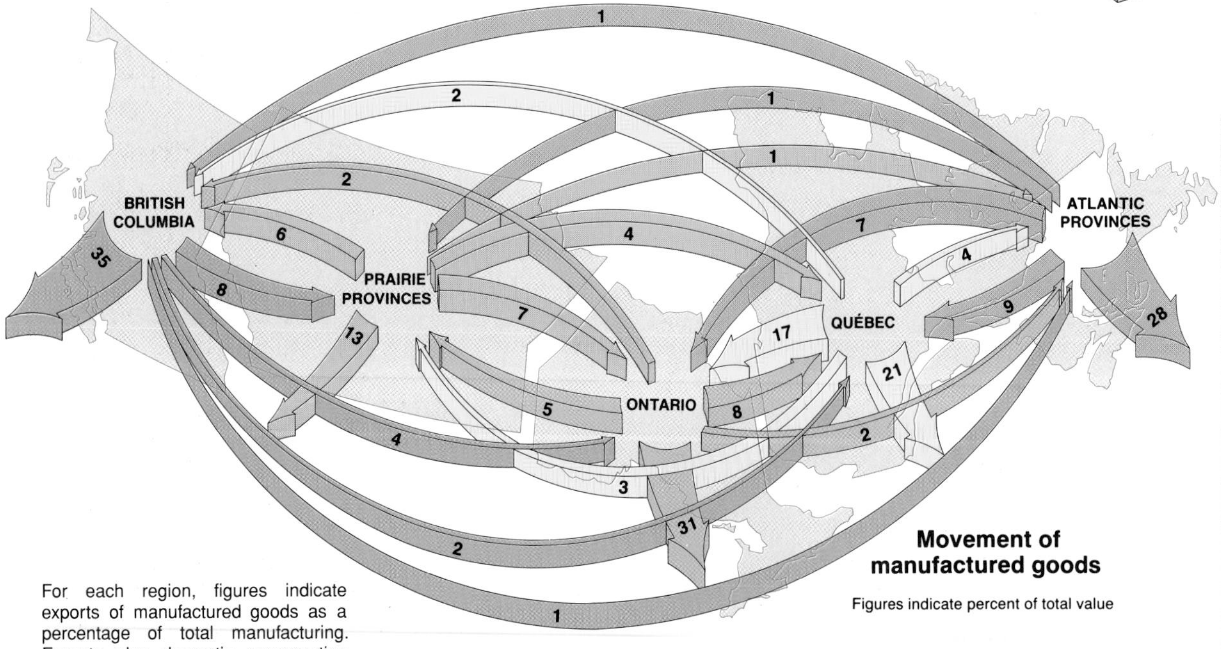

Movement of manufactured goods

Figures indicate percent of total value

For each region, figures indicate exports of manufactured goods as a percentage of total manufacturing. Exports plus domestic consumption equal 100 percent.

Canada's largest companie

Company and product

- **General Motors of Canada** $1
  automobiles, railway equipment
- **Ford Motor Company of Canada** 1
  automobiles
- **Imperial Oil**
  integrated oil
- **Alcan Aluminum**
  mining and processing
- **Noranda Inc.**
  lead, zinc, pulp and paper
- **Chrysler Canada**
  automobiles
- **Seagram**
  alcoholic beverages, juices
- **John Labatt Brewing**
  brewing, foods
- **Petro-Canada**
  oil and gas
- **Shell Canada**
  oil and gas
- **Amoco Canada**
  oil and gas
- **Mitsui and Company**
  dry goods, chemicals
- **Bombardier**
  aircraft, vehicles, parts
- **Maple Leaf Foods**
  prepared meat, food
- **Inco**
  nickel

Manufacturing accounts for approximately 20 percent of Canada's total output of goods and services, employing about two million people. Manufacturing is good for Canada's economy. For every three new jobs created in this sector, one additional job is also created in manufacturing and one each in the resource and services sectors. Much of Canada's manufacturing is resource-based, and over one third of manufactured goods are exported.

Today, Ontario and Québec lead in manufacturing, especially of consumer goods, but the rapid growth of manufacturing in the Western provinces will likely continue over the next decade. Canada is currently trying to become more competitive in this area, while dealing with global trends such as free trade.

Total manufacturing by type

Total (1992): $283 754 000 000

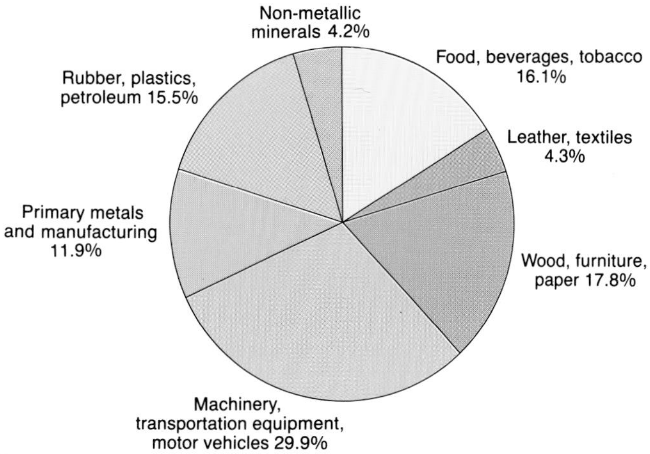

Resource dependency* of selected countries

Canada	33%
United States	29%
United Kingdom	21%
Korea	16%
Italy	14%
Sweden	14%
Germany	10%
Japan	3%

* Unprocessed and semi-processed resource exports as a percent of total exports

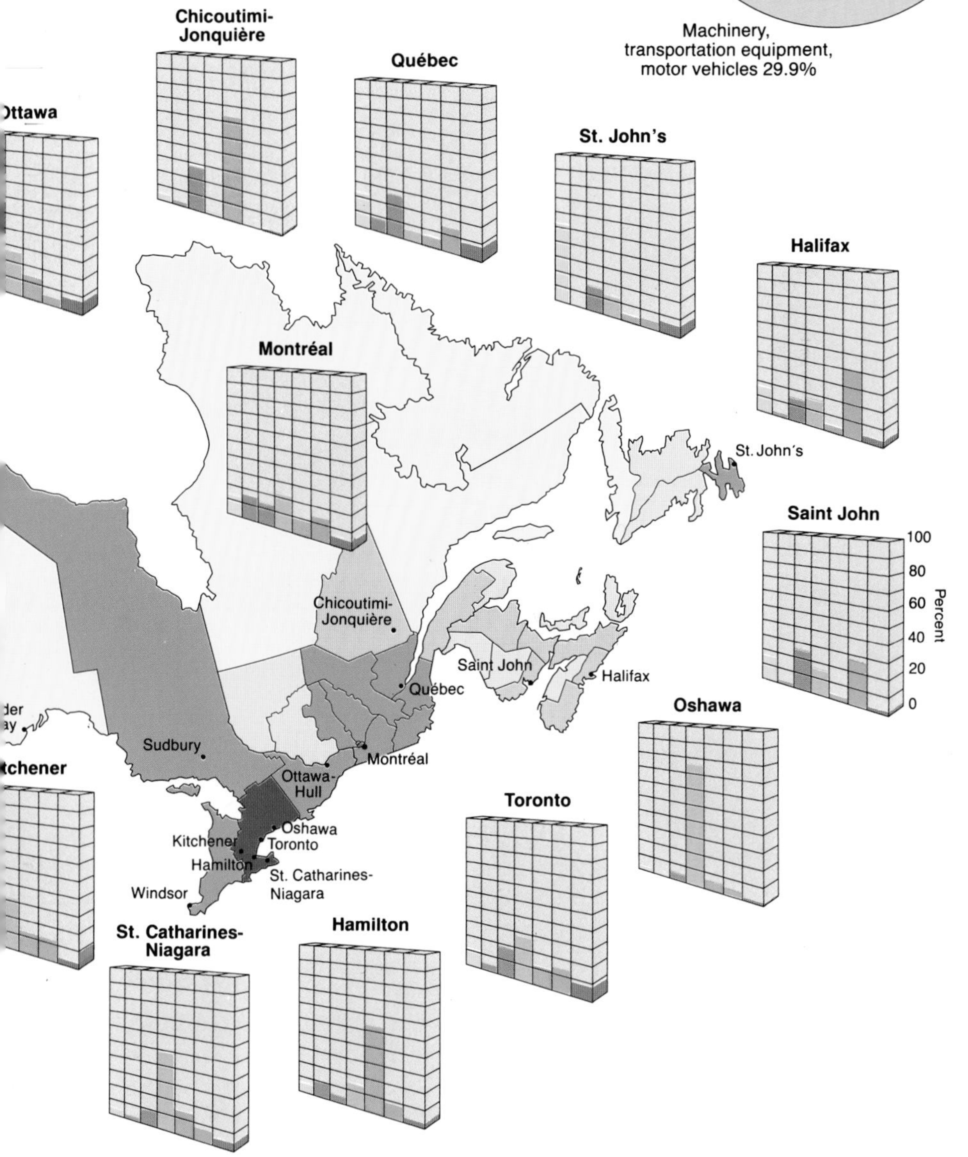

Advanced manufacturing technologies* in Canada and the U.S.

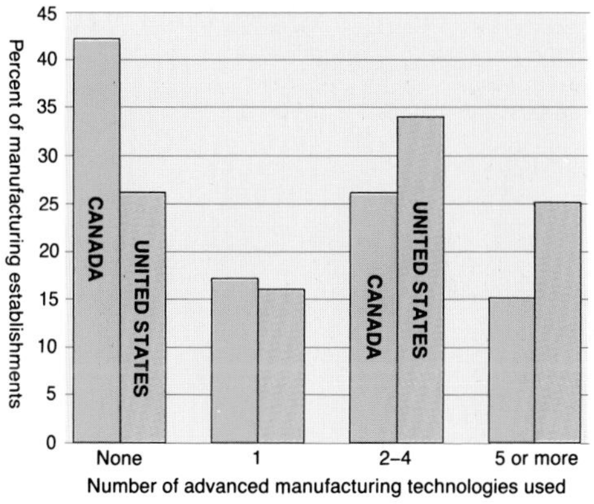

*Technologies include: Computer-aided design, robotics, automatic measuring devices, computer networks, computer-based inspections and quality testing, automatic process control

Exports of manufactured goods

Total (1992): $128 563 000 000

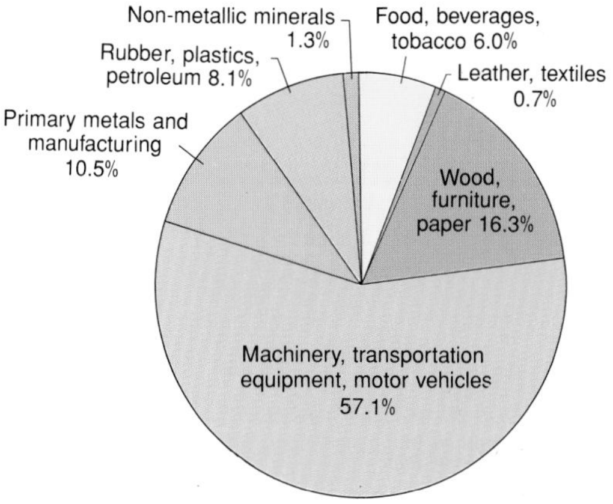

Manufacturing output per person per hour

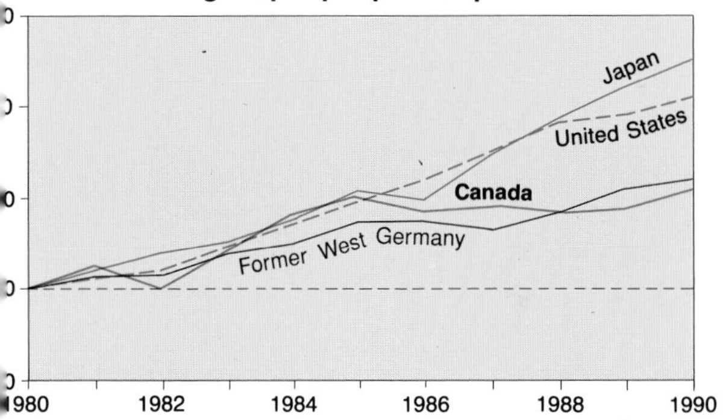

Expenditure on research and development as a percent of GDP

	1981	1989
Germany	2.42%	2.88%
Switzerland	2.29	2.86
Japan	2.14	2.85
United States	2.45	2.82
Sweden	2.30	2.76
France	1.97	2.32
Netherlands	1.88	2.26
United Kingdom	2.41	2.20
Austria	1.17	1.40
Canada	1.21	1.33
Italy	0.87	1.29
Australia	1.01	1.24
Average	**1.84**	**2.18**

Manufacturing in Canada and in other nations is changing. The use of high technology reduces the need for labour. The growth of world markets means that there are many more customers—but many more competitors. Free trade allows Canada to sell freely in markets abroad—but as well other nations can also sell here. Some people will lose their jobs because some Canadian industries may not be able to compete on those terms.

Canada depends on the export of resource-based products to global markets. At the same time it must also try to market itself as an attractive place for both foreign and domestic businesses to invest in. This investment will help create jobs and provide a tax base for social services.

CANADA-THEMATIC

Over the last century, Canada has been transformed from a predominantly agricultural nation into an affluent, urban industrial society. The high standard of living most Canadians enjoy is based primarily on exports of rich natural resources, industrial development and a skilled labour force. Recent years have seen rapid growth in the service sector of the economy.

Among the different regions of Canada, there are significant differences in GDP per capita, average family income (see pages 49–50) and unemployment rates. A system of equalization payments and development programs have helped to reduce these differences, but significant regional disparity still exists.

Unemployment has risen over the past 20 years. Those workers with less education and training are often the most affected by downturns in the economy. To face the constantly changing marketplace of the 21st century with confidence, Canada's work force needs to be educated and adaptable.

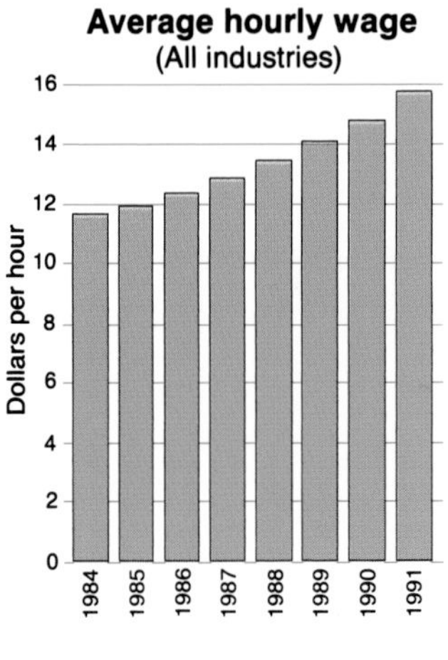

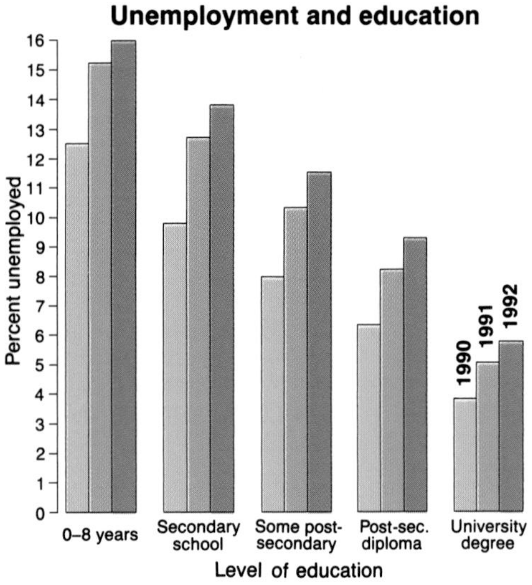

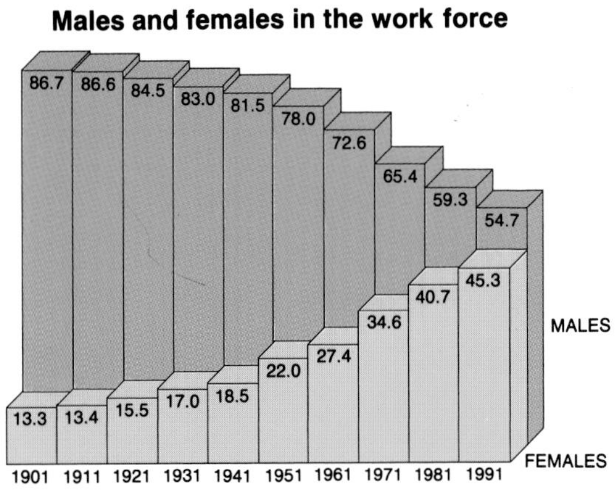

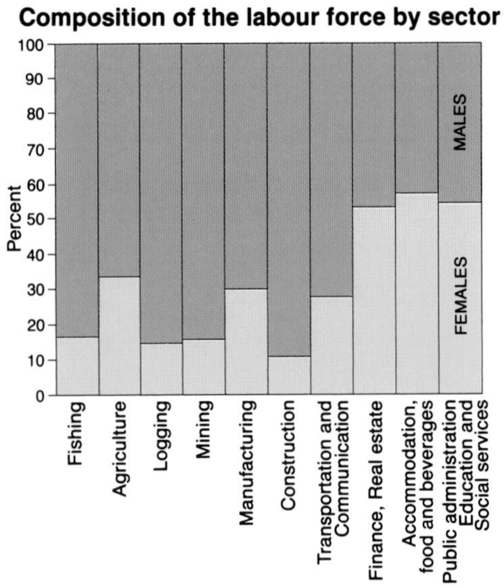

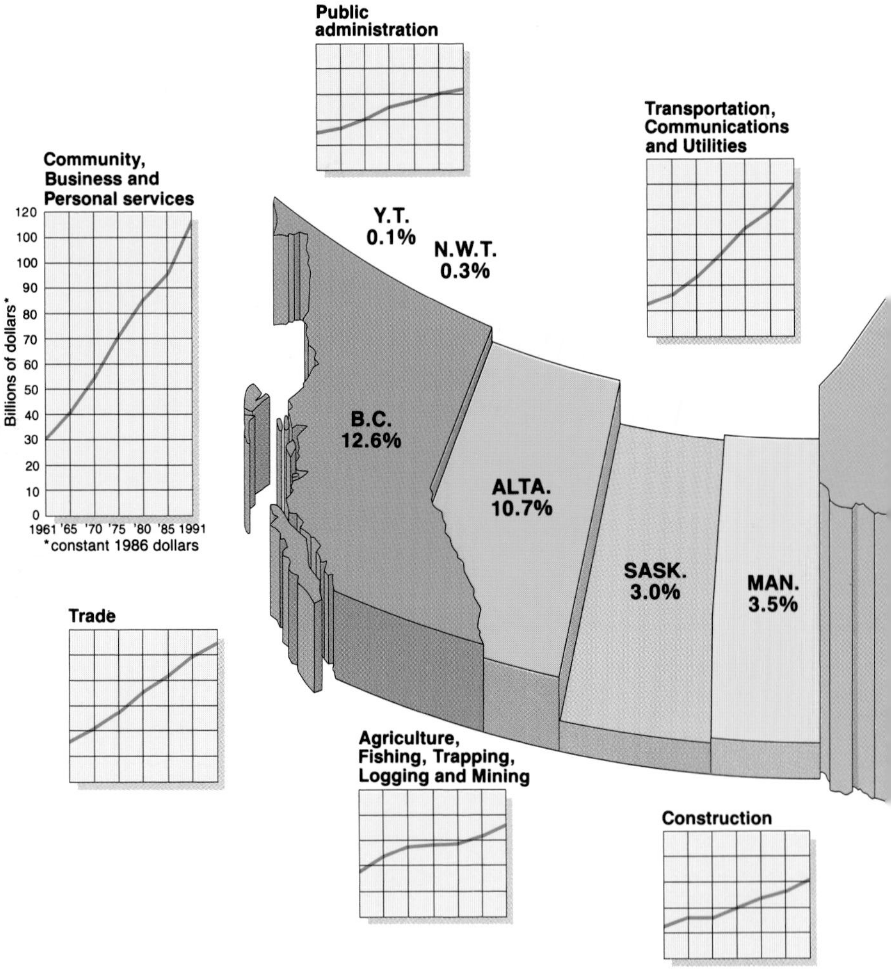

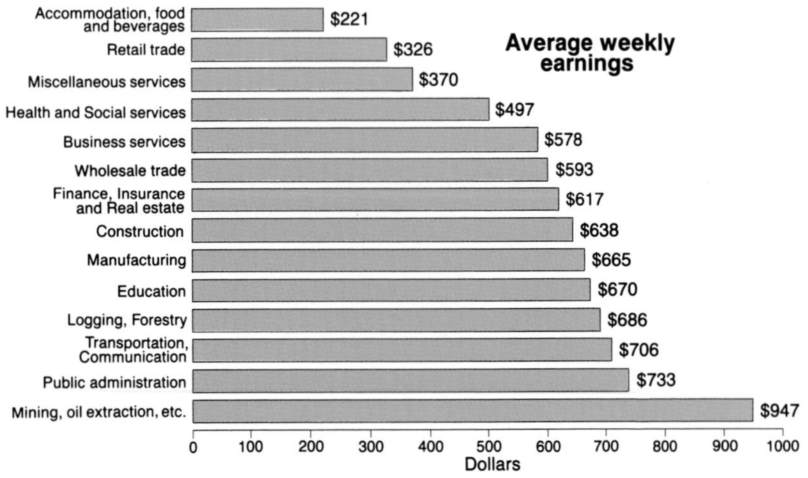

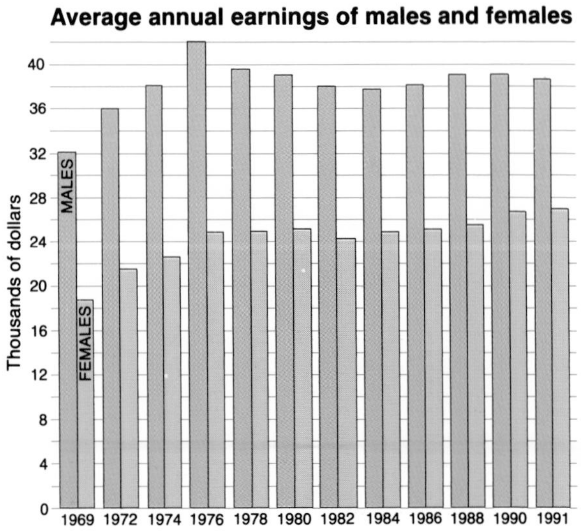

tudent employment by occupation

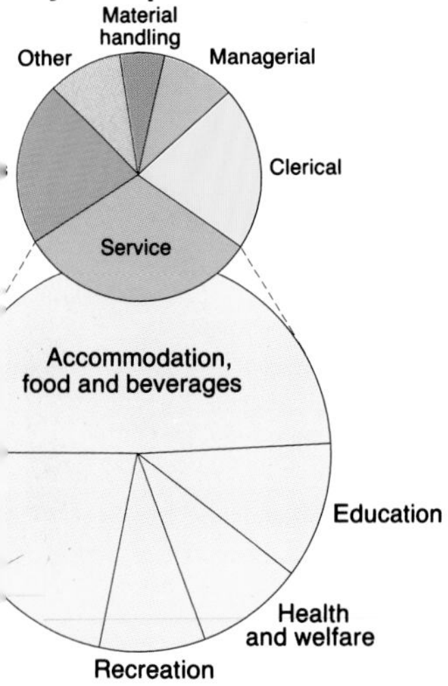

Student employment*

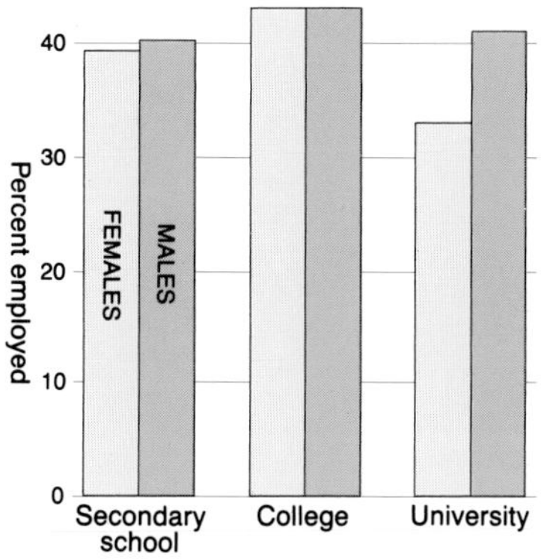

* working more than 10 hours a week

Employment by industry

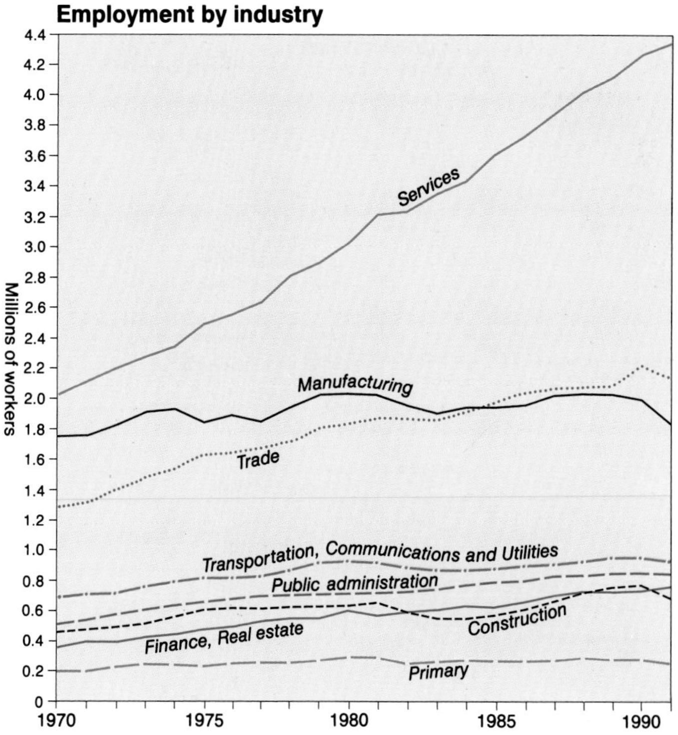

rovincial share of the oss Domestic Product

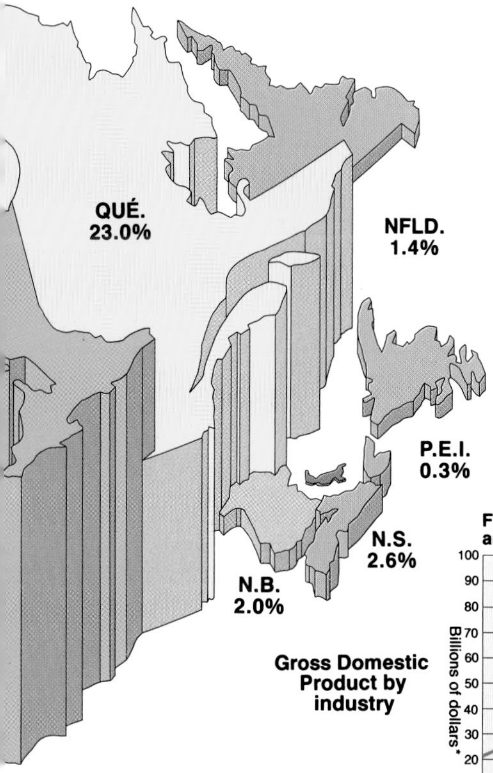

ada: $671 668 000 000

Manufacturing

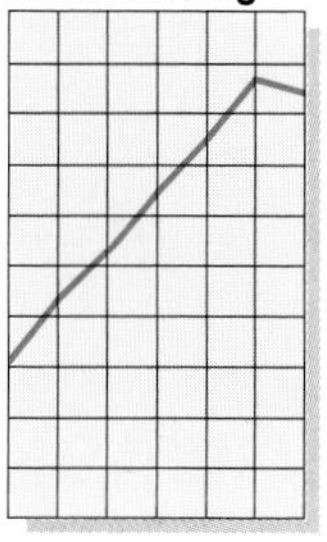

Gross Domestic Product by industry

Finance, Insurance and Real estate

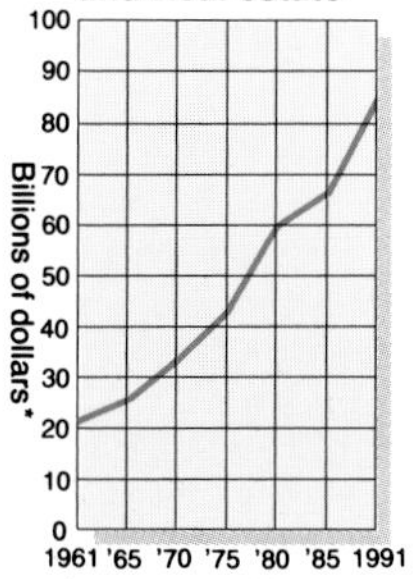

*constant 1986 dollars

Part-time work and unemployment

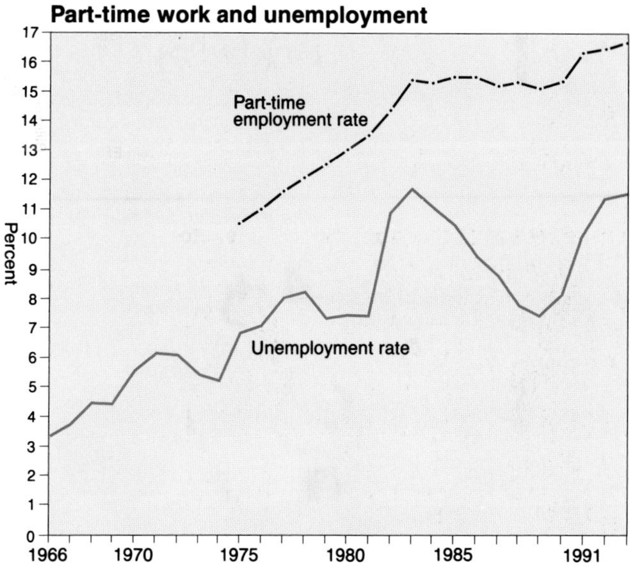

oss Domestic Product per capita

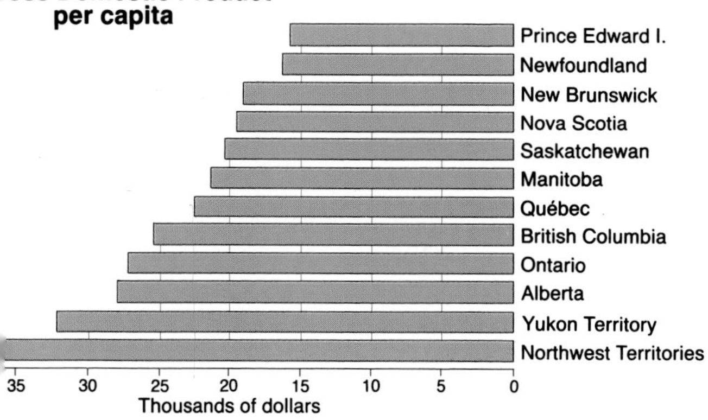

Gross Domestic Product
(Selected industries)

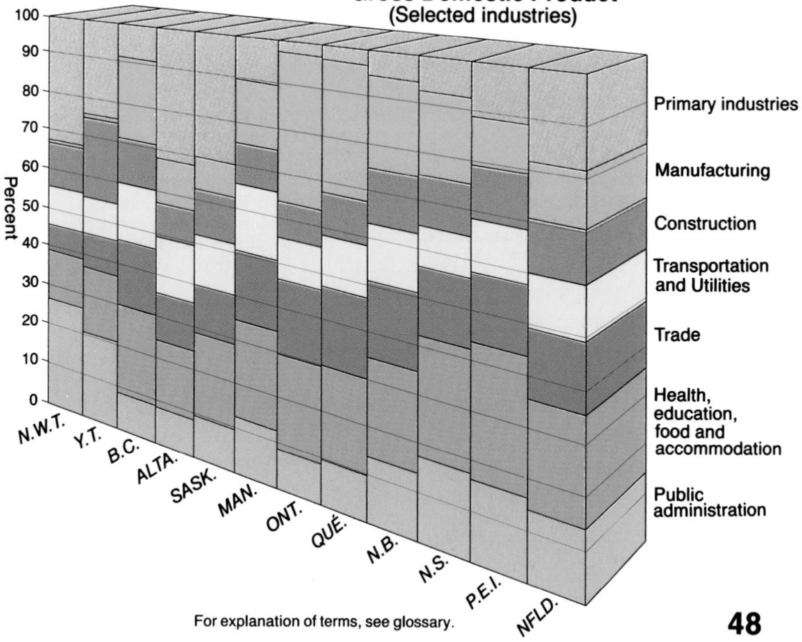

For explanation of terms, see glossary.

CANADA-THEMATIC

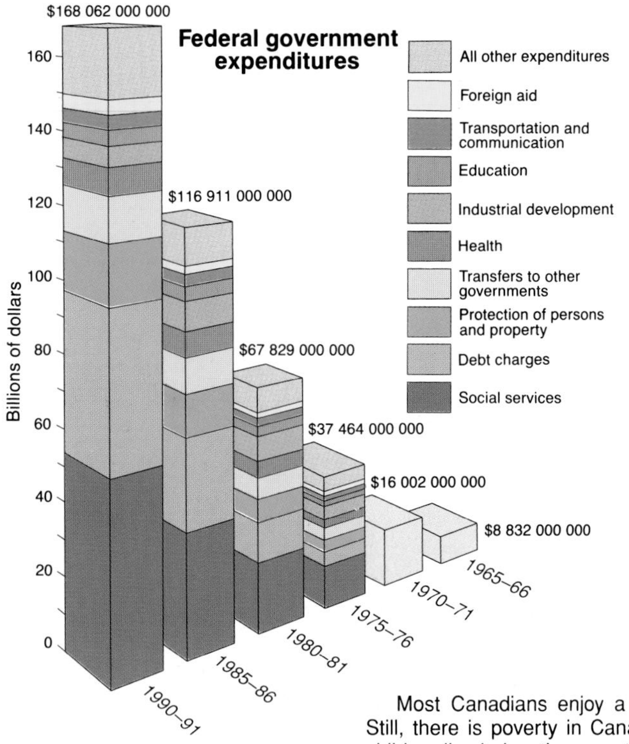

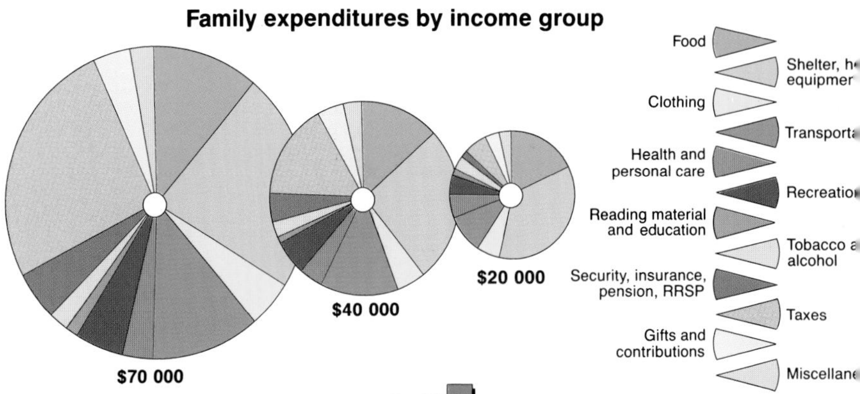

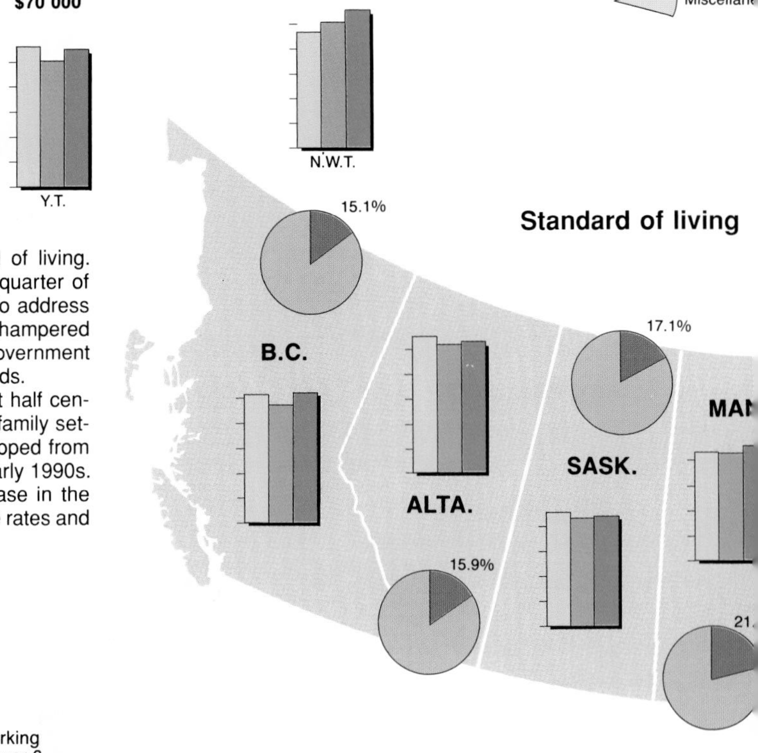

Most Canadians enjoy a high standard of living. Still, there is poverty in Canada; over one-quarter of children live below the poverty line. Efforts to address this and other social problems have been hampered by growing government debt. All levels of government are re-examining how they spend public funds.

Canadian society is changing. In the last half century, the proportion of people not living in a family setting has increased. Average family size dropped from 3.9 persons in 1960 to 3.0 persons in the early 1990s. This reflects an aging population, an increase in the number of people living alone, higher divorce rates and the increasing average age at marriage.

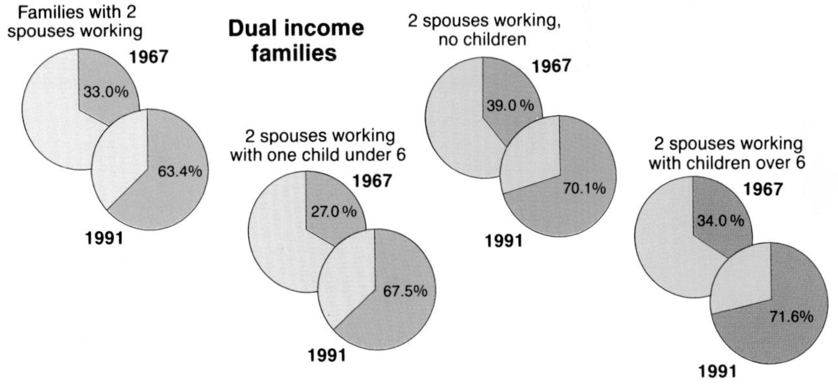

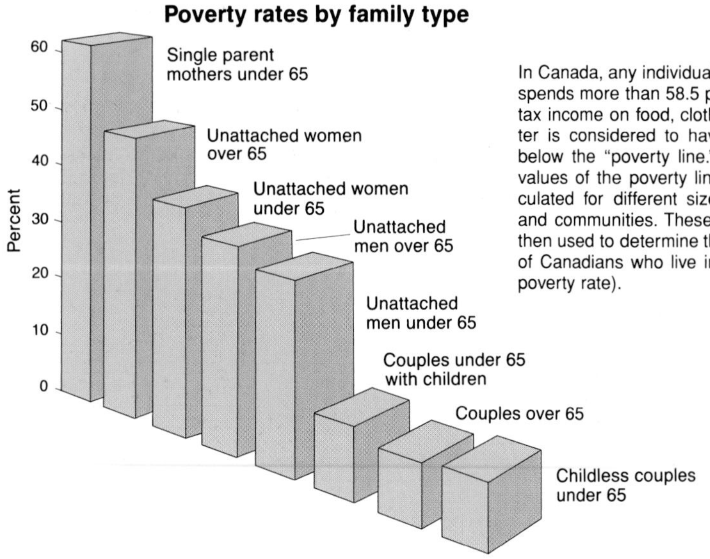

In Canada, any individual or family that spends more than 58.5 percent of pre-tax income on food, clothing and shelter is considered to have an income below the "poverty line." Approximate values of the poverty line can be calculated for different sizes of families and communities. These numbers are then used to determine the percentage of Canadians who live in poverty (the poverty rate).

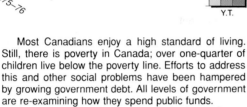

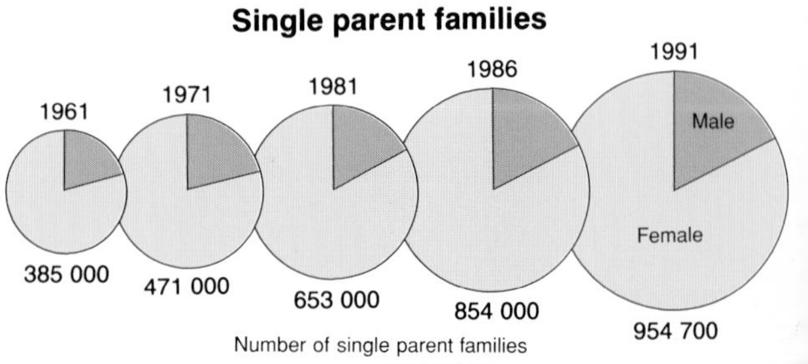

Reading and numeracy skills

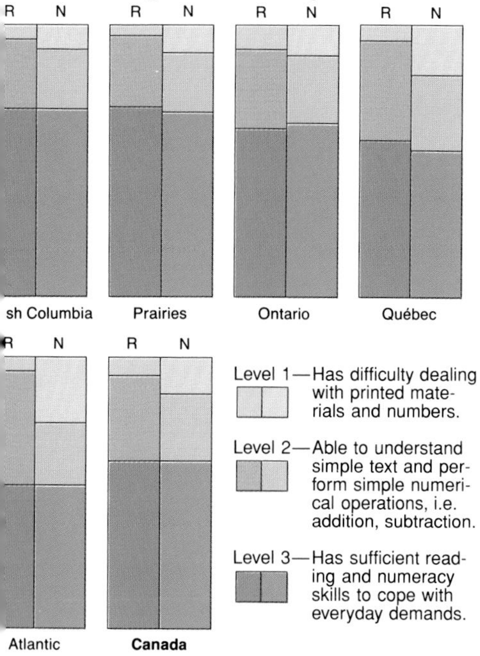

Homicide rate, Canada and United States

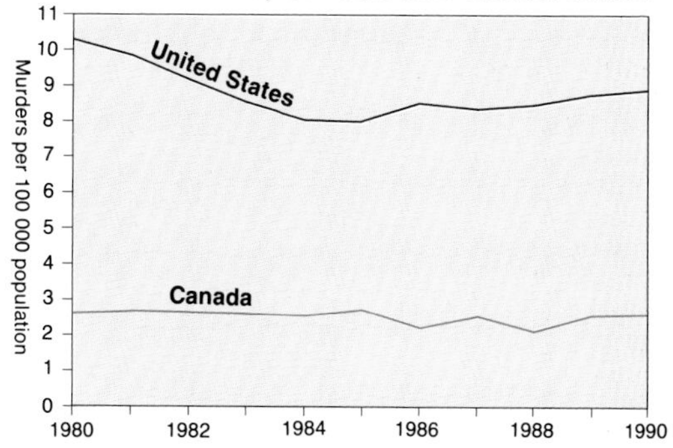

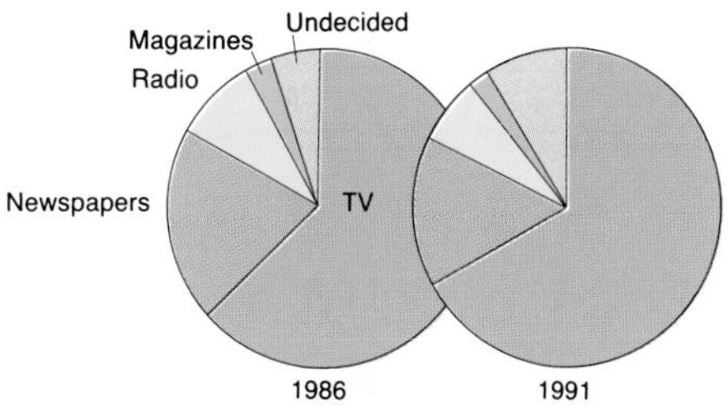

Where we get the news

Crime

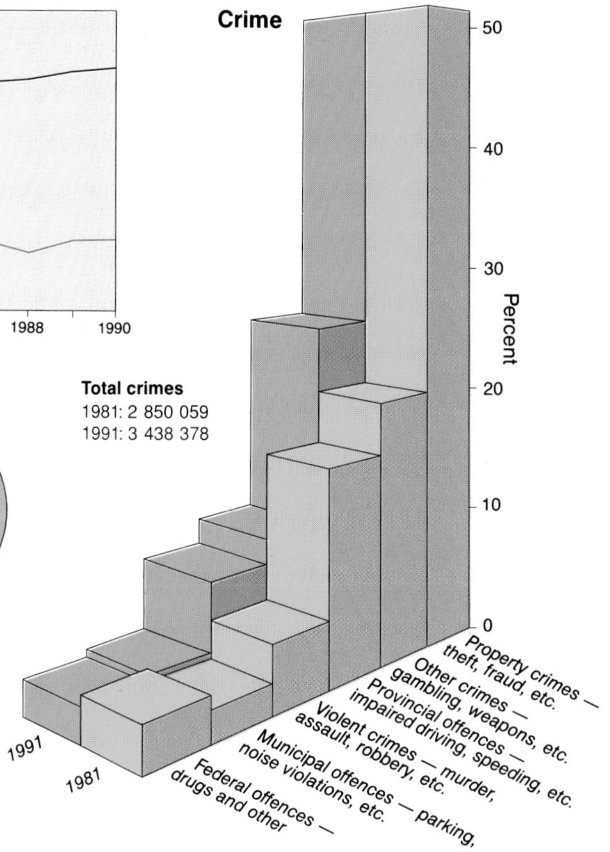

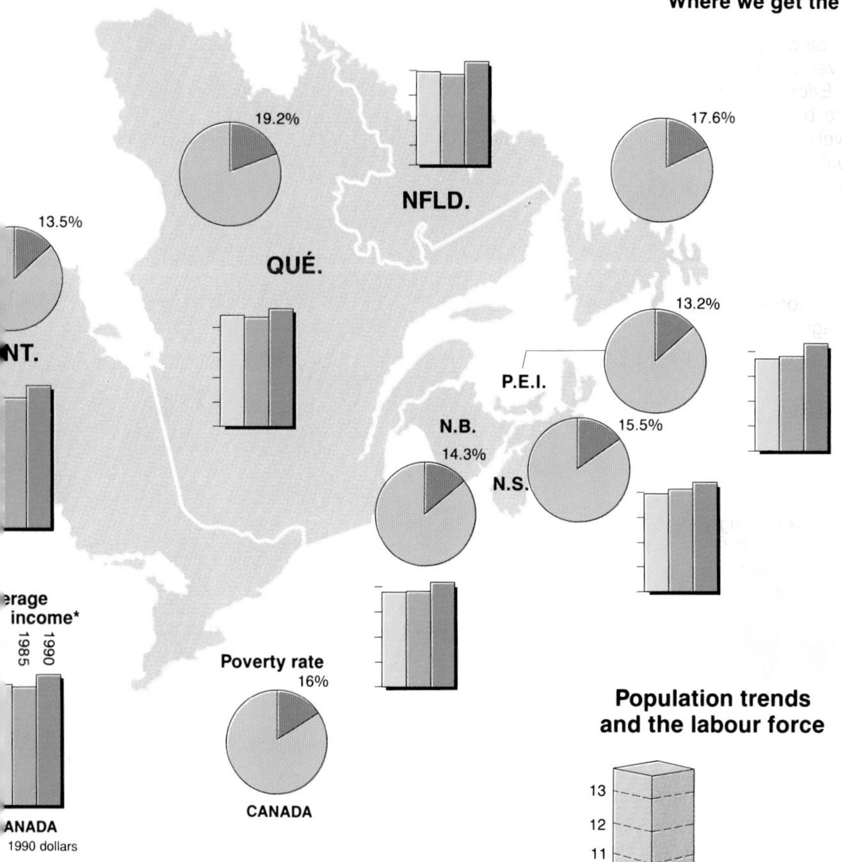

The technology revolution

Percent of families owning these products

	1980	1985	1991
Microwave ovens	8.0%	23.0%	68.2%
Television (cable)	54.8	62.5	71.4
Video cassette recorders	—	23.5	66.3
Compact disc players	—	7.9	15.4
Camcorders	—	2.8	5.6
Home computers	—	12.6	16.3

Population trends and the labour force

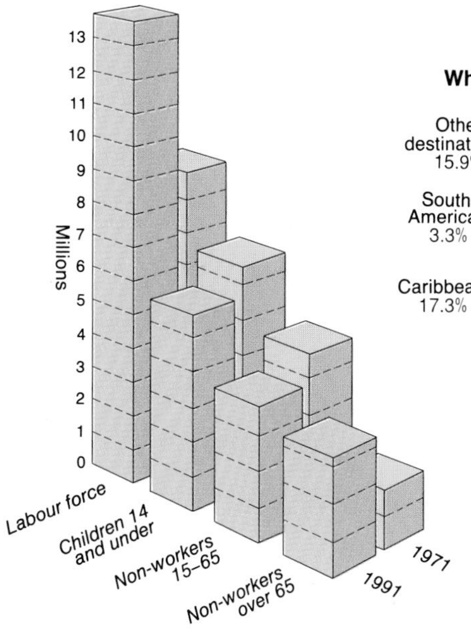

Composition of families

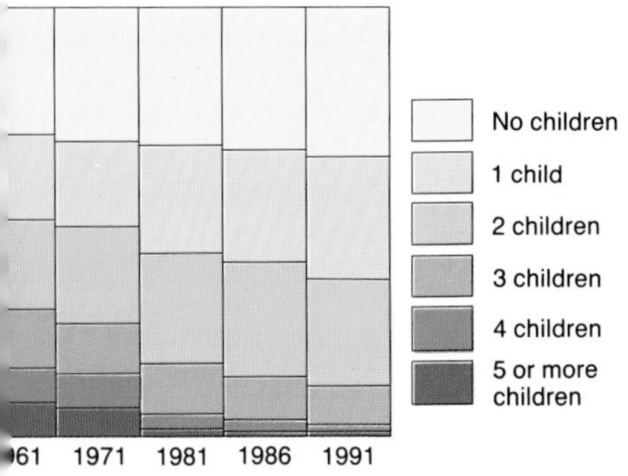

Canadians travelling abroad

(not including United States)
Total non-U.S. trips (1990): 3 153 000
Total U.S. trips (1990): 70 433 000

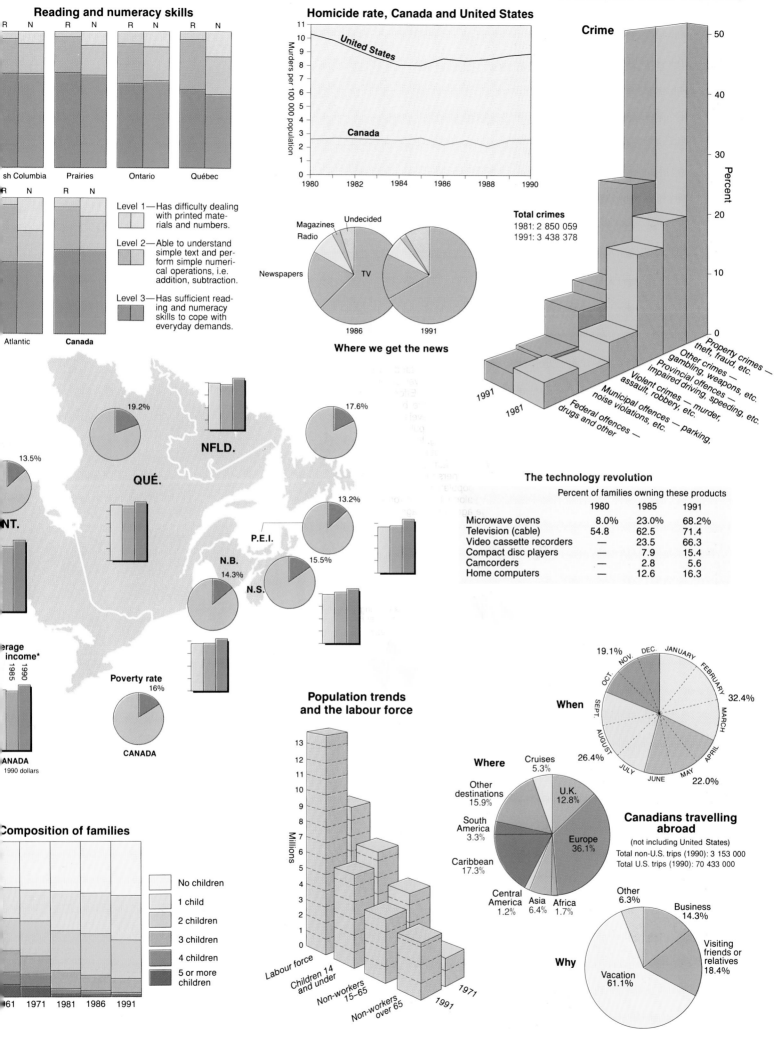

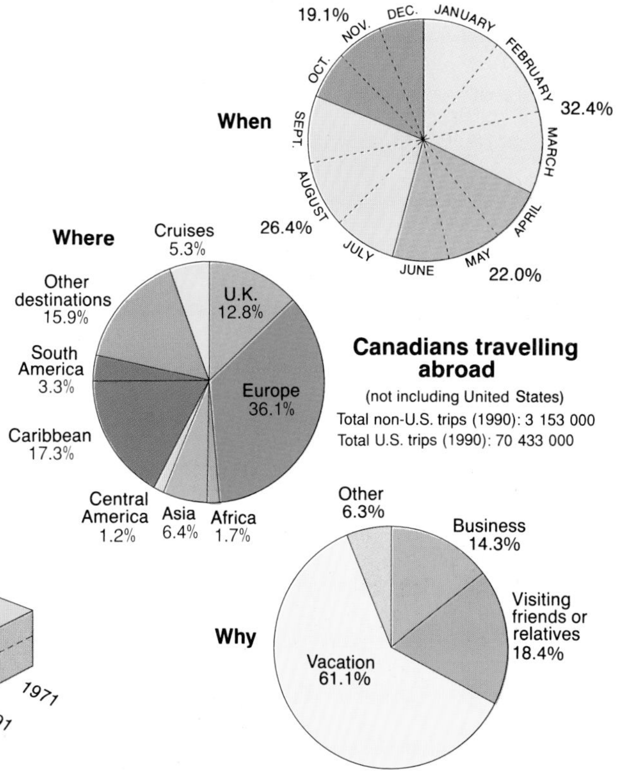

For explanation of terms, see glossary.

CANADA-THEMATIC

The transportation of people and goods has been critical to Canada's growth. New transportation routes opened up new territories, linked communities and created opportunities for economic development. Canada's lakes and rivers provided fur traders with access to many areas of Canada; the transcontinental railway stimulated the western economy and linked Canada politically. Today, strong transportation links with the rest of the world are crucial to Canada's success as a trading nation.

Transportation produces 4.4 percent of Canada's Gross Domestic Product (GDP). Nearly 5 percent of the labour force is directly employed in this industry, and an additional 2 percent work in the manufacturing and servicing of transportation equipment.

Recent trends in the transportation industry include an increase in the movement of goods by air and truck. Passenger train and inter-city bus travel has decreased as automobile and air travel has increased.

Domestic freight transported

(by weight)

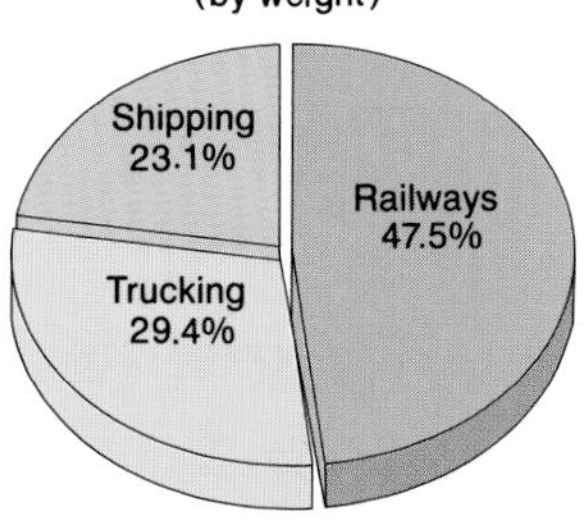

Accidents involving dangerous goods, 199

	Road	Rail	Shipping	Air	Tot
Compressed or liquefied gases	44	105	—	—	14
Flammable and combustible liquids	194	67	3	7	27
Corrosive materials	94	39	2	1	13
Poisonous substances	42	9	1	2	5
Other	52	14	1	2	6
Total	426	234	7	12	67

Fuel use in transportation

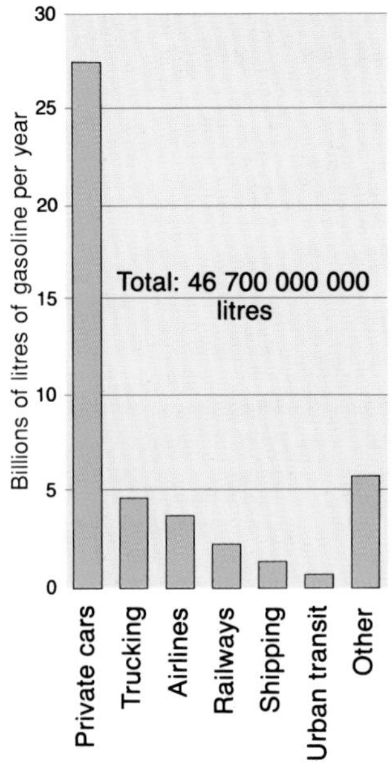

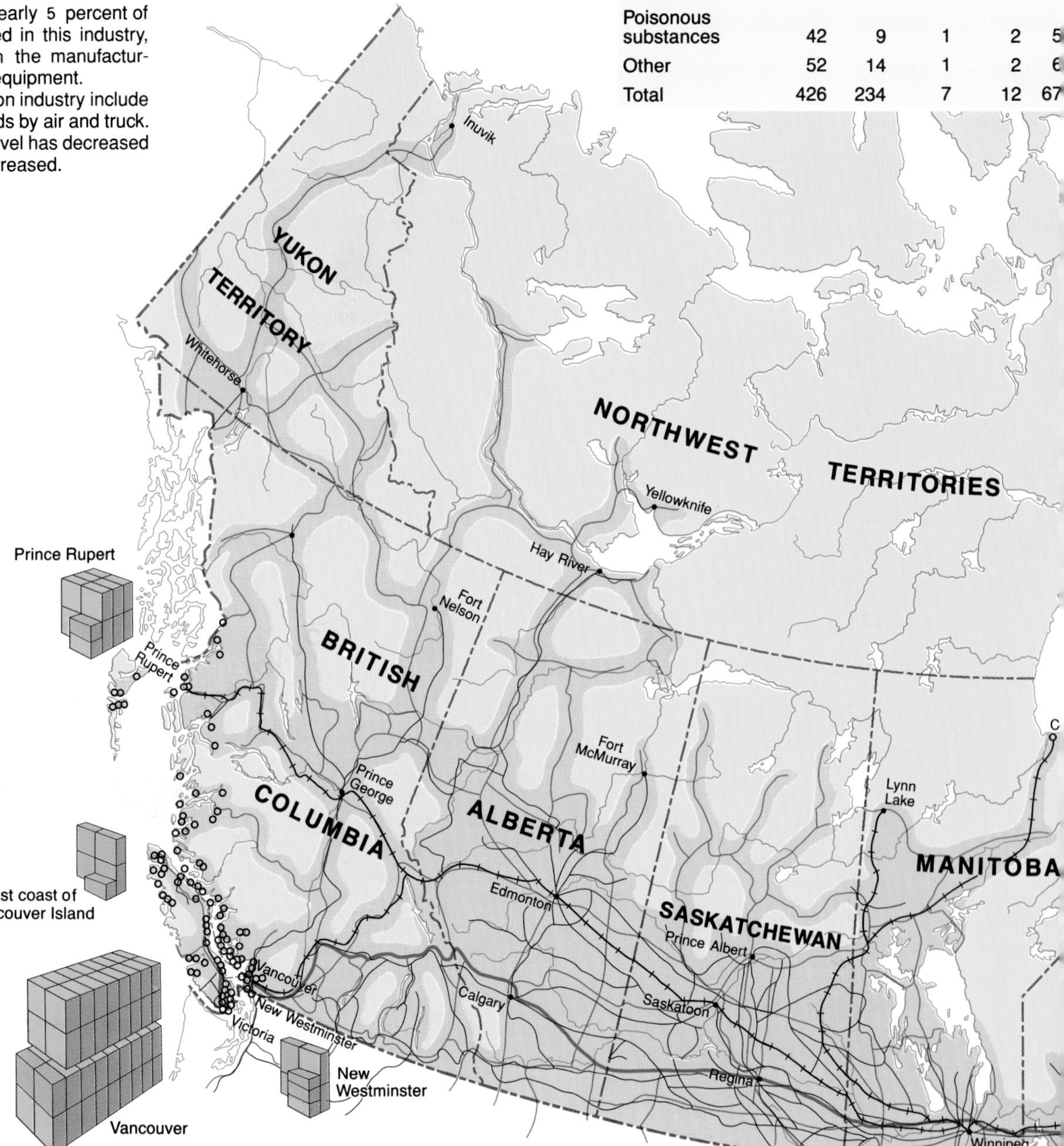

About 76 percent of commuters drive to work alone, while only 18 percent use public transportation. Although cars have become more fuel efficient over the last decade, they are also being driven greater distances, so the total amount of fuel used by cars has remained fairly constant.

Use of public transit

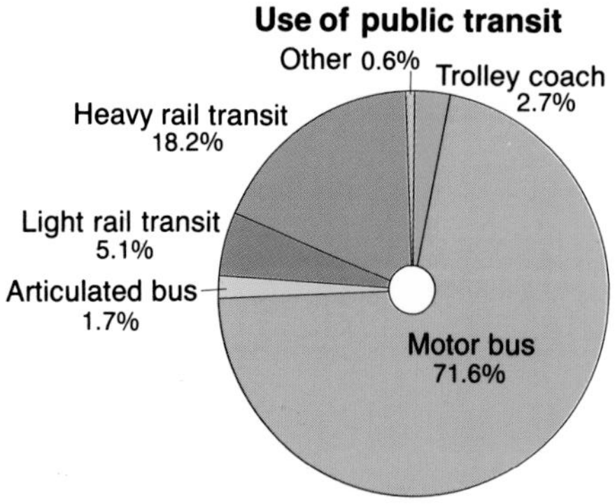

Energy efficiency

Average energy use per person per kilometre

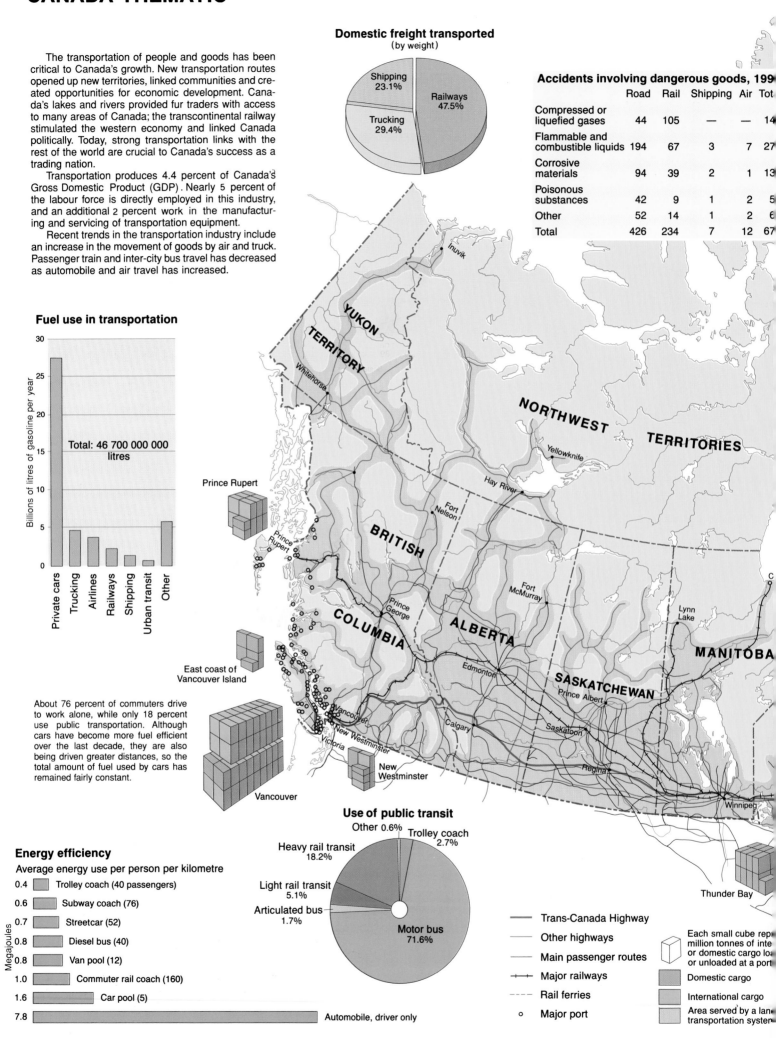

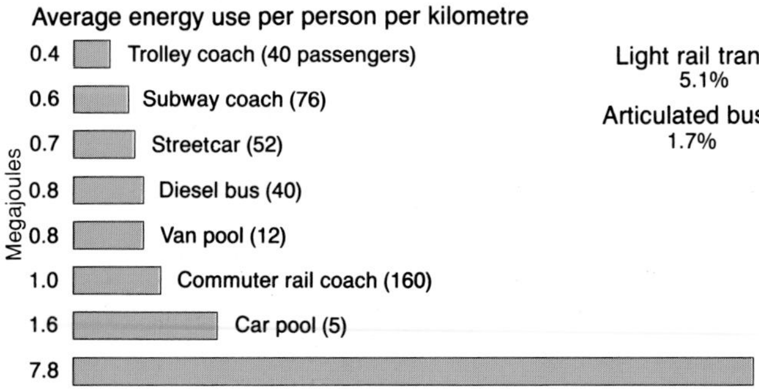

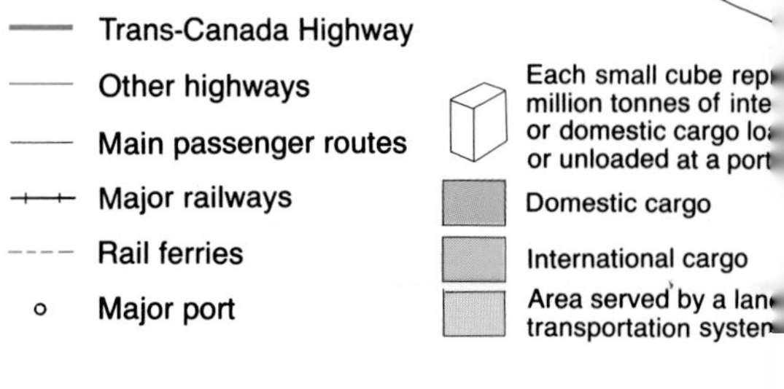

Transportation revenue

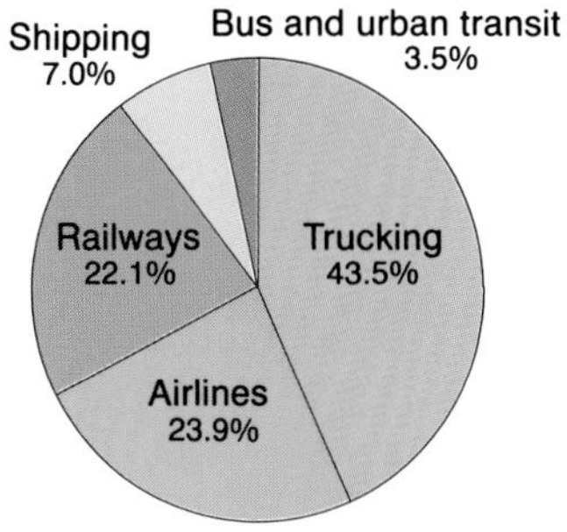

Canada's busiest airports

(Number of flights)

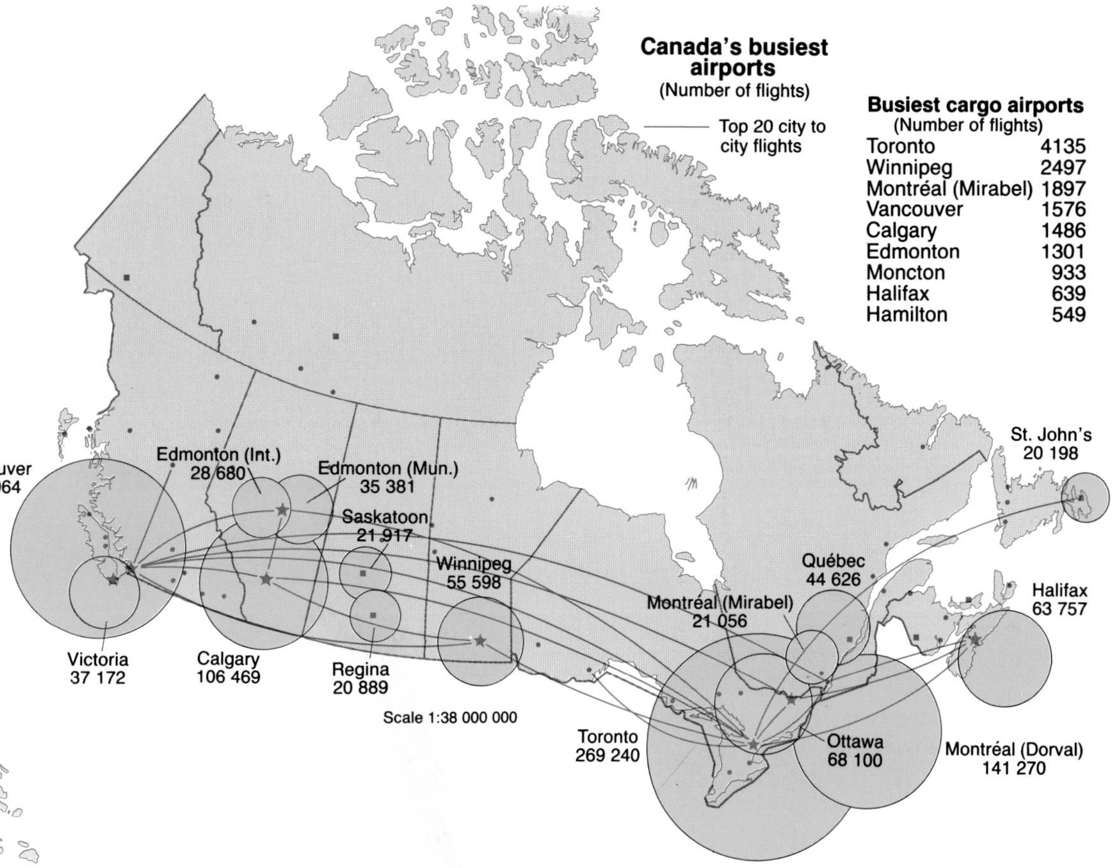

Busiest cargo airports

(Number of flights)

Toronto	4135
Winnipeg	2497
Montréal (Mirabel)	1897
Vancouver	1576
Calgary	1486
Edmonton	1301
Moncton	933
Halifax	639
Hamilton	549

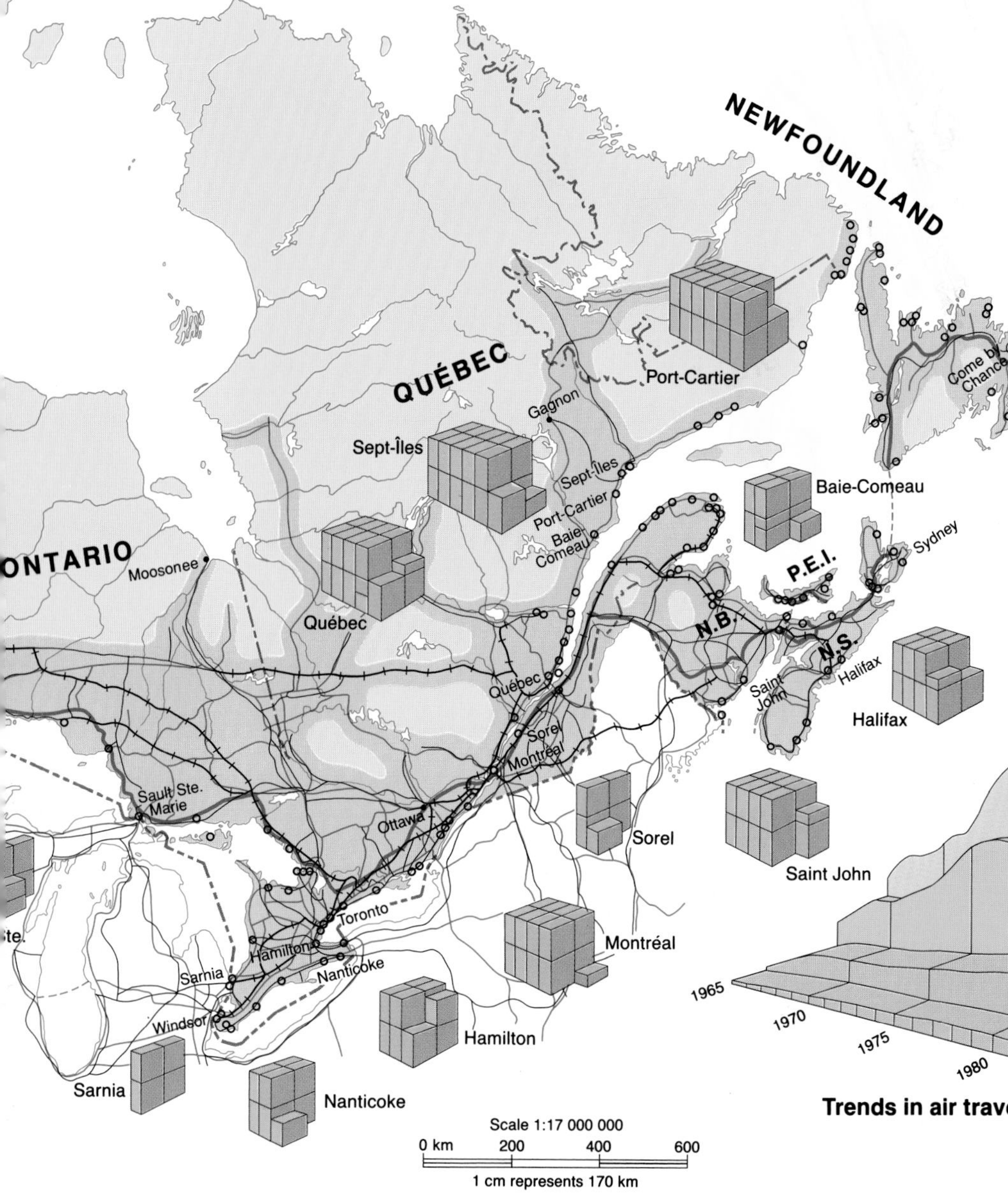

Busiest passenger airports

(Number of passengers)

Toronto	18 489 392
Vancouver	8 996 027
Montréal (Dorval)	5 590 476
Calgary	4 573 569
Ottawa	2 420 548
Halifax	2 292 370
Montréal (Mirabel)	2 254 313
Winnipeg	2 072 599
Edmonton (Int.)	1 810 020
Edmonton (Mun.)	922 589
Québec	683 810
Victoria	666 543
St.John's	613 277
Saskatoon	562 423
Regina	556 549

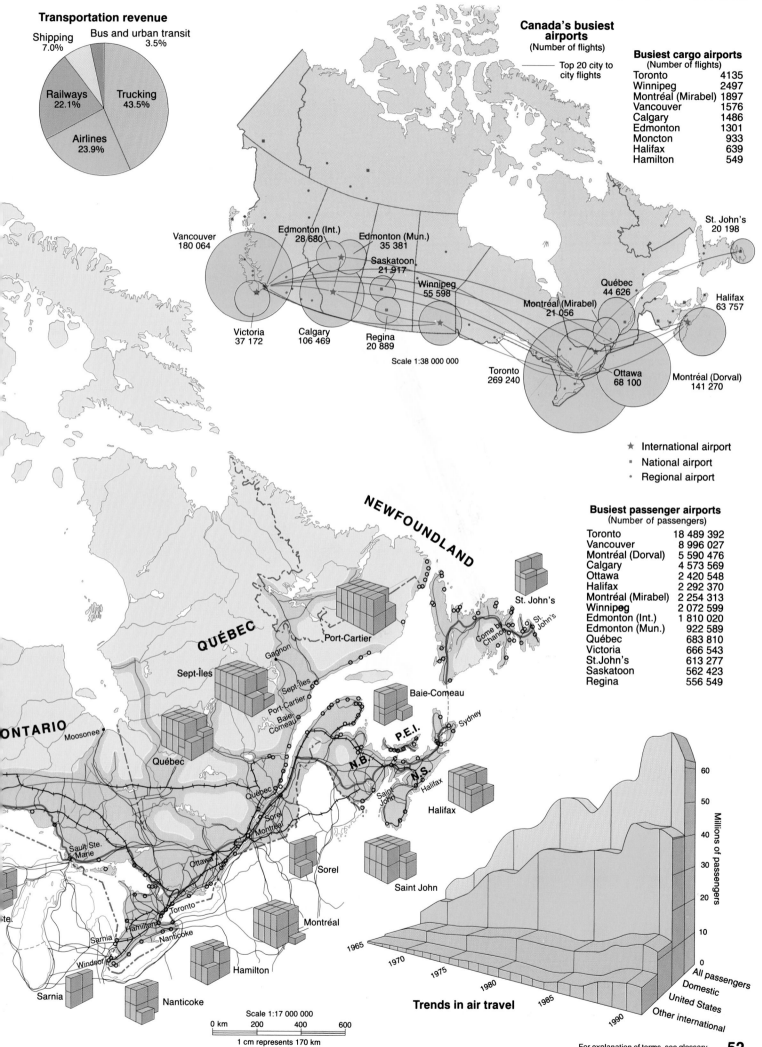

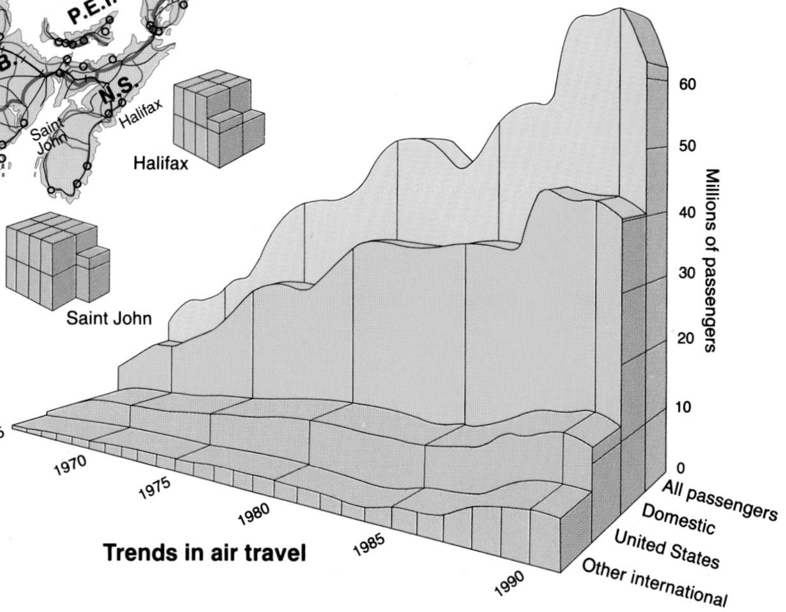

Trends in air travel

For explanation of terms, see glossary.

Canada's population has become increasingly urbanized over the last century. Today, at least three out of every four Canadians live in an urban environment. More than 61 percent live in the 25 largest cities (also known as Census Metropolitan Areas or CMAs). The trend towards increasing urbanization is levelling off, however. Although the decrease in the rural farming population continues, there is a movement of urban residents to the rural fringes of cities that keeps the percentage of rural population roughly constant.

People are drawn to cities for a variety of reasons. Because manufacturing and service industries tend to be concentrated in urban areas, cities and towns offer the widest variety of job opportunities. This makes cities attractive to both rural job-seekers and new immigrants. Cities also offer many recreational and cultural activities. The negative effects of urban growth include the loss of good farmland, water and air pollution, overcrowding and crime.

The fastest growing urban centres in Canada

		1986	1991	% change
1	Keswick-Elmhurst Beach, Ont.	8 473	13 713	61.8%
2	Bradford, Ont.	8 825	12 800	45.0
3	Varennes, Qué.	9 741	13 951	43.2
4	Port Hammond, B.C.	31 126	43 226	38.9
5	Kanata, Ont.	30 174	39 851	32.1
6	Matsqui, B.C.	70 635	92 975	31.6
7	Barrie, Ont.	59 841	78 447	31.1
8	Bolton, Ont.	7 165	9 291	29.7
9	Okotoks, Alta.	5 226	6 720	28.6
10	Westbank, B.C.	10 553	13 551	28.4
11	Stony Plain, Alta.	5 505	6 962	26.5
12	Fergus, Ont.	6 275	7 940	26.5
13	Whitchurch-Stouffville, Ont.	7 186	9 085	26.4
14	White Rock, B.C.	41 166	51 712	25.6

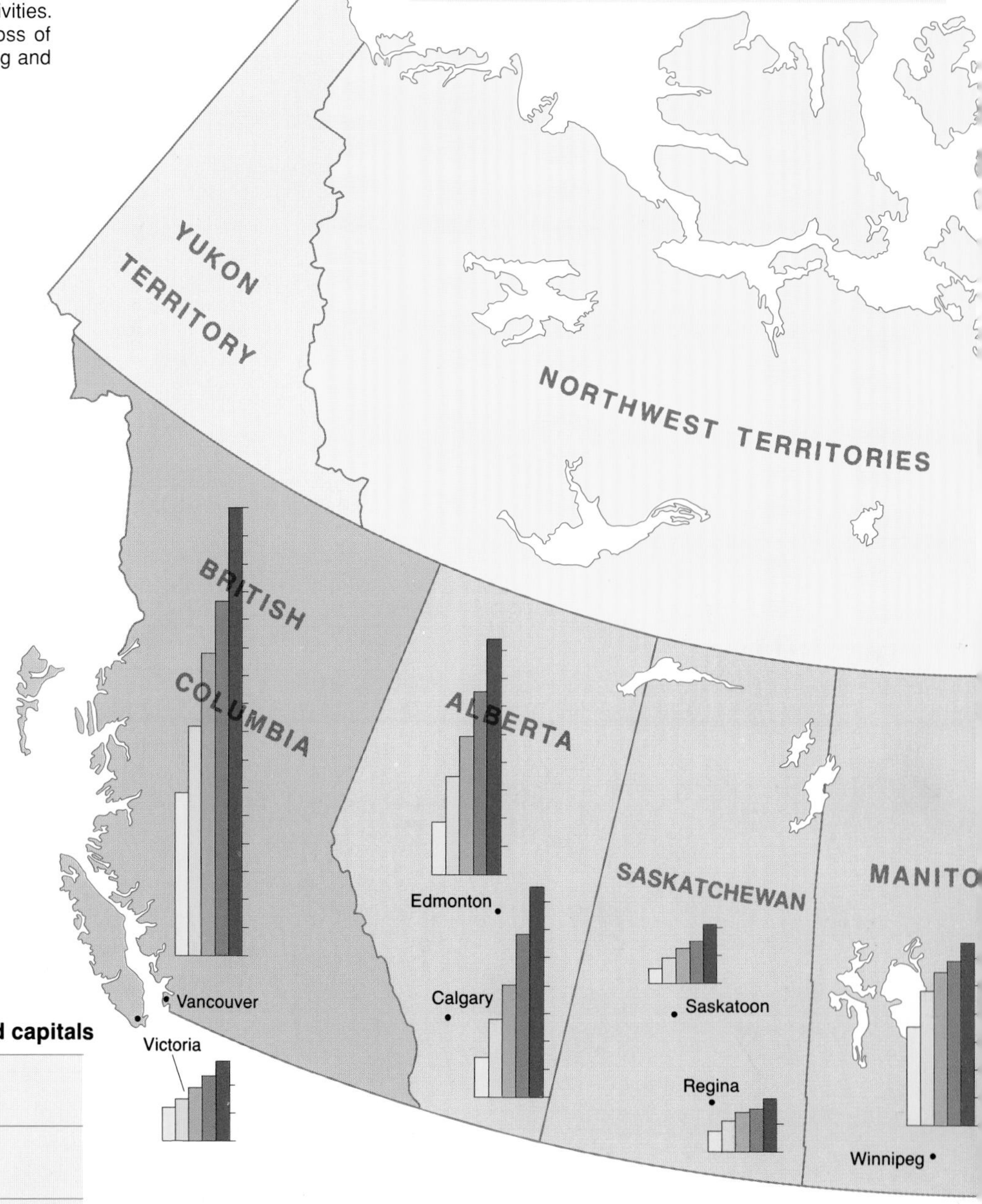

Rural to urban land conversion

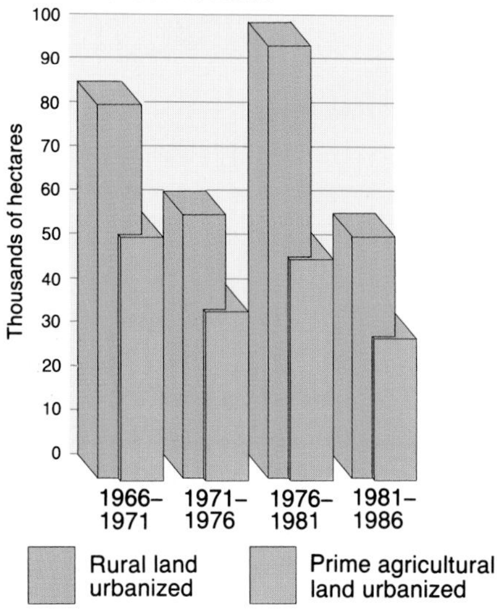

Urban population growth, provinces and capitals 1986-1991

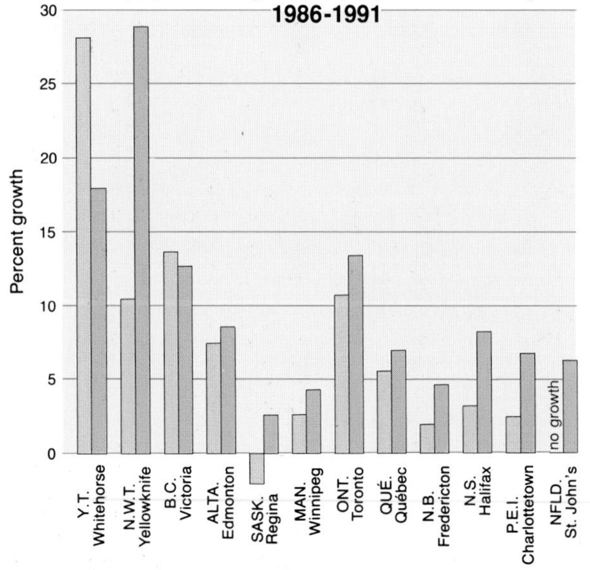

Urban population comparisons, 1990

Israel	91.6%
Sweden	84.0%
Canada	76.6%
United States	75.0%
Mexico	72.6%
Nigeria	35.2%
China	33.4%
India	30.5%
Ethiopia	12.9%
Nepal	9.6%

Urban and rural population

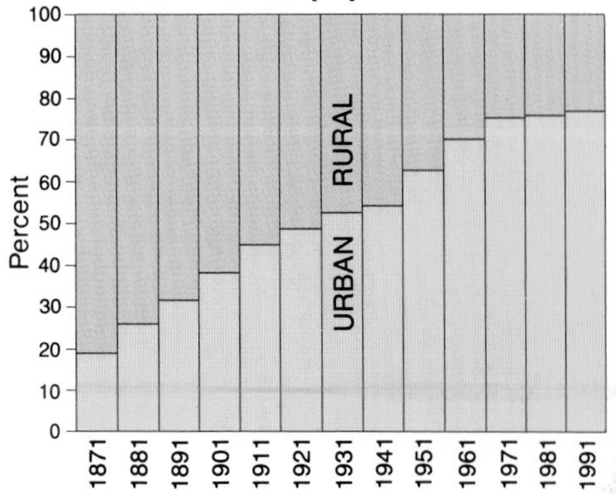

Urban air quality, selected cities

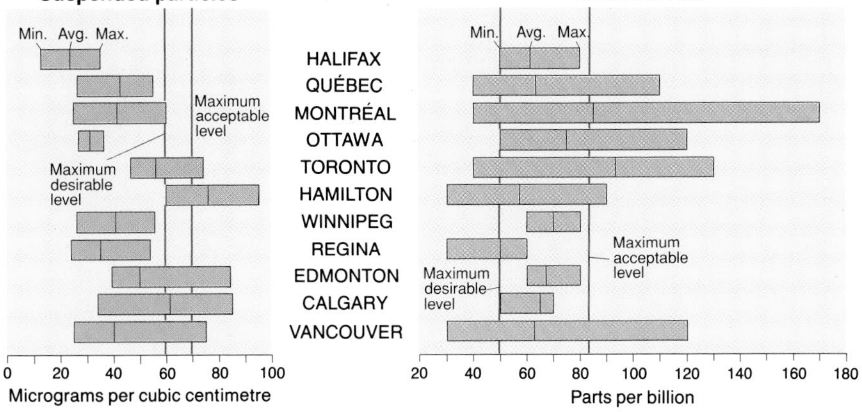

Municipal waste disposal methods

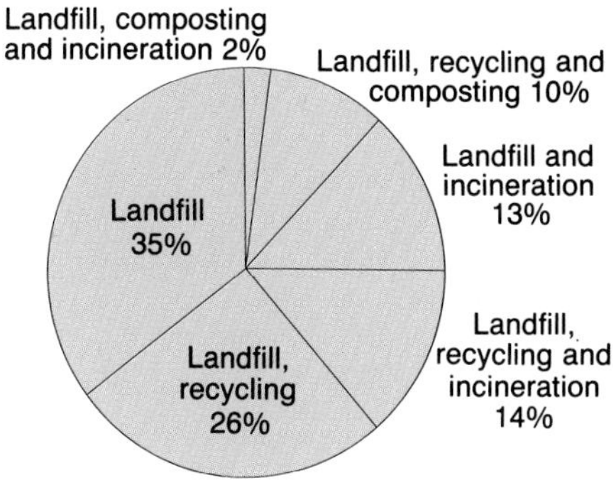

Municipal sewage treatment, selected cities

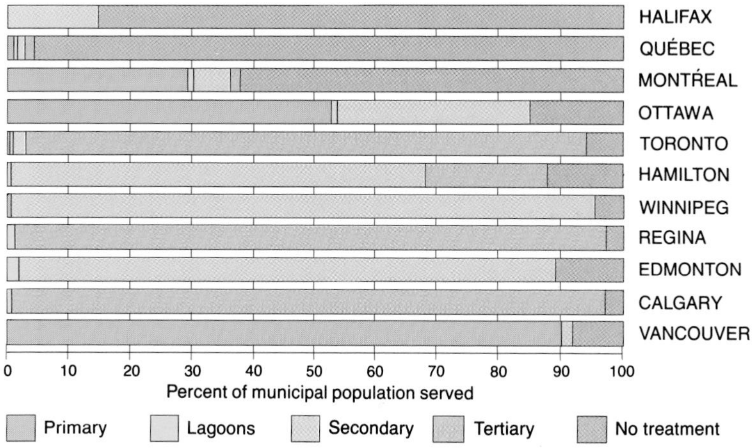

Urban development causes great stress on the environment. Natural habitats are destroyed by the construction of roads and buildings. Air and water pollution are common urban problems. Recycling and composting programs have reduced the amount of solid waste generated, but many urban areas are still running out of landfill space.

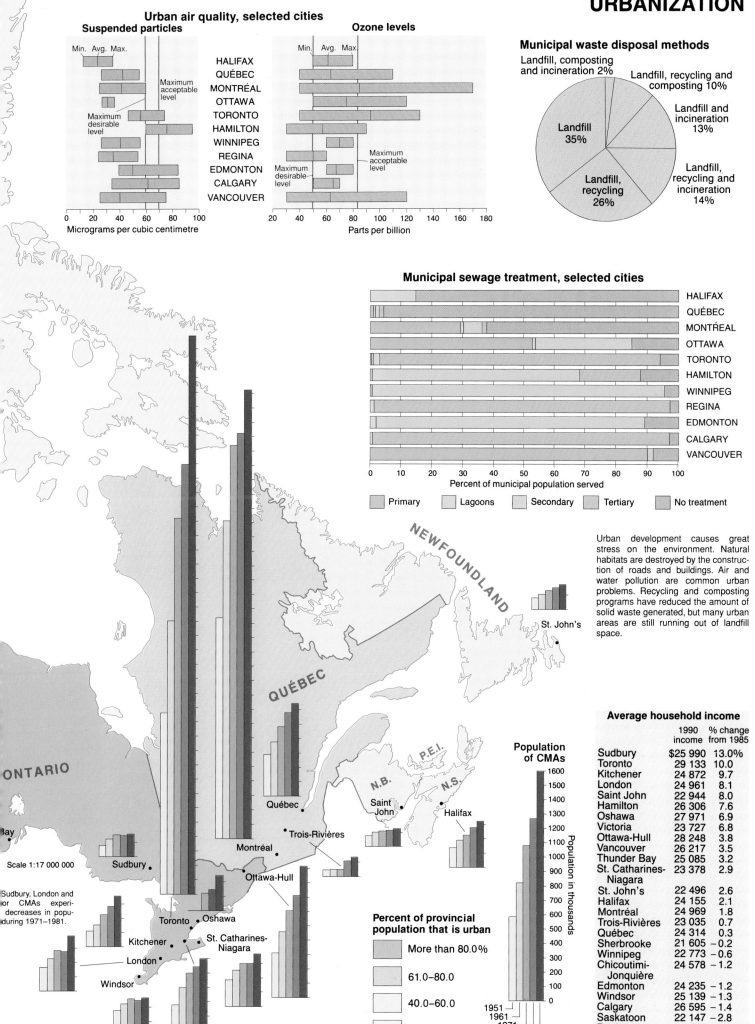

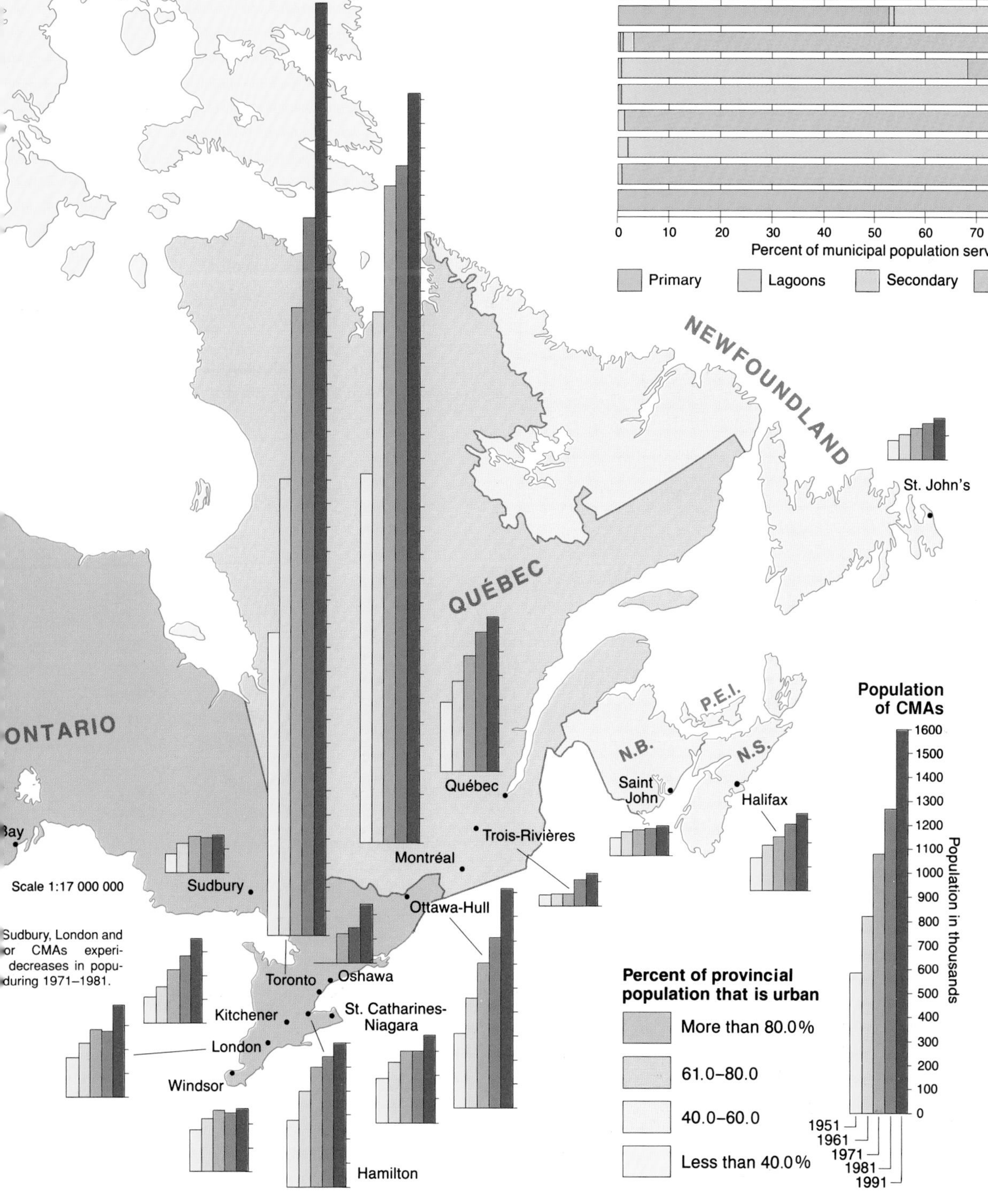

Average household income

	1990 income	% change from 1985
Sudbury	$25 990	13.0%
Toronto	29 133	10.0
Kitchener	24 872	9.7
London	24 961	8.1
Saint John	22 944	8.0
Hamilton	26 306	7.6
Oshawa	27 971	6.9
Victoria	23 727	6.8
Ottawa-Hull	28 248	3.8
Vancouver	26 217	3.5
Thunder Bay	25 085	3.2
St. Catharines-Niagara	23 378	2.9
St. John's	22 496	2.6
Halifax	24 155	2.1
Montréal	24 969	1.8
Trois-Rivières	23 035	0.7
Québec	24 314	0.3
Sherbrooke	21 605	−0.2
Winnipeg	22 773	−0.6
Chicoutimi-Jonquière	24 578	−1.2
Edmonton	24 235	−1.2
Windsor	25 139	−1.3
Calgary	26 595	−1.4
Saskatoon	22 147	−2.8
Regina	23 794	−2.9
Canada	**$24 328**	**4.3%**

CANADA-THEMATIC

Major influences on forest ecosystems

ECOZONE	Fire	Insects, Disease	Harvest-ing	Change to other use	Transport. corridors	Pollution
Boreal Cordillera	✓			✓	✓	
Pacific Maritime		✓	✓	✓	✓	✓
Montane Cordillera	✓	✓	✓	✓	✓	
Boreal Plain	✓		✓	✓	✓	
Taiga Plain	✓				✓	
Taiga Shield	✓					
Boreal Shield	✓	✓	✓	✓		✓
Mixed-Wood Plain		✓		✓		✓
Atlantic Maritime	✓	✓	✓	✓		✓

Major influences on farmland

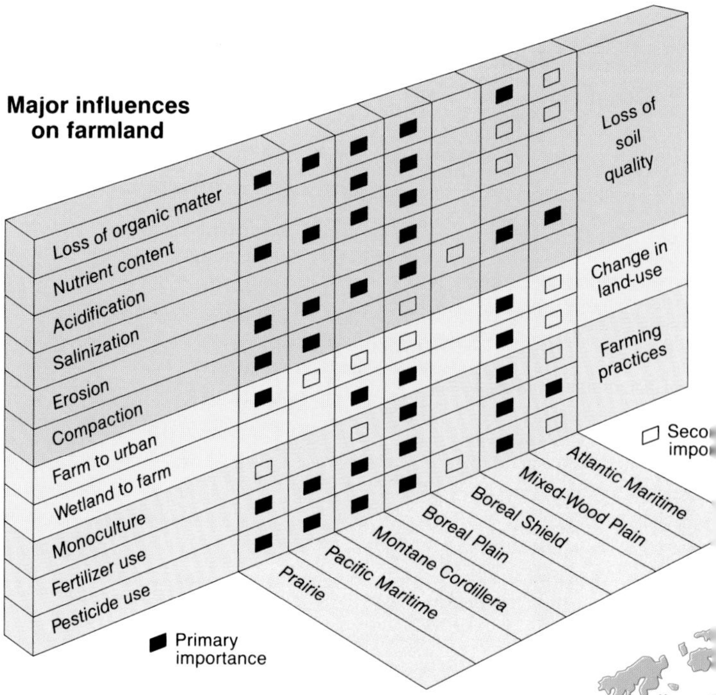

Wetlands near Winnipeg, Manitoba

Acid precipitation, or acid rain, is formed when sulphur dioxide and nitrogen oxides mix with moisture in the upper atmosphere. Fossil-fuelled power plants, metal smelters and automobile exhaust fumes are the main sources of these gases. Acid rain damages crops and forests, and seriously disrupts some aquatic ecosystems. Below a pH of 5, reproduction of aquatic species is affected; below a pH of 4, fish start to die. The worst damage occurs in aquatic ecosystems in environments lacking natural buffers, such as large areas of surface limestone, that can neutralize the acid. An estimated 14 000 lakes have been killed in eastern Canada as a result of acid rain. By 2000, both Canada and the United States aim to reduce acid-rain causing emissions by approximately one-half from 1987 totals. The situation is presently improving but much remains to be done to repair the environmental damage that has already occurred.

Causes of wetland habitat decline

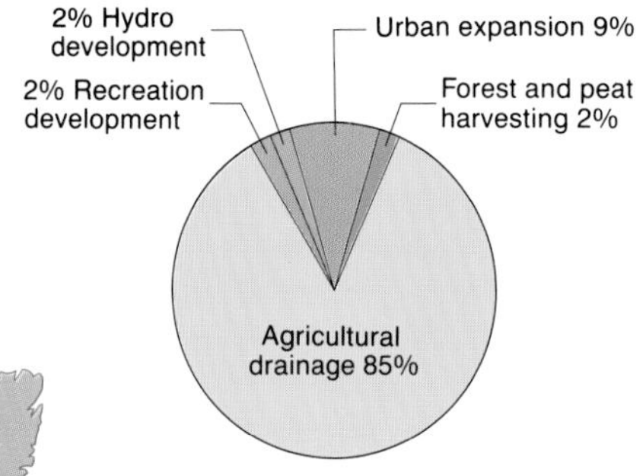

Acid rain

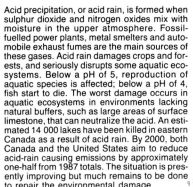

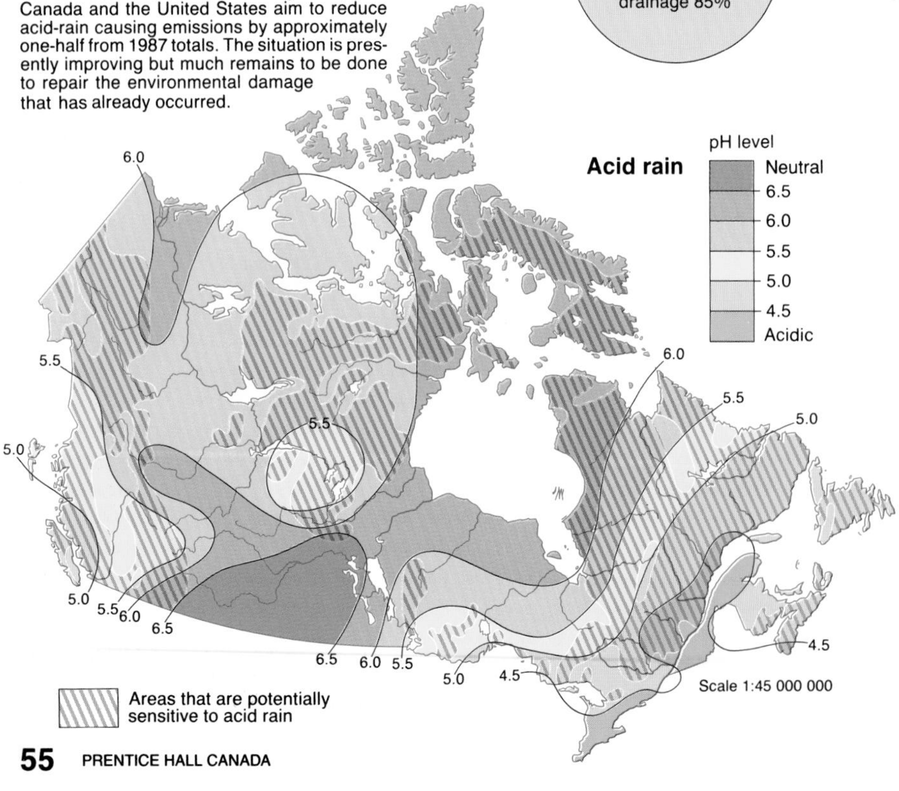

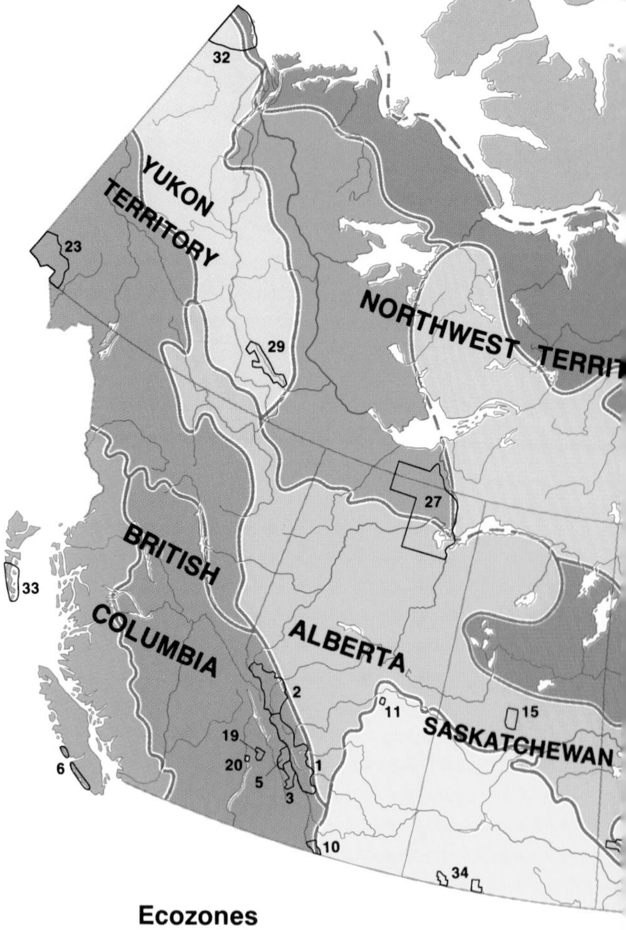

Ecozones

Tundra Cordillera	
Boreal Cordillera	Boreal Shield
Pacific Maritime	Hudson Bay P
Montane Cordillera	Mixed-Wood P
Boreal Plain	Atlantic Maritir
Taiga Plain	Southern Arcti
Prairie	Northern Arcti
Taiga Shield	Arctic Cordille

ENVIRONMENTAL ISSUES

The environment is made up of both the physical environment ·hich includes such elements as landforms, minerals, climates, er, air and soil — and living things such as plants and animals. of these elements are interdependent. The interaction of these erent elements in any one region is called an ecosystem. An system can be as small as a woodlot or pond, or as large as the eal forest or the oceans.

All ecosystems have a certain capacity to support animal and t life. However, the natural balance of an ecosystem can be upted, either by natural events or human activities. In recent rs, sophisticated technology and an increasing population have ant that humans are having more impact on the environment ever before.

Today, Canadians recognize that all life is interdependent and attempting to minimize the adverse effects of their activities. ough much has been accomplished, it will take continuing effort n all Canadians to make sure a healthy environment is preserved legacy for future generations.

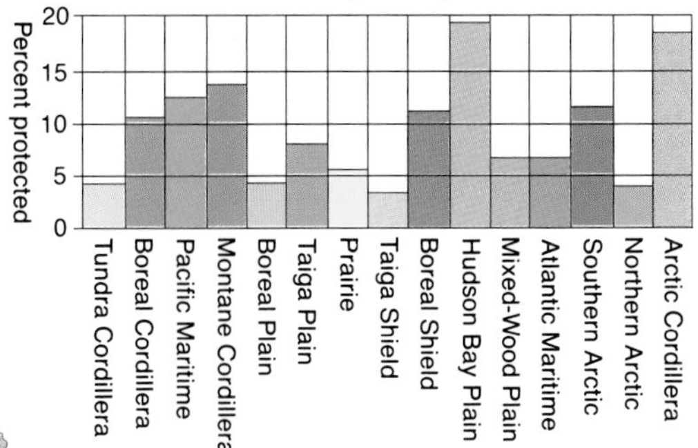

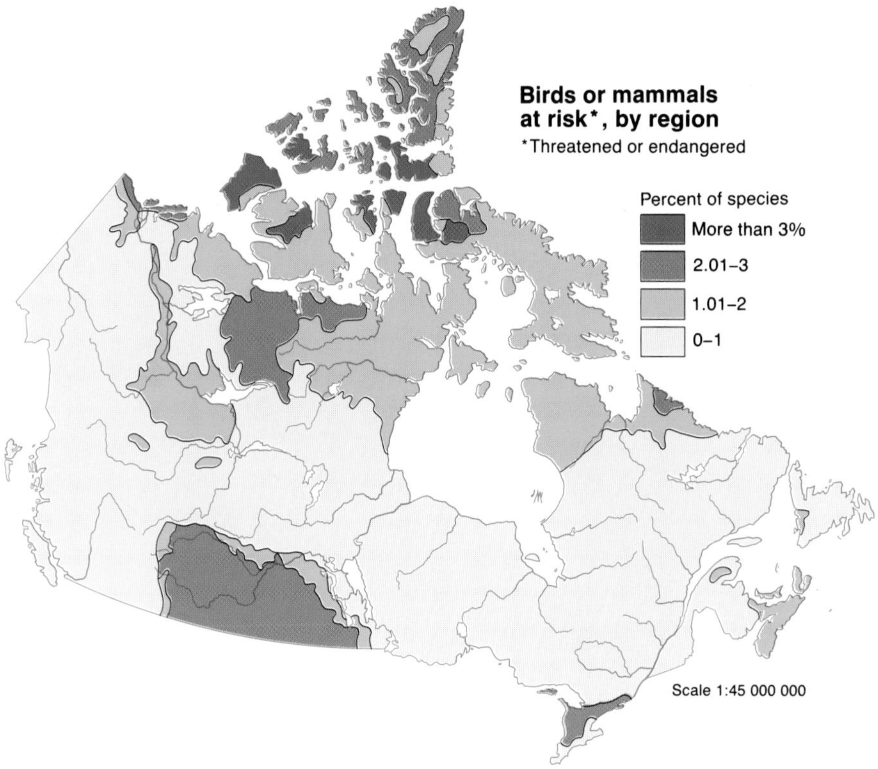

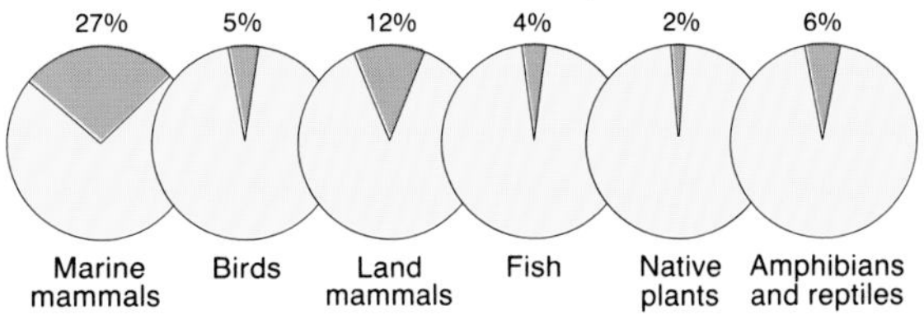

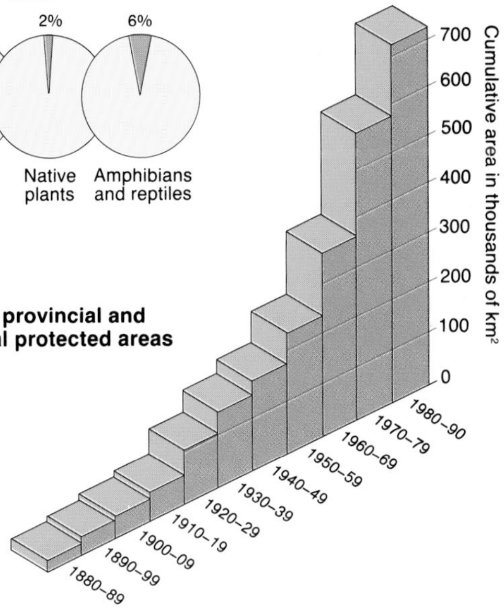

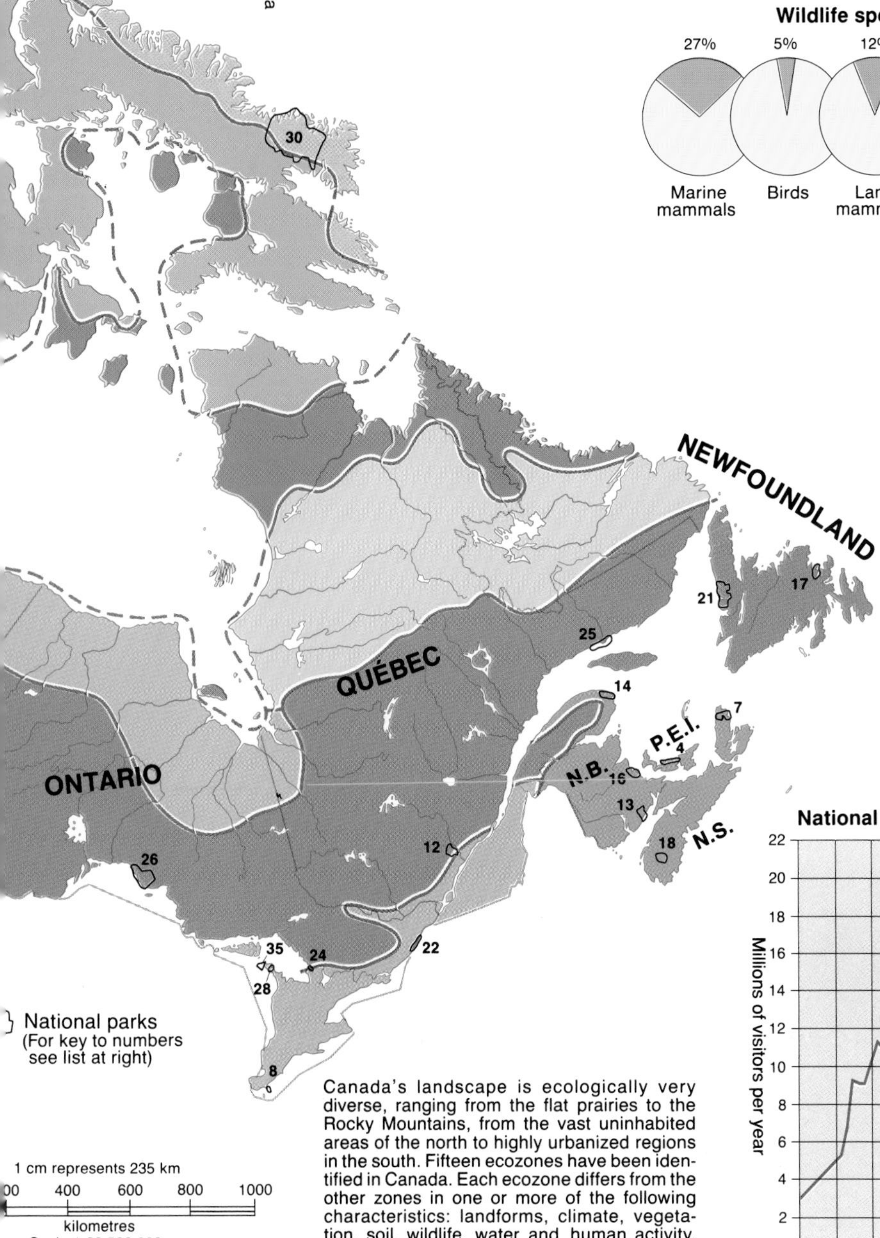

Canada's landscape is ecologically very diverse, ranging from the flat prairies to the Rocky Mountains, from the vast uninhabited areas of the north to highly urbanized regions in the south. Fifteen ecozones have been identified in Canada. Each ecozone differs from the other zones in one or more of the following characteristics: landforms, climate, vegetation, soil, wildlife, water and human activity. The zones vary in size and often cross provincial boundaries.

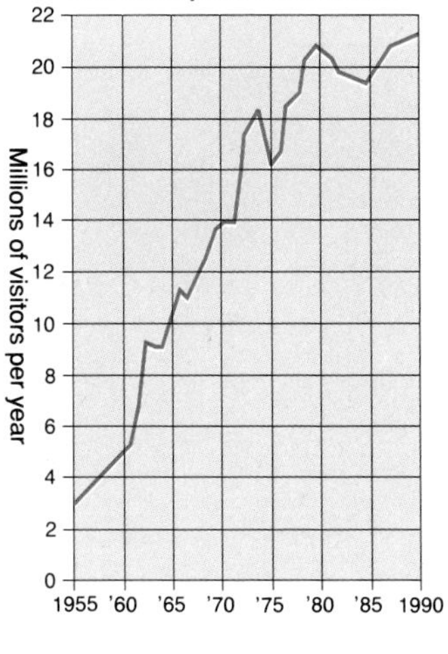

Popularity of parks, 1990

National park	Attendance	Percent of total
1 Banff, Alta.	4 030 000	19.78
2 Jasper, Alta.	1 310 000	6.44
3 Kootenay, B.C.	1 160 000	5.66
4 Prince Edward Island, P.E.I.	810 000	3.97
5 Yoho, B.C.	680 000	3.33
6 Pacific Rim, B.C.	600 000	2.95
7 Cape Breton Highlands, N.S.	570 000	2.79
8 Point Pelee, Ont.	460 000	2.26
9 Riding Mountain, Man.	390 000	1.92
10 Waterton Lakes, Alta.	350 000	1.73
11 Elk Island, Alta.	300 000	1.46
12 La Mauricie, Qué.	280 000	1.38
13 Fundy, N.B.	230 000	1.11
14 Forillon, Qué.	210 000	1.03
15 Prince Albert, Sask.	190 000	0.92
16 Kouchibouguac, N.B.	180 000	0.89
17 Terra Nova, Nfld.	170 000	0.84
18 Kejimkujik, N.S.	170 000	0.83
19 Glacier, B.C.	160 000	0.80
20 Mount Revelstoke, B.C.	160 000	0.80
21 Gros Morne, Nfld.	98 000	0.48
22 St. Lawrence Islands, Ont.	85 000	0.42
23 Kluane, Y.T.	69 000	0.34
24 Georgian Bay Islands, Ont.	49 000	0.24
25 Mingan Archipelago, Qué.	25 000	0.12
26 Pukaskwa, Ont.	15 000	0.17
27 Wood Buffalo, Alta.-N.W.T.	6600	0.03
28 Bruce Peninsula, Ont.	3500	0.02
29 Nahanni, N.W.T.	1300	0.01
30 Auyuittuq, N.W.T.	350	0.01
31 Ellesmere Island, N.W.T.	230	0.01
32 Northern Yukon, Y.T.	99	0.01
33 South Moresby, B.C.	no data available	
34 Grasslands, Sask.	no data available	
35 Fathom Five, Ont.	no data available	

For explanation of terms, see glossary.

Exports from Canada

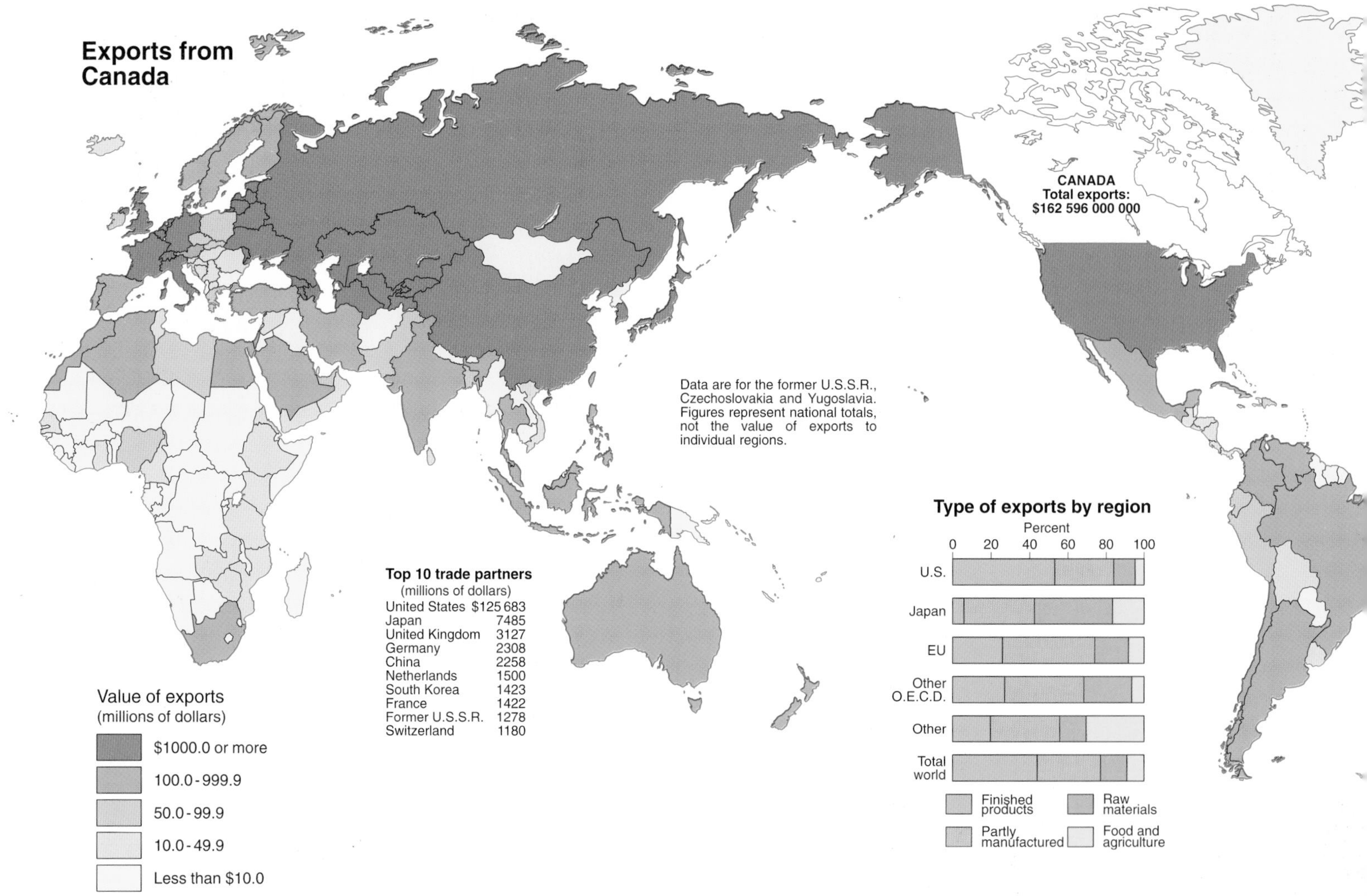

Top 10 trade partners
(millions of dollars)

United States	$125 683
Japan	7485
United Kingdom	3127
Germany	2308
China	2258
Netherlands	1500
South Korea	1423
France	1422
Former U.S.S.R.	1278
Switzerland	1180

Imports to Canada

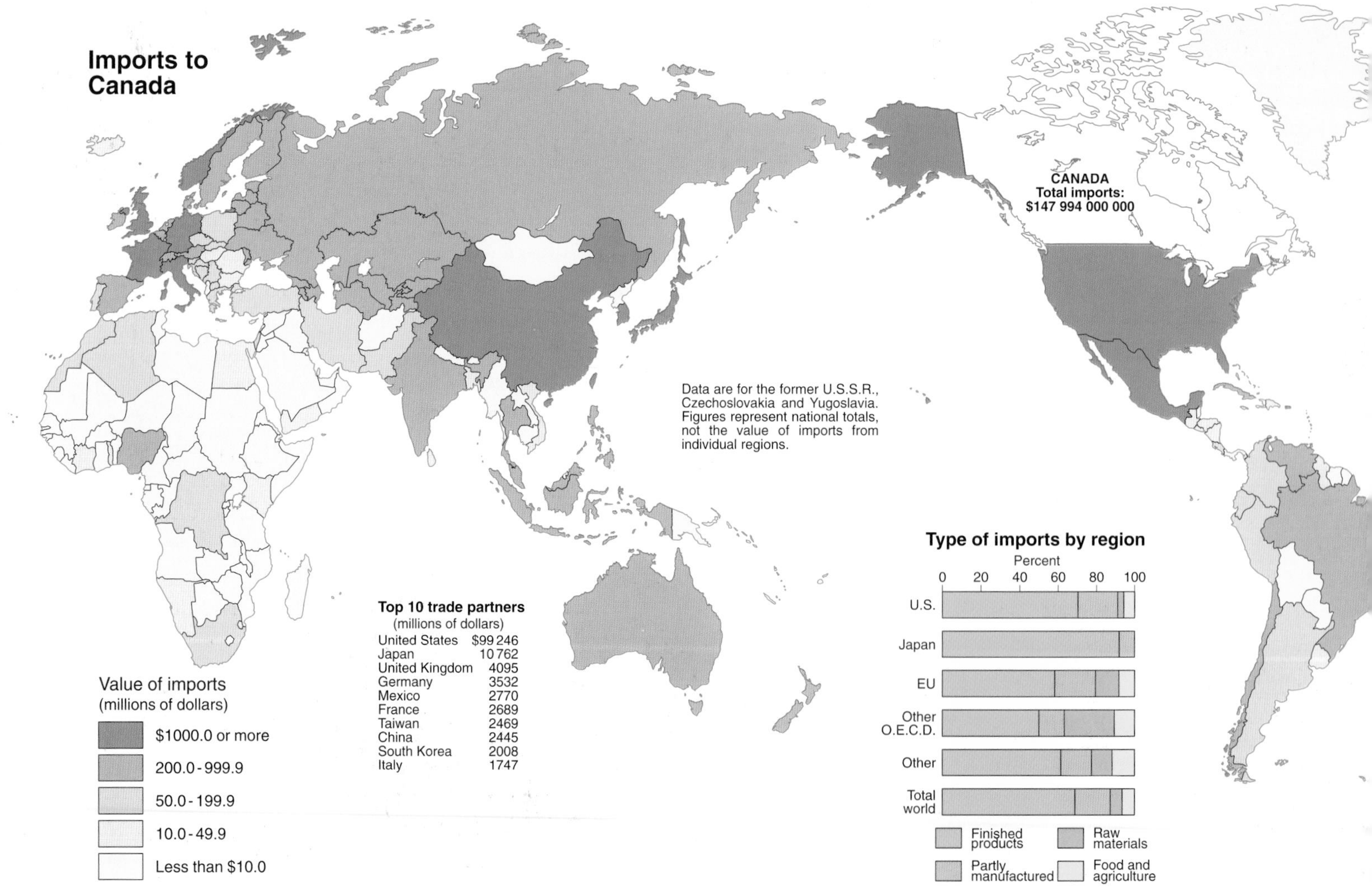

Top 10 trade partners
(millions of dollars)

United States	$99 246
Japan	10 762
United Kingdom	4095
Germany	3532
Mexico	2770
France	2689
Taiwan	2469
China	2445
South Korea	2008
Italy	1747

ernational trade has shaped the development of la. Exchanges of raw materials, goods and es have made this country a major trading nation. er to compete globally, Canadians must maintain international relations. With the longest ected border in the world stretching between la and the United States, it is no wonder that cans are Canada's main trading partners. In years, Canada's trade with Pacific Rim countries ilso increased. Trading blocs such as the ean Union (EU), and agreements such as the al Agreement on Tariffs and Trade (GATT) and orth American Free Trade Agreement (NAFTA), a major impact on Canada's economic relations ther countries.

chinery and equipment and automobiles account largest shares of imports and exports. Forestry cts, energy products and agricultural and fishing cts are significant exports. Consumer goods, ially from the Pacific Rim, are significant imports.

Canadian expenditures on foreign aid

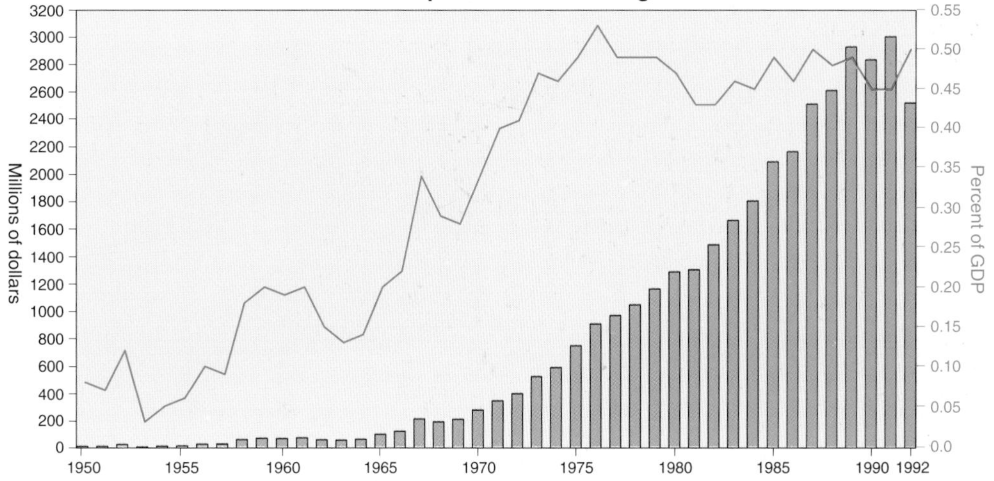

Foreign aid

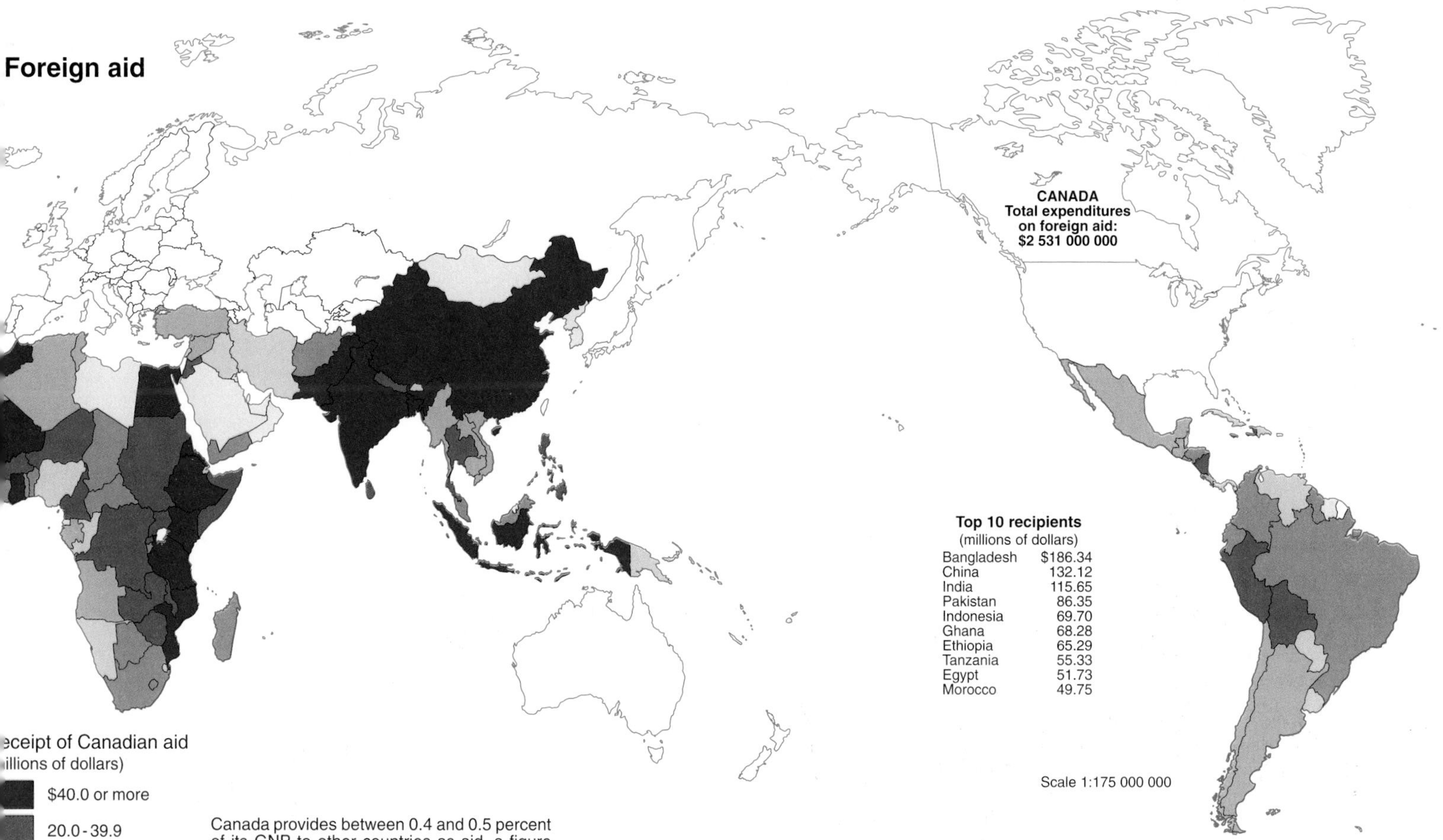

Top 10 recipients
(millions of dollars)

Bangladesh	$186.34
China	132.12
India	115.65
Pakistan	86.35
Indonesia	69.70
Ghana	68.28
Ethiopia	65.29
Tanzania	55.33
Egypt	51.73
Morocco	49.75

eceipt of Canadian aid
illions of dollars)

- $40.0 or more
- 20.0-39.9
- 10.0-19.9
- 5.0-9.9
- 1.0-4.9
- Less than $1.0
- No aid

Canada provides between 0.4 and 0.5 percent of its GNP to other countries as aid, a figure below the United Nations target of 0.7 percent. Aid takes different forms—short-term emergency aid for disasters such as floods, and long-term aid to improve local living conditions. Canadian branches of non-governmental organizations (NGOs) such as Save the Children, Oxfam and Rotary International, as well as various religious organizations, also provide assistance in the form of financial, material or human resources.

Canadian aid by region, 1991-92

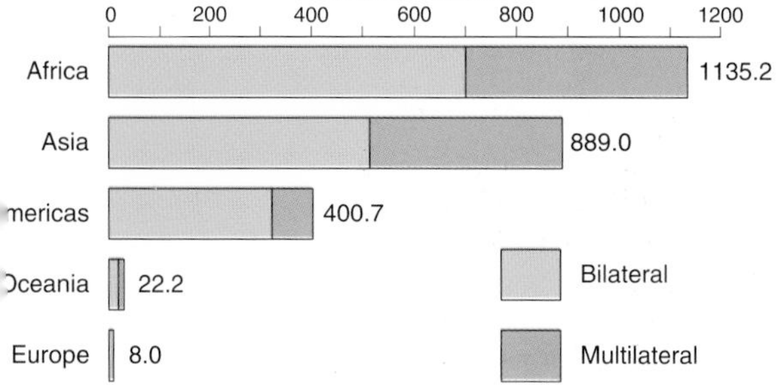

Canadian aid programs

(ODA – Official Development Assistance)

National initiatives
The Canadian government works directly with recipient country governments to plan programs and projects.

Partnership programs
Funding is provided to national and international organizations running their own programs and projects.

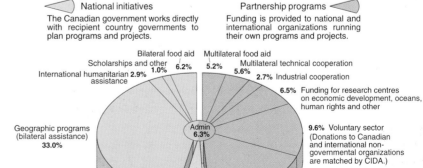

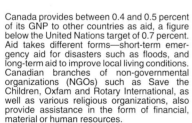

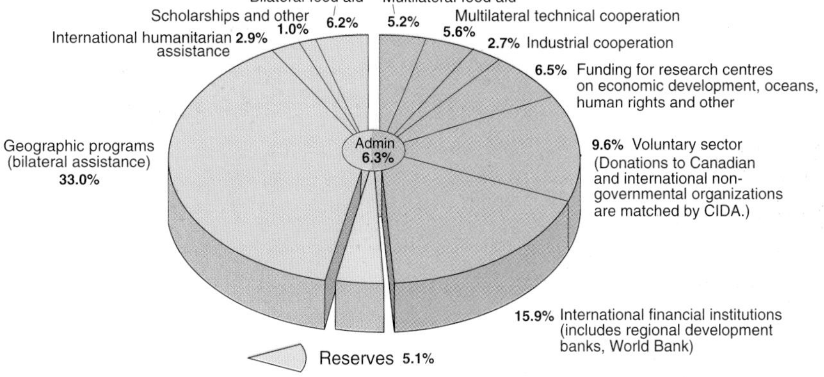

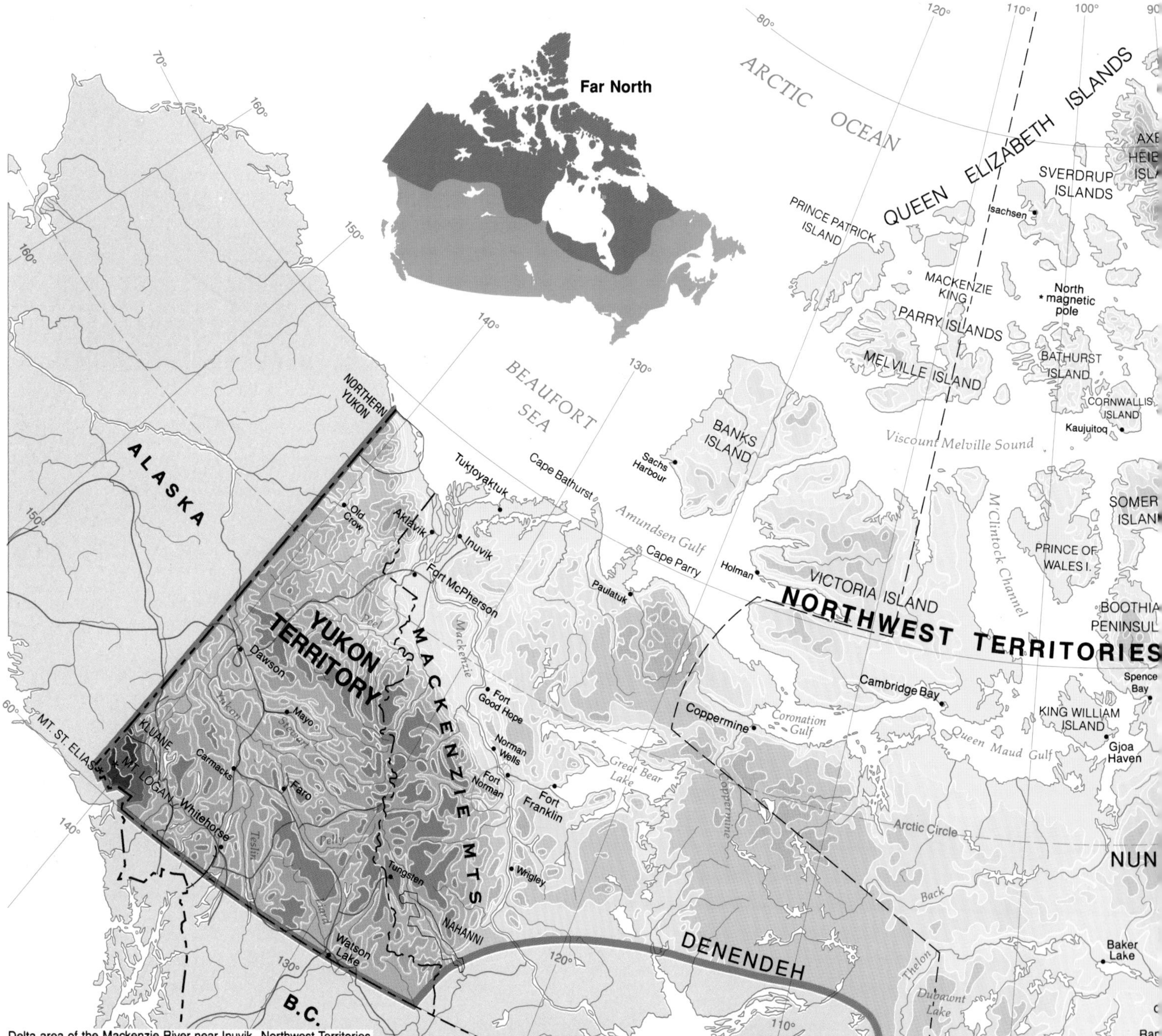
Far North
ARCTIC OCEAN
QUEEN ELIZABETH ISLANDS
SVERDRUP ISLANDS
PRINCE PATRICK ISLAND
Isachsen
MACKENZIE KING I.
North magnetic pole
PARRY ISLANDS
MELVILLE ISLAND
BATHURST ISLAND
CORNWALLIS ISLAND
Kaujuitoq
BEAUFORT SEA
BANKS ISLAND
Sachs Harbour
Viscount Melville Sound
NORTHERN YUKON
ALASKA
Tuktoyaktuk
Cape Bathurst
Amundsen Gulf
Cape Parry
Holman
Paulatuk
M'Clintock Channel
PRINCE OF WALES I.
VICTORIA ISLAND
NORTHWEST TERRITORIES
BOOTHIA PENINSUL
Old Crow
Aklavik
Inuvik
Fort McPherson
YUKON TERRITORY
MACKENZIE MTS
Peel
Mackenzie
Dawson
Mayo
Fort Good Hope
Cambridge Bay
Spence Bay
Coppermine
Coronation Gulf
Queen Maud Gulf
KING WILLIAM ISLAND
Gjoa Haven
MT. ST. ELIAS
KLUANE
MT. LOGAN
Carmacks
Faro
Whitehorse
Teslin
Pelly
Norman Wells
Fort Norman
Fort Franklin
Great Bear Lake
Coppermine
Arctic Circle
Wrigley
Tungsten
NAHANNI
Watson Lake
B.C.
DENENDEH
Back
Thelon
Dubawnt Lake
Baker Lake
MANITOBA
Tadoule Lake
SASK.
120°
110°
100°
130°
140°
150°
160°
70°
60°
80°

Delta area of the Mackenzie River near Inuvik, Northwest Territories

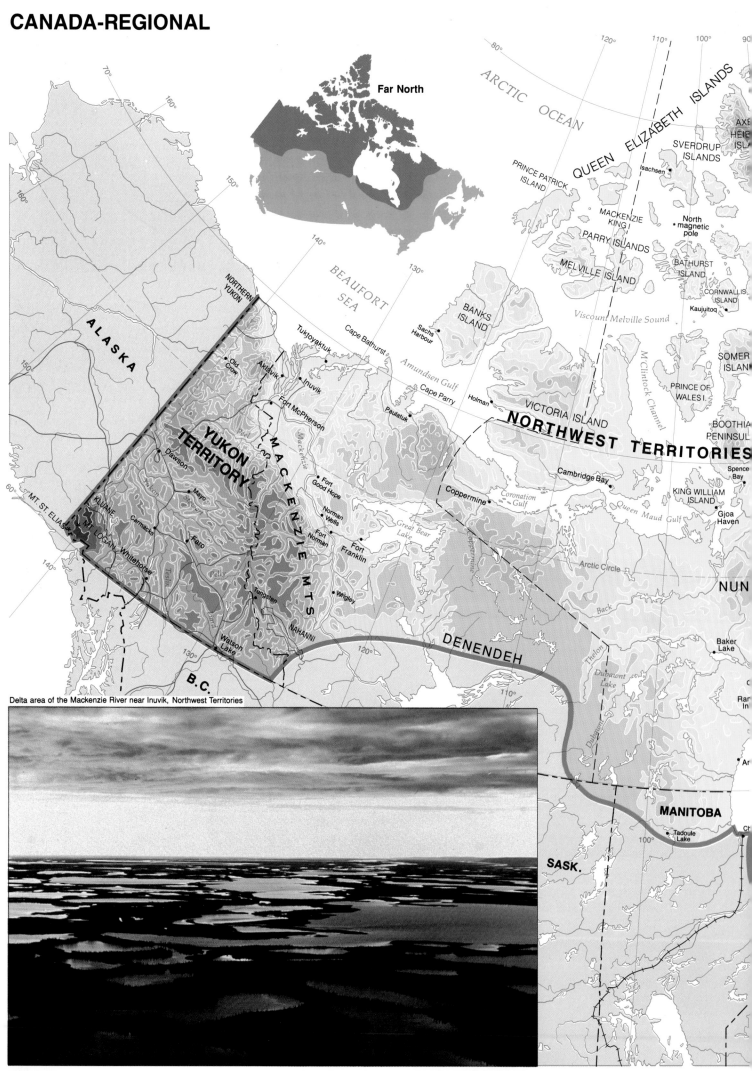

FAR NORTH

The Far North, Canada's largest region, has a unique character because of its physical environment and variety of cultures. The climate is harsh, particularly in the northern part, with long, cold winters, extended periods of darkness, a lack of vegetation and permanently frozen ground.

At the time of European arrival, Indian cultural groups lived in the south and Inuit groups in the north. Europeans came to the area initially in hopes of finding the Northwest Passage and later for whales, white fox and seals. Today, the focus of development is on minerals, gas and oil.

The Inuit and other Native peoples have been greatly affected by these changes, as has the environment. Some changes have been beneficial—education, medical services and employment opportunities. Other changes are less welcome—pollution, substance abuse, tuberculosis and potential loss of cultural identity. Through the settlement of land claims, Native groups are gaining more control over both government and resource development in the region.

Life in the Far North can be isolated, costly and dangerous. To encourage workers to migrate from the south, employers pay well. However, many residents of the Far North enjoy the rugged beauty of the landscape and the challenge of working in a frontier community.

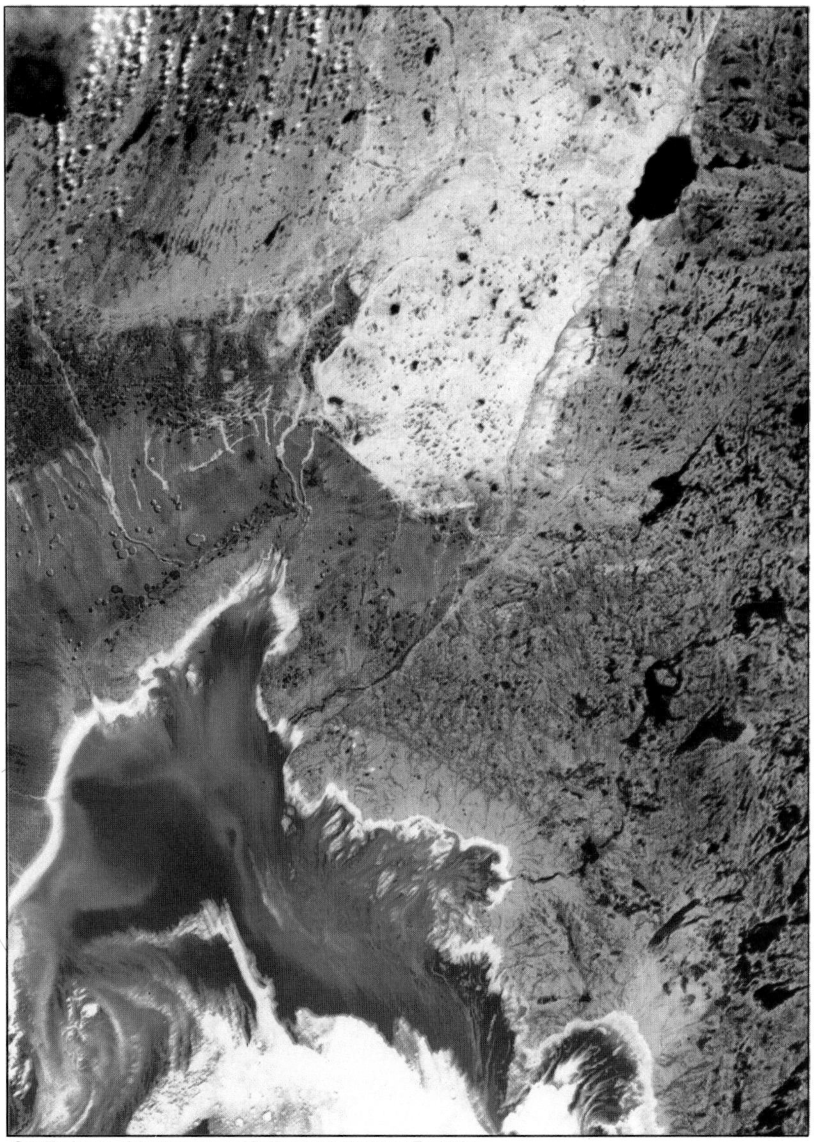

Satellite image of the west coast of Baffin Island, Northwest Territories

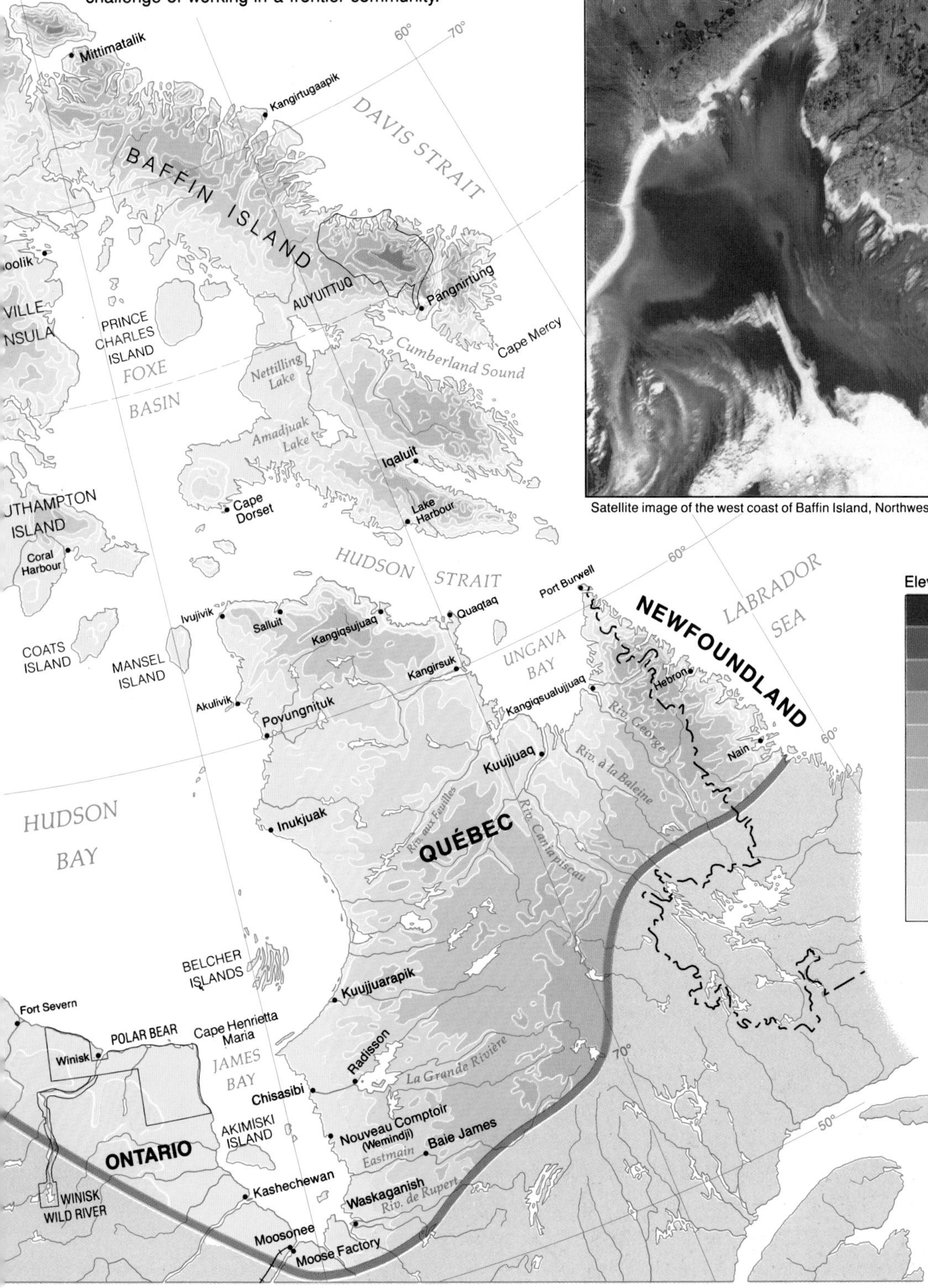

Population of major urban centres

Whitehorse	Y.T.	17 925
Iqaluit	N.W.T.	3552
Inuvik	N.W.T.	3206
Baie James	Qué.	3157
Chisasibi	Qué.	2306
Rankin Inlet	Nfld.	1706
Kuujjuaq	Qué.	1405
Waskaganish	Qué.	1355
Arviat	N.W.T.	1323
Faro	Y.T.	1221
Moosonee	Ont.	1213
Pangnirtung	N.W.T.	1135
Cambridge Bay	N.W.T.	1116
Povungnituk	Qué.	1091
Inukjuak	Qué.	1044

1 cm represents 125 km

0 100 200 300 400 500
kilometres
Scale 1: 12 500 000

CANADA-REGIONAL

Day and night in the Far North

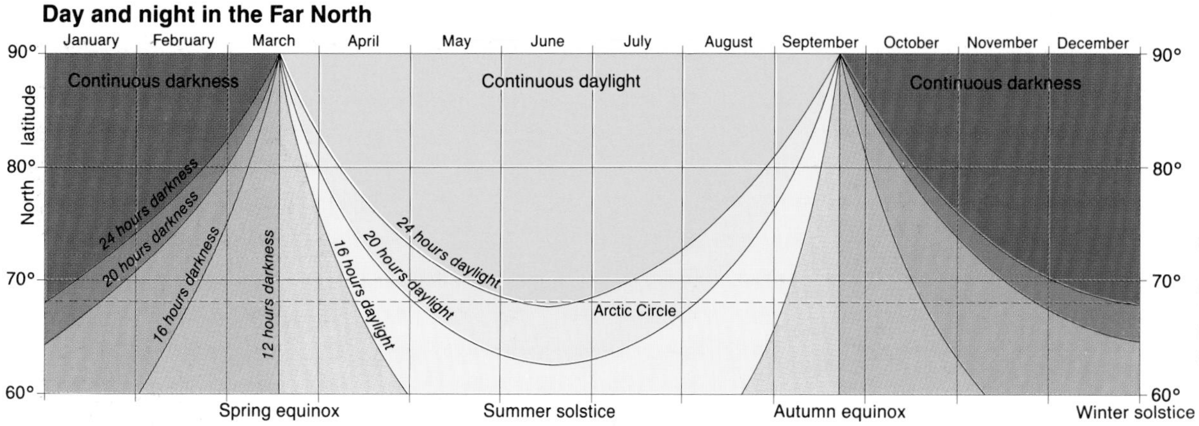

Most of the Far North region has mafrost. Although the upper layer c (called the active layer) melts in the mer, the permanently frozen gr beneath it prevents water drainage. areas, this creates boggy condition

Permafrost creates obstacle human development. Most surface rials become unstable during the mer thaw. Structures that are h must be separated from the fr ground by insulation or by supp Paved roads and airstrips must ha insulating layer of sand or gravel u neath them. Permafrost also makes ing and drilling for oil and gas difficult.

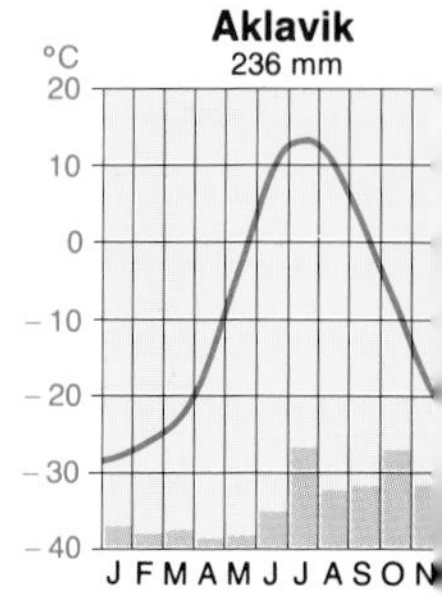

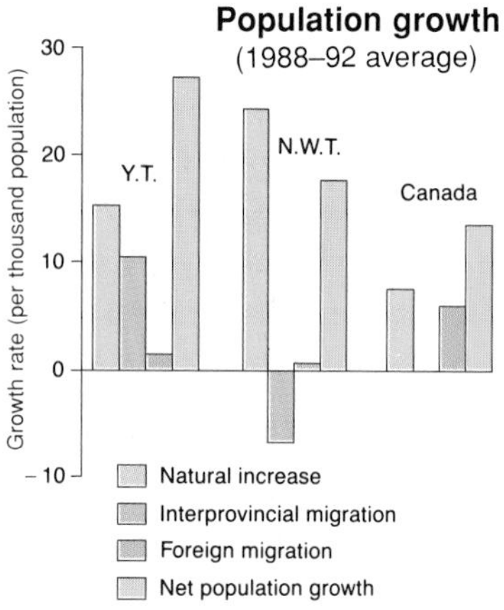

Ethnicity

(single responses only)

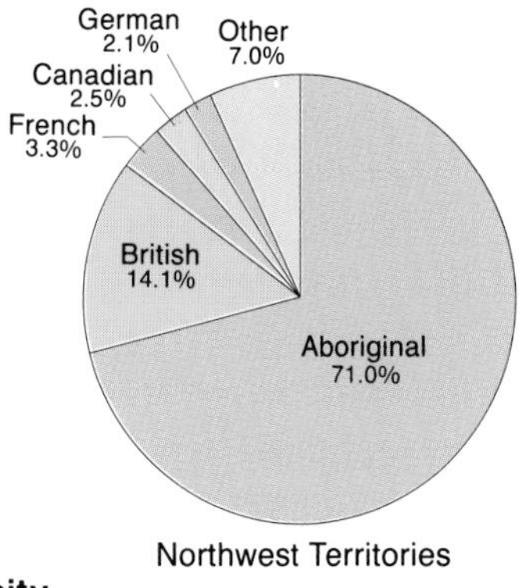

Northwest Territories

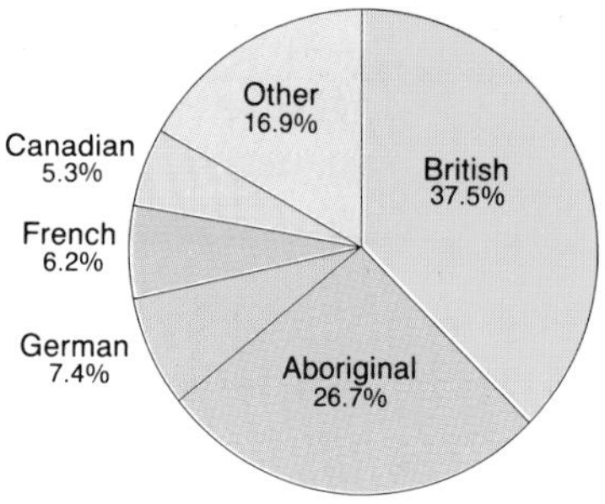

Yukon Territory

The Far North has a large aboriginal population. Land claims settlements are giving these groups increasing power in government and the economic development of the region. In 1992, the division of the Northwest Territories into two areas called Denendeh and Nunavut was ratified. The population of the western area, Denendeh, is largely Dene, Métis and non-Native, while Nunavut is mostly Inuit.

Arctic Bay, Baffin Island, in winter

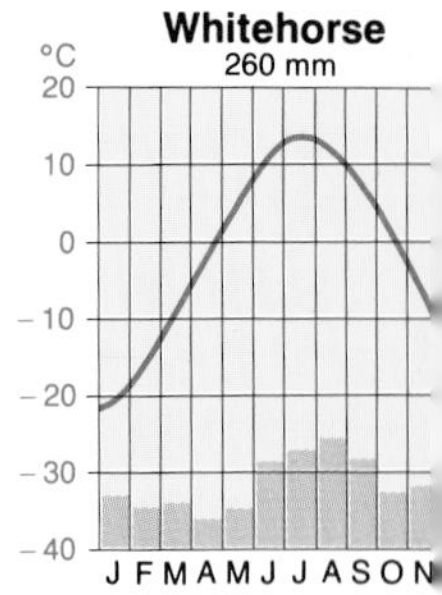

Many residents of the Far North still get much of their food from hunting and fishing.

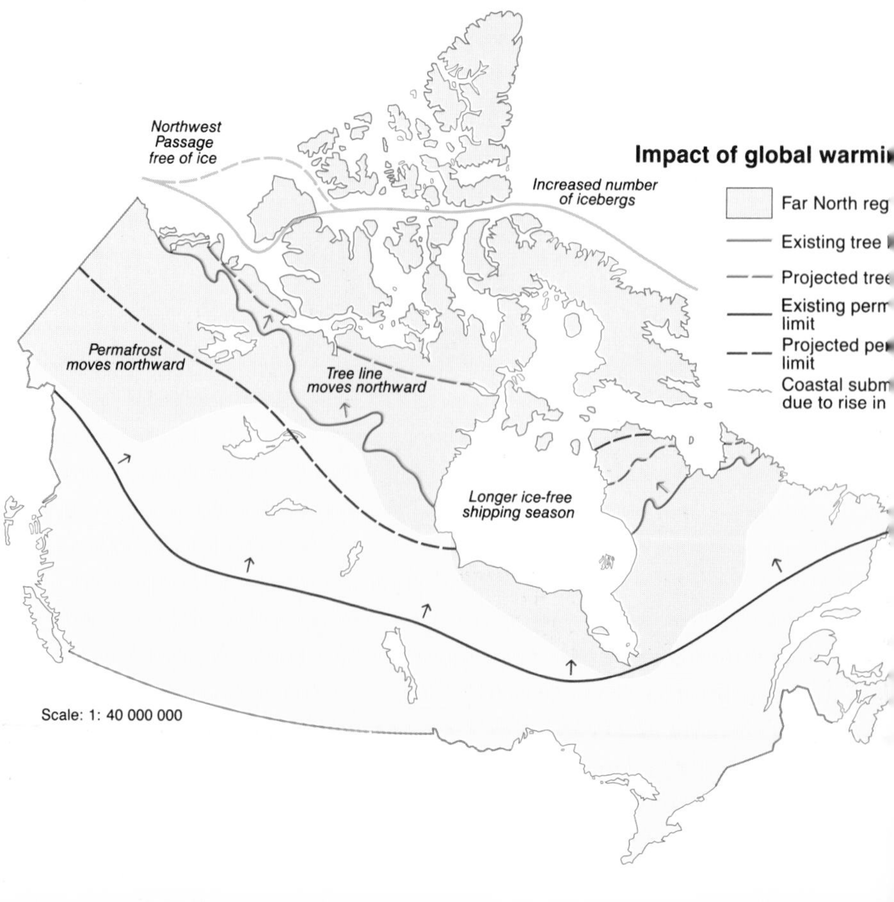

r sea ice

50–100% coverage

10–49%

Less than 10%

Forest use

Tundra (no trees)

Local

Moderate commercial

resources

Gold

Silver

Lead

Zinc

Gas field

Oil field

Oil pipeline

Hydroelectric dam

Weather station

Climograph station

Glacier

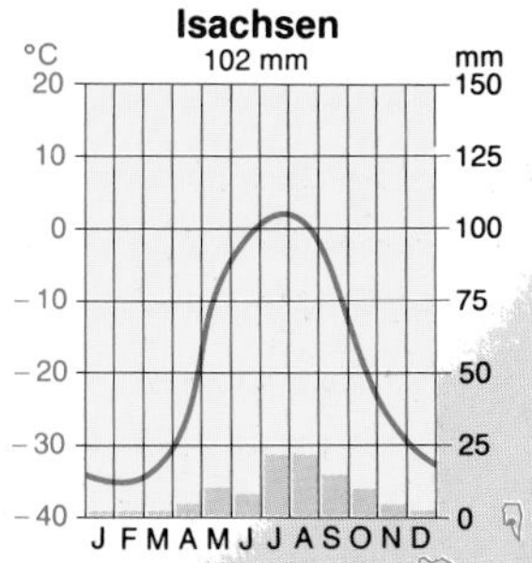

The community of Iqaluit, Baffin Island

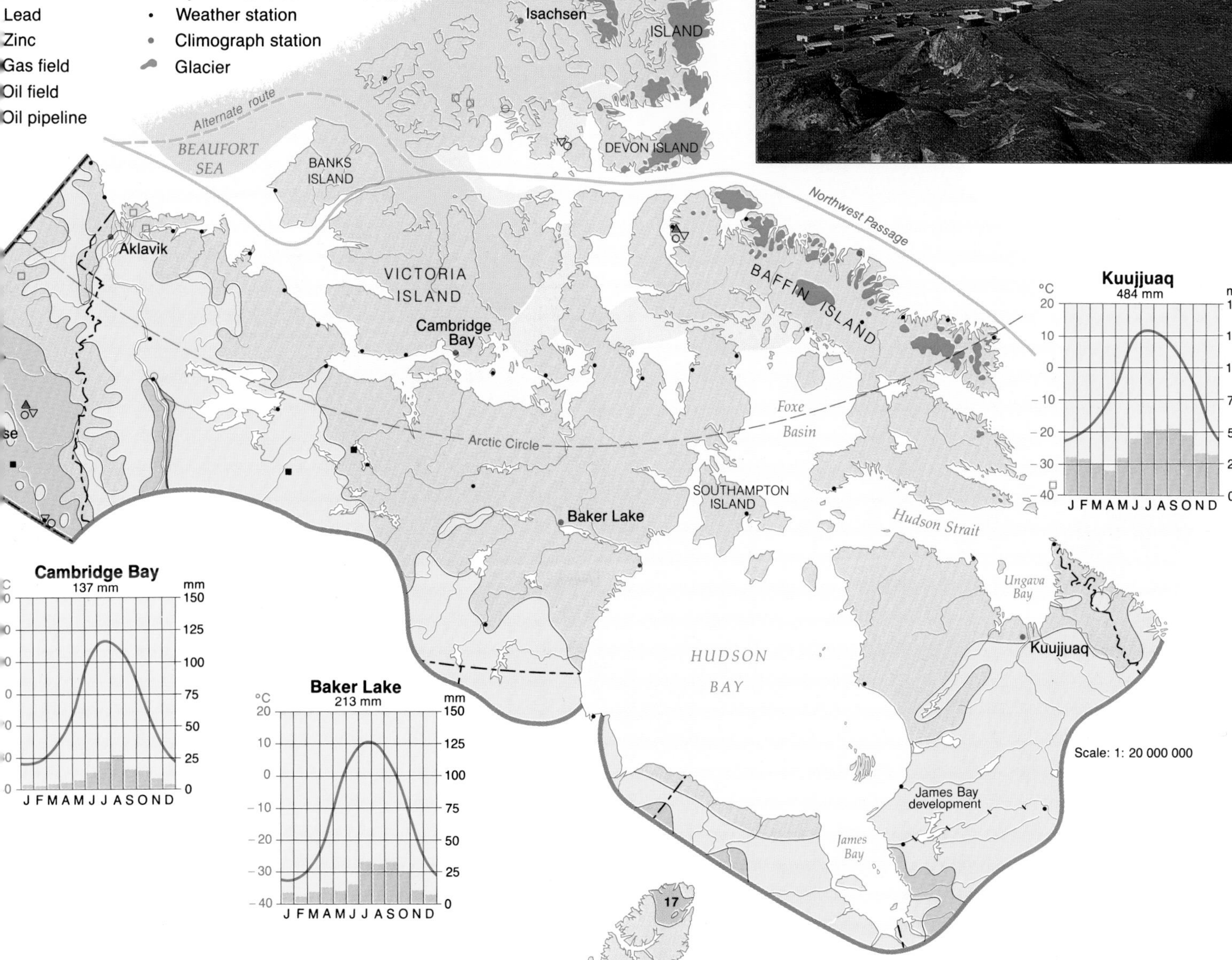

ease in global temperatures
ed with the greenhouse effect
ed to be greatest in the polar
n particular, winter tempera-
the Far North could increase
as 8 to 10°C. This would
ermafrost, glaciers and per-
e in the Arctic Ocean to melt,
in a rise in world ocean levels.
ng of permafrost would also
roads, buildings and other
s that are now resting on per-
frozen ground. Warming
o cause the tree line to move
d and wildlife to be dislocated.
mber of icebergs would
creating more problems for
d ships. However, the melting
rmanent ice would open up
west Passage and make it a
ipping route for the summer
Heavy shipping traffic and
foreign interest in the area
ve implications for Canada's
nty.

Protected areas and endangered species

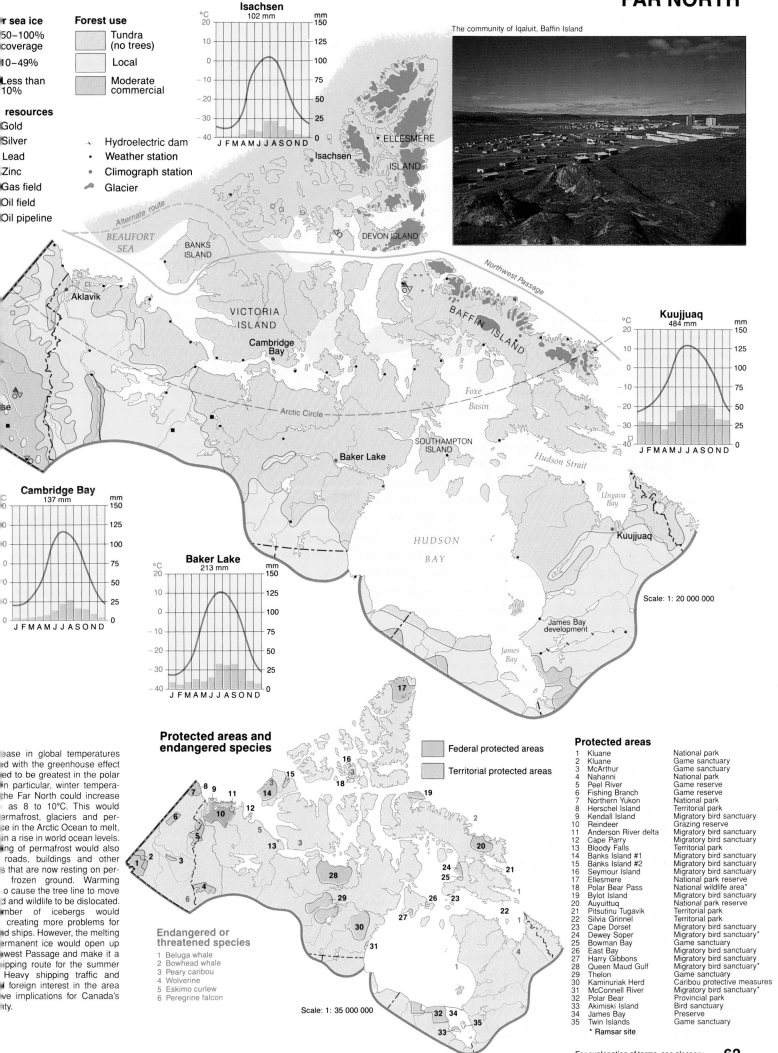

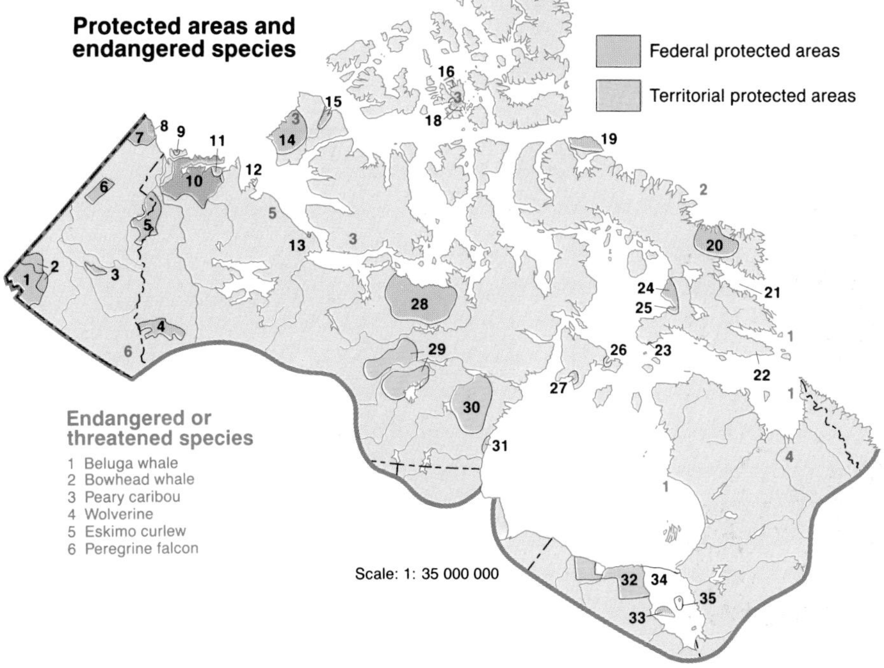

Protected areas

1	Kluane	National park
2	Kluane	Game sanctuary
3	McArthur	Game sanctuary
4	Nahanni	National park
5	Peel River	Game reserve
6	Fishing Branch	Game reserve
7	Northern Yukon	National park
8	Herschel Island	Territorial park
9	Kendall Island	Migratory bird sanctuary
10	Reindeer	Grazing reserve
11	Anderson River delta	Migratory bird sanctuary
12	Cape Parry	Migratory bird sanctuary
13	Bloody Falls	Territorial park
14	Banks Island #1	Migratory bird sanctuary
15	Banks Island #2	Migratory bird sanctuary
16	Seymour Island	Migratory bird sanctuary
17	Ellesmere	National park reserve
18	Polar Bear Pass	National wildlife area*
19	Bylot Island	Migratory bird sanctuary
20	Auyuittuq	National park reserve
21	Pitsutinu Tugavik	Territorial park
22	Silvia Grinnel	Territorial park
23	Cape Dorset	Migratory bird sanctuary
24	Dewey Soper	Migratory bird sanctuary*
25	Bowman Bay	Game sanctuary
26	East Bay	Migratory bird sanctuary
27	Harry Gibbons	Migratory bird sanctuary
28	Queen Maud Gulf	Migratory bird sanctuary*
29	Thelon	Game sanctuary
30	Kaminuriak Herd	Caribou protective measures
31	McConnell River	Migratory bird sanctuary*
32	Polar Bear	Provincial park
33	Akimiski Island	Bird sanctuary
34	James Bay	Preserve
35	Twin Islands	Game sanctuary

* Ramsar site

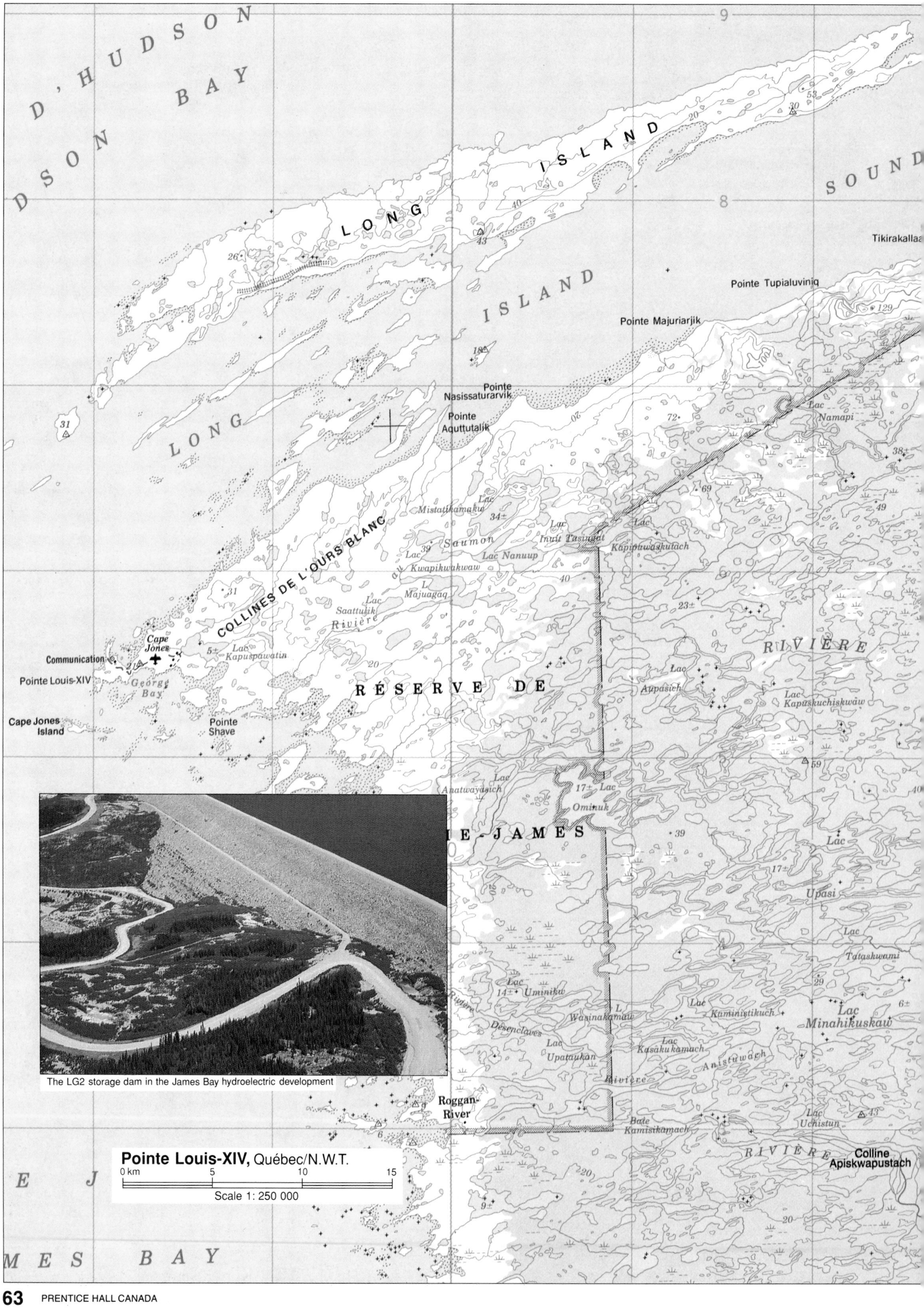
D'HUDSON
BAY
DSON
LONG ISLAND
LONG ISLAND
LONG
SOUND
Tikirakallaa
Pointe Tupialuviniq
Pointe Majuriarjik
Pointe Nasissaturarvik
Pointe Aqutlutalik
COLLINES DE L'OURS BLANC
Lac Mistatikamakw
Lac Saumon
Lac Inuit Tasingat
Lac Kapiputwaskutach
Lac Nanuup
Lac Kwapikwakwaw
L. Majuagaq
Lac Saattulik
Rivière
Cape Jones
Communication
Pointe Louis-XIV
George Bay
Cape Jones Island
Pointe Shave
Lac Kapuspawatin
RÉSERVE DE
RIVIÈRE
Lac Aupasich
Lac Kapaskuchiskwaw
Lac Namapi
Lac Anatwayasich
Lac Ominuk
E-JAMES
Lac Upasi
Lac Tataskwami
Lac Uminikw
L. Wasinakamaw
Lac Kuministikuch
Lac Minahikuskaw
Lac Upataukan
Lac Kasakukamach
Anistuwagh
Rivière
Roggan River
Baie Kamisikamach
Lac Uchistun
RIVIÈRE
Colline Apiskwapustach
Pointe Louis-XIV, Québec/N.W.T.
0 km
5
10
15
Scale 1: 250 000
E J
MES BAY

The LG2 storage dam in the James Bay hydroelectric development

The contact of the Canadian Shield with the Great Lakes–St. Lawrence Lowlands near Ottawa, Ontario

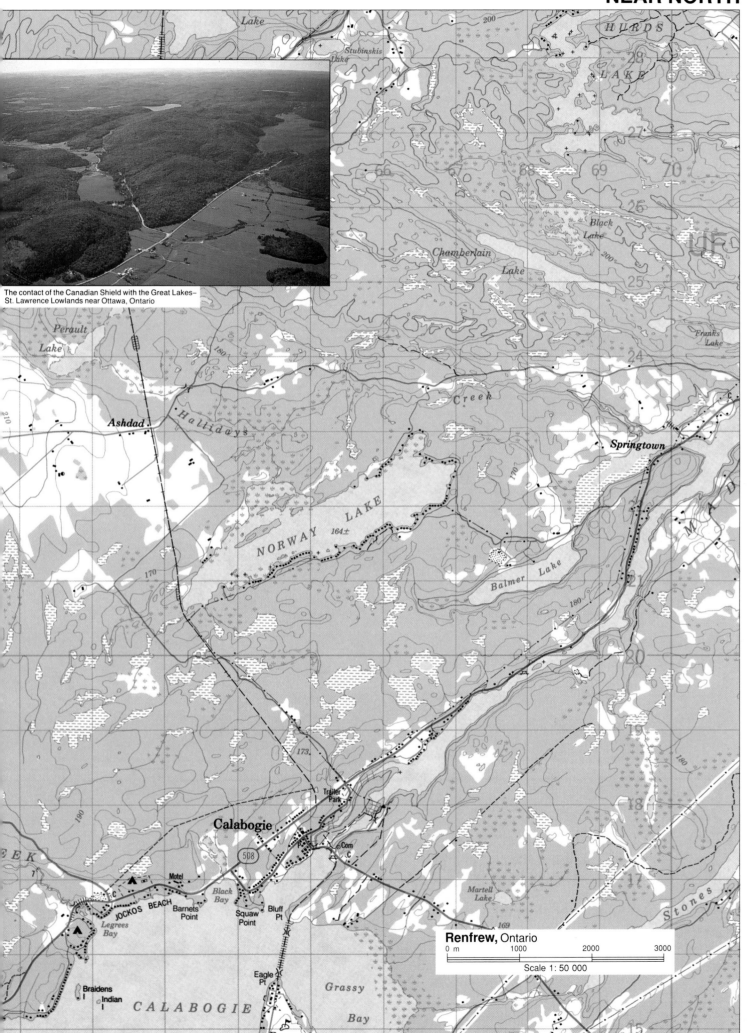

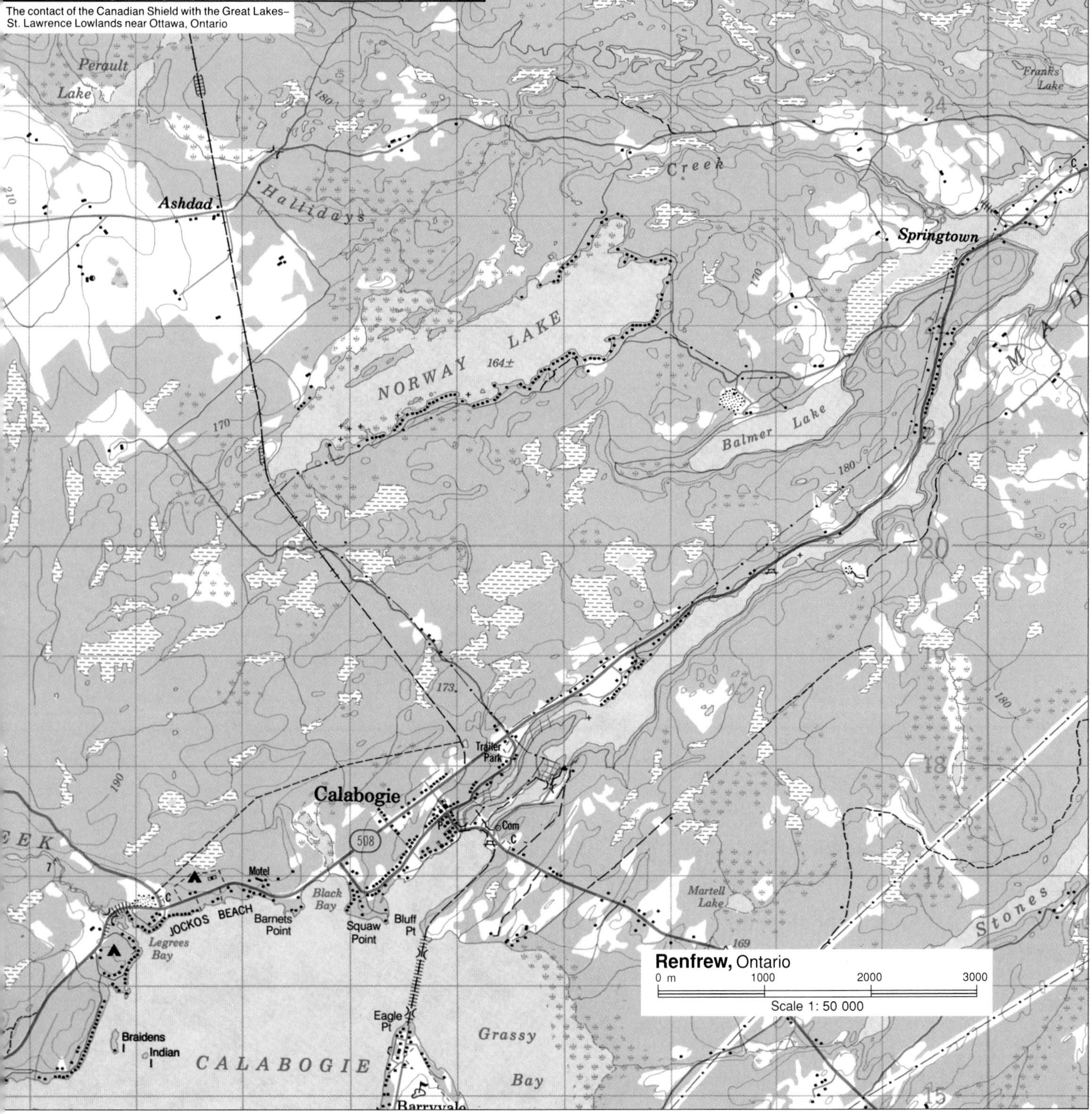

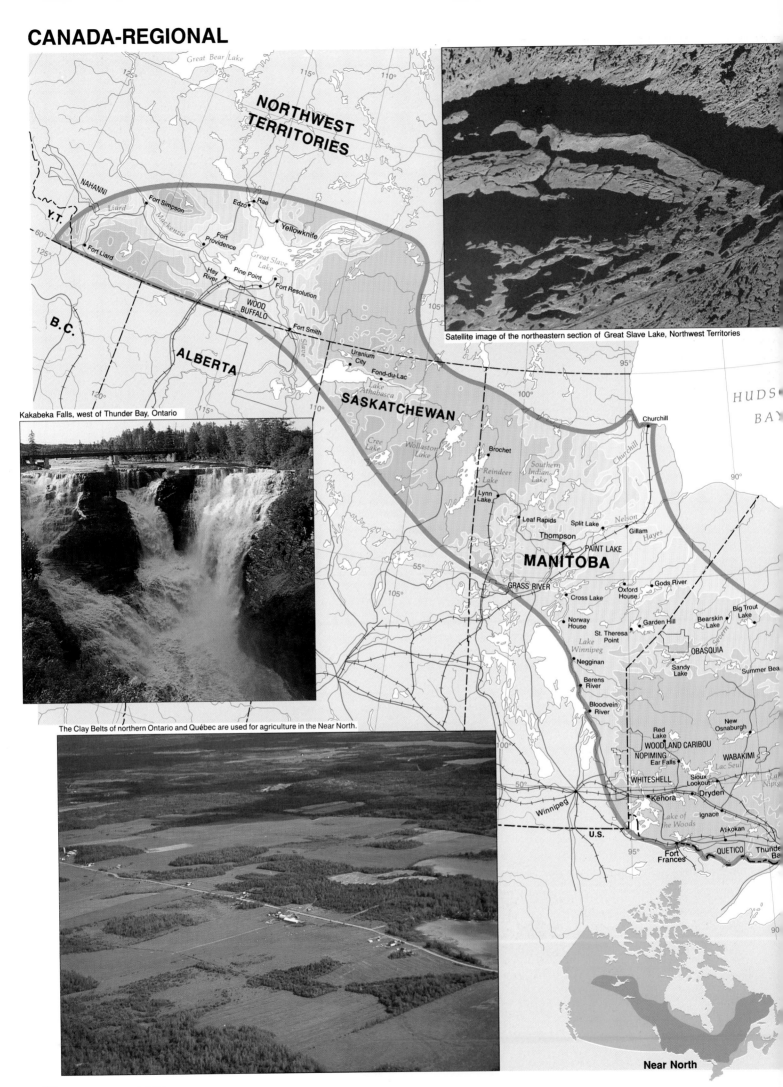

CANADA-REGIONAL

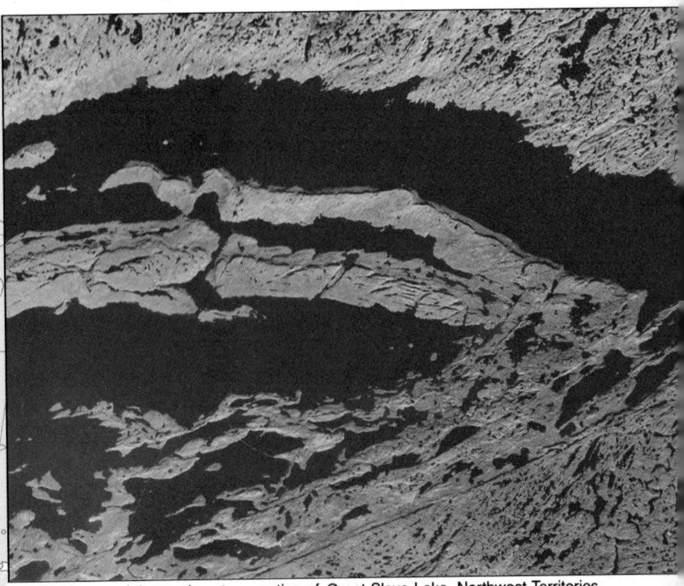

Satellite image of the northeastern section of Great Slave Lake, Northwest Territories

Kakabeka Falls, west of Thunder Bay, Ontario

The Clay Belts of northern Ontario and Québec are used for agriculture in the Near North.

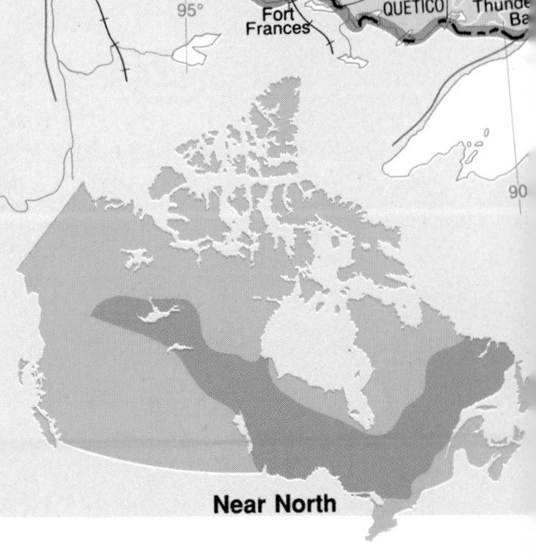

Near North

NEAR NORTH

The Near North, as its name suggests, is a transitional area between the industrialized south and the more isolated Far North. It stretches across Canada in an arc which includes much of the Canadian Shield.

Originally, First Nations groups, primarily Algonquian and Athapaskan, lived nomadically by hunting and fishing. The methods of transportation that they developed—such as the canoe, toboggan, and snowshoe — were also adopted by Europeans, who were attracted to the region by its rich natural resources. The First Nations and the Europeans often worked together, trapping and trading furs. Conflicts later arose between the two groups when the Europeans tapped lumber, mineral and water resources.

Today, the population and the economy are still closely tied to natural resources. Urban centres such as Thompson, Chibougamau and Sudbury have developed because of nearby mineral resources. Forests, particularly in the southern and eastern parts, are harvested for pulp and paper. Agriculture is practised in a few areas. National and provincial parks preserve the scenic beauty of many areas in the Near North.

o smelter in Thompson, Manitoba

Population of major urban centres

Chicoutimi-Jonquière CMA	Qué.	161 000
Sudbury CMA	Ont.	158 000
Thunder Bay CMA	Ont.	124 000
Sault Ste. Marie	Ont.	82 000
North Bay	Ont.	55 500
Timmins	Ont.	47 500
Rouyn, Noranda	Qué.	26 000
Alma	Qué.	26 000
Baie-Comeau	Qué.	26 000
Sept-Îles	Qué.	25 000
Val-d'Or	Qué.	24 000
La Baie	Qué.	21 000
Yellowknife	N.W.T.	15 500
Thompson	Man.	15 000
Elliot Lake	Ont.	15 000
Huntsville	Ont.	15 000

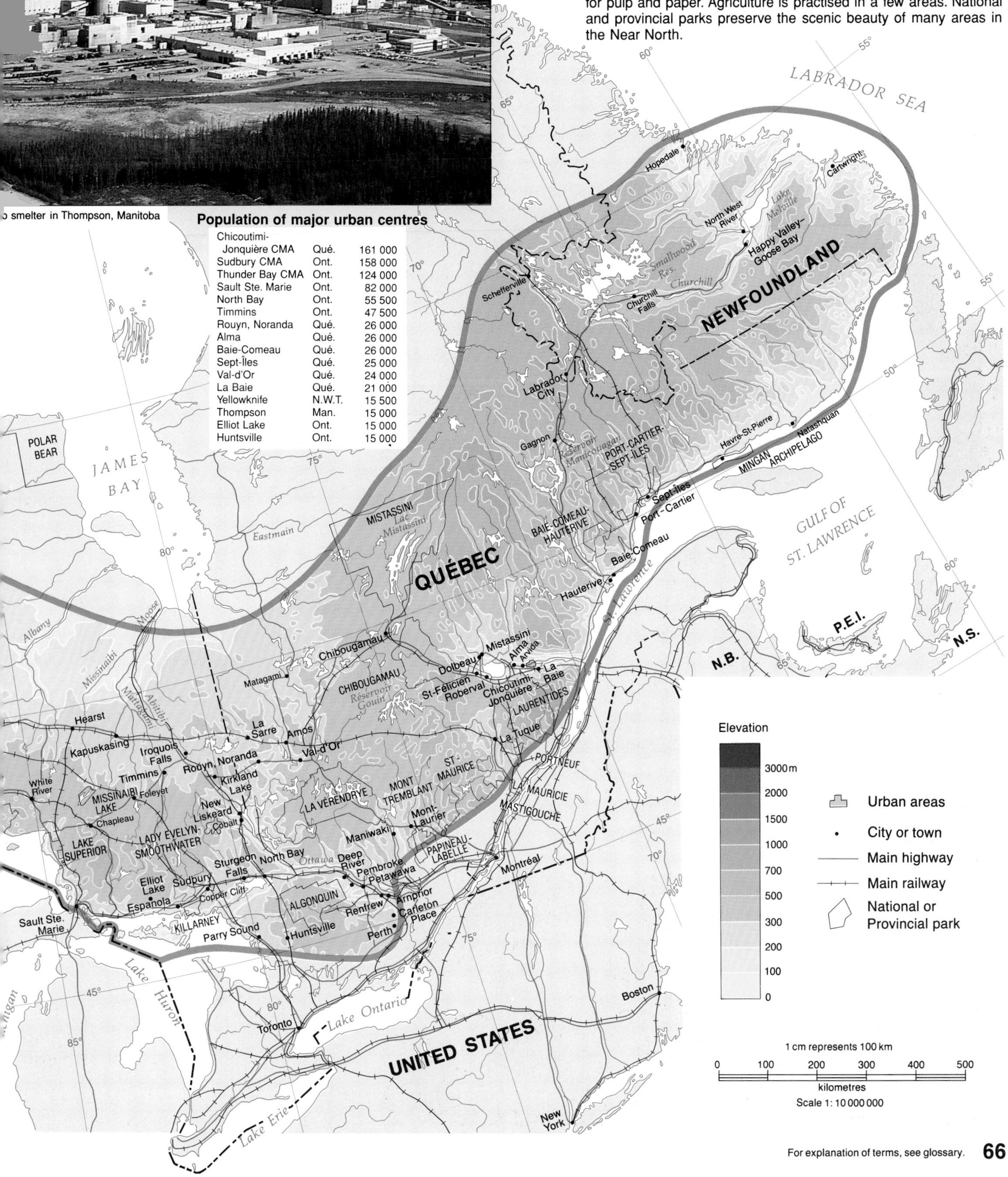

CANADA-REGIONAL

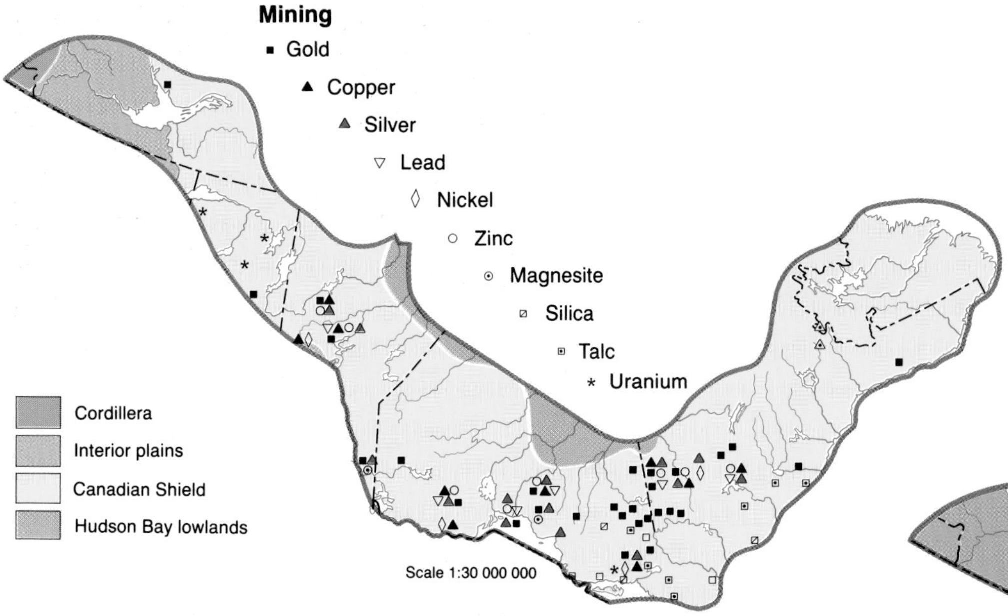

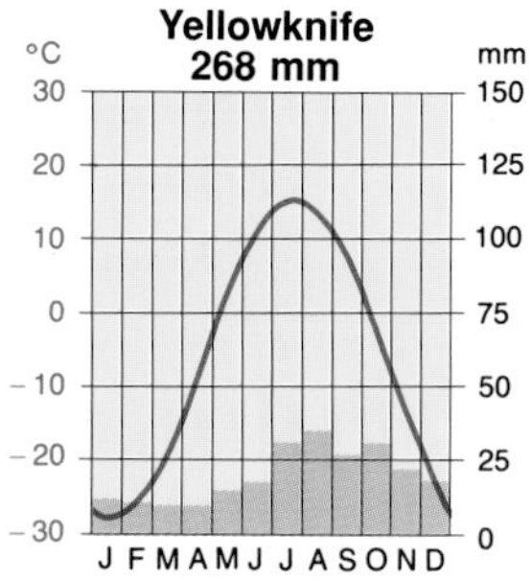

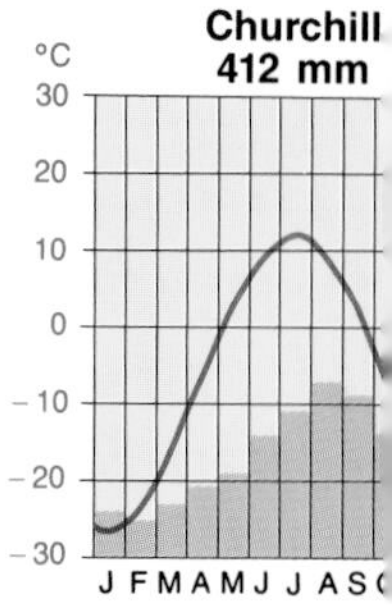

One of the important activities in the Near North is mining. The Kidd Creek operation north of Timmins uses both open pit and shaft methods.

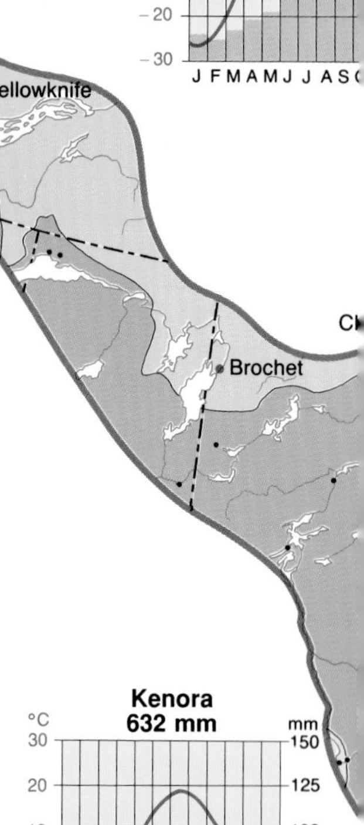

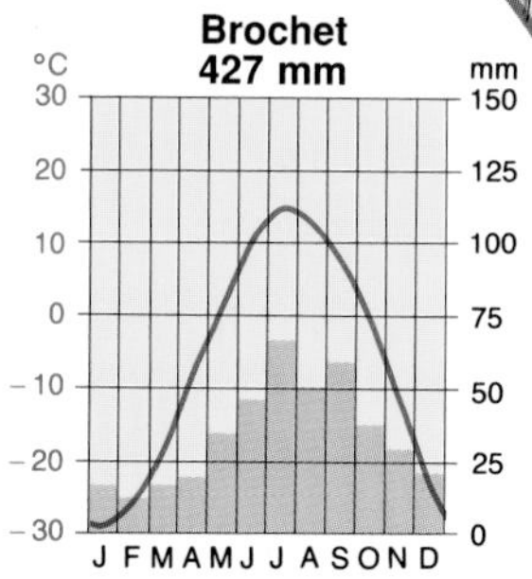

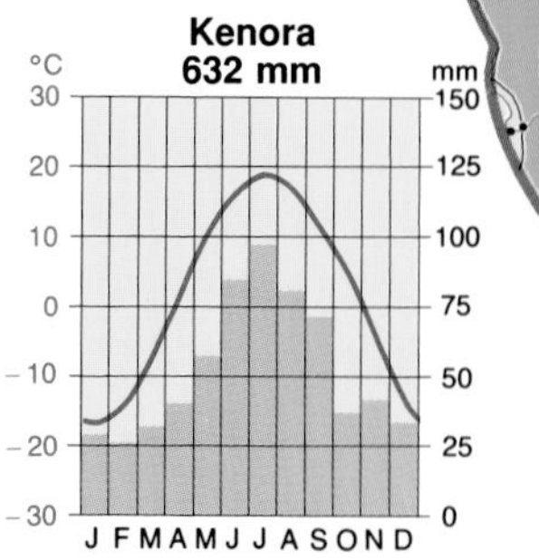

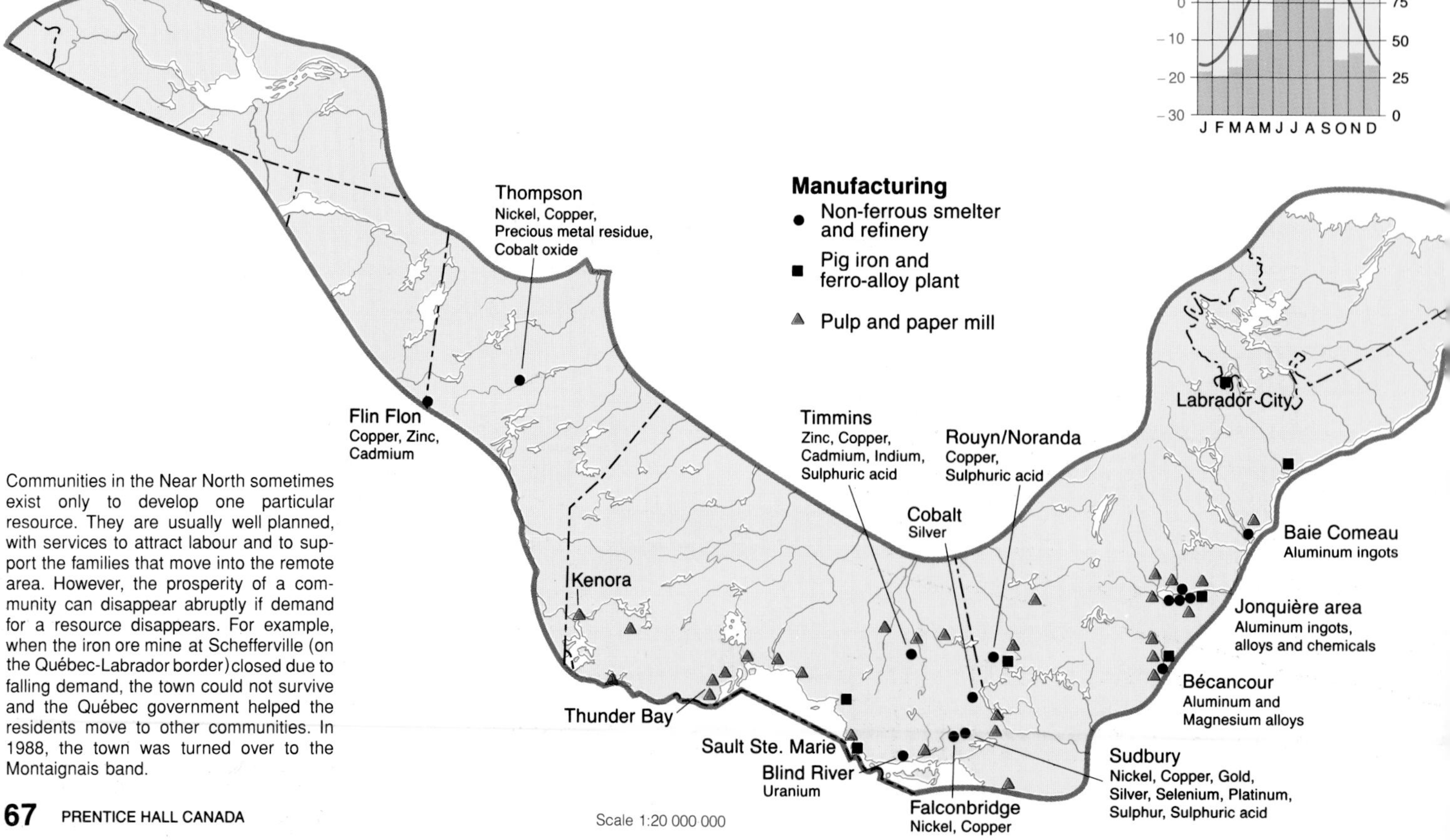

Communities in the Near North sometimes exist only to develop one particular resource. They are usually well planned, with services to attract labour and to support the families that move into the remote area. However, the prosperity of a community can disappear abruptly if demand for a resource disappears. For example, when the iron ore mine at Schefferville (on the Québec-Labrador border) closed due to falling demand, the town could not survive and the Québec government helped the residents move to other communities. In 1988, the town was turned over to the Montagnais band.

e Near North is rich in natural resources — c mineral deposits, forests and water. The main ries are pulp and paper, smelting and refining, e production of hydroelectric power. However, tural resources that have brought prosperity also made the region vulnerable to changes in orld marketplace. The effect of resource devel- t on the environment must also be considered. ample, smelters and pulp mills create air and pollution, and large power developments may e extensive flooding and changes in natural g patterns.

The boreal forest is an important resource. A feller buncher is being used to fell and delimb trees.

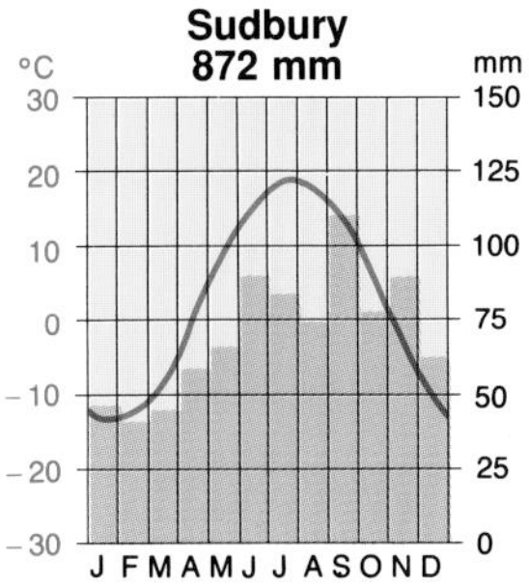

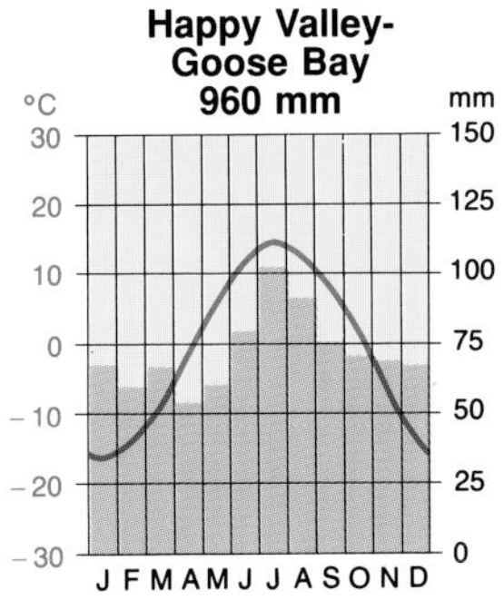

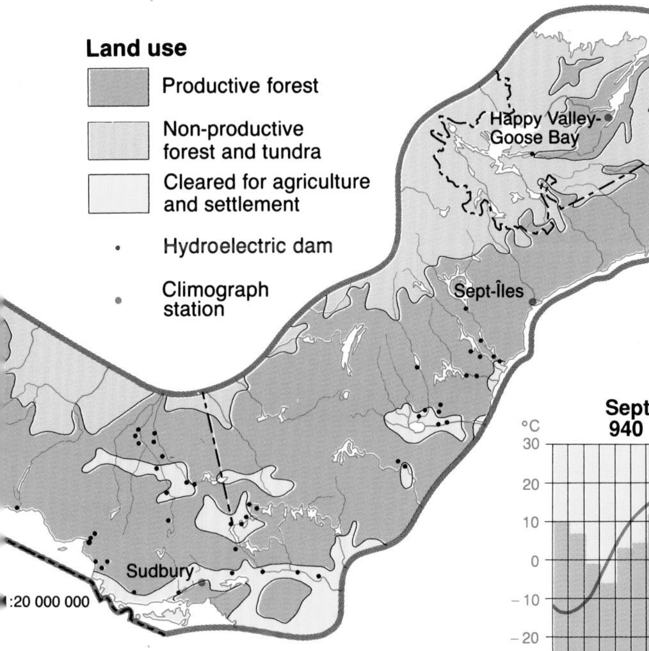

Ethnicity
(single responses only)

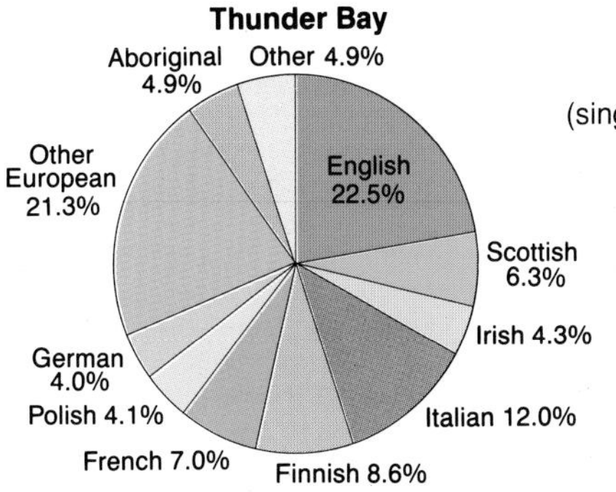

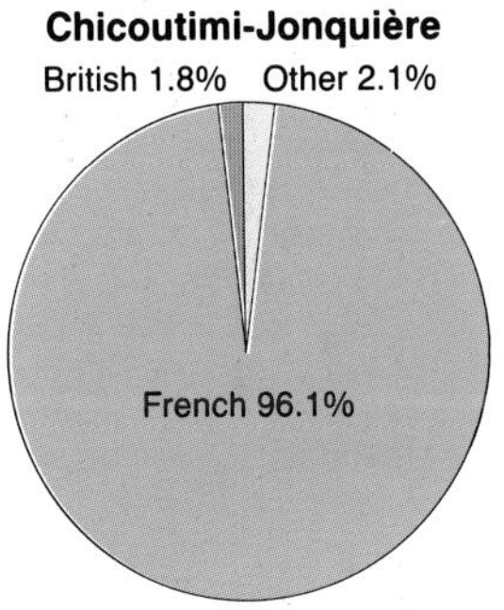

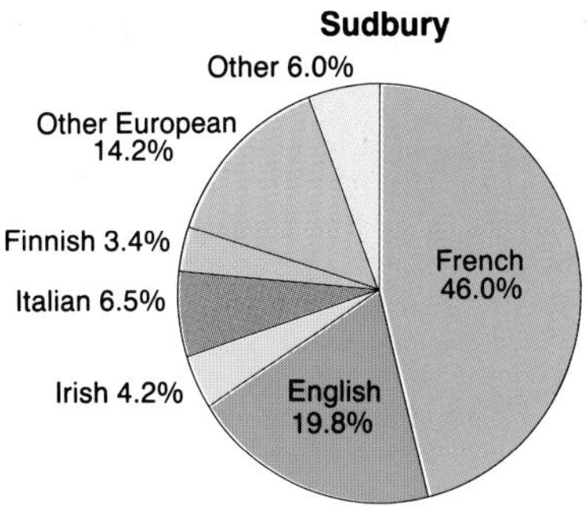

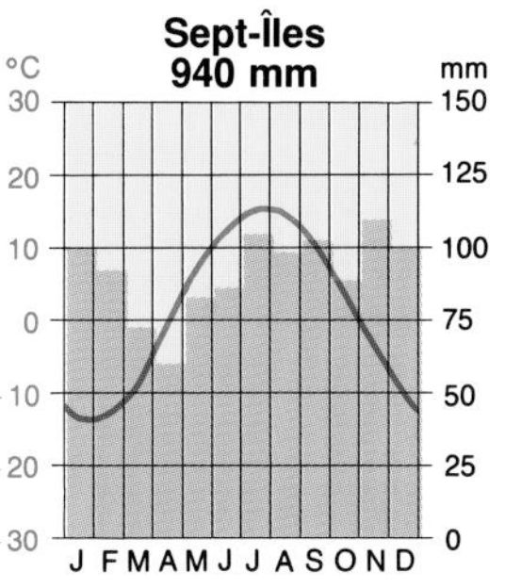

Acid rain in Eastern Canada

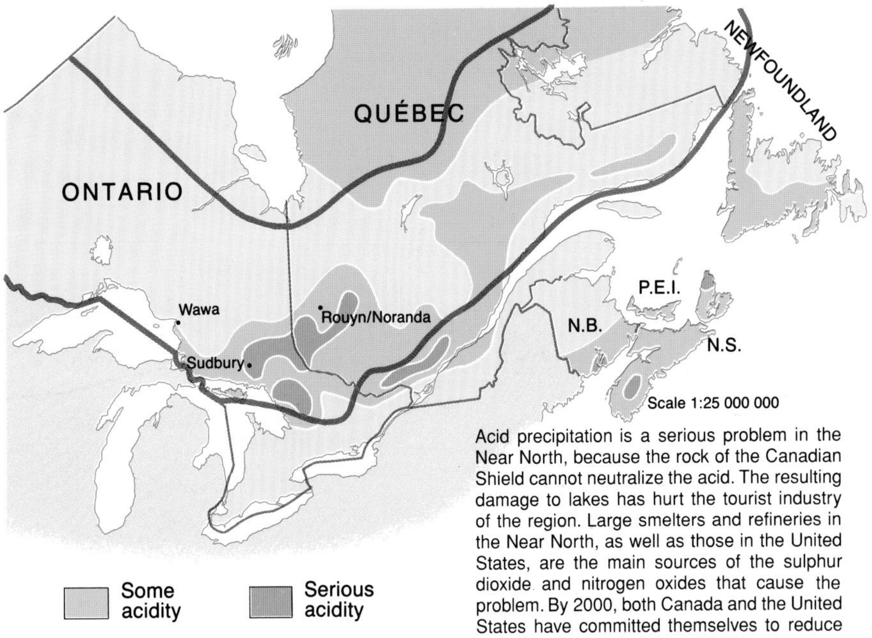

Acid precipitation is a serious problem in the Near North, because the rock of the Canadian Shield cannot neutralize the acid. The resulting damage to lakes has hurt the tourist industry of the region. Large smelters and refineries in the Near North, as well as those in the United States, are the main sources of the sulphur dioxide and nitrogen oxides that cause the problem. By 2000, both Canada and the United States have committed themselves to reduce emissions by one half from 1987 levels.

hing attracts many tourists to the Near North

CANADA-REGIONAL

Western Mountains

The Western Mountains region is also known as *The Cordillera*, a Spanish name meaning a "series of parallel ridges." Two major mountain ranges cover most of the region. Most people live in the southern coastal areas around Vancouver and Victoria where the climate is mild. Other smaller centres, including Cranbrook and Revelstoke, are located in southern valleys.

The physiography of the region prevented much interaction among the First Nations groups who traditionally fished in the coastal waters and hunted for deer and moose. Europeans came to this region later than to other parts of Canada. They were first interested in obtaining furs, then timber, and later minerals.

Today the rich natural resources—forests, fish, minerals, water and fertile land—provide the basis for a high standard of living. Most industries are directly related to these natural resources. Value-added manufacturing of natural resources is increasing in importance. There is a growing and significant base of "high tech" and computer/software industry-related firms that has had an impact on the region's economy. Tourism is an important service industry.

The Western Mountains region is heavily involved in international trade, particularly with Pacific Rim countries. Vancouver is the major port. While Canadian wheat and coal are exported to Northeast Asia, Japanese electronics and cars, Korean textiles, and Chinese footwear and clothing are distributed throughout Canada. These international connections have greatly increased the strength and diversity of the region's economy.

Satellite image of Vancouver and the delta of the Fraser River

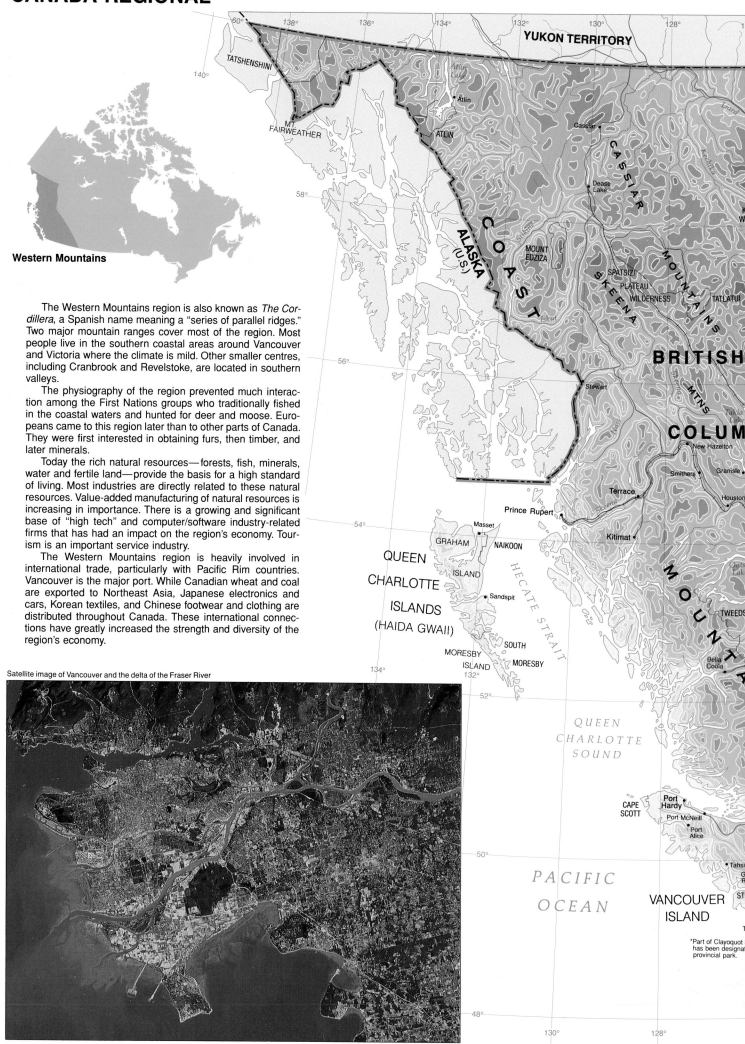

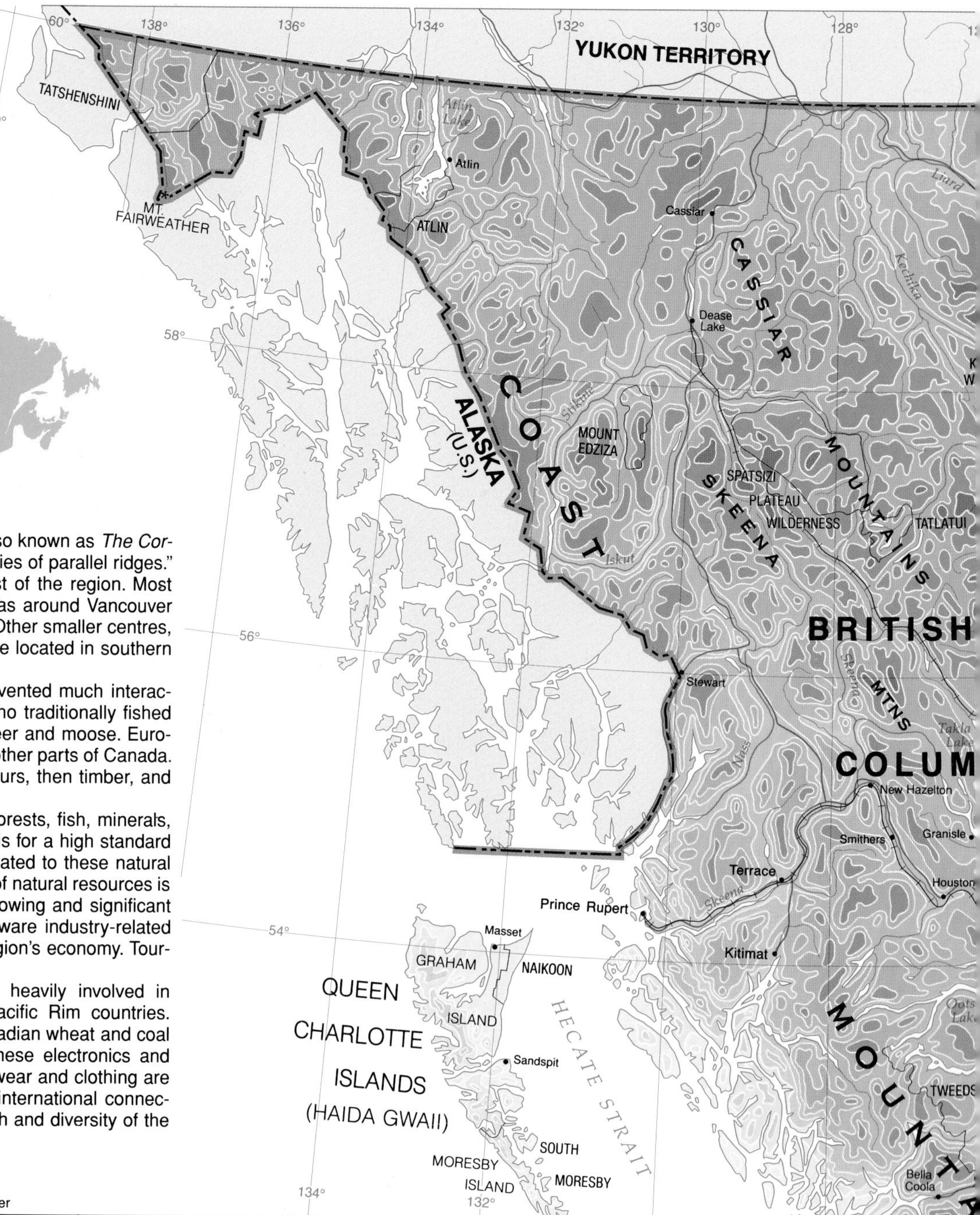

The Coast Mountains and Garibaldi Lake

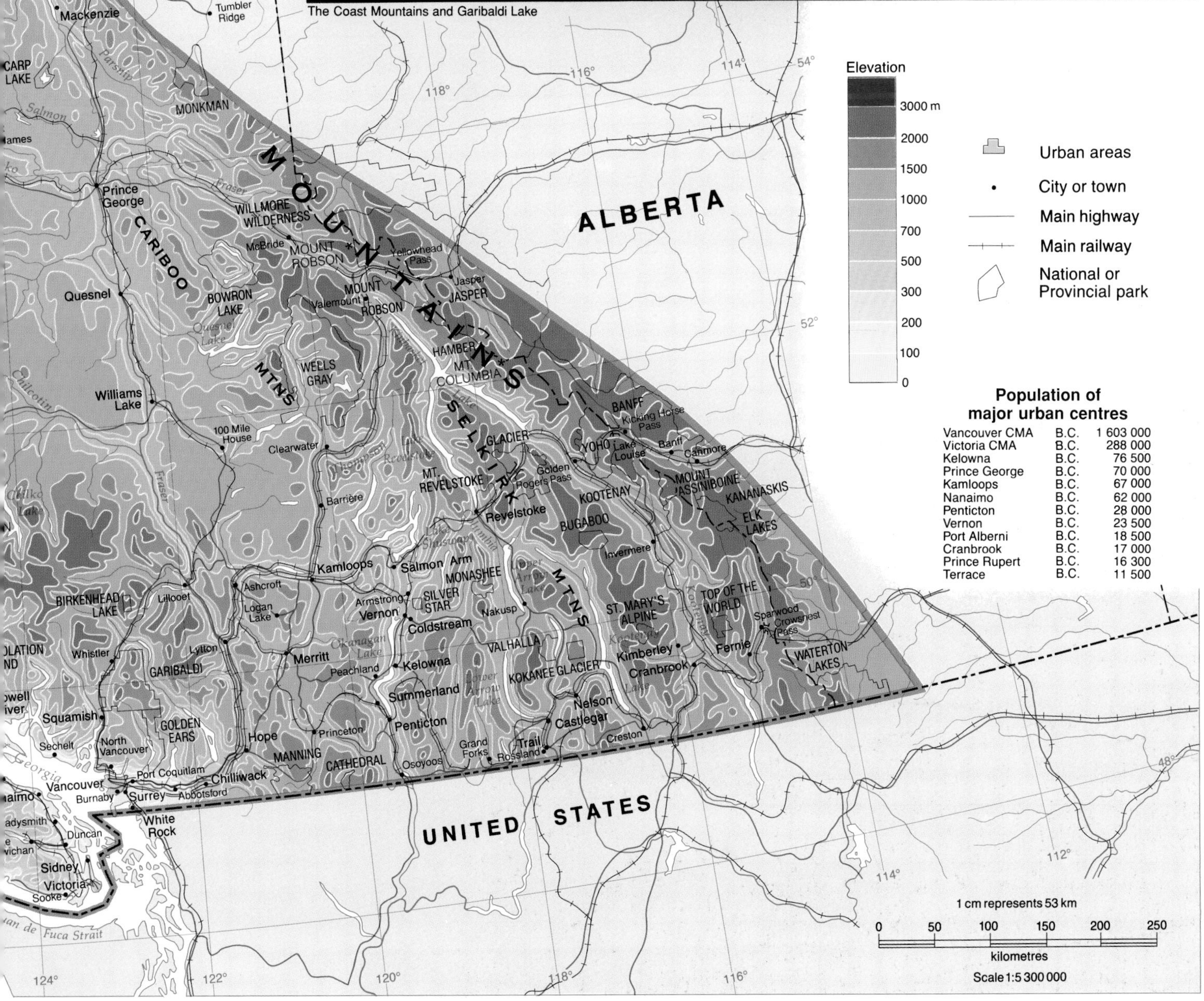

Population of major urban centres

Vancouver CMA	B.C.	1 603 000
Victoria CMA	B.C.	288 000
Kelowna	B.C.	76 500
Prince George	B.C.	70 000
Kamloops	B.C.	67 000
Nanaimo	B.C.	62 000
Penticton	B.C.	28 000
Vernon	B.C.	23 500
Port Alberni	B.C.	18 500
Cranbrook	B.C.	17 000
Prince Rupert	B.C.	16 300
Terrace	B.C.	11 500

CANADA-REGIONAL

Containers waiting for loading, Vancouver

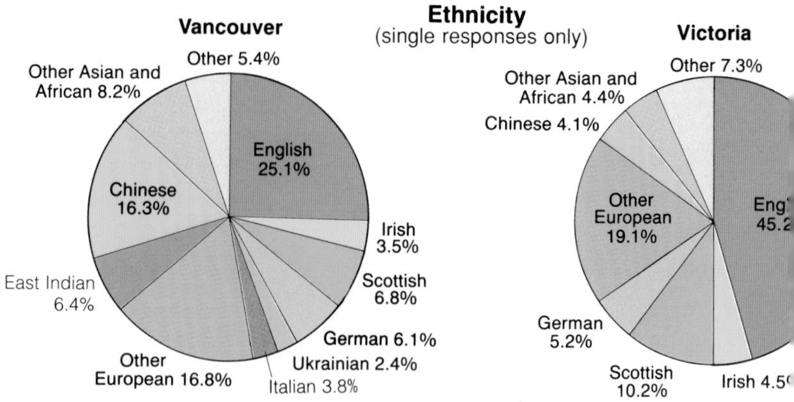

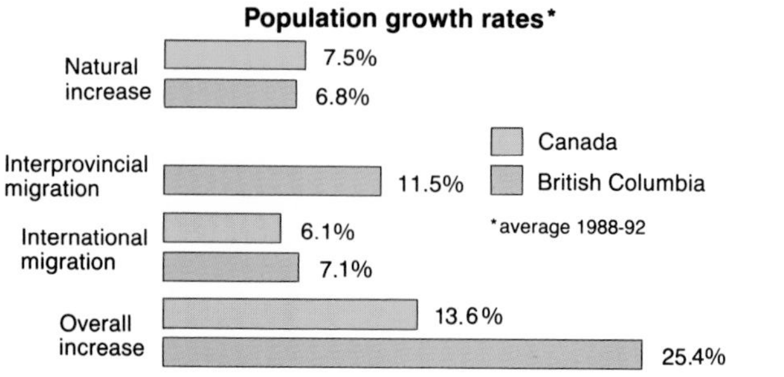

The port of Vancouver

International cargo	66 200 000 tonnes
Domestic cargo	2 200 000 tonnes
Share of Canadian shipping	19.5%
Top commodities shipped:	coal, wheat, sulphur, potash, lumber, wood pulp
Top foreign markets:	Japan, China, South Korea

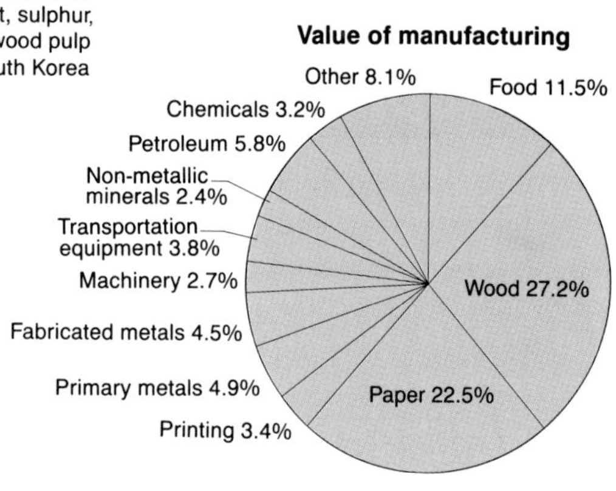

Pulp mill at Squamish, British Columbia

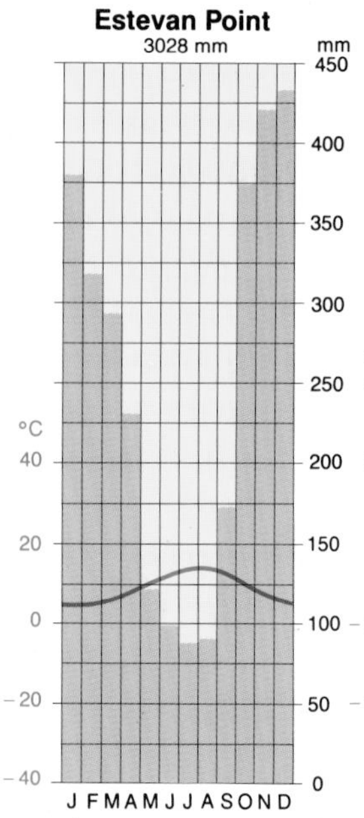

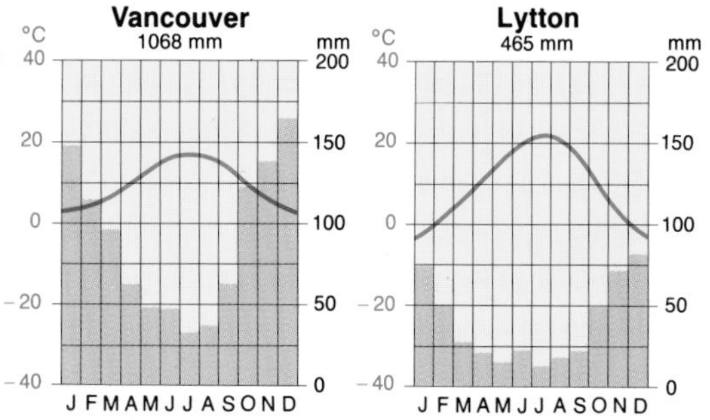

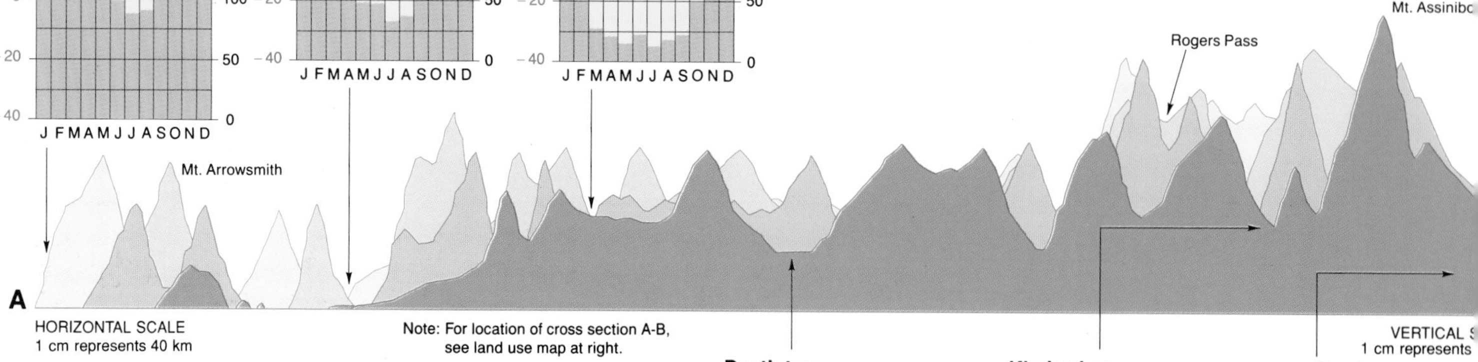

The mountains of British Columbia

The Western Mountains region extends more than 1600 km from north to south and from 118° W to 141° W. Generally, near the Pacific coast, temperatures are milder and there is more precipitation. However, local weather conditions can vary from this pattern; whether a town is located in a valley, at a high altitude or in a rain shadow has a great influence on typical weather conditions.

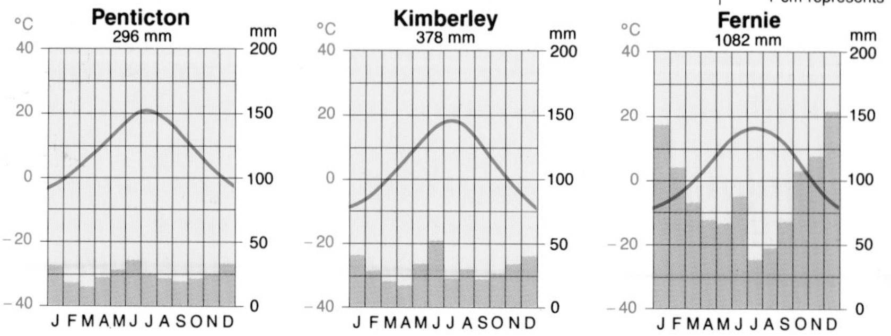

WESTERN MOUNTAINS

he economy of the Western Mountains is ly based on natural resources: forests, miner- nd fish. However, public concerns have been d about damage to the natural environment may result from removing trees, developing te areas with fragile ecosystems, or permitting water pollution. Proposals for clearcutting old th forests in this region have been very con- rsial. (See pages 41/42 for the location of old th forests on southern Vancouver Island.) In ion, the tourist trade, which is also an impor- industry, depends on retaining the natural ty of the region. The debate over the balance een protecting the environment and protecting in the resource industries will continue into the century.

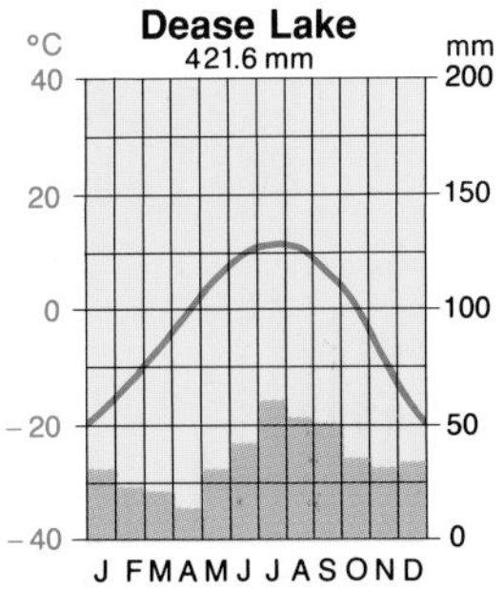

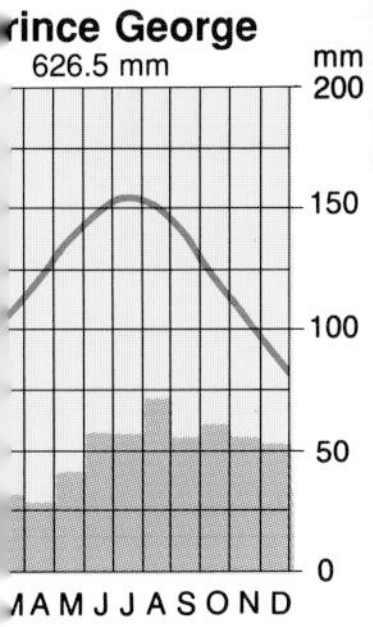

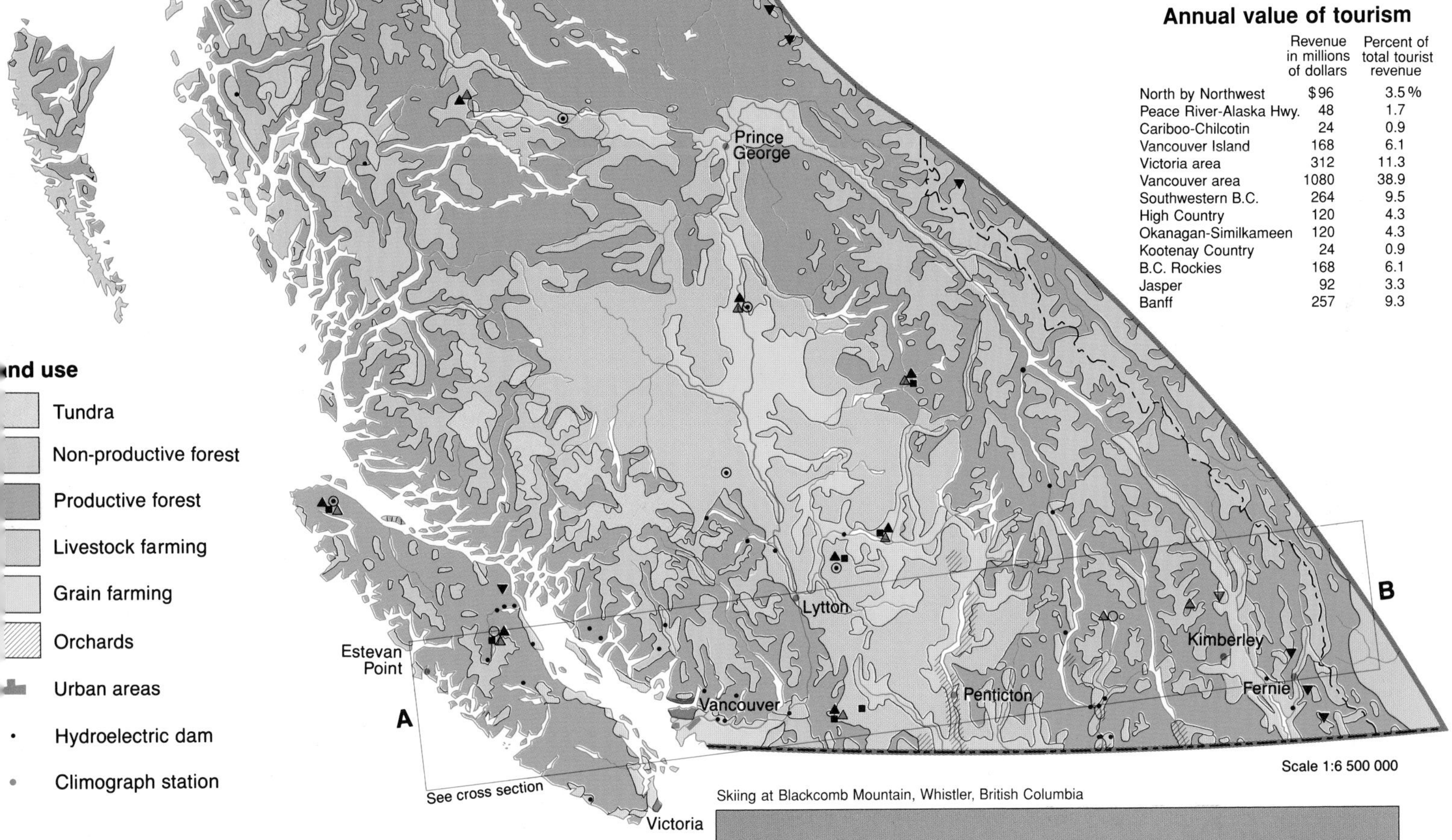

nd use

- Tundra
- Non-productive forest
- Productive forest
- Livestock farming
- Grain farming
- Orchards
- Urban areas
- Hydroelectric dam
- Climograph station

inerals

- ■ Gold
- △ Silver
- ▲ Copper
- ○ Zinc
- ⊙ Molybdenum
- ▽ Gypsum
- ▼ Coal

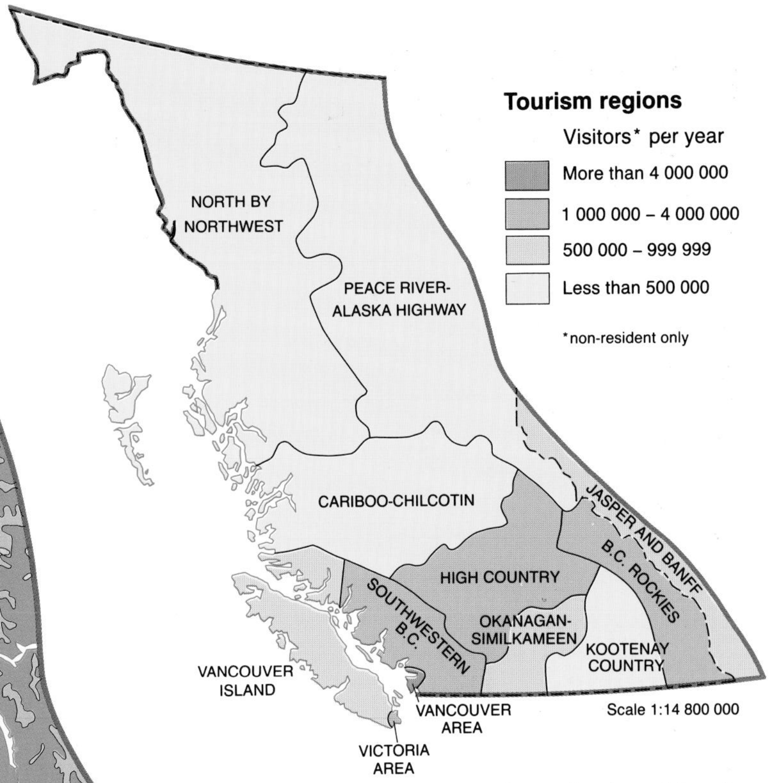

Tourism regions

Visitors* per year

- More than 4 000 000
- 1 000 000 – 4 000 000
- 500 000 – 999 999
- Less than 500 000

*non-resident only

Annual value of tourism

	Revenue in millions of dollars	Percent of total tourist revenue
North by Northwest	$96	3.5%
Peace River-Alaska Hwy.	48	1.7
Cariboo-Chilcotin	24	0.9
Vancouver Island	168	6.1
Victoria area	312	11.3
Vancouver area	1080	38.9
Southwestern B.C.	264	9.5
High Country	120	4.3
Okanagan-Similkameen	120	4.3
Kootenay Country	24	0.9
B.C. Rockies	168	6.1
Jasper	92	3.3
Banff	257	9.3

Over 90 percent of the land area is this region is mountainous—scenic but poor for farming. Half of the available agricultural land is suitable only for grazing cattle. Most urban centres and much of the arable land (suitable for crops) are at the lower elevations. Agricultural Land Reserves have been formed to prevent the growth of cities from reducing farmland. These reserves protect land from development other than agriculture. For example, some forested land may be cleared for farming but not for other purposes.

Skiing at Blackcomb Mountain, Whistler, British Columbia

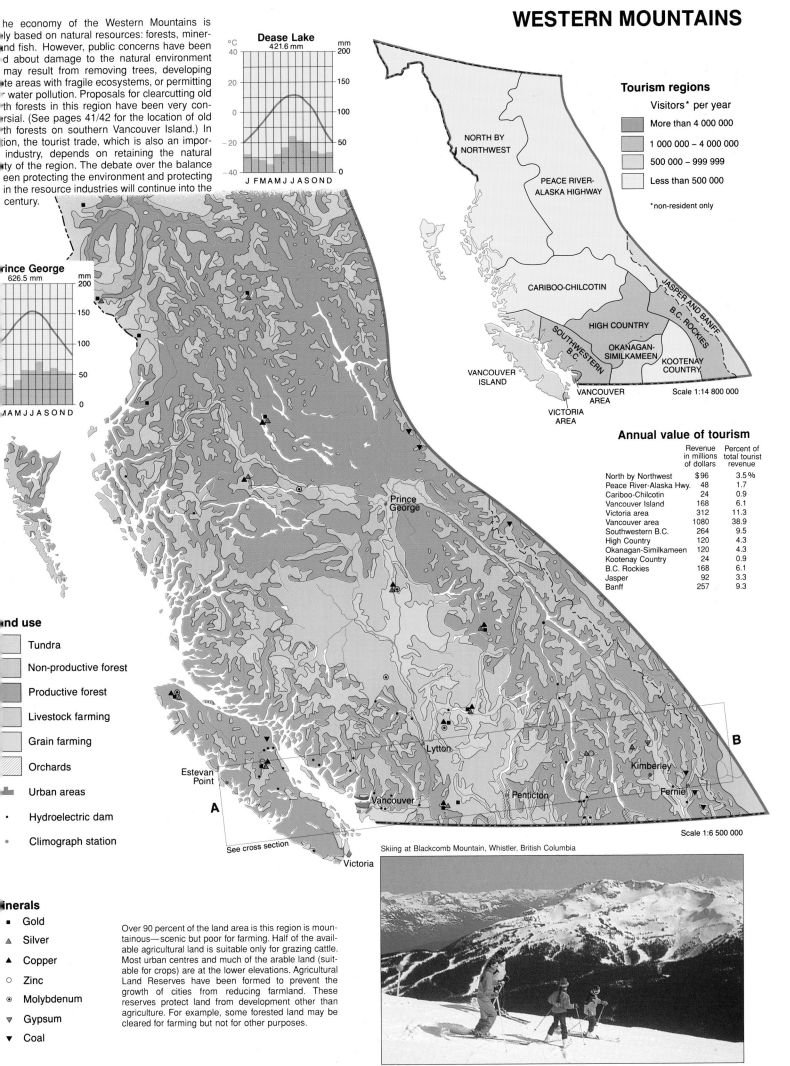

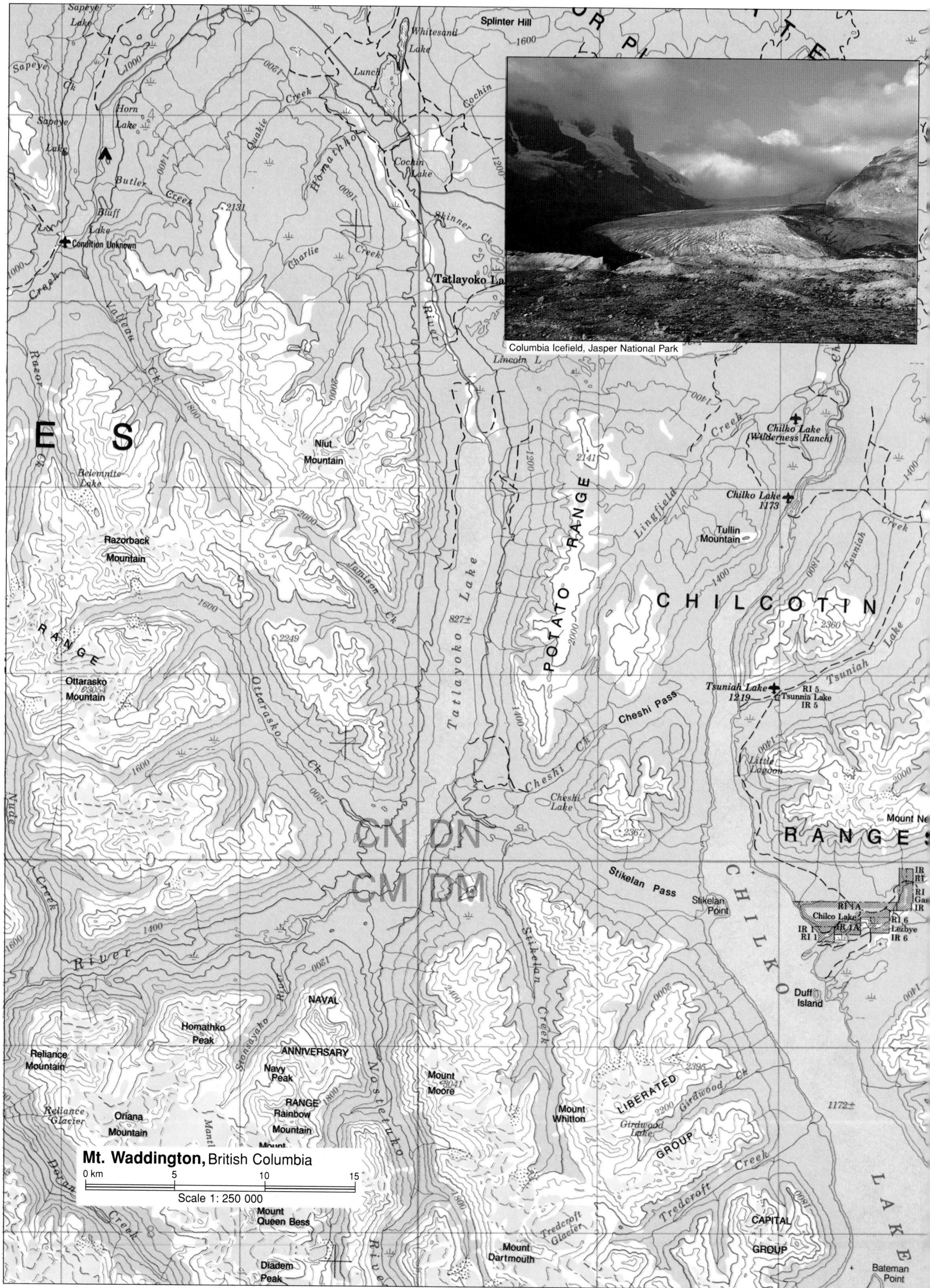

Columbia Icefield, Jasper National Park

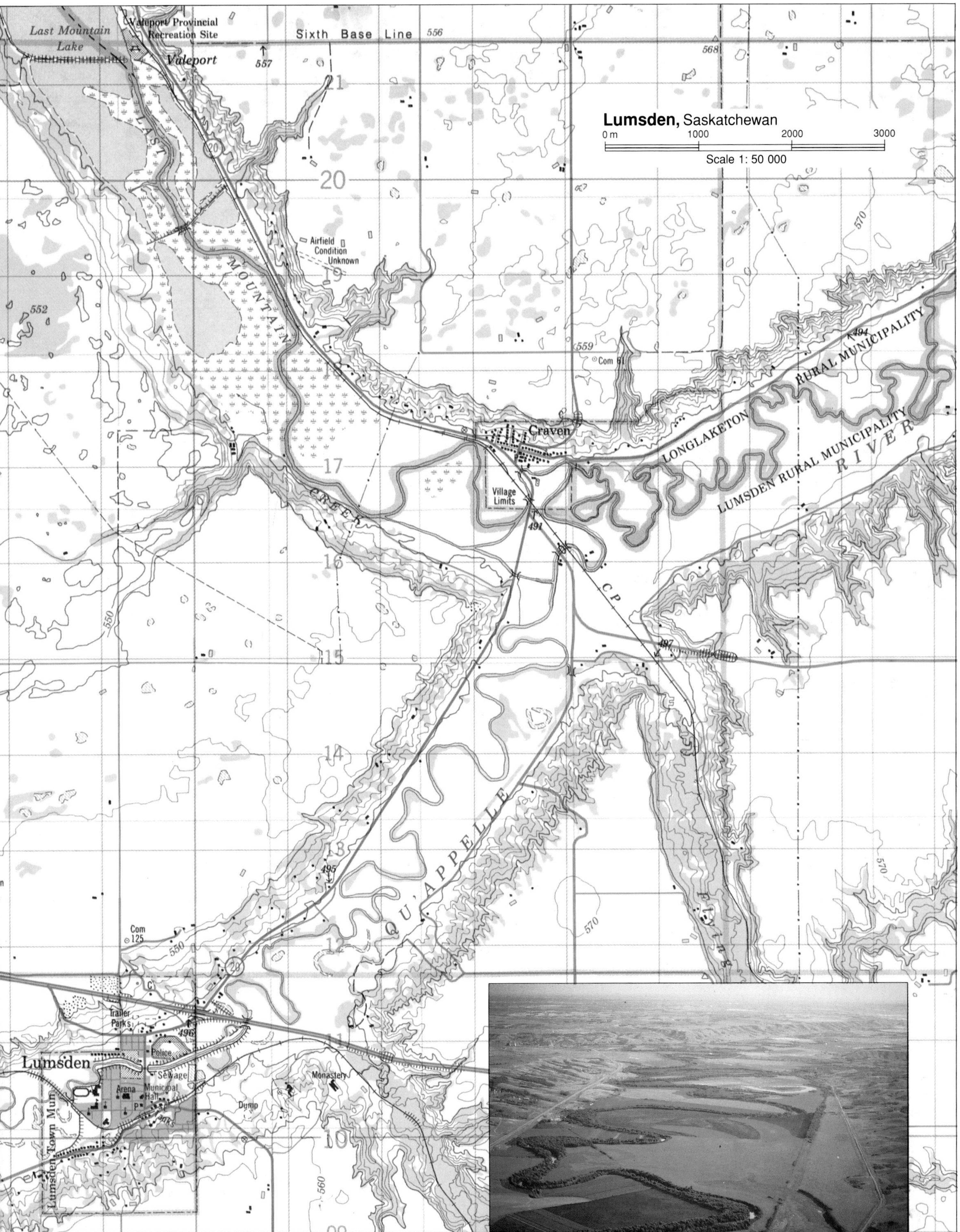

Qu'Appelle Valley looking east from Lumsden, Saskatchewan

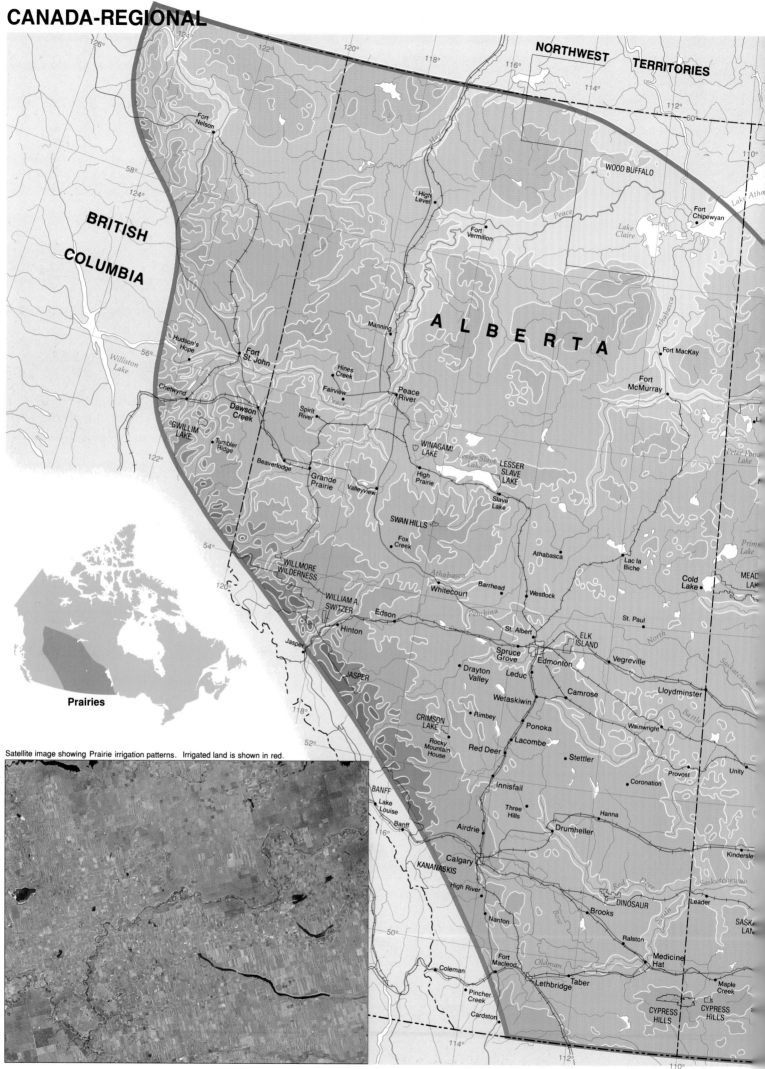

CANADA-REGIONAL

Satellite image showing Prairie irrigation patterns. Irrigated land is shown in red.

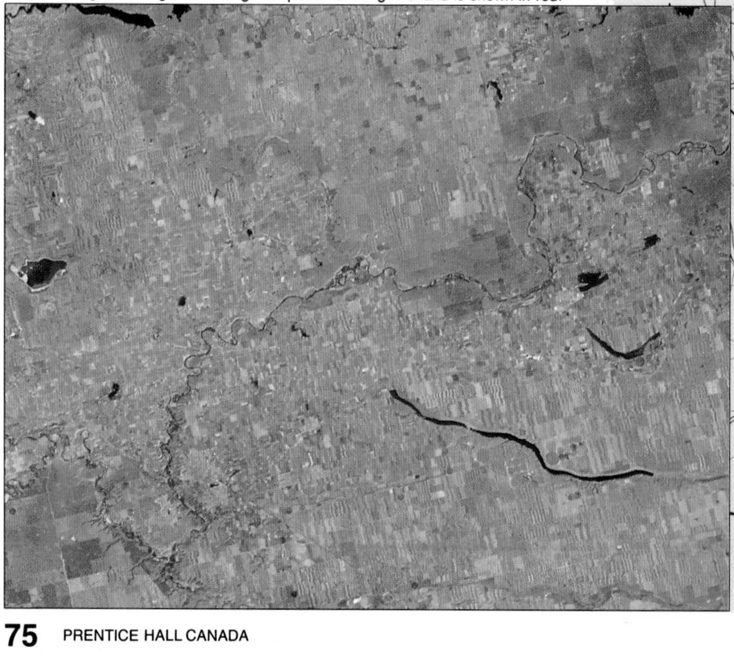

The Prairies are often known as Canada's breadbasket and agricultural heartland. No other region in Canada has as much uniformity in its physical character. Ethnically, however, the region is diverse. Although people of British ancestry are in the majority, there are many small communities founded by immigrants from countries such as Russia, the Ukraine, Finland, Poland and Germany. Native peoples, such as the Blackfoot, Peigan, Blood, Sarcee, Cree, Saulteaux and Assiniboine, represent 3 percent of the population.

The First Nations of this region originally hunted buffalo for food, clothing, tools and shelter. After the arrival of Europeans, the fur trade became important, with the First Nations trading furs and pemmican (dried buffalo meat) for firearms, metal goods and textiles from the Europeans. From the 1890s to the eve of World War I, vast numbers of settlers moved to the Prairies.

Today the land is used mostly for agriculture. Wheat is the main crop in Saskatchewan, while ranching dominates much of Alberta. Irrigation has been introduced to make drier areas more fertile. The discovery of oil and natural gas has stimulated the economic growth of the region. Manufacturing activity in the region has also increased, making the economy more diversified.

Agricultural area west of Lethbridge, Alberta

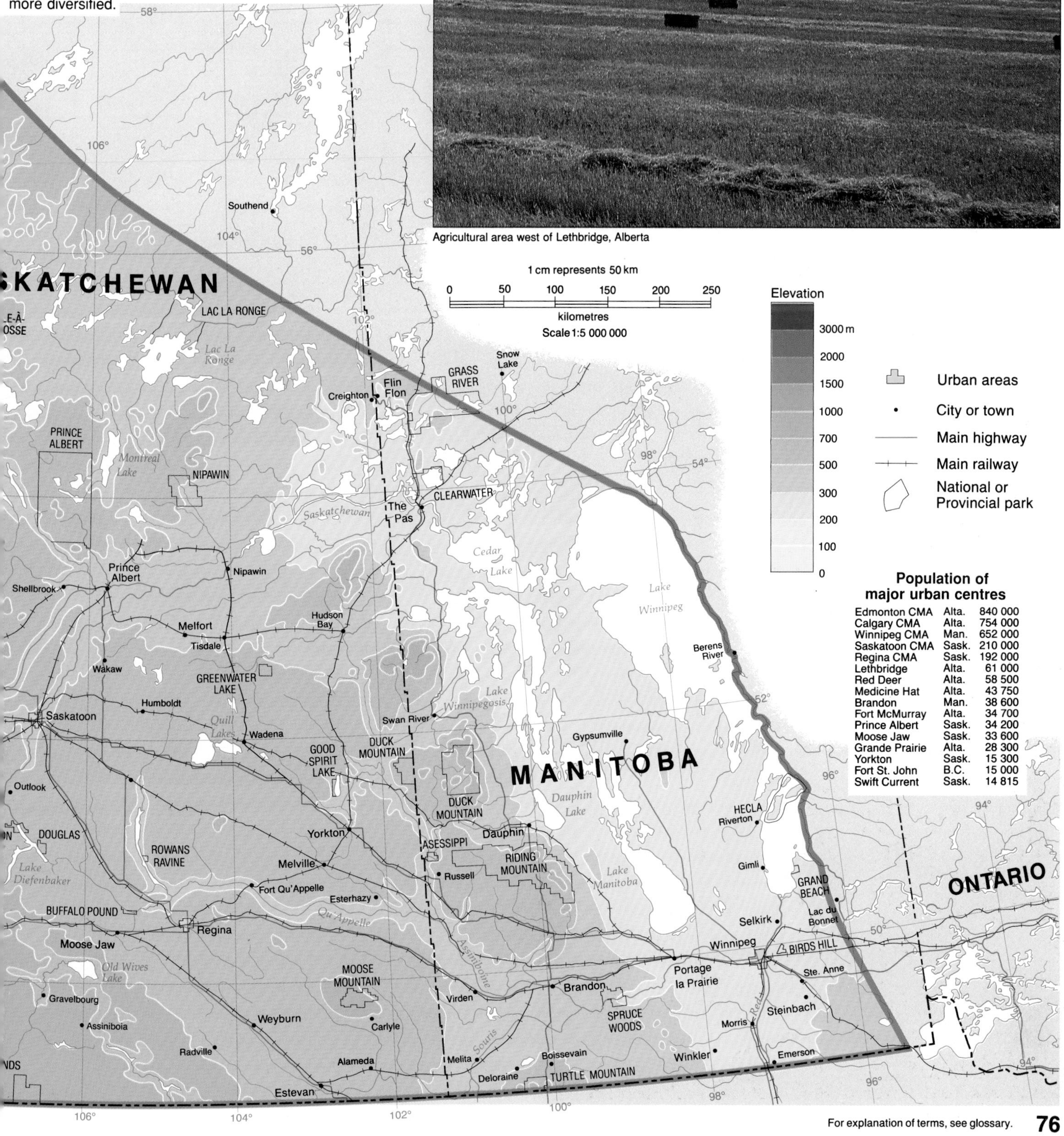

Population of major urban centres

Edmonton CMA	Alta.	840 000
Calgary CMA	Alta.	754 000
Winnipeg CMA	Man.	652 000
Saskatoon CMA	Sask.	210 000
Regina CMA	Sask.	192 000
Lethbridge	Alta.	61 000
Red Deer	Alta.	58 500
Medicine Hat	Alta.	43 750
Brandon	Man.	38 600
Fort McMurray	Alta.	34 700
Prince Albert	Sask.	34 200
Moose Jaw	Sask.	33 600
Grande Prairie	Alta.	28 300
Yorkton	Sask.	15 300
Fort St. John	B.C.	15 000
Swift Current	Sask.	14 815

For explanation of terms, see glossary.

CANADA-REGIONAL

Over the past 30 years, the economy of the Prairies has become increasingly diversified. Although agriculture, particularly wheat and cattle, provides the economic base for the region, the proportion of the population employed in this industry has fallen. Because farms are increasing both in size and in mechanization, fewer workers are needed. In the past 40 years, the proportion of the population in the rural areas of the prairies has declined drastically. Many small towns that existed in part to provide services to the rural population have also declined quite significantly.

Many people have left Saskatchewan and Manitoba to seek jobs in other provinces. In Alberta, oil and gas production has formed the basis for the development of manufacturing. As a result, this province has a more diversified economy and more varied job opportunities.

An aerial view of Calgary

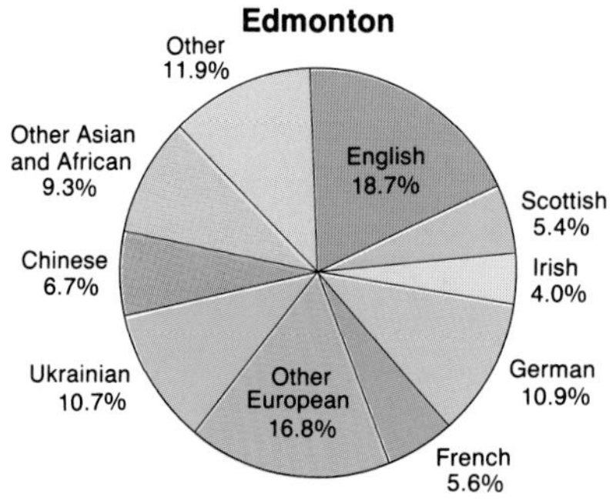

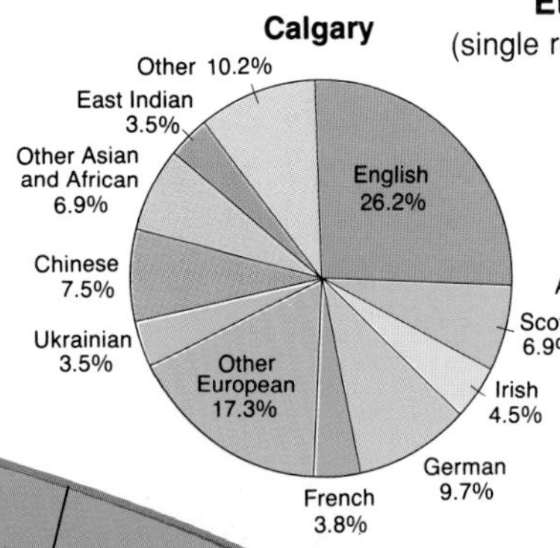

Ethnicity
(single responses only)

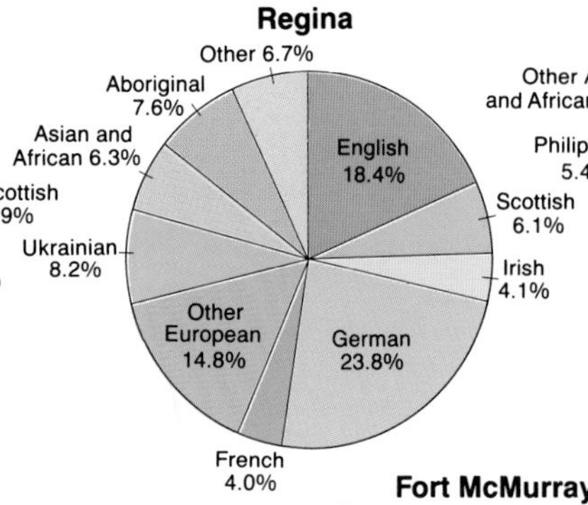

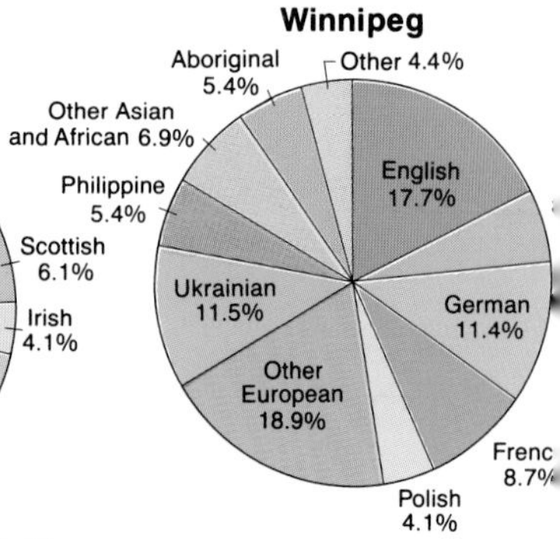

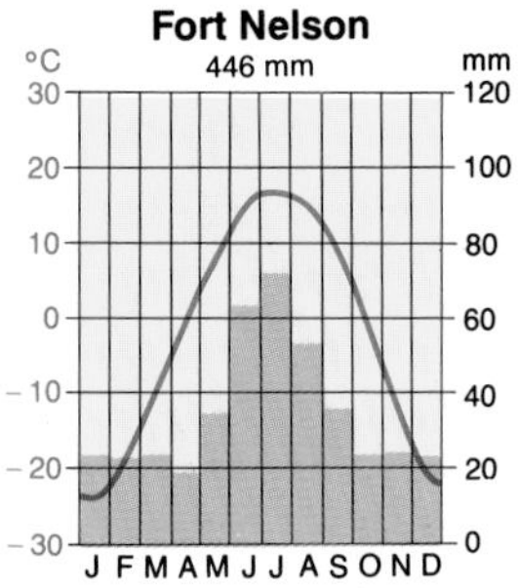

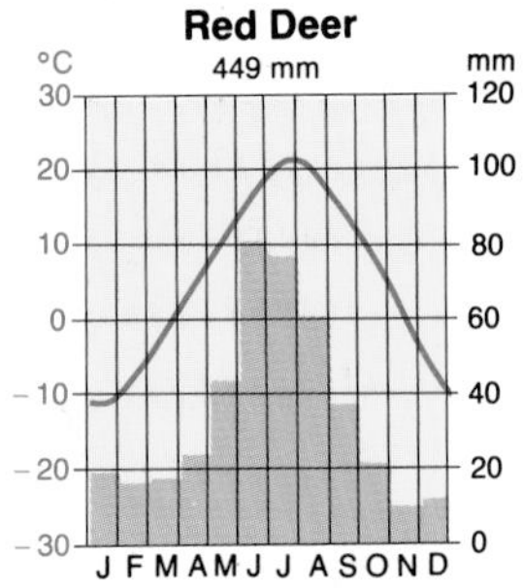

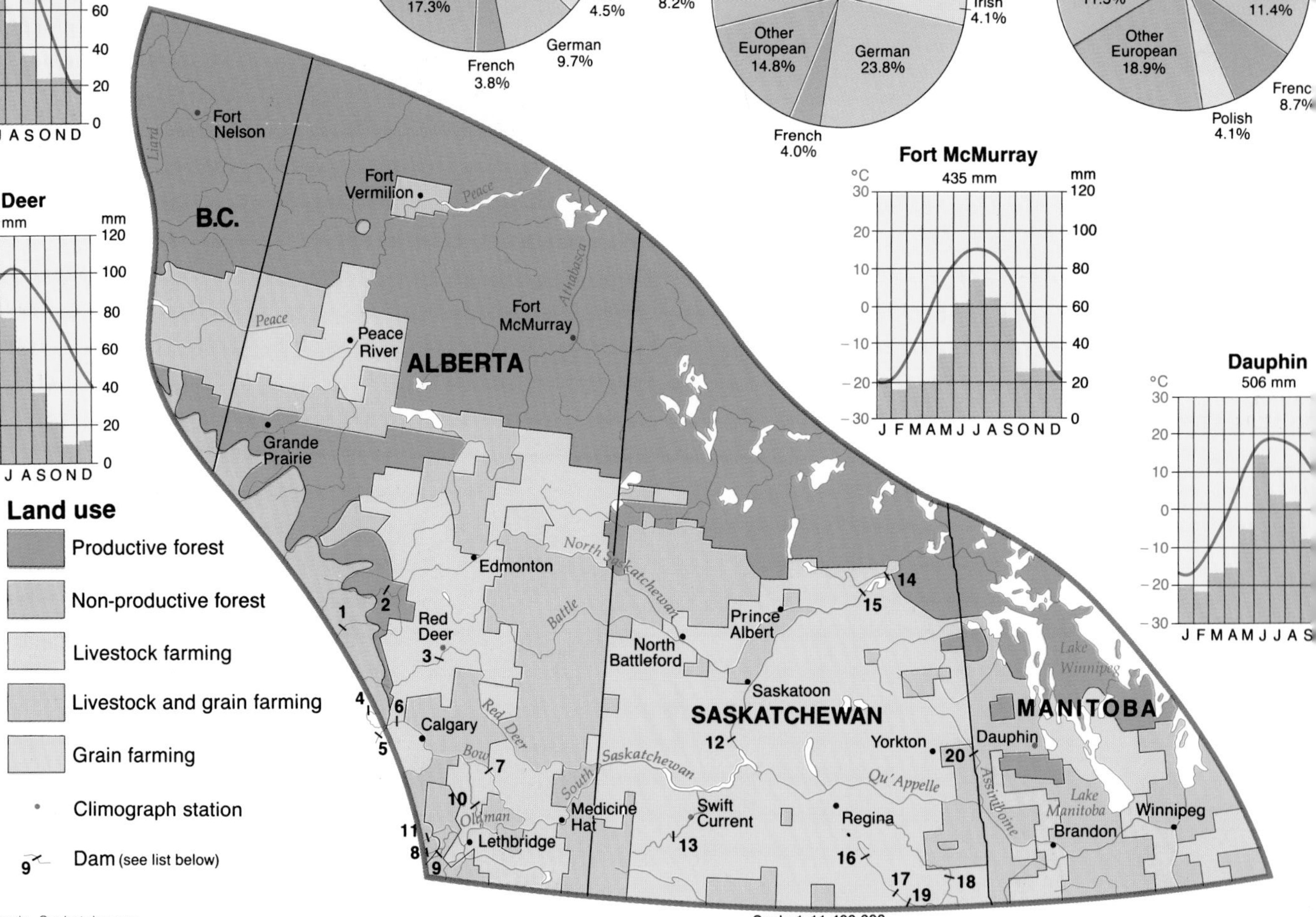

An oil well near Alameda, Saskatchewan

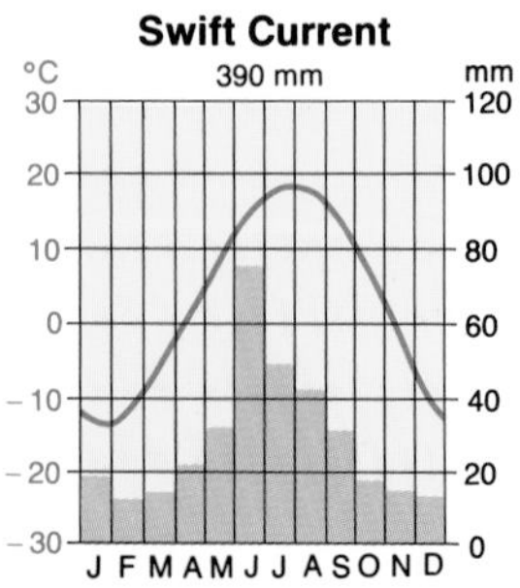

Major dams in the Prairies

1. Bighorn
2. Brazeau
3. Dickson
4. Cascade
5. Three Sisters
6. Ghost
7. Bassano
8. Waterton
9. St. Mary
10. Travers
11. Oldman
12. Gardiner
13. Duncairn
14. E.B. Campbell
15. Francois-Finlay
16. Weyburn
17. Rafferty
18. Alameda
19. Boundary
20. Shellmouth

Farms and income

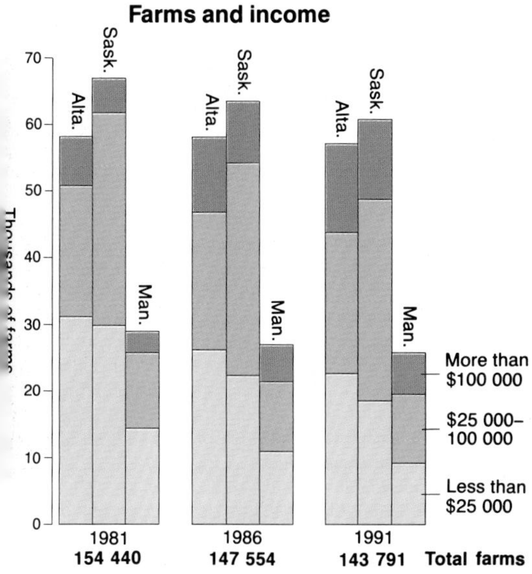

Wheat exports

Total (1991): 21 911 261 tonnes

Port of shipment

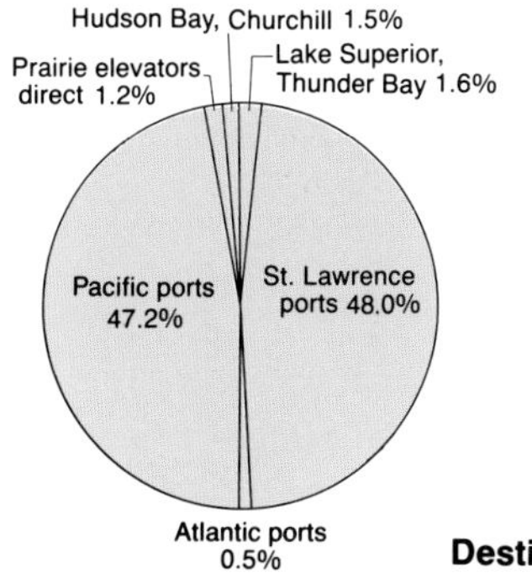

Destination

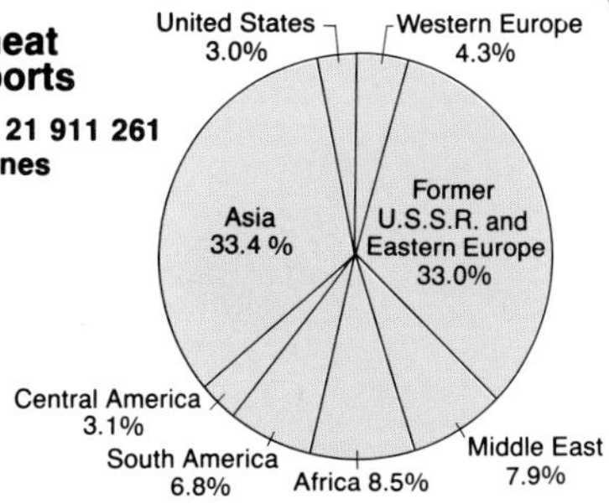

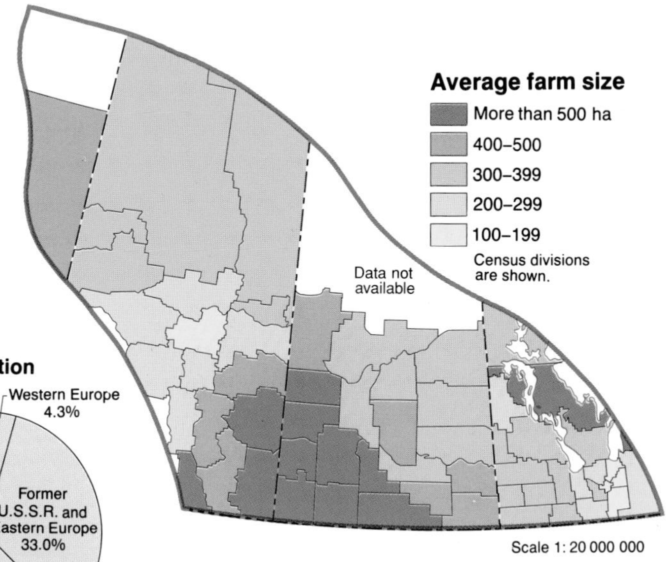

Population change by province

(1988–92 average)

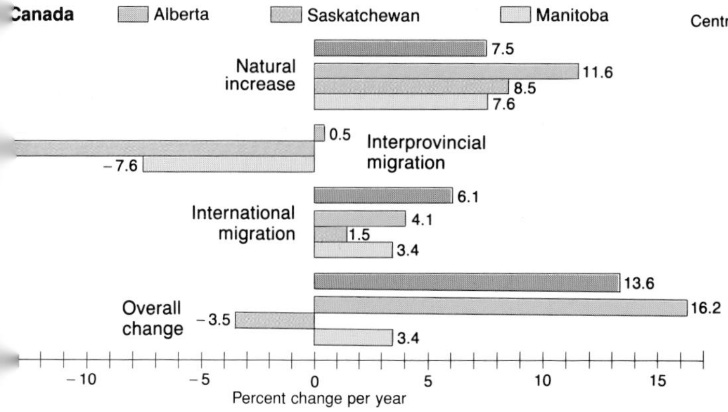

Mean annual moisture deficiency

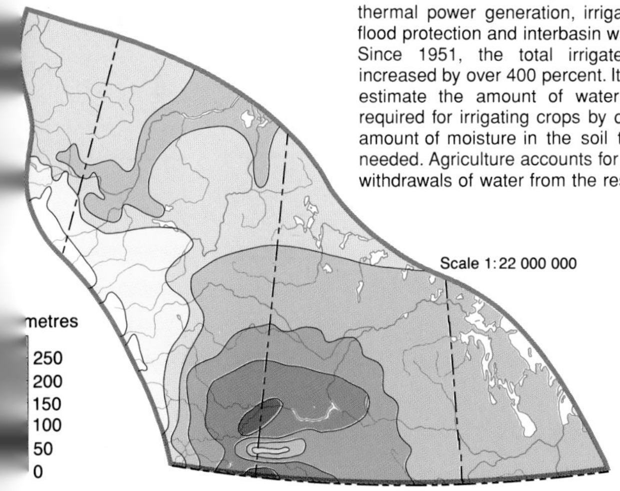

Because the Prairies are a relatively dry region, water management is important. Reservoirs are used to store water for hydro and thermal power generation, irrigation projects, flood protection and interbasin water transfers. Since 1951, the total irrigated area has increased by over 400 percent. It is possible to estimate the amount of water that will be required for irrigating crops by comparing the amount of moisture in the soil to the amount needed. Agriculture accounts for 46 percent of withdrawals of water from the reservoirs.

A Prairie grain elevator and silos

Crude oil production

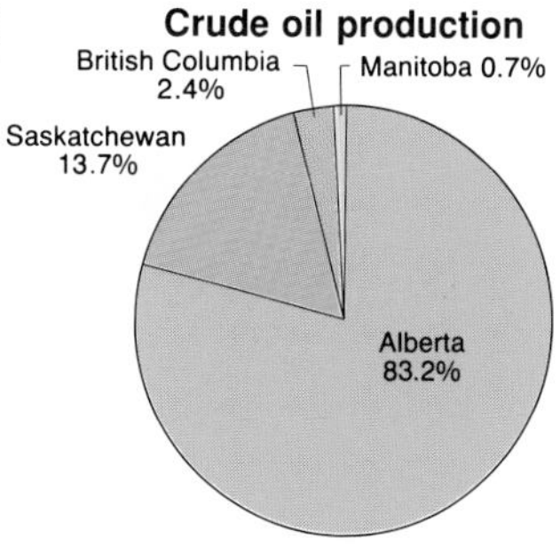

Total (1992): 98 234 000 m³

Major industries in the Prairies

Value of shipments of goods of own manufacture

		Millions of dollars
1	Meat and meat products	$3290.7
2	Non-ferrous smelting	818.9
3	Livestock feed	634.5
4	Plastics and synthetics	618.5
5	Agricultural implements	607.2
6	Newspapers, magazines	541.0
7	Sawmill products	540.1
8	Construction/mining mach.	535.2
9	Printing	506.7
10	Fluid milk	433.3
11	Inorganic chemicals	345.2
12	Other food products	329.7
13	Concrete	300.0
14	Soft drinks	287.3
15	Pressed metal products	259.1

Note: Some industries are not included in this list because the data are confidential.

Effect of the Gardiner dam on the South Saskatchewan river flow

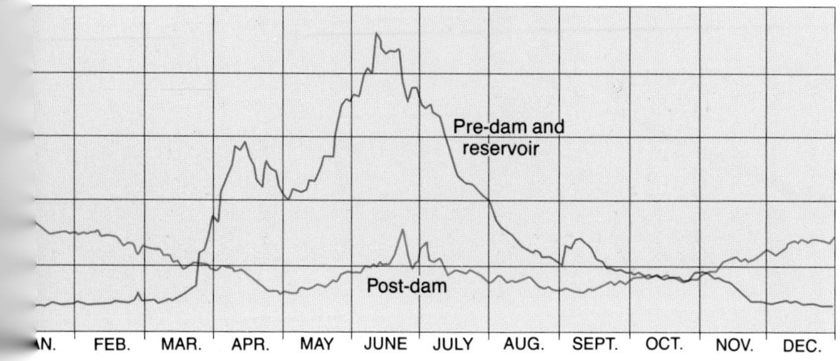

Oil and gas reserves

	Petroleum		Natural gas	
	Millions of m³		Billions of m³	
B.C.	17.7	2.4%	229.2	12.4%
Alta.	810.7	83.2%	1678.6	82.2%
Sask.	110.3	13.7%	71.2	5.4%
Man.	7.8	0.7%	—	—
Prairies	**946.5**	**100%**	**1979.0**	**100%**

Natural gas production

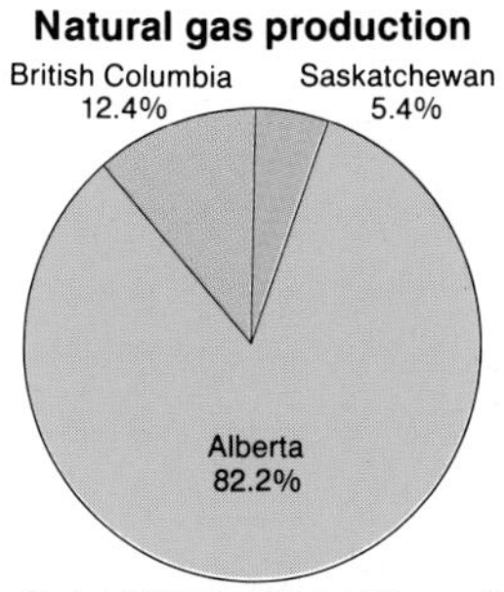

Total (1992): 115.1 billion m³

CANADA-REGIONAL

Aerial view of Québec City

The Great Lakes–St. Lawrence region has b
described as Canada's "main street," its heartl
Approximately 50 percent of the country's popula
is found in this southern strip of Québec and Ont
The region is culturally and ethnically diverse; Fre
is mainly spoken in Québec and English in Ontario
both areas include speakers of many other langua
The area attracts many immigrants, as well as
seekers from across Canada.

The mild climate and fertile soils of this region h
always made it suitable for human settlement. W
the Europeans arrived, Algonquian and Iroq
nations lived here, growing corn, beans, and squ
as well as hunting and trapping. The French and
lish were first drawn to the area by furs but later es
lished farms and villages. Population growth, comb
with superior water transportation routes and prox
to the industrial heartland of the United States,
moted the development of industry in this area.

Today, this wealthy region has a significant
nomic and political influence on the rest of Car
Over 70 percent of Canada's manufactured goods
produced in the Great Lakes–St. Lawrence re
Many investment and insurance companies, banks
corporations have their head offices here.

Population of major urban centres

Toronto CMA	Ont.	3 893 000
Montréal CMA	Qué.	3 127 000
Ottawa-Hull CMA	Ont./Qué.	921 000
Québec CMA	Qué.	646 000
Hamilton CMA	Ont.	600 000
London CMA	Ont.	382 000
St. Catharines-Niagara CMA	Ont.	365 000
Kitchener CMA	Ont.	356 000
Windsor CMA	Ont.	262 000
Oshawa CMA	Ont.	240 000
Sherbrooke CMA	Qué.	139 000
Trois-Rivières CMA	Qué.	136 000
Oakville	Ont.	115 000
Brantford	Ont.	82 000
Cambridge	Ont.	80 000
Guelph	Ont.	79 000
Barrie	Ont.	63 000
Peterborough	Ont.	61 000
Kingston	Ont.	55 000

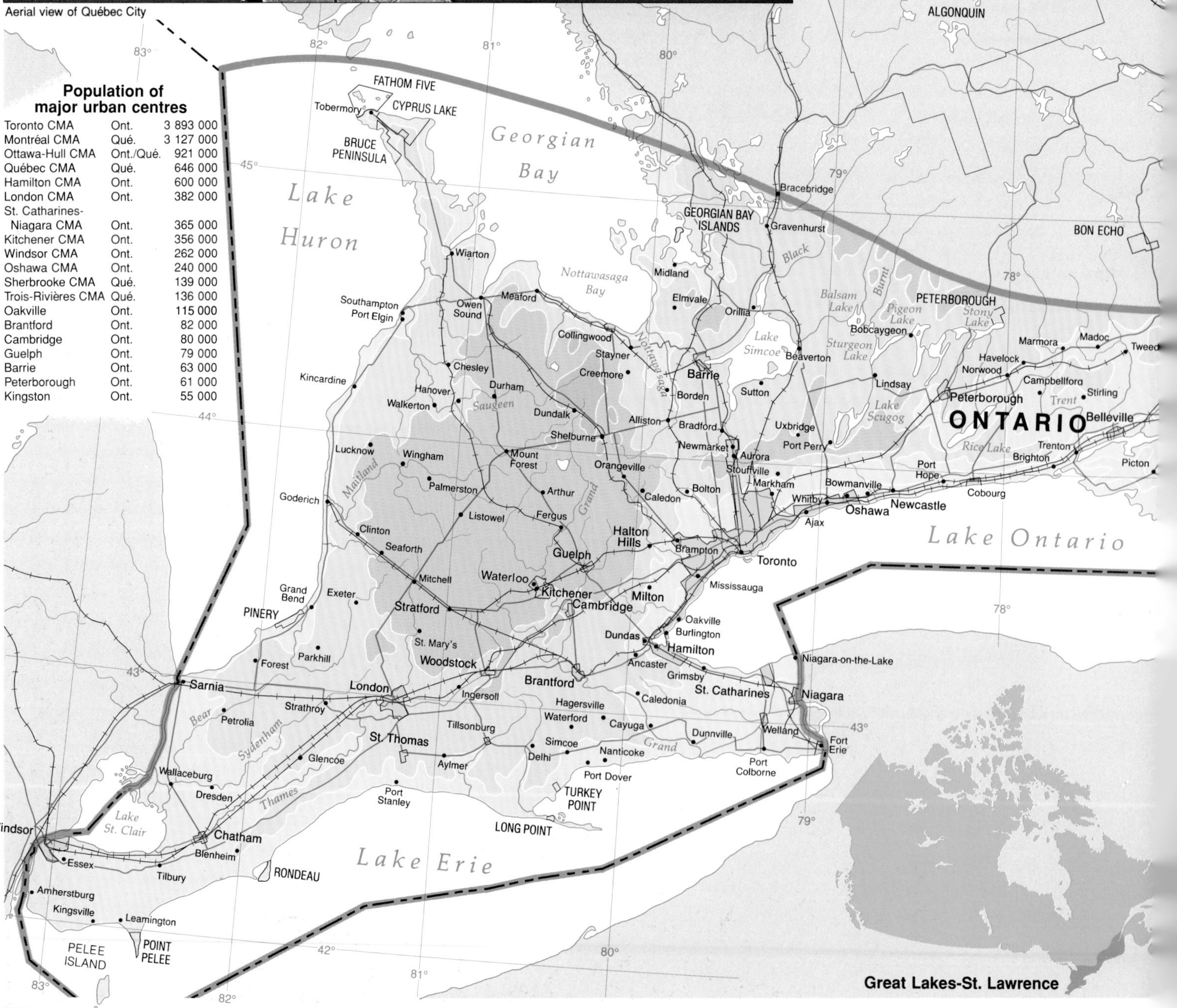

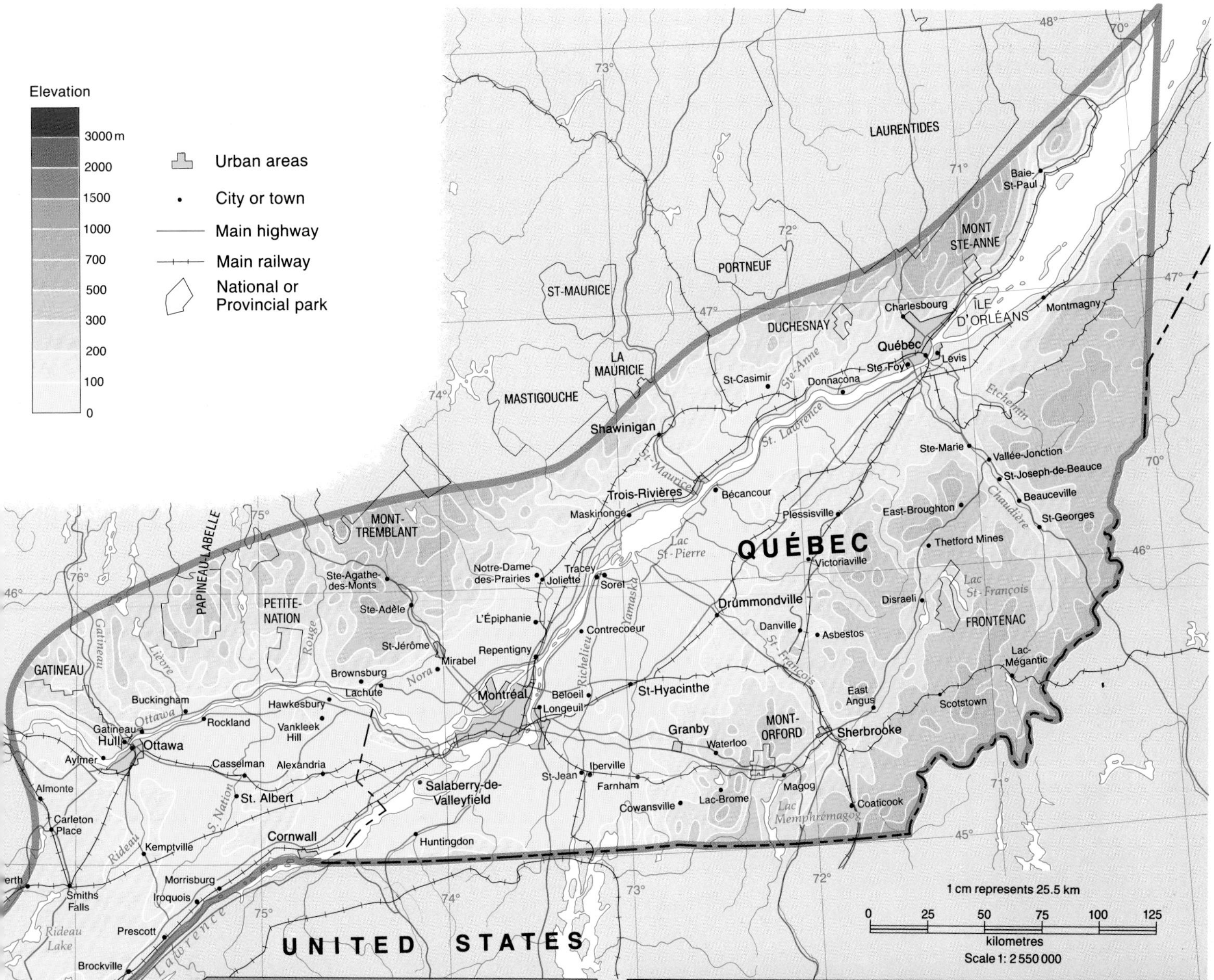
Elevation
3000 m
2000
1500
1000
700
500
300
200
100
0
Urban areas
City or town
Main highway
Main railway
National or
Provincial park
QUÉBEC
UNITED STATES
LAURENTIDES
MONT STE-ANNE
PORTNEUF
ST-MAURICE
DUCHESNAY
LA MAURICIE
MASTIGOUCHE
MONT-TREMBLANT
PAPINEAU-LABELLE
PETITE-NATION
GATINEAU
FRONTENAC
MONT-ORFORD
ÎLE D'ORLÉANS
Baie-St-Paul
Charlesbourg
Montmagny
Québec
Lévis
Ste-Foy
St-Casimir
Donnacona
Shawinigan
Trois-Rivières
Bécancour
Maskinongé
Plessisville
East-Broughton
Ste-Marie
Vallée-Jonction
St-Joseph-de-Beauce
Beauceville
St-Georges
Thetford Mines
Victoriaville
Disraeli
Drummondville
Danville
Asbestos
Lac-Mégantic
Notre-Dame-des-Prairies
Joliette
Tracey
Sorel
Ste-Agathe-des-Monts
Ste-Adèle
L'Épiphanie
Contrecoeur
St-Jérôme
Repentigny
Mirabel
Brownsburg
Lachute
Montréal
Beloeil
Longeuil
St-Hyacinthe
Buckingham
Hawkesbury
Rockland
Vankleek Hill
Gatineau
Hull
Ottawa
Aylmer
Casselman
Alexandria
Almonte
St. Albert
Salaberry-de-Valleyfield
Granby
Waterloo
Sherbrooke
East Angus
Scotstown
St-Jean
Iberville
Farnham
Cowansville
Lac-Brome
Magog
Coaticook
Carleton Place
Kemptville
Cornwall
Huntingdon
Morrisburg
Iroquois
Smiths Falls
Prescott
Brockville
St. Lawrence
Ottawa
Rideau
Rideau Lake
Lac St-Pierre
Lac St-François
Lac Memphrémagog
Chaudière
Etchemin
Richelieu
Yamaska
St-François
St-Maurice
Ste-Anne
Gatineau
Lièvre
Rouge
Nord
S. Nation
1 cm represents 25.5 km
0 25 50 75 100 125
kilometres
Scale 1: 2 550 000

Downtown Toronto, looking north-east

Satellite image of the western end of Lake Ontario, also known as the Golden Horseshoe

For explanation of terms, see glossary.

CANADA-REGIONAL

The Great Lakes-St. Lawrence region is densely populated and continues to attract people despite economic downturns. Urban sprawl has overtaken some of the most productive agricultural land in the region. In addition to air and water pollution (see pages 53/54), waste disposal has become a hot political controversy. Large urban centres lack landfill sites; other communities nearby do not want to be alternative deposit sites.

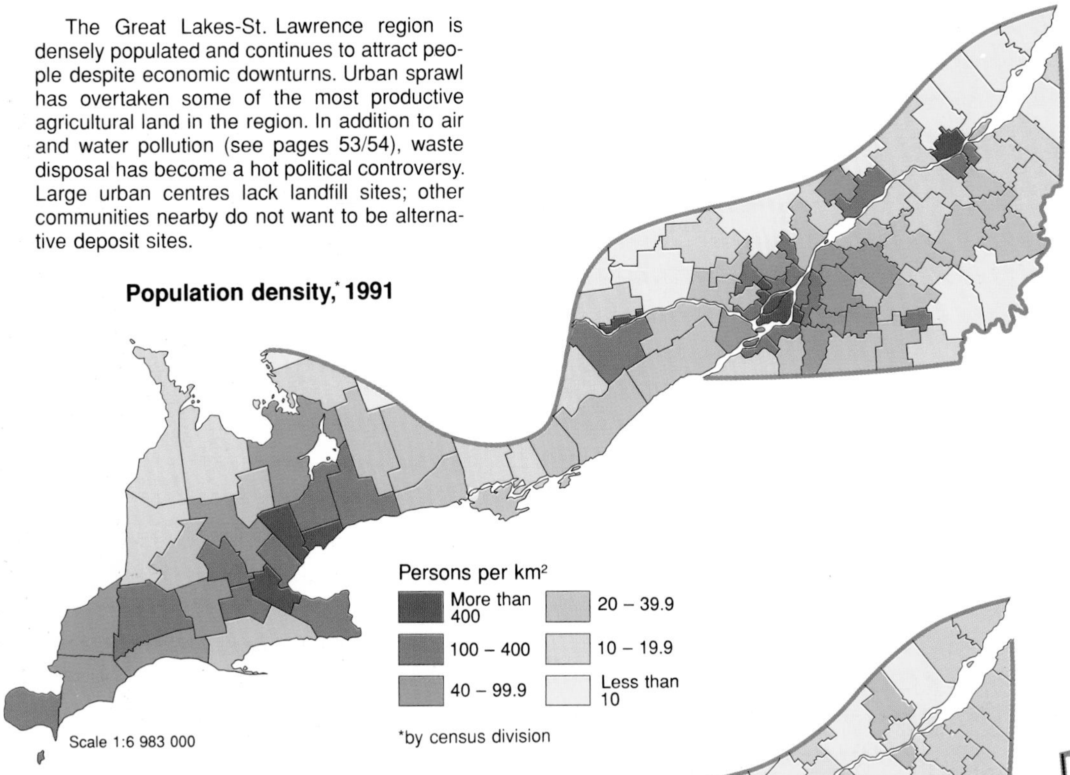

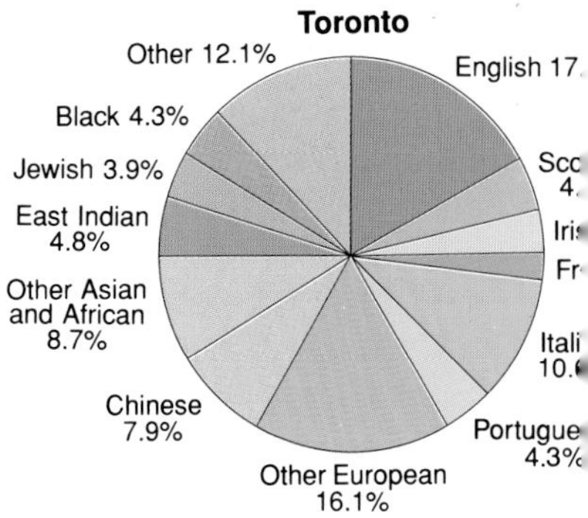

Over half of all immigrants to Canada settle in the Metropolitan Toronto area, along with many people from other parts of Canada. This influx of population has given the region a great ethnic diversity. Tourists are also attracted to the region because of its major league sports, world class theatre, theme parks and shopping.

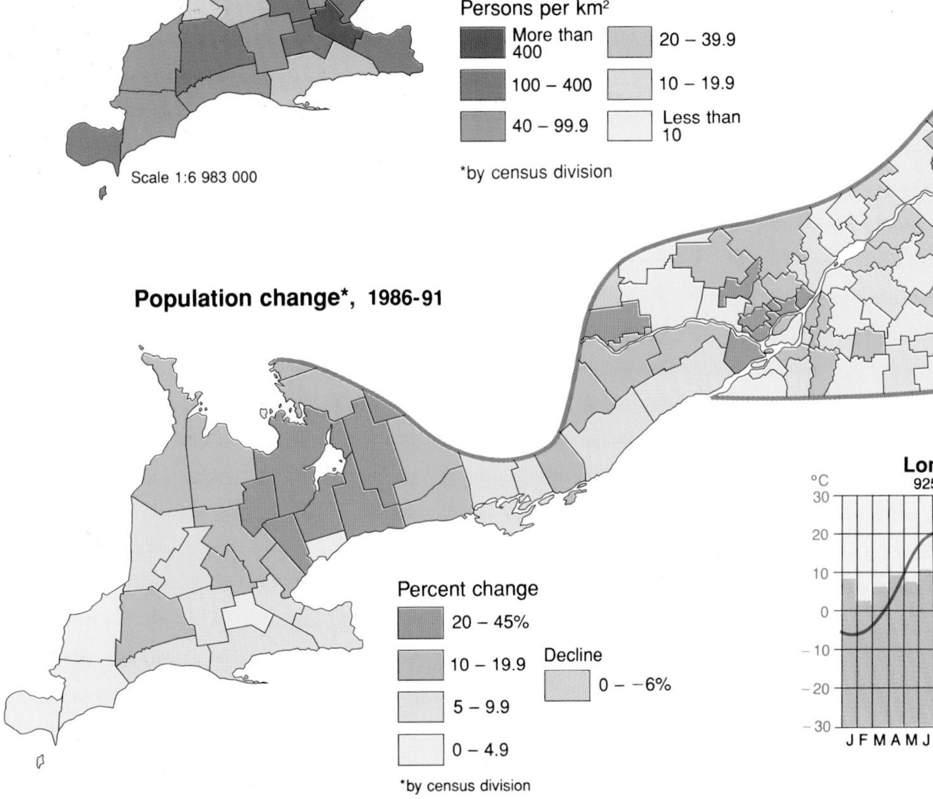

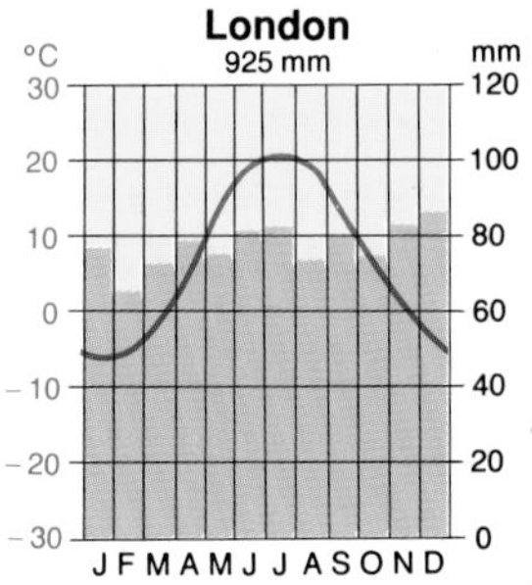

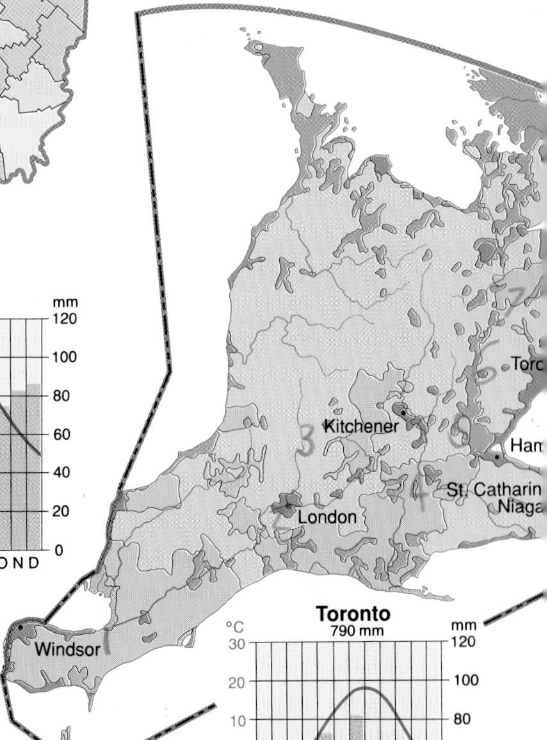

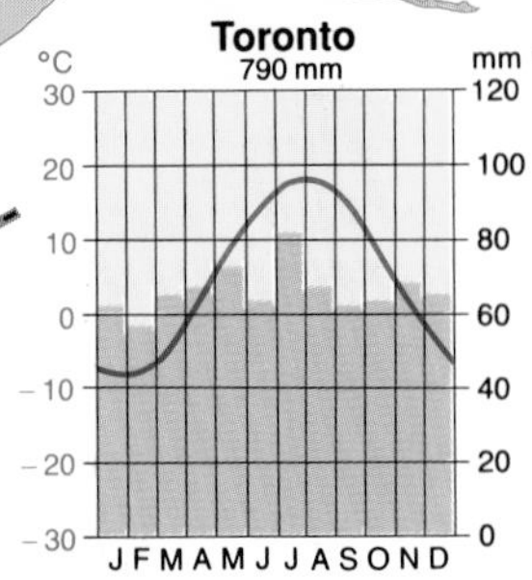

Agricultural workers in the Holland Marsh, a very fertile area near Bradford, Ontario

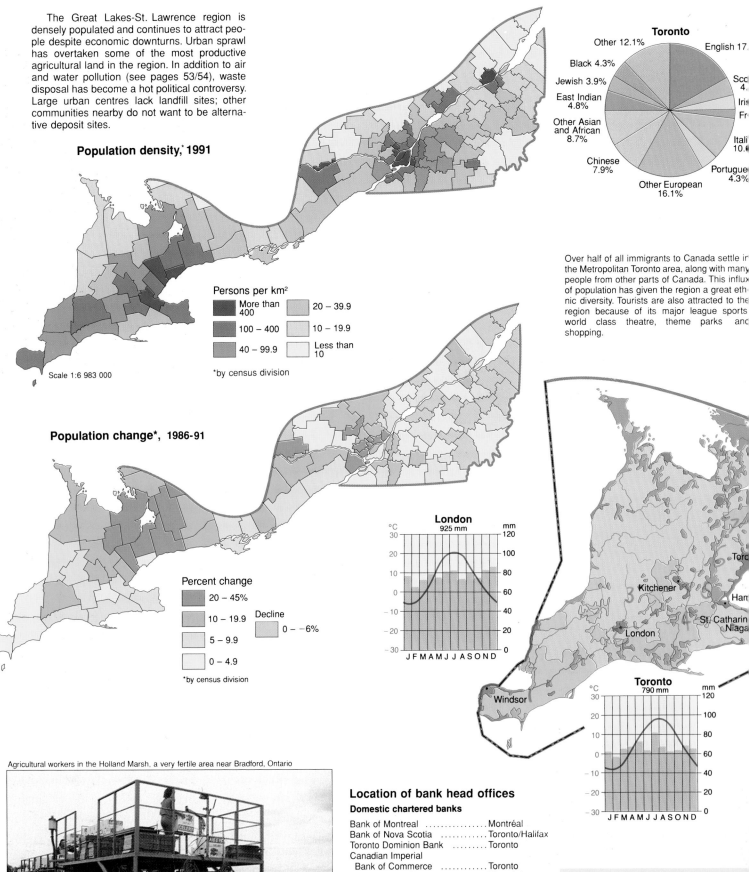

Location of bank head offices

Domestic chartered banks

Bank of Montreal	Montréal
Bank of Nova Scotia	Toronto/Halifax
Toronto Dominion Bank	Toronto
Canadian Imperial Bank of Commerce	Toronto
Royal Bank of Canada	Montréal
National Bank of Canada	Montréal
Laurentian Bank of Canada	Montréal
Canadian Western Bank	Edmonton
Manulife Bank of Canada	Orillia, Ontario

Foreign chartered banks

45 in Toronto
6 in Montréal
3 in Vancouver

Trading on Canadian Stock Exchange
Total 1991 trading: $90 115 613 501

	Dollar value
Toronto	$67 749 538 37
Montréal	18 332 759 95
Vancouver	3 486 268 04
Alberta	586 877 05
Winnipeg	1 170 07

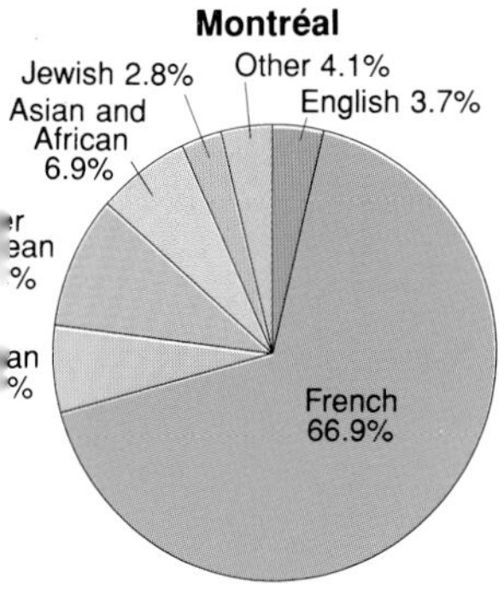

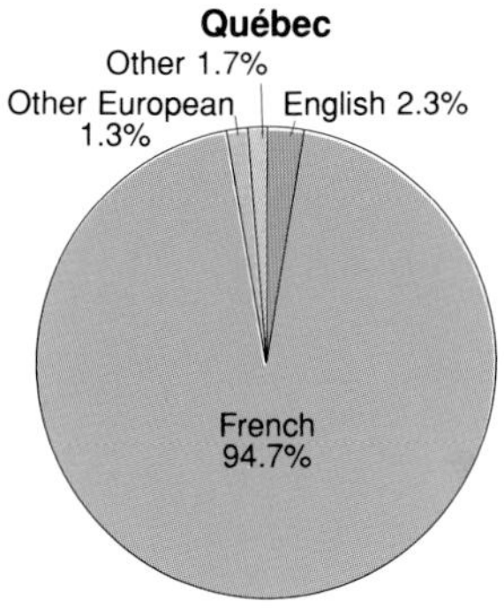

Montréal
Jewish 2.8%
Other 4.1%
English 3.7%
Asian and African 6.9%
French 66.9%
Québec
Other 1.7%
Other European 1.3%
English 2.3%
French 94.7%
Ethnicity
(single responses only)
Québec
1089 mm
°C
mm
J F M A M J J A S O N D
Tourists in old Québec City
FUNICULAIRE
Haliburton
928 mm
Québec
Trois-Rivières
Montréal
Sherbrooke
Ottawa-Hull
Scale 1:4 050 000
Sherbrooke
928 mm
St. Lawrence Seaway cargo
Other processed products and misc. 11.5%
Other grains and soybeans 6.9%
Corn 5.6%
Wheat 18.6%
Iron ore 24.0%
Coal, coke, salt and other mine products 23.9%
Iron and steel 9.5%
Value of manufacturing by CMA
Toronto 19.9%
Other areas of Canada 35.6%
Montréal 12.5%
Windsor 3.7%
Hamilton 3.7%
Kitchener 2.2%
St. Catharines-Niagara 2.3%
Vancouver 3.8%
Edmonton 2.9%
Winnipeg 1.6%
Other CMAs excluding Oshawa 11.7%
The St. Lawrence Seaway opened in 1959. The United States and Canada jointly constructed the canals, dams and locks that allow ocean-going ships to rise 177 m over 3769 km. The Seaway is still important to the region despite a decline in shipping volume, which peaked in 1977. Changes in manufacturing in the surrounding area and new methods of shipping grain have contributed to the decline.
use
Urban areas
Cropland
Improved pasture and cropland
Orchards/vineyards
Climograph station
Grazing
Recreation
Woodland
Marsh/swamp
Types of manufacturing by CMA
London
Hamilton
Toronto
Ottawa-Hull
Trois-Rivières
Québec
Windsor
Kitchener
St. Cath.-Niagara
Oshawa
Montréal
Sherbrooke
Percent
Industrial grouping
Non-metallic minerals and miscellaneous
Rubber, plastic, petroleum, coal and chemicals
Primary metals and metal fabricating
Machinery, transport equipment and electrical products
Wood, furniture, paper, publishing and printing
Leather, textiles, knitting and clothing
Food and beverages
ndustry in Hamilton, Ontario

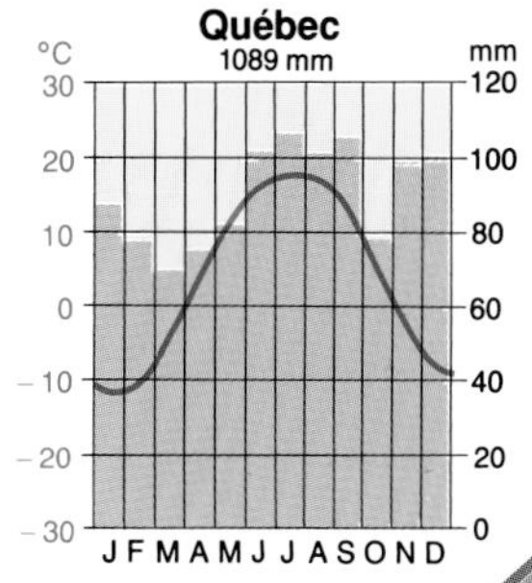

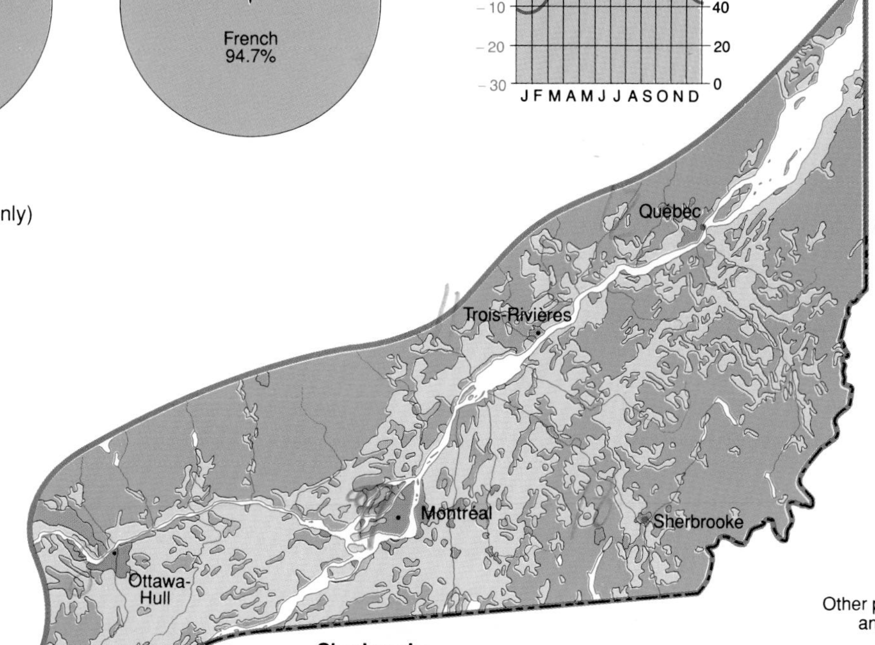

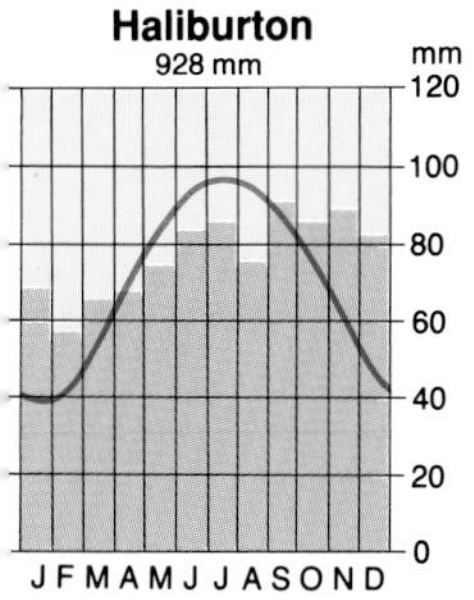

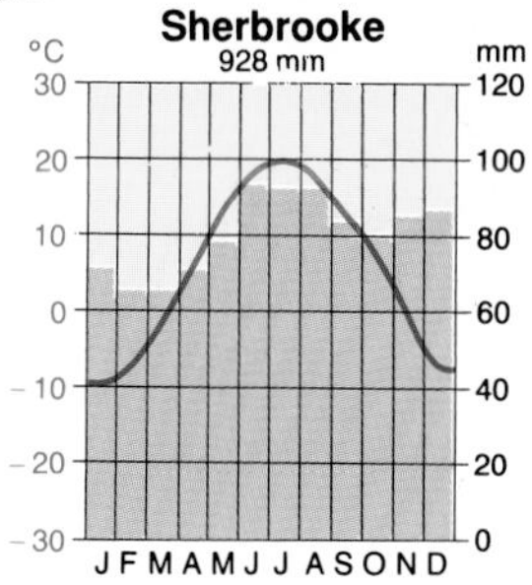

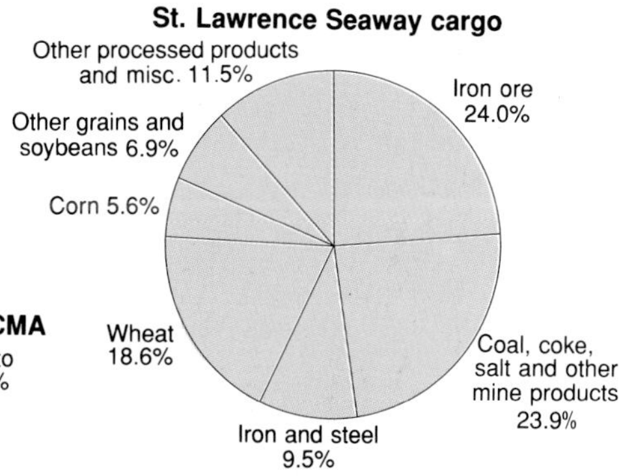

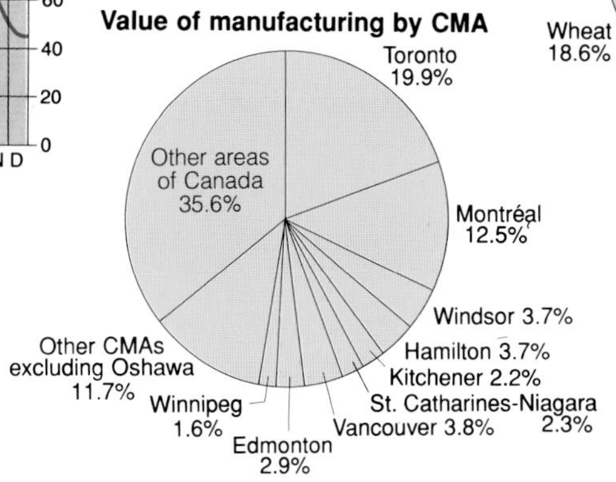

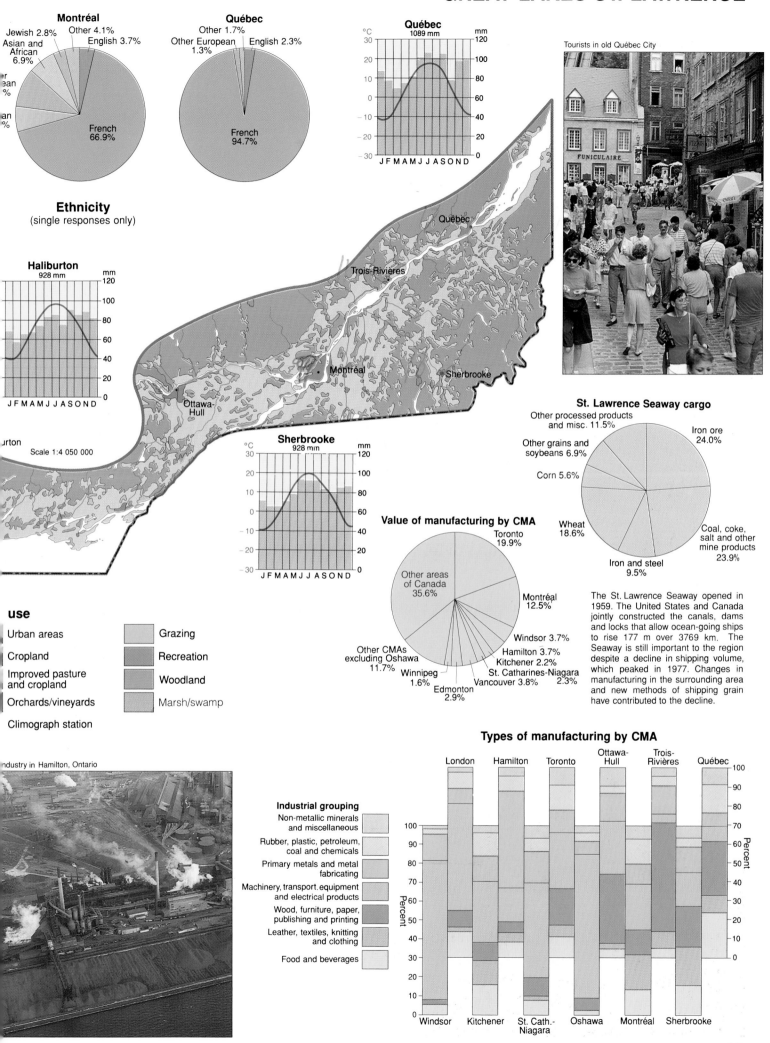

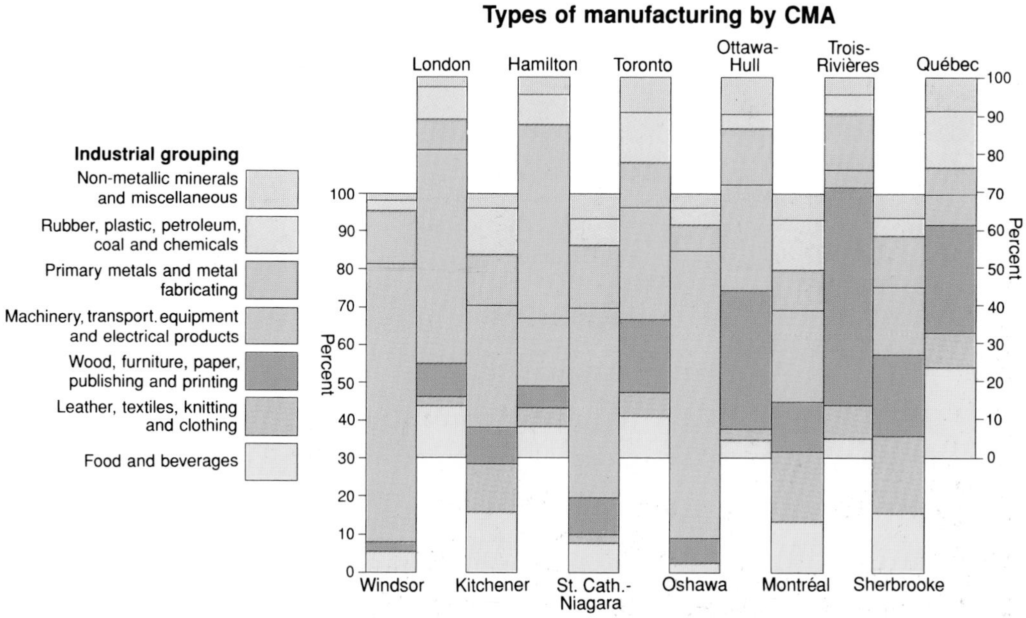

GREAT LAKES-ST. LAWRENCE

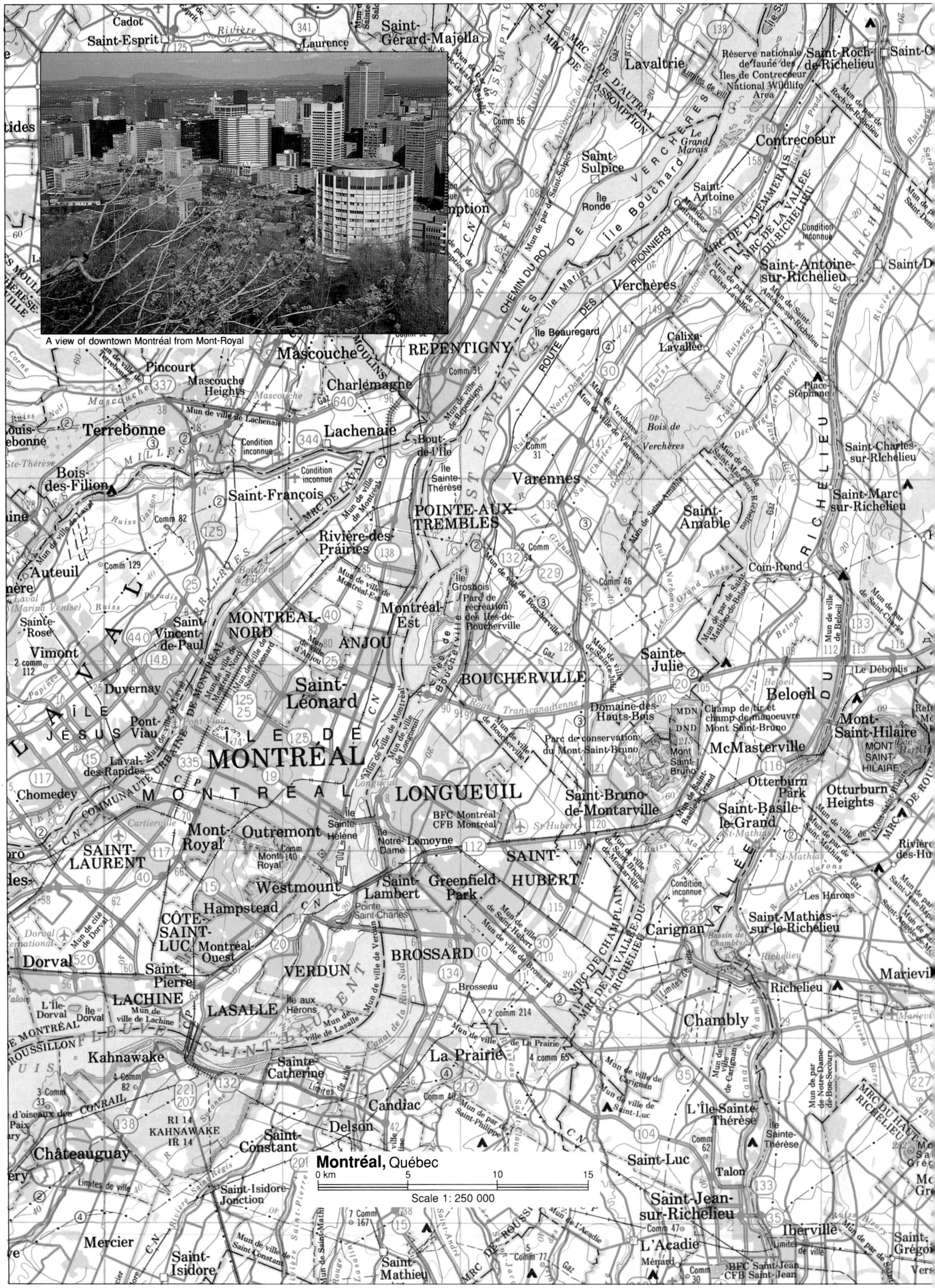

A view of downtown Montréal from Mont-Royal

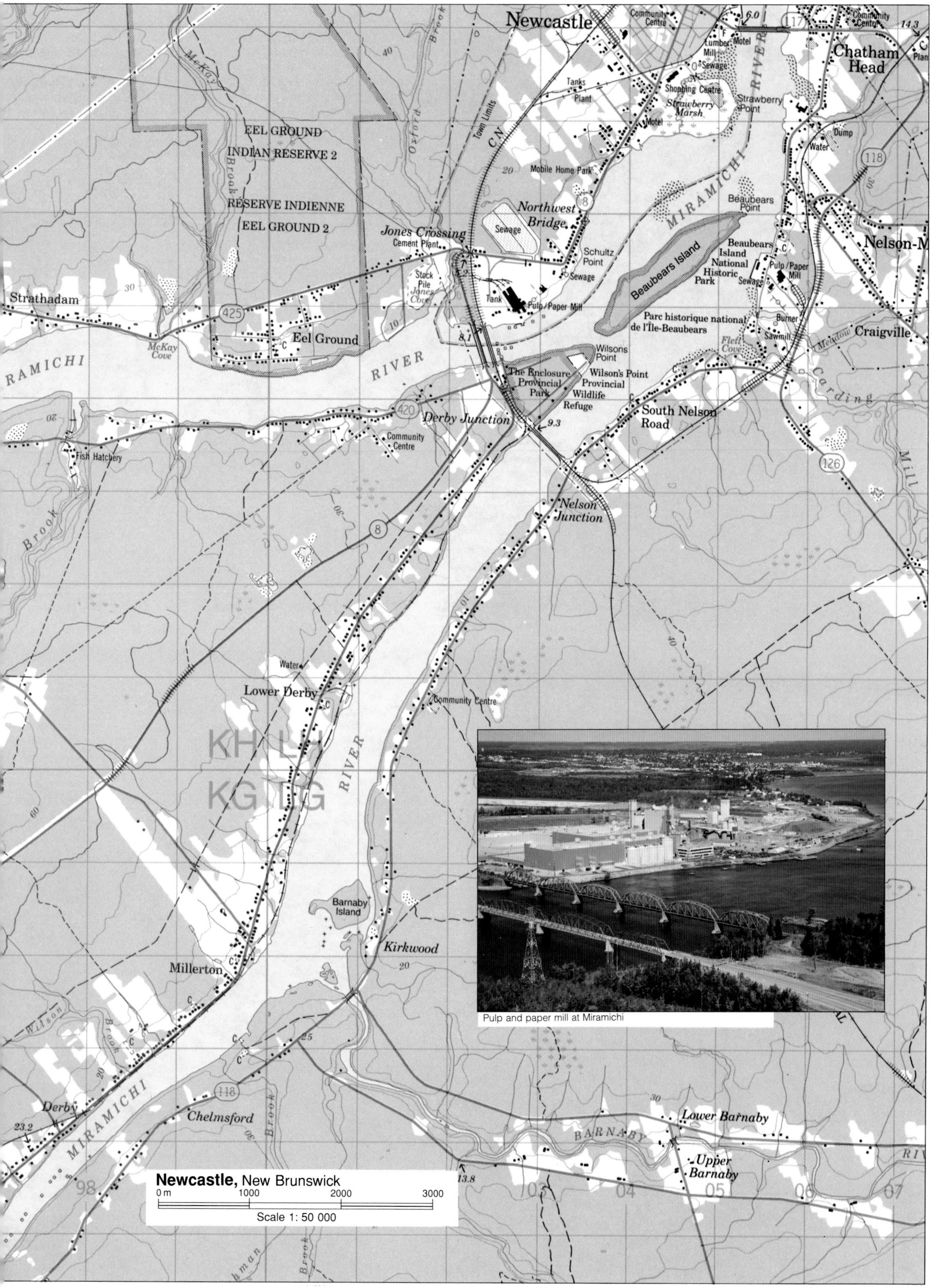
Newcastle
Community Centre
Lumber Mill
Motel
Sewage
Shopping Centre
Strawberry Marsh
Strawberry Point
Chatham Head
Dump
Water
Tanks
Plant
Motel
Town Limits
CN
Mobile Home Park
MIRAMICHI
RIVER
Beaubears Point
EEL GROUND
INDIAN RESERVE 2
RÉSERVE INDIENNE
EEL GROUND 2
Jones Crossing
Cement Plant
Northwest Bridge
Sewage
Schultz Point
Beaubears Island
Beaubears Island National Historic Park
Pulp/Paper Mill
Nelson-M
Stock Pile
Jones Cove
Tank
Pulp/Paper Mill
Burner
Sawmill
Parc historique national de l'Île-Beaubears
Strathadam
Eel Ground
McKay Cove
Craigville
Wilsons Point
The Enclosure Provincial Park
Wilson's Point Provincial Wildlife Refuge
Fleet Cove
Carding
Meadow
South Nelson Road
Derby Junction
Community Centre
Fish Hatchery
Nelson Junction
Mill
Oxford
Brook
McKay
Water
Lower Derby
Community Centre
KH LH
KG LG
Barnaby Island
Kirkwood
Millerton
Wilson
Derby
Chelmsford
Lower Barnaby
BARNABY
Upper Barnaby
98
03
04
05
06
07
Newcastle, New Brunswick
0 m
1000
2000
3000
Scale 1: 50 000

Pulp and paper mill at Miramichi

CANADA-REGIONAL

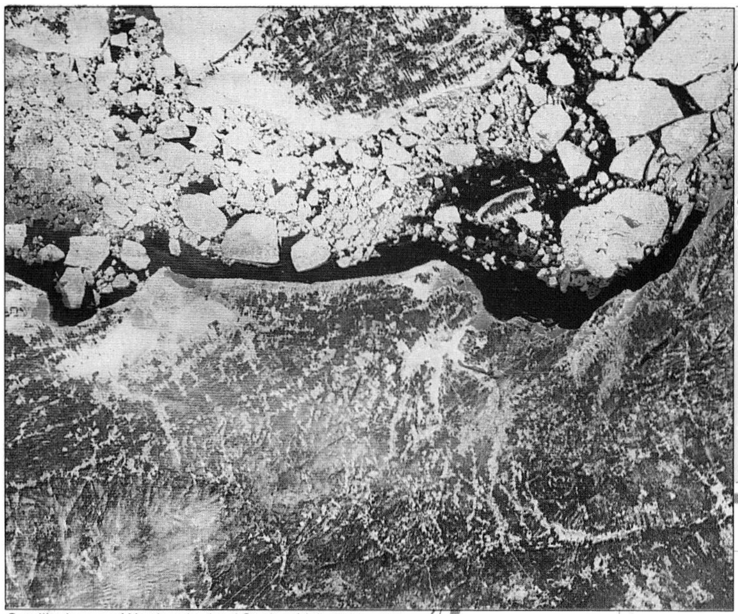

Satellite image of Northumberland Strait, which divides Prince Edward Island from Nova Scotia and New Brunswick

The Atlantic region includes all or parts of five provinces. A variety of physical features gives the region a special beauty: miles of coastline, low, rounded mountains, rugged uplands and fertile lowlands.

Before Europeans arrived, the Atlantic region was sparsely settled by Algonquian (mainly Micmacs and Malecites) and Beothuk First Nations. While the Micmacs hunted and fished along the coast, the Malecites hunted and farmed in the Saint John River valley. The Beothuks, a separate cultural group in Newfoundland, were destroyed by disease and several years of conflict with the new settlers and the Algonquian.

Due to its strategic location, the Atlantic region was settled early by Europeans. French colonists named part of the region Acadie, but lost the territory to the British at the end of the Seven Years' War. Today, there is still a large francophone population in New Brunswick, and St-Pierre et Miquelon is a self-governing department of France.

Most people in the Atlantic region are urban dwellers, employed in manufacturing and service industries. Tourism is of growing importance in the region. However, recent economic growth has been slow. The fishing industry, especially in Newfoundland, is suffering from declining fish stocks. Offshore oil and gas developments such as Hibernia may provide more money and jobs for the Atlantic economy.

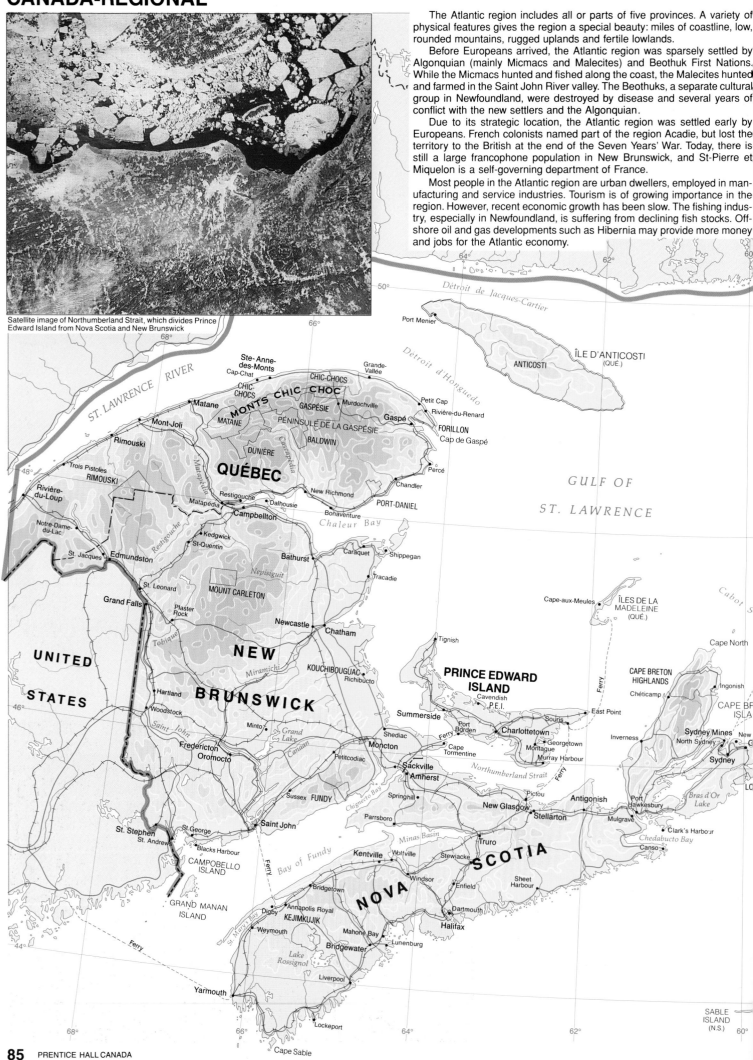

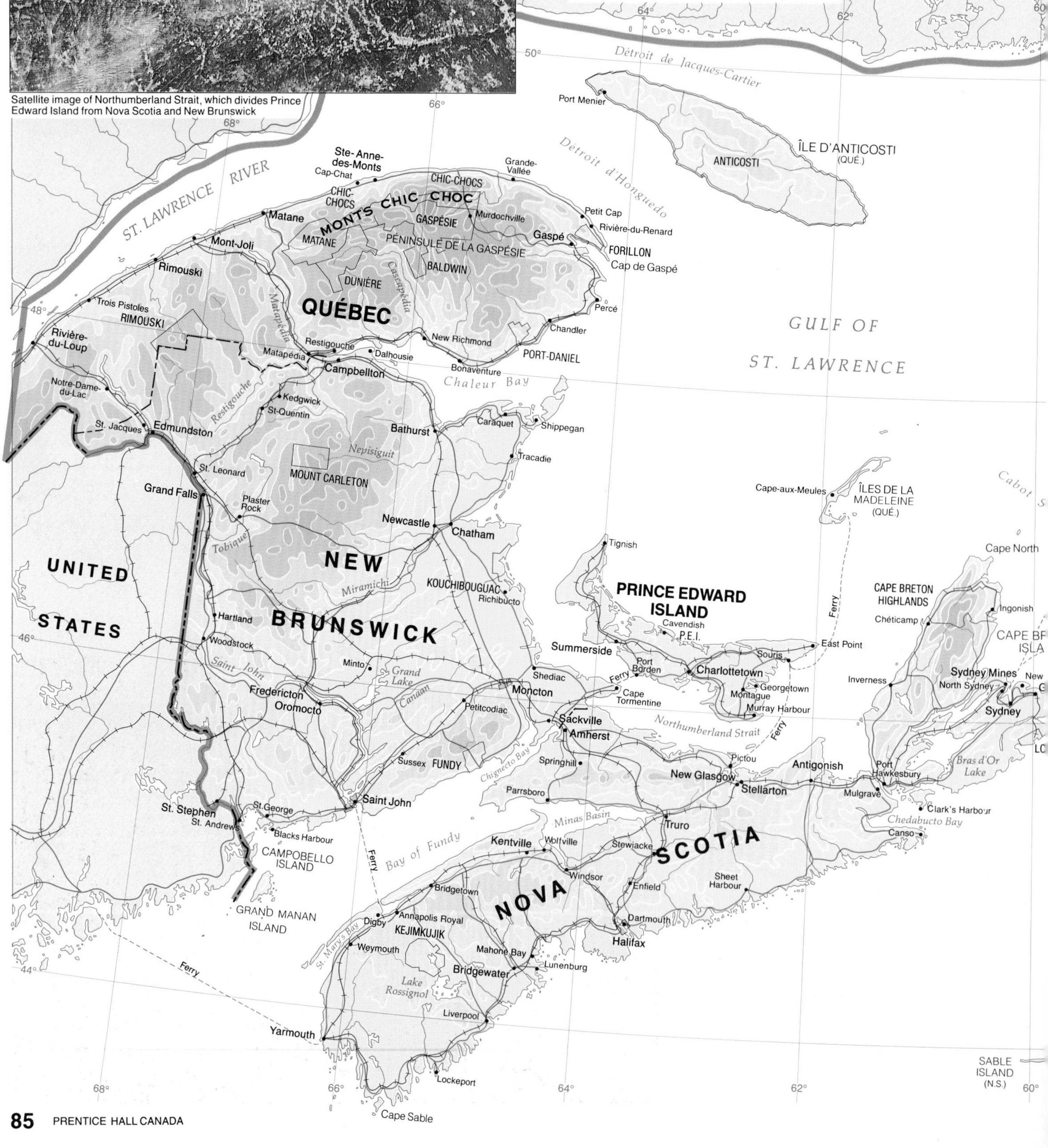

Halifax, Nova Scotia

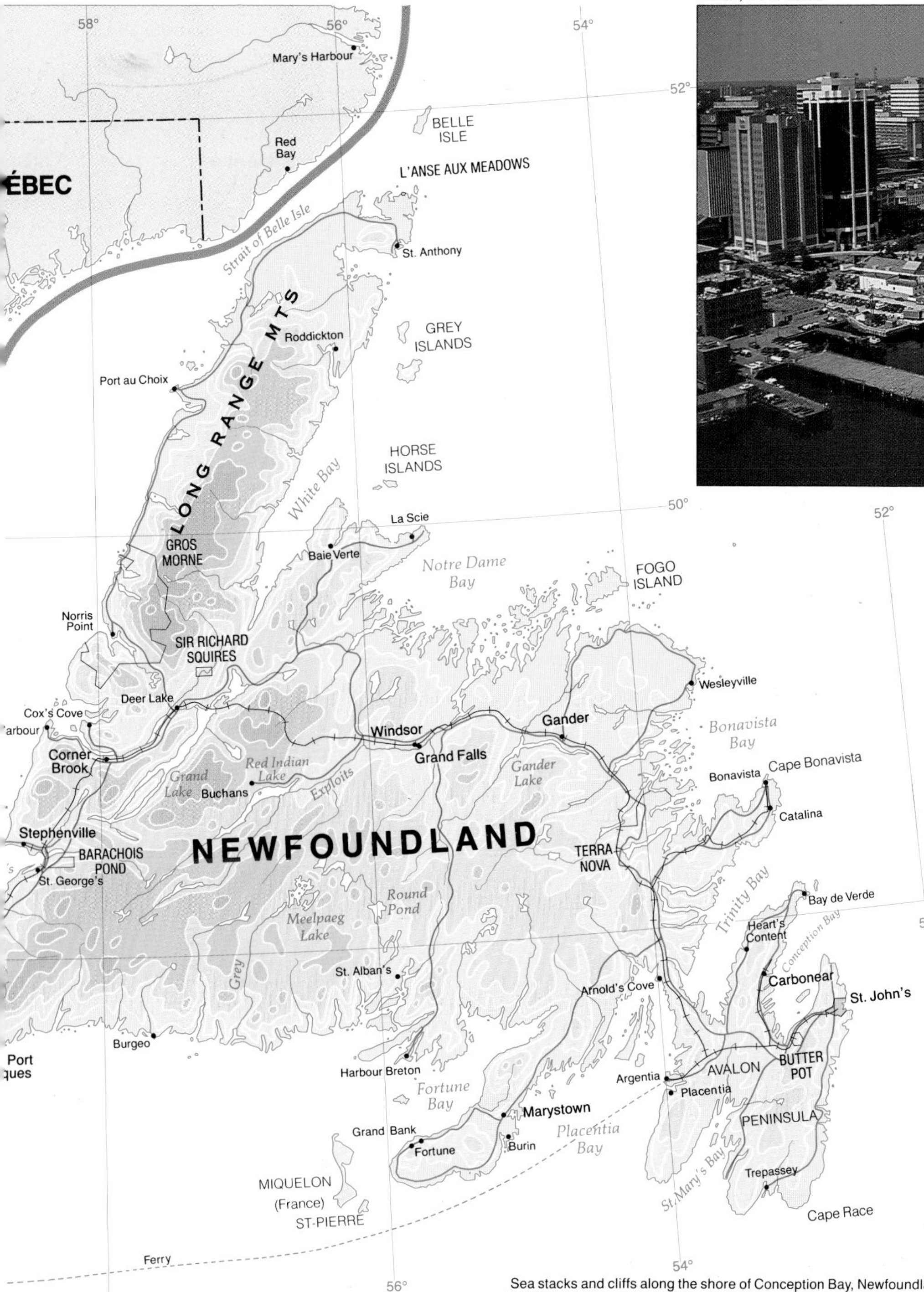

Population of major urban centres

Halifax CMA	N.S.	321 000
St. John's CMA	Nfld.	172 000
Saint John CMA	N.B.	125 000
Moncton	N.B.	57 000
Fredericton	N.B.	46 500
Rimouski	Qué.	31 000
Sydney	N.S.	26 100
Corner Brook	Nfld.	22 500
Glace Bay	N.S.	19 500
Grand Falls-Windsor	Nfld.	15 500
Charlottetown	P.E.I.	15 400
Bathurst	N.B.	14 400
Rivière-du-Loup	Qué.	13 400
Matane	Qué.	12 700
Truro	N.S.	11 700
Edmundston	N.B.	10 900

Elevation

3000 m
2000
1500
1000
700
500
300
200
100
0

Urban areas

City or town

Main highway

Main railway

National or Provincial park

Scale 1:3 500 000

0 50 100 150

kilometres

1 cm represents 35 km

Sea stacks and cliffs along the shore of Conception Bay, Newfoundland

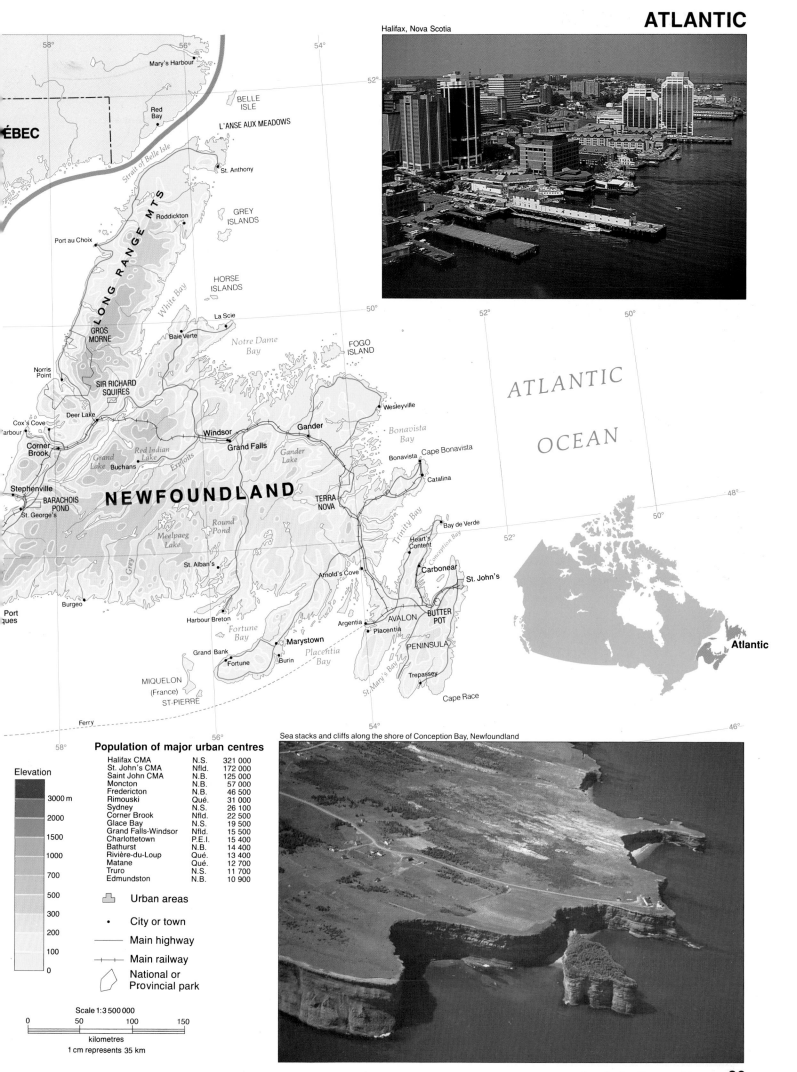

For explanation of terms, see glossary.

CANADA-REGIONAL

The Atlantic economy is based largely on fishing, forestry, farming and mining. Because the economy depends so much on supply, demand and prices for the resources of the region, it is vulnerable to both sudden downturns and long recessions. The decline in the northern cod stocks, in particular, has caused a serious crisis. High levels of unemployment and a regional per capita income that is the lowest in Canada have caused many Atlantic Canadians to move to other parts of Canada to find work.

Government incentives and development programs have attempted to create new industries to strengthen the economy. Tourists from the northern United States are an important source of seasonal income during the summer. In addition, the region is becoming attractive to people of retirement age, both for its scenic beauty and relatively lower housing costs.

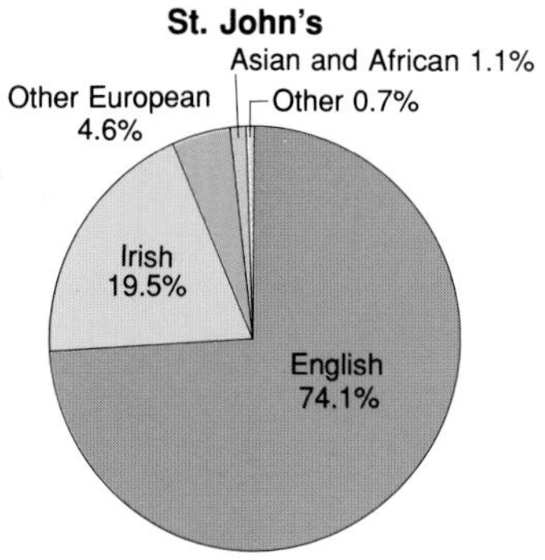

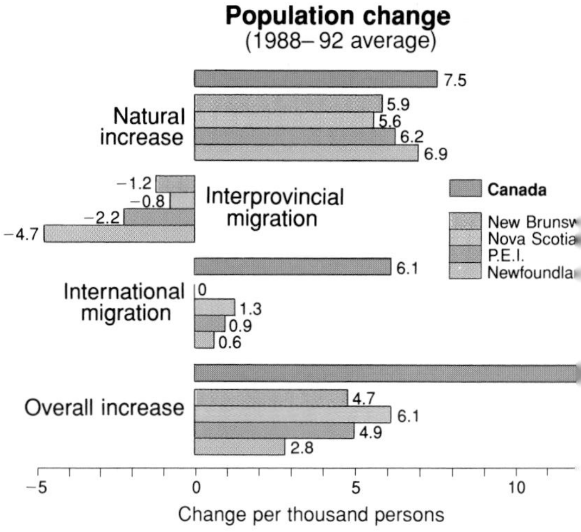

Ethnicity
(single responses only)

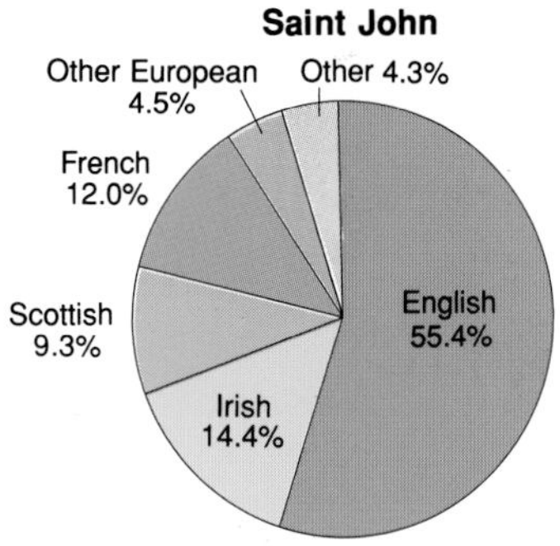

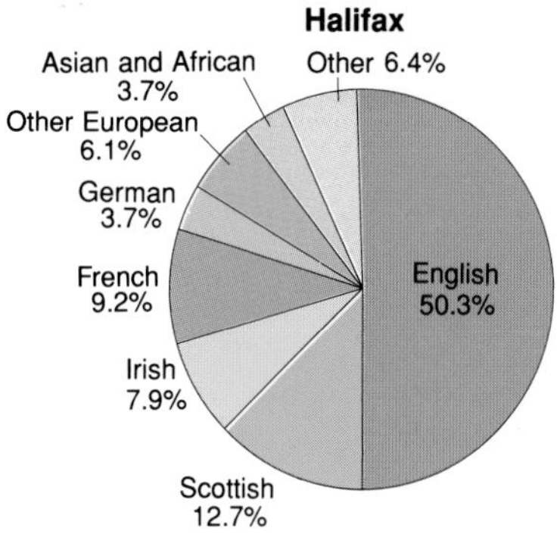

Fish processing plants are spread out all over the Atlantic region (see pages 43/44). Some are large plants, open year round, while others have only a few employees and operate only during the summer. The crisis in fish stocks has forced many plants to shut down or slow down, putting many people out of work.

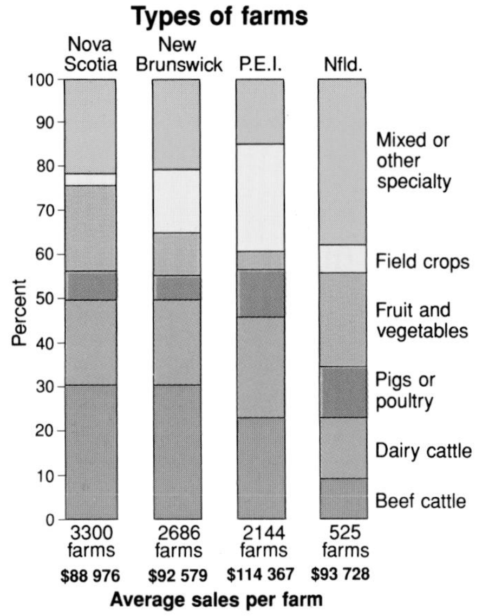

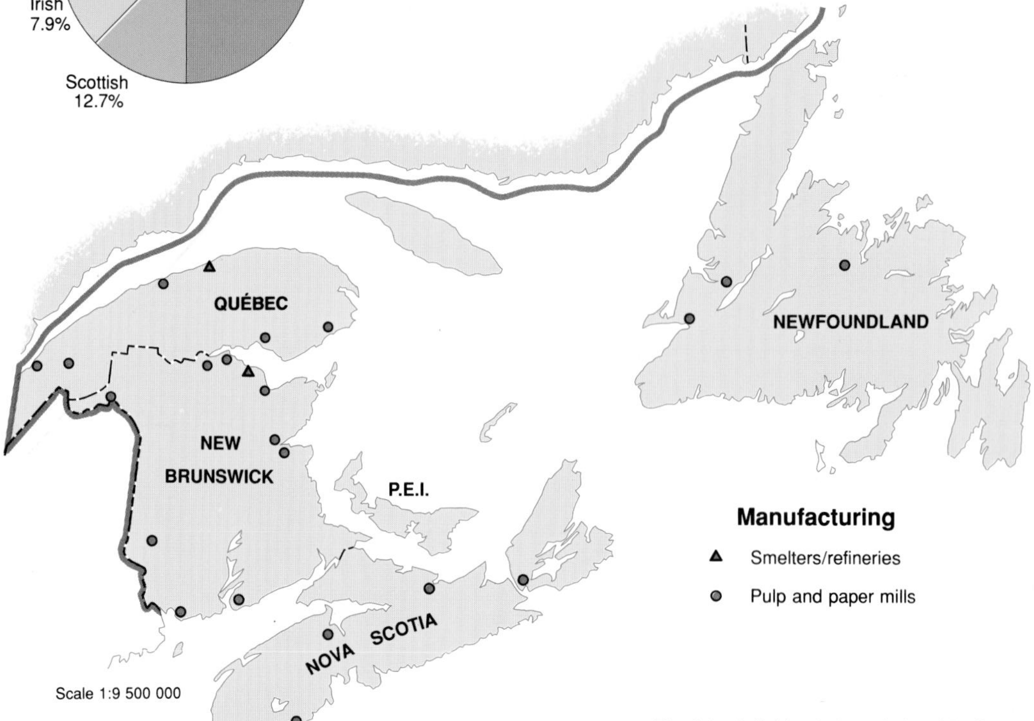

The Atlantic fishing industry is in crisis. Seven out of its 11 most important species are in decline, particularly northern cod. In 1992, a moratorium was declared on fishing of northern cod to give stocks a chance to recover. The nearly 1000 communities that are largely or entirely dependent upon the fisheries have been severely affected by these fish shortages.

The outport fishing village of Bay de Verde, Newfoundland

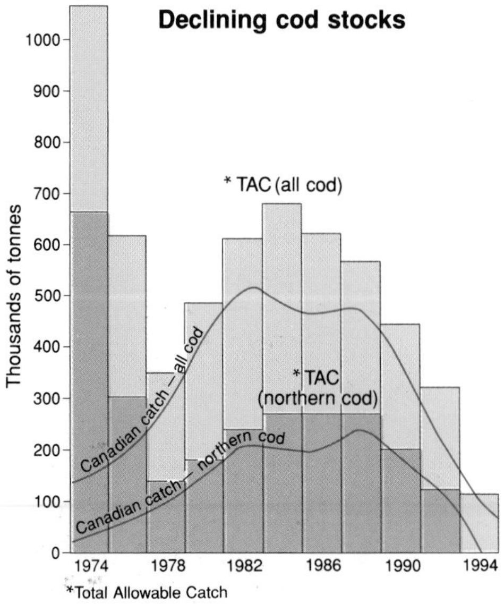

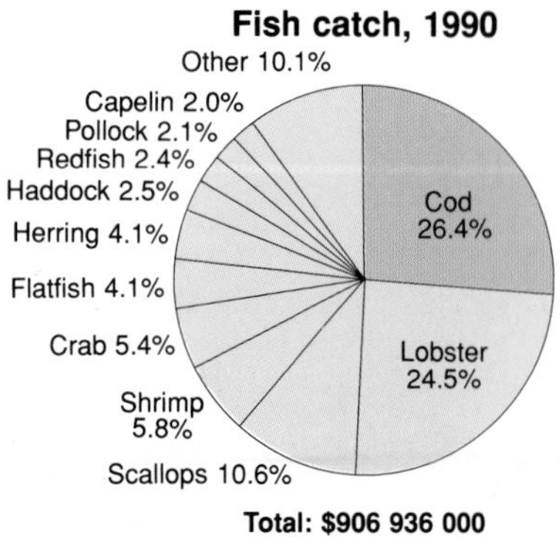

ice conditions and Hibernia

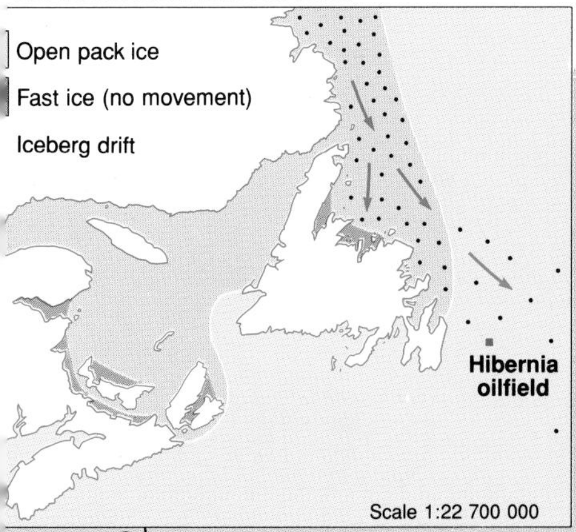
Open pack ice
Fast ice (no movement)
Iceberg drift
Hibernia oilfield
Scale 1:22 700 000

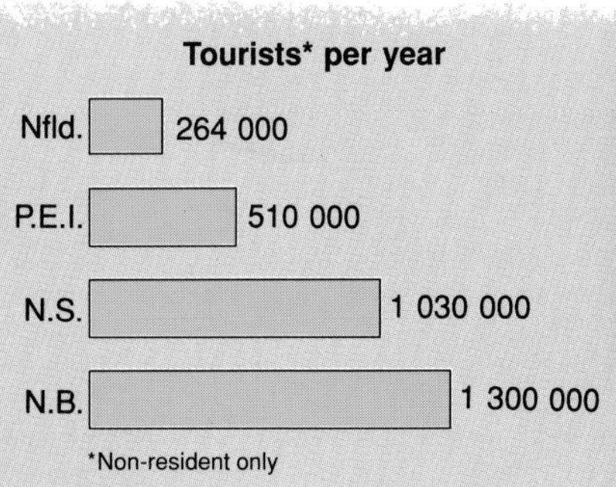
Tourists* per year
Nfld. 264 000
P.E.I. 510 000
N.S. 1 030 000
N.B. 1 300 000
*Non-resident only

Brackley Beach in Prince Edward Island National Park

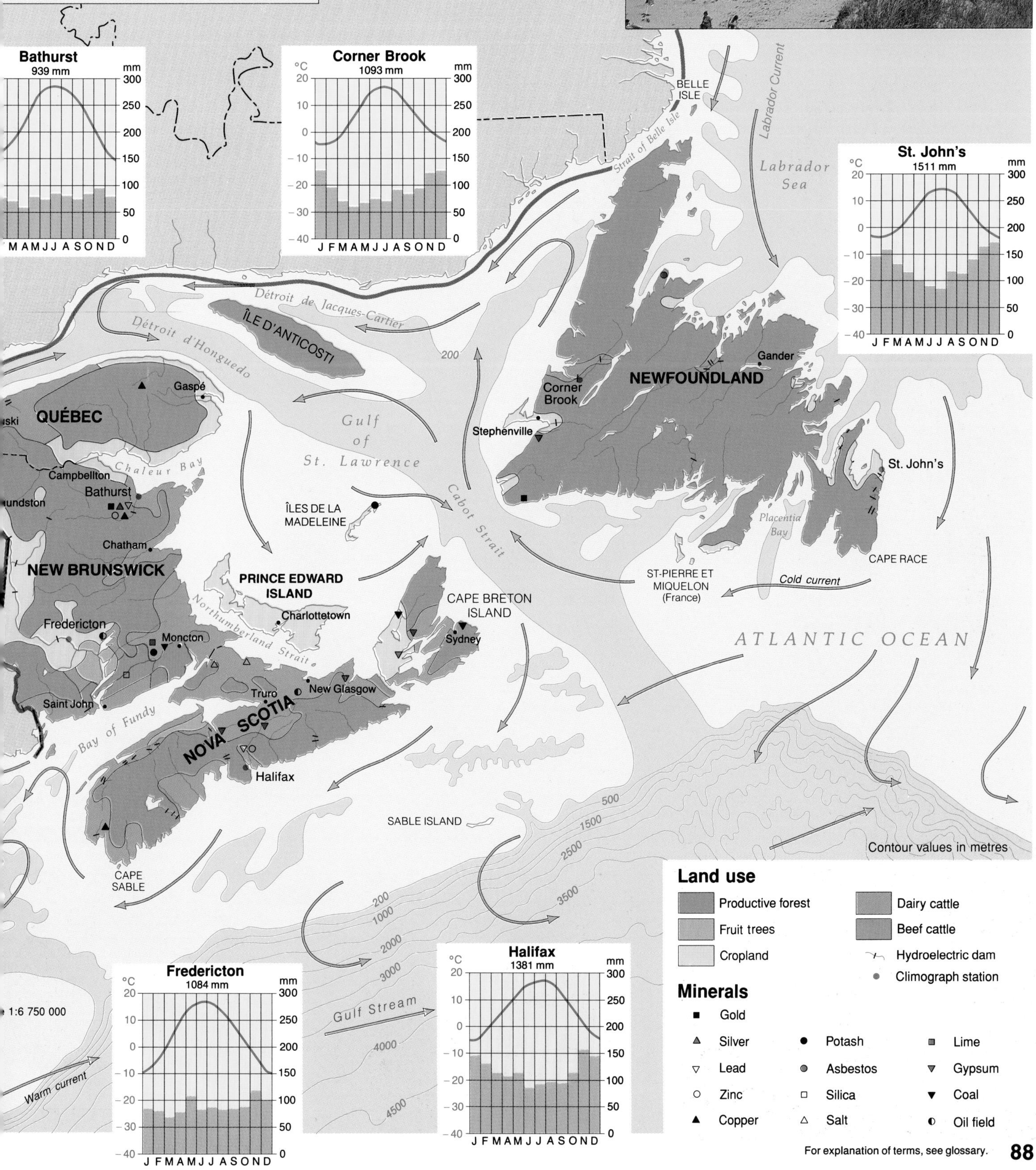
Bathurst
939 mm
Corner Brook
1093 mm
St. John's
1511 mm
Fredericton
1084 mm
Halifax
1381 mm
°C
mm
J F M A M J J A S O N D
BELLE ISLE
Strait of Belle Isle
Labrador Current
Labrador Sea
Détroit de Jacques-Cartier
ÎLE D'ANTICOSTI
Détroit d'Honguedo
Gaspé
QUÉBEC
Chaleur Bay
Campbellton
Bathurst
Chatham
NEW BRUNSWICK
Fredericton
Moncton
Saint John
Gulf of St. Lawrence
ÎLES DE LA MADELEINE
PRINCE EDWARD ISLAND
Charlottetown
Northumberland Strait
Cabot Strait
CAPE BRETON ISLAND
Sydney
Truro
New Glasgow
NOVA SCOTIA
Halifax
Bay of Fundy
CAPE SABLE
SABLE ISLAND
Corner Brook
Stephenville
NEWFOUNDLAND
Gander
St. John's
Placentia Bay
CAPE RACE
ST-PIERRE ET MIQUELON (France)
Cold current
ATLANTIC OCEAN
Gulf Stream
Warm current
1:6 750 000
Contour values in metres
Land use
Productive forest
Fruit trees
Cropland
Dairy cattle
Beef cattle
Hydroelectric dam
Climograph station
Minerals
Gold
Silver
Lead
Zinc
Copper
Potash
Asbestos
Silica
Salt
Lime
Gypsum
Coal
Oil field

For explanation of terms, see glossary.

WORLD-THEMATIC

Continental drift

According to the theory of continental drift, the continents used to be united in a supercontinent named Pangaea. About 200 million years ago, Pangaea started to break up and the shape of the present-day continents began to appear. The plates are still moving today. Sea floor spreading continues to widen the Atlantic and Indian Oceans. Australia is moving northwards and the Arabian Peninsula is slowly separating from northeastern Africa.

200 000 000 years ago

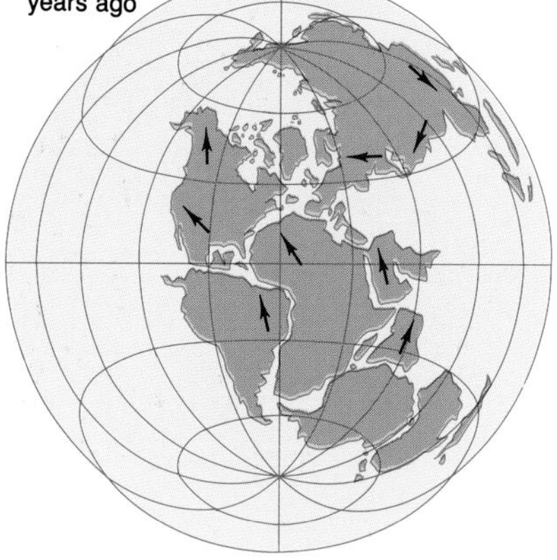

150 000 000 years ago

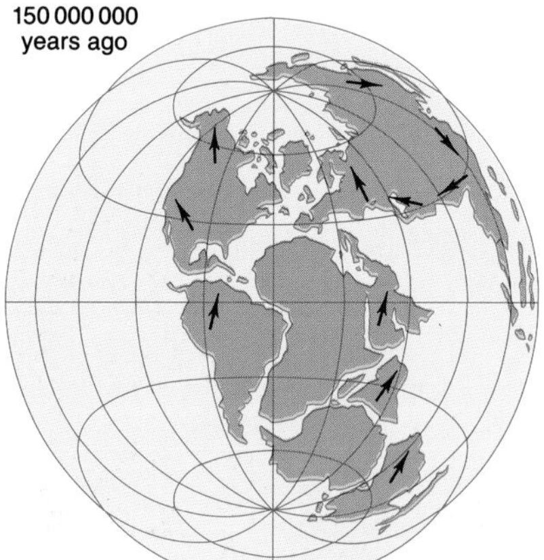

100 000 000 years ago

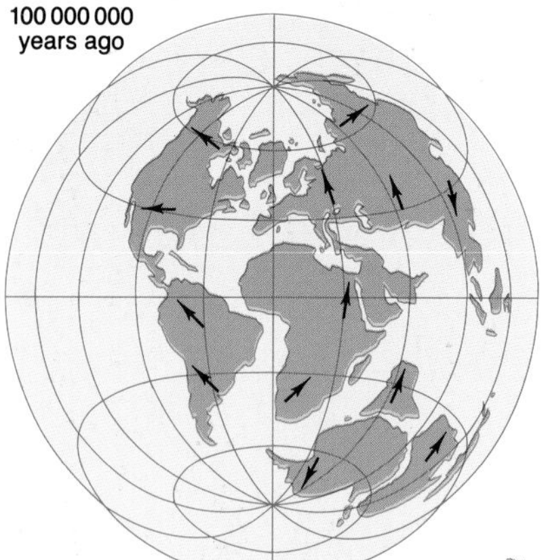

50 000 000 years ago

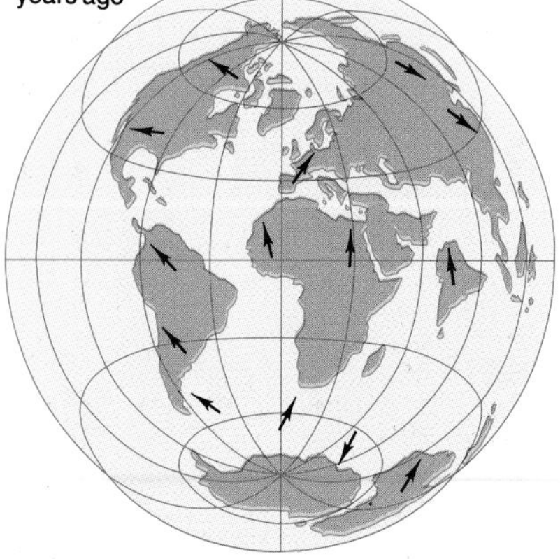

Plate tectonics

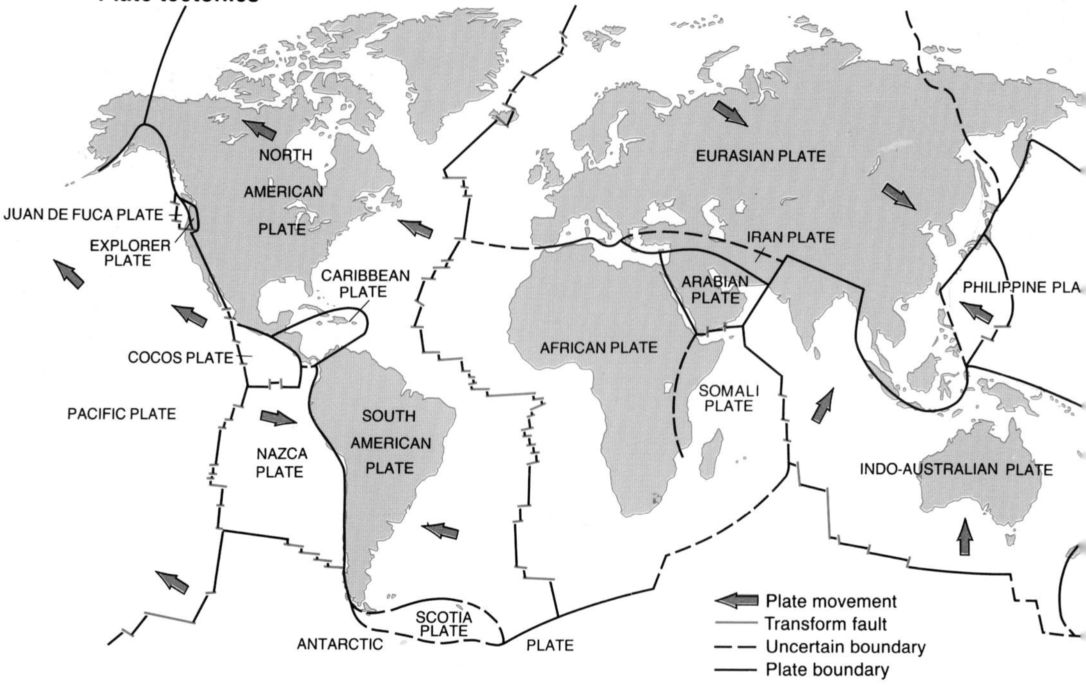

Earthquakes and volcanoes

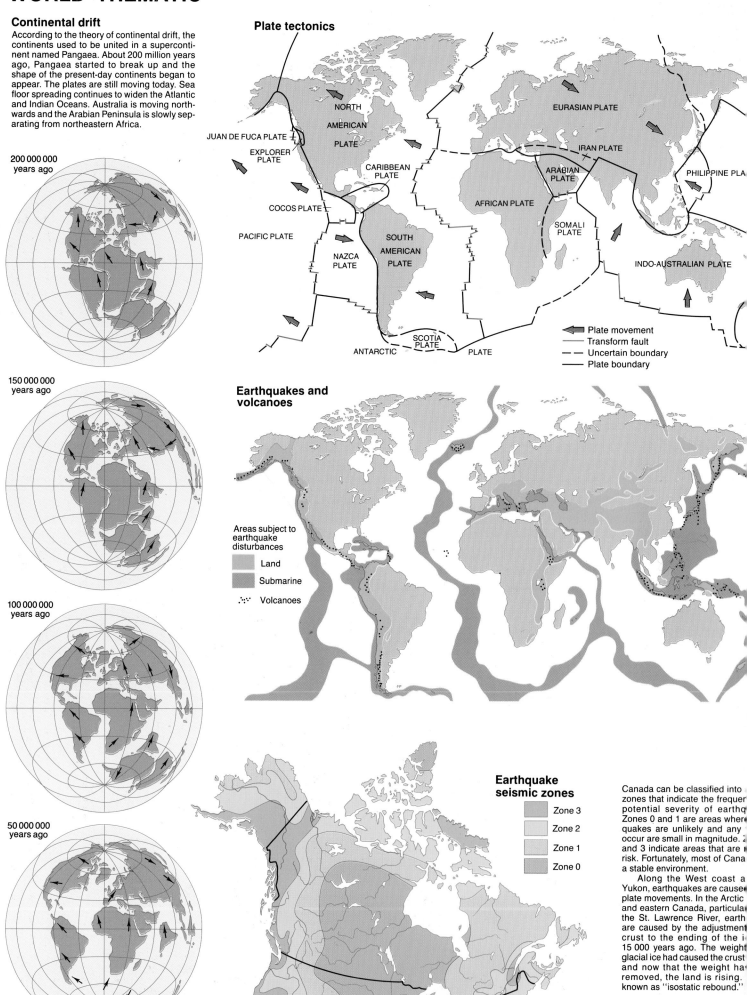

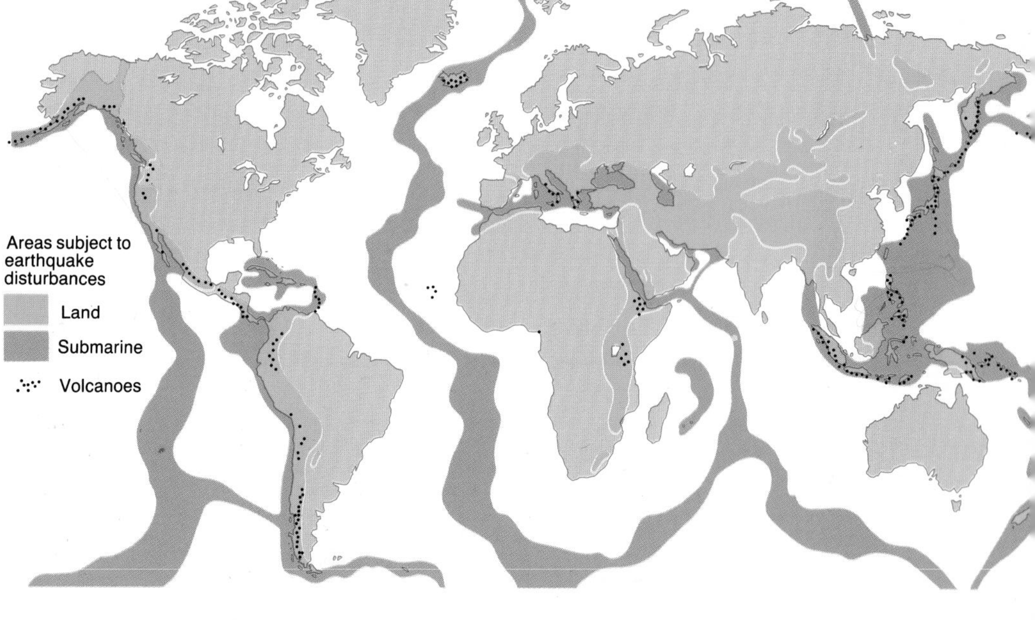

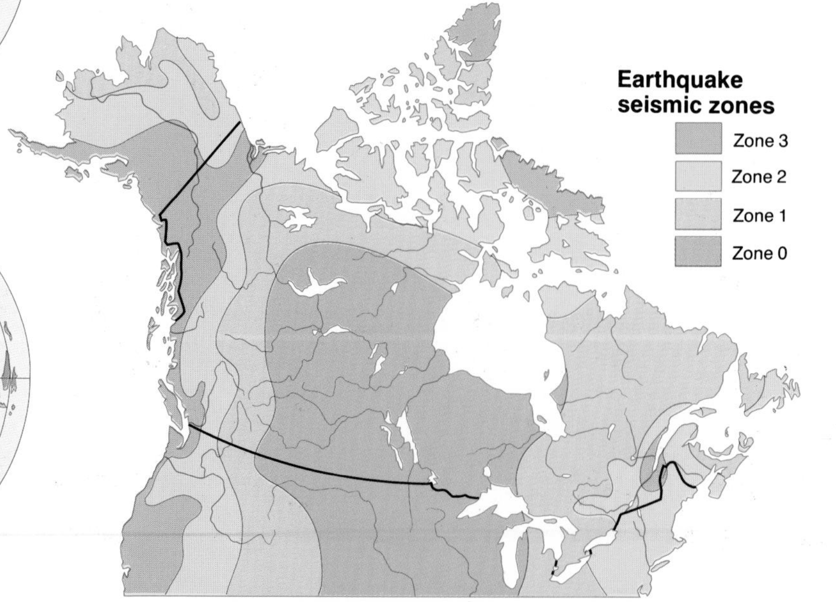

Canada can be classified into zones that indicate the frequer potential severity of earthq Zones 0 and 1 are areas wher quakes are unlikely and any occur are small in magnitude. and 3 indicate areas that are risk. Fortunately, most of Cana a stable environment.

Along the West coast a Yukon, earthquakes are cause plate movements. In the Arctic and eastern Canada, particula the St. Lawrence River, earth are caused by the adjustment crust to the ending of the i 15 000 years ago. The weight glacial ice had caused the crust and now that the weight ha removed, the land is rising. known as "isostatic rebound."

According to the latest theories of continental ·t, the earth's lithosphere or crust is made up of ge, rigid sections called plates. Convection cur- ıts flow in the asthenosphere and cause the tes above them to move. Along the boundaries ween plates, several activities may occur. A con- ental plate can override an oceanic plate, result- g in the formation of deep oceanic trenches, canoes or folded mountain ranges such as the des. Plates can also collide, causing mountains develop along the line of collision. Where plates parate, ocean ridges may develop, such as the d-Atlantic Ridge shown below. Plates may move rallel to one another, sliding past to create trans- m faults such as the San Andreas fault in Cali- nia. As plates move past each ıer, stresses often build up at te boundaries. Earthquakes cur when the crust shifts sud- nly and the energy is released.

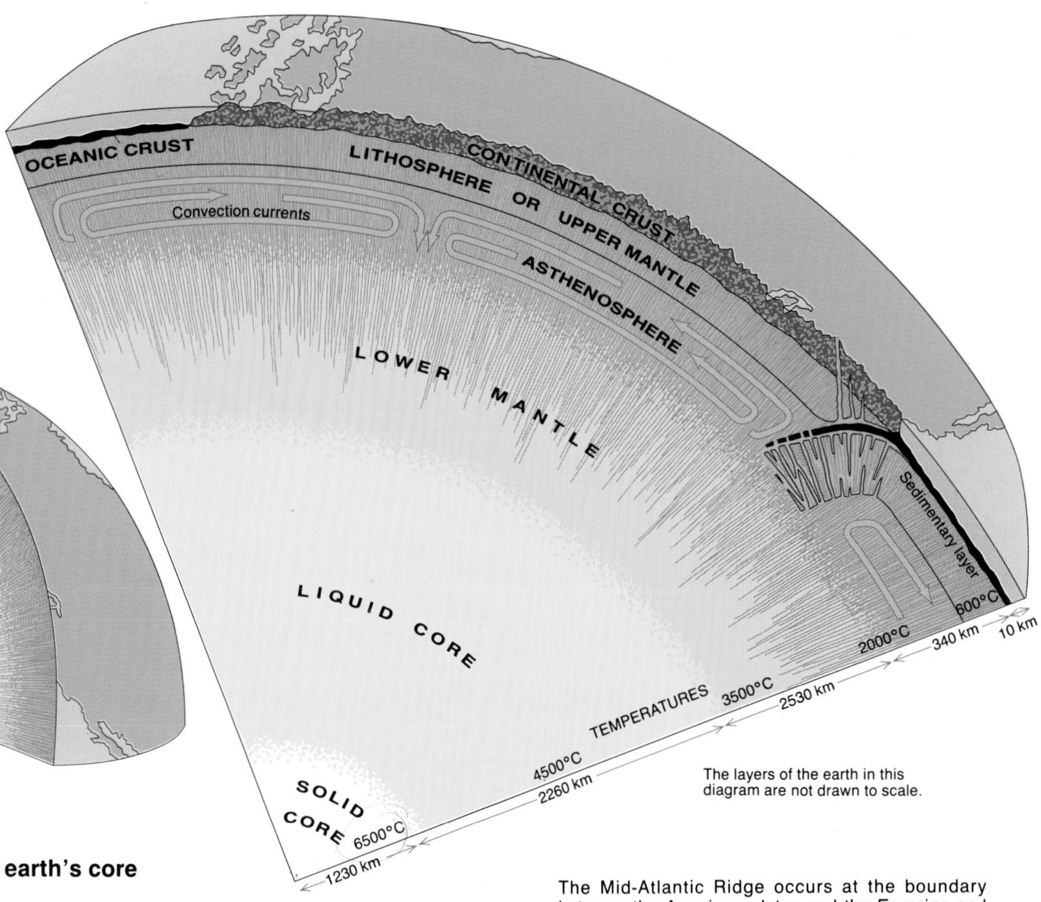

The layers of the earth in this diagram are not drawn to scale.

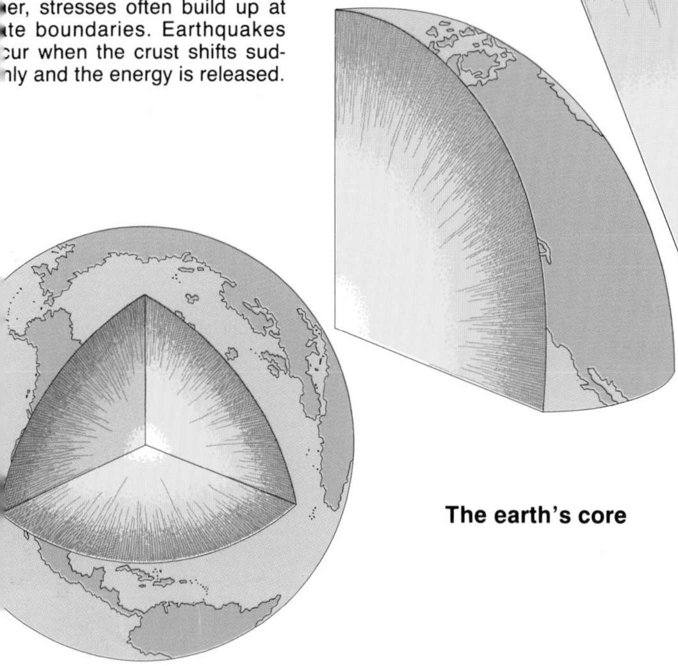

The earth's core

The Mid-Atlantic Ridge occurs at the boundary between the American plates and the Eurasian and African plates in the Atlantic Ocean. As the plates move away from each other, volcanic material rises up in the centre of the split and forms new sea floor as it cools. This is known as sea floor spreading. The rate of spreading can be determined by scientists using radioactive dating and magnetic patterns in the rock.

ʌtlantic Ocean floor

ıt centre of diagram is 1:26 600 000

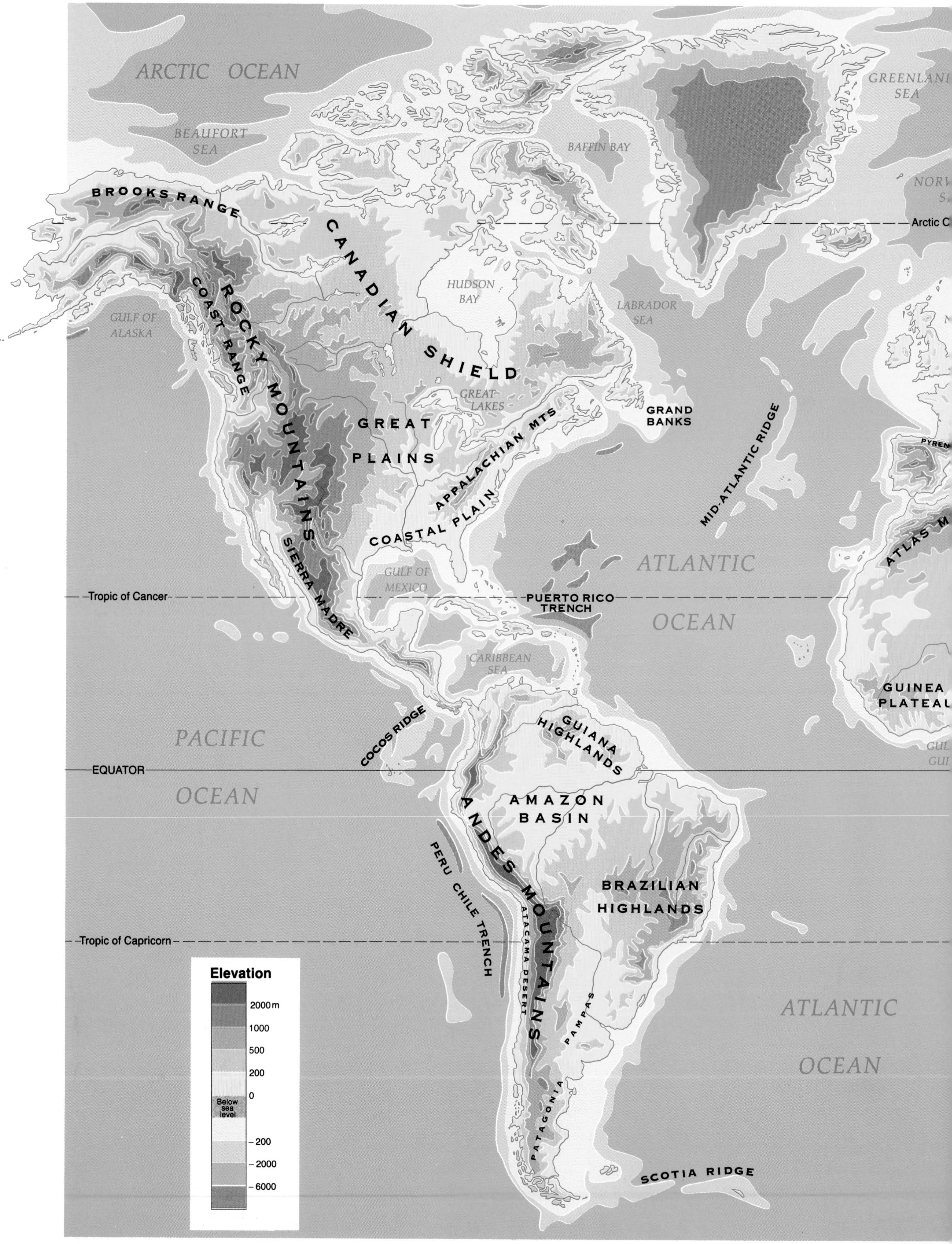
ARCTIC OCEAN
BEAUFORT SEA
BAFFIN BAY
GREENLAND SEA
NORW S
Arctic C
BROOKS RANGE
CANADIAN SHIELD
HUDSON BAY
LABRADOR SEA
GULF OF ALASKA
COAST RANGE
ROCKY MOUNTAINS
GREAT LAKES
GRAND BANKS
MID-ATLANTIC RIDGE
GREAT PLAINS
APPALACHIAN MTS
COASTAL PLAIN
PYREN
ATLAS M
SIERRA MADRE
GULF OF MEXICO
ATLANTIC
Tropic of Cancer
PUERTO RICO TRENCH
OCEAN
CARIBBEAN SEA
GUINEA PLATEAU
PACIFIC
COCOS RIDGE
GUIANA HIGHLANDS
GUL GUI
EQUATOR
OCEAN
AMAZON BASIN
ANDES MOUNTAINS
PERU CHILE TRENCH
BRAZILIAN HIGHLANDS
ATACAMA DESERT
Tropic of Capricorn
Elevation
2000 m
1000
500
200
0
Below sea level
-200
-2000
-6000
PAMPAS
ATLANTIC
OCEAN
PATAGONIA
SCOTIA RIDGE

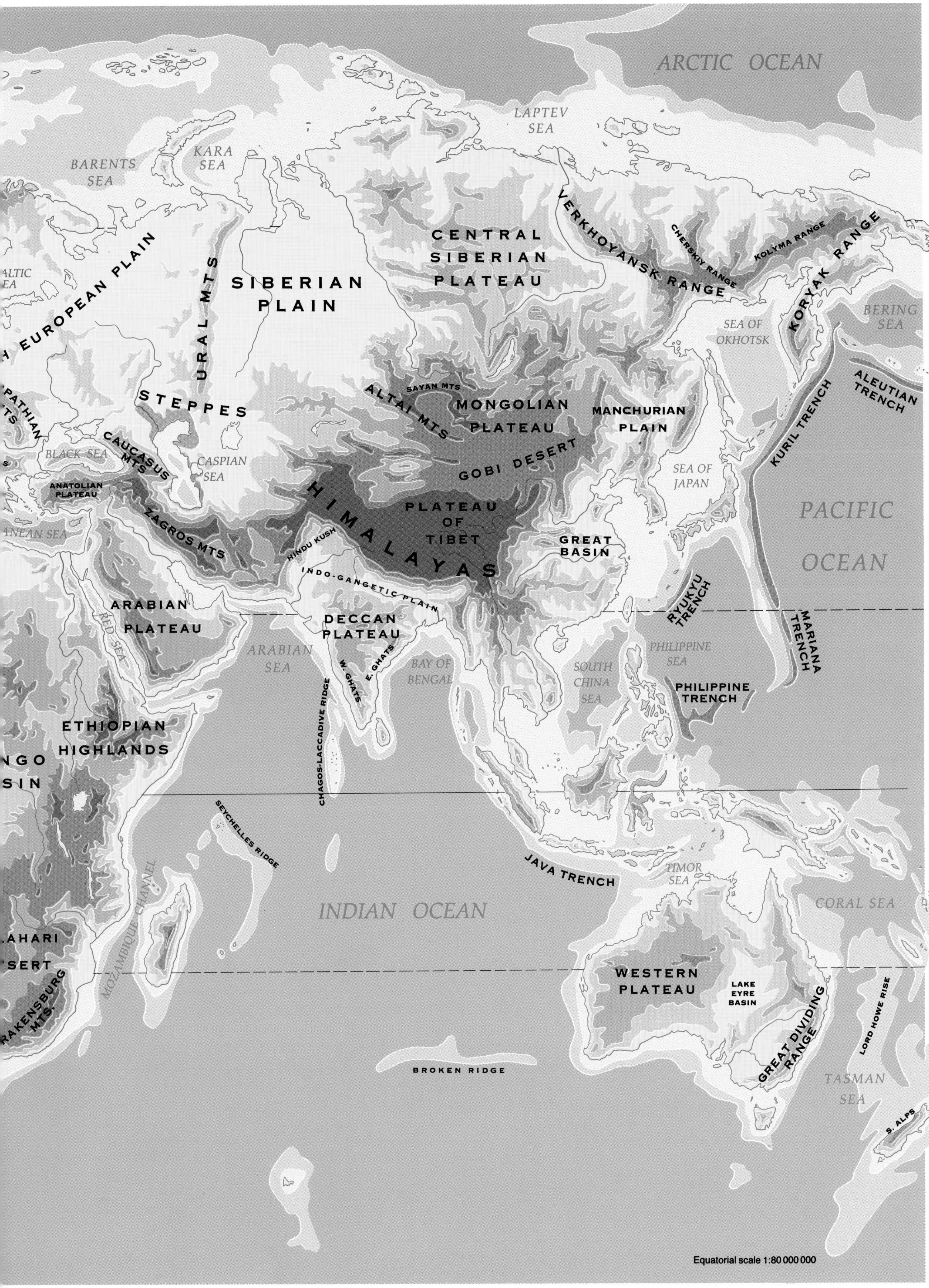
ARCTIC OCEAN
LAPTEV SEA
KARA SEA
BARENTS SEA
CENTRAL SIBERIAN PLATEAU
VERKHOYANSK RANGE
CHERSKIY RANGE
KOLYMA RANGE
KORYAK RANGE
SIBERIAN PLAIN
URAL MTS
EUROPEAN PLAIN
SEA OF OKHOTSK
BERING SEA
STEPPES
SAYAN MTS
ALTAI MTS
MONGOLIAN PLATEAU
MANCHURIAN PLAIN
ALEUTIAN TRENCH
KURIL TRENCH
CAUCASUS MTS
BLACK SEA
CASPIAN SEA
GOBI DESERT
SEA OF JAPAN
ANATOLIAN PLATEAU
HIMALAYAS
PLATEAU OF TIBET
PACIFIC OCEAN
ZAGROS MTS
HINDU KUSH
GREAT BASIN
INDO-GANGETIC PLAIN
RYUKYU TRENCH
ARABIAN PLATEAU
DECCAN PLATEAU
MARIANA TRENCH
RED SEA
ARABIAN SEA
E. GHATS
W. GHATS
BAY OF BENGAL
SOUTH CHINA SEA
PHILIPPINE SEA
PHILIPPINE TRENCH
ETHIOPIAN HIGHLANDS
CHAGOS-LACCADIVE RIDGE
SEYCHELLES RIDGE
JAVA TRENCH
TIMOR SEA
INDIAN OCEAN
CORAL SEA
MOZAMBIQUE CHANNEL
WESTERN PLATEAU
LAKE EYRE BASIN
GREAT DIVIDING RANGE
LORD HOWE RISE
TASMAN SEA
BROKEN RIDGE
S. ALPS
Equatorial scale 1:80 000 000

January temperature

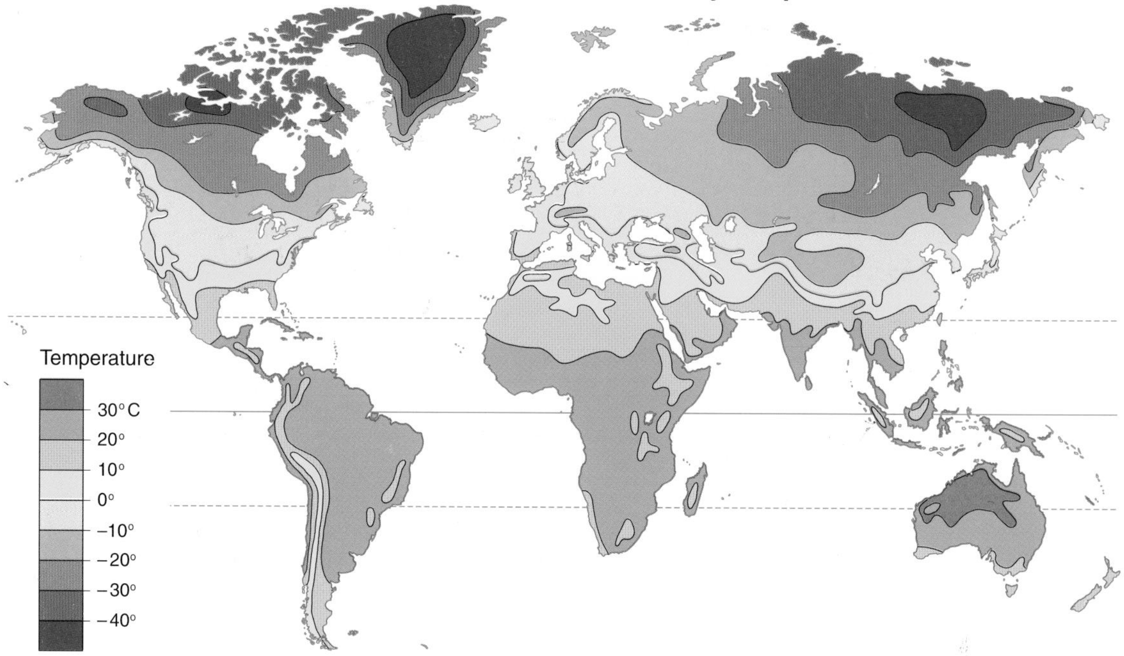

Angle of the sun's rays according to latitude

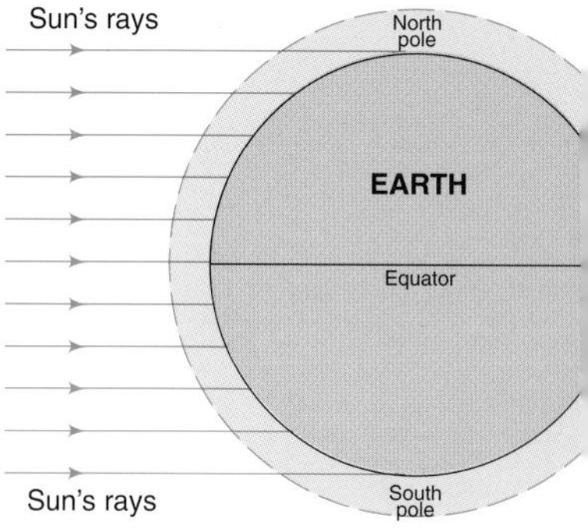

January precipitation

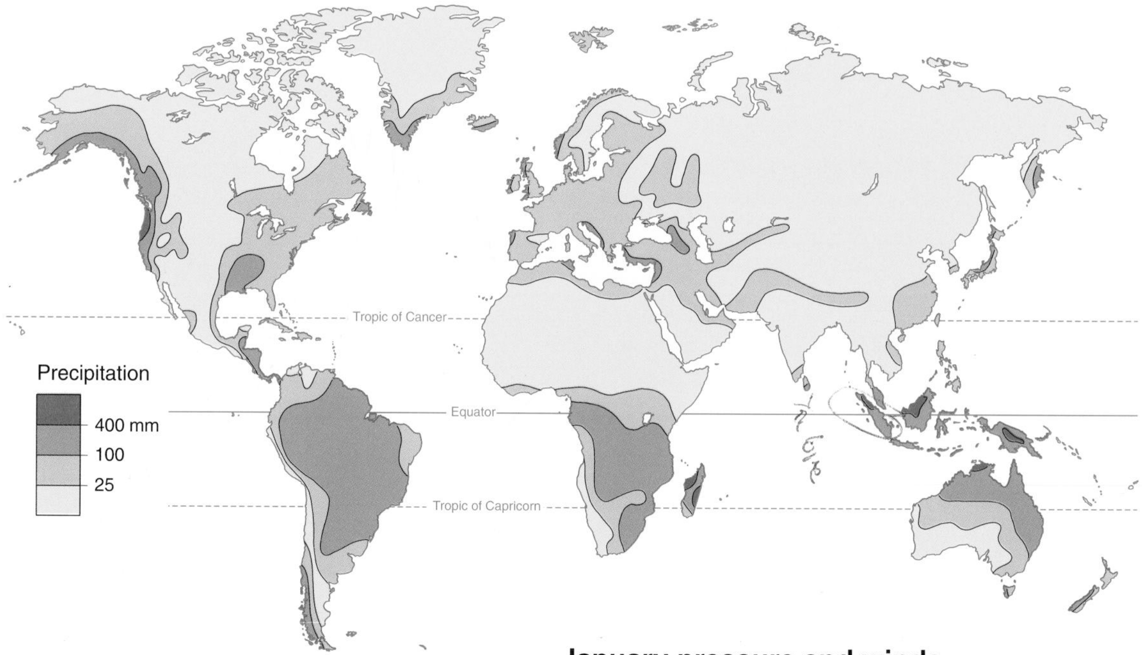

Relief precipitation, which is often foun
coastline ranges, is caused by moist, warn
moving over mountains. The mountains ca
the air to rise, and, as it rises, it expands
cools. Because cooler air cannot hold as m
water vapour, the water begins to conde
creating clouds and precipitation. After pas
over the mountains, the air descends
grows warmer. The water vapour in the clo
evaporates and the precipitation stops.
warm, dry wind that blows down the mount
is called a chinook in Canada and a föh
Europe. A classic example of this patter
precipitation occurs along the west coast
North and South America.

January pressure and winds

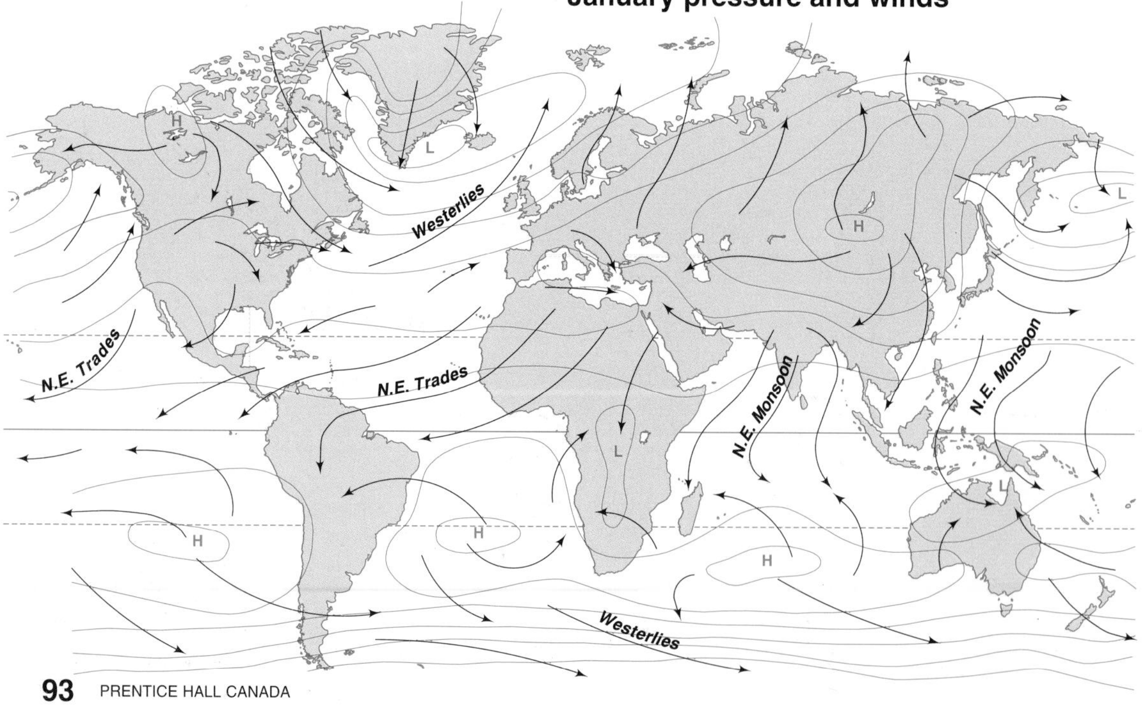

Effect of elevation on temperature

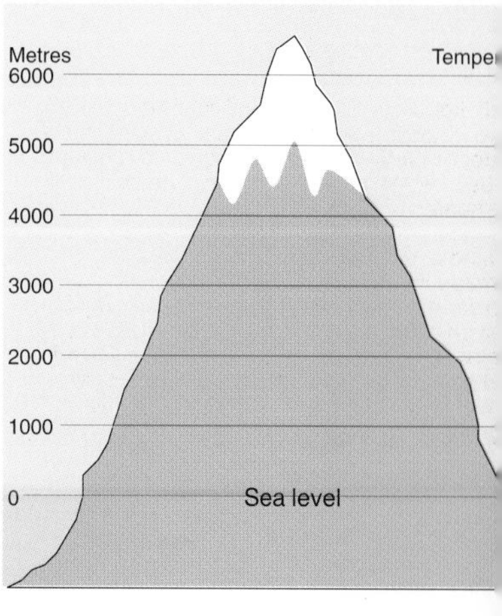

CLIMATE

July temperature

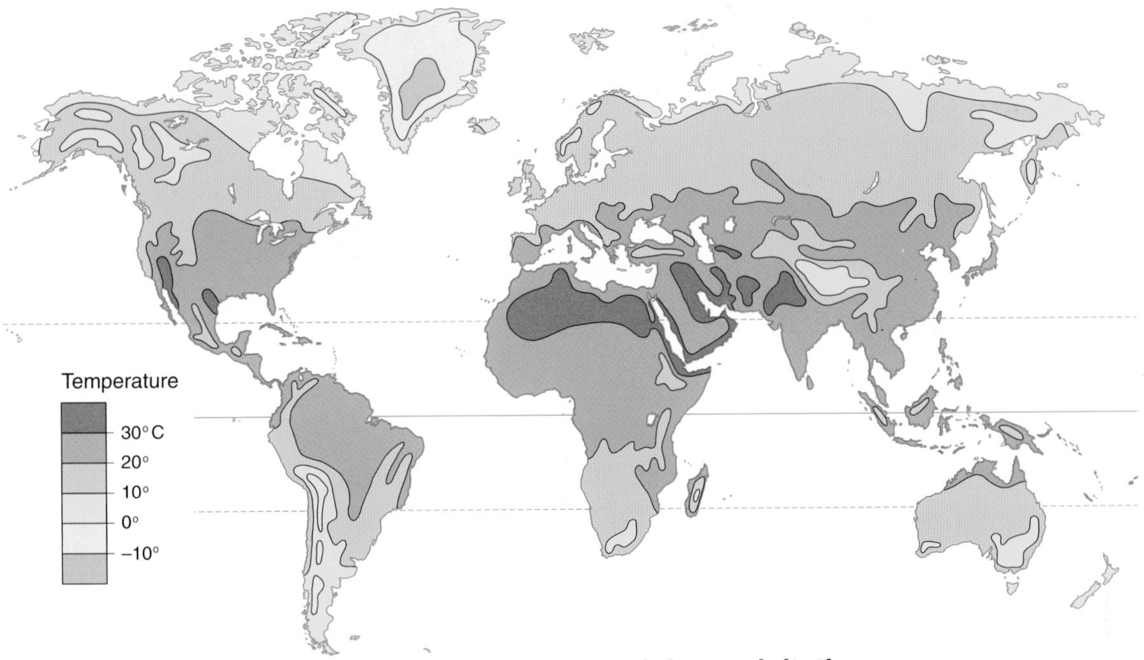

sun's energy creates the earth's climate erns. Different locations receive varying ounts of solar radiation, depending on their ude and the season. These variations se vastly different climates. At low latitudes se to the equator), the noontime sun is ost directly overhead year round, and solar ation is at maximum strength. At higher udes in the winter, the sun's rays strike the th at an angle, spreading the solar radiation r a larger area. As the rays must also travel eater distance through the atmosphere, the ation is more scattered and less intense. s means that less energy per unit area of the th's surface can be turned into heat.

July precipitation

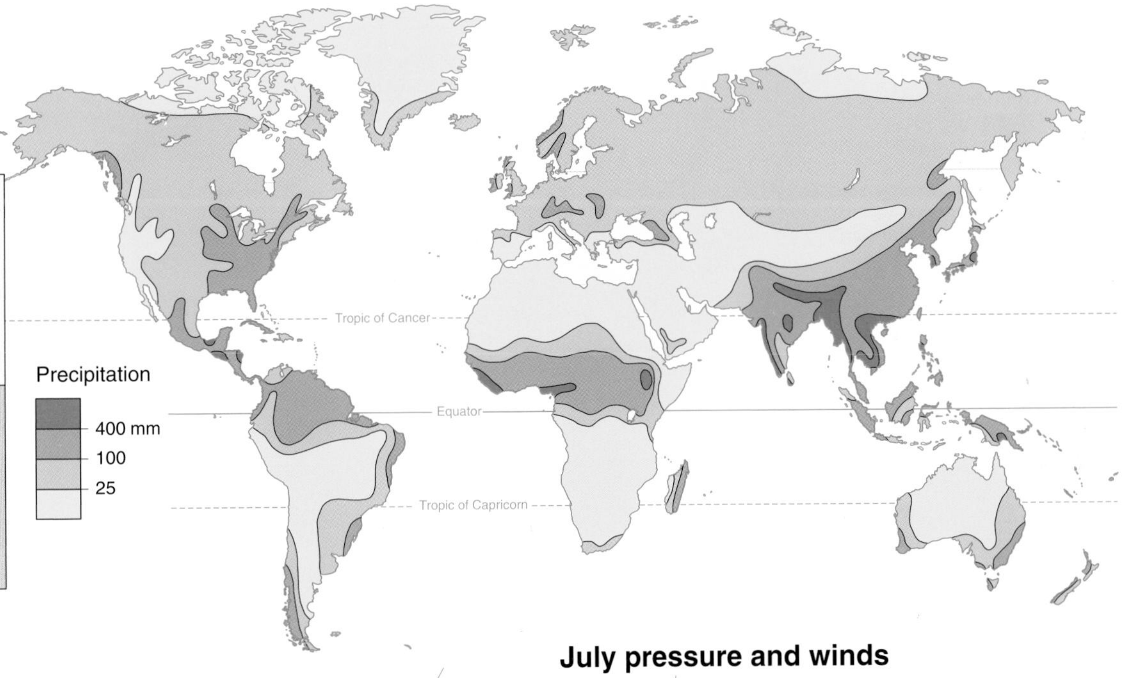

ffect of mountains on precipitation

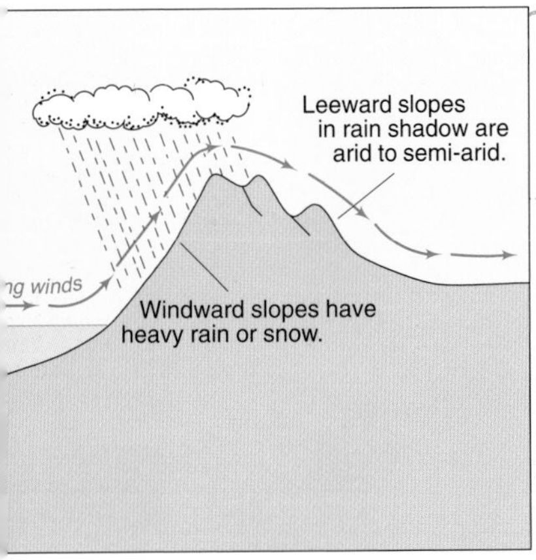

July pressure and winds

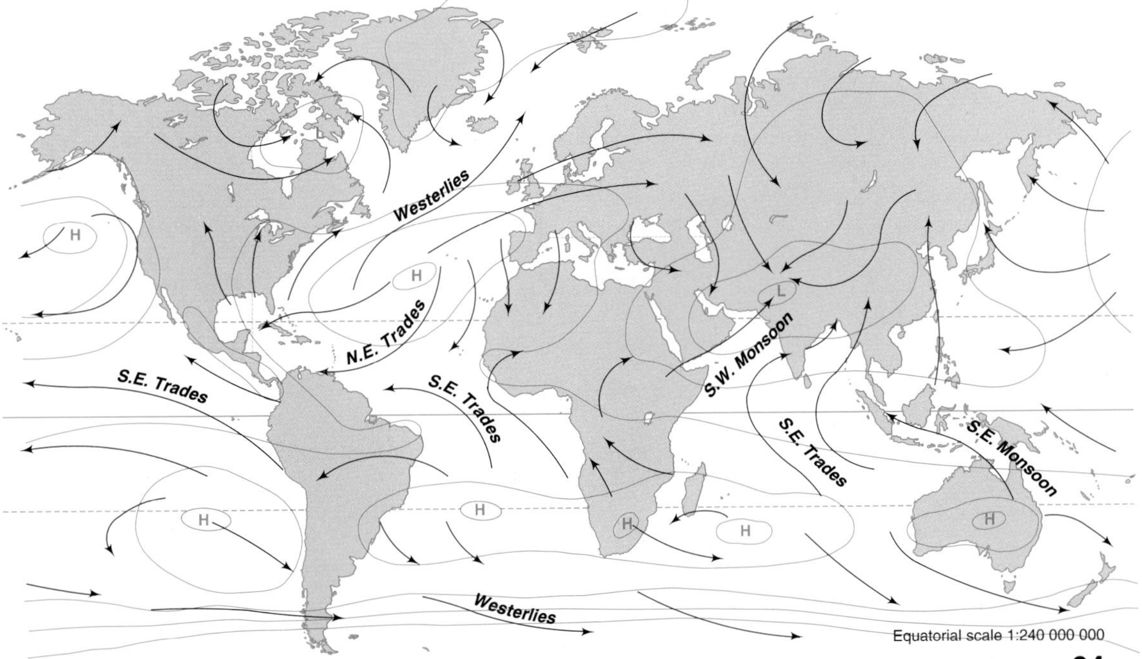

diation from the sun is converted into long ve radiation when it meets a solid mass ater or land). When this long wave radiation absorbed by the atmosphere, it heats the air. e heated air rises and then expands at higher itudes, where air pressure is lower. As it pands, the air loses heat. Therefore, an crease in altitude will normally result in a crease in temperature. For example, the nyan cities of Nairobi and Mombasa are at nilar latitudes, but different altitudes; Nairobi almost 2000 m higher. As a result, the mean nual temperature in Nairobi is 6 Celsius grees lower than in Mombasa.

For explanation of terms, see glossary.

WORLD-THEMATIC

Climate zones

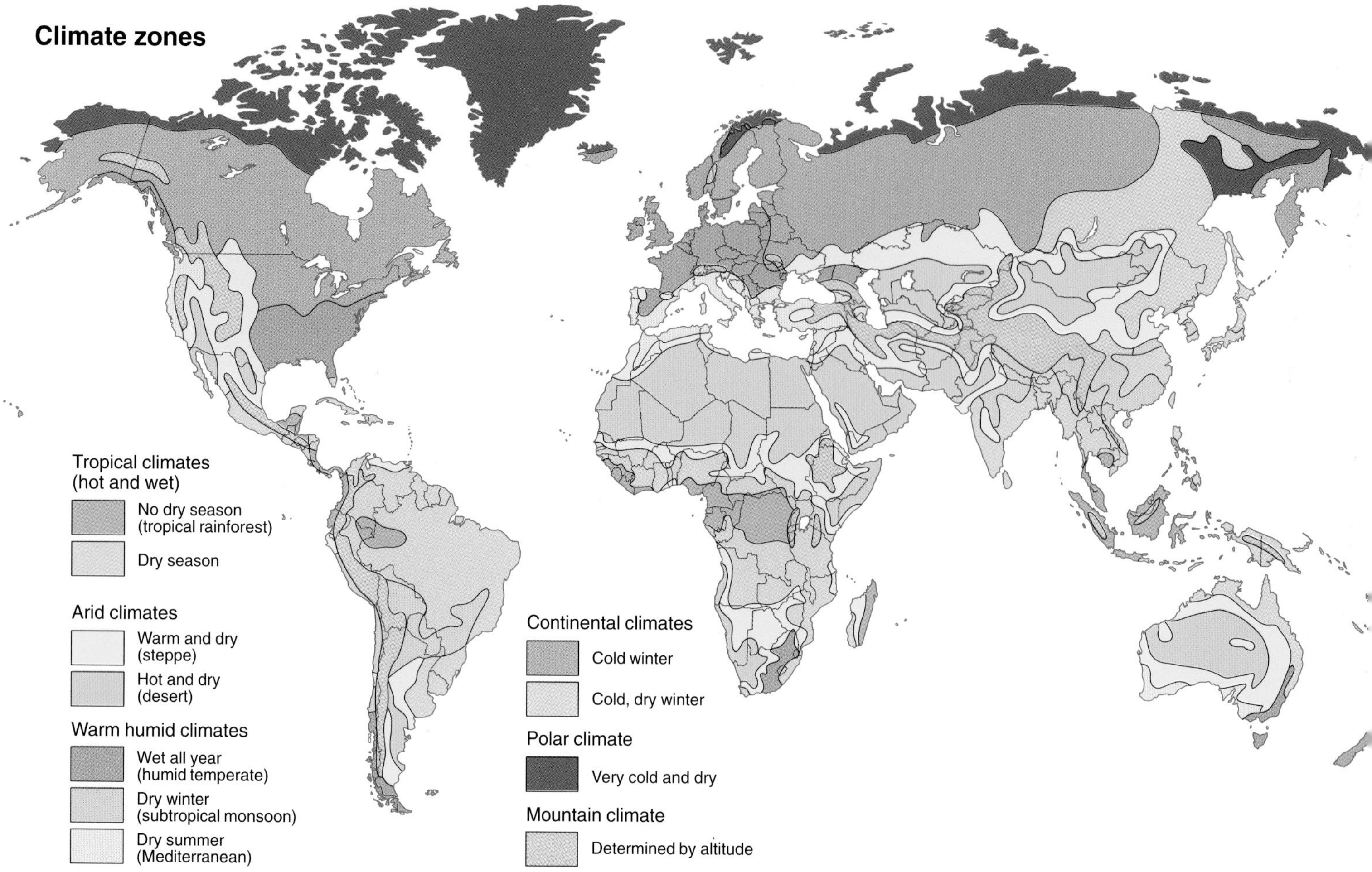
Tropical climates
(hot and wet)
No dry season
(tropical rainforest)
Dry season
Arid climates
Warm and dry
(steppe)
Hot and dry
(desert)
Warm humid climates
Wet all year
(humid temperate)
Dry winter
(subtropical monsoon)
Dry summer
(Mediterranean)
Continental climates
Cold winter
Cold, dry winter
Polar climate
Very cold and dry
Mountain climate
Determined by altitude

Soils

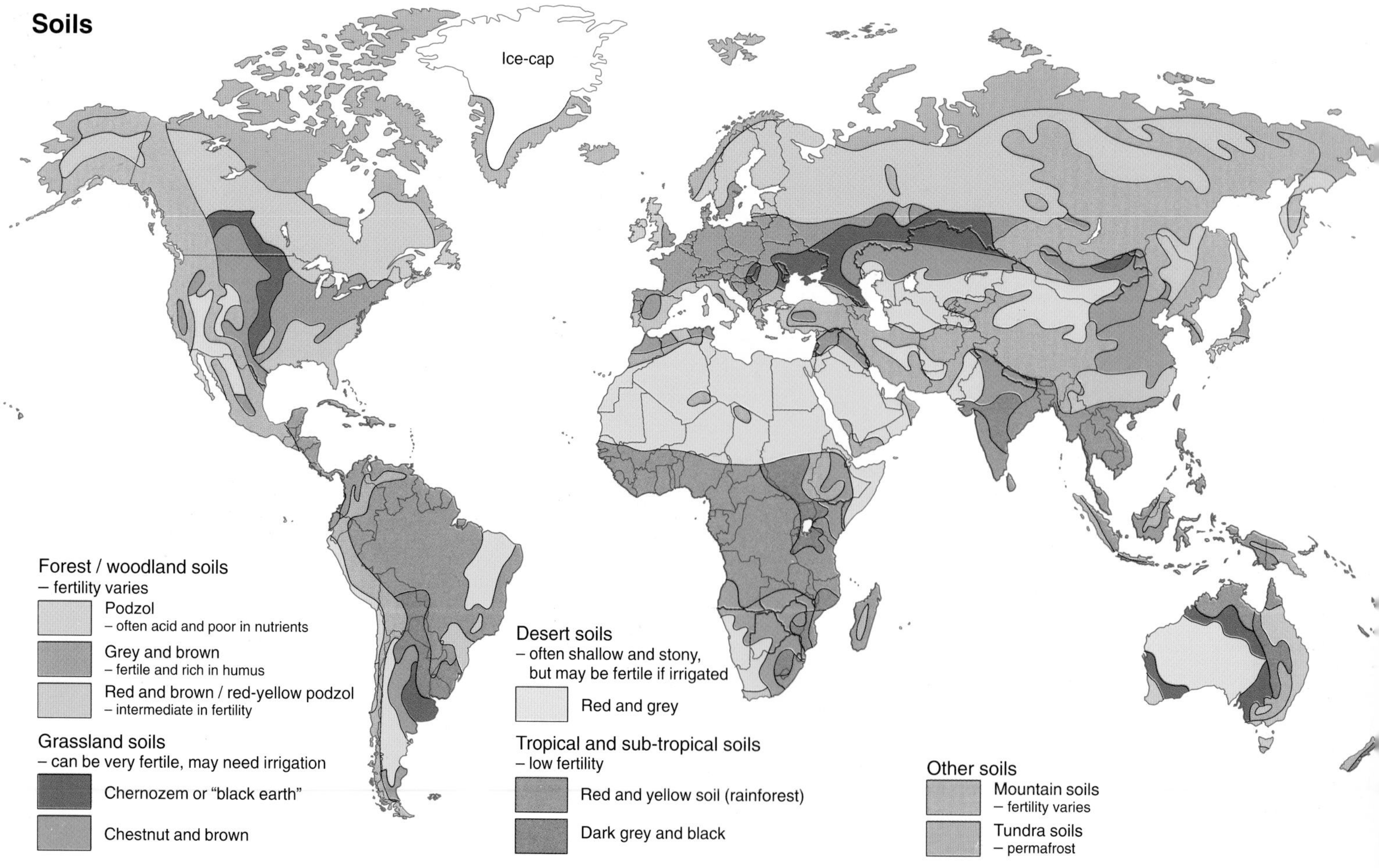
Ice-cap
Forest / woodland soils
– fertility varies
Podzol
– often acid and poor in nutrients
Grey and brown
– fertile and rich in humus
Red and brown / red-yellow podzol
– intermediate in fertility
Grassland soils
– can be very fertile, may need irrigation
Chernozem or "black earth"
Chestnut and brown
Desert soils
– often shallow and stony,
but may be fertile if irrigated
Red and grey
Tropical and sub-tropical soils
– low fertility
Red and yellow soil (rainforest)
Dark grey and black
Other soils
Mountain soils
– fertility varies
Tundra soils
– permafrost

ıter deficiency d surplus

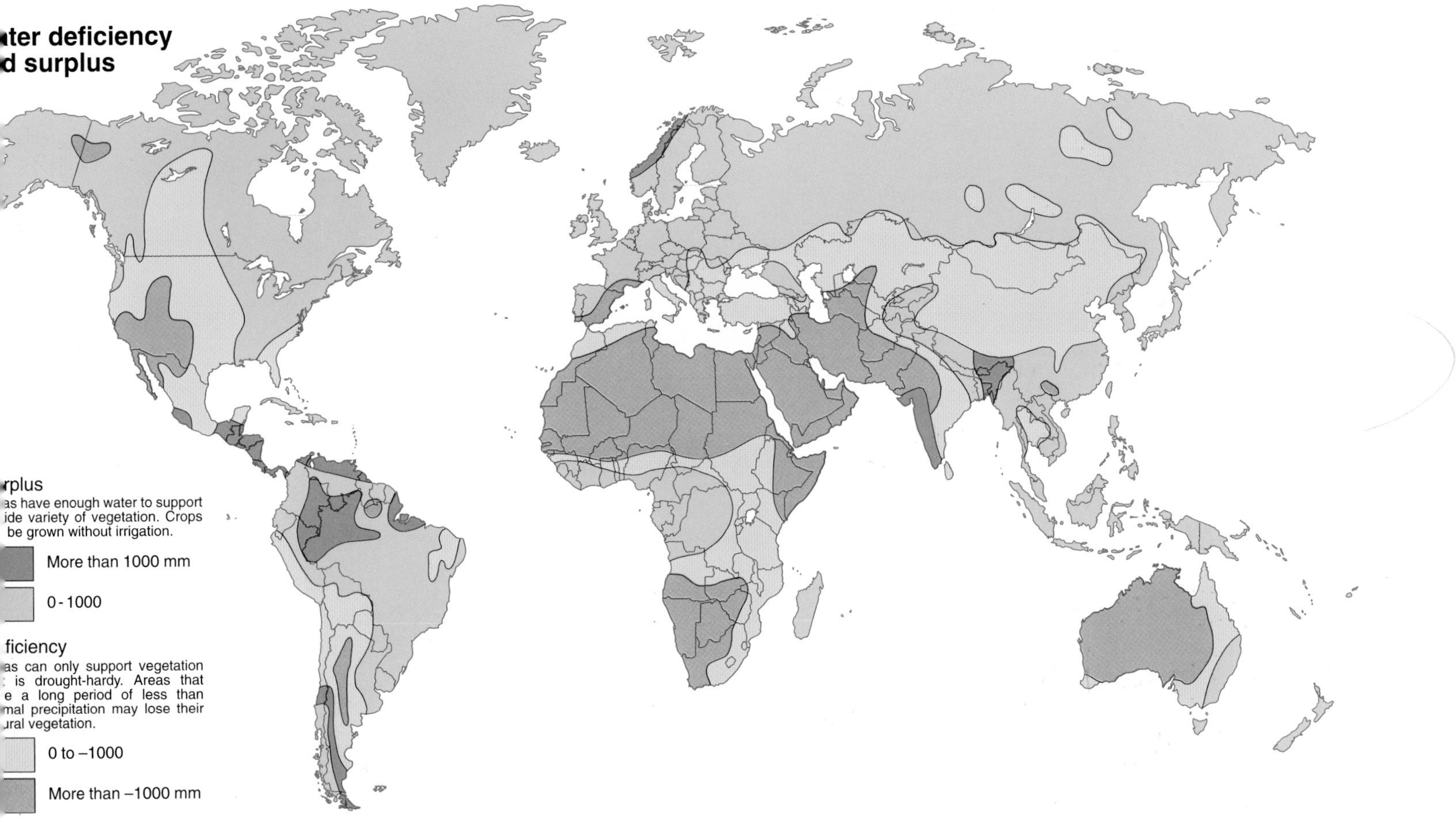

·getation

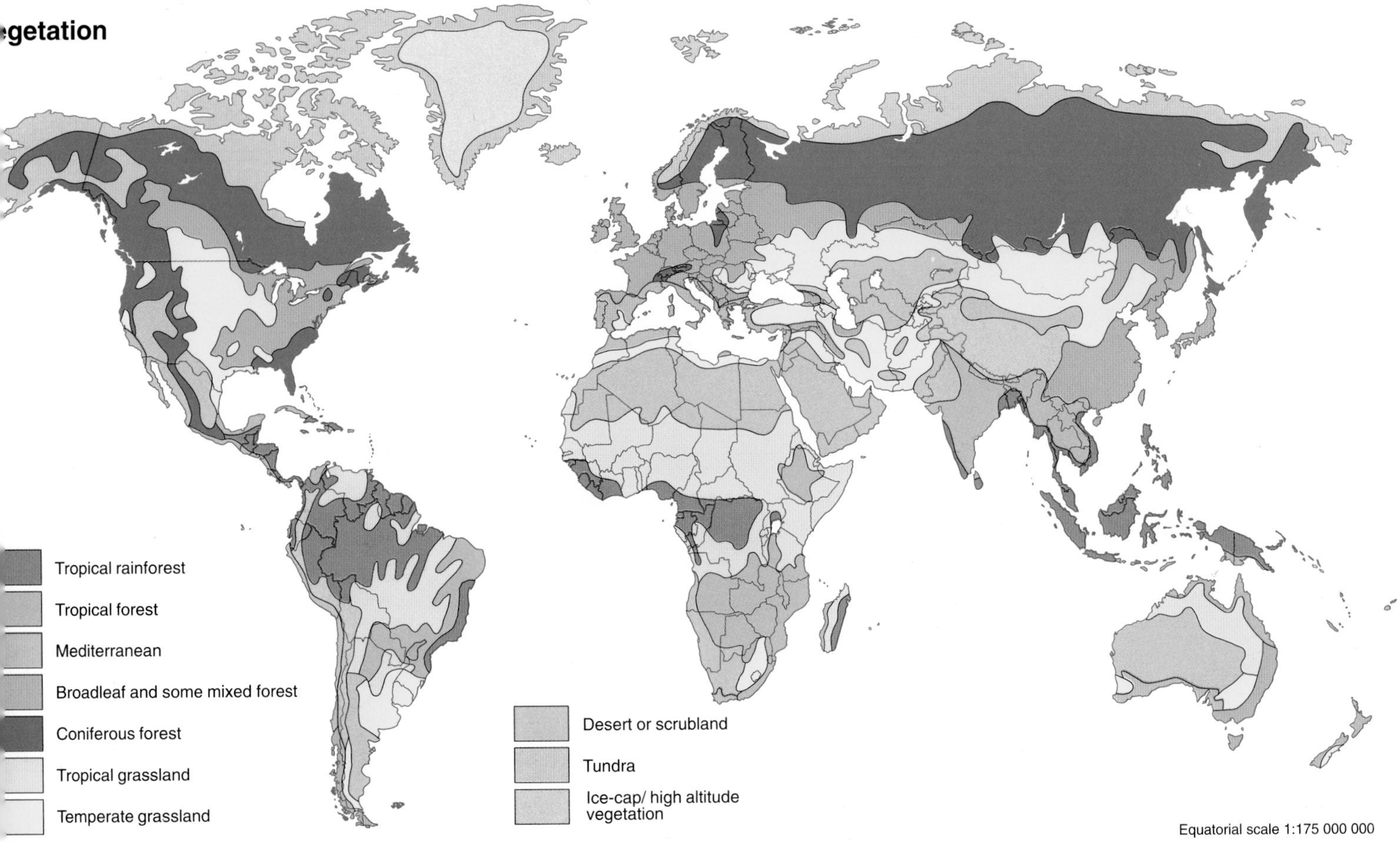

Equatorial scale 1:175 000 000

WORLD-THEMATIC

EU (European Union)

Belgium	Italy
Denmark	Luxembourg
France	Netherlands
Germany	Portugal
Greece	Spain
Ireland	United Kingdom

O.P.E.C. (Organization of Petroleum Exporting Countries)

Algeria	Libya
Ecuador	Nigeria
Gabon	Qatar
Indonesia	Saudi Arabia
Iran	United Arab Emirates
Iraq	Venezuela
Kuwait	

United Nations membership

Year	Number of countries
1945	50
1950	54
1960	91
1970	116
1980	140
1990	159
1993	184

Type of government

Monarchy
A ruling individual or family can remove the government; the position of leader is inherited.

Single party
One political party dominates and has so much power that no other party has an opportunity to be elected. Opposition parties may be banned completely in some countries.

Democracy
Countries in which it is possible to elect alternative parties and where all adult citizens can participate in the process of electing a government.

Conditional democracy
A democratic government where there is political repression, censorship, or civil or guerilla warfare.

Military rule
The military runs the country; power was seized and the people do not have the opportunity to elect a government.

Central Europe

Scale 1:58 40

ASEAN (Association of Southeast Asian Nations)

Brunei
Indonesia
Malaysia
Philippines
Singapore
Thailand

Commonwealth

Antigua and Barbuda
Australia
Bahamas
Bangladesh
Barbados
Belize
Botswana
Brunei
Canada
Cyprus
Dominica
Fiji
Gambia
Ghana
Grenada
Guyana
India
Jamaica
Kenya
Kiribati
Lesotho
Malawi
Malaysia
Maldives
Malta
Mauritius
Namibia
Nauru
New Zealand
Nigeria
Pakistan
Papua New Guinea
St. Kitts and Nevis
St. Lucia
St. Vincent and the Grenadines
Seychelles
Sierra Leone
Singapore
Solomon Islands
Sri Lanka
Swaziland
Tanzania
Tonga
Trinidad and Tobago
Tuvalu
Uganda
United Kingdom
Vanuatu
Western Samoa
Zambia
Zimbabwe

For explanation of terms, see glossary.

WORLD-THEMATIC

Birth rate

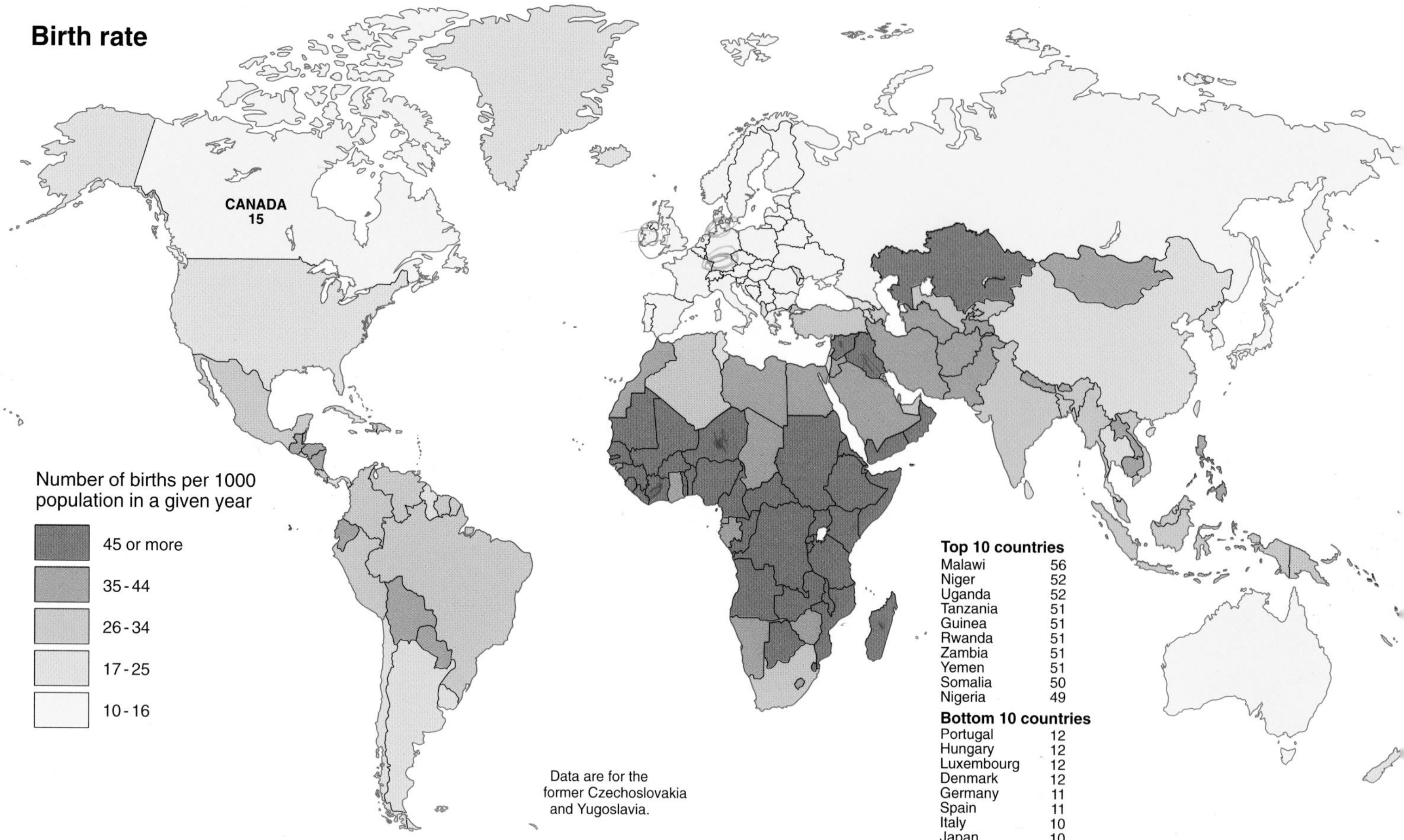

Top 10 countries

Malawi	56
Niger	52
Uganda	52
Tanzania	51
Guinea	51
Rwanda	51
Zambia	51
Yemen	51
Somalia	50
Nigeria	49

Bottom 10 countries

Portugal	12
Hungary	12
Luxembourg	12
Denmark	12
Germany	11
Spain	11
Italy	10
Japan	10
Greece	10
San Marino	10

Population density (total area)

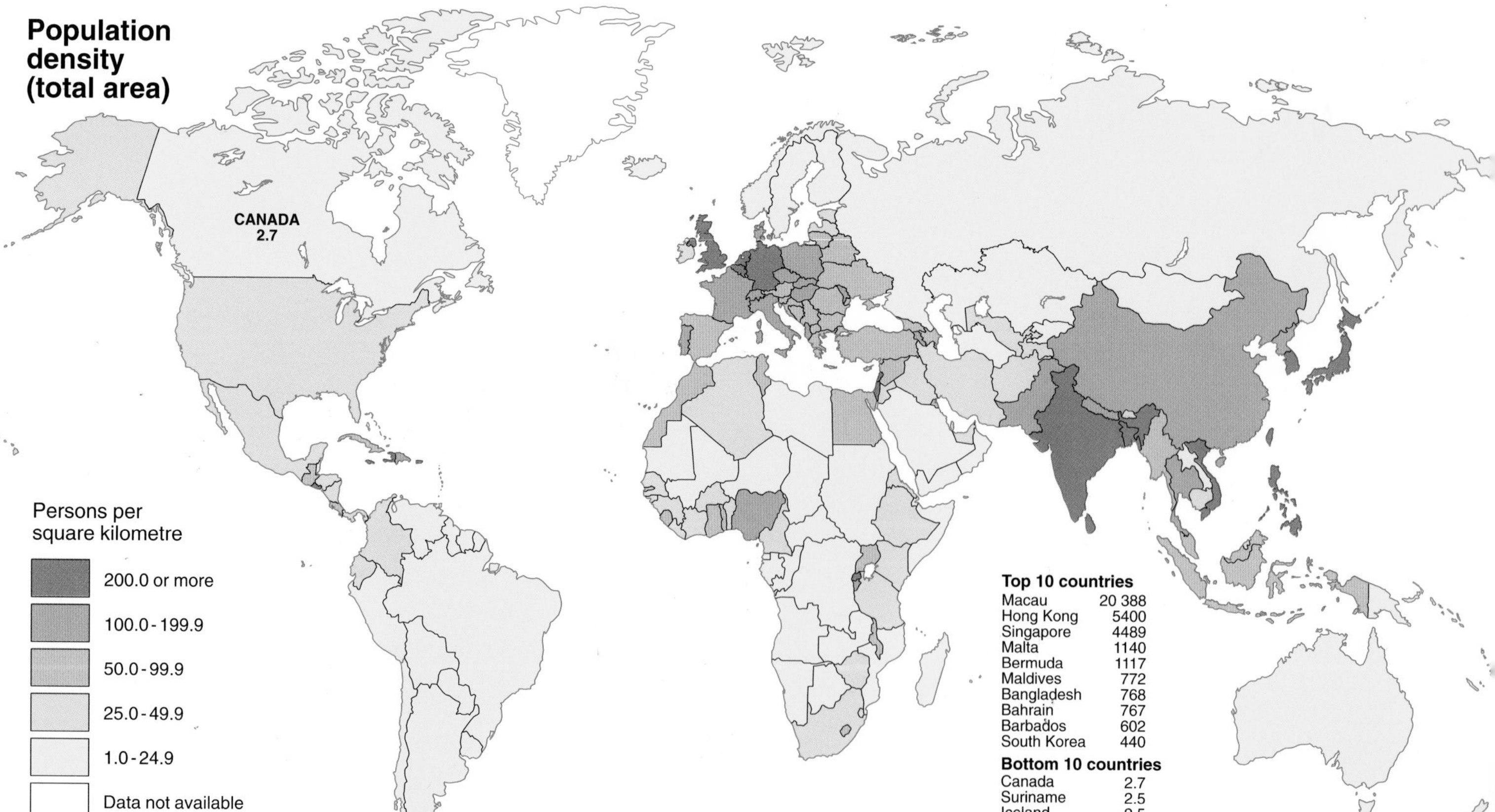

Top 10 countries

Macau	20 388
Hong Kong	5400
Singapore	4489
Malta	1140
Bermuda	1117
Maldives	772
Bangladesh	768
Bahrain	767
Barbados	602
South Korea	440

Bottom 10 countries

Canada	2.7
Suriname	2.5
Iceland	2.5
Libya	2.5
Australia	2.3
Botswana	2.3
Mauritania	2.0
Namibia	1.7
Mongolia	1.4
French Guiana	1.4

pulation owth rate

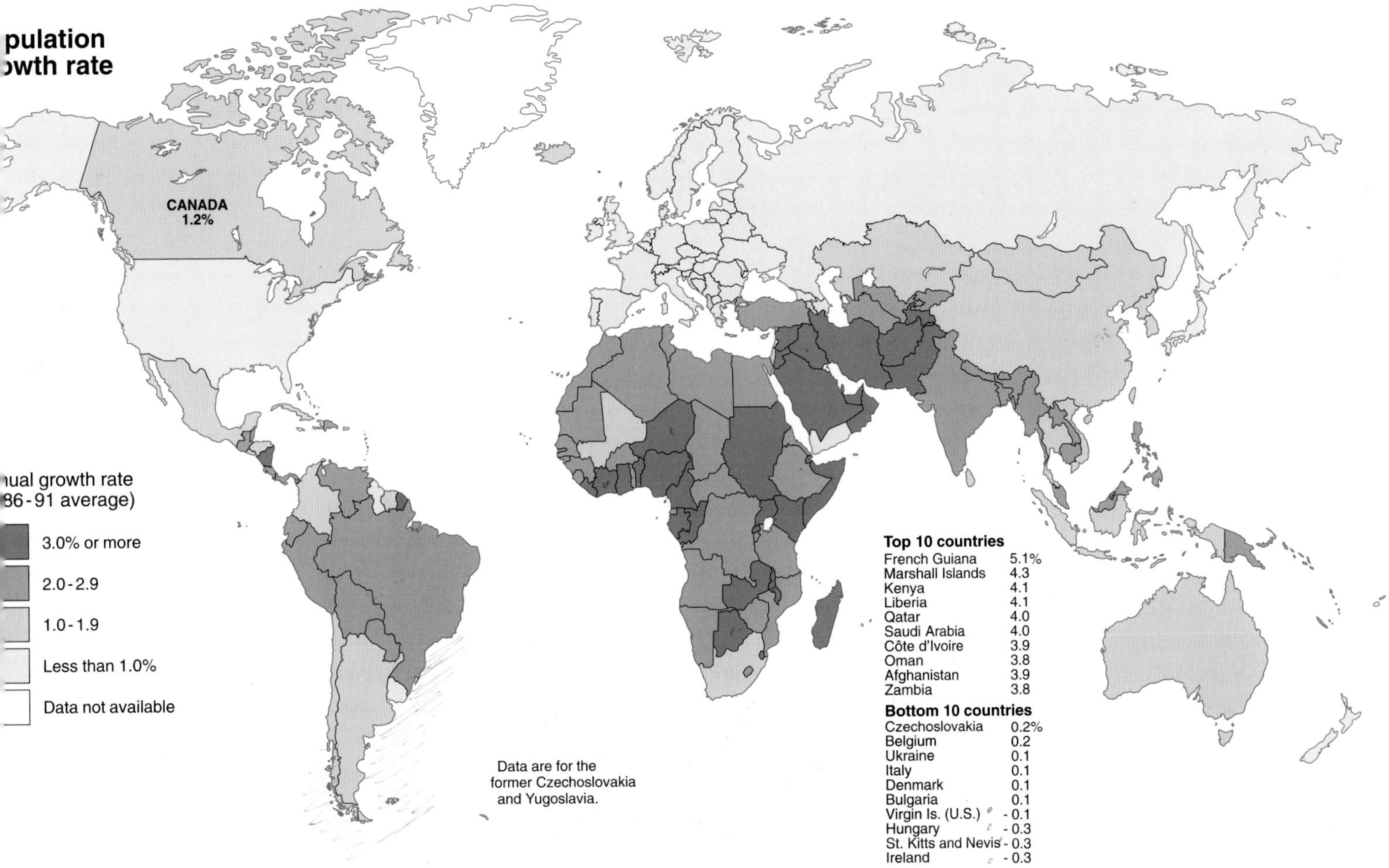

opulation stribution

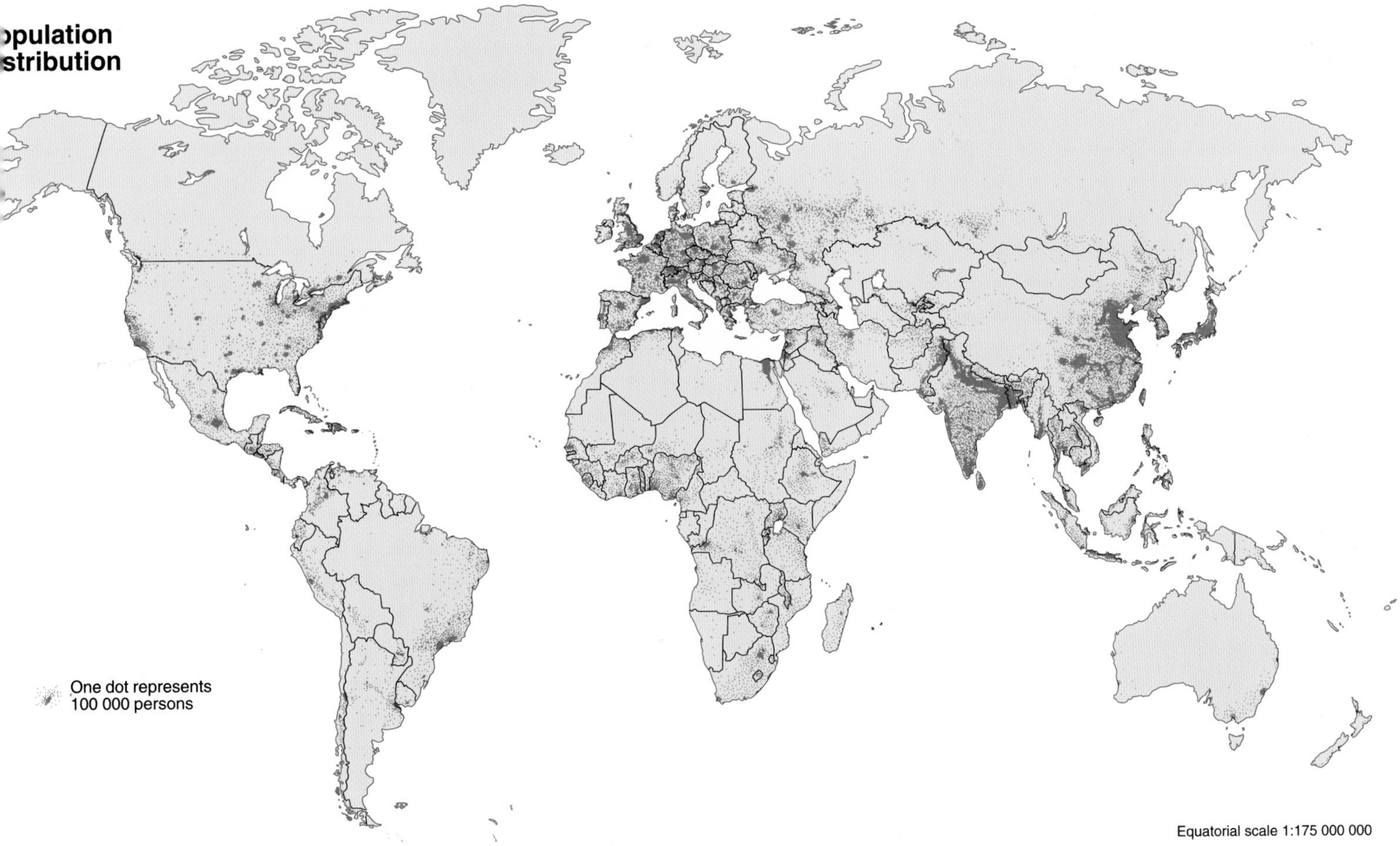

Equatorial scale 1:175 000 000

Top 10 countries
French Guiana	5.1%
Marshall Islands	4.3
Kenya	4.1
Liberia	4.1
Qatar	4.0
Saudi Arabia	4.0
Côte d'Ivoire	3.9
Oman	3.8
Afghanistan	3.9
Zambia	3.8
Bottom 10 countries
Czechoslovakia	0.2%
Belgium	0.2
Ukraine	0.1
Italy	0.1
Denmark	0.1
Bulgaria	0.1
Virgin Is. (U.S.)	- 0.1
Hungary	- 0.3
St. Kitts and Nevis	- 0.3
Ireland	- 0.3

WORLD-THEMATIC

Over the last 200 years, the urban population of the world has increased more than 40 times, to a total of two billion people. By the year 2010, it is expected that there will be more people living in cities—nearly three billion—than in the countryside. The main cause of urban growth is the migration of rural poor to cities, drawn by the chance of employment and access to services such as health and education. There are many problems associated with rapid urban growth: pollution, high cost of living, crime, expansion onto agricultural land, and stress on surrounding areas for food, water and fuel. However, in spite of these problems, urban growth is likely to continue unless the issue of rural poverty is addressed.

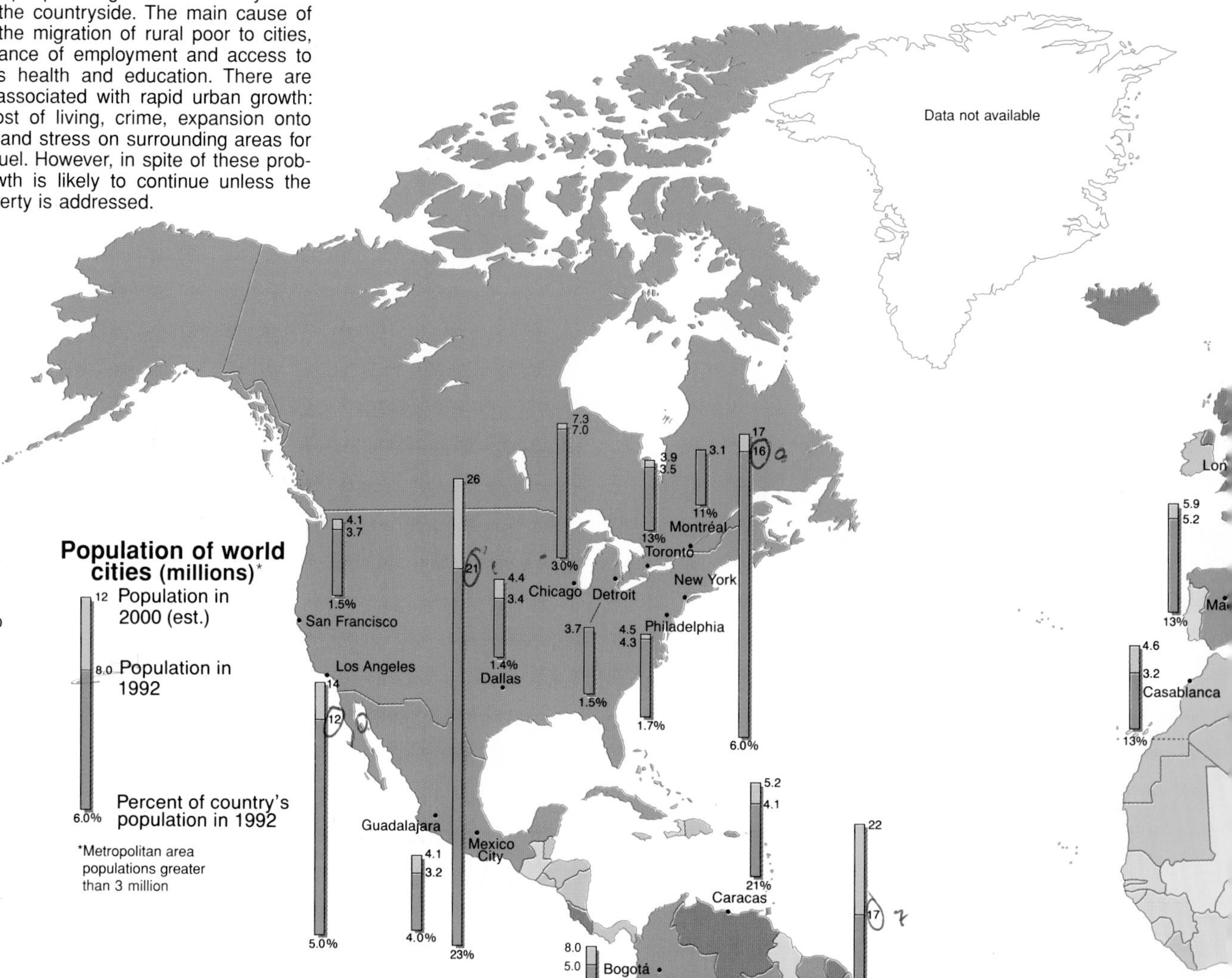
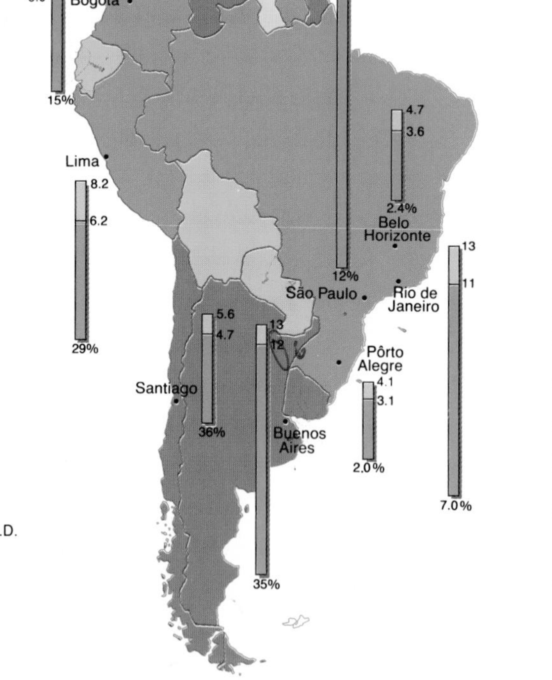

All cities with metropolitan area populations greater than three million are shown. The populations of Detroit, Montréal, Rome and London are not projected to increase between 1992 and 2000.

Early urban societies

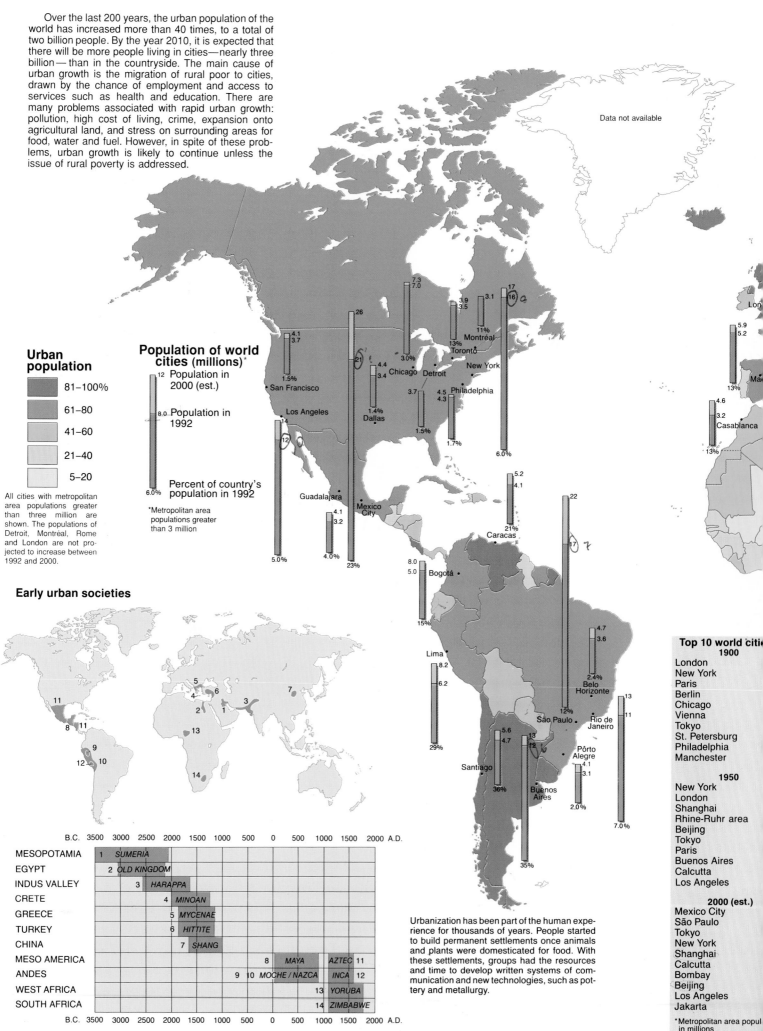

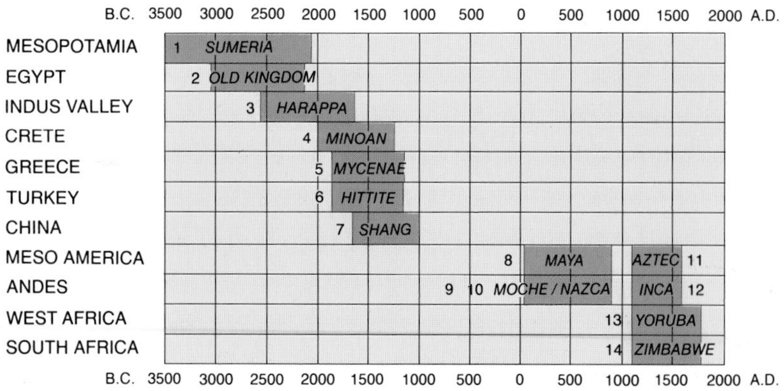

Urbanization has been part of the human experience for thousands of years. People started to build permanent settlements once animals and plants were domesticated for food. With these settlements, groups had the resources and time to develop written systems of communication and new technologies, such as pottery and metallurgy.

Top 10 world citi

1900
London
New York
Paris
Berlin
Chicago
Vienna
Tokyo
St. Petersburg
Philadelphia
Manchester

1950
New York
London
Shanghai
Rhine-Ruhr area
Beijing
Tokyo
Paris
Buenos Aires
Calcutta
Los Angeles

2000 (est.)
Mexico City
São Paulo
Tokyo
New York
Shanghai
Calcutta
Bombay
Beijing
Los Angeles
Jakarta

*Metropolitan area popul in millions

The colour of each country indicates the percentage of the population that is urban. The locations of the urban areas are not shown, other than the cities indicated. Urban areas are generally found in regions of high population density, but there are exceptions. Some cities are found in otherwise sparsely settled areas, while other densely populated areas have no big cities.

The fastest growing cities, 1950–2000 (est.)

	1950	1975	2000
Nouakchott, Mauritania	5000	25 000	1 100 000
'Ammān, Jordan	30 000	800 000	1 500 000
Lagos, Nigeria	270 000	4 000 000	13 000 000
Nairobi, Kenya	140 000	800 000	2 900 000
Baghdād, Iraq	650 000	4 000 000	5 000 000
Santa Cruz, Bolivia	60 000	250 000	1 000 000
Bogotá, Columbia	600 000	2 900 000	5 000 000
Mexico City, Mexico	2 500 000	16 000 000	26 000 000
Jakarta, Indonesia	1 400 000	7 000 000	14 000 000
São Paulo, Brazil	2 500 000	15 000 000	22 000 000
Manaus, Brazil	110 000	500 000	1 100 000
Delhi, India	1 750 000	6 200 000	13 000 000
Cairo, Egypt	2 500 000	8 500 000	12 000 000
Bombay, India	3 300 000	9 000 000	15 000 000

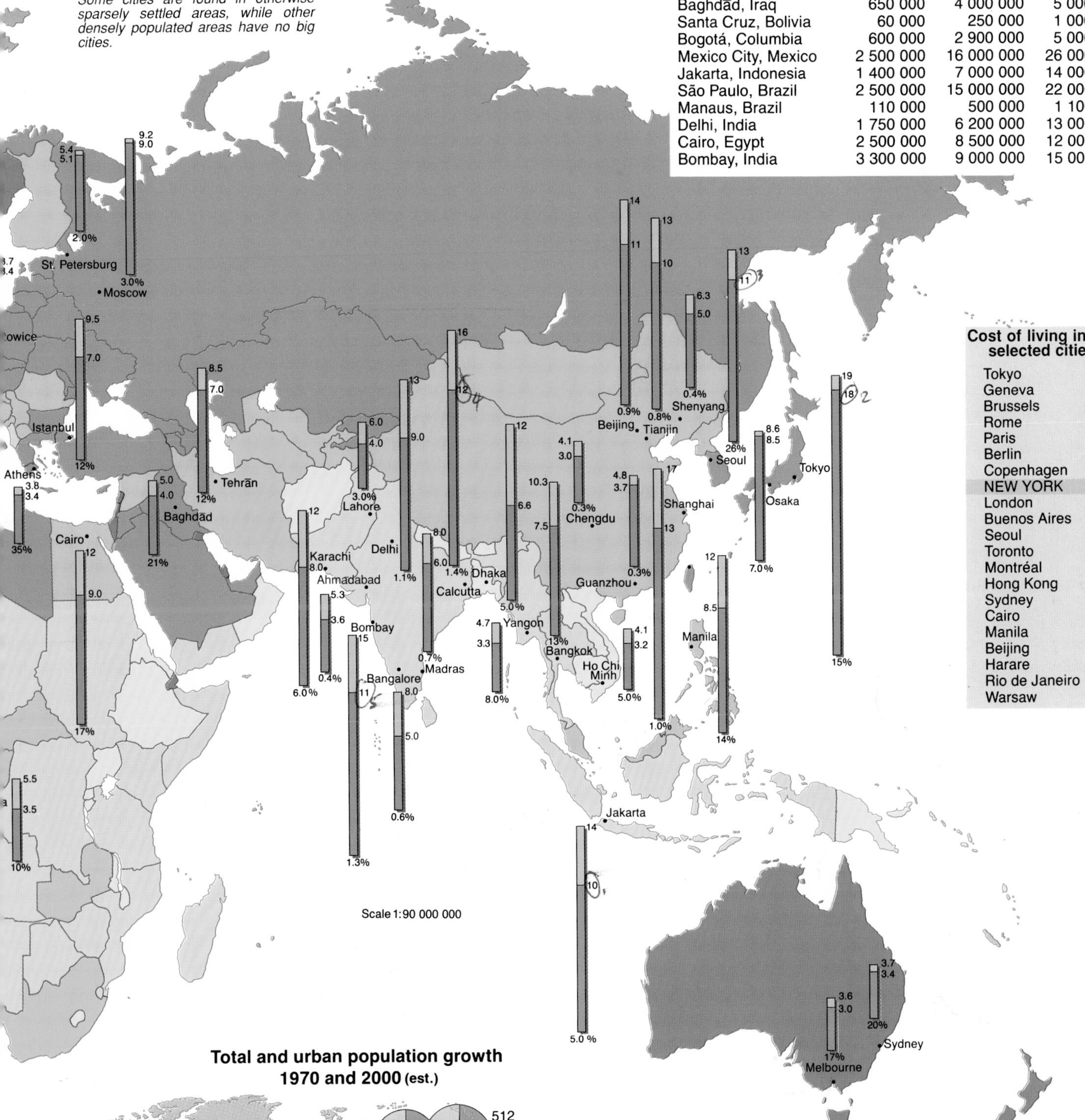

Cost of living index, selected cities

Tokyo	148
Geneva	135
Brussels	123
Rome	110
Paris	108
Berlin	103
Copenhagen	102
NEW YORK	100
London	97
Buenos Aires	93
Seoul	91
Toronto	89
Montréal	87
Hong Kong	87
Sydney	85
Cairo	81
Manila	73
Beijing	70
Harare	65
Rio de Janeiro	63
Warsaw	51

Total and urban population growth 1970 and 2000 (est.)

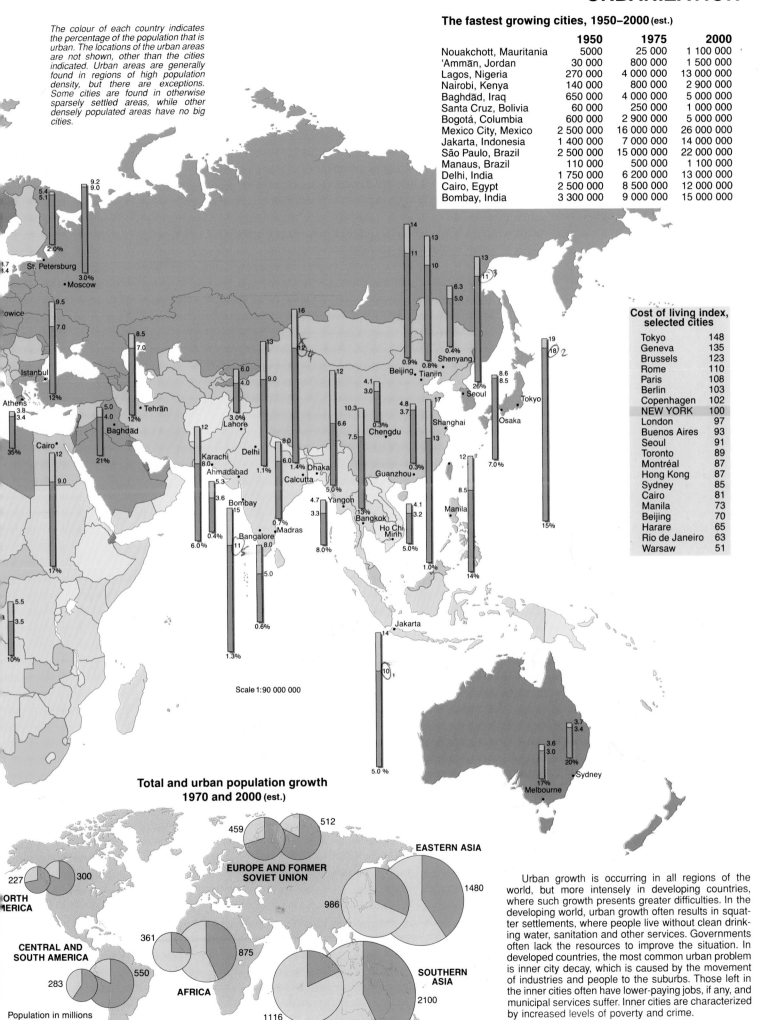

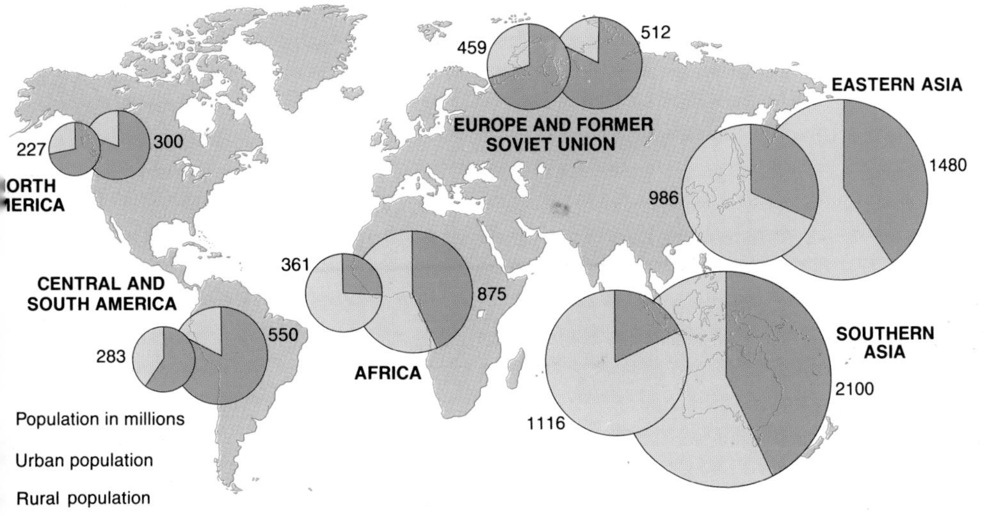

Urban growth is occurring in all regions of the world, but more intensely in developing countries, where such growth presents greater difficulties. In the developing world, urban growth often results in squatter settlements, where people live without clean drinking water, sanitation and other services. Governments often lack the resources to improve the situation. In developed countries, the most common urban problem is inner city decay, which is caused by the movement of industries and people to the suburbs. Those left in the inner cities often have lower-paying jobs, if any, and municipal services suffer. Inner cities are characterized by increased levels of poverty and crime.

For explanation of terms, see glossary.

WORLD-THEMATIC

Agriculture

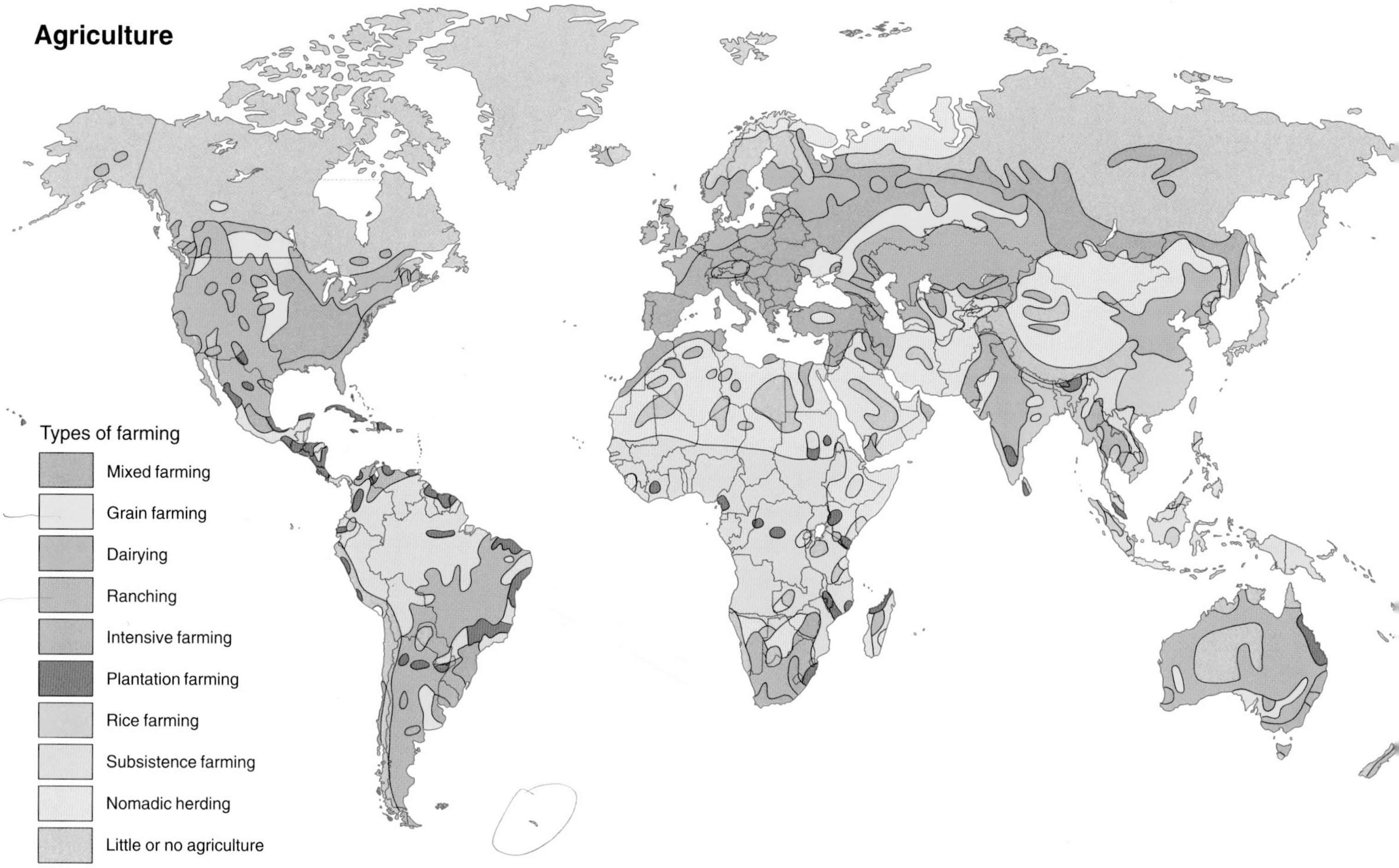

Food intake

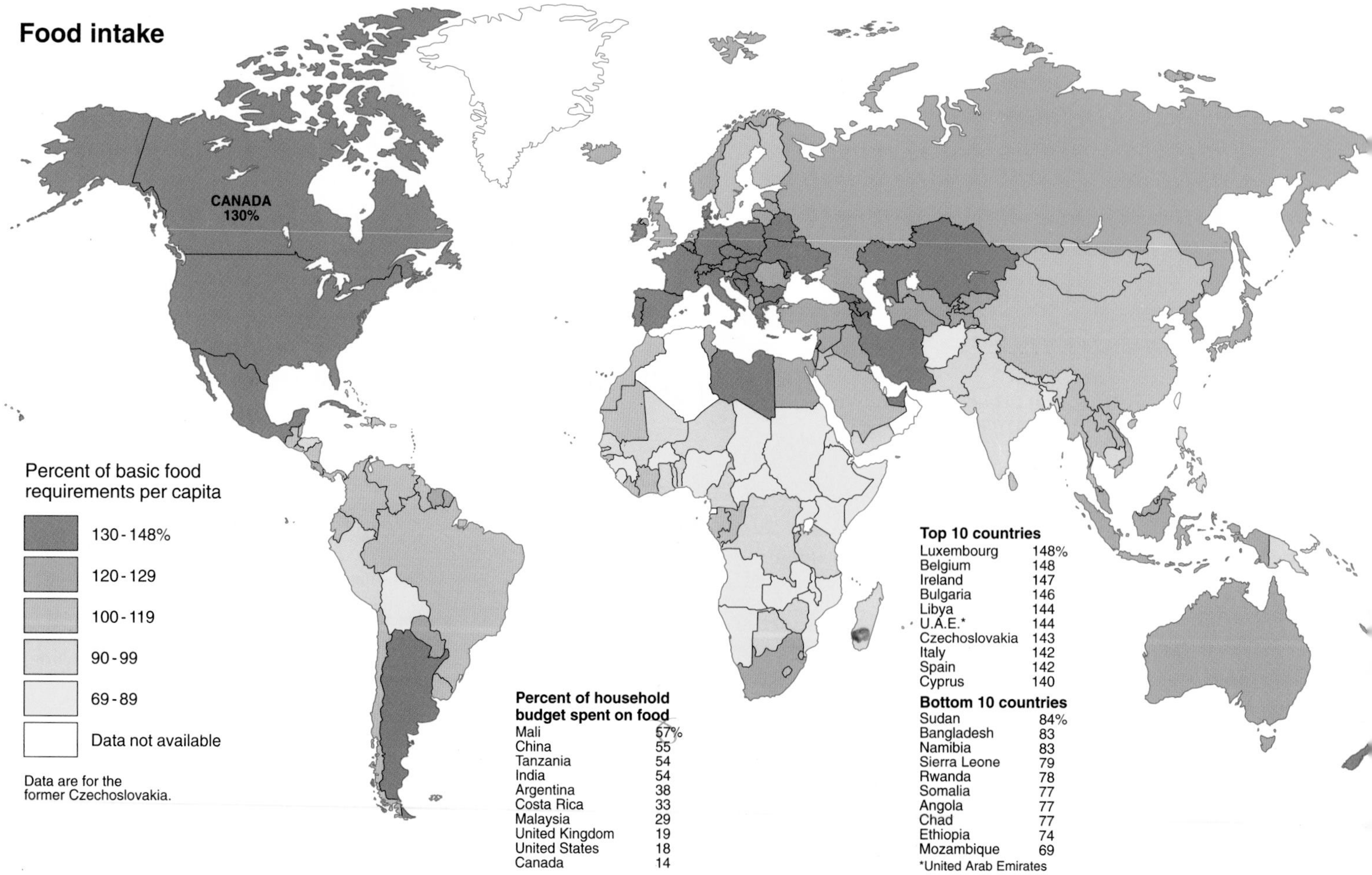

Data are for the former Czechoslovakia.

Percent of household budget spent on food

Mali	57%
China	55
Tanzania	54
India	54
Argentina	38
Costa Rica	33
Malaysia	29
United Kingdom	19
United States	18
Canada	14

Top 10 countries

Luxembourg	148%
Belgium	148
Ireland	147
Bulgaria	146
Libya	144
U.A.E.*	144
Czechoslovakia	143
Italy	142
Spain	142
Cyprus	140

Bottom 10 countries

Sudan	84%
Bangladesh	83
Namibia	83
Sierra Leone	79
Rwanda	78
Somalia	77
Angola	77
Chad	77
Ethiopia	74
Mozambique	69

*United Arab Emirates

shing

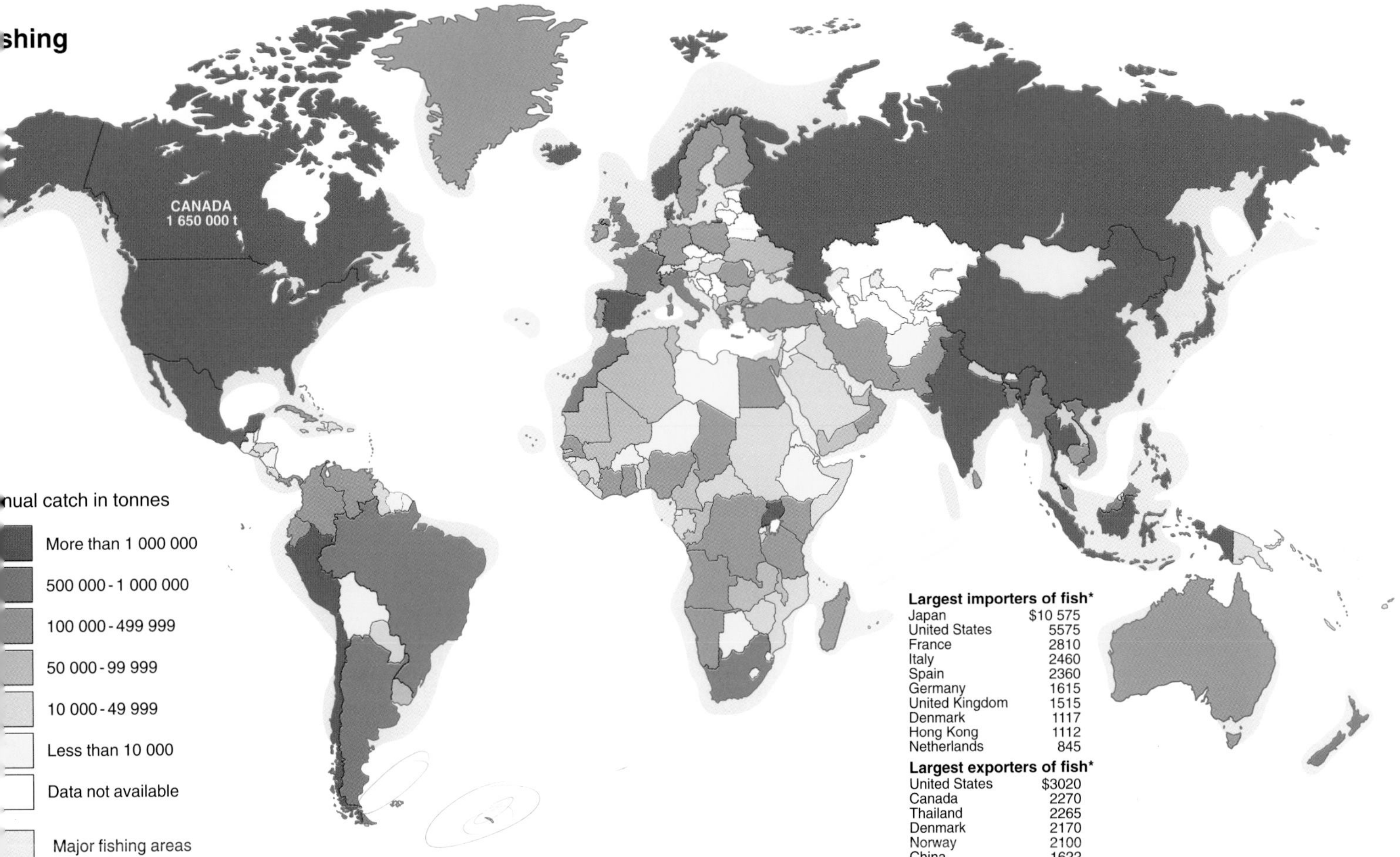

nual catch in tonnes

- More than 1 000 000
- 500 000 - 1 000 000
- 100 000 - 499 999
- 50 000 - 99 999
- 10 000 - 49 999
- Less than 10 000
- Data not available
- Major fishing areas

Largest importers of fish*

Japan	$10 575
United States	5575
France	2810
Italy	2460
Spain	2360
Germany	1615
United Kingdom	1515
Denmark	1117
Hong Kong	1112
Netherlands	845

Largest exporters of fish*

United States	$3020
Canada	2270
Thailand	2265
Denmark	2170
Norway	2100
China	1622
South Korea	1370
Netherlands	1333
Iceland	1245
France	935

*millions of U.S. dollars

nported food

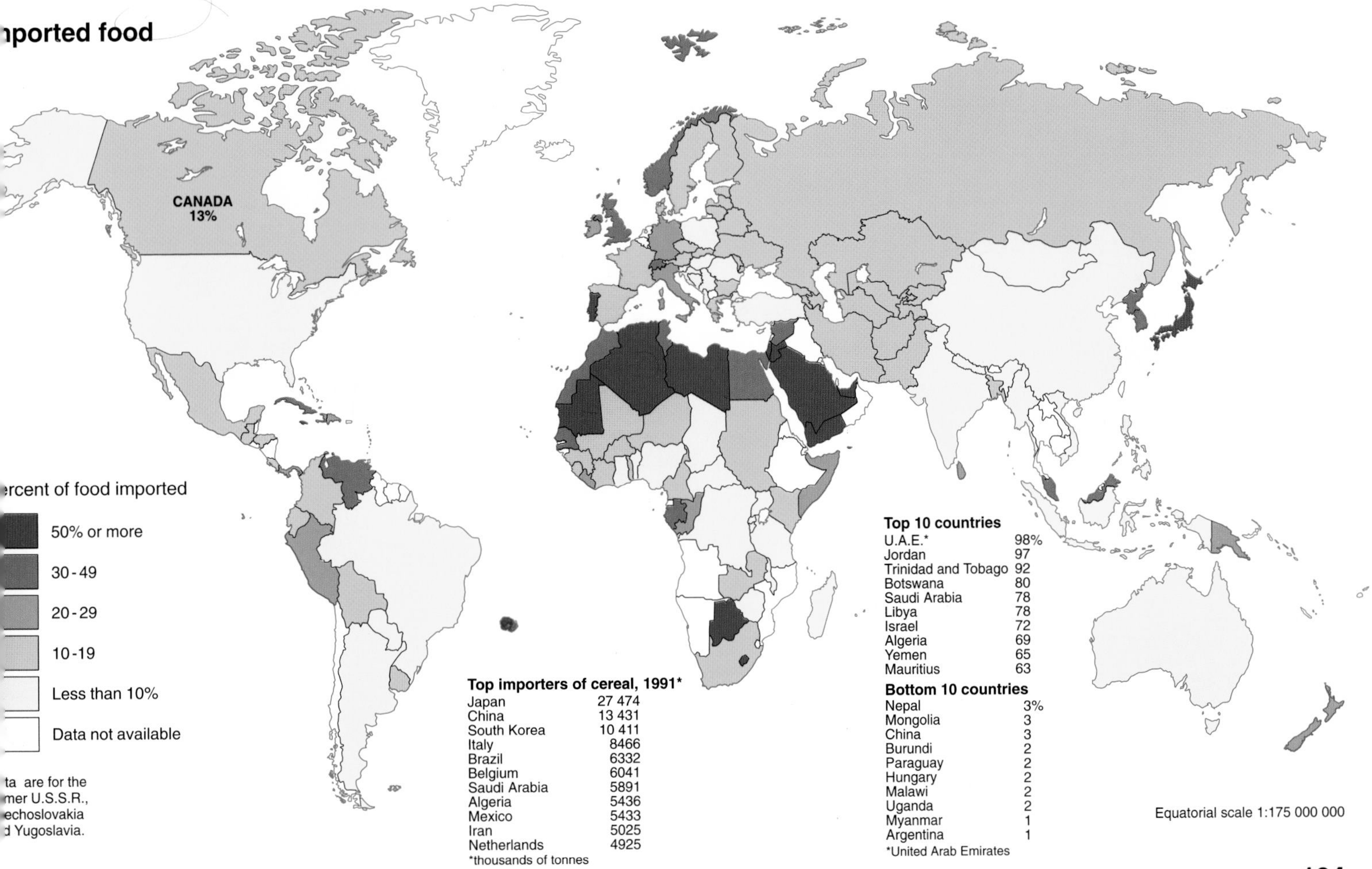

ercent of food imported

- 50% or more
- 30 - 49
- 20 - 29
- 10 - 19
- Less than 10%
- Data not available

ta are for the
mer U.S.S.R.,
echoslovakia
d Yugoslavia.

Top importers of cereal, 1991*

Japan	27 474
China	13 431
South Korea	10 411
Italy	8466
Brazil	6332
Belgium	6041
Saudi Arabia	5891
Algeria	5436
Mexico	5433
Iran	5025
Netherlands	4925

*thousands of tonnes

Top 10 countries

U.A.E.*	98%
Jordan	97
Trinidad and Tobago	92
Botswana	80
Saudi Arabia	78
Libya	78
Israel	72
Algeria	69
Yemen	65
Mauritius	63

Bottom 10 countries

Nepal	3%
Mongolia	3
China	3
Burundi	2
Paraguay	2
Hungary	2
Malawi	2
Uganda	2
Myanmar	1
Argentina	1

*United Arab Emirates

Equatorial scale 1:175 000 000

For explanation of terms, see glossary.

GNP per capita

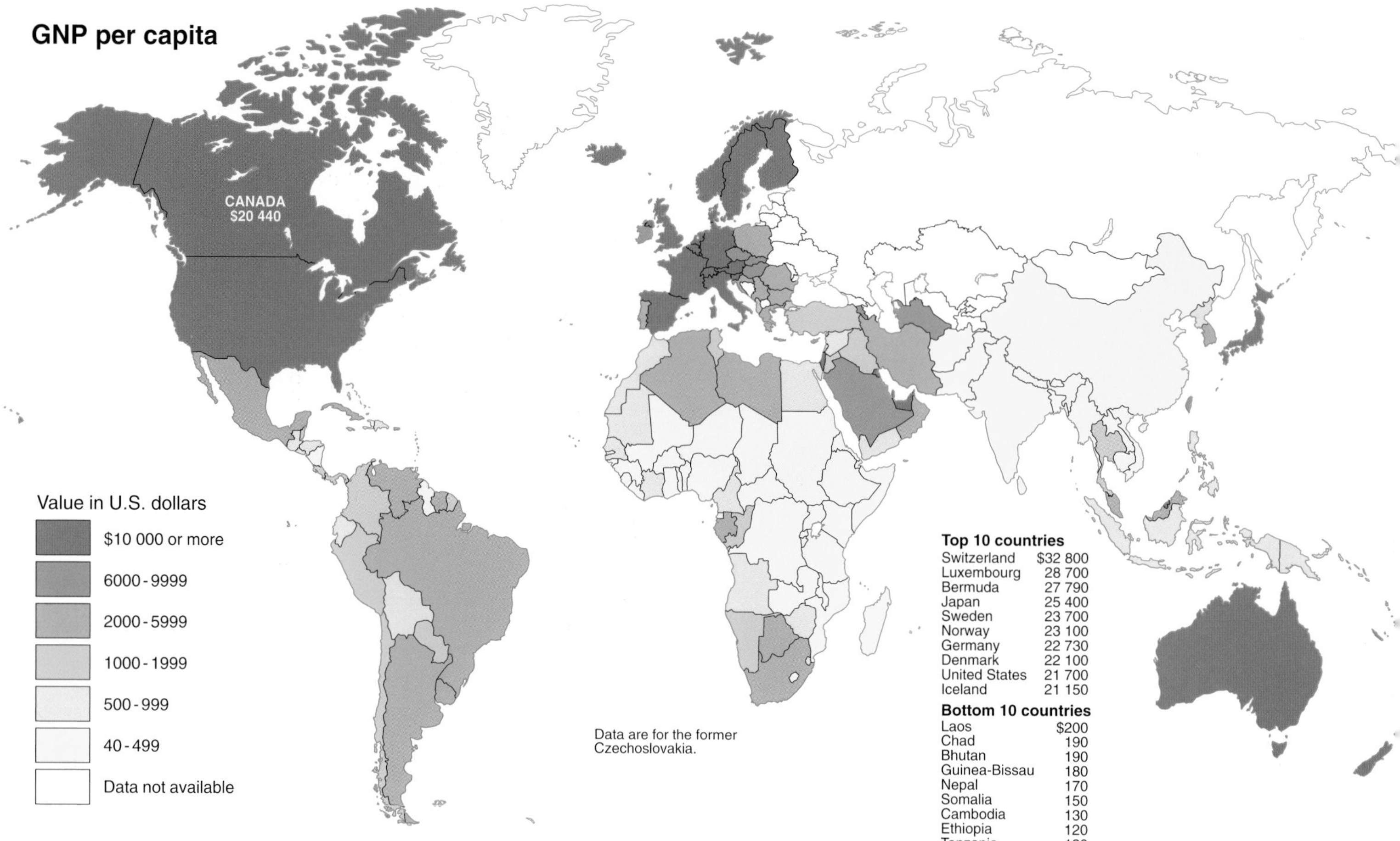

Data are for the former Czechoslovakia.

Top 10 countries

Switzerland	$32 800
Luxembourg	28 700
Bermuda	27 790
Japan	25 400
Sweden	23 700
Norway	23 100
Germany	22 730
Denmark	22 100
United States	21 700
Iceland	21 150

Bottom 10 countries

Laos	$200
Chad	190
Bhutan	190
Guinea-Bissau	180
Nepal	170
Somalia	150
Cambodia	130
Ethiopia	120
Tanzania	120
Mozambique	80

The GNP (Gross National Product the value of all goods and servi produced domestically in one year, p income earned by foreigners f domestic production.

Manufacturing

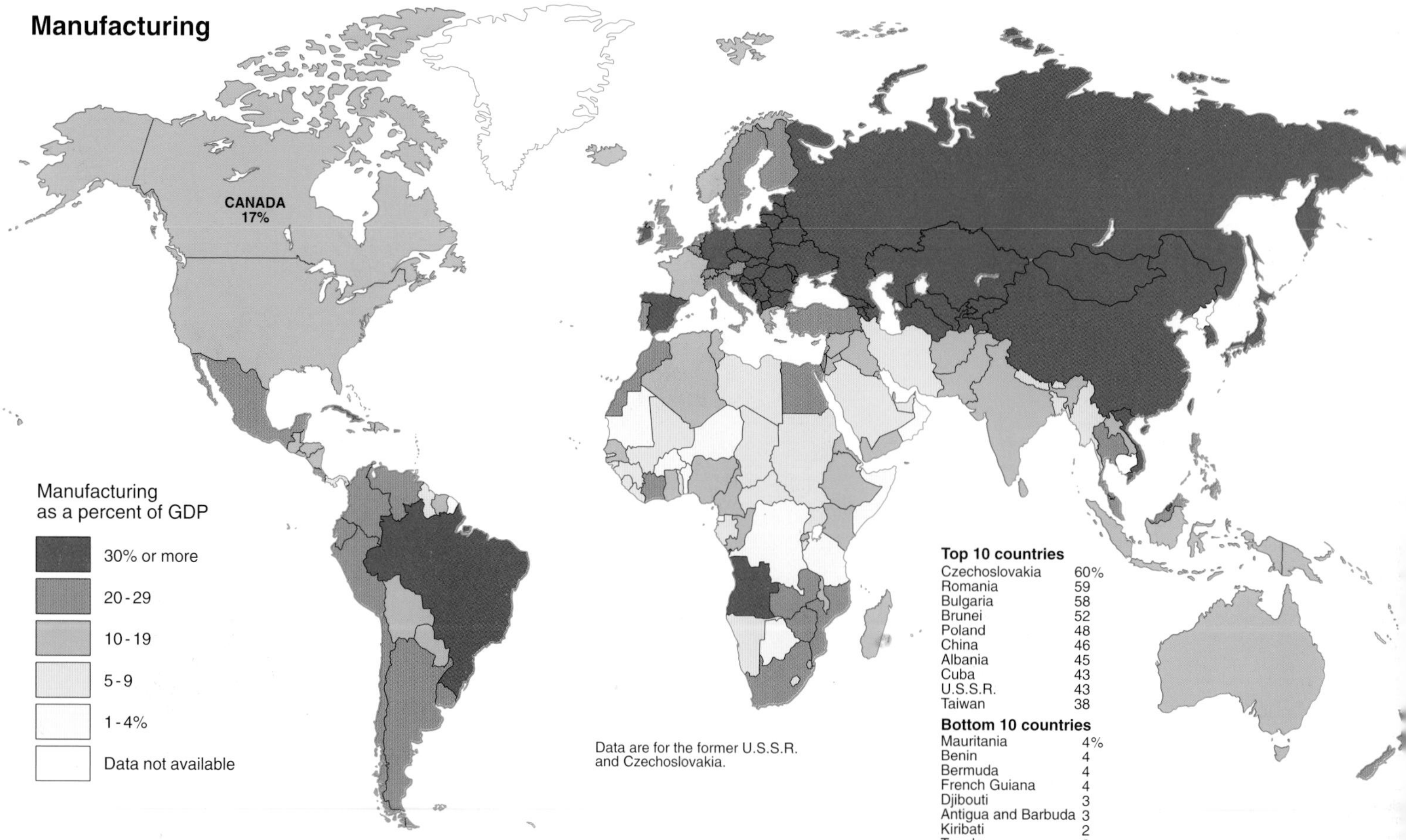

Data are for the former U.S.S.R. and Czechoslovakia.

Top 10 countries

Czechoslovakia	60%
Romania	59
Bulgaria	58
Brunei	52
Poland	48
China	46
Albania	45
Cuba	43
U.S.S.R.	43
Taiwan	38

Bottom 10 countries

Mauritania	4%
Benin	4
Bermuda	4
French Guiana	4
Djibouti	3
Antigua and Barbuda	3
Kiribati	2
Tuvalu	2
Equatorial Guinea	2
Zaire	1

riculture

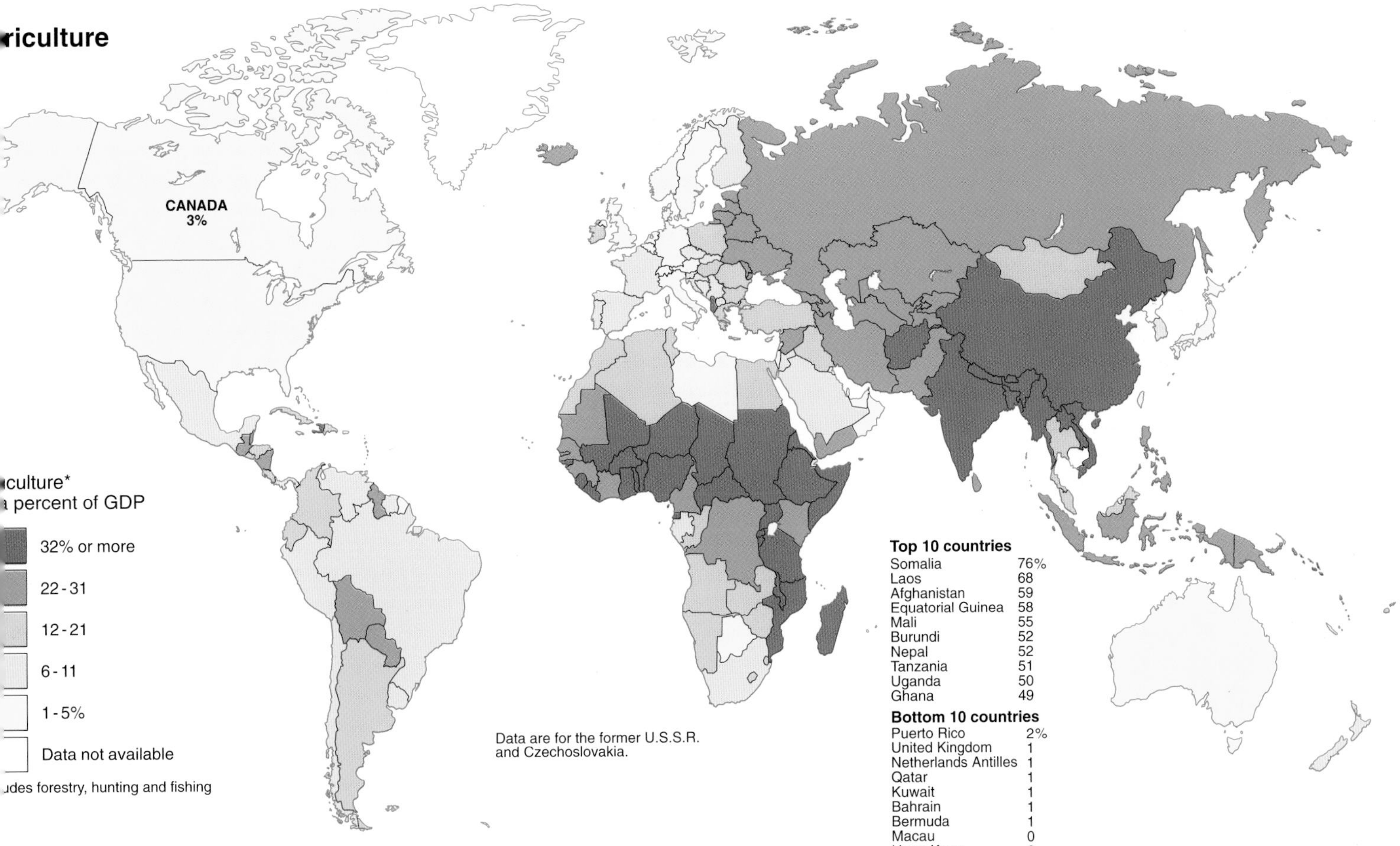

culture*
a percent of GDP

- 32% or more
- 22-31
- 12-21
- 6-11
- 1-5%
- Data not available

udes forestry, hunting and fishing

Top 10 countries

Somalia	76%
Laos	68
Afghanistan	59
Equatorial Guinea	58
Mali	55
Burundi	52
Nepal	52
Tanzania	51
Uganda	50
Ghana	49

Bottom 10 countries

Puerto Rico	2%
United Kingdom	1
Netherlands Antilles	1
Qatar	1
Kuwait	1
Bahrain	1
Bermuda	1
Macau	0
Hong Kong	0
Singapore	0

GDP (Gross Domestic Product) is value of all goods and services luced domestically in one year. For y countries, the GDP and GNP are close in value.

rvices

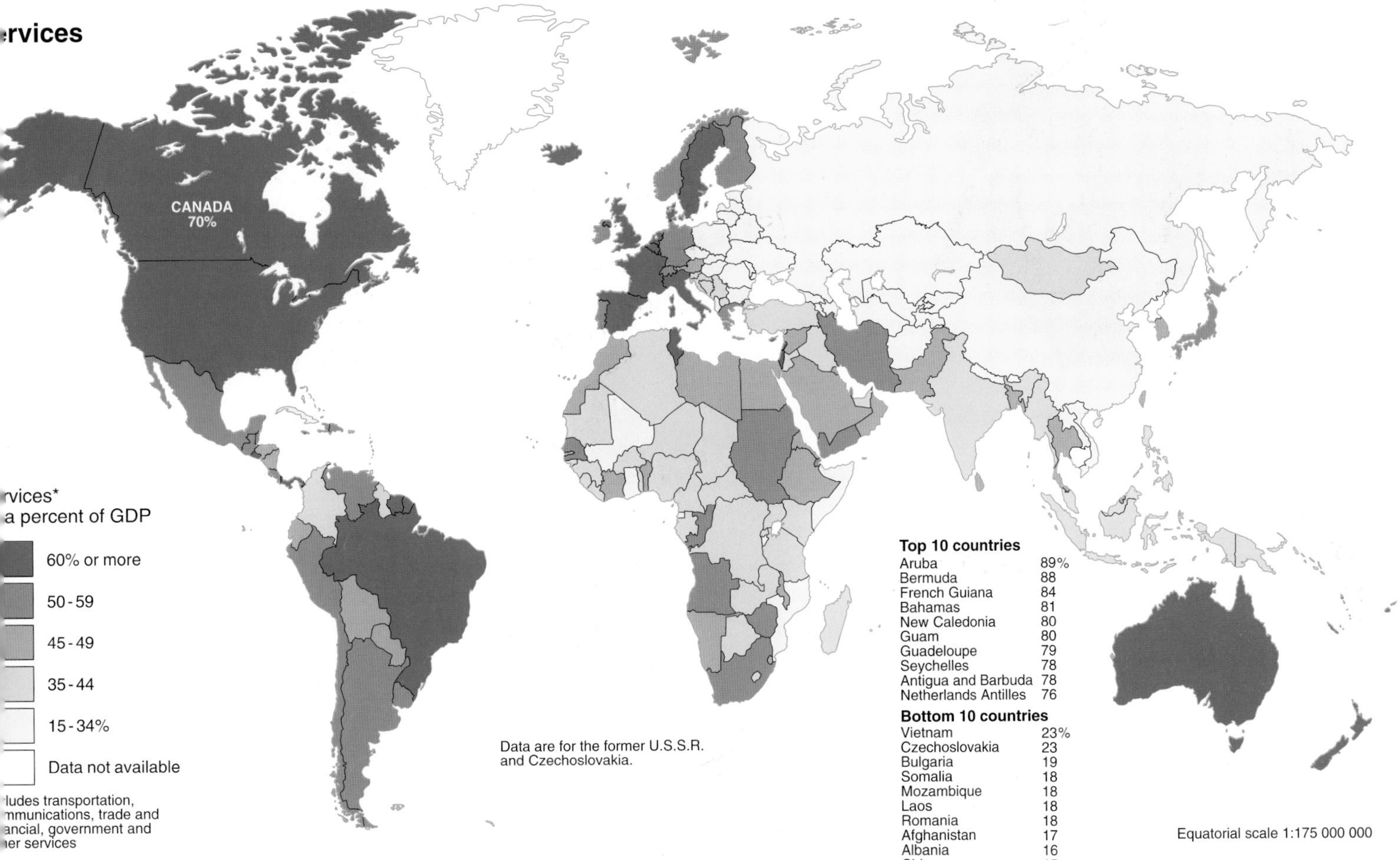

rvices*
a percent of GDP

- 60% or more
- 50-59
- 45-49
- 35-44
- 15-34%
- Data not available

ludes transportation, mmunications, trade and ancial, government and ner services

Top 10 countries

Aruba	89%
Bermuda	88
French Guiana	84
Bahamas	81
New Caledonia	80
Guam	80
Guadeloupe	79
Seychelles	78
Antigua and Barbuda	78
Netherlands Antilles	76

Bottom 10 countries

Vietnam	23%
Czechoslovakia	23
Bulgaria	19
Somalia	18
Mozambique	18
Laos	18
Romania	18
Afghanistan	17
Albania	16
China	15

WORLD-THEMATIC

Exports

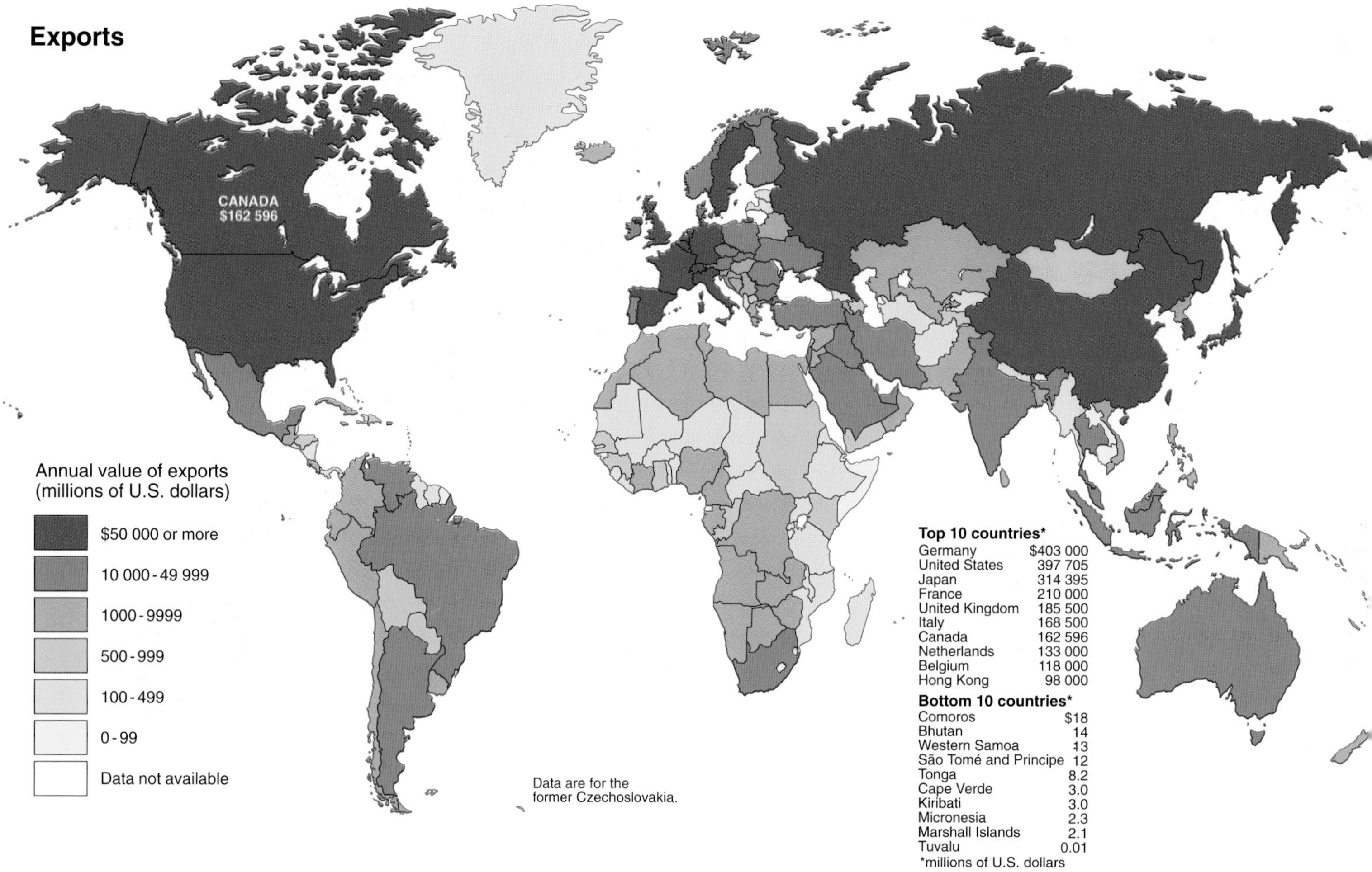

CANADA
$162 596
Annual value of exports (millions of U.S. dollars)
$50 000 or more
10 000 - 49 999
1000 - 9999
500 - 999
100 - 499
0 - 99
Data not available
Data are for the former Czechoslovakia.
Top 10 countries*
Germany $403 000
United States 397 705
Japan 314 395
France 210 000
United Kingdom 185 500
Italy 168 500
Canada 162 596
Netherlands 133 000
Belgium 118 000
Hong Kong 98 000
Bottom 10 countries*
Comoros $18
Bhutan 14
Western Samoa 13
São Tomé and Principe 12
Tonga 8.2
Cape Verde 3.0
Kiribati 3.0
Micronesia 2.3
Marshall Islands 2.1
Tuvalu 0.01
*millions of U.S. dollars

Manufacturing

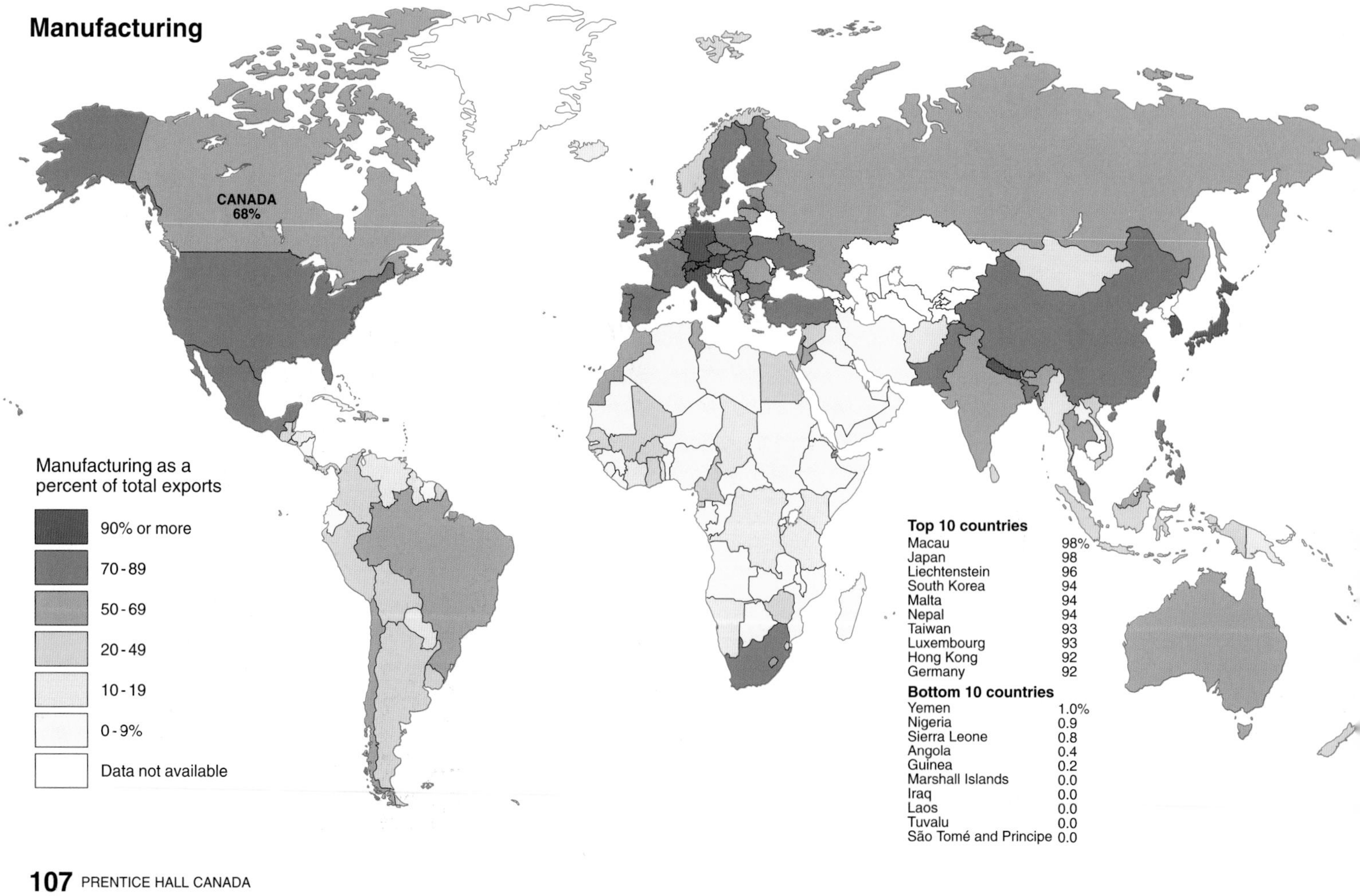

CANADA
68%
Manufacturing as a percent of total exports
90% or more
70 - 89
50 - 69
20 - 49
10 - 19
0 - 9%
Data not available
Top 10 countries
Macau 98%
Japan 98
Liechtenstein 96
South Korea 94
Malta 94
Nepal 94
Taiwan 93
Luxembourg 93
Hong Kong 92
Germany 92
Bottom 10 countries
Yemen 1.0%
Nigeria 0.9
Sierra Leone 0.8
Angola 0.4
Guinea 0.2
Marshall Islands 0.0
Iraq 0.0
Laos 0.0
Tuvalu 0.0
São Tomé and Principe 0.0

ports

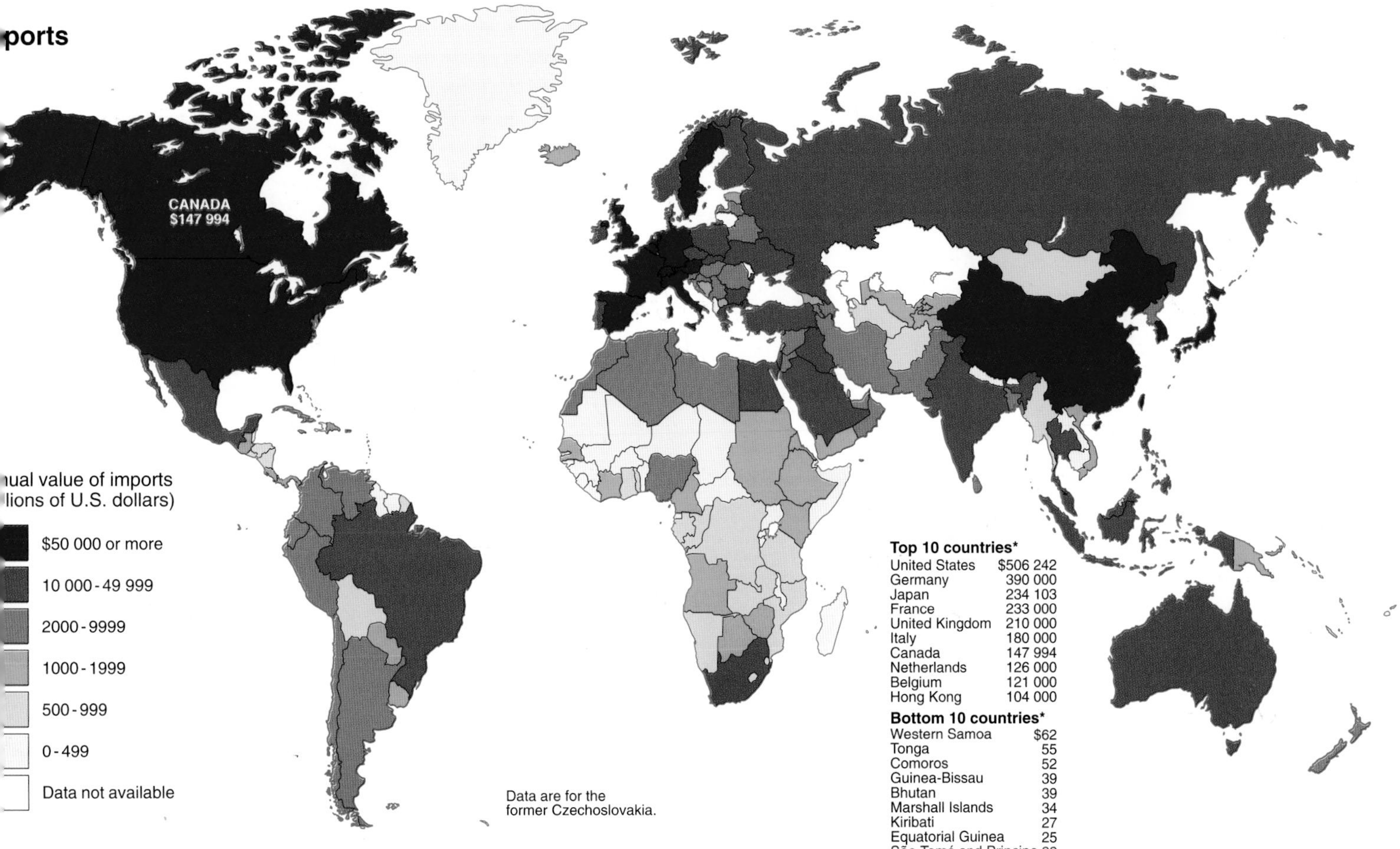

Top 10 countries*

United States	$506 242
Germany	390 000
Japan	234 103
France	233 000
United Kingdom	210 000
Italy	180 000
Canada	147 994
Netherlands	126 000
Belgium	121 000
Hong Kong	104 000

Bottom 10 countries*

Western Samoa	$62
Tonga	55
Comoros	52
Guinea-Bissau	39
Bhutan	39
Marshall Islands	34
Kiribati	27
Equatorial Guinea	25
São Tomé and Principe	22
Tuvalu	4

*millions of U.S. dollars

urism

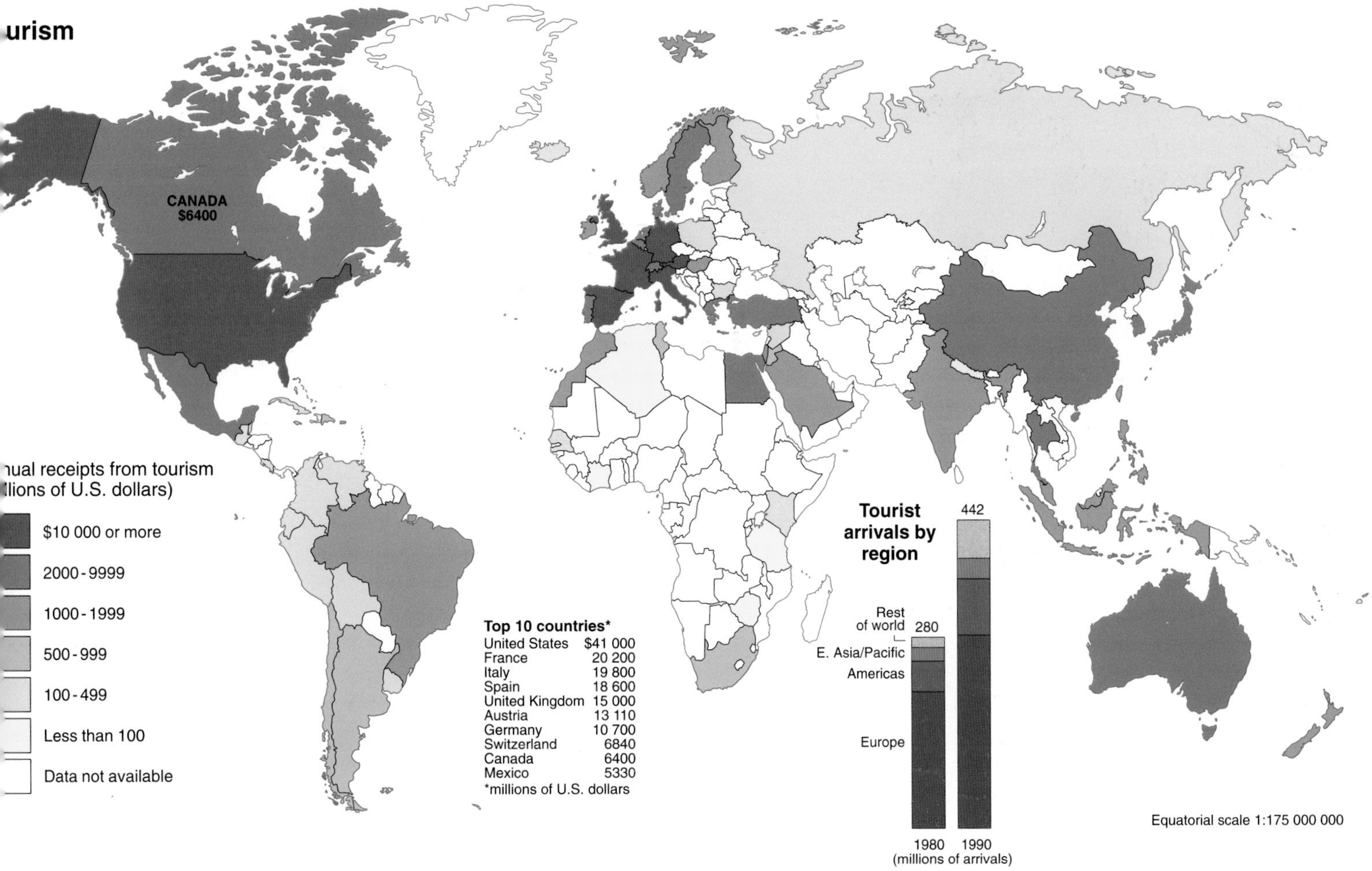

Top 10 countries*

United States	$41 000
France	20 200
Italy	19 800
Spain	18 600
United Kingdom	15 000
Austria	13 110
Germany	10 700
Switzerland	6840
Canada	6400
Mexico	5330

*millions of U.S. dollars

Tourist arrivals by region

Equatorial scale 1:175 000 000

WORLD-THEMATIC

Energy consumption

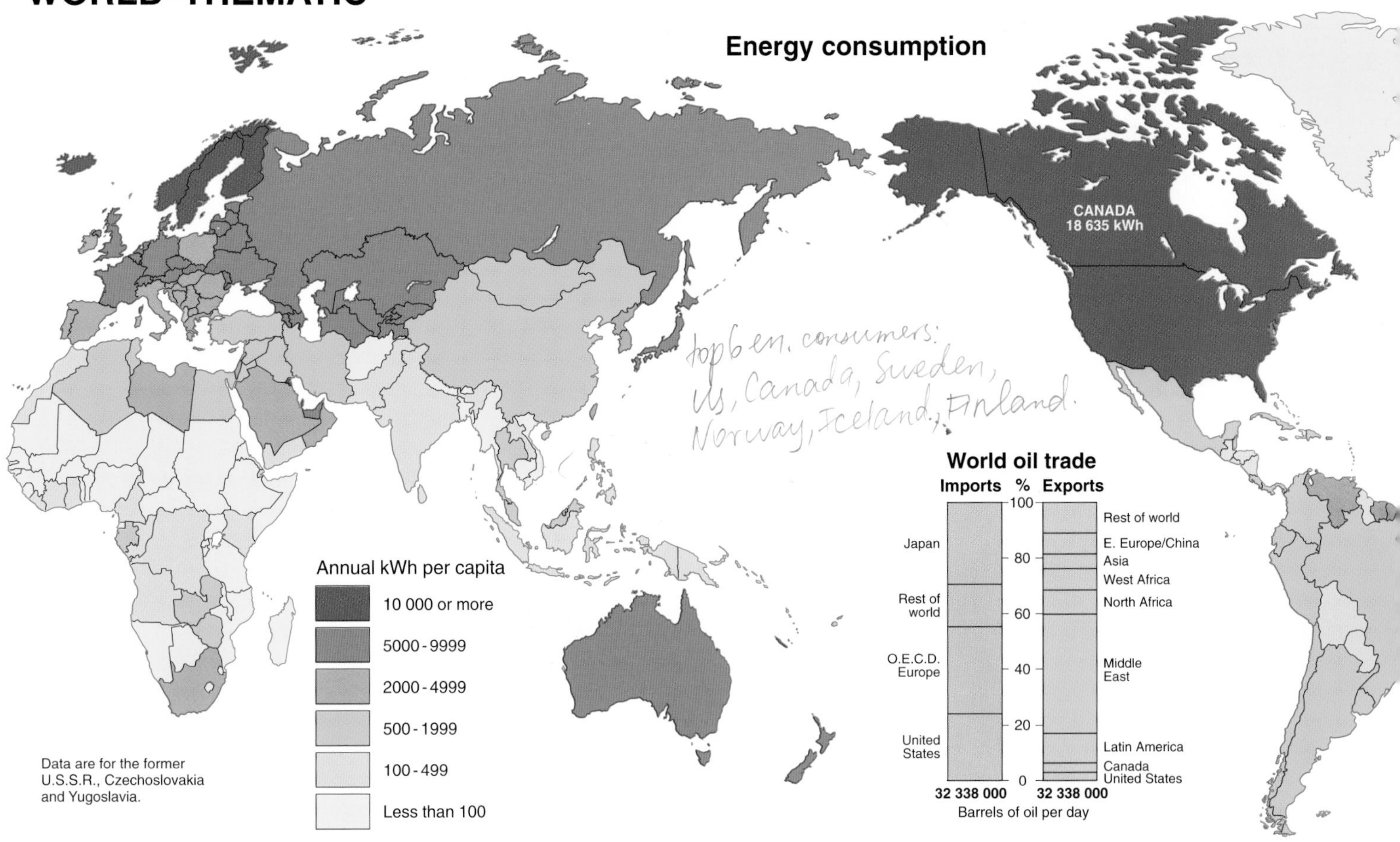

Energy consumption by type of fuel

Tonnes of oil equivalent per year

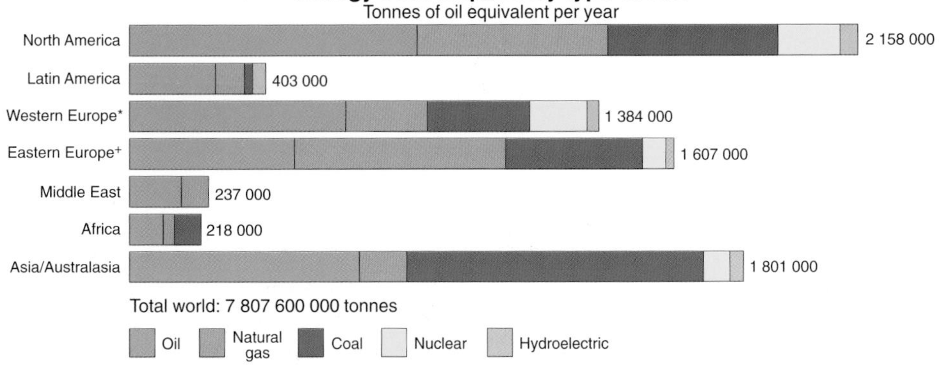

* Organization for Economic Cooperation and Development (O.E.C.D.) countries only
+ Includes former U.S.S.R.

The global energy budget has neither social nor environmental balance. Some countries consume disproportionate amounts of energy; some countries produce disproportionate amounts. Canada and the United States, with 5 percent of the world's population, consume nearly 28 percent of the global energy supply.

Many developed nations have relied on abundant energy resources for their prosperity, but energy consumption and production cause enormous environmental stress. Scientists estimate that half of the greenhouse gases in the atmosphere are linked to energy use. In North America, energy conservation measures have had some effect. Although the economy has grown by about 2.5 percent in recent years, energy needs have remained fairly constant.

Economic and political factors have also encouraged research into alternative energy sources. The oil embargo in the 1970s, the Gulf War and political instability in the former Soviet Union have shown that fossil fuel supplies are an effective weapon in international conflict. Scientists are studying other energy sources such as wind, tidal power, biomass, geothermal and solar photovoltaic cells.

Oil reserves

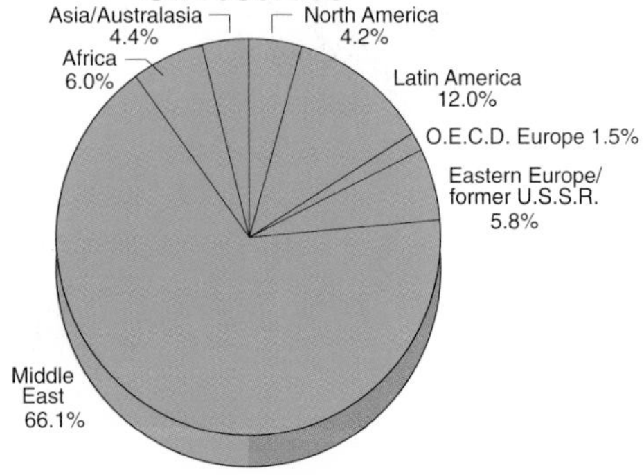

Estimated years of oil reserves remaining: 43

Top producers of oil

Country	% of world total
Former U.S.S.R.	16.5
United States	13.6
Saudi Arabia	13.4
Iran	5.2
Mexico	4.8
China	4.5
Venezuela	4.3
U.A.E.*	3.2
Canada	3.0
Nigeria	3.0

*United Arab Emirates

Gas reserves

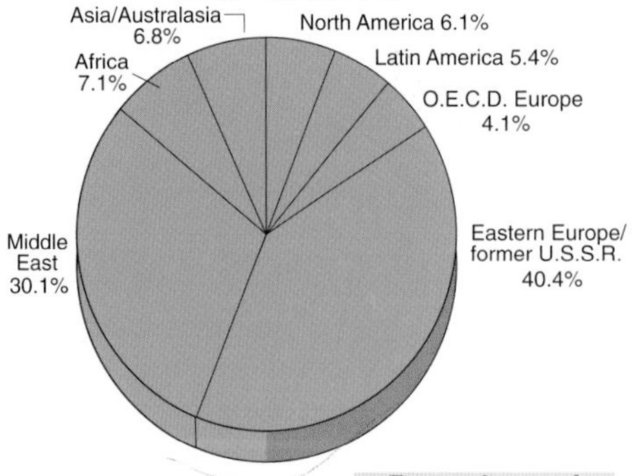

Estimated years of natural gas reserves remaining: 59

Top producers of natural gas

Country	% of world total
Former U.S.S.R.	35.9
United States	25.1
Canada	5.2
Netherlands	3.4
United Kingdom	2.7
Indonesia	2.5
Algeria	2.5
Saudi Arabia	1.6
Mexico	1.4
Iran	1.4

Coal reserves

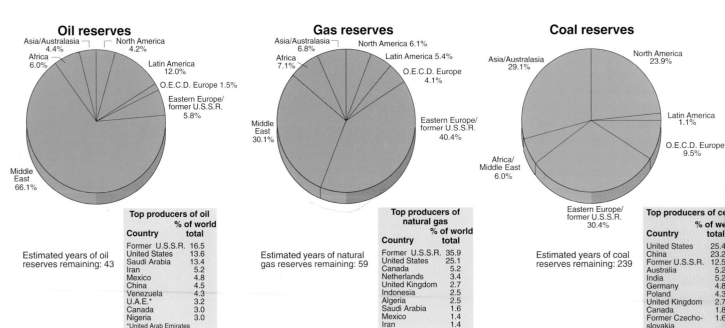

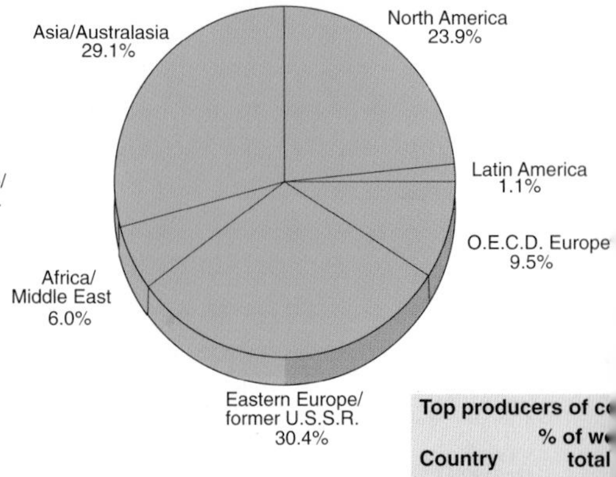

Estimated years of coal reserves remaining: 239

Top producers of c

Country	% of w total
United States	25.4
China	23.2
Former U.S.S.R.	12.5
Australia	5.2
India	5.2
Germany	4.8
Poland	4.3
United Kingdom	2.7
Canada	1.8
Former Czecho-slovakia	1.6

Nearly half of the world's population depends on ood for warmth, light and cooking. But the supply of elwood is decreasing at an alarming rate. Already illions of people struggle to meet their minimum ergy needs, and the United Nations' Food and griculture Organization predicts that, by the year)00, one billion more people will face chronic fuelwood ortages.

Although this crisis in the wood supply is partly used by the growing population of rural poor, there e other causes as well. Heavy logging and slash-d-burn cultivation have deforested many areas, and e growing urban demand in developing countries for th fuelwood and charcoal is another serious factor.

Reforestation is critical. Fast-growing species may lp to sustain the wood supply, but trees must be anted in community lots near people who depend on elwood, rather than in plantations. More efficient od-burning stoves and improved fuelwood distribu-n networks would also ease the escalating shortage.

Fuelwood's share of total energy consumption

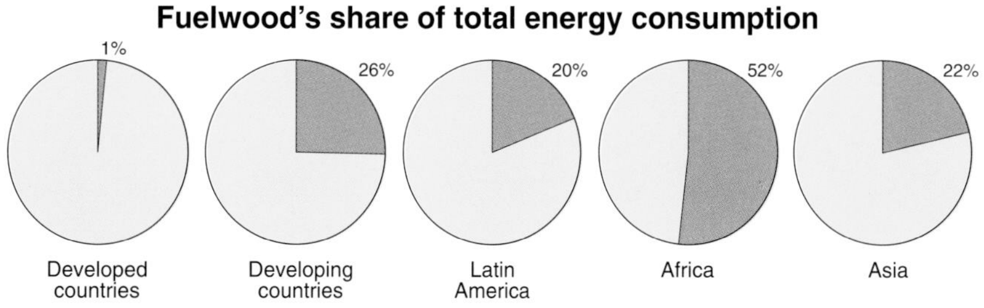

Top producers of nuclear energy

untry	% of world total
ted States	32.3
nce	12.7
mer U.S.S.R.	10.6
an	10.5
rmany	7.4
nada	4.3
eden	3.6
ted Kingdom	3.0
th Korea	2.7
ain	2.5

Top producers of hydroelectric power

Country	% of world total
Canada	13.5
United States	13.3
Former U.S.S.R.	10.3
Brazil	9.4
China	5.4
Japan	4.4
Norway	4.2
Mexico	4.1
Sweden	2.7
France	2.1

Top producers of fuelwood

Country	% of world total
India	14.2
China	10.3
Brazil	8.4
Indonesia	7.6
United States	6.7
Nigeria	5.8
Former U.S.S.R.	4.7
Ethiopia	2.2
Thailand	2.1
Philippines	1.9

Population affected by fuelwood shortages

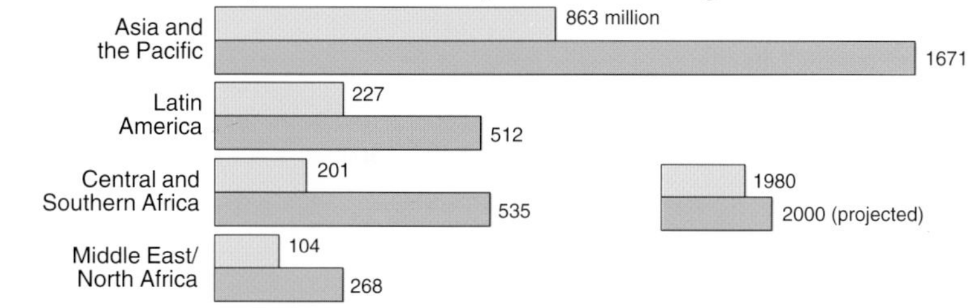

Top producers of geothermal energy

Megawatts per year

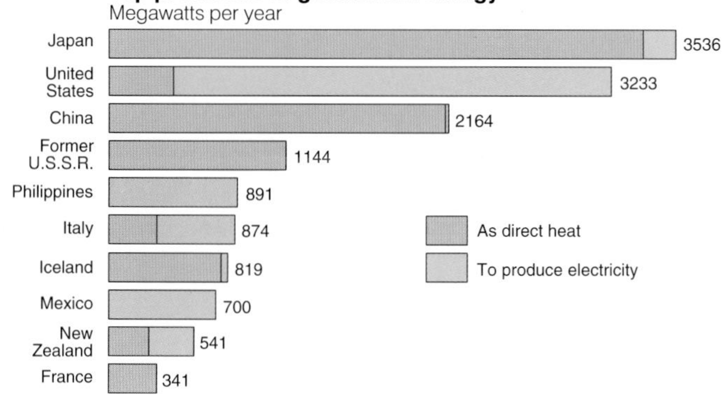

Wind power generation

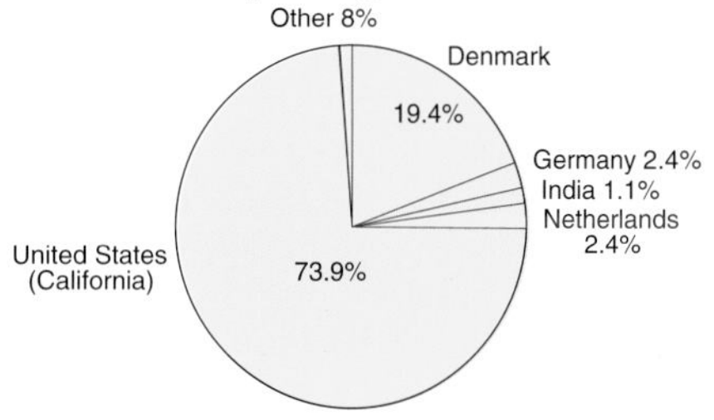

1991: 3 757 000 000 kWh

Tidal power

Country	Average annual energy output (million kWh)	Capacity (megawatts)
Canada	50.0	17.8
China	10.0	3.4
France	540.0	240.0
Russia	0.8	0.4

Electricity generation

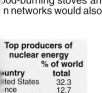

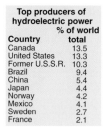

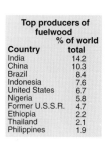

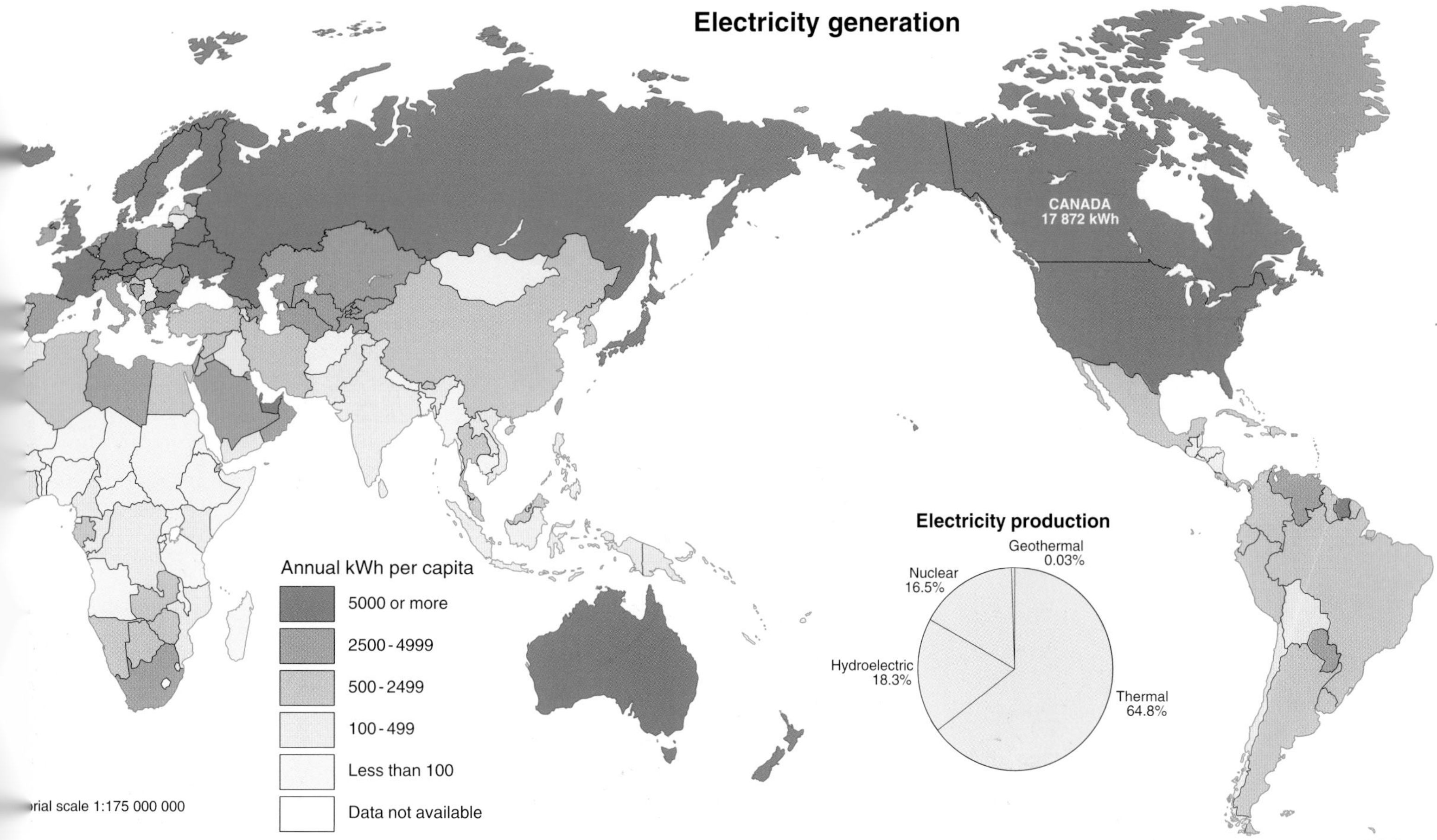

rial scale 1:175 000 000

Infant mortality

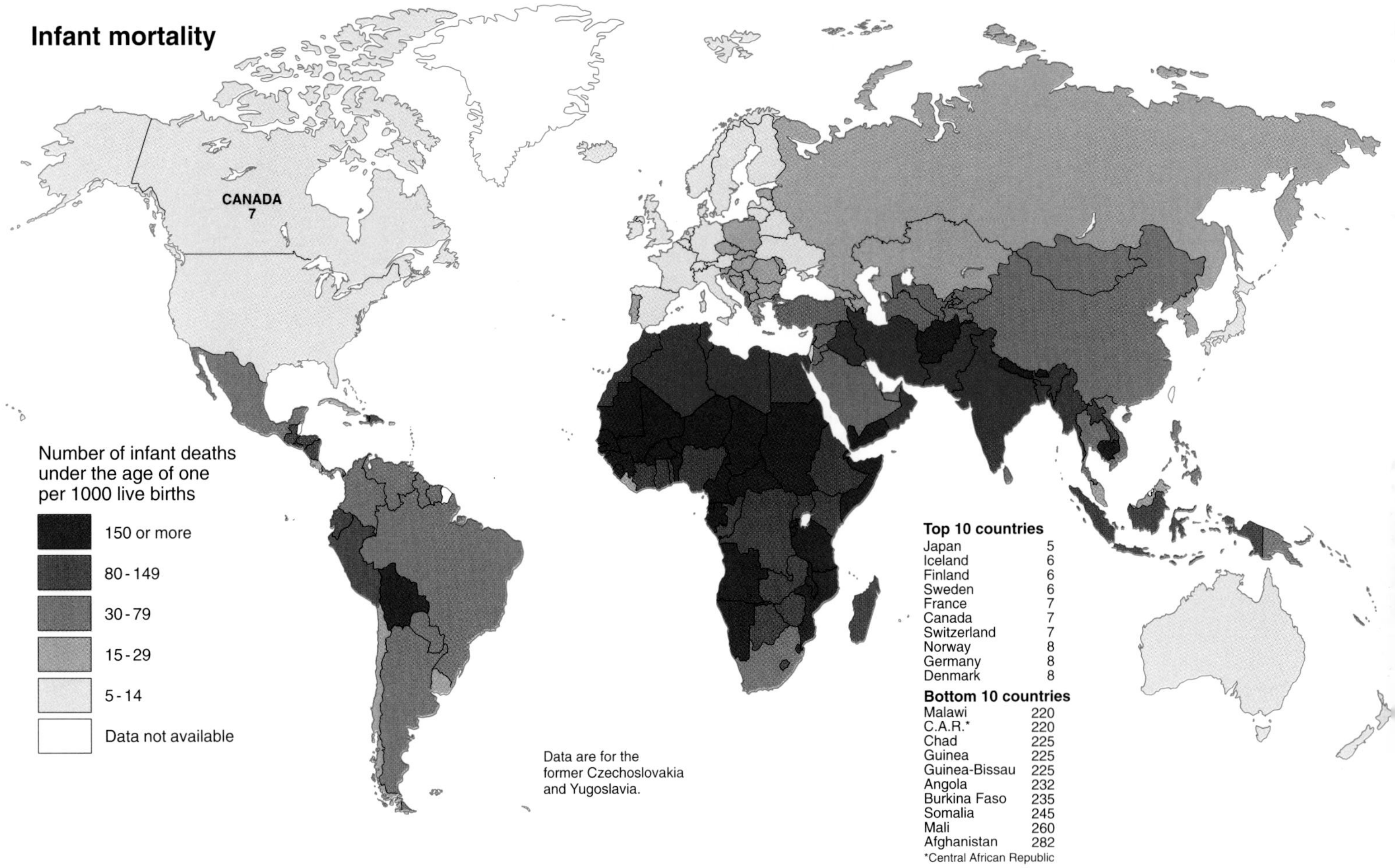
CANADA
7
Number of infant deaths under the age of one per 1000 live births
150 or more
80-149
30-79
15-29
5-14
Data not available
Data are for the former Czechoslovakia and Yugoslavia.
Top 10 countries
Japan 5
Iceland 6
Finland 6
Sweden 6
France 7
Canada 7
Switzerland 7
Norway 8
Germany 8
Denmark 8
Bottom 10 countries
Malawi 220
C.A.R.* 220
Chad 225
Guinea 225
Guinea-Bissau 225
Angola 232
Burkina Faso 235
Somalia 245
Mali 260
Afghanistan 282
*Central African Republic

People per physician

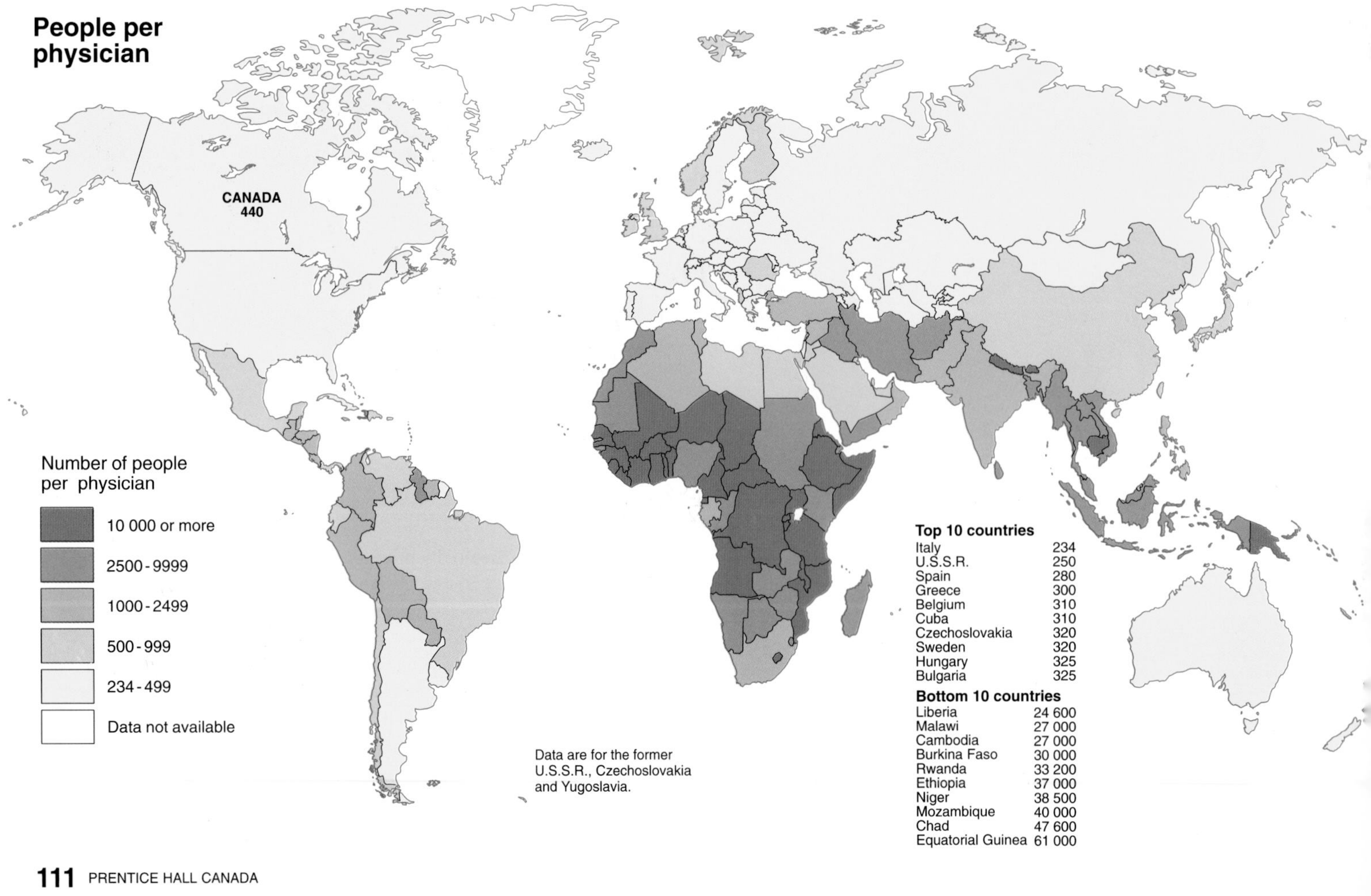
CANADA
440
Number of people per physician
10 000 or more
2500-9999
1000-2499
500-999
234-499
Data not available
Data are for the former U.S.S.R., Czechoslovakia and Yugoslavia.
Top 10 countries
Italy 234
U.S.S.R. 250
Spain 280
Greece 300
Belgium 310
Cuba 310
Czechoslovakia 320
Sweden 320
Hungary 325
Bulgaria 325
Bottom 10 countries
Liberia 24 600
Malawi 27 000
Cambodia 27 000
Burkina Faso 30 000
Rwanda 33 200
Ethiopia 37 000
Niger 38 500
Mozambique 40 000
Chad 47 600
Equatorial Guinea 61 000

fe expectancy
birth

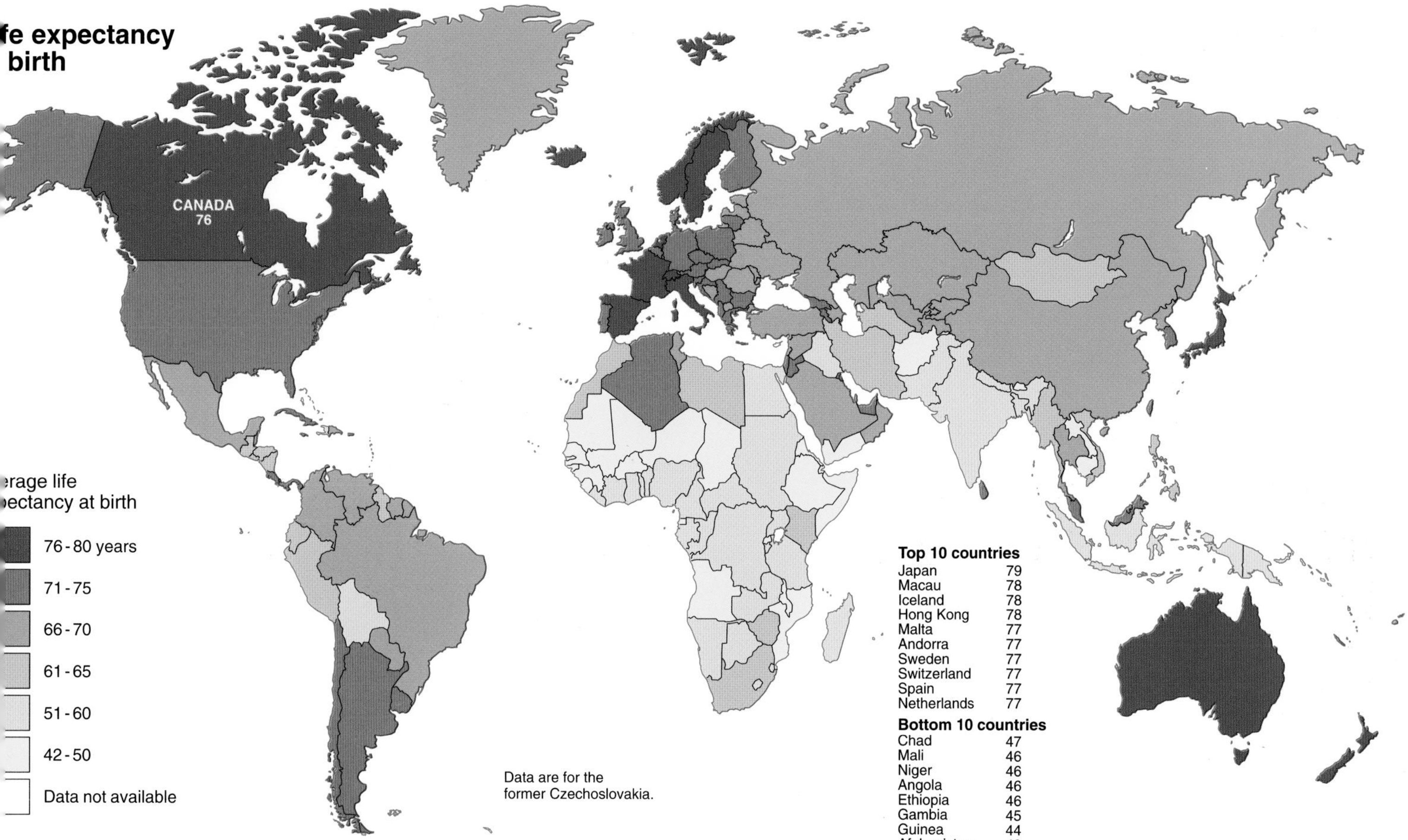

Data are for the former Czechoslovakia.

Top 10 countries

Japan	79
Macau	78
Iceland	78
Hong Kong	78
Malta	77
Andorra	77
Sweden	77
Switzerland	77
Spain	77
Netherlands	77

Bottom 10 countries

Chad	47
Mali	46
Niger	46
Angola	46
Ethiopia	46
Gambia	45
Guinea	44
Afghanistan	43
Guinea-Bissau	43
Sierra Leone	42

ccess to
ean water

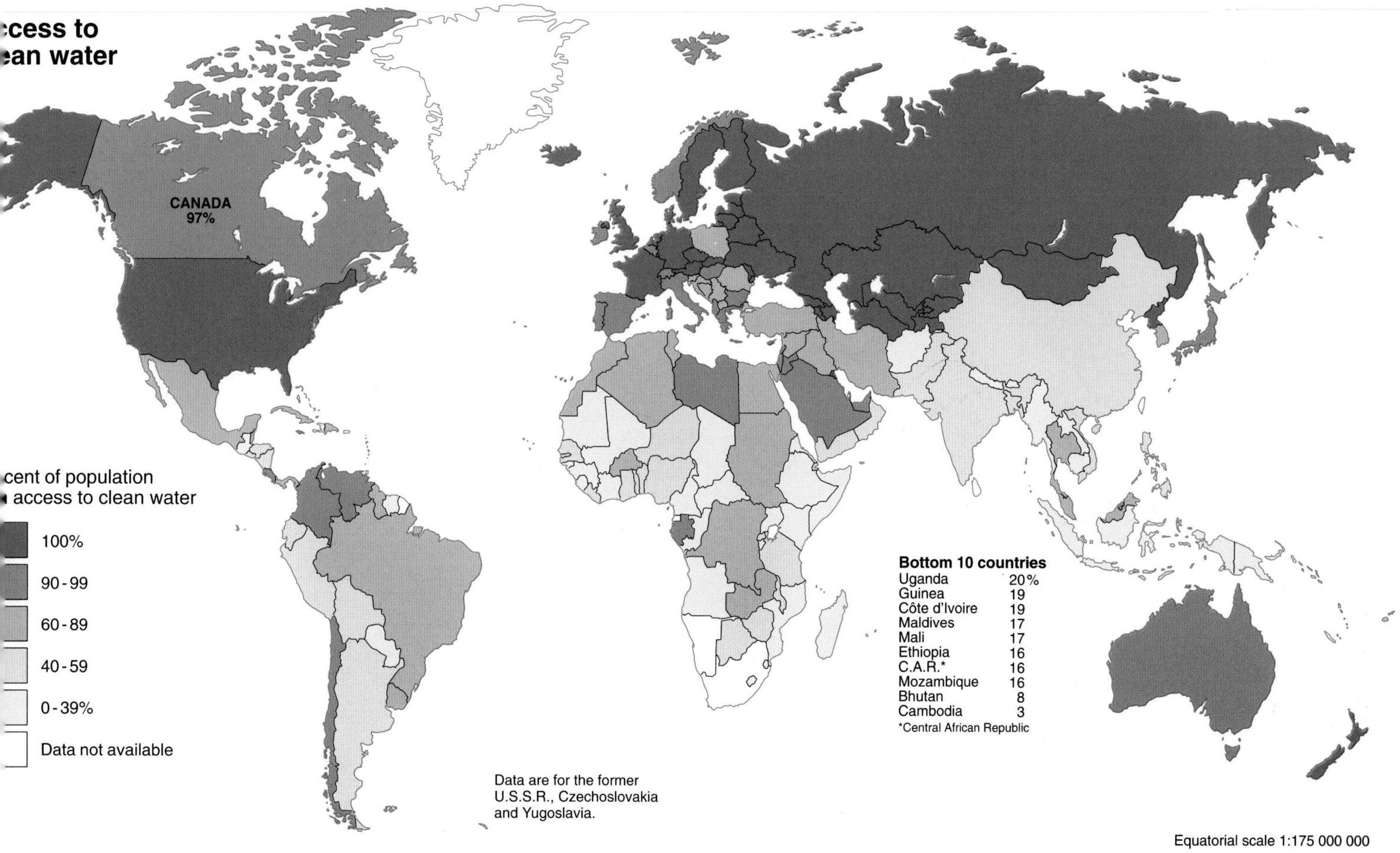

Bottom 10 countries

Uganda	20%
Guinea	19
Côte d'Ivoire	19
Maldives	17
Mali	17
Ethiopia	16
C.A.R.*	16
Mozambique	16
Bhutan	8
Cambodia	3

*Central African Republic

Data are for the former U.S.S.R., Czechoslovakia and Yugoslavia.

Equatorial scale 1:175 000 000

WORLD-THEMATIC

Language

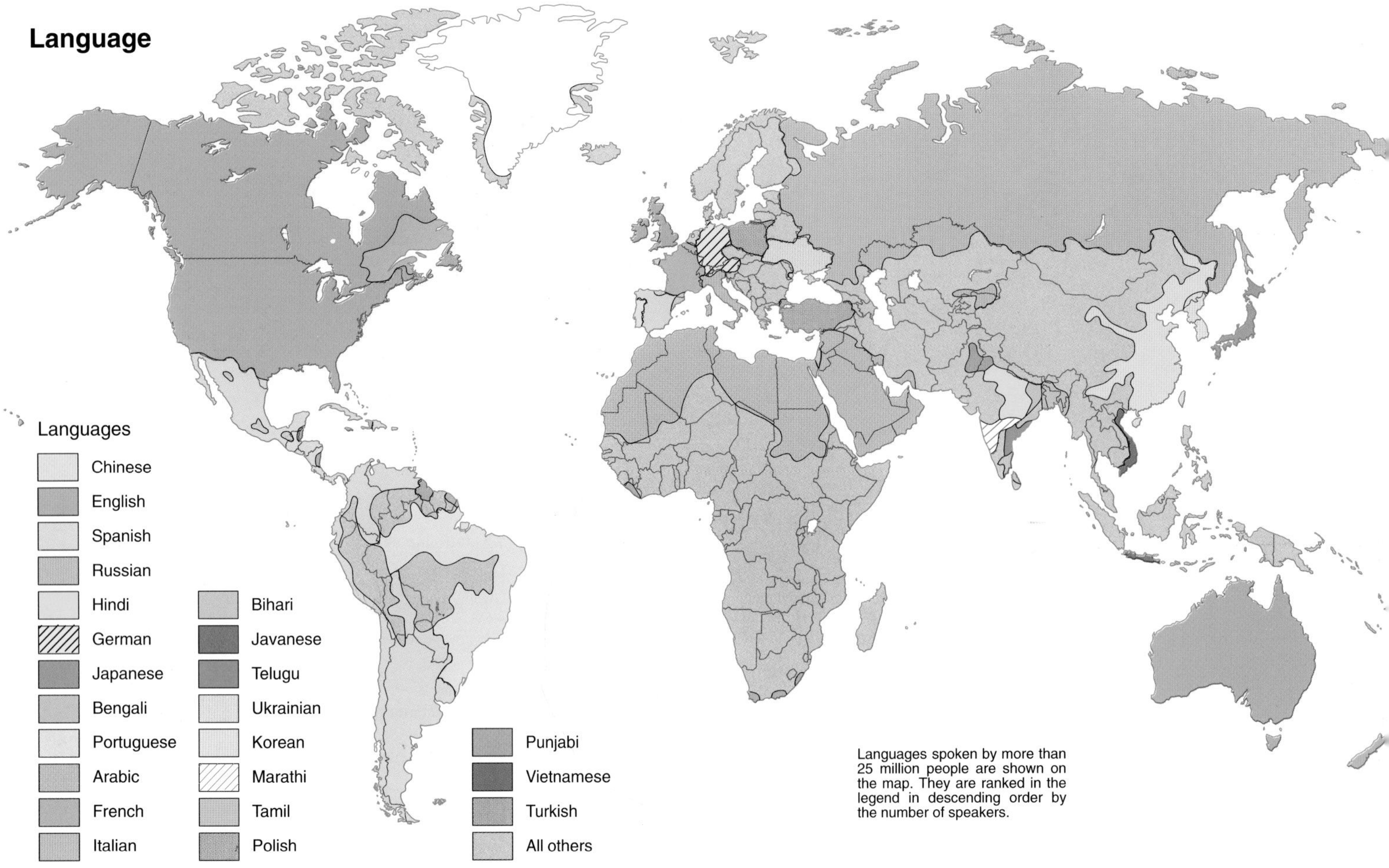

Languages spoken by more than 25 million people are shown on the map. They are ranked in the legend in descending order by the number of speakers.

Literacy

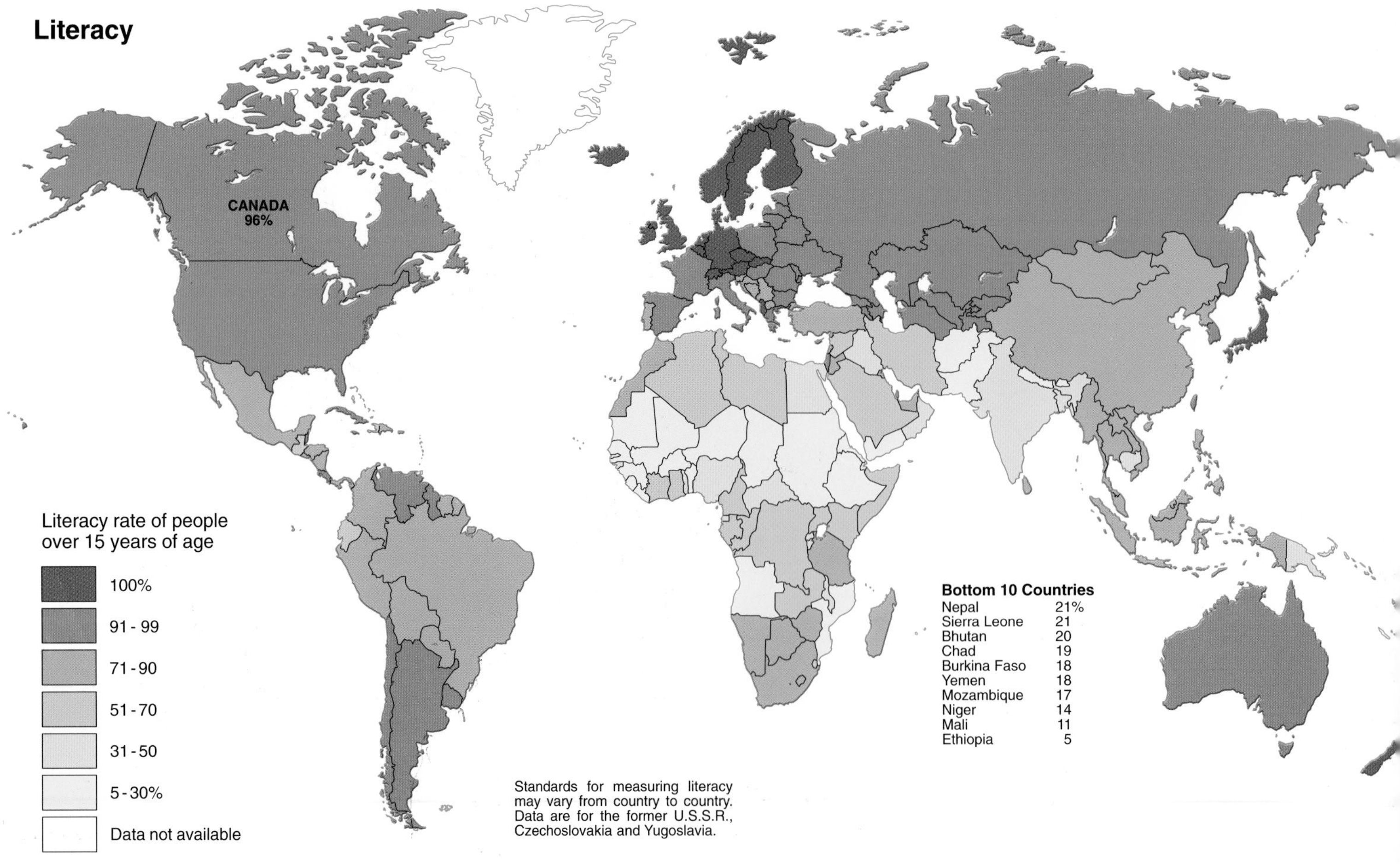

Bottom 10 Countries

Nepal	21%
Sierra Leone	21
Bhutan	20
Chad	19
Burkina Faso	18
Yemen	18
Mozambique	17
Niger	14
Mali	11
Ethiopia	5

Standards for measuring literacy may vary from country to country. Data are for the former U.S.S.R., Czechoslovakia and Yugoslavia.

ligion

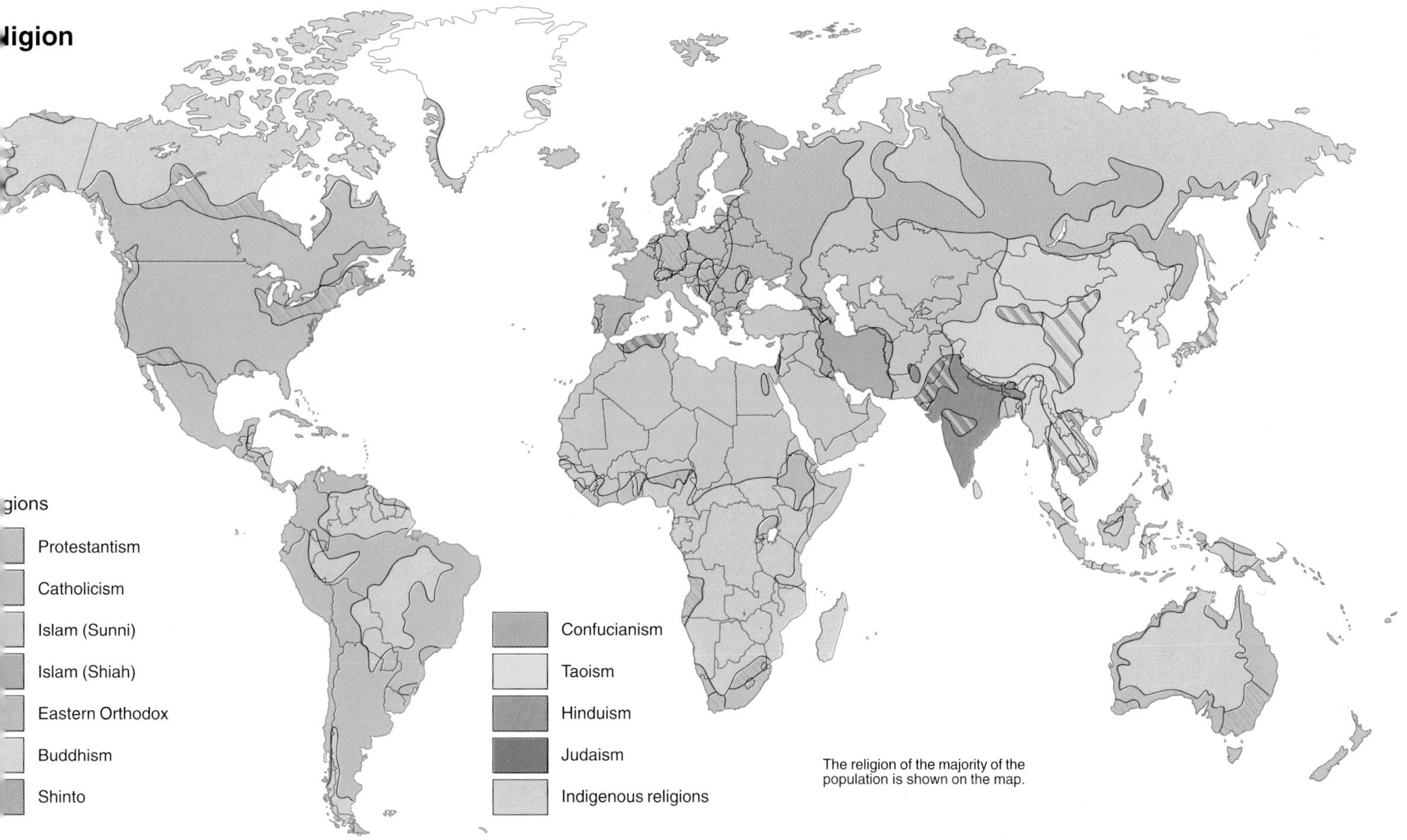

The religion of the majority of the population is shown on the map.

men in the our force

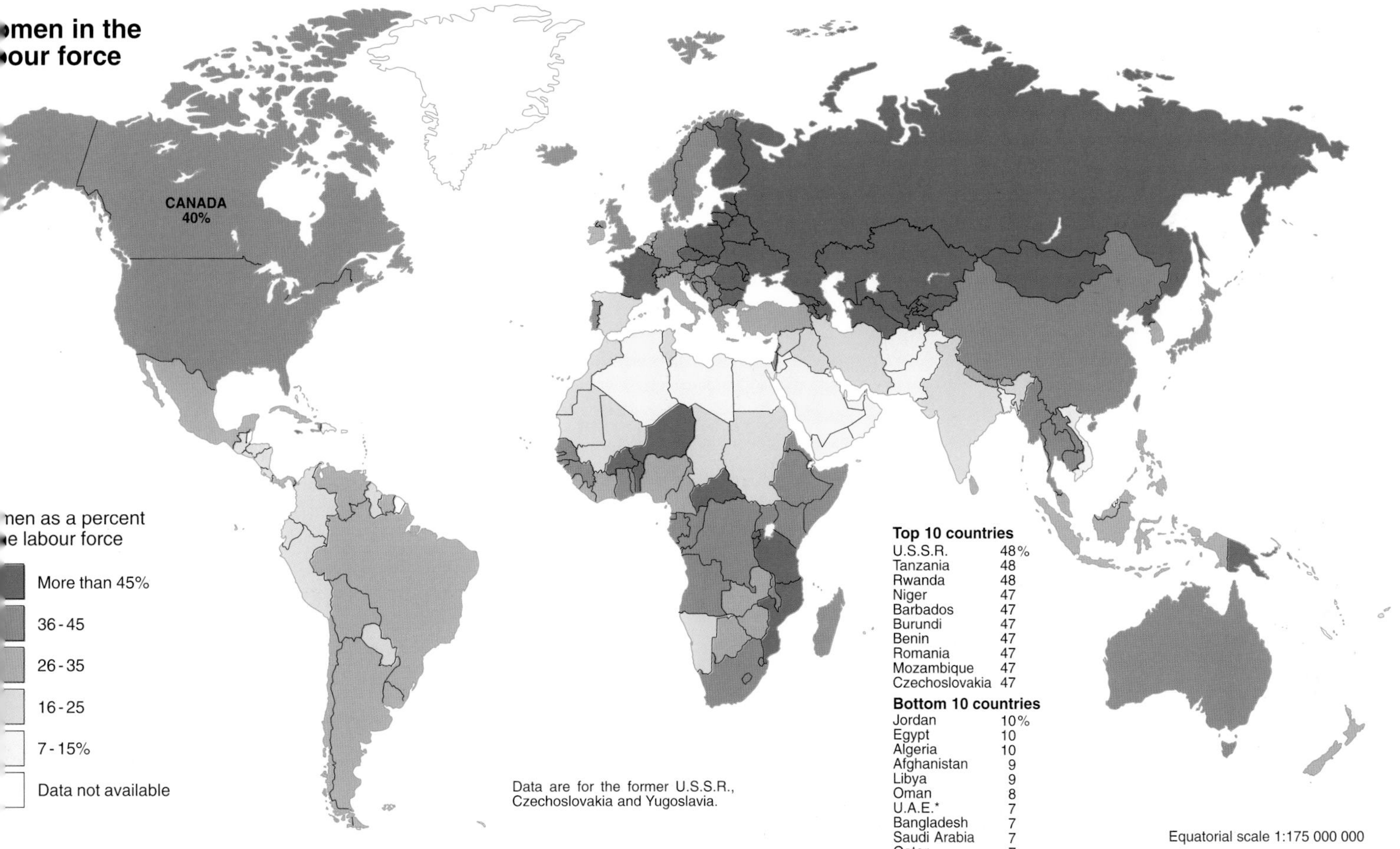

Data are for the former U.S.S.R., Czechoslovakia and Yugoslavia.

Top 10 countries

U.S.S.R.	48%
Tanzania	48
Rwanda	48
Niger	47
Barbados	47
Burundi	47
Benin	47
Romania	47
Mozambique	47
Czechoslovakia	47

Bottom 10 countries

Jordan	10%
Egypt	10
Algeria	10
Afghanistan	9
Libya	9
Oman	8
U.A.E.*	7
Bangladesh	7
Saudi Arabia	7
Qatar	7

*United Arab Emirates

Equatorial scale 1:175 000 000

WORLD-THEMATIC

Military expenditures

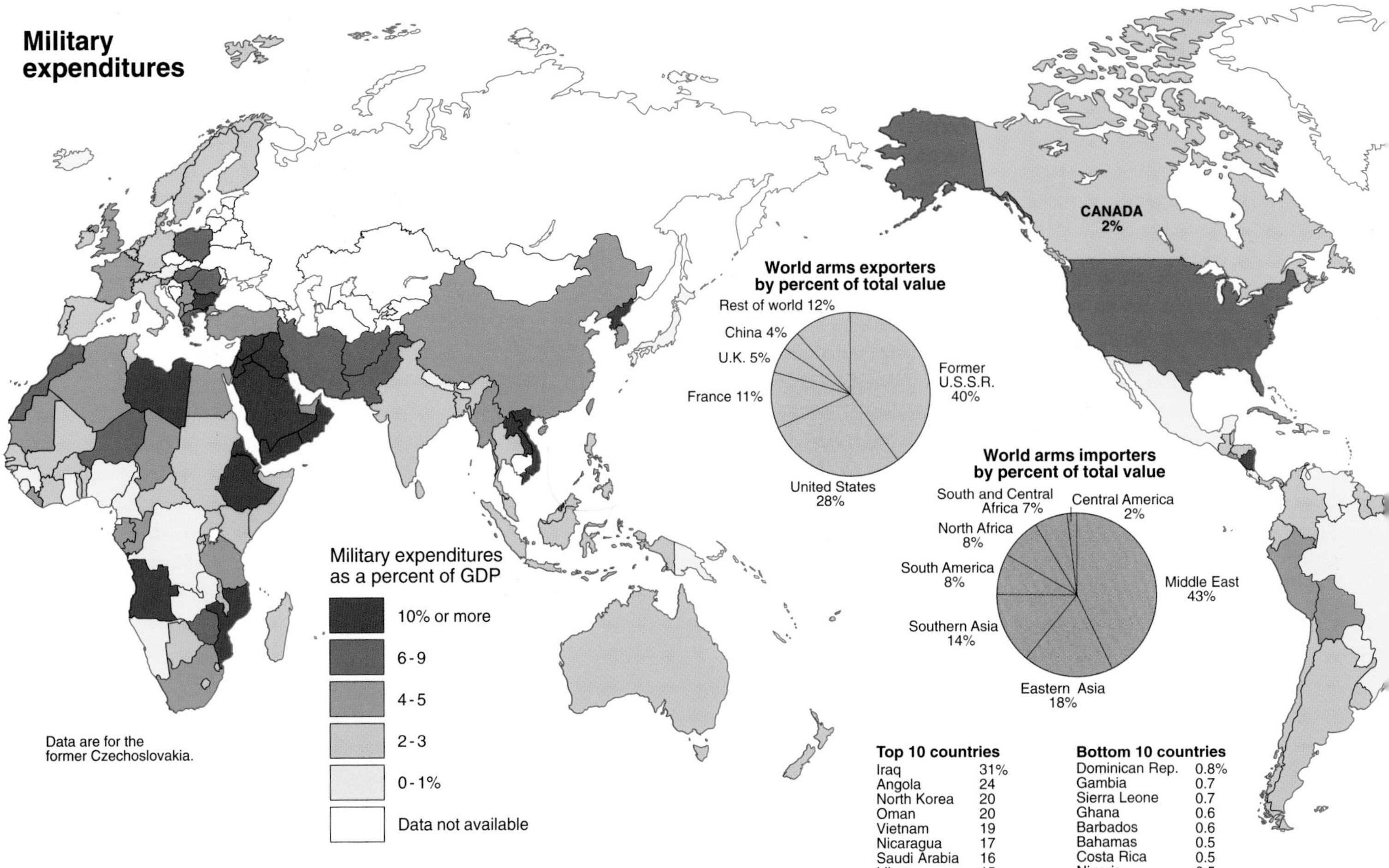

Top 10 countries	
Iraq	31%
Angola	24
North Korea	20
Oman	20
Vietnam	19
Nicaragua	17
Saudi Arabia	16
Libya	15
Jordan	13
Israel	13

Bottom 10 countries	
Dominican Rep.	0.8%
Gambia	0.7
Sierra Leone	0.7
Ghana	0.6
Barbados	0.6
Bahamas	0.5
Costa Rica	0.5
Nigeria	0.5
Mexico	0.5
Mauritius	0.2

Twenty countries have no military expenditures.

Foreign aid

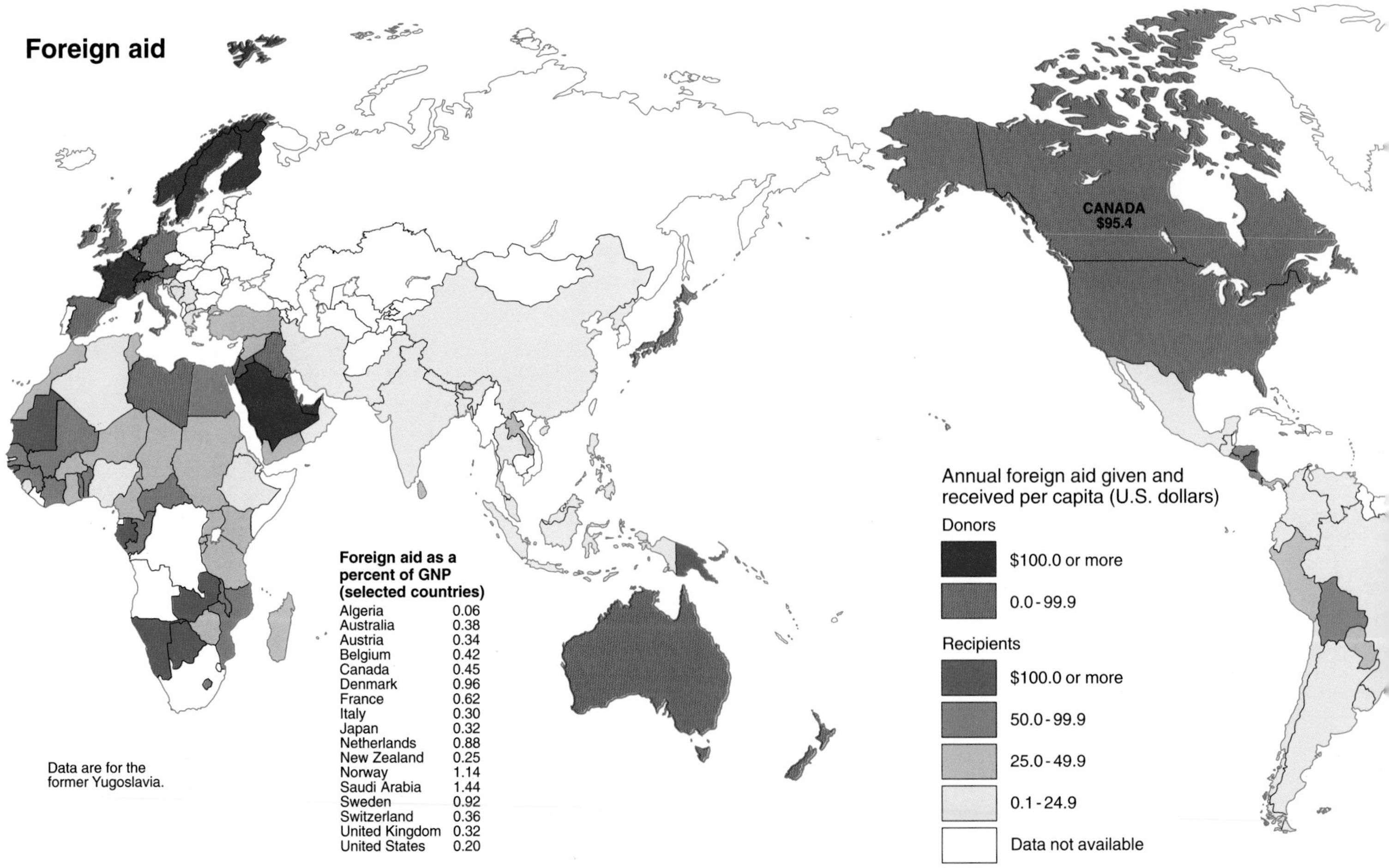

Foreign aid as a percent of GNP (selected countries)

Algeria	0.06
Australia	0.38
Austria	0.34
Belgium	0.42
Canada	0.45
Denmark	0.96
France	0.62
Italy	0.30
Japan	0.32
Netherlands	0.88
New Zealand	0.25
Norway	1.14
Saudi Arabia	1.44
Sweden	0.92
Switzerland	0.36
United Kingdom	0.32
United States	0.20

ugees

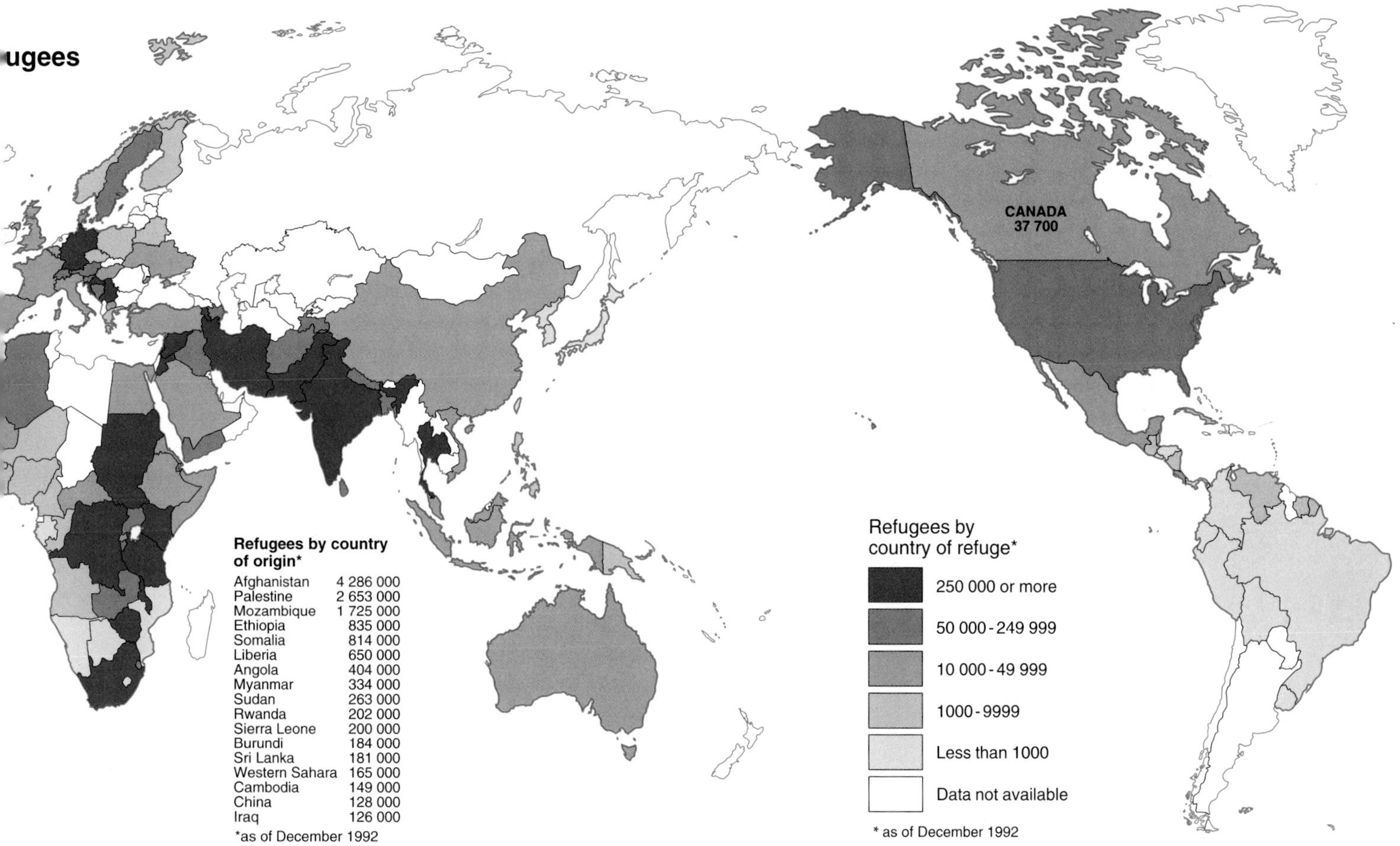

Refugees by country of origin*

Afghanistan	4 286 000
Palestine	2 653 000
Mozambique	1 725 000
Ethiopia	835 000
Somalia	814 000
Liberia	650 000
Angola	404 000
Myanmar	334 000
Sudan	263 000
Rwanda	202 000
Sierra Leone	200 000
Burundi	184 000
Sri Lanka	181 000
Western Sahara	165 000
Cambodia	149 000
China	128 000
Iraq	126 000

*as of December 1992

nan
velopment
ex*

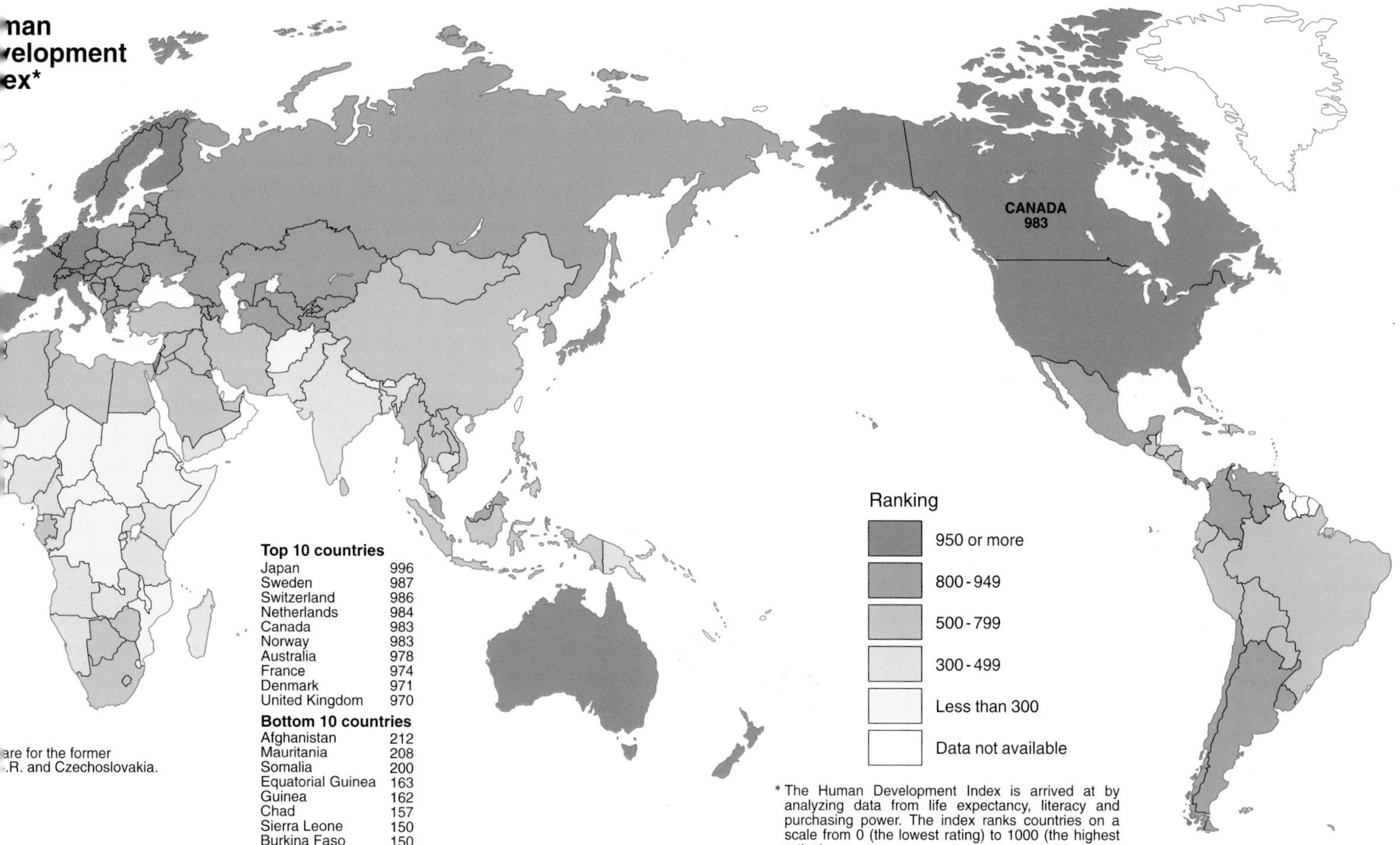

Top 10 countries

Japan	996
Sweden	987
Switzerland	986
Netherlands	984
Canada	983
Norway	983
Australia	978
France	974
Denmark	971
United Kingdom	970

Bottom 10 countries

Afghanistan	212
Mauritania	208
Somalia	200
Equatorial Guinea	163
Guinea	162
Chad	157
Sierra Leone	150
Burkina Faso	150
Mali	143
Niger	116

are for the former
.R. and Czechoslovakia.

* The Human Development Index is arrived at by analyzing data from life expectancy, literacy and purchasing power. The index ranks countries on a scale from 0 (the lowest rating) to 1000 (the highest rating).

Equatorial scale 1:175 000 000

Pollution

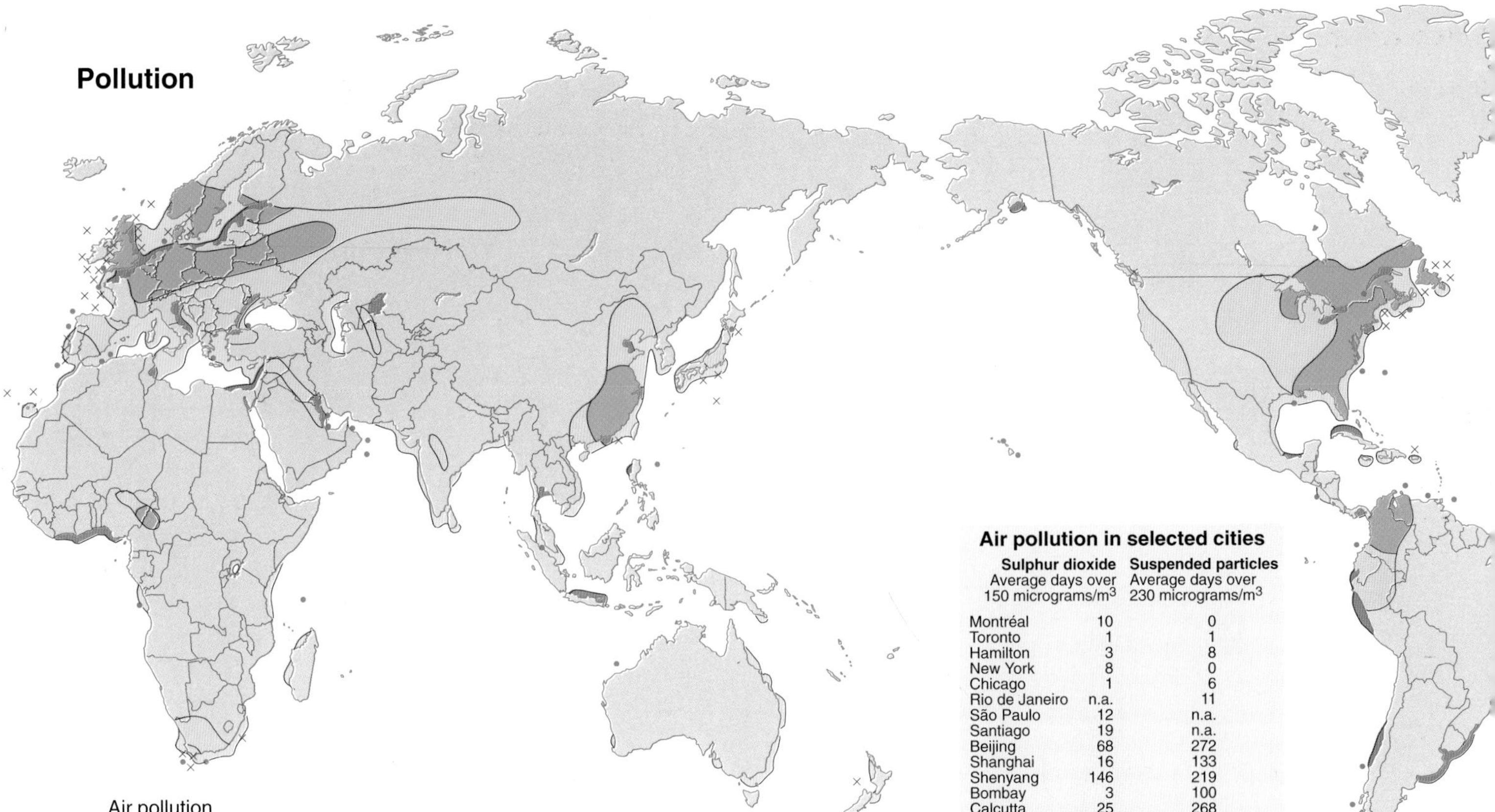

Air pollution causing acid rain

- Areas of air pollution
- Present acid rain problem areas

Water pollution

- Severe coastal pollution
- Persistent coastal pollution
- Significant oil spill
- Deep-sea dump sites

Air pollution in selected cities

	Sulphur dioxide Average days over 150 micrograms/m^3	Suspended particles Average days over 230 micrograms/m^3
Montréal	10	0
Toronto	1	1
Hamilton	3	8
New York	8	0
Chicago	1	6
Rio de Janeiro	n.a.	11
São Paulo	12	n.a.
Santiago	19	n.a.
Beijing	68	272
Shanghai	16	133
Shenyang	146	219
Bombay	3	100
Calcutta	25	268
Tehran	104	174
Tokyo	0	2
Manila	24	14
Bangkok	0	97
Helsinki	2	19
Frankfurt	20	0
Milan	29	n.a.
Warsaw	10	n.a.
Madrid	35	n.a.
London	7	n.a.
Sydney	2	3

The 1992 conference on the environment in Rio de Janeiro focused on international agreements that protect the integrity of the global environmental system. The first principle of the conference dealt with sustainable development, stating that "human beings are entitled to a healthy and productive life in harmony with nature." Unfortunately, human activity is damaging many areas of the world. Climatic change has also had an effect on specific areas—nearly one third of the African continent is now threatened by desertification.

Population growth and energy usage have the greatest impact on the world's environment. Although the population of developed countries such as Canada and the United States may not grow substantially in the next ten years, their highly developed, heavily consumer-oriented economies will have devastating effects on global warming, unless patterns of consumption and energy use change.

Waste generated per year

Country	Total waste tonnes	Municipal waste kg per capita	Industrial waste kg per capita	Nuclear waste kg* per capita
United States	986 800 000	864	3145	7.9
Japan	360 600 000	394	2548	6.3
Former W. Germany	81 630 000	331	1005	5.9
Canada	77 400 000	632	2351	50.1
U.K.	67 770 000	353	997	17.9
France	67 000 000	304	894	16.9
Italy	61 000 000	301	760	–
Australia	30 000 000	681	1362	–
Spain	17 660 000	322	131	6.9
Finland	15 700 000	608	2573	15.6
Austria	14 990 000	228	1748	–
Netherlands	13 590 000	467	453	1.0
Belgium	11 080 000	313	813	2.4
Portugal	8 970 000	231	651	–
Greece	7 450 000	314	429	–
Sweden	6 650 000	317	478	28.7
Denmark	4 800 000	469	469	–
Norway	4 190 000	475	520	–
Switzerland	2 850 000	427	–	2.7
Ireland	2 680 000	311	447	–
New Zealand	2 410 000	662	94	–

* of heavy metal

The "throw-away" mentality of developed nations generates the largest amount of garbage. Developing nations, however, have difficulty disposing of hazardous chemicals and can rarely afford pollution controls. Every year, the earth's population produces millions of tonnes of household and toxic industrial waste, and the world is running out of disposal sites.

Habitat destruction

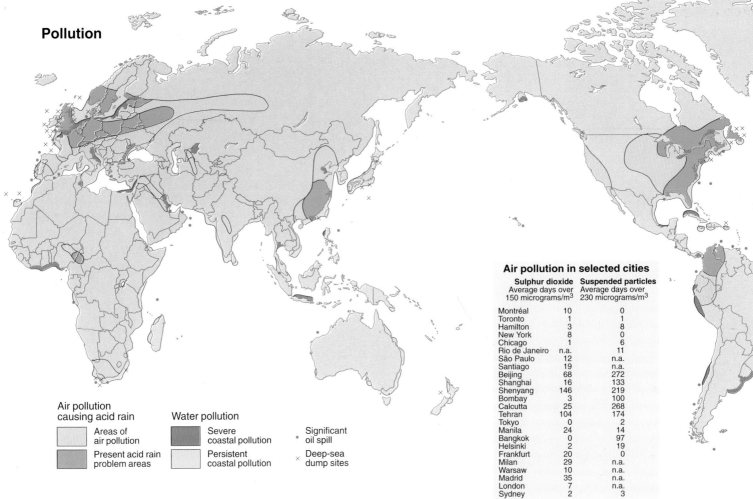

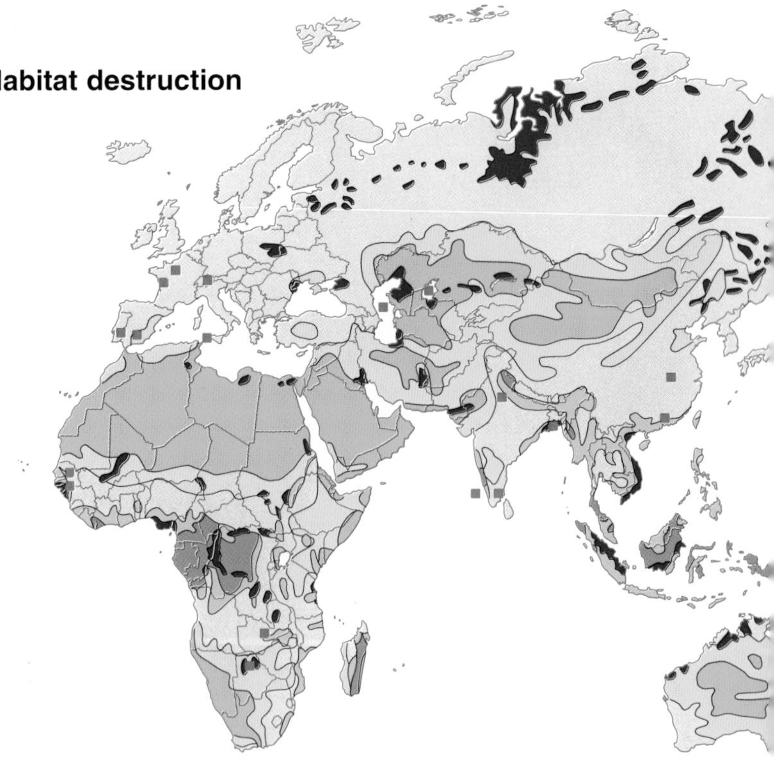

Desertification

- Existing deserts
- Areas at risk of desertification

Deforestation

- Existing tropical forests
- Forests destroyed since 1940
- Large areas of wetlands
- Wetlands under threat

rotected land

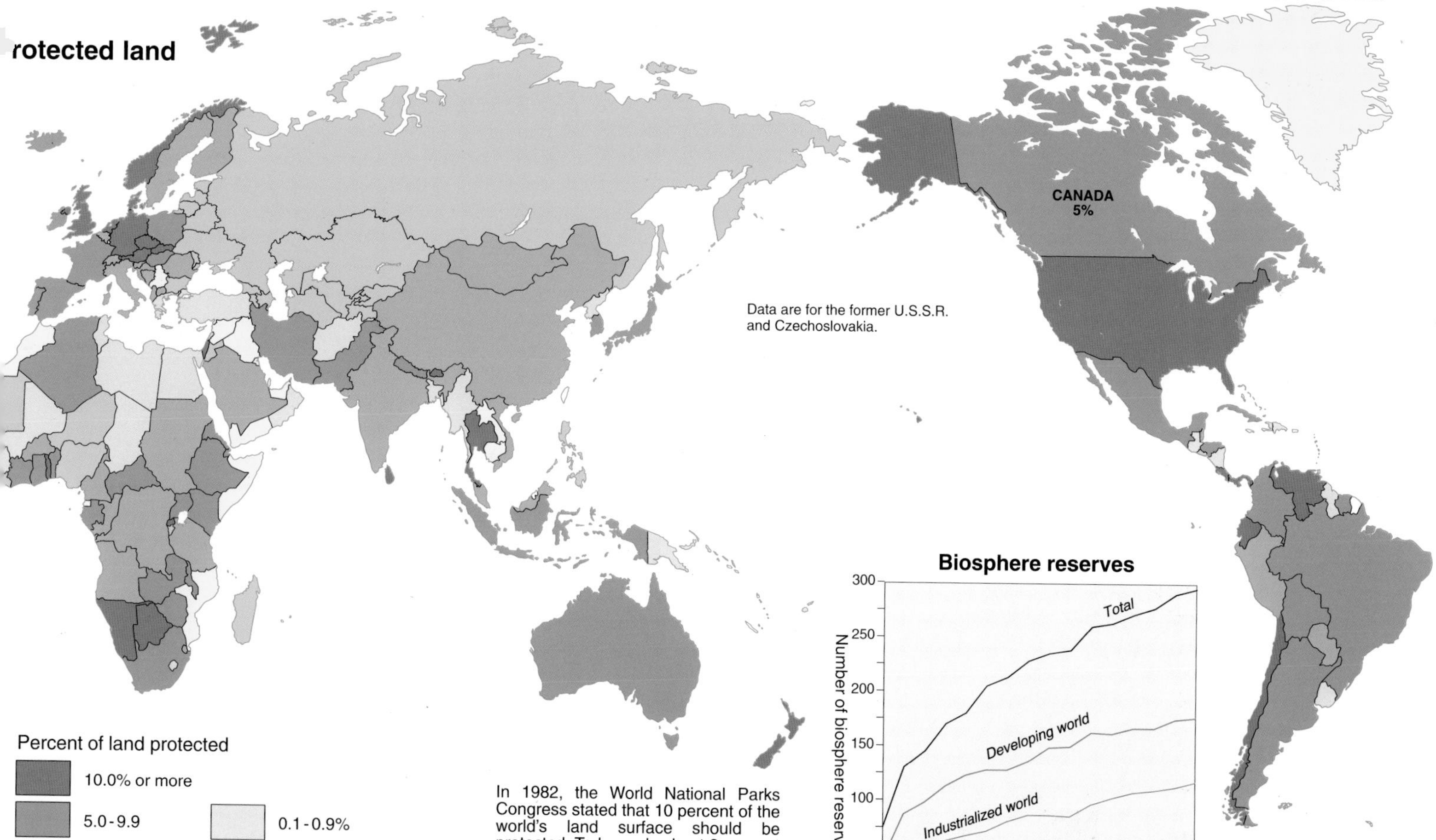

Percent of land protected

- 10.0% or more
- 5.0 - 9.9
- 2.0 - 4.9
- 1.0 - 1.9
- 0.1 - 0.9%
- No land protected
- Data not available

In 1982, the World National Parks Congress stated that 10 percent of the world's land surface should be protected. Today, only about 3 percent is protected in the form of national parks and wildlife and nature reserves. Some of these areas are left undisturbed; others are used for a variety of activities, including the sustainable production of natural resources.

Deforestation

	Forest area (ha) 1990	Average annual deforestation (ha) 1981-90	Average annual rate of change 1981-90
Latin America	**839 900 000**	**8 300 000**	**−0.9**%
Central America and Mexico	63 500 000	1 400 000	−1.8
Caribbean subregion	47 100 000	200 000	−0.4
Tropical South America	729 300 000	6 800 000	−0.8
Asia	**274 900 000**	**3 600 000**	**−1.2**
South Asia	66 200 000	400 000	−0.6
Continental Southeast Asia	69 700 000	1 300 000	−1.6
Southeast Asia islands	138 900 000	1 800 000	−1.2
Africa	**600 100 000**	**5 000 000**	**−0.8**
West Sahelian Africa	38 000 000	400 000	−0.9
East Sahelian Africa	85 300 000	700 000	−0.8
West Africa	43 400 000	1 200 000	−2.1
Central Africa	215 400 000	1 500 000	−0.6
Tropical Southern Africa	206 300 000	1 100 000	−0.5
African islands	11 700 000	200 000	−1.2

Percent of world species found in rainforests

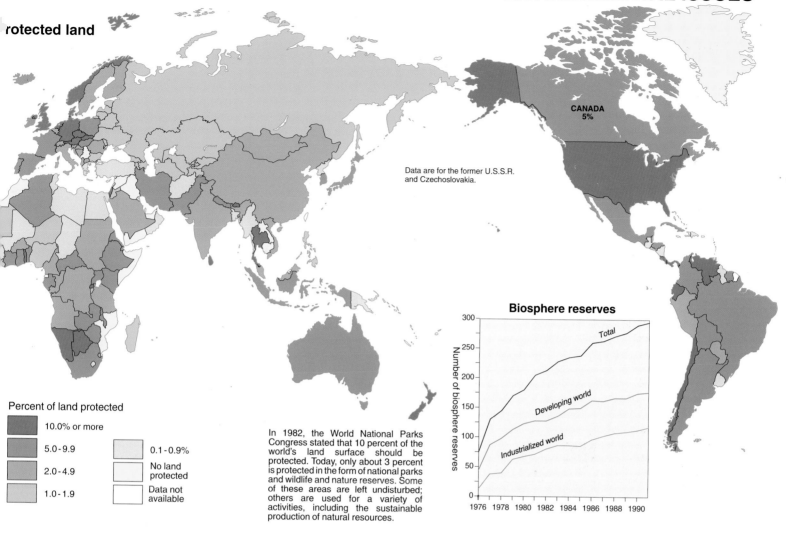

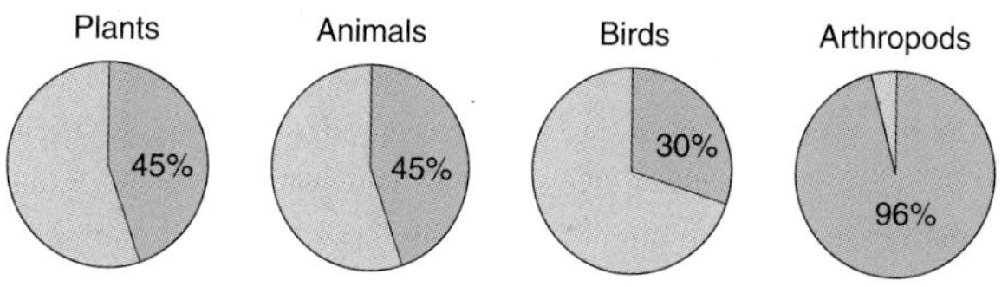

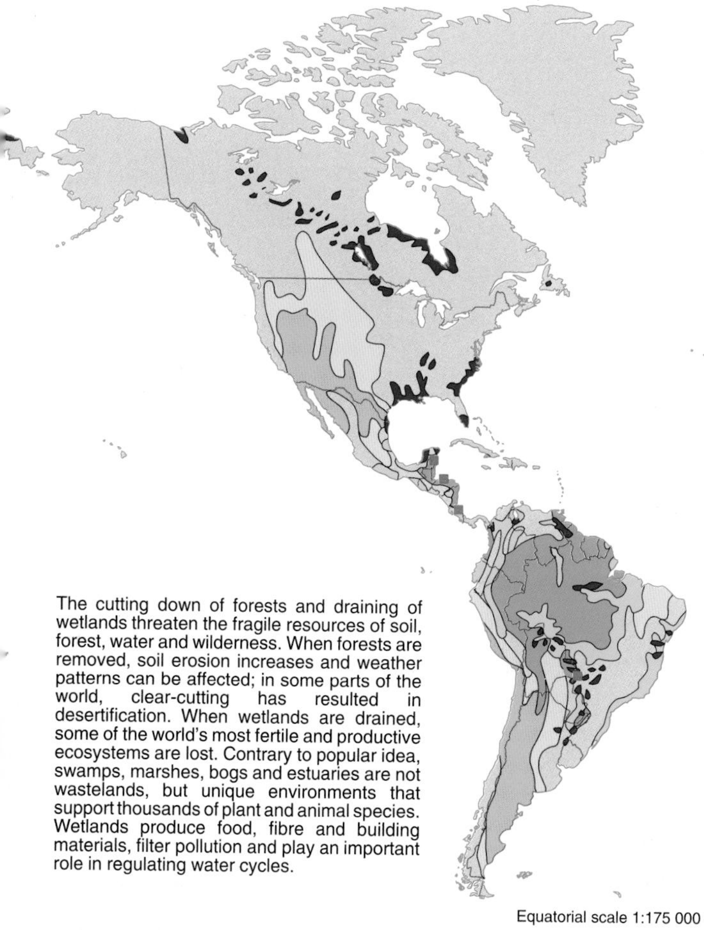

The cutting down of forests and draining of wetlands threaten the fragile resources of soil, forest, water and wilderness. When forests are removed, soil erosion increases and weather patterns can be affected; in some parts of the world, clear-cutting has resulted in desertification. When wetlands are drained, some of the world's most fertile and productive ecosystems are lost. Contrary to popular idea, swamps, marshes, bogs and estuaries are not wastelands, but unique environments that support thousands of plant and animal species. Wetlands produce food, fibre and building materials, filter pollution and play an important role in regulating water cycles.

Equatorial scale 1:175 000 000

The rainforest is home to over 60 percent of identified plant species and 90 percent of all primates; 20 percent of all bird species are found in the Amazon forests alone. Scientists estimate that the destruction of the tropical rainforest eliminates at least one species a day, and that 10 percent of the world's species could become extinct by the year 2000.

Estimated number of species

	Already identified	Estimated percent of total
Micro-organisms	5760	3 - 27%
Invertebrates	1 020 561	
Plants	322 311	67 - 100
Fish	19 056	83 - 100
Birds	9040	94 - 100
Reptiles and amphibians	10 484	90 - 95
Mammals	4000	
TOTAL	**1 392 485**	

Estimated number of world species: 4.5 to 33.5 million
Percent of species yet to be identified: 69 to 96%

For explanation of terms, see glossary.

Ozone (O_3) is a pungent-smelling colourless gas. In the earth's upper atmosphere, ozone is produced by natural processes; at ground level, it can also be formed from air pollutants. The region of the upper atmosphere with the highest concentration of ozone is called the ozone layer. It protects life on earth from the damaging ultraviolet rays of the sun.

A large thin area in the ozone layer, or a "hole," was first discovered over Antarctica in 1985. Scientists have also detected significant thinning over the Arctic, as well as a world-wide decrease in average ozone levels.

Today, atmospheric scientists think that emissions of ozone-depleting chemicals are responsible for this loss of ozone. The most significant chemicals are chlorofluorocarbons (CFCs), although other substances are also thought to be involved. CFCs are especially dangerous because the molecules remain intact for many years and can rise to the upper atmosphere.

The ozone layer

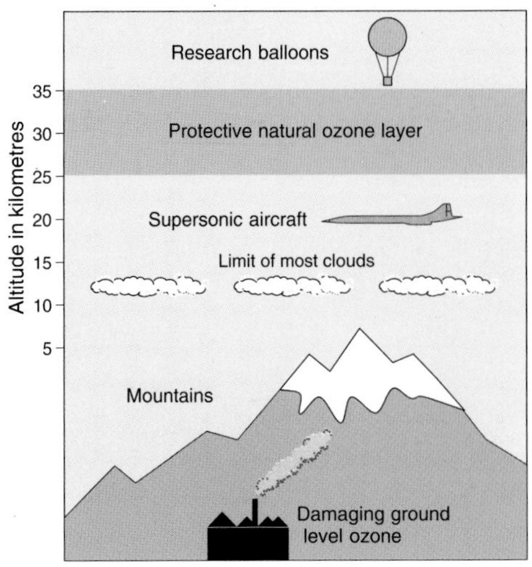

Vertical distribution of ozone

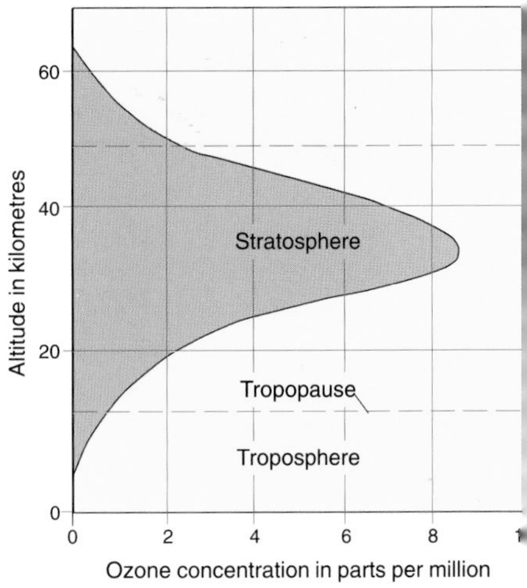

Ozone loss

Northern hemisphere, April 1993

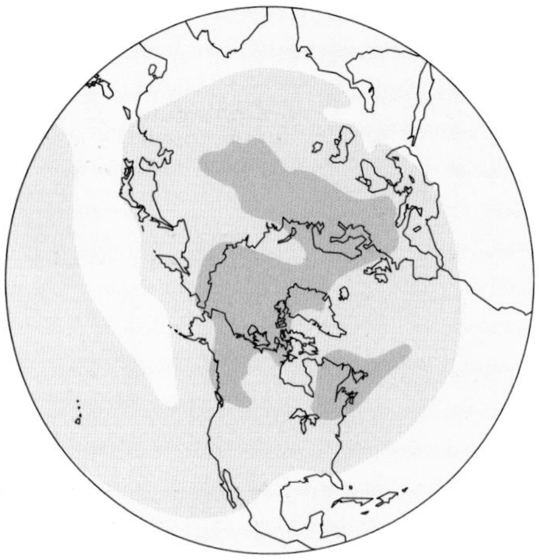

Southern hemisphere, October 1992

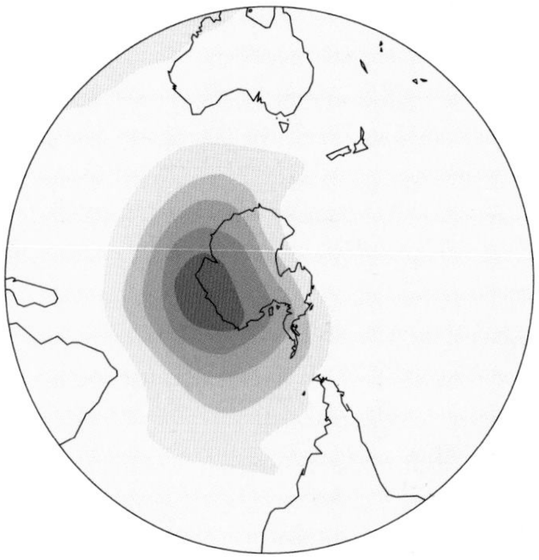

Percent decrease in ozone
(from 1979-82 average)

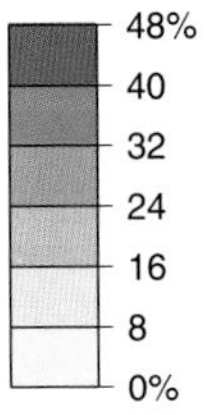

The depletion of the ozone layer is especially marked over the north and south poles. In Antarctica, the depletion is worst from September to December. In the Arctic, thinning is most pronounced in March and April. Scientists have found that the area of ozone depletion and the amount of depletion are increasing, although Canadian ozone thinning varies greatly from year to year. In addition to these ozone "holes," global levels of ozone have also decreased from earlier levels.

The destruction of ozone

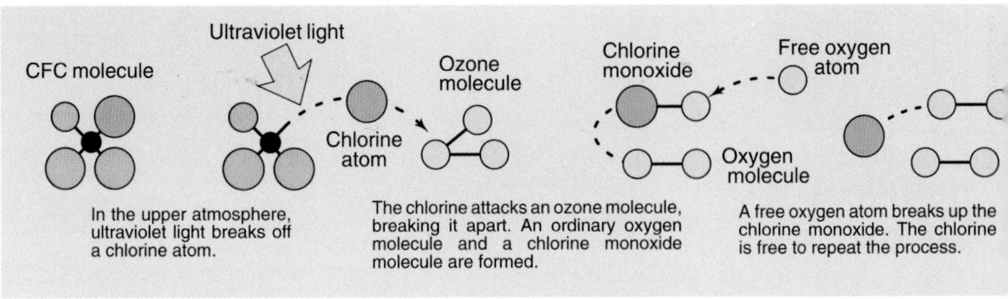

Emissions of CFCs

Percent of world total

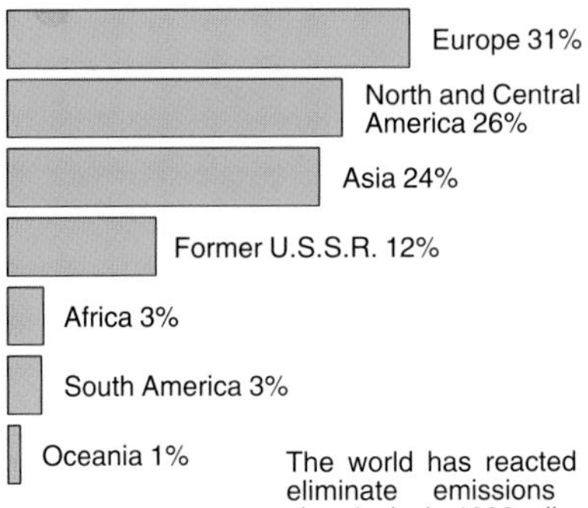

Contribution of gases to ozone destruc

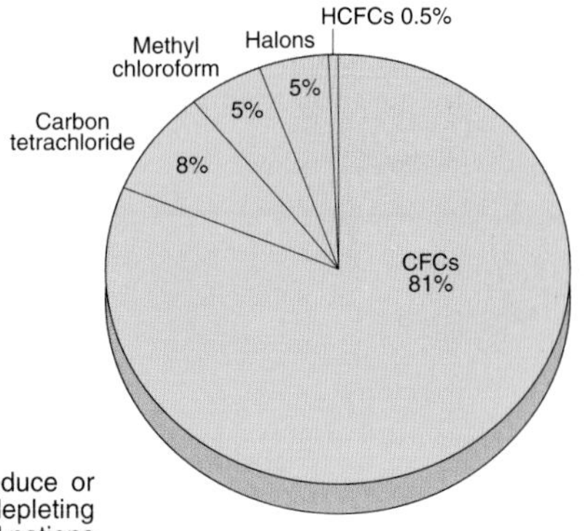

The world has reacted quickly to reduce or eliminate emissions of ozone-depleting chemicals. In 1990, all major industrial nations agreed to eliminate the production and trade of CFCs by the year 2000. This target date was later moved up to 1996. As well, an international fund has been established to help developing countries avoid a dependence on these chemicals. If all ozone-depleting chemicals are strictly controlled, the ozone layer will eventually heal itself. However, because these chemicals will remain in the atmosphere for many years, ozone depletion will continue before the situation improves.

Ozone-depleting chemicals

Chemical	Use	Lifetime in atmosphere in years	Year to cease production
CFCs	Refrigerators, car air conditioners, foam cushioning and insulation, solvent, sterilizing agent	50-110	1996
HCFCs	Home air conditioners, plastic insulation and packaging foam, some aerosols	15	2020-2030
Halons	Fire extinguishers, fire suppressant systems	25-110	1994
Carbon tetrachloride	Solvents, production of CFCs, chlorine and pesticides	50	1996
Methyl chloroform	Industrial solvent for cleaning metal and electronics	6	1996

OZONE LAYER / GLOBAL WARMING

The earth is warmed by a natural greenhouse effect: gases in the atmosphere absorb infra-red radiation emitted from the earth's surface. Some gases, such as water vapour, carbon dioxide, methane and CFCs, absorb this radiation particularly well. Scientists now believe that increases in concentrations of carbon dioxide, methane and CFCs since preindustrial times are causing the temperature of the earth's surface to rise.

Although the temperature increase is estimated to be small (perhaps one or two Celsius degrees by 2020), even this change will have severe effects on life on earth. Sea levels will rise as polar ice caps melt and ocean waters expand, flooding low-lying coastal areas. Precipitation patterns will change, disrupting harvests. Wildlife will have trouble adjusting to rapid changes in habitat. Global warming can best be controlled by conserving energy, halting deforestation and planting more trees. To achieve this goal, international cooperation is essential.

Atmospheric gases and the greenhouse effect

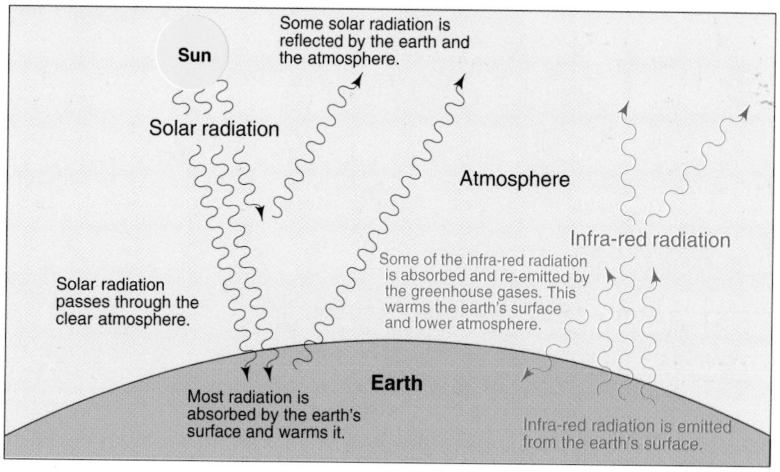

Top producers of greenhouse gases
(percent of global emissions)

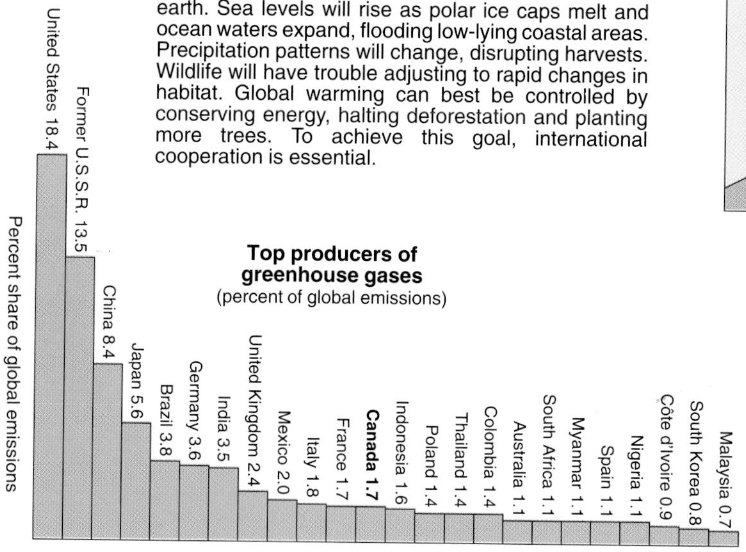

Contribution of greenhouse gases to global warming

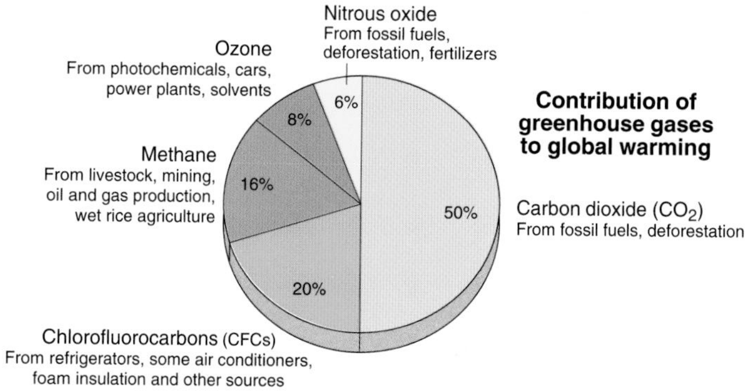

Emissions of CO_2 and Methane

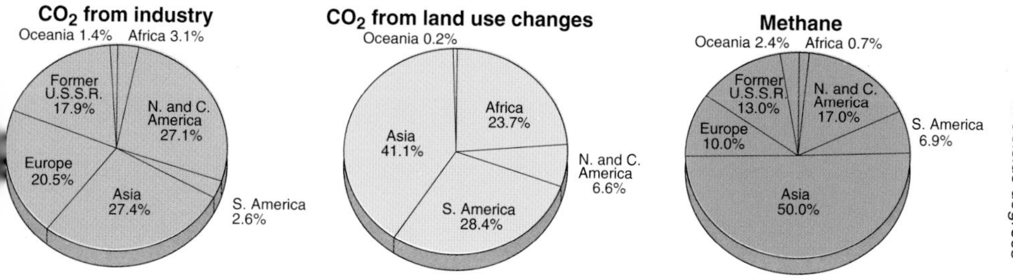

Temperature change due to greenhouse effect

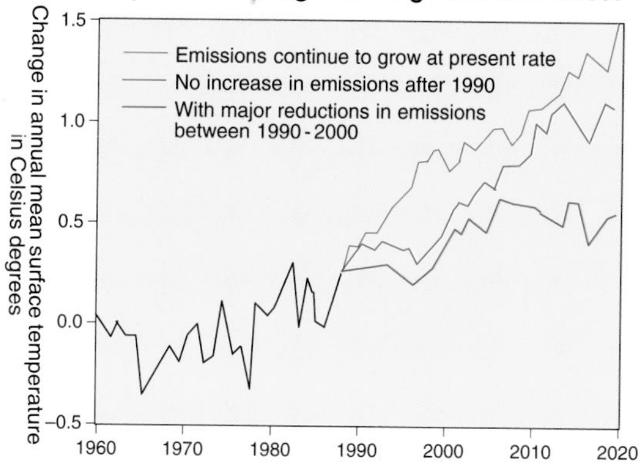

Mean annual temperature change from 1930-59 to 1960-89

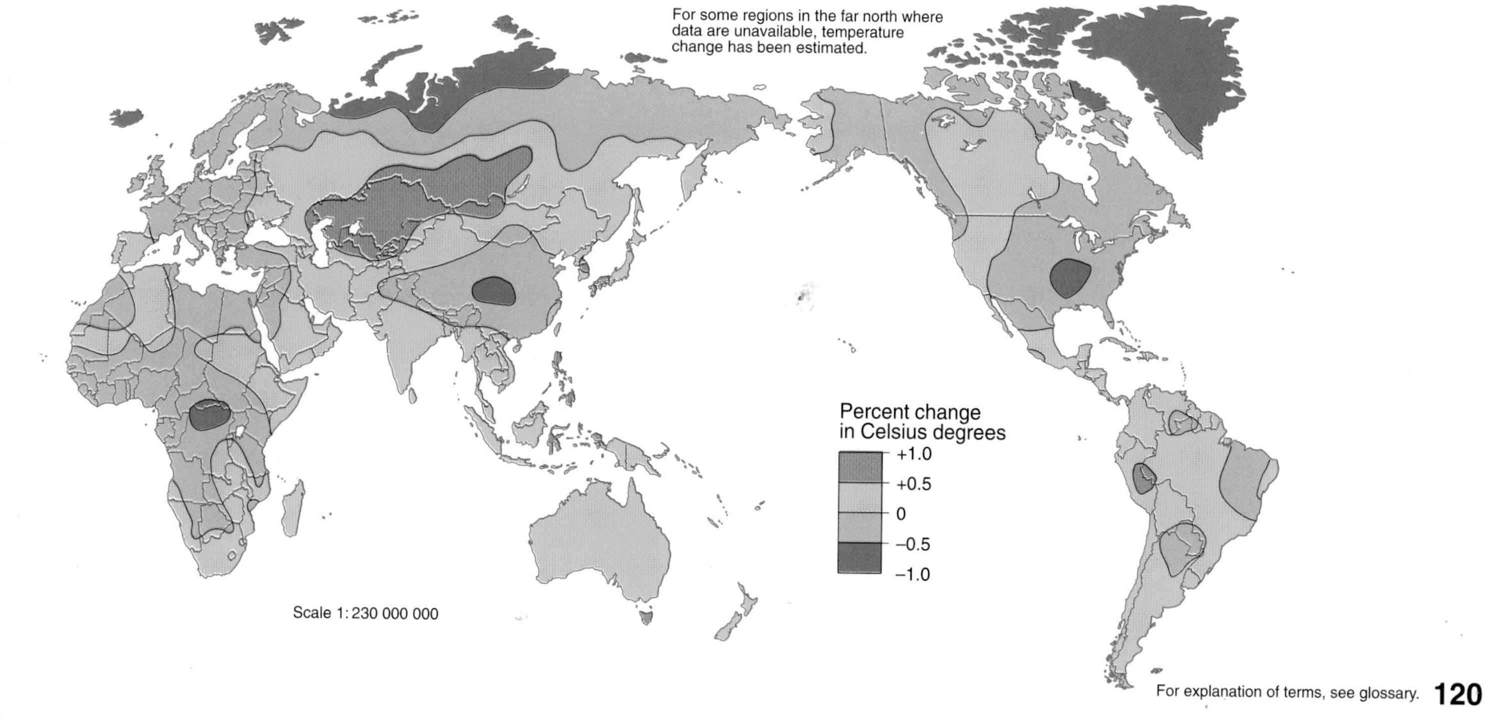

NORWAY
ICELAND
Reykjavík
Arctic Circle
DENMARK STRAIT
GREENLAND SEA
Greenwich Meridian
KALAALLIT NUNAAT
(Denmark)
Nuuk
(Godthåb)
LABRADOR SEA
DAVIS STRAIT
BAFFIN BAY
BAFFIN ISLAND
HUDSON STRAIT
HUDSON BAY
Iqaluit
Kuujjuaq
Inukjuak
Kuujjuarapik
Goose Bay
Smallwood Reservoir
Churchill
Sept-Îles
Gander
St. John's
ST-PIERRE ET MIQUELON
(France)
Charlottetown
Moosonee
Albany
ARCTIC OCEAN
North Pole
ELLESMERE ISLAND
DEVON I.
North Magnetic Pole
Resolute
QUEEN ELIZABETH ISLANDS
VICTORIA ISLAND
BANKS ISLAND
BEAUFORT SEA
CANADA
Churchill
Lake Winnipeg
Saskatchewan
Saskatoon
Fort McMurray
Athabasca
Peace
Yellowknife
Great Slave Lake
Great Bear Lake
Mackenzie
Tuktoyaktuk
Inuvik
Edmonton
Calgary
Prince George
Fraser
Columbia
Kamloops
Vancouver
Victoria
VANCOUVER ISLAND
QUEEN CHARLOTTE ISLANDS (HAIDA GWAII)
Juneau
Whitehorse
Dawson
Prudhoe Bay
Fairbanks
ALASKA
(UNITED STATES)
Yukon
Valdez
Anchorage
GULF OF ALASKA
Nome
BERING STRAIT
BERING SEA
RUSSIA

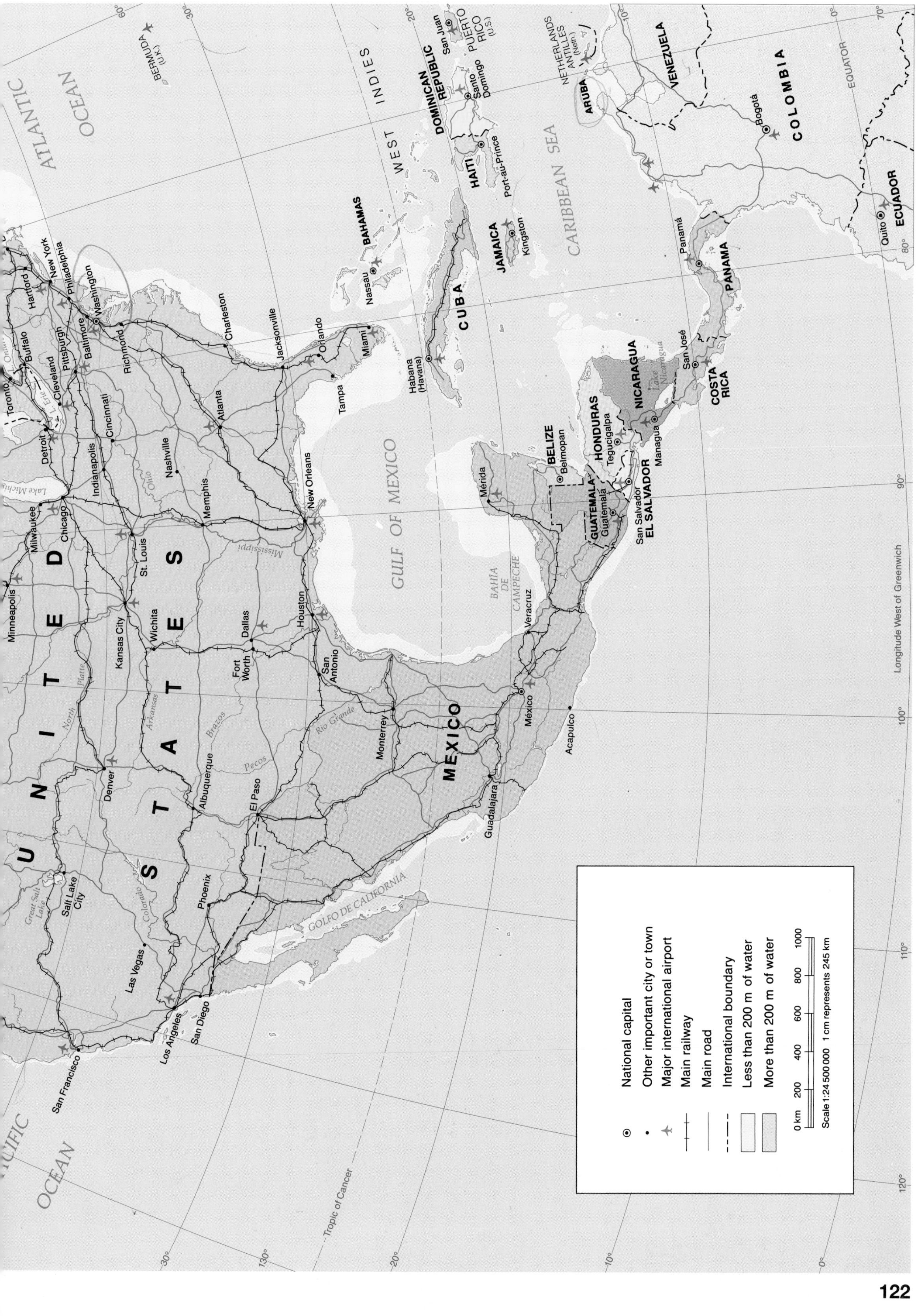

ATLANTIC OCEAN
PACIFIC OCEAN
BERMUDA (U.K.)
WEST INDIES
GULF OF MEXICO
CARIBBEAN SEA
BAHÍA DE CAMPECHE
GOLFO DE CALIFORNIA
UNITED STATES
MEXICO
BAHAMAS
CUBA
JAMAICA
HAITI
DOMINICAN REPUBLIC
PUERTO RICO (U.S.)
NETHERLANDS ANTILLES (Neth.)
ARUBA
VENEZUELA
COLOMBIA
ECUADOR
BELIZE
GUATEMALA
HONDURAS
EL SALVADOR
NICARAGUA
COSTA RICA
PANAMA
Toronto
Buffalo
Hartford
New York
Philadelphia
Washington
Baltimore
Pittsburgh
Cleveland
Detroit
Richmond
Charleston
Jacksonville
Orlando
Miami
Tampa
Atlanta
Cincinnati
Indianapolis
Nashville
Memphis
New Orleans
Chicago
Milwaukee
Minneapolis
St. Louis
Kansas City
Wichita
Dallas
Fort Worth
Houston
San Antonio
Denver
Albuquerque
El Paso
Salt Lake City
Phoenix
Las Vegas
Los Angeles
San Diego
San Francisco
Monterrey
Guadalajara
México
Acapulco
Veracruz
Mérida
Belmopan
Guatemala
San Salvador
Tegucigalpa
Managua
San José
Panamá
Habana (Havana)
Nassau
Kingston
Port-au-Prince
Santo Domingo
San Juan
Bogotá
Quito
Lake Michigan
L. Erie
Ohio
Mississippi
Arkansas
North Platte
Brazos
Pecos
Rio Grande
Colorado
Great Salt Lake
Lake Nicaragua
Tropic of Cancer
EQUATOR
Longitude West of Greenwich
National capital
Other important city or town
Major international airport
Main railway
Main road
International boundary
Less than 200 m of water
More than 200 m of water
0 km 200 400 600 800 1000
Scale 1:24 500 000 1 cm represents 245 km

CANADA
MONTANA
NORTH DAKO
SOUTH DAKO
WASHINGTON
OREGON
IDAHO
WYOMING
NEBRASK
NEVADA
UTAH
COLORADO
KANS
CALIFORNIA
ARIZONA
NEW MEXICO
OK
TEXA
MEXICO
ROCKY MOUNTAINS
CASCADE RANGE
COAST RANGE
WASSACH RANGE
GREAT PLAINS
BLACK HILLS
GRAND CANYON
PACIFIC OCEAN
GOLFO DE CALIFORNIA
Victoria
Vancouver
Bellingham
Seattle
Olympia
Tacoma
Spokane
Coeur d'Alene
Portland
Salem
Eugene
Walla Walla
Great Falls
Helena
Butte
Billings
Regina
Bismarck
Aberde
Rapid City
Pierre
Boise
Pocatello
Casper
Eureka
Reno
Carson City
Ogden
Salt Lake City
Provo
Cheyenne
Greeley
Napa
San Francisco
Oakland
Sacramento
San Jose
Monterey
Salinas
Fresno
Steamboat Springs
Vail
Aspen
Denver
Aurora
Colorado Springs
Pueblo
Dodge City
San Luis Obispo
Bakersfield
Las Vegas
Santa Barbara
Los Angeles
San Bernardino
San Diego
Tijuana
Mexicali
Yuma
Phoenix
Flagstaff
Gallup
Santa Fe
Albuquerque
Amarillo
Oklaho
Lawton
Lubbock
Tucson
El Paso
Ciudad Juarez
Carlsbad
Abilene
Odessa
Chihuahua
Laredo
Monterrey
Torreón
Columbia
Missouri
Fort Peck Reservoir
Lake Sakakawea
Yellowstone
Snake
Great Salt Lake
Humboldt
North Platte
Platte
Green
Colorado
Lake Powell
Rio Grande
Lake Mead
Arkansas
Lake Havasu
Gila
Canadian
Brazos
Pecos
Rio Bravo Del Norte
Conchos
HAWAIIAN ISLANDS
KAUAI
OAHU
Waipahu
Honolulu
HAWAII
MAUI
Hilo
125°
120°
115°
110°
105°
100°
50°
45°
40°
35°
30°
25°
20°
160°

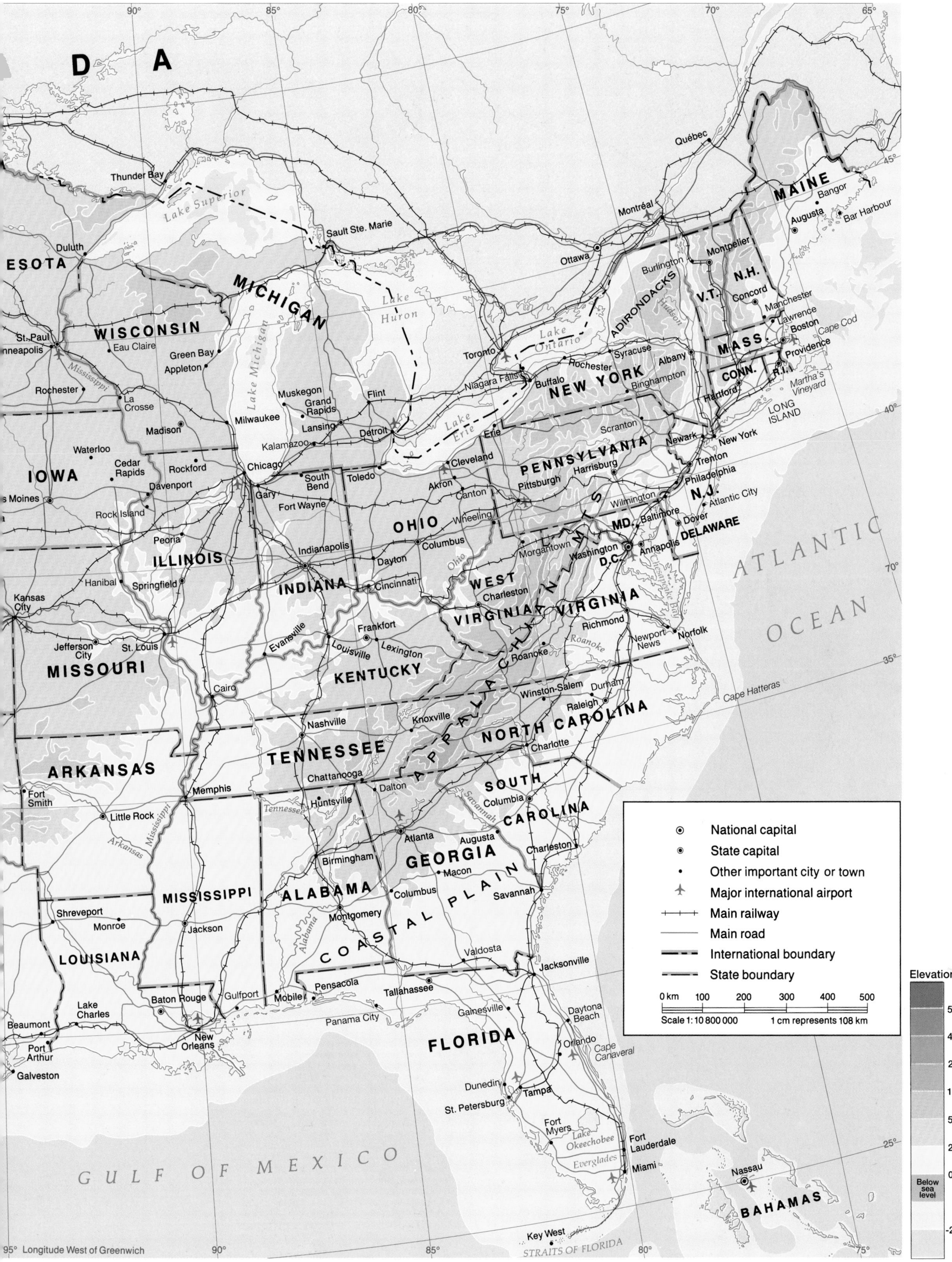

90°
85°
80°
75°
70°
65°
45°
40°
35°
25°
70°
D A
ESOTA
MICHIGAN
WISCONSIN
IOWA
ILLINOIS
INDIANA
OHIO
MISSOURI
KENTUCKY
WEST VIRGINIA
VIRGINIA
PENNSYLVANIA
NEW YORK
MAINE
N.H.
V.T.
MASS.
CONN.
R.I.
N.J.
MD.
DELAWARE
D.C.
TENNESSEE
NORTH CAROLINA
SOUTH CAROLINA
ARKANSAS
MISSISSIPPI
ALABAMA
GEORGIA
LOUISIANA
FLORIDA
BAHAMAS
ADIRONDACKS
APPALACHIAN MTS
COASTAL PLAIN
ATLANTIC OCEAN
GULF OF MEXICO
STRAITS OF FLORIDA
Lake Superior
Lake Huron
Lake Michigan
Lake Ontario
Lake Erie
Lake Okeechobee
Everglades
Mississippi
Ohio
Tennessee
Arkansas
Alabama
Savannah
Roanoke
Hudson
Chesapeake Bay
LONG ISLAND
Martha's Vineyard
Cape Cod
Cape Hatteras
Cape Canaveral
Thunder Bay
Sault Ste. Marie
Duluth
St. Paul
Minneapolis
Eau Claire
Green Bay
Appleton
Rochester
La Crosse
Madison
Milwaukee
Muskegon
Grand Rapids
Flint
Lansing
Detroit
Kalamazoo
Waterloo
Cedar Rapids
Rockford
Chicago
South Bend
Toledo
Davenport
Des Moines
Gary
Fort Wayne
Rock Island
Peoria
Indianapolis
Hanibal
Springfield
Kansas City
Jefferson City
St. Louis
Cairo
Evansville
Louisville
Frankfort
Lexington
Cincinnati
Dayton
Columbus
Cleveland
Akron
Canton
Wheeling
Erie
Pittsburgh
Harrisburg
Scranton
Morgantown
Charleston
Washington
Baltimore
Annapolis
Wilmington
Dover
Philadelphia
Trenton
Newark
New York
Atlantic City
Richmond
Newport News
Norfolk
Roanoke
Winston-Salem
Durham
Raleigh
Charlotte
Knoxville
Nashville
Chattanooga
Dalton
Memphis
Huntsville
Fort Smith
Little Rock
Birmingham
Atlanta
Augusta
Columbia
Charleston
Macon
Columbus
Savannah
Montgomery
Shreveport
Monroe
Jackson
Valdosta
Jacksonville
Baton Rouge
Gulfport
Mobile
Pensacola
Tallahassee
Panama City
Gainesville
Daytona Beach
Orlando
Dunedin
Tampa
St. Petersburg
Fort Myers
Fort Lauderdale
Miami
Key West
Nassau
Lake Charles
Beaumont
Port Arthur
Galveston
New Orleans
Toronto
Niagara Falls
Buffalo
Rochester
Syracuse
Binghampton
Albany
Hartford
Providence
Boston
Lawrence
Manchester
Concord
Montpelier
Burlington
Augusta
Bangor
Bar Harbour
Québec
Montréal
Ottawa
National capital
State capital
Other important city or town
Major international airport
Main railway
Main road
International boundary
State boundary
0 km 100 200 300 400 500
Scale 1: 10 800 000
1 cm represents 108 km
Elevation
5000 m
4000
2000
1000
500
200
0
Below sea level
-200
95° Longitude West of Greenwich

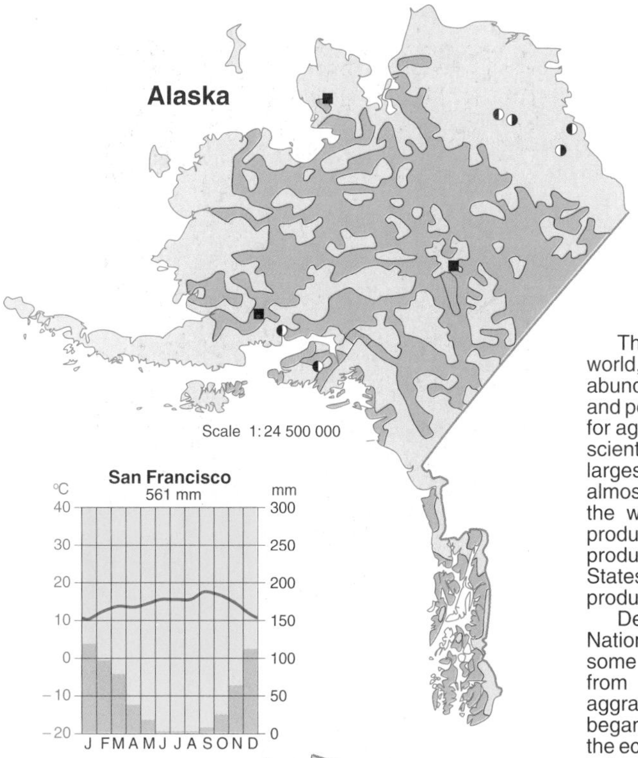

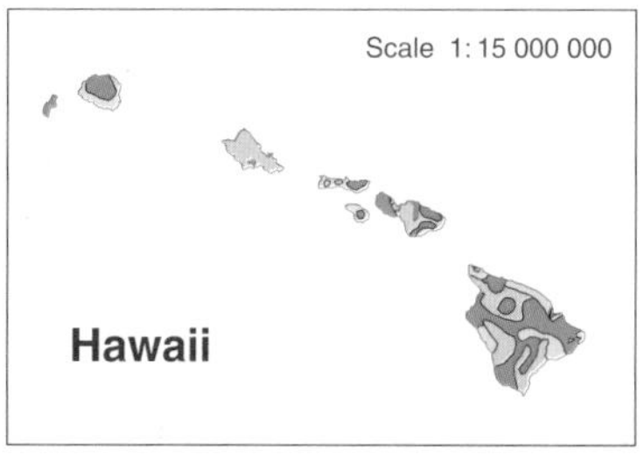

Land use

- Urban
- Cropland
- Cropland and woodland
- Cropland and grazing
- Grassland and grazing
- Temperate forest
- Tropical forest
- Scrubland and grassland, or desert
- Swamp and marsh
- Tundra
- ○ Climograph station

Minerals

- ■ Gold
- ▲ Silver
- ▼ Lead
- ● Zinc
- ◆ Copper
- ✚ Iron
- Tungster
- Molybder
- Coal
- Oil
- Gas
- Uranium
- S Sulphur
- P Phospha
- B Borax

The United States is the fourth largest country in the world, in both area and population. Blessed with abundant resources, it has developed into an economic and political leader. Over 40 percent of the land is used for agriculture, a sector that is largely mechanized and scientifically controlled. The United States is the world's largest exporter of grain. The country also produces almost 14 percent of the world's oil, and 25 percent of the world's natural gas and coal. It is the leading producer of nuclear power and the second largest producer of hydroelectric power. However, the United States consumes even more natural resources than it produces.

Despite having one of the world's highest Gross National Products, the United States has experienced some economic problems in recent years. A shift away from heavy industry towards electronics has aggravated unemployment. Large budget deficits began in 1985. The recession in the early 1990s, after the economic expansion in the 1980s, has forced many industries to cut staff and close facilities.

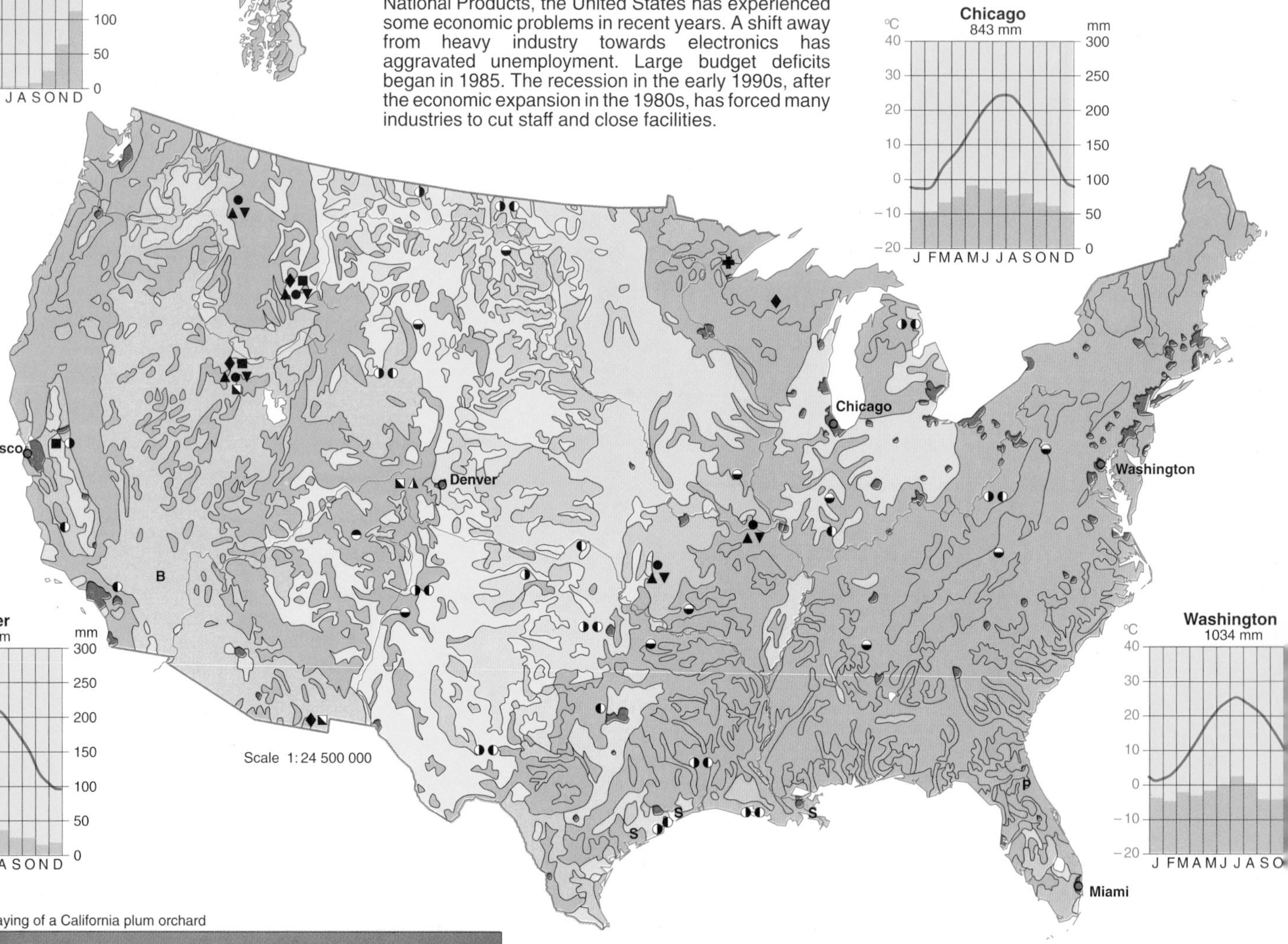

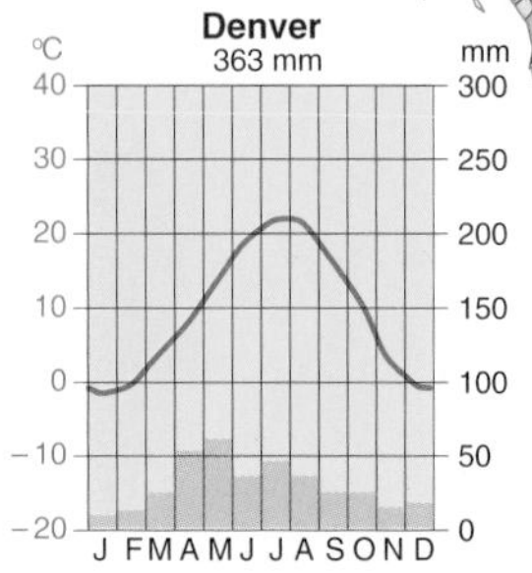

Aerial spraying of a California plum orchard

The United States has diverse landscapes, ranging from the low-lying Florida swamps to the majestic Rocky Mountains. Most of the country has a temperate climate, but extremes such as Death Valley, Alaska and Key West illustrate the variety of conditions. A world conservation pioneer, the United States has protected 400 areas, more than 10 percent of its land area. However, Americans' heavy consumption of non-recycled goods and toxic chemicals has caused serious environmental problems. Runoff of excess pesticides and fertilizers continues to threaten lakes and rivers. As the largest single source of greenhouse gases, the United States also contributes significantly to global warming. Another serious problem, acid rain, concerns both Canadians and Americans. The U.S. government has recently tightened air pollution regulations in order to reduce the emissions that cause acid precipitation.

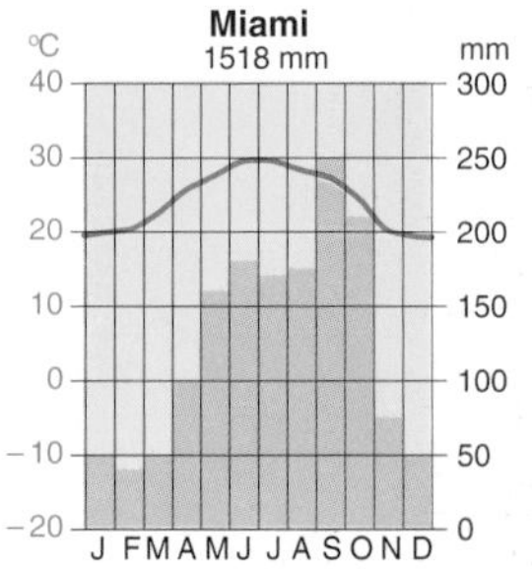

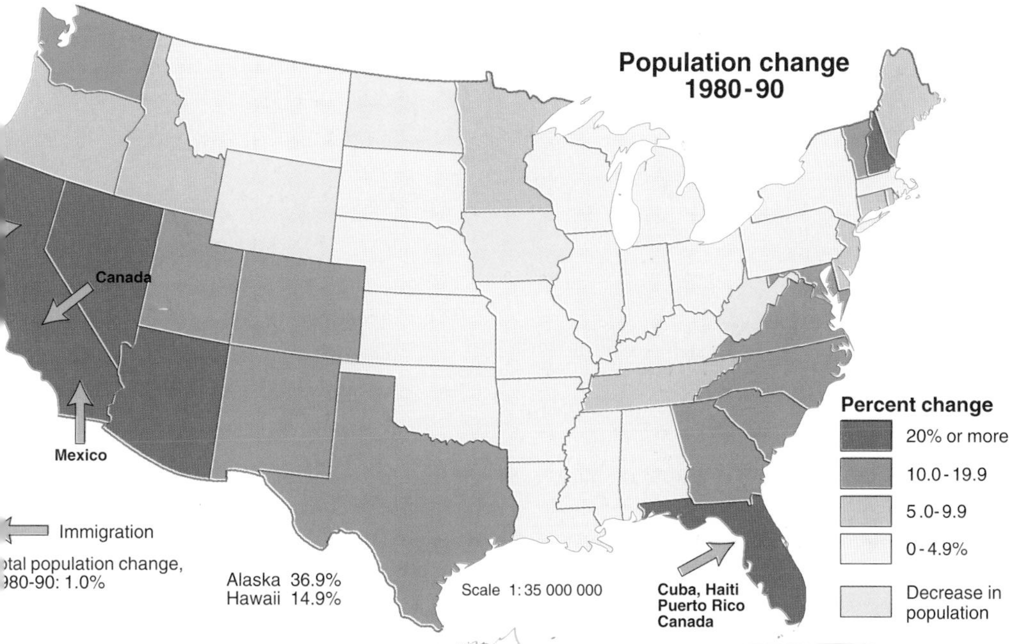

A view of New York City at sunset

Imports

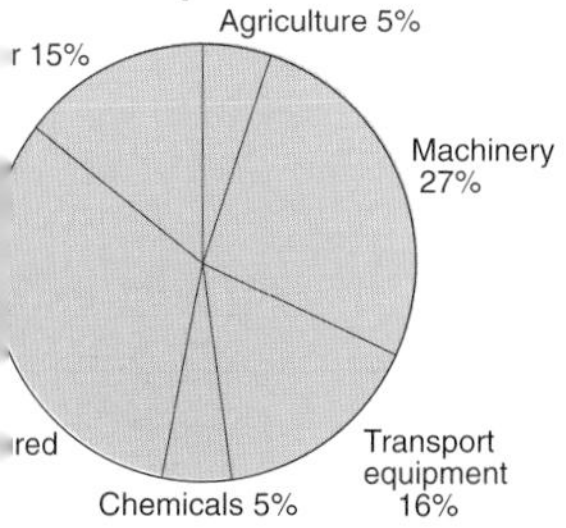

…in imports (billions of U.S. dollars)
…or vehicles $50.0
…$47.0
…nputers and parts $19.1
…ctronic circuits $8.2
…cks $8.1
…ts for tractors and motor vehicles $7.8
…twear $6.2
…l oils $4.8
…vsprint $4.3

Exports

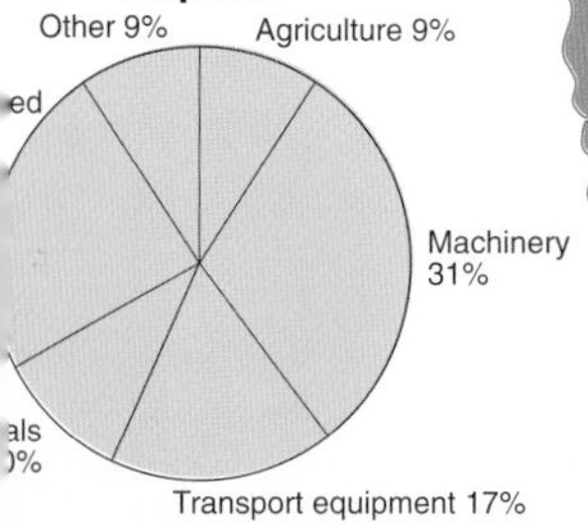

…in exports (billions of U.S. dollars)
…planes $18.0
…ce machines, computers and parts $14.0
…or vehicles $10.0
…ts for airplanes $8.3
…ts for tractors and motor vehicles $7.6
…ctronic circuits $7.0
…ze $6.0
…arettes $4.7
…l $4.5

Population change is often linked to economic change. People pursuing jobs move to areas where the economy is strong. However, retirees can also stimulate an area's economy, as they have in Florida and the Southwest. Generally, the American population continues to move west and southwest. California and Texas are becoming the centres for the fast-growing high-tech industry. The Northeast still features heavy industry.

Culturally, the United States of the 1990s is more of a melting pot than ever. When the immigration laws changed in 1968, people flowed into the country from Eastern Europe, the Caribbean and Central America, and the Pacific Rim.

GDP by sector

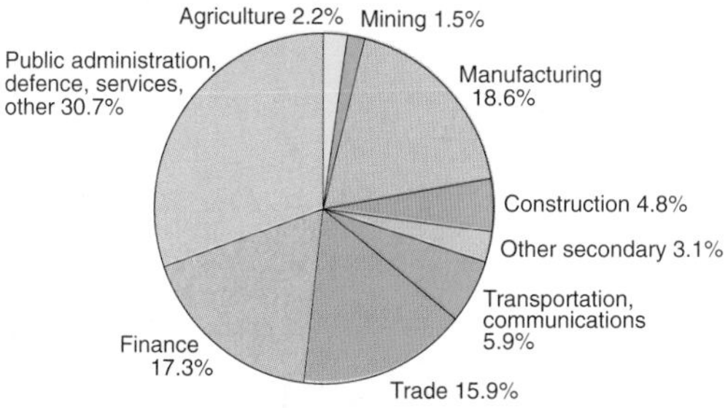

Total: $5181.0 billion (U.S.)
Annual growth rate, 1980-90: 3.2%

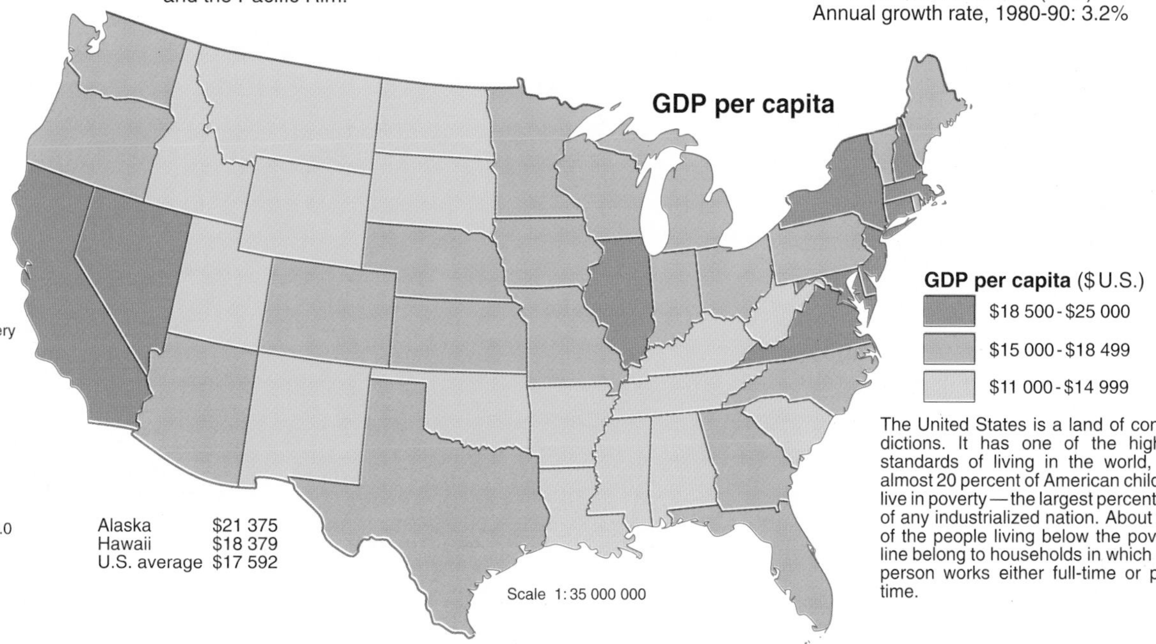

The United States is a land of contradictions. It has one of the highest standards of living in the world, but almost 20 percent of American children live in poverty — the largest percentage of any industrialized nation. About half of the people living below the poverty line belong to households in which one person works either full-time or part-time.

Trade

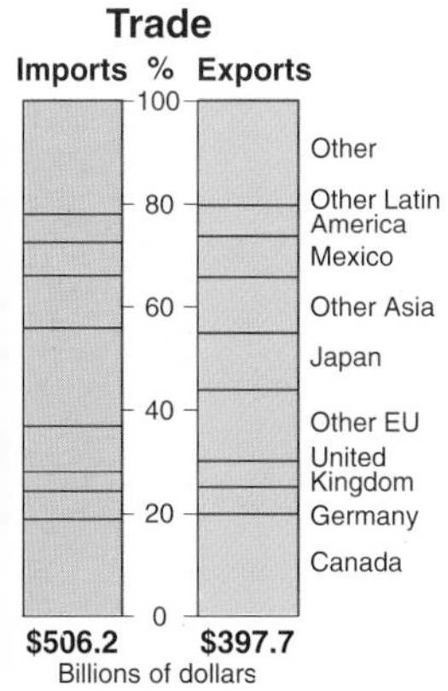

Rapidly changing industries, 1979-89

Growing Industry	Employment index*	Declining Industry	Employment index*
Computers, services	282	Iron ores	38
Outpatient care facilities	281	Combined real estate, insurance, etc.	40
Personnel supply services	257	Watches, clocks and watchcases	41
Mortgage bankers, brokers	239	Copper ores	42
Correspondence and vocational schools	234	Blast furnaces and basic steel products	48
Business credit institutions	222	Rubber, plastic footwear	49
Individual and family services	219	Handbags and personal leather goods	49
Mailing, reproduction, stenographic services	217	Railroad equipment	53
Residential care	212	Railroad transportation	53
Sanitary services	206	Footwear, except rubber	54
Guided missiles, space vehicles, parts	205	Bituminous coal and lignite mining	55
Air transportation services	205	Musical instruments	55
Security brokers, dealers	204	Operative builders	55
Legal services	195	Taxicabs	59
Holding and other investment offices	191	Cement, hydraulic	59

***1979 = 100** The employment index measures the change in the number of employees.

Testing for HIV in a southern U.S. laboratory

UNITED STA
MEXICO
GUATEMALA
BELIZE
EL SALVADOR
PACIFIC OCEAN
GULF OF MEXICO
GOLFO DE CALIFORNIA
BAJA CALIFORNIA
BAHIA DE CAMPECHE
GOLFO DE TEHUANTEPEC
SIERRA MADRE OCCIDENTAL
SIERRA MADRE ORIENTAL
SIERRA MADRE DEL SUR
ISLAS TRES MARIAS
ISLAS DE REVILLAGIGEDO
GUADALUPE
Los Angeles
San Diego
Tijuana
Ensenada
Mexicali
Phoenix
Tucson
Nogales
Albuquerque
Amarillo
Oklahoma City
Little Rock
Memphi
El Paso
Ciudad Juárez
Fort Worth
Dallas
Jackson
Austin
San Antonio
Houston
Baton Rouge
New Orleans
Mobile
Hermosillo
Guaymas
Ciudad Obregón
Chihuahua
Hidalgo del Parral
Los Mochis
La Paz
Culiacan
Torreón
Durango
Mazatlán
Ciudad Acuna
Nuevo Laredo
Laredo
Corpus Christi
Reynosa
Brownsville
Matamoros
Monterrey
Saltillo
Ciudad Victoria
Zacatecas
Aguascalientes
Tepic
San Luis Potosi
Ciudad Madero
Tampico
Guadalajara
Léon
Irapuato
Manzanillo
Uruapan
Morelia
México
Cuernavaca
Xalapa
Puebla
Veracruz
Orizaba
Acapulco
Oaxaca
Mérida
Campeche
Ciudad del Carmen
Coatzacoalcos
Villahermosa
Tuxtla Gutiérrez
Juchitan
Tapachula
Huehuetenango
Guatemala
Quezaltenango
Santa Ana
San Salvador
Belm
Colorado
Brazos
Red
Pecos
Rio Grande
Bravo del Norte
Mississippi
Balsas
Yaqui
National capital
Other important city or town
Major international airport
Main railway
Main road
International boundary
0 km 250 500 750
Scale 1:14 300 000
1 cm represents 143 km
120° 110° 105° 100° 95° 90°
35° 30° 25° 20° 15° 10° 5°
115°
Longitude West of Greenwich

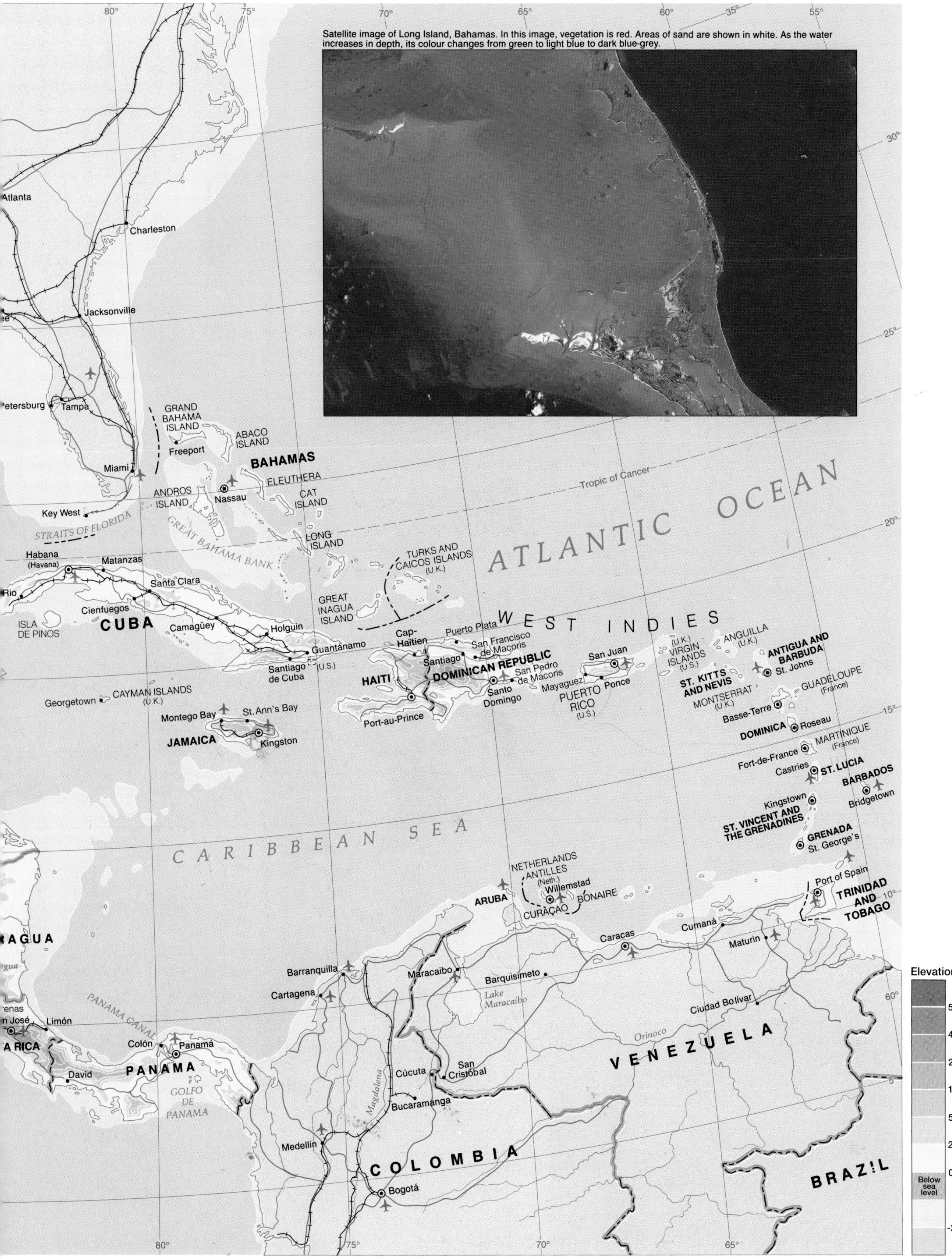

Satellite image of Long Island, Bahamas. In this image, vegetation is red. Areas of sand are shown in white. As the water increases in depth, its colour changes from green to light blue to dark blue-grey.

WORLD-REGIONAL

Although small in size and population, this region is physically and culturally diverse. The area is covered with folded and volcanic mountains, more than 30 of which are active. Inhabited for at least 13 000 years, the Central American countries were dominated in historical times by the Mayan and Toltec civilizations, which domesticated plants important for agriculture. The Spanish then ruled for three centuries until the region regained independence in 1823. Today, the descendants of the Mayans and Toltecs form the basis of the Central American population. Most countries in the region are Spanish-speaking, but some islands are French or English.

Rich natural resources—minerals, energy fuels, forests, fertile soils—promise the area great economic potential. But disparities of wealth, huge national debts and dependence on foreign markets test the political and economic stability.

The tropical climate of the Caribbean attracts many tourists.

Value of tourism
Millions of U.S. dollars

Mexico	5324	2.
Puerto Rico	1367	6.
Bahamas	1333	46.
Dominican Republic	750	12.
Jamaica	740	21.
U.S. Virgin Islands	700	70.
Barbados	500	30
Bermuda	500	30
Netherlands Antilles	480	30
Aruba	353	49.
Cayman Islands	325	70.
Costa Rica	275	5.
Cuba	268	1.
Antigua and Barbuda	250	69.
Guadeloupe	250	20.
Martinique	240	1.
Guatemala	185	2.
Panama	167	3.
St. Lucia	154	56.
Trinidad and Tobago	95	2.
Belize	91	24.
Haiti	75	3.
El Salvador	69	1.
Grenada	38	20.
Honduras	29	0.
Nicaragua	12	0.

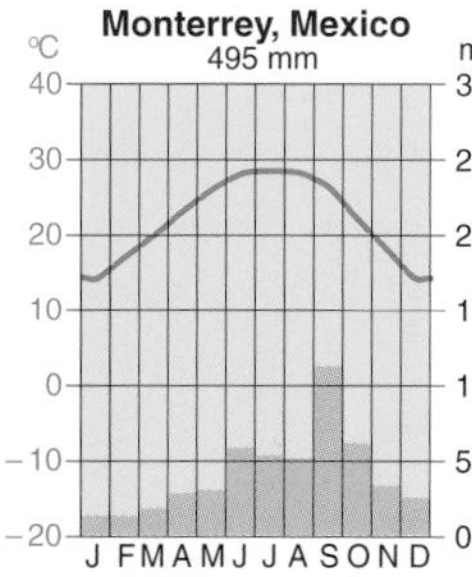

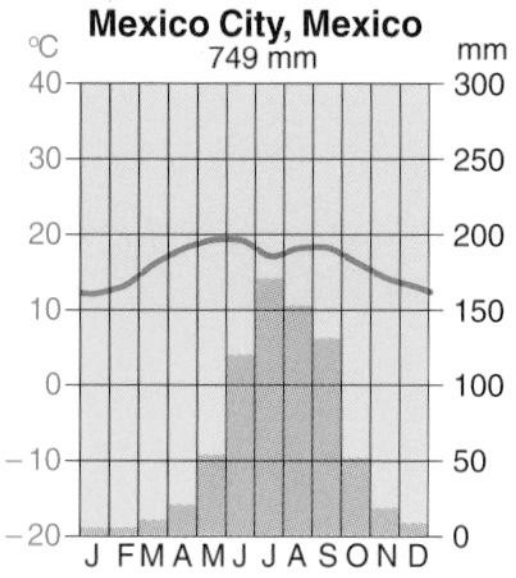

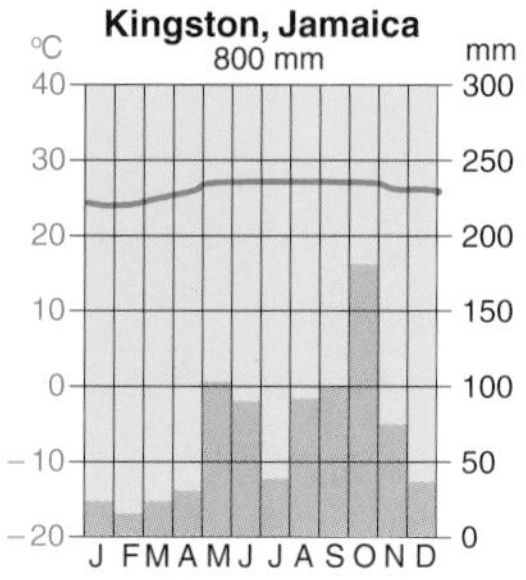

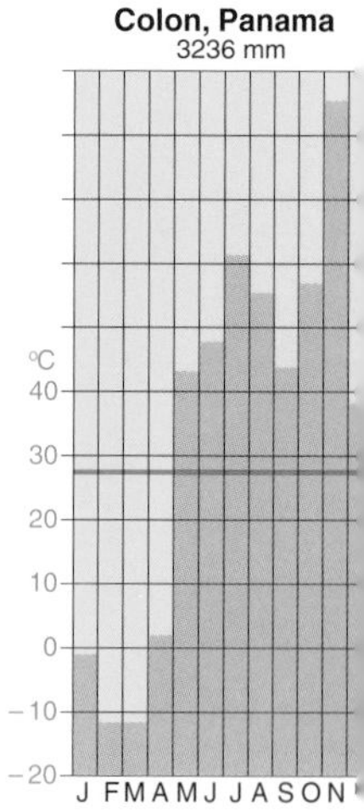

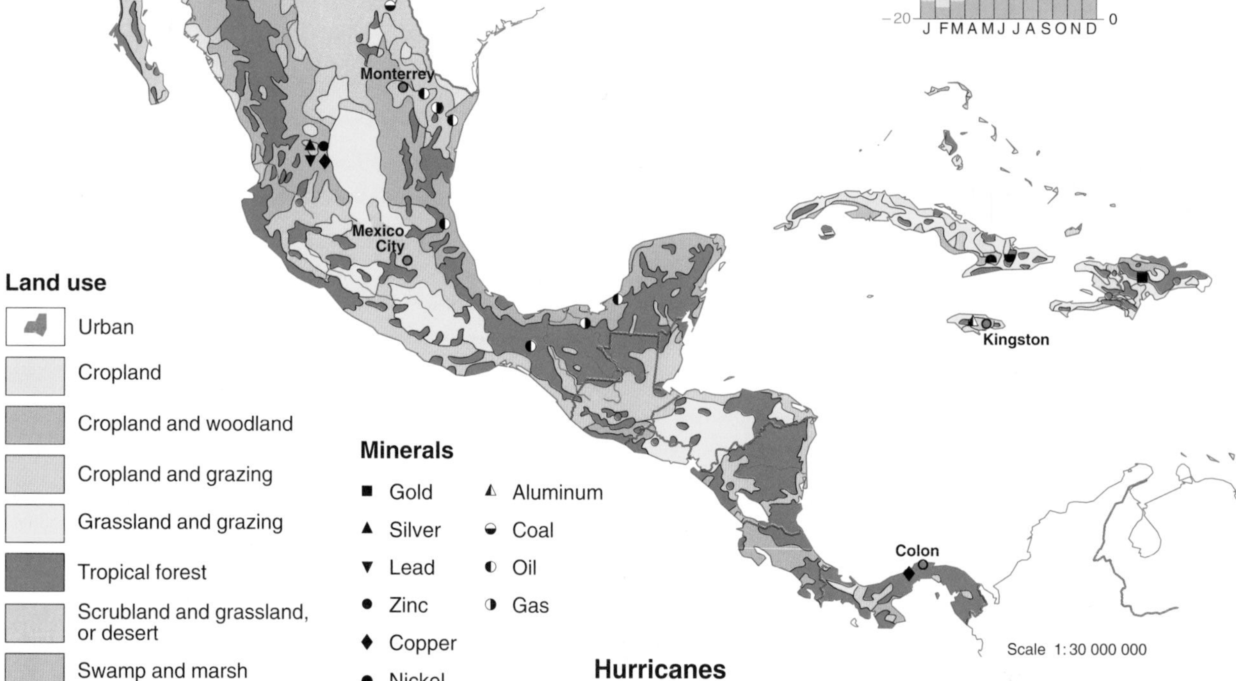

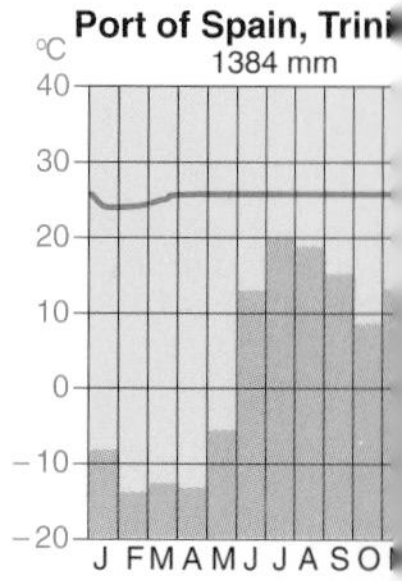

Hurricanes

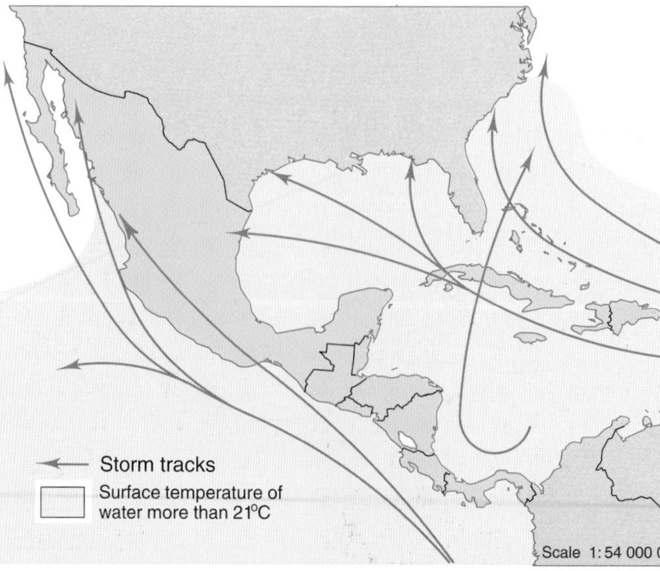

Gas pipelines in Tabasco, Mexico

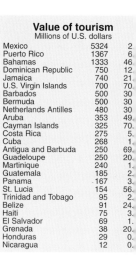

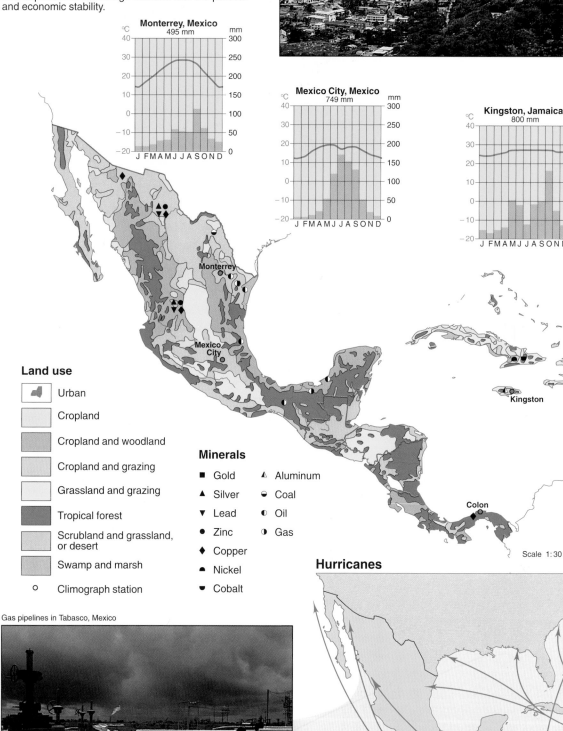

a Canal

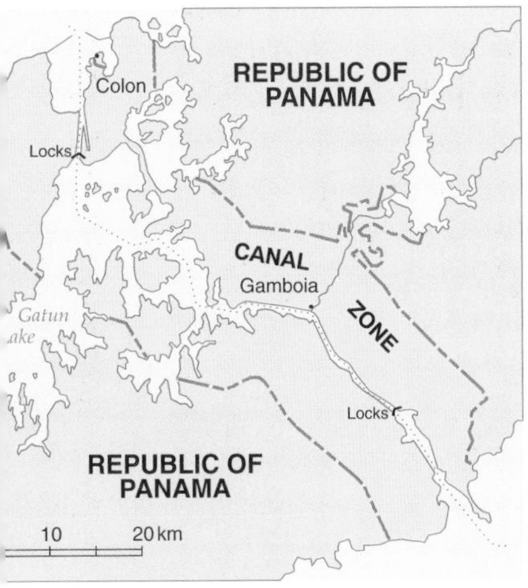

nama Canal, cut through the s of Panama, is a strategic and rcial link between the Atlantic cific Oceans that eliminates the 5 000 km journey around South a. After ten years of construc- hich cost the lives of many ean workers, the 82 km canal in 1914. Although the Canal as been administered by the States during this century, under nama Canal Treaty of 1977, a is to take control by the year

Political conflict

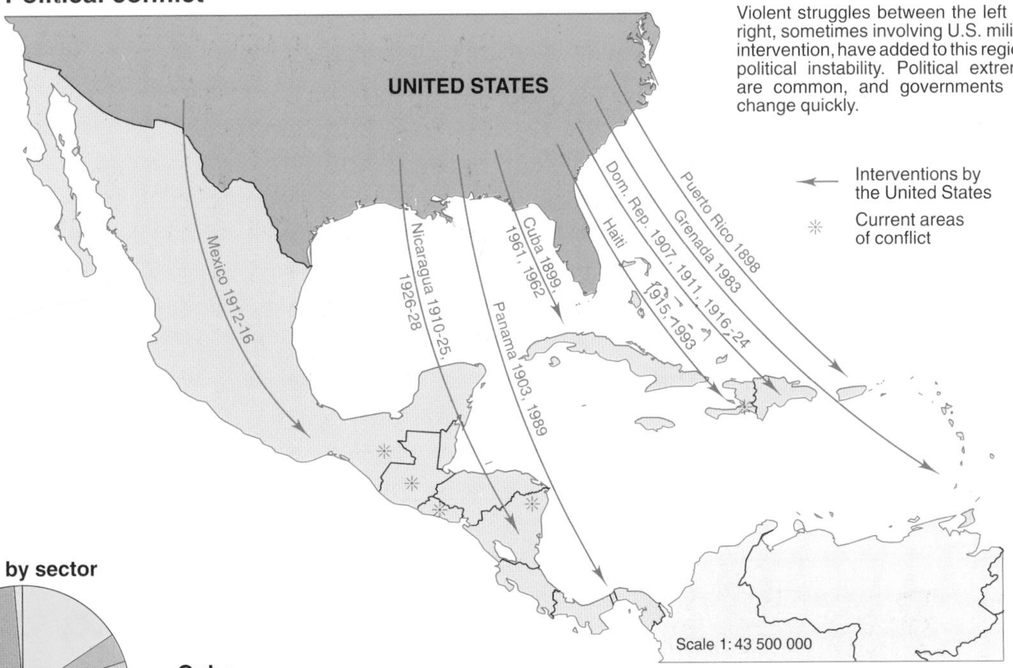

Violent struggles between the left and right, sometimes involving U.S. military intervention, have added to this region's political instability. Political extremes are common, and governments can change quickly.

Interventions by the United States

Current areas of conflict

Cuba

GDP by sector

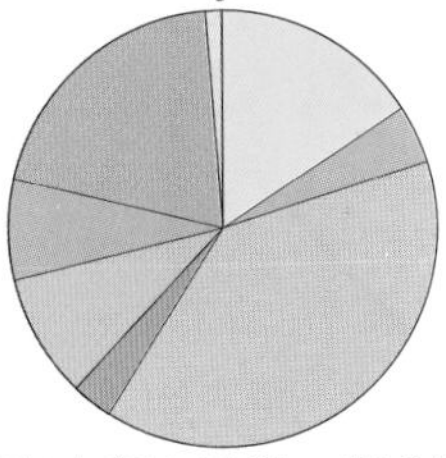

Total: $21.0 billion (U.S.)

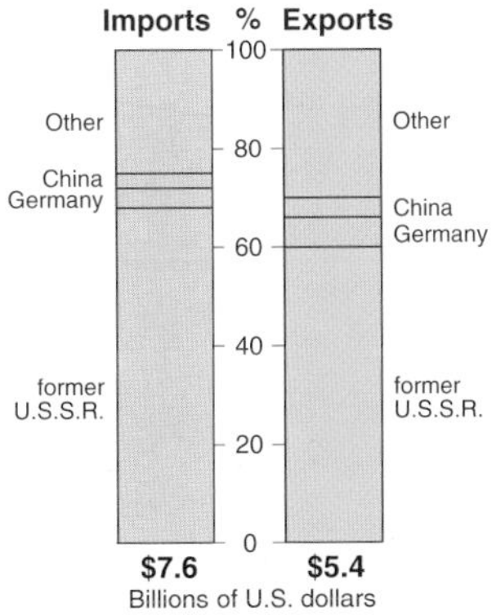

Main imports:
Mineral fuels 33%
Machinery and transport equipment 31%
Food 12%

Main exports:
Sugar 73%
Mineral fuels 9%
Agricultural goods 4%

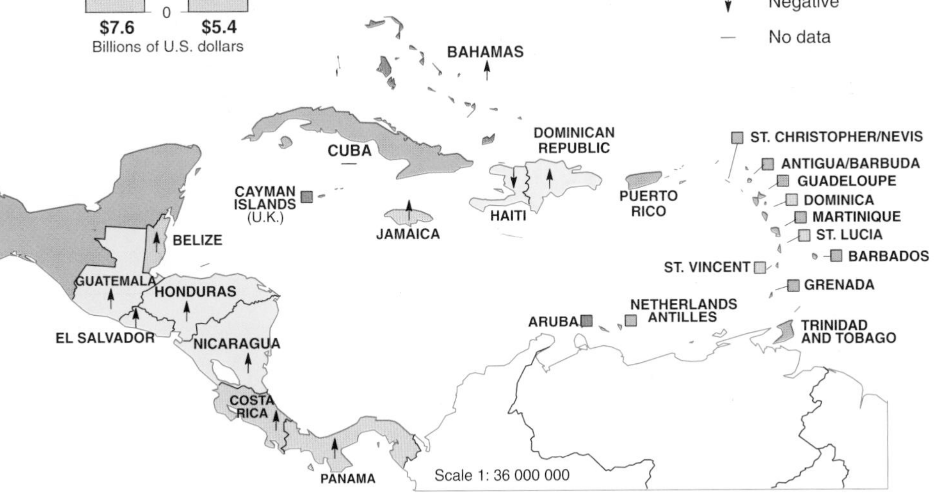

GDP per capita ($U.S.)
- $10 000 or more
- 2000 - 9999
- 1000 - 1999
- Less than $1000

Annual change in GDP (1980-90 average)
- Positive
- Negative
- No data

Mexico

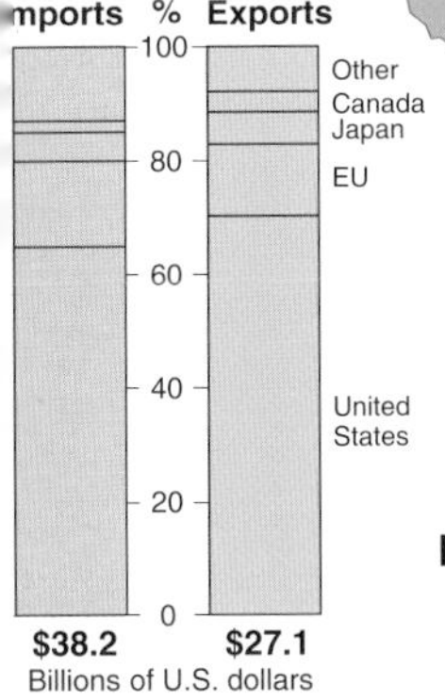

Main imports:
Metallic products, machinery, equipment 53%
Chemical products 9%

Main exports:
Metallic products, machinery, equipment 28%
Petroleum 26%
Chemical products 6.3%

GDP by sector

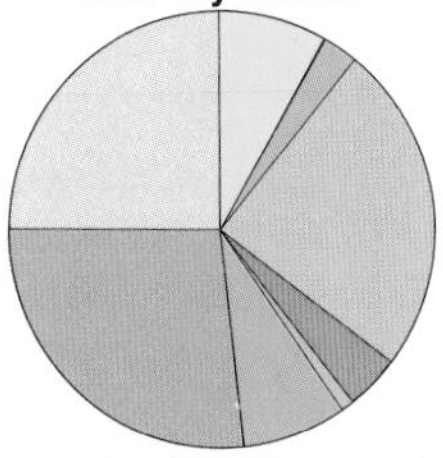

Total: $215.0 billion (U.S.)

plantation in Costa Rica

Costa Rica

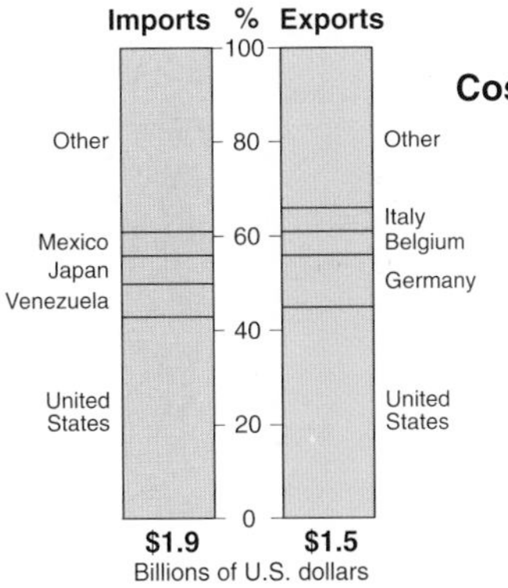

GDP by sector

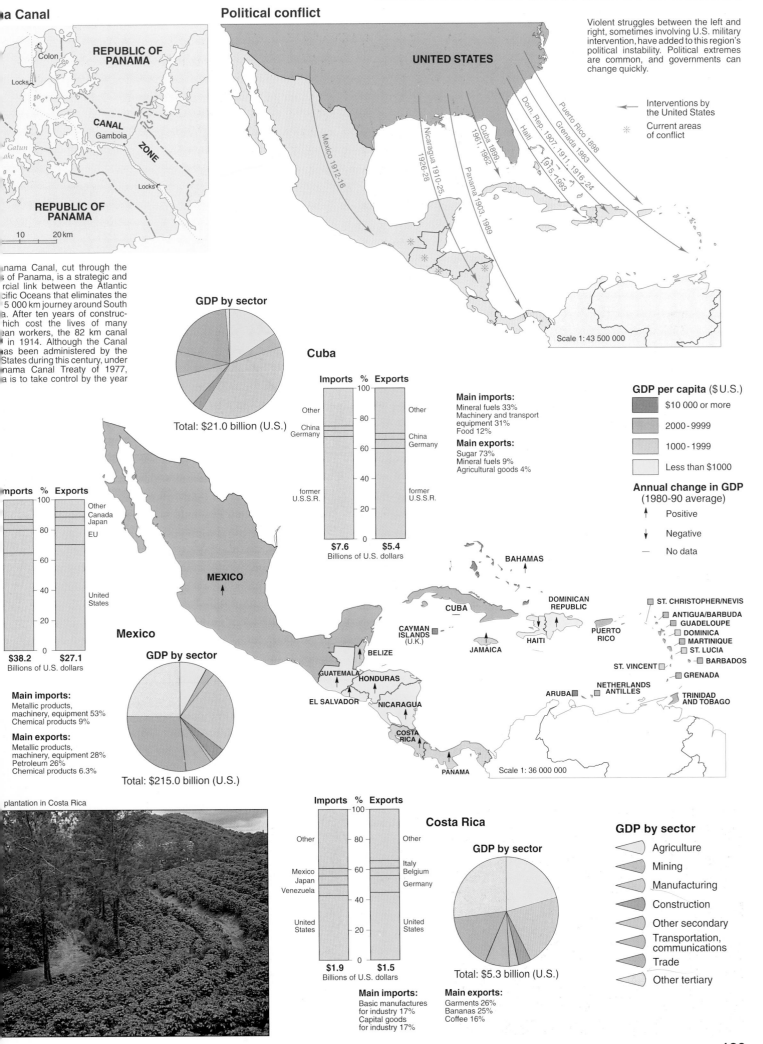

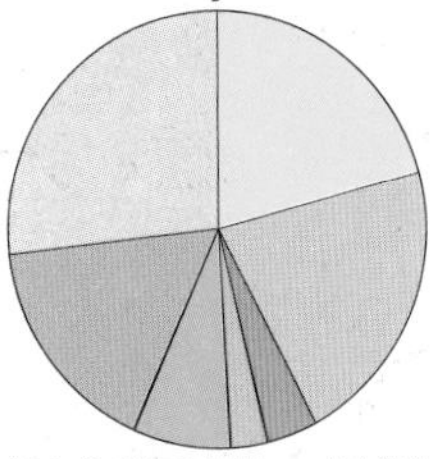

Total: $5.3 billion (U.S.)

Main imports:
Basic manufactures for industry 17%
Capital goods for industry 17%

Main exports:
Garments 26%
Bananas 25%
Coffee 16%

GDP by sector
- Agriculture
- Mining
- Manufacturing
- Construction
- Other secondary
- Transportation, communications
- Trade
- Other tertiary

CARIBBEAN SEA
ST. LUCIA
ST. VINCENT AND THE GRENADINES
GRENADA
NETHERLANDS ANTILLES
ARUBA
Willemstad
BONAIRE
CURAÇAO
Santa Marta
Barranquilla
Cartagena
Valledupar
Maracaibo
Cabimas
Coro
Caracas
Valencia
Maracay
Barquisimeto
Cumaná
Barcelona
Port of Spain
TRINIDAD AND TOBAGO
Maturín
Mouths of the Orinoco
PANAMA
COSTA RICA
Panama Canal
Colón
Panamá
Golfo de Panama
Sincelejo
Montería
Lake Maracaibo
Valera
Mérida
Barinas
Acarigua
San Fernando de Apure
Orinoco
Ciudad Guayana
Ciudad Bolívar
Tumeremo
Cúcuta
San Cristóbal
Arauca
Atrato
Bucaramanga
Barrancabermeja
Floridablanca
VENEZUELA
GUYANA
Bello
Medellín
Itagüí
Magdalena
ANDES
LLANOS
GUIANA HIGHLANDS
Caura
Manizales
Pereira
Armenia
Bogotá
Ibagué
Villavicencio
Meta
Guaviare
Buenaventura
Cali
Palmira
COLOMBIA
Boa Vista
Neiva
Popayán
Guainía
Tumaco
Pasto
Içana
Branco
Jauaperi
Esmeraldas
Negro
Caquetá
Apaporis
Quito
0°—EQUATOR
Portoviejo
ECUADOR
Ambato
Japurí
Arica
Putumayo
Napo
Pastaza
Guayaquil
Golfo de Guayaquil
Cuenca
Machala
Amazonas
Tefé
Manaus
Iquitos
Juruá
Purus
Madeira
Marañón
Javari
Sullana
Piura
SELVAS
Ucayali
Juruá
Chiclayo
Cajamarca
Cruzeiro do Sul
Pôrto Velho
Jiparaná
Trujillo
Pucallpa
Purus
Chimbote
Rio Branco
Abunã
Madre de Dios
Mamoré
Cerro de Pasco
PERU
Beni
Callao
Lima
Huancayo
Apurimac
Machu Picchu
Cuzco
Ayacucho
Guaporé
Itenez
Paraguá
Vilhena
Ica
PACIFIC OCEAN
ANDES MOUNTAINS
Juliaca
Lake Titicaca
BOLIVIA
Arequipa
La Paz
Mamoré
Cochabamba
Santa Cruz
Tacna
Oruro
Arica
Lake Poopó
Sucre
Potosí
15°
10°
5°
0°
5°
10°
15°
85°
80°
75°
70°
65°
60°
Longitude West of Greenwich

ATLANTIC OCEAN
National capital
Other important city or town
Major international airport
Main railway
Main road
International boundary
0 km 200 400 600 800
Scale 1:17 000 000
1 cm represents 140 km
Paramaribo
Cayenne
FRENCH GUIANA
RINAME
Maroni
Cumina
Araguari
Mouths of the Amazon
Jari
Macapá
Paru
Amazonas
Pará
Belém
Xingu
Santarém
São Luís
Gurupi
Pindaré
Tocantins
Bacabal
Parnaíba
Fortaleza
Grajaú
Imperatriz
Teresina
Iriri
São Félix do Xingu
Itapecuru
Jaguaribe
Mossoró
Natal
Floriano
PLATEAU OF BORBOREMA
Juàzeiro do Norte
Campina Grande
João Pessoa
BRAZIL
Caruaru
Olinda
Recife
São Francisco
Garanhuns
Juàzeiro
Araguaia
Maceió
Gurupi
Grande
Aracaju
PLATEAU OF MATO GROSSO
BRAZILIAN HIGHLANDS
Fiera de Santana
Paraguaçu
Salvador (Bahia)
Contas
Cuiabá
Vitória da Conquista
Itabuna
Brasília
Anápolis
Goiânia
Jequitinhonha
Montes Claros
Paranaíba
Uberlândia
Governador Valadares
EQUATOR
55°
50°
45°
40°
35°
30°
15°
10°
5°
0°
Elevation
5000 m
4000
2000
1000
500
200
0
Below sea level
-200

PACIFIC OCEAN
CHILE
ARGENTINA
BOLIVIA
PARA
ANDES MOUNTAINS
GRAN CHACO
PAMPAS
Tropic of Capricorn
Sucre
Iquique
Uyuni
Tarija
Mariscal Estigarribia
Calama
Antofagasta
San Salvador de Jujuy
Salta
San Miguel de Tucuman
Santiago del Estero
Catamarca
La Rioja
Vallenar
La Serena
Coquimbo
San Juan
Córdoba
Mendoza
Viña del Mar
Valparaíso
Santiago
Godoy Cruz
San Bernardo
San Luis
Rio Cuarto
Rancagua
San Rafael
Curico
Talca
Santa Rosa
Talcahuano
Concepción
Chillan
Los Angeles
Temuco
Neuquén
Valdivia
Osorno
San Carlos de Bariloche
San Antonio Oeste
Puerto Montt
ISLA DE CHILOÉ
Trelew
Golfo Corcovado
ARCHIPIÉLAGO DE LOS CHONOS
Puerto Aisén
Perito Moreno
Comodoro Rivadavia
Golfo San Jorge
Puerto Deseado
Gobernad
Puerto Santo Cruz
Bahia Grande
Puerto Natales
Rio Gallegos
Punta Arenas
TIERRA DEL FUEGO
ISLA DE LOS ESTADOS
FALKLAND IS
ISLAS JUAN FERNÁNDEZ (Chile)
Pres. Roque Saénz Peña
Resistencia
San Justo
Santa Fe
Paraná
Rosario
Junin
San Isidro
Buenos Aires
Olavarría
Bahia Blanca
Golfo San Matias
Mar del
Asu
Form
Goy
Conco
Verde
Bermejo
Pilcomayo
Teuco
Salado
Paraná
Diamante
Chadileuvu
Colorado
Negro
Limay
Chubut
Chico
Deseado
Santa Cruz
Estrecho de Magallanes
Longitude West of Greenwich
National capital
Other important city or town
Major international airport
Main railway
Main road
International boundary
0 km 200 400 600 800
Scale 1:15 000 000
1 cm represents 150 km

Satellite image of the Andes Mountains, west of La Rioja, Argentina.
Snow is shown in blue, while vegetation is green. Recent volcanoes are shown in orange.

WORLD-REGIONAL

The countries of South America differ greatly in resources, level of economic development and physical geography, as well as in culture and traditions. The "spine" of the continent is the Andes, made up of mountains and high plateaus. On a long, thin strip of land bordering the Pacific Ocean lies one of the world's driest areas, the Atacama Desert. Yet to the east is the tropical rainforest of Amazonia. Tropical conditions prevail over half the continent, but genuinely cold climates are found in southern Chile and Argentina.

The history of South America includes the remarkable Inca civilization in Peru, which flourished between the 12th and early 16th centuries. At its widest extent, the Inca empire comprised modern Peru, Ecuador, Bolivia, northern Argentina and northern Chile. In 1911, the ruins of the Incan city of Machu Picchu were discovered. In South America's more recent history, extremes of wealth and poverty have led to political instability, resulting in military rule or dictatorships. But most countries are moving slowly towards democracy and greater social justice.

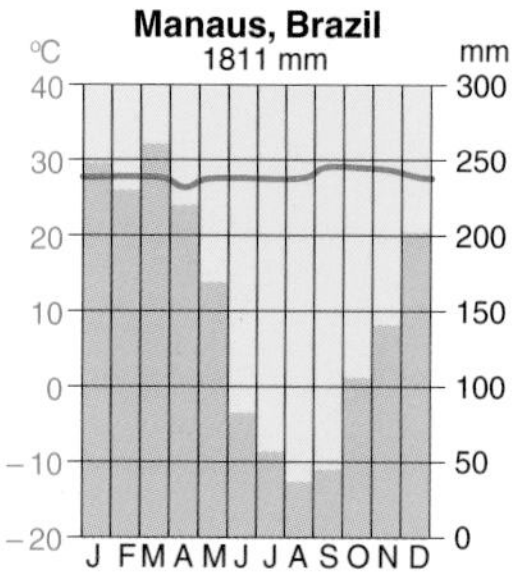

Much of Brazil's rainforest is being turned into pasture land.

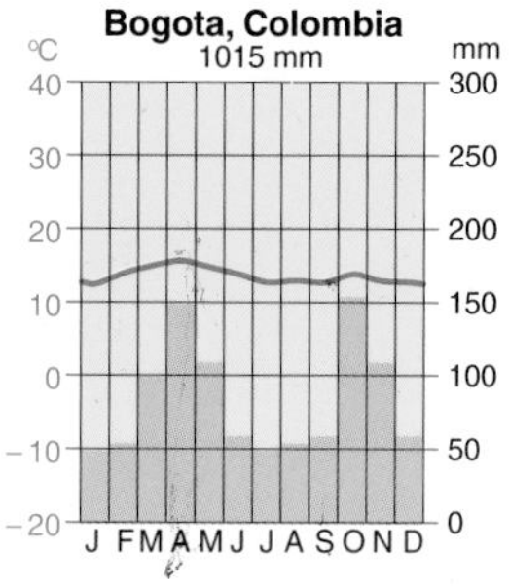

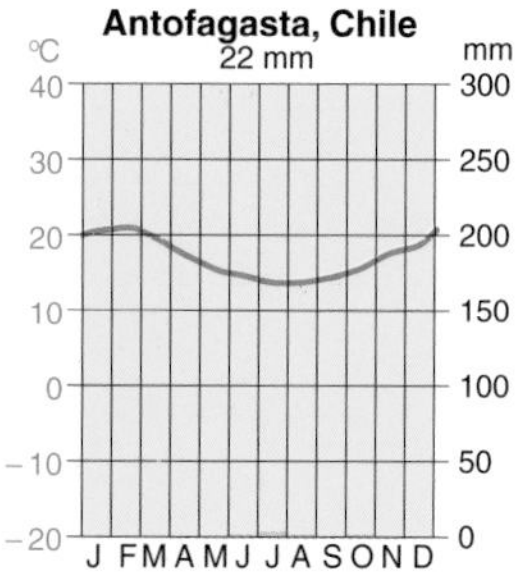

South America's tropical rainforests have become focal point for a new global awareness of environment. Brazil has almost one third of the wo rainforests, and this precious resource is threatene agriculture and economic activity. To many develo countries, the forests represent a desperately nee source of both export products, such as timber and p and agricultural land. Once logging companies have through an area of rainforest, peasant families slash-and-burn methods to clear the land subsistence crops or cattle. But the destruction of forest is felt locally and around the world. In Brazil, lives and cultures of many indigenous peoples threatened, and thousands of plant and animal spe have been wiped out. World-wide, the enorm quantities of carbon dioxide released into the air h contributed to global warming.

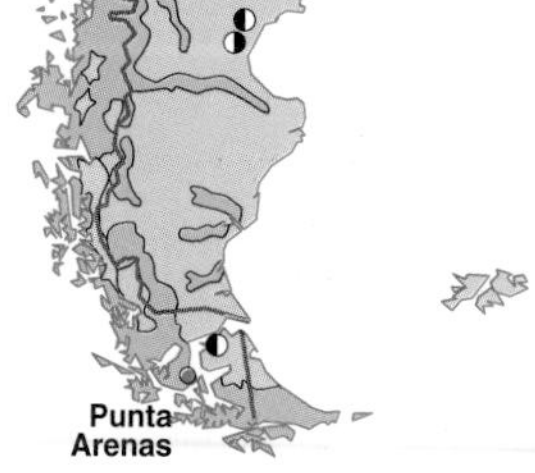

Scale 1:39 000 000

Land use

- Urban
- Cropland
- Cropland and woodland
- Cropland and grazing
- Grassland and grazing
- Tropical forest
- Temperate forest
- Scrubland and grassland, or desert
- Swamp and marsh
- Tundra
- o Climograph station

Minerals

- ▲ Silver
- ▼ Lead
- ● Zinc
- ♦ Copper
- ✚ Iron
- Tin
- Aluminum
- Tungsten
- Coal
- Oil
- Gas
- Uranium
- Mn Manganese
- P Phosphate

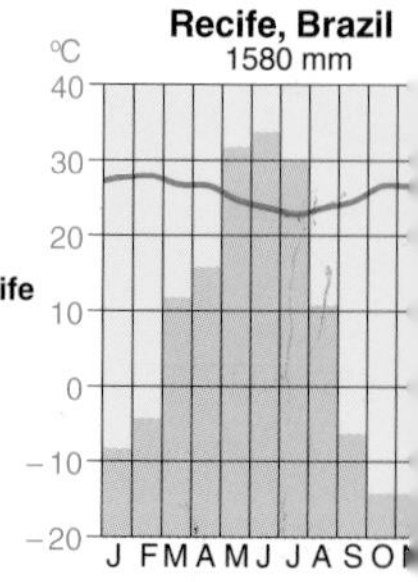

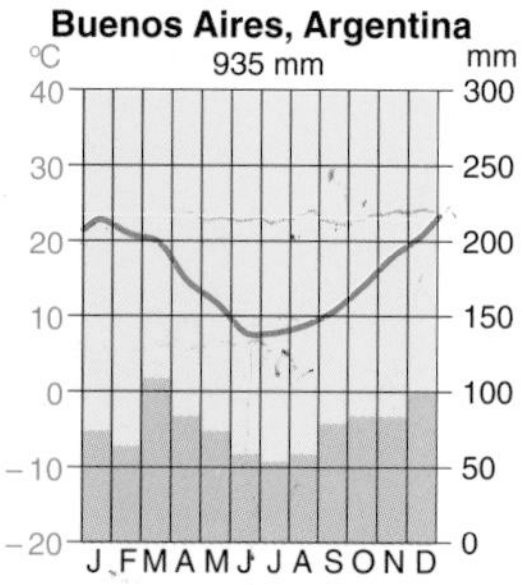

Cattle ranching provides a source of farm income in Colombia.

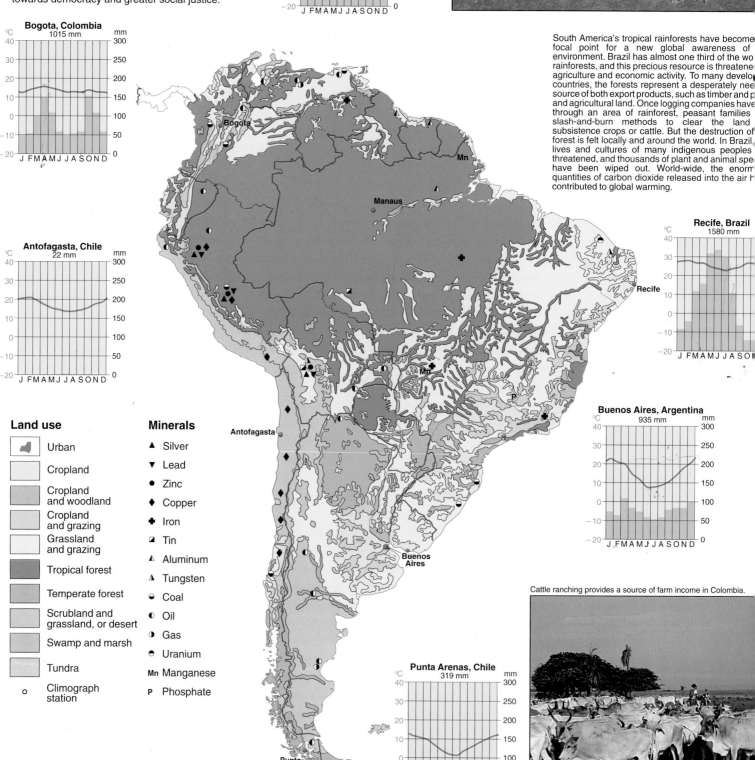

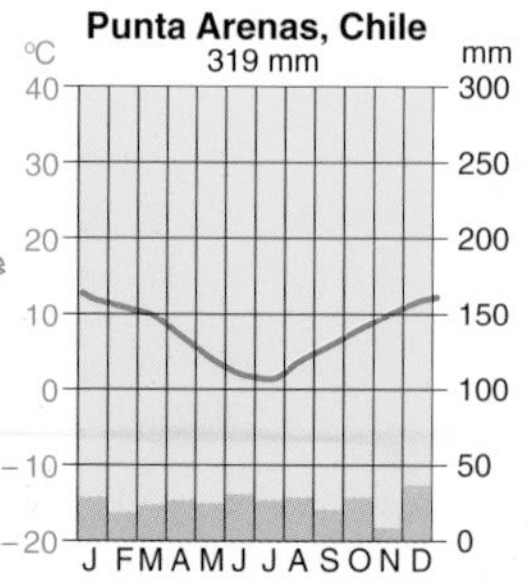

…w of Rio de Janeiro shows the favelas of the poorer citizens in the foreground.

Colonization circa 1650

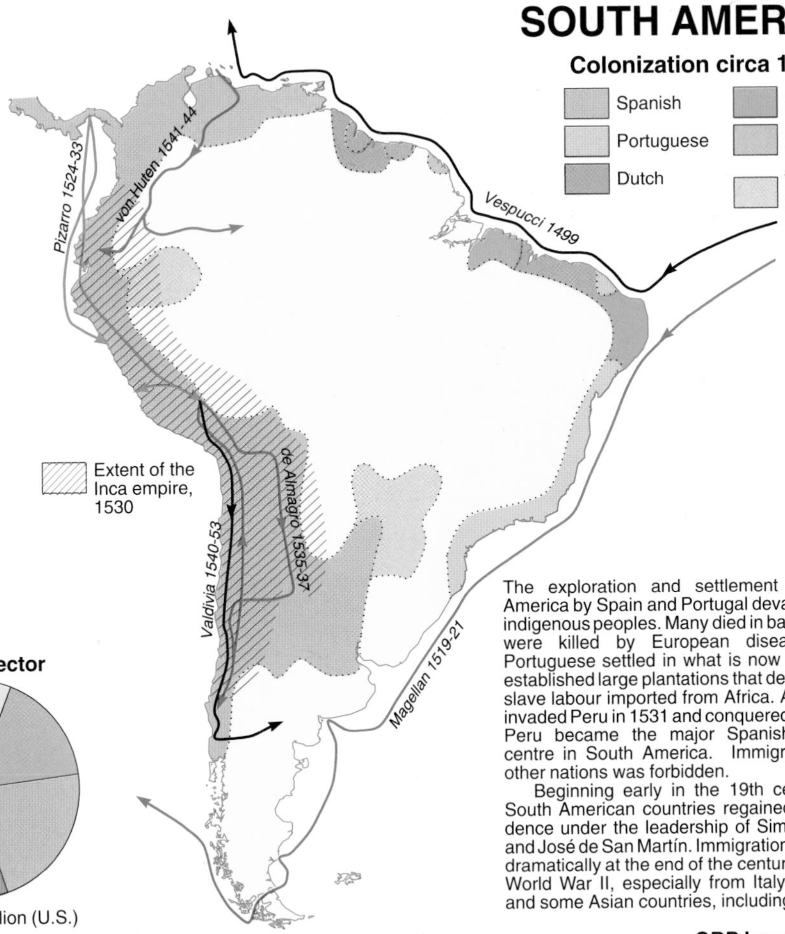

The exploration and settlement of South America by Spain and Portugal devastated the indigenous peoples. Many died in battle; others were killed by European diseases. The Portuguese settled in what is now Brazil and established large plantations that depended on slave labour imported from Africa. After Spain invaded Peru in 1531 and conquered the Incas, Peru became the major Spanish-speaking centre in South America. Immigration from other nations was forbidden.

Beginning early in the 19th century, the South American countries regained independence under the leadership of Simón Bolívar and José de San Martín. Immigration increased dramatically at the end of the century and after World War II, especially from Italy, Germany and some Asian countries, including Japan.

Venezuela

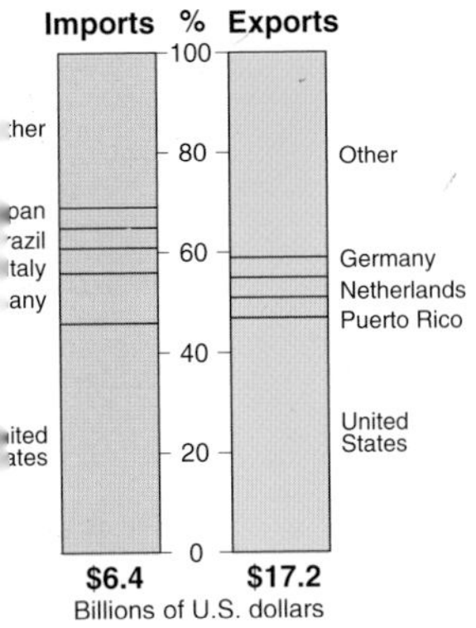

Main imports:
Machinery and transport equipment 33%
Manufactured metals 10%
Vegetables and vegetable products 6%
Minerals 6%

Main exports:
Crude petroleum 80%
Iron ore 2%

GDP by sector

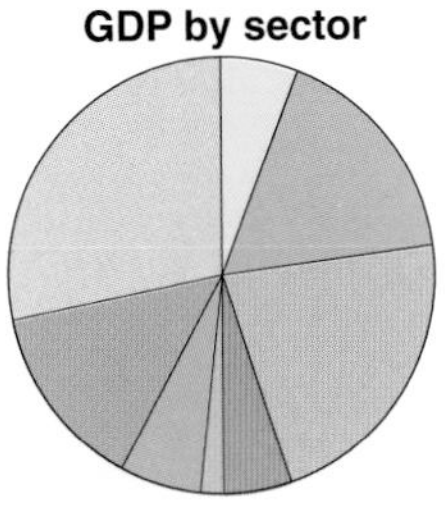

Total: $51.0 billion (U.S.)

Brazil

GDP by sector

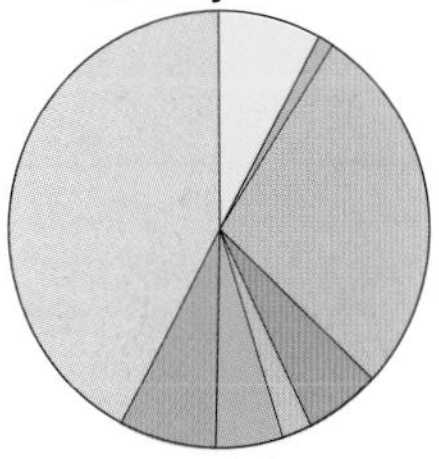

Total: $405.0 billion (U.S.)

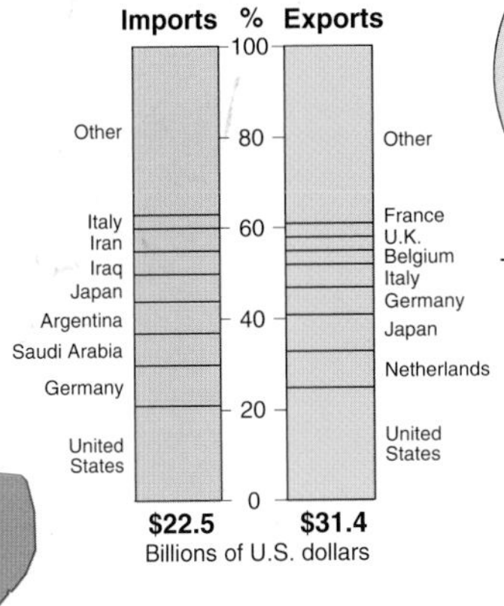

Main imports:
Fuel 30%
Machinery and transport equipment 29%
Chemicals and related products 17%
Food and agricultural raw materials 8%

Main exports:
Food and agricultural raw materials 31%
of which coffee beans 4%
Machinery and transport equipment 18%
Mineral ores and concentrates 10%
Paper and cellulose 4%

Data not available for Falkland Islands.

Chile

Main imports:
Machinery and transport equipment 43%
Fuel 16%
Chemicals 12%
Food 7%

Main exports:
Minerals 48%
Food and agricultural products 33%
Paper and paper products 5%
Chemicals 4%

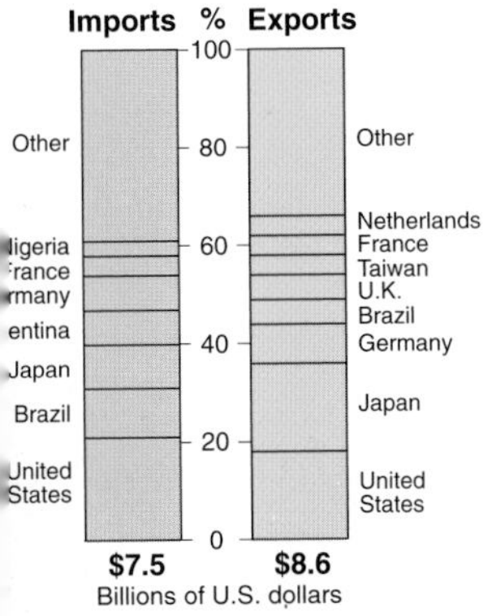

GDP by sector

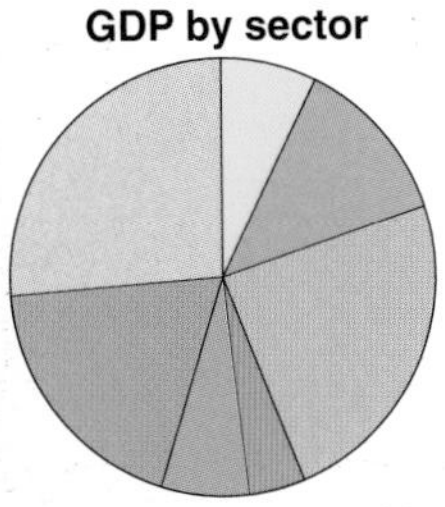

Total: $26.0 billion (U.S.)

External debt

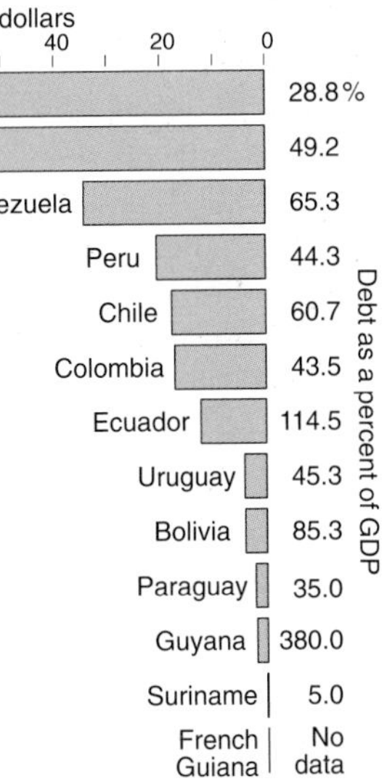

GDP by sector

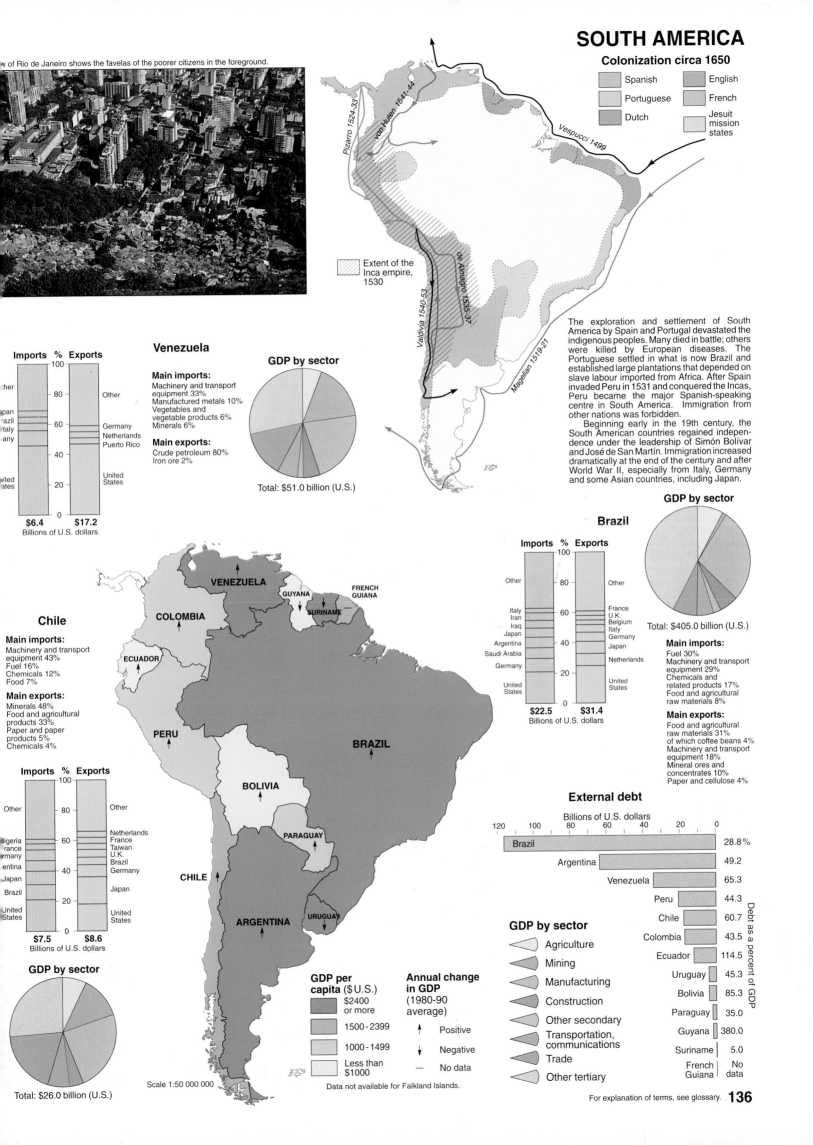

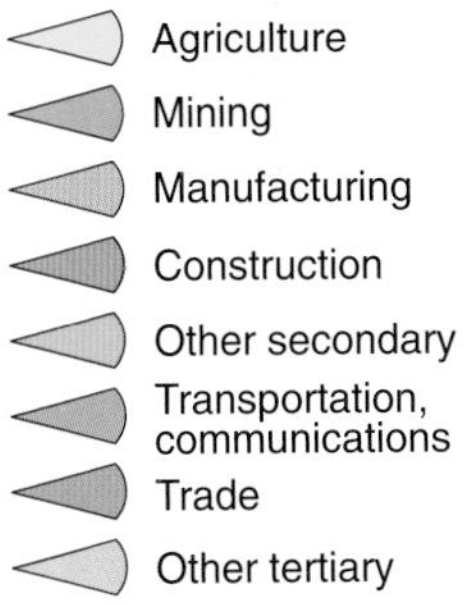

CARIBBEAN SEA
GUADELOUPE (France)
DOMINICA
MARTINIQUE (France)
ST. LUCIA
BARBADOS
ST. VINCENT AND THE GRENADINES
GRENADA
NETHERLANDS ANTILLES (Neth.)
ARUBA
TRINIDAD AND TOBAGO
ATLANTIC OCEAN
PANAMA
Panama
Cartagena
Maracaibo
Lake Maracaibo
Caracas
Maturin
Orinoco
VENEZUELA
GUYANA
Georgetown
Paramaribo
Cayenne
SURINAME
FRENCH GUIANA (France)
Bogotá
COLOMBIA
Negro
Quito
ECUADOR
Guayaquil
Iquitos
Manaus
Amazonas
Belém
EQUATOR
Fortaleza
Teresina
Natal
Recife
Juruá
Purus
Ucayali
Madeira
Xingu
Pôrto Velho
BRAZIL
Chiclayo
PERU
Lima
Cuzco
Mamoré
BOLIVIA
Cuiabá
Araguaia
São Francisco
Salvador
Brasília
Goiânia
Arequipa
Lake Titicaca
La Paz
Sucre
Santa Cruz
Campo Grande
Paraguay
PARAGUAY
Paraná
Belo Horizonte
Antofagasta
Salta
Asunción
São Paulo
Rio de Janeiro
Tropic of Capricorn
CHILE
Resistencia
Salado
ARGENTINA
Pôrto Alegre
Córdoba
Mendoza
Rosario
URUGUAY
Rio Grande
PACIFIC OCEAN
Valparaiso
Santiago
Buenos Aires
Montevideo
Concepción
Colorado
Bahia Blanca
Negro
Valdivia
Chubut
Comodoro Rivadavia
FALKLAND ISLANDS (U.K.)
Stanley
TIERRA DEL FUEGO
Punta Arenas
SOUTH GEORGIA (U.K.)
0 km 500 1000
Scale 1 : 29 800 000 1 cm represents 298 km
Longitude West of Greenwich
80° 70° 60° 50° 40° 30° 20°
10° 0° 10° 20° 30° 40°

FRANCE
SPAIN
PORTUGAL
CORSICA
SARDINIA
ITALY
GREECE
TURKEY
BLACK SEA
SICILY
MALTA
CRETE
CYPRUS
SYRIA
LEBANON
ISRAEL
JORDAN
MEDITERRANEAN SEA
Tanger
Rabat
Casablanca
Marrakech
MADEIRA ISLANDS (Portugal)
CANARY ISLANDS (Spain)
MOROCCO
Alger (Algiers)
Tunis
Sfax
TUNISIA
Ghardaïa
Tarabulus (Tripoli)
Banghāzī (Benghazi)
ALGERIA
LIBYA
El Qâhira (Cairo)
Nile
EGYPT
Aswân
Lake Nasser
RED SEA
SAUDI ARABIA
Tropic of Cancer
Nouadhibou
MAURITANIA
Nouakchott
MALI
Tombouctou (Timbuktu)
NIGER
Agadez
CHAD
Lake Chad
Wadi Halfa
Port Sudan
SUDAN
Khartoum
Asmera
ERITREA
YEMEN
SENEGAL
Bamako
Niamey
BURKINA FASO
Ouagadougou
Kano
Kaduna
N'Djamena
DJIBOUTI
Djibouti
GUINEA
Conakry
Freetown
SIERRA LEONE
BENIN
NIGERIA
Addis Ababa
Hargeisa
COTE D'IVOIRE
TOGO
GHANA
Lagos
Accra
Lomé
Porto Novo
Benue
CENTRAL AFRICAN REPUBLIC
ETHIOPIA
SOMALIA
Monrovia
LIBERIA
Greenville
Abidjan
CAMEROON
Yaoundé
Bangui
Blue Nile
White Nile
Malabo
EQUATORIAL GUINEA
São Tomé
SAO TOME AND PRINCIPE
Libreville
GABON
CONGO
Zaïre
Kisangani
UGANDA
Kampala
Lake Turkana
KENYA
Mogadishu (Muqdisho)
EQUATOR
ZAÏRE
RWANDA
Kigali
Nairobi
Lake Victoria
BURUNDI
Bujumbura
Brazzaville
Kinshasa
Congo
CABINDA (Angola)
Kananga
TANZANIA
Dodoma
Mombasa
INDIAN OCEAN
Zanzibar
Dar es Salaam
Lake Tanganyika
SEYCHELLES
Luanda
ANGOLA
Likasi
MALAWI
Lake Malawi
COMOROS
Huambo
Kitwe
Lilongwe
ZAMBIA
Lusaka
Zambeze
MOZAMBIQUE
Mozambique
Moçâmedes
Harare
ZIMBABWE
MADAGASCAR
Antananarivo
NAMIBIA
Beira
Bulawayo
Limpopo
ATLANTIC OCEAN
Walvis Bay
Windhoek
BOTSWANA
Gaberone
Toliary
Tropic of Capricorn
Pretoria
Johannesburg
Maputo
Mbabane
SWAZILAND
Lüderitz
SOUTH AFRICA
Vaal
Orange
Maseru
LESOTHO
Durban
Cape Town
Port Elizabeth
Longitude East of Greenwich
Longitude West of Greenwich
20°
10°
0°
30°
40°
50°
Scale 1 : 3 480 000
1 cm represents 348 km
0 km
500
1000
National capital
Other important city
Major international airport
Main railway
Main road
International boundary
Less than 200 metres of water
More than 200 metres of water

WORLD-REGIONAL

Africa is the probable homeland of all homo sapiens. The civilizations of the upper and lower Nile are some of the world's oldest. This huge continent straddling the equator represents one fifth of the world's landmass. Africa has mineral wealth, vast agricultural potential and rich human resources, but it is also home to the world's poorest nations with the least developed economies.

Although African nations suffered colonialization in recent times, most liberated themselves within a century. However, during the Cold War, the United States and U.S.S.R. armed political factions, which destabilized many emerging nations and resulted in political turmoil. More recently, there has been a trend towards more democratic government.

Rainfall variability in the Sahel

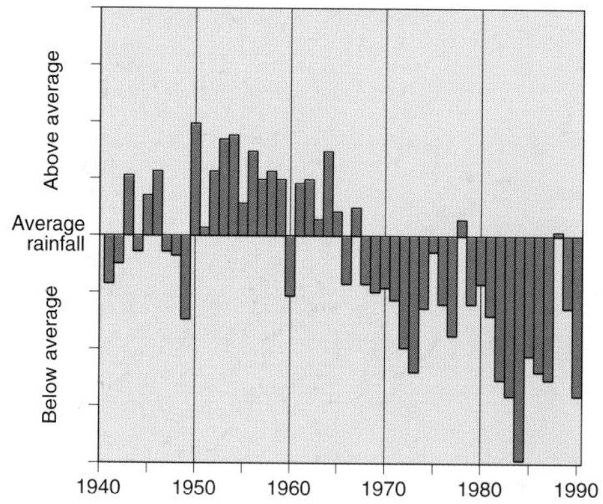

Many villages in the Sahel struggle against desertification.

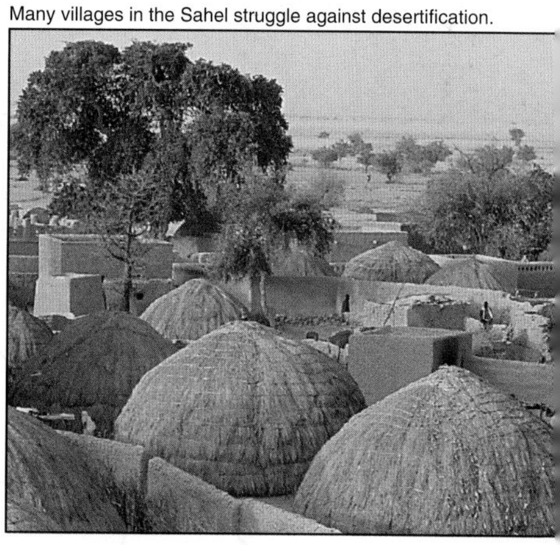

The area to the south of the Sahara Desert, ca Sahel, has been undergoing a period of d cation—the spreading of desert into adjacent Scientists estimate that nearly 700 000 km² south of the Sahara (an area larger than Albert been affected in the last 100 years. In Mali, the has moved south 350 km in only two decade growing population in this region puts inc pressure on the land. Farmers are forced to gr in marginal areas, thereby depleting the soil of n and making the land less productive in the future regions will benefit from the development of water resources. The Senegal River, for examp great potential for irrigation and energy product

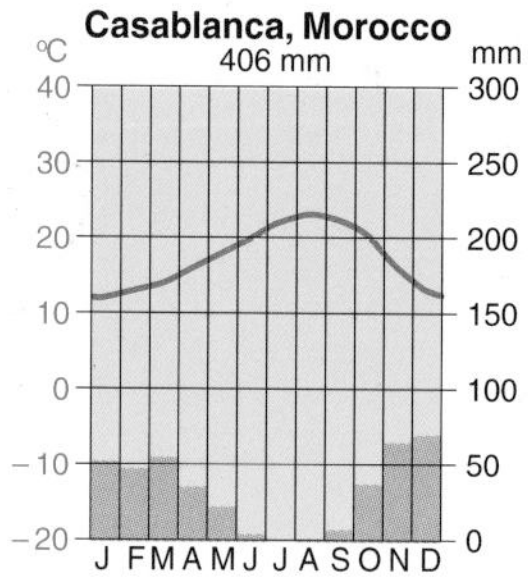

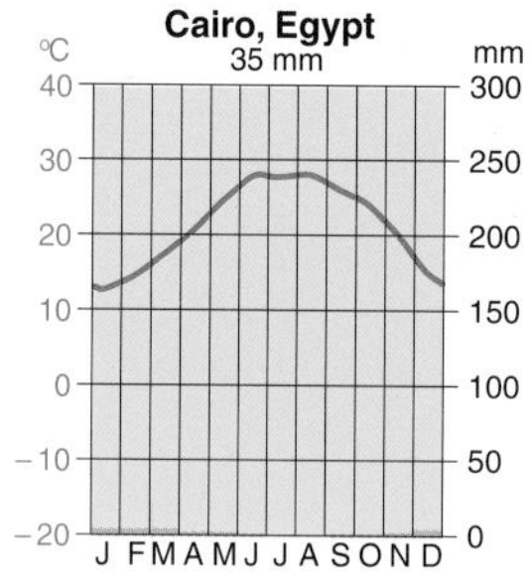

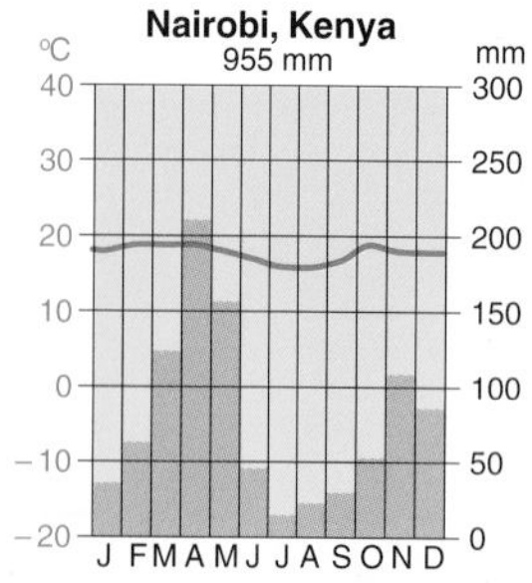

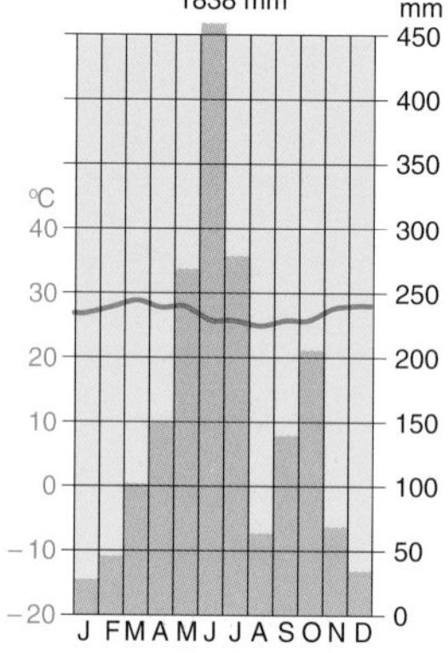

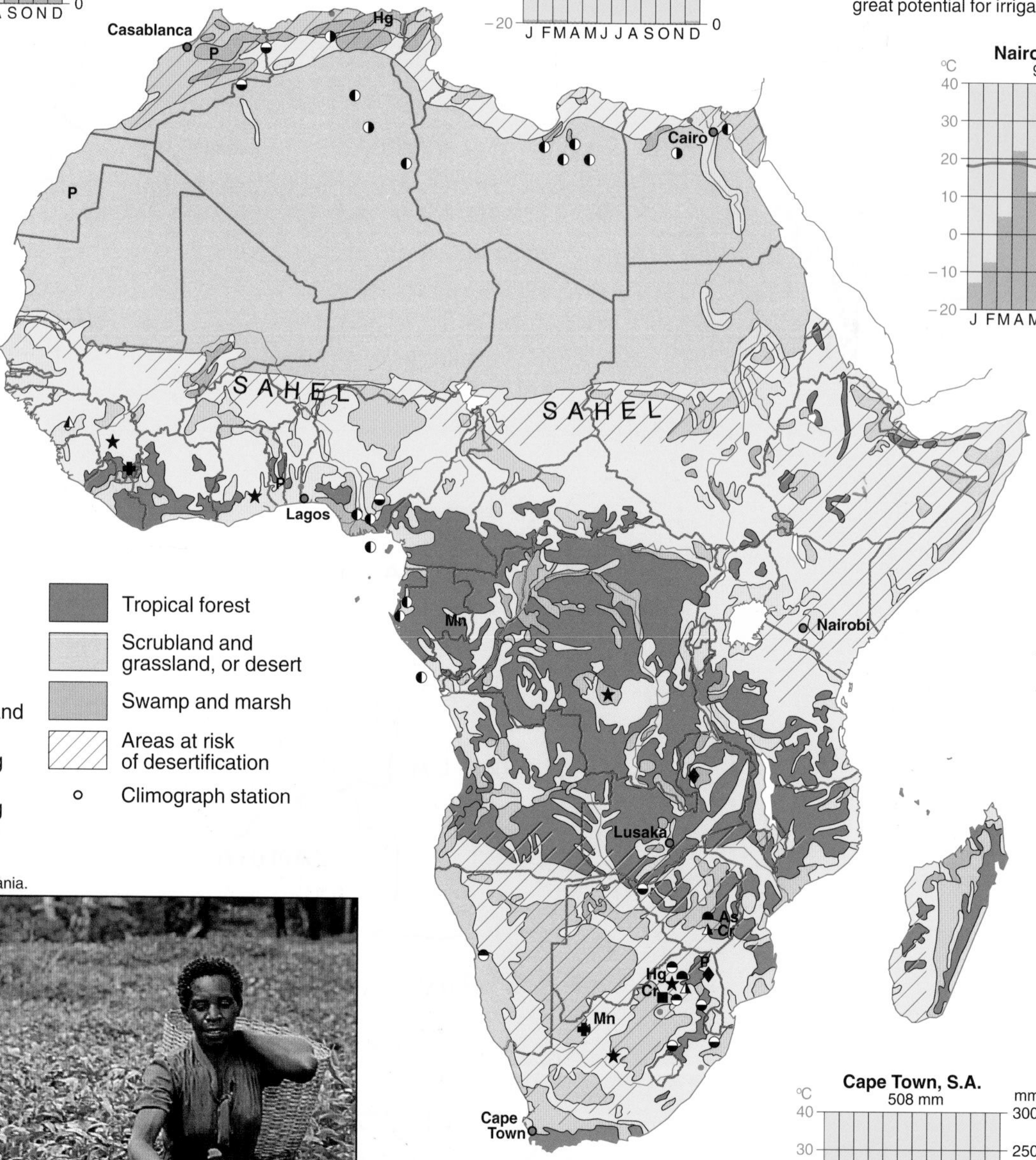

Scale 1:55 000 000

Land use

- Urban
- Cropland
- Cropland and woodland
- Cropland and grazing
- Grassland and grazing
- Tropical forest
- Scrubland and grassland, or desert
- Swamp and marsh
- Areas at risk of desertification
- Climograph station

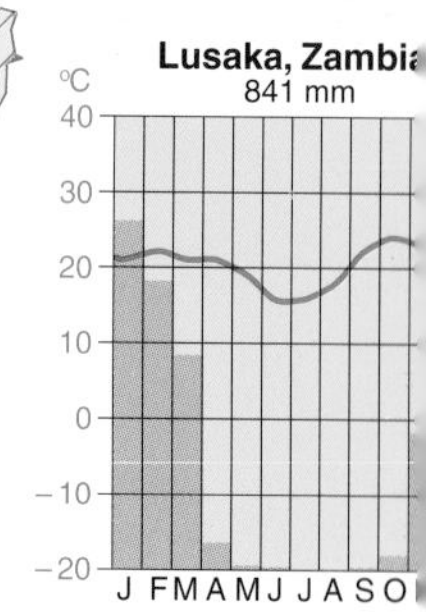

Minerals

- Gold
- Diamond
- Copper
- Iron
- Nickel
- Aluminum
- Tungsten
- Coal
- Oil
- Gas
- Uranium
- Cr Chromium
- Mn Manganese
- Hg Mercury
- As Asbestos
- P Phosphate

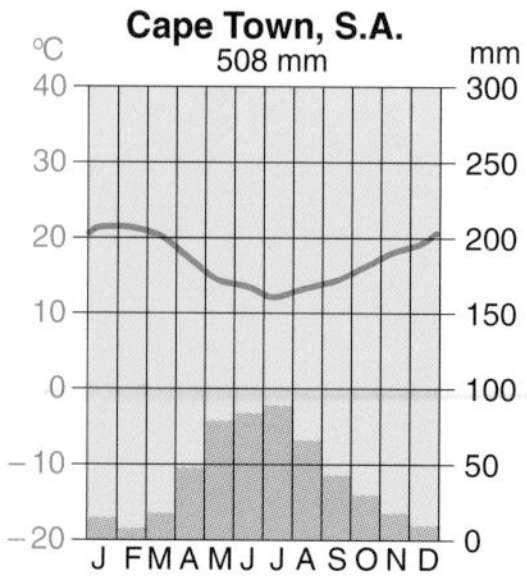

Tea is an important cash crop in Tanzania.

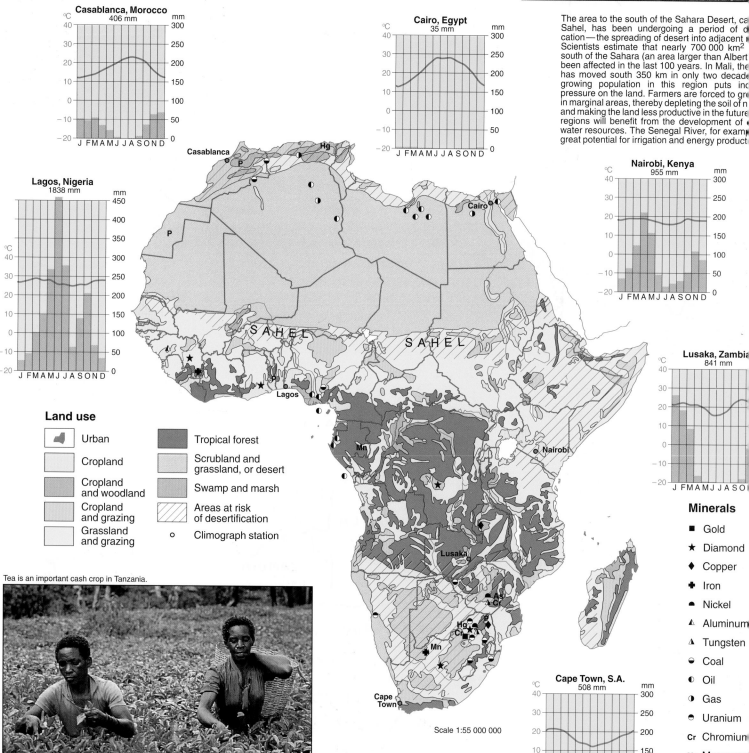

AFRICA

Independence and conflict

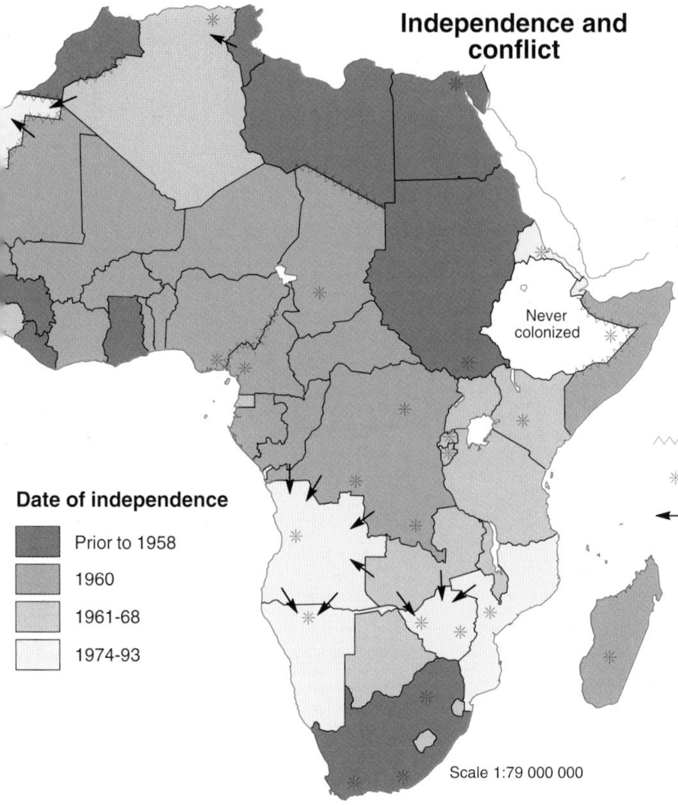

Date of independence

- Prior to 1958
- 1960
- 1961-68
- 1974-93

- Current border conflicts
- Violent insurrection or war
- Guerilla movements

Before the last two decades of the 19th century, the African continent, particularly south of the Sahara Desert, consisted of hundreds of small states and tribal kingdoms. Along the east coast and in the north, Africans traded with Arab nations, but the mountains and deserts south of the Mediterranean limited contact with Europeans. Only in South Africa were settlements established by the British and Dutch. The European conquest was part of 19th-century colonial expansion by the French, British, Germans, Portuguese and Dutch. The continent was divided into colonies with little regard for either tribal boundaries or African cultures. In the 20th century, almost all African nations have regained their independence. Egypt was the first in 1922, and Eritrea the most recent in 1993.

Downtown Harare, Zimbabwe.
A growing number of Africans live and work in urban areas.

Mali

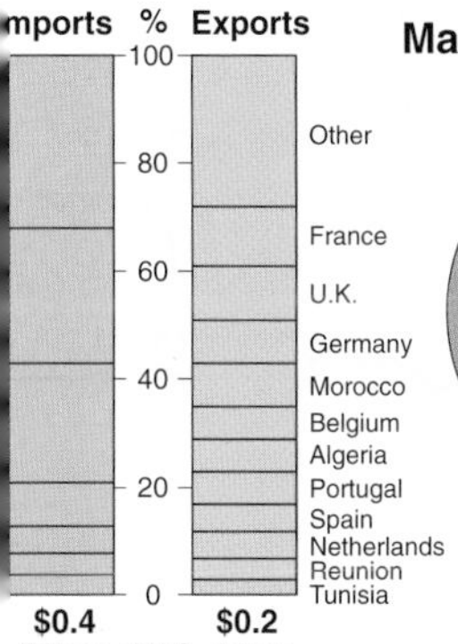

GDP by sector

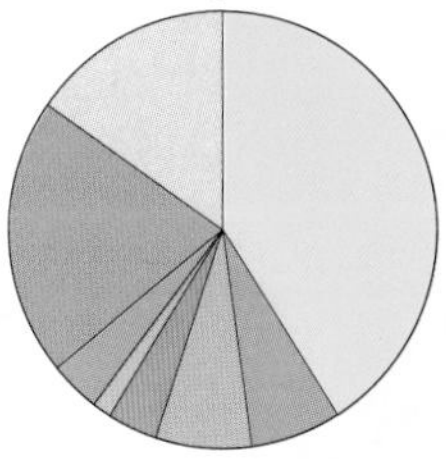

Total: $2.3 billion (U.S.)

Main imports:
nery, transport
nent and
nces 30%
13%
cals 10%
eum 10%

Main exports:
45%
ls 24%
nd diamonds 13%

GDP per capita ($U.S.)

- $1000 or more
- 500-999
- 300-499
- Less than $300

GDP by sector

- Agriculture
- Mining
- Manufacturing
- Construction
- Other secondary
- Transportation, communications
- Trade
- Other tertiary

Egypt

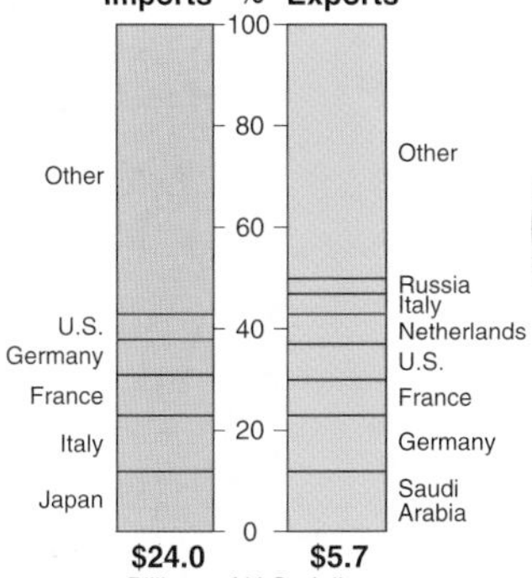

GDP by sector

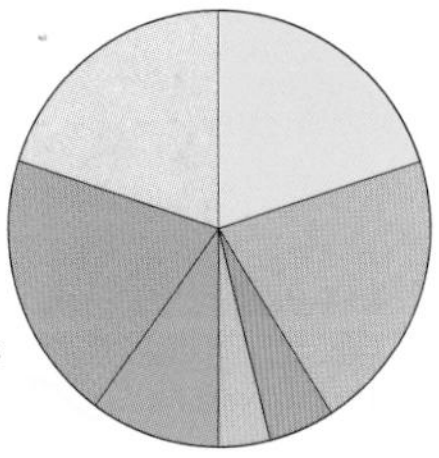

Total: $32.0 billion (U.S.)

Main imports:
Machinery and transport equipment 20%
Food 14%
Chemicals 10%
Base metals 7%

Main exports:
Petroleum products 50%
Cotton yarn, textiles 14%
Raw cotton 2%

Annual change in GDP (1980-90 average)

- ↑ Positive
- ↓ Negative
- — No data

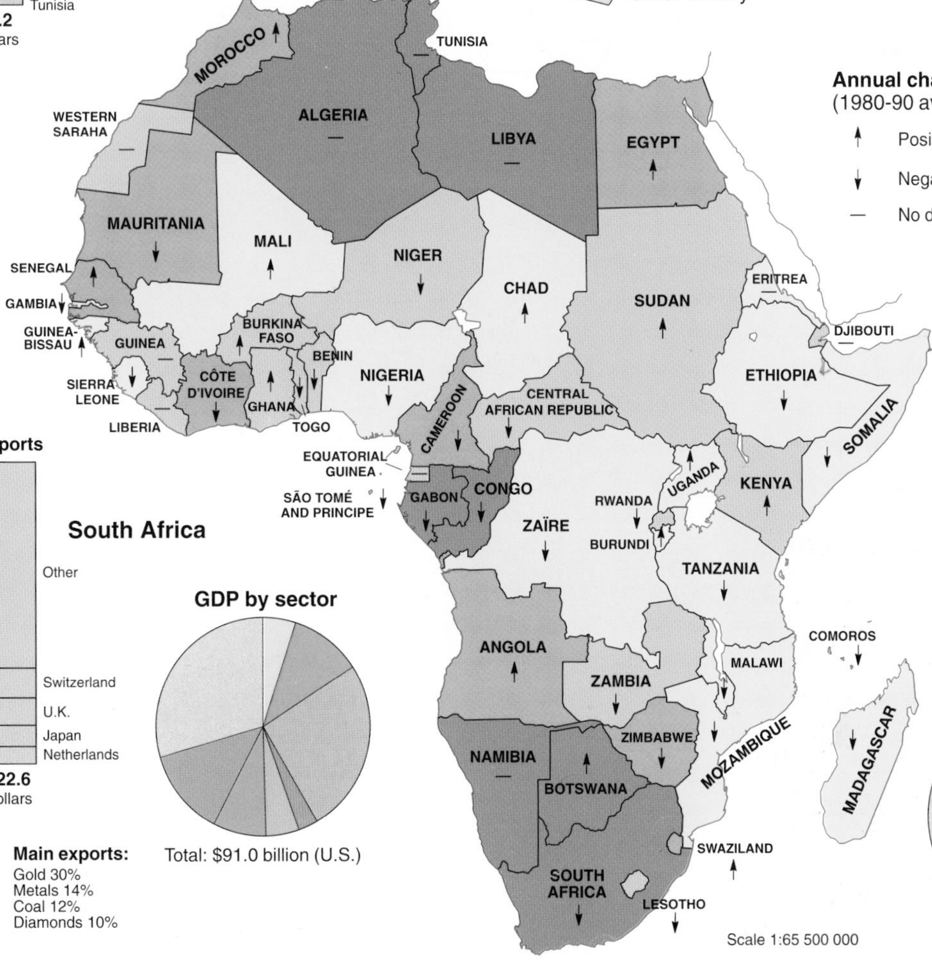

South Africa

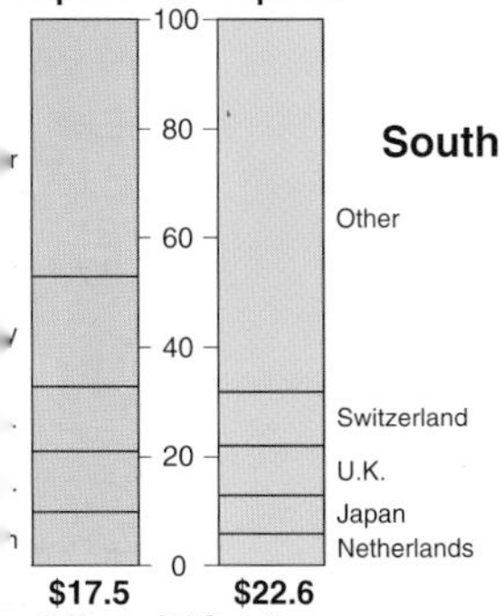

GDP by sector

Total: $91.0 billion (U.S.)

Main imports:
ery 15%
ort equipment 15%
cals 11%
s and clothing 5%

Main exports:
Gold 30%
Metals 14%
Coal 12%
Diamonds 10%

Kenya

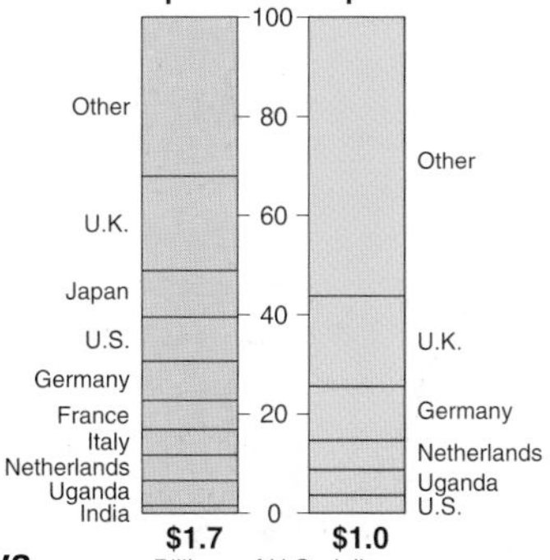

GDP by sector

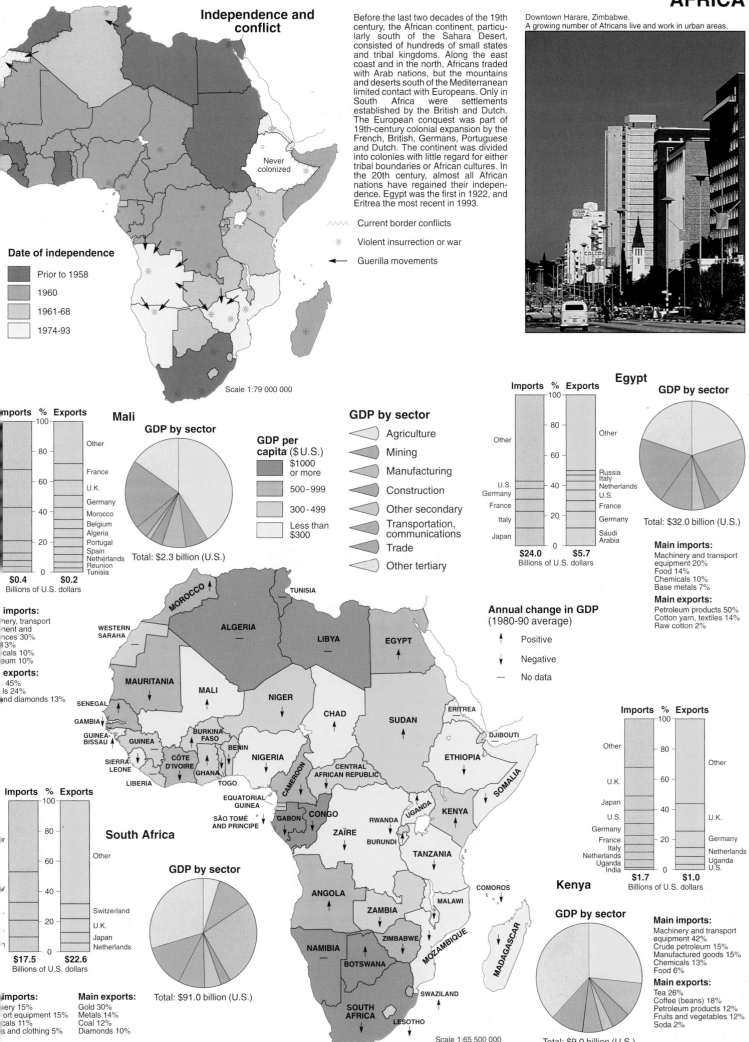

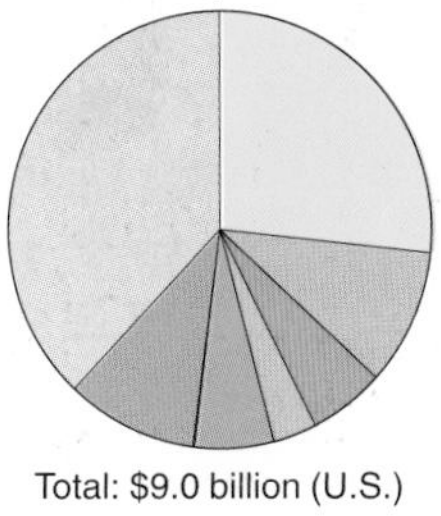

Total: $9.0 billion (U.S.)

Main imports:
Machinery and transport equipment 42%
Crude petroleum 15%
Manufactured goods 15%
Chemicals 13%
Food 6%

Main exports:
Tea 26%
Coffee (beans) 18%
Petroleum products 12%
Fruits and vegetables 12%
Soda 2%

For explanation of terms, see glossary.

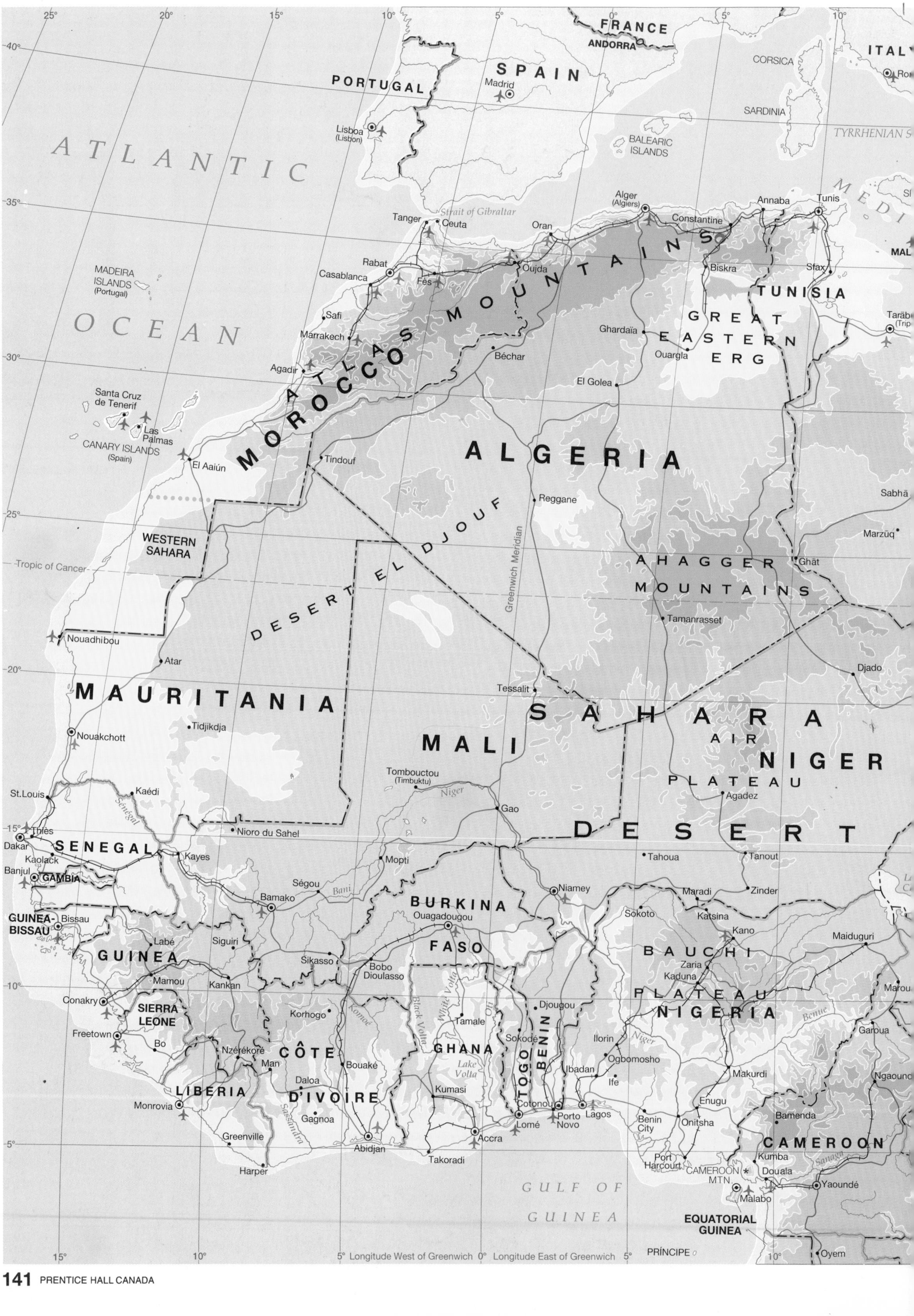
FRANCE
ANDORRA
SPAIN
Madrid
PORTUGAL
Lisboa
(Lisbon)
CORSICA
ITALY
SARDINIA
BALEARIC
ISLANDS
TYRRHENIAN S
ATLANTIC
OCEAN
MADEIRA
ISLANDS
(Portugal)
Strait of Gibraltar
Tanger
Ceuta
Alger
(Algiers)
Oran
Annaba
Tunis
Constantine
Rabat
Casablanca
Fès
Oujda
Biskra
Sfax
TUNISIA
Safi
Marrakech
Agadir
Ghardaïa
Ouargla
GREAT
EASTERN
ERG
Béchar
El Golea
ATLAS MOUNTAINS
MOROCCO
Santa Cruz
de Tenerif
Las
Palmas
CANARY ISLANDS
(Spain)
El Aaiún
Tindouf
ALGERIA
Reggane
Sabhā
Marzūq
Ghāt
WESTERN
SAHARA
Tropic of Cancer
DESERT EL DJOUF
Greenwich Meridian
AHAGGER
MOUNTAINS
Tamanrasset
Nouadhibou
Atar
Djado
MAURITANIA
Tessalit
SAHARA
Nouakchott
Tidjikdja
MALI
AIR
PLATEAU
NIGER
Agadez
Tombouctou
(Timbuktu)
Niger
Gao
St.Louis
Kaédi
Sénégal
Thiès
Dakar
Nioro du Sahel
DESERT
SENEGAL
Kaolack
Kayes
Mopti
Tahoua
Tanout
Banjul
GAMBIA
Ségou
Bani
Bamako
Niamey
Maradi
Zinder
GUINEA-
BISSAU
Bissau
BURKINA
FASO
Ouagadougou
Sokoto
Katsina
Kano
Maiduguri
Labé
GUINEA
Siguiri
Sikasso
Bobo
Dioulasso
BAUCHI
PLATEAU
Zaria
Kaduna
Mamou
Kankan
Conakry
SIERRA
LEONE
Freetown
Bo
Korhogo
Komoé
Black Volta
White Volta
Oti
Tamale
Djougou
Sokodé
NIGERIA
Benue
Garoua
Ilorin
Ogbomosho
Nzérékoré
Man
CÔTE
D'IVOIRE
Bouaké
GHANA
Lake
Volta
TOGO
BENIN
Ibadan
Ife
Makurdi
Ngaound
LIBERIA
Monrovia
Daloa
Kumasi
Cotonou
Porto
Novo
Lagos
Enugu
Gagnoa
Sassandra
Lomé
Benin
City
Onitsha
Bamenda
Greenville
Abidjan
Accra
CAMEROON
Kumba
Harper
Takoradi
Port
Harcourt
CAMEROON
MTN
Douala
Sanaga
Yaoundé
GULF OF
GUINEA
Malabo
EQUATORIAL
GUINEA
PRÍNCIPE
Oyem
5° Longitude West of Greenwich 0° Longitude East of Greenwich 5°

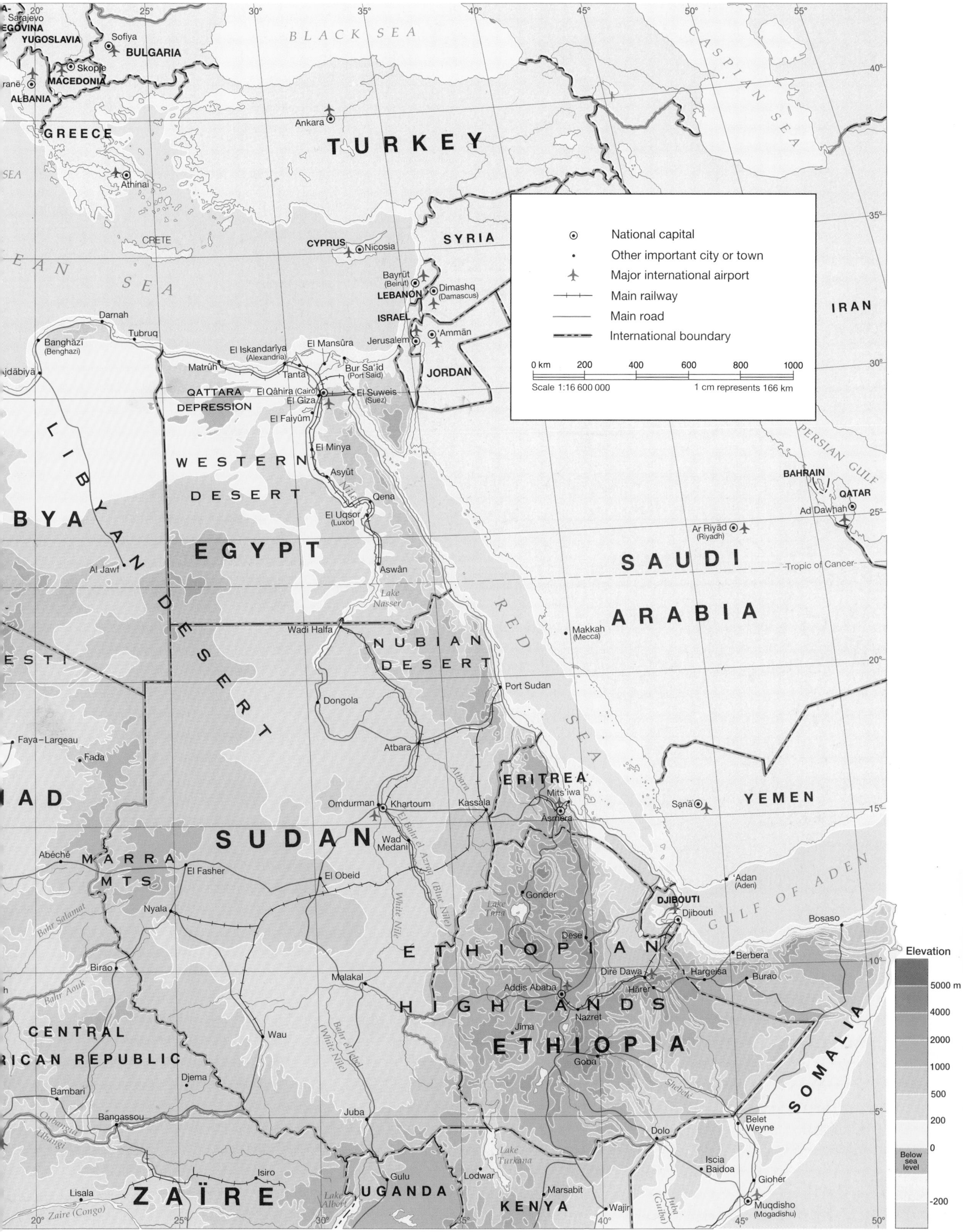
National capital
Other important city or town
Major international airport
Main railway
Main road
International boundary
0 km 200 400 600 800 1000
Scale 1:16 600 000
1 cm represents 166 km
Elevation
5000 m
4000
2000
1000
500
200
0
Below sea level
-200
BLACK SEA
CASPIAN SEA
RED SEA
GULF OF ADEN
PERSIAN GULF
Tropic of Cancer
TURKEY
GREECE
BULGARIA
YUGOSLAVIA
MACEDONIA
ALBANIA
CYPRUS
SYRIA
LEBANON
ISRAEL
JORDAN
IRAN
SAUDI ARABIA
BAHRAIN
QATAR
YEMEN
DJIBOUTI
SOMALIA
ETHIOPIA
ERITREA
KENYA
UGANDA
ZAÏRE
CENTRAL AFRICAN REPUBLIC
SUDAN
EGYPT
LIBYA
WESTERN DESERT
LIBYAN DESERT
NUBIAN DESERT
ETHIOPIAN HIGHLANDS
MARRA MTS
QATTARA DEPRESSION
Sarajevo
Sofiya
Skopje
Ankara
Athinai
CRETE
Nicosia
Bayrūt (Beirut)
Dimashq (Damascus)
Jerusalem
'Ammān
Darnah
Tubruq
Banghāzī (Benghazi)
Matrûh
El Iskandarîya (Alexandria)
El Mansûra
Tanta
Bur Sa'îd (Port Said)
El Qâhira (Cairo)
El Gîza
El Suweis (Suez)
El Faiyûm
El Minya
Asyût
Qena
El Uqsor (Luxor)
Aswân
Lake Nasser
Nile
Al Jawf
Wadi Halfa
Dongola
Port Sudan
Atbara
Faya-Largeau
Fada
Omdurman
Khartoum
Kassala
Mits'iwa
Asmera
Wad Medani
Abéché
El Fasher
El Obeid
Nyala
Birao
Malakal
Wau
Djema
Bambari
Bangassou
Juba
Gulu
Lodwar
Marsabit
Wajir
Isiro
Lisala
Lake Albert
Lake Turkana
Lake Tana
Gonder
Dese
Djibouti
Dire Dawa
Harer
Hargeisa
Berbera
Burao
Bosaso
Addis Ababa
Nazret
Jima
Goba
Dolo
Belet Weyne
Iscia Baidoa
Gioher
Muqdisho (Mogadishu)
'Adan (Aden)
Sanā
Makkah (Mecca)
Ar Riyād (Riyadh)
Ad Dawhah
Atbara
Bahr el Azraq (Blue Nile)
White Nile
Bahr el Jebel (White Nile)
Bahr Salamat
Bahr Aouk
Oubangui
Zaïre (Congo)
Shebelē
Juba (Giuba)
20°
25°
30°
35°
40°
45°
50°
55°

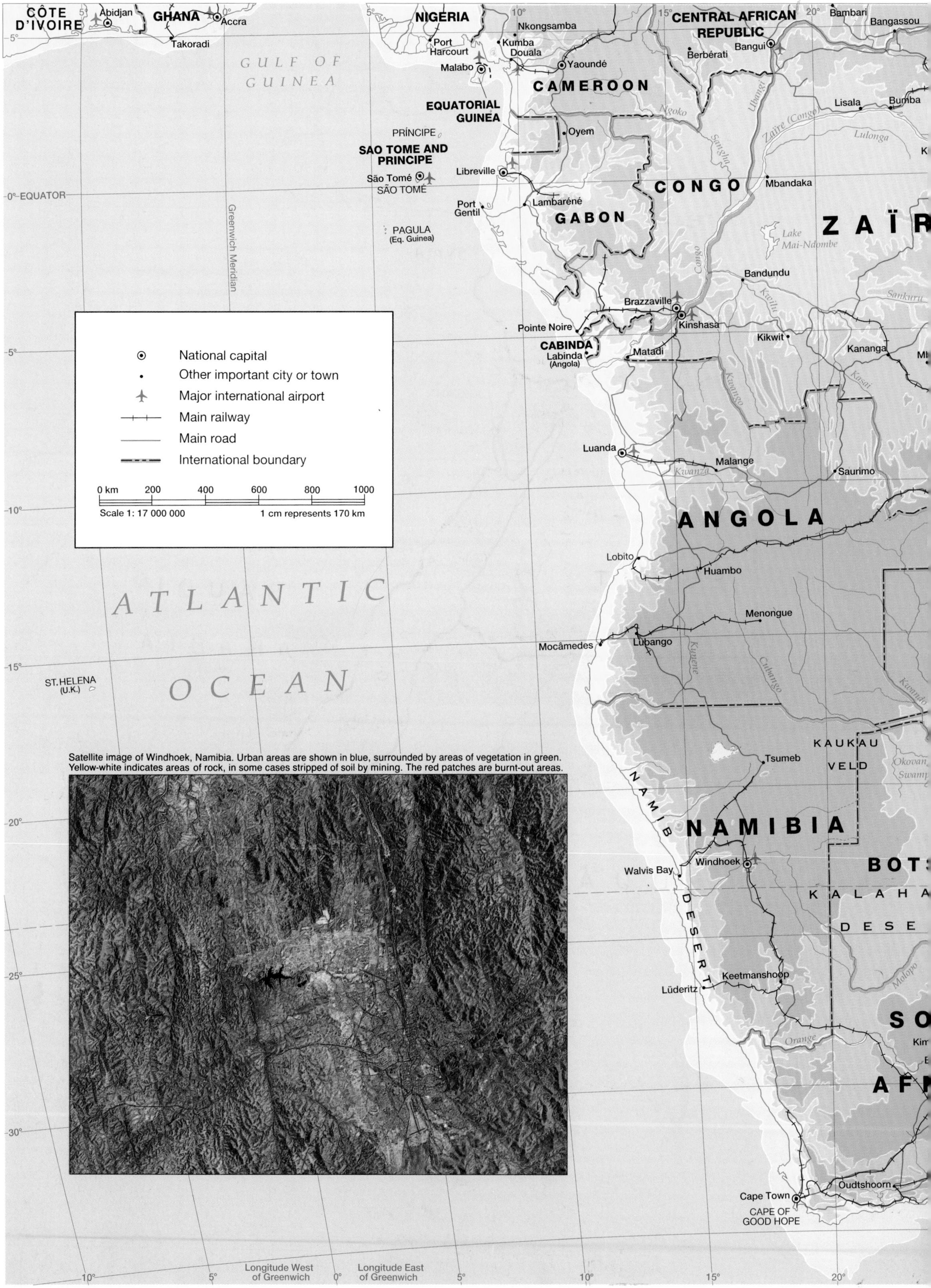

Satellite image of Windhoek, Namibia. Urban areas are shown in blue, surrounded by areas of vegetation in green. Yellow-white indicates areas of rock, in some cases stripped of soil by mining. The red patches are burnt-out areas.

SUDAN
ETHIOPIA
SOMALIA
UGANDA
KENYA
RWANDA
BURUNDI
TANZANIA
MALAWI
MOZAMBIQUE
MADAGASCAR
SWAZILAND
COMOROS
SEYCHELLES
MAURITIUS
RÉUNION
INDIAN
OCEAN
MOZAMBIQUE CHANNEL
SERENGETI
MT. KENYA
MT. KILIMANJARO
MUNCHINGA MTS
DRAKENSBERG MTS
PEMBA ISLAND
ZANZIBAR ISLAND
MAFIA ISLAND
Juba
Gulu
Lodwar
Marsabit
Wajir
Belet Weyne
Iscia Baidoa
Gioher
Muqdisho (Mogadishu)
Kismayu
Lake Turkana
Lake Albert
Lake Kyoga
Fort Portal
Kampala
Jinja
Tororo
Entebbe
Kisumu
Nakuru
Nanyuki
Nyeri
Nairobi
Lake Edward
Lake Victoria
Bukoba
Kigali
Bukavu
Mwanza
Arusha
Lake Eyasi
Lamu
Malindi
Mombasa
Tanga
Tana
Gitega
Tabora
Kalemie
Dodoma
Morogoro
Zanzibar
Dar es Salaam
Lake Tanganyika
Lake Rukwa
Iringa
Rufiji
Mbeya
Lake Mweru
Lindi
Mtwara
Ruvuma
Mzuzu
Lubumbashi
Lake Malawi
Lichinga
Ndola
Lilongwe
Kabwe
Pemba
Moroni
Antsiranana
Mahajanga
Zambezi
Nampula
Mozambique
Blantyre
Tete
Lake Kariba
Harare
Quelimane
Toamasina
Antananarivo
Mamica
Gwelo
Bulawayo
Beira
Save
Limpopo
Mangoky
Fianarantsoa
Toliary
Tropic of Capricorn
Xai-Xai
Maputo
Springs
Mbabane
Maseru
Pietermaritzburg
Durban
East London
EQUATOR
Elevation
5000
4000
2000
1000
500
200
0
Below sea level
-200

WORLD-REGIONAL

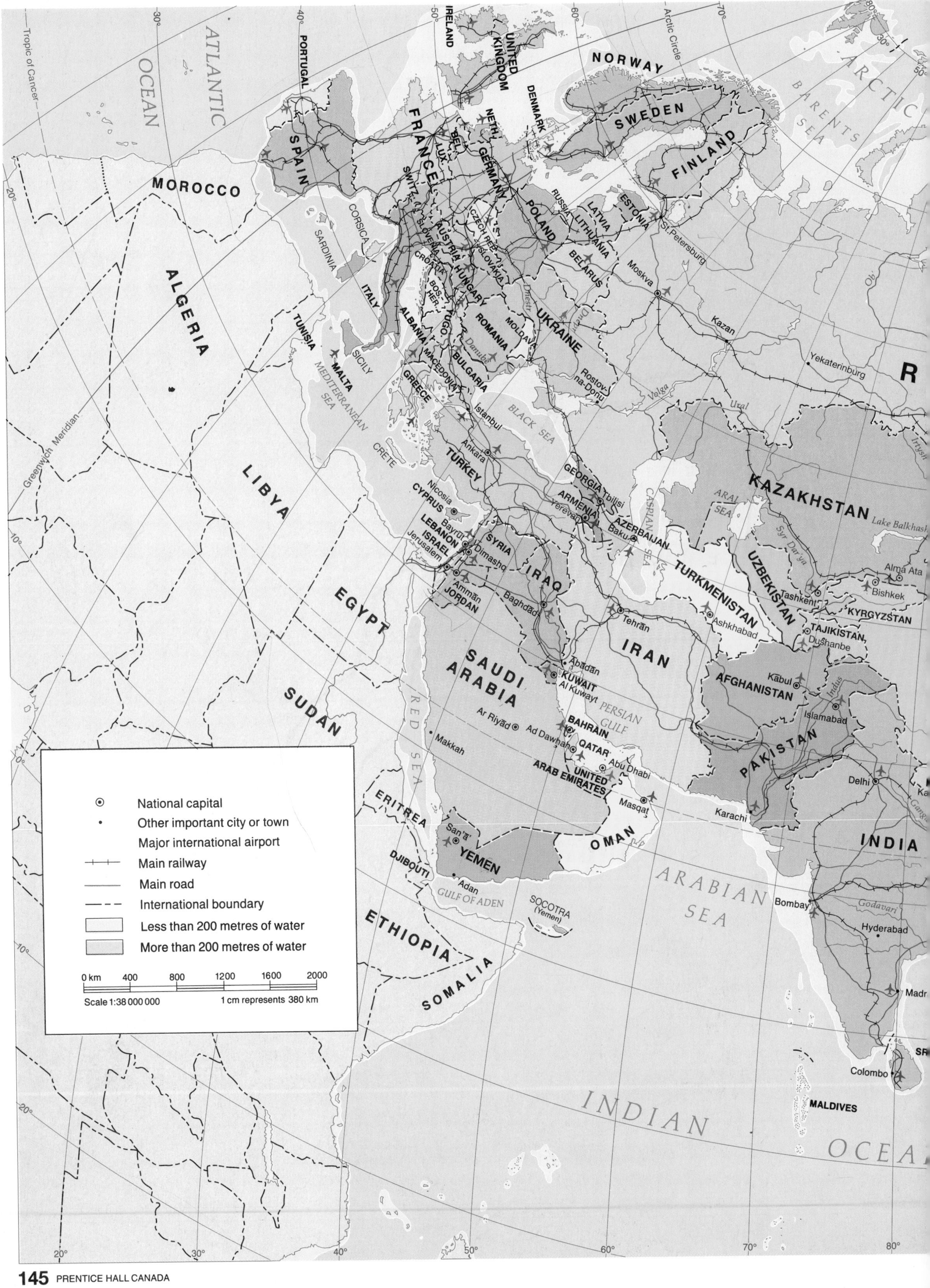

National capital
Other important city or town
Major international airport
Main railway
Main road
International boundary
Less than 200 metres of water
More than 200 metres of water

0 km 400 800 1200 1600 2000
Scale 1:38 000 000 1 cm represents 380 km

OCEAN
BERING SEA
International date line
Lena
Okhotsk
SEA OF OKHOTSK
Amur
S I A
Lake Baykal
JAPAN
Vladivostok
Ulaanbaatar
NORTH KOREA
SEA OF JAPAN
MONGOLIA
Shenyang
P'yŏngyang
Tōkyō
PACIFIC OCEAN
Beijing
Sŏul
Huang (Yellow)
SOUTH KOREA
Pusan
YELLOW SEA
CHINA
Shanghai
EAST CHINA SEA
Chang (Yangtze)
Brahmaputra
T'aipei
TAIWAN
BHUTAN
Guangzhou
Victoria
HONG KONG (U.K.)
Mekong
Dhaka
Salween
Hanoi
MYANMAR
SOUTH CHINA SEA
PHILIPPINES
EQUATOR 0°
Irrawaddy
LAOS
Manila
Viangchan
Yangon
VIETNAM
THAILAND
Krung Thep
CAMBODIA
Phnum Penh
Ho Chi Minh
NEW GUINEA
PAPUA-NEW GUINEA
(India)
BRUNEI
IRIAN JAYA
MALAYSIA
Port Moresby
Kuala Lumpur
SINGAPORE
BORNEO
SULAWESI
CORAL SEA
SUMATRA
INDONESIA
TIMOR
AUSTRALIA
Jakarta
JAVA
Longitude East of Greenwich 100°
110°
120°
130°
140°
150°

WORLD-REGIONAL

RUSSIA
FINLAND
ESTONIA
LATVIA
BELARUS
UKRAINE
MOLDOVA
ROMANIA
BULGARIA
GEORGIA
ARMENIA
AZERBAIJAN
AZER.
KAZAKHSTAN
UZBEKISTAN
TURKMENISTAN
TURKEY
SYRIA
IRAQ
IRAN
CYPRUS
WHITE SEA
BLACK SEA
SEA OF AZOV
CASPIAN SEA
ARAL SEA
AEGEAN SEA
CRETE
RHODES
Lake Ladoga
Lake Onega
Eastern extent of Europe
Murmansk
Arkhangel'sk
Oulu
Petrozavodsk
Vyborg
Helsinki
St Petersburg
Tartu
Vilnius
Minsk
Moskva (Moscow)
Tula
Nizhniy Novgorod
Kazan'
Kirov
Perm'
Yekaterinburg
Samara
Saratov
Volgograd
Astrakhan
Kiyev (Kiev)
Khar'kov
L'vov
Dnepropetrovsk
Kishinev
Odessa
Rostov na Donu
Sevastopol'
Groznyy
Tbilisi
Yerevan
Baku
Cluj-Napoca
Bucureşti (Bucharest)
Constanţa
Sofiya (Sofia)
Plovdiv
İstanbul
Thessaloníki
Ankara
İzmir
Athínai (Athens)
Nicosia
Halab
Baghdād
Tehrān
Esfahān
Pechora
Severnaya Dvina
Vychegda
Onega
Sukhona
Volga
Oka
Don
Dnepr
Dnestr
Bug
Daugava
Dunărea (Danube)
Ural
Ob'
Irtysh
Tobol
Surgut
30°
35°
40°
45°
50°
55°
60°
65°
70°
75°
25°

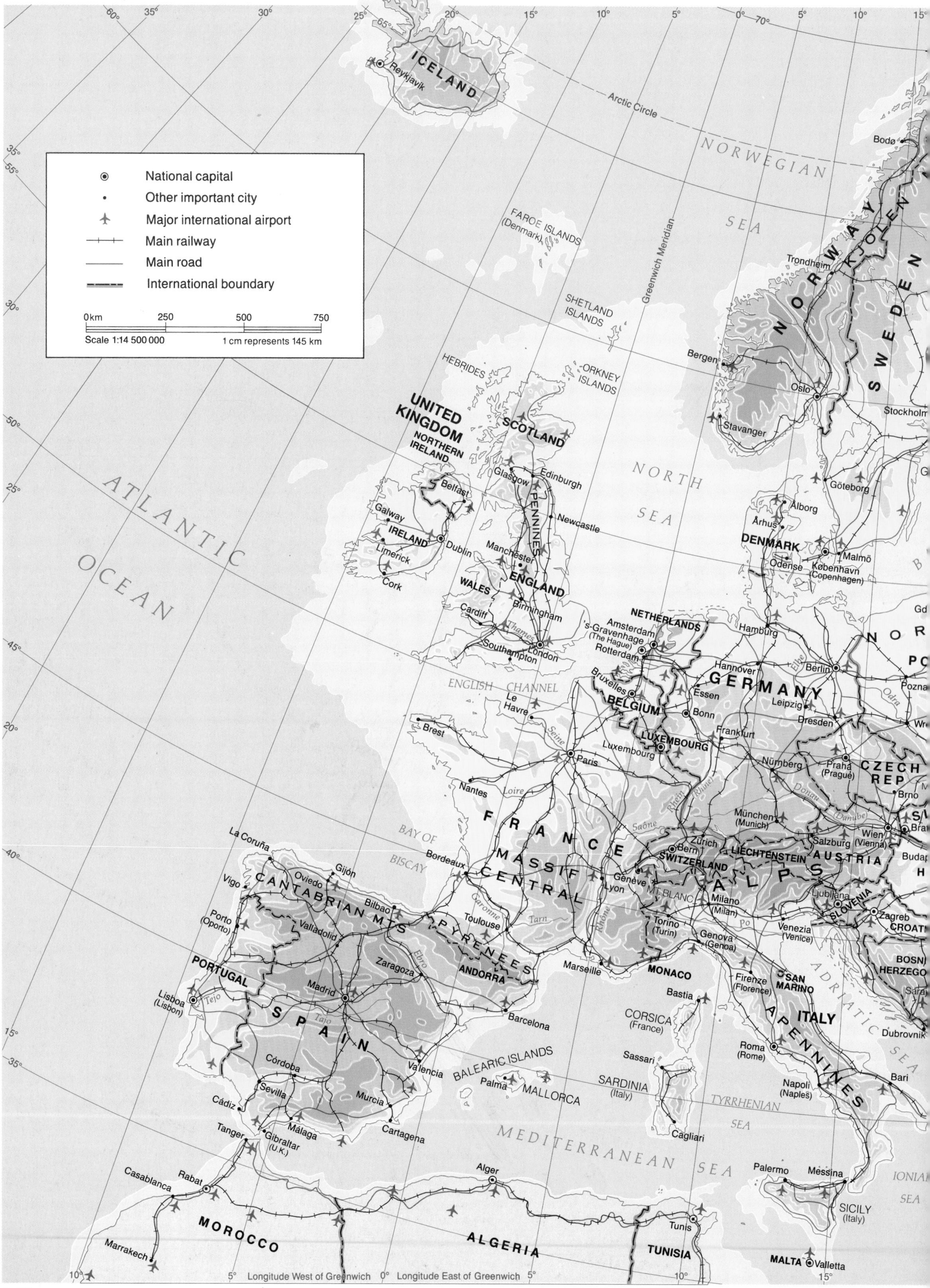
National capital
Other important city
Major international airport
Main railway
Main road
International boundary
0km
250
500
750
Scale 1:14 500 000
1 cm represents 145 km
ICELAND
Reykjavik
Arctic Circle
NORWEGIAN SEA
FAROE ISLANDS (Denmark)
SHETLAND ISLANDS
ORKNEY ISLANDS
HEBRIDES
Greenwich Meridian
UNITED KINGDOM
NORTHERN IRELAND
SCOTLAND
Glasgow
Edinburgh
PENNINES
Newcastle
Belfast
Galway
IRELAND
Dublin
Limerick
Cork
Manchester
ENGLAND
WALES
Cardiff
Birmingham
Thames
Southampton
London
ATLANTIC OCEAN
NORTH SEA
NORWAY
KJOLEN
SWEDEN
Bodø
Trondheim
Bergen
Oslo
Stavanger
Stockholm
Göteborg
Ålborg
Århus
DENMARK
Odense
København (Copenhagen)
Malmö
NETHERLANDS
Amsterdam
's-Gravenhage (The Hague)
Rotterdam
Hamburg
Hannover
Berlin
GERMANY
Essen
Bruxelles
BELGIUM
Bonn
Leipzig
Dresden
LUXEMBOURG
Luxembourg
Frankfurt
Nürnberg
Praha (Prague)
CZECH REP
Brno
Elbe
Odra
Poznań
ENGLISH CHANNEL
Le Havre
Brest
Seine
Paris
Nantes
Loire
FRANCE
MASSIF CENTRAL
BAY OF BISCAY
Bordeaux
Rhein (Rhine)
Donau (Danube)
München (Munich)
Wien (Vienna)
Salzburg
AUSTRIA
Zürich
Bern
SWITZERLAND
LIECHTENSTEIN
ALPS
Saône
Genève
Lyon
MT BLANC
Milano (Milan)
Ljubljana
SLOVENIA
Zagreb
CROATIA
Torino (Turin)
Genova (Genoa)
Venezia (Venice)
Po
Rhône
Tarn
Garonne
Toulouse
PYRENEES
ANDORRA
Marseille
MONACO
La Coruña
Gijón
Oviedo
Vigo
CANTABRIAN MTS
Bilbao
Porto (Oporto)
Valladolid
Ebro
Zaragoza
PORTUGAL
Madrid
Tejo
Tajo
Lisboa (Lisbon)
SPAIN
Barcelona
Córdoba
Sevilla
Valencia
Murcia
Cádiz
Málaga
Gibraltar (U.K.)
Cartagena
Tanger
BALEARIC ISLANDS
Palma
MALLORCA
CORSICA (France)
Bastia
SARDINIA (Italy)
Sassari
Cagliari
Firenze (Florence)
SAN MARINO
ITALY
APENNINES
ADRIATIC SEA
Roma (Rome)
Napoli (Naples)
Bari
Dubrovnik
TYRRHENIAN SEA
MEDITERRANEAN SEA
Palermo
Messina
SICILY (Italy)
IONIAN SEA
MALTA
Valletta
Alger
Casablanca
Rabat
MOROCCO
Marrakech
ALGERIA
Tunis
TUNISIA
Longitude West of Greenwich
Longitude East of Greenwich

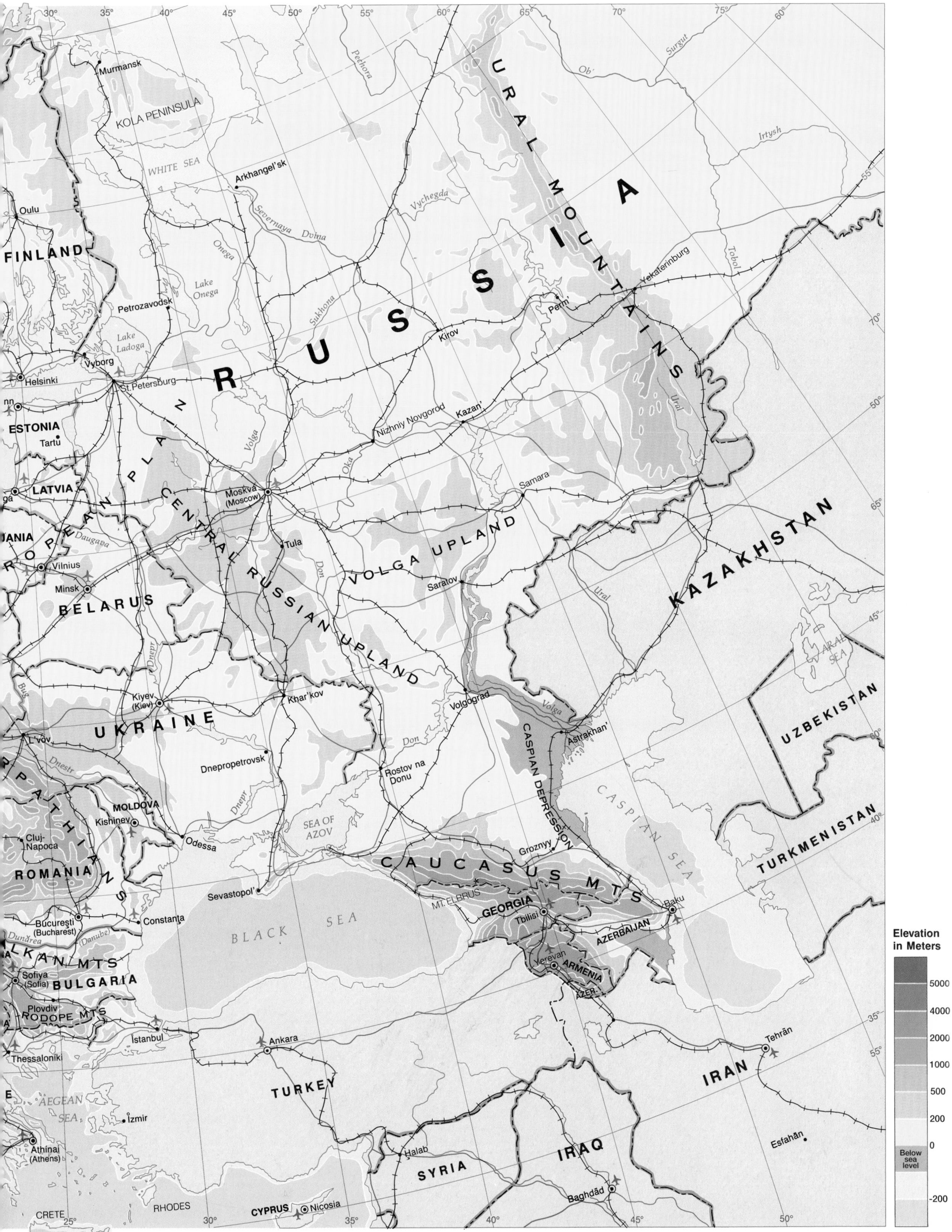
RUSSIA
URAL MOUNTAINS
KOLA PENINSULA
WHITE SEA
Murmansk
Arkhangel'sk
Oulu
FINLAND
Petrozavodsk
Lake Onega
Lake Ladoga
Vyborg
Helsinki
St. Petersburg
ESTONIA
Tartu
LATVIA
Vilnius
Minsk
BELARUS
EUROPEAN PLAIN
CENTRAL RUSSIAN UPLAND
VOLGA UPLAND
Moskva (Moscow)
Tula
Kirov
Perm'
Yekaterinburg
Nizhniy Novgorod
Kazan'
Samara
Saratov
Volgograd
Astrakhan'
KAZAKHSTAN
UZBEKISTAN
TURKMENISTAN
ARAL SEA
CASPIAN DEPRESSION
CASPIAN SEA
Kiyev (Kiev)
Khar'kov
UKRAINE
L'vov
Dnepropetrovsk
Rostov na Donu
SEA OF AZOV
MOLDOVA
Kishinev
Odessa
Cluj-Napoca
CARPATHIANS
ROMANIA
Sevastopol'
Constanta
București (Bucharest)
BALKAN MTS
Sofiya (Sofia)
BULGARIA
Plovdiv
RODOPE MTS
BLACK SEA
CAUCASUS MTS
MT. ELBRUS
Groznyy
GEORGIA
Tbilisi
AZERBAIJAN
Baku
Yerevan
ARMENIA
AZER.
İstanbul
Thessaloníki
Ankara
TURKEY
İzmir
AEGEAN SEA
Athínai (Athens)
CRETE
RHODES
CYPRUS
Nicosia
Halab
SYRIA
IRAQ
Baghdād
IRAN
Tehrān
Esfahān
Pechora
Ob'
Surgut
Irtysh
Tobol
Ural
Vychegda
Severnaya Dvina
Onega
Sukhona
Volga
Oka
Don
Dnepr
Daugava
Dnestr
Bug
Danube
Dunărea
Elevation in Meters
5000
4000
2000
1000
500
200
0
Below sea level
-200

WORLD-REGIONAL

Europeans are never more than 600 km from a major body of water; they are tied both physically and economically to the oceans that surround them. Europe was the first major world region to construct a modern economy (commercial agriculture combined with industrial development). However, in the 20th century, Europe has also undergone great political turmoil, including two world wars that cost millions of lives. The collapse of communism in the late 1980s resulted in the reunification of East and West Germany, but also led to the terrible civil war between the states of the former Yugoslavia. As well, forced resettlements of ethnic populations have increased tensions and caused outbreaks of violence. Generally, however, Europe is progressing towards economic integration. It remains an economic powerhouse, with average per capita incomes among the world's highest.

The changing face of Europe

Prelude to war 1914

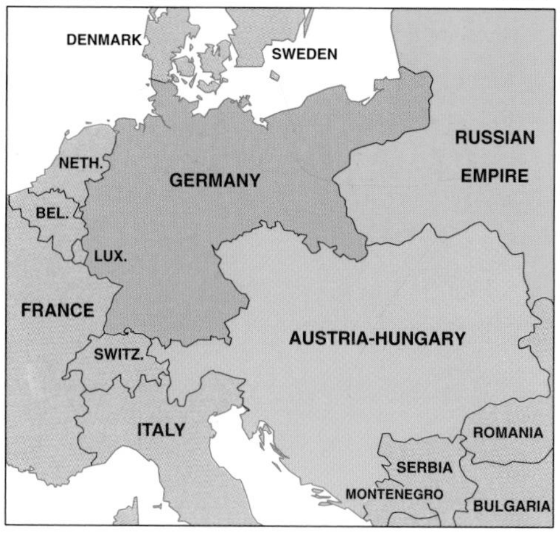

Aftermath o

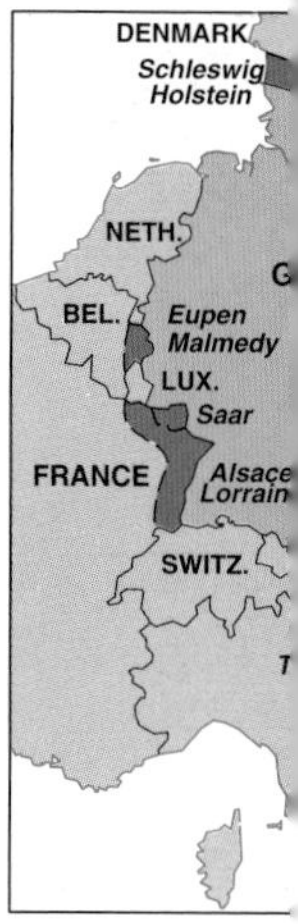

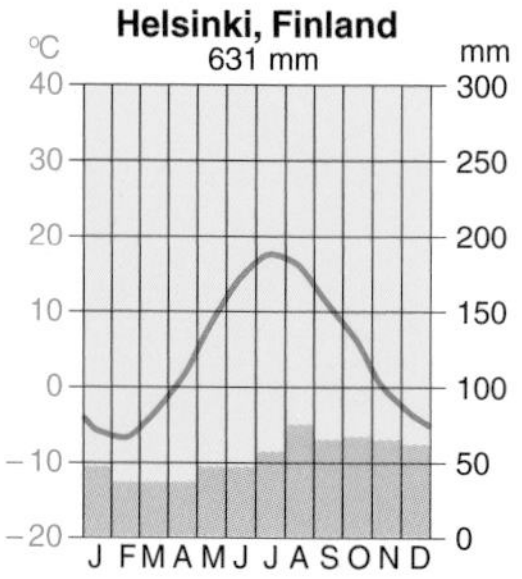

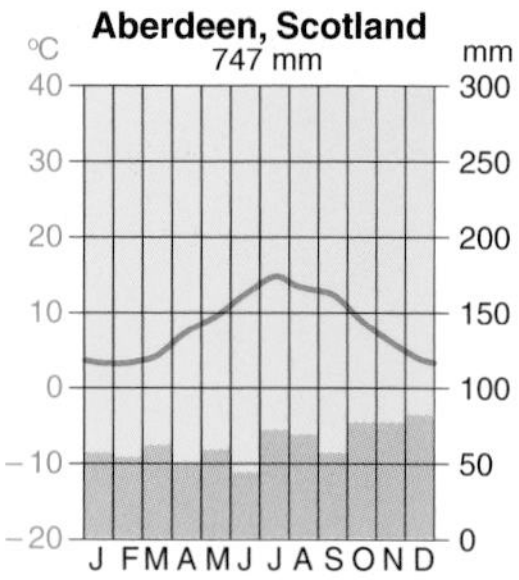

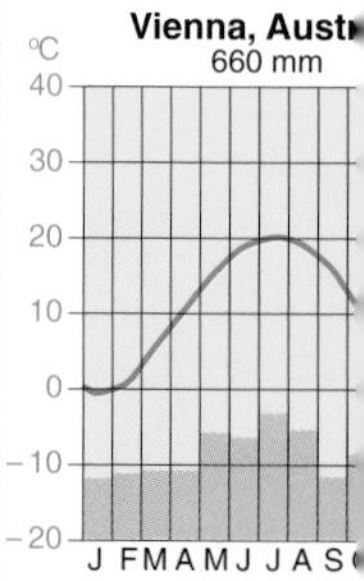

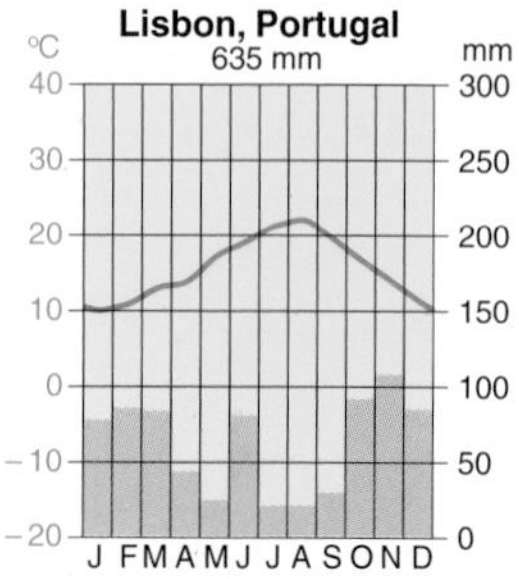

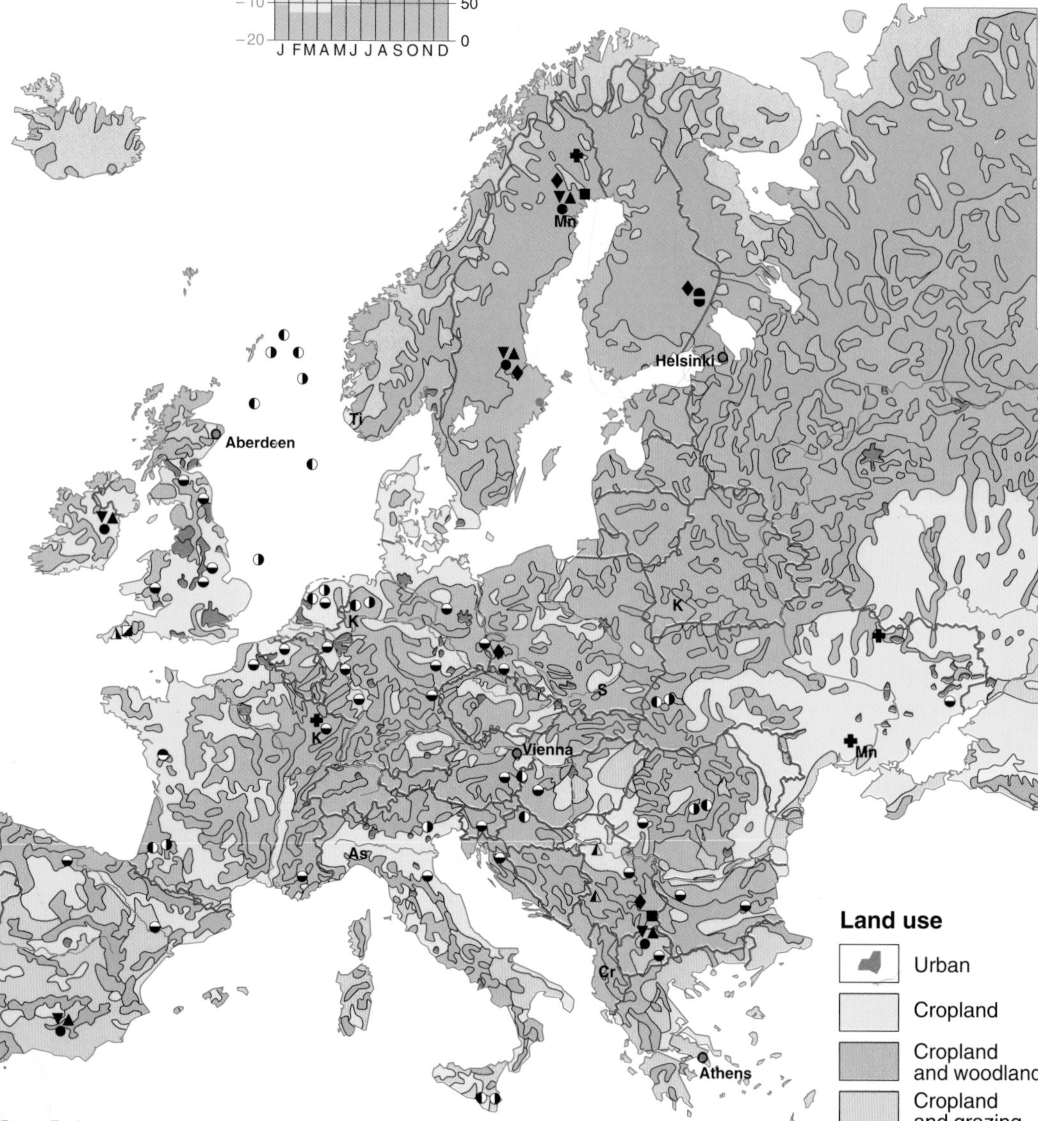

Scale 1: 27 000 000

Land use

- Urban
- Cropland
- Cropland and woodland
- Cropland and grazing
- Grassland and grazing
- Tempera forest
- Scrublar and gras
- Swamp marsh
- Tundra
- Climogra station

Minerals

- ■ Gold
- ▲ Silver
- ▼ Lead
- ● Zinc
- ◆ Copper
- ✚ Iron
- Nickel
- Tin
- Cobalt
- Aluminum
- Tungsten
- Coal
- Oil
- Gas
- Uraniu
- Cr Chron
- Mn Mang
- Ti Titani
- S Sulph
- K Potas
- As Asbes

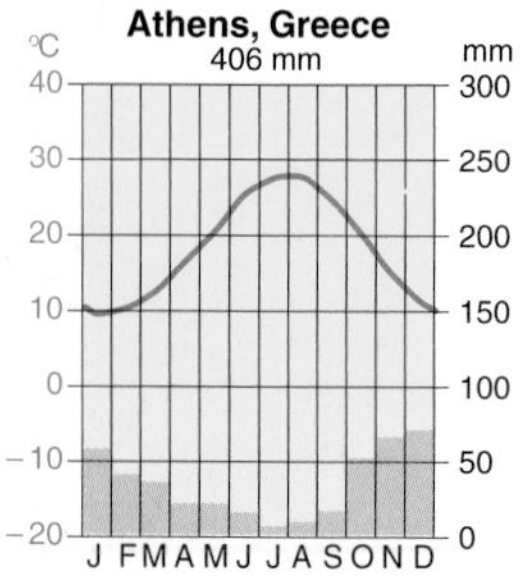

The worksite of the Channel Tunnel in Dover, England

Aftermath of WWII

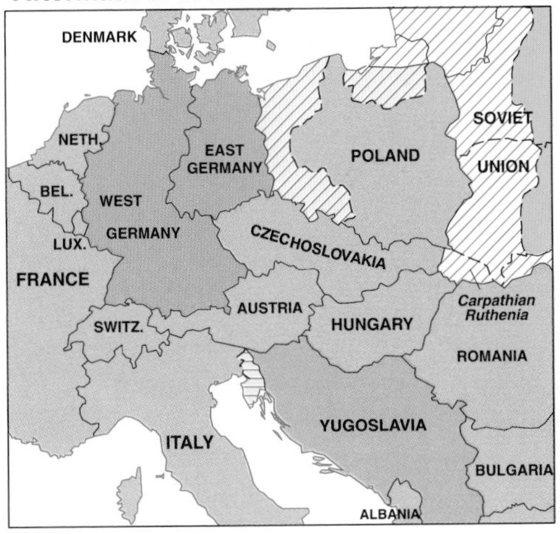

1920
Territory lost by
- Soviet Union
- Germany
- Austria-Hungary
- Bulgaria

1945
Territory annexed by
- Soviet Union
- Yugoslavia

Territory lost
- Germany

Scale 1: 27 500 000

Vineyards near Martigny, Switzerland

United Kingdom

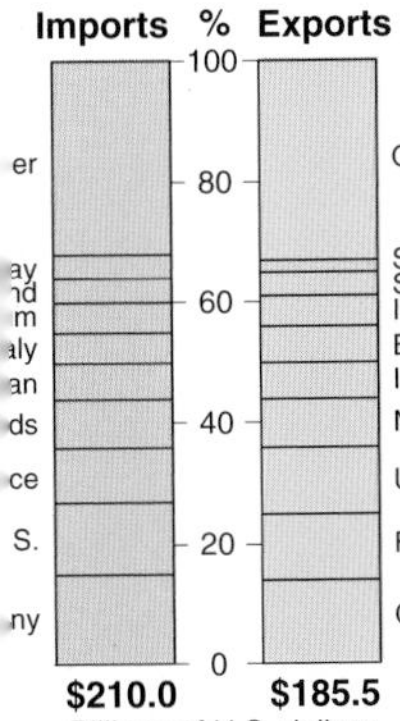

Main imports:
Machinery and transport equipment 36%
Data processing equipment 7%
Petroleum 10%
Chemicals 10%
Food 8.5%

Main exports:
Machinery and transport equipment 41%
Data processing equipment 6%
Chemicals 13%
Petroleum 7%
Iron and steel 3%

GDP by sector

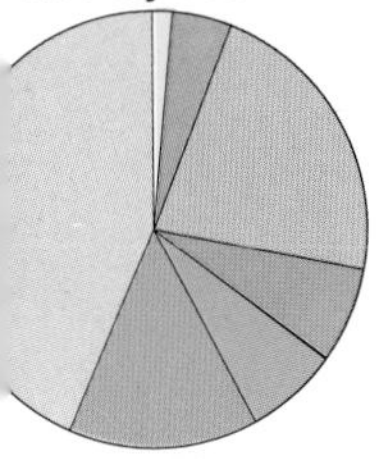

Total: $925.0 billion (U.S.)

GDP by sector
- Agriculture
- Mining
- Manufacturing
- Construction
- Other secondary
- Transportation, communications
- Trade
- Other tertiary

GDP per capita ($U.S.)
- More than $20 000
- 15 000 - 20 000
- 5000 - 14 999
- Less than $5000
- No data

Annual change in GDP (1980-90 average)
- ↑ Positive
- ↓ Negative
- — No data

Poland

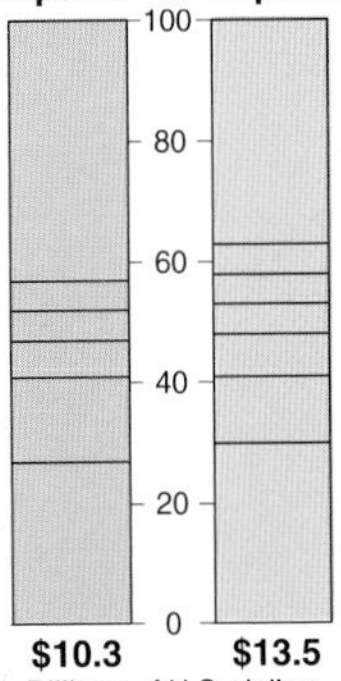

GDP by sector

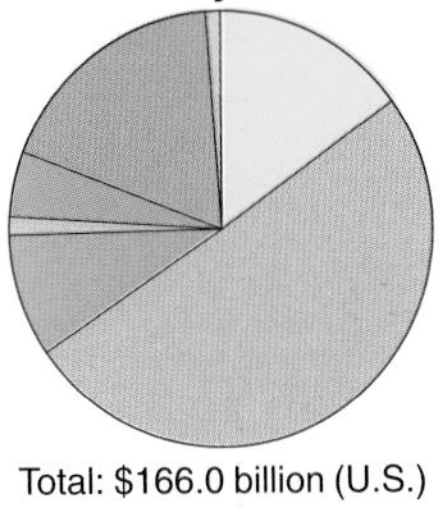

Total: $166.0 billion (U.S.)

Main exports:
Machinery and transport equipment 22%
Iron and steel products 16%
Chemicals 12%
Fuel 10%
Food 10%

Main imports:
Machinery and transport equipment 38%
Fuel 19%
Chemicals 13%
Food 11%
Manufactured consumer goods 6%

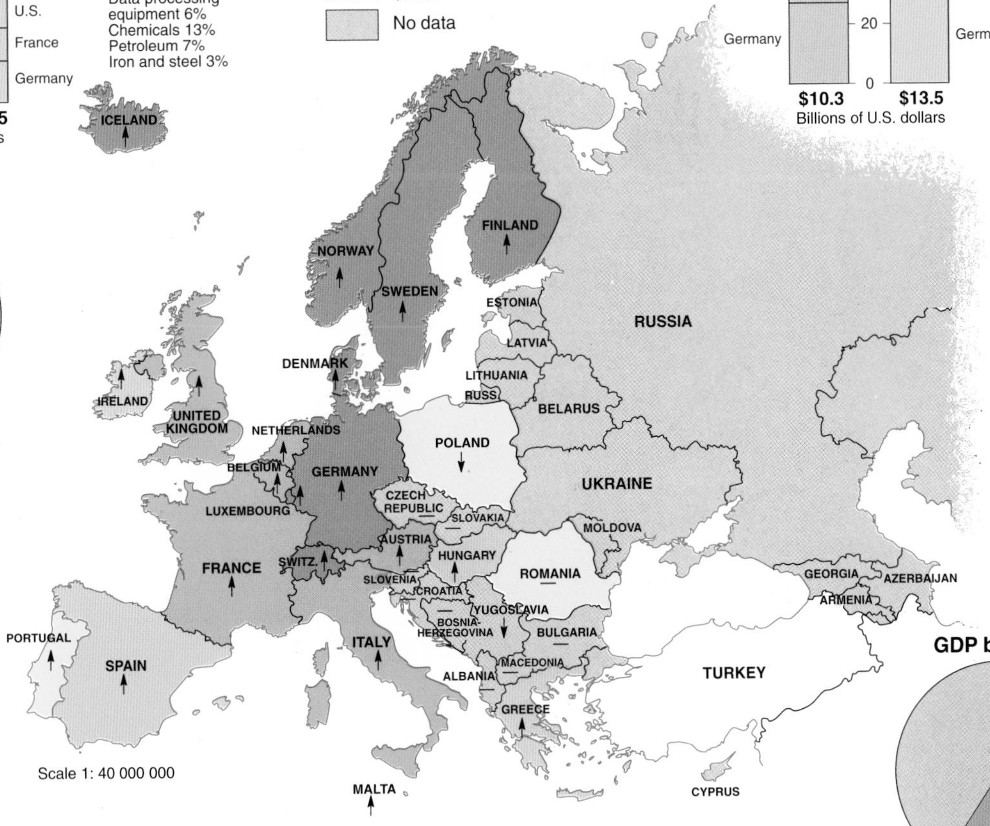

Scale 1: 40 000 000

Germany

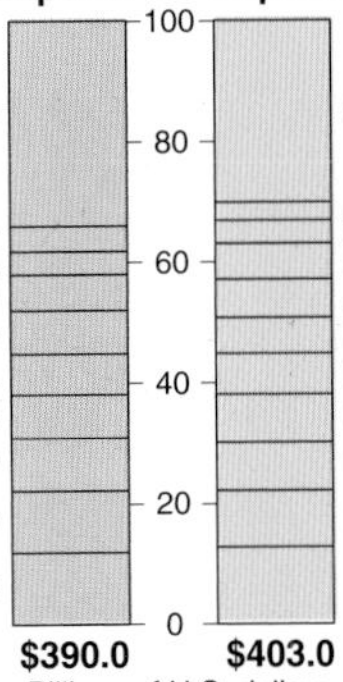

GDP by sector

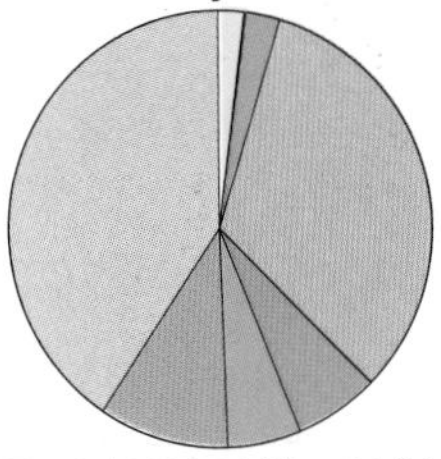

Total: $1420.0 billion (U.S.)

Main imports:
Machinery and transport equipment 35%
Office equipment 5%
Chemical products 8%
Food 8%
Petroleum 7%
Clothing 6%

Main exports:
Machinery and transport equipment 49%
Chemicals 13%

Spain

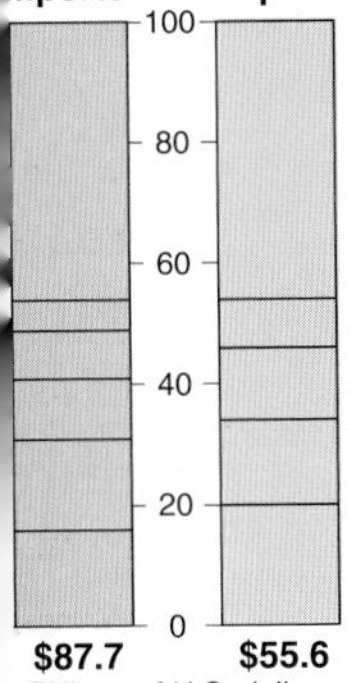

GDP by sector

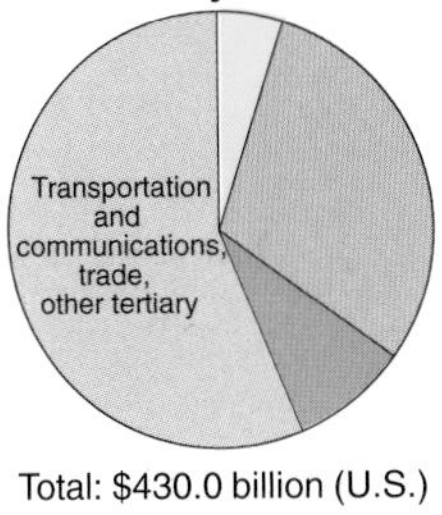

Total: $430.0 billion (U.S.)

Main imports:
Machinery 14%
Petroleum 11%
Agricultural products 10%
Transportation equipment 10%

Main exports:
Transportation equipment 20%
Agricultural products 15%
Machinery 8%

EU: A major economic power

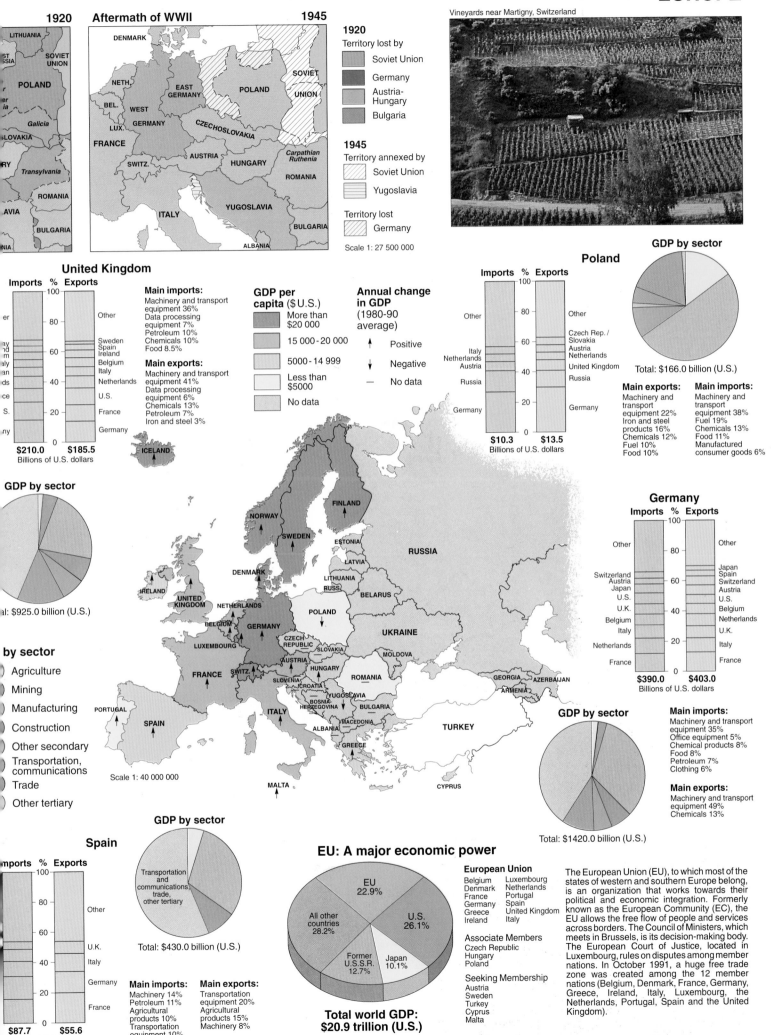

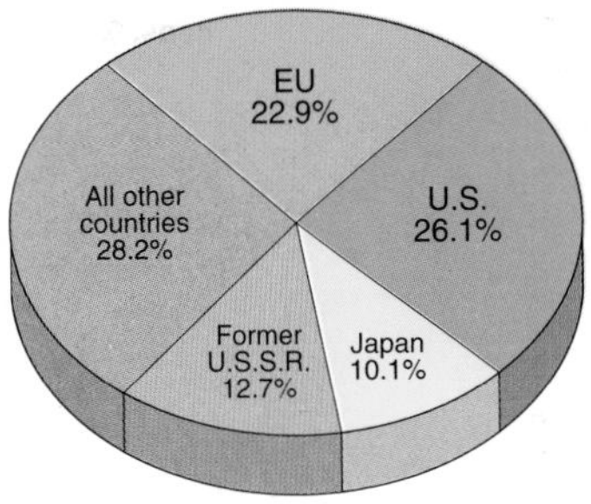

Total world GDP: $20.9 trillion (U.S.)

European Union
Belgium, Denmark, France, Germany, Greece, Ireland, Luxembourg, Netherlands, Portugal, Spain, United Kingdom, Italy

Associate Members
Czech Republic
Hungary
Poland

Seeking Membership
Austria
Sweden
Turkey
Cyprus
Malta

The European Union (EU), to which most of the states of western and southern Europe belong, is an organization that works towards their political and economic integration. Formerly known as the European Community (EC), the EU allows the free flow of people and services across borders. The Council of Ministers, which meets in Brussels, is its decision-making body. The European Court of Justice, located in Luxembourg, rules on disputes among member nations. In October 1991, a huge free trade zone was created among the 12 member nations (Belgium, Denmark, France, Germany, Greece, Ireland, Italy, Luxembourg, the Netherlands, Portugal, Spain and the United Kingdom).

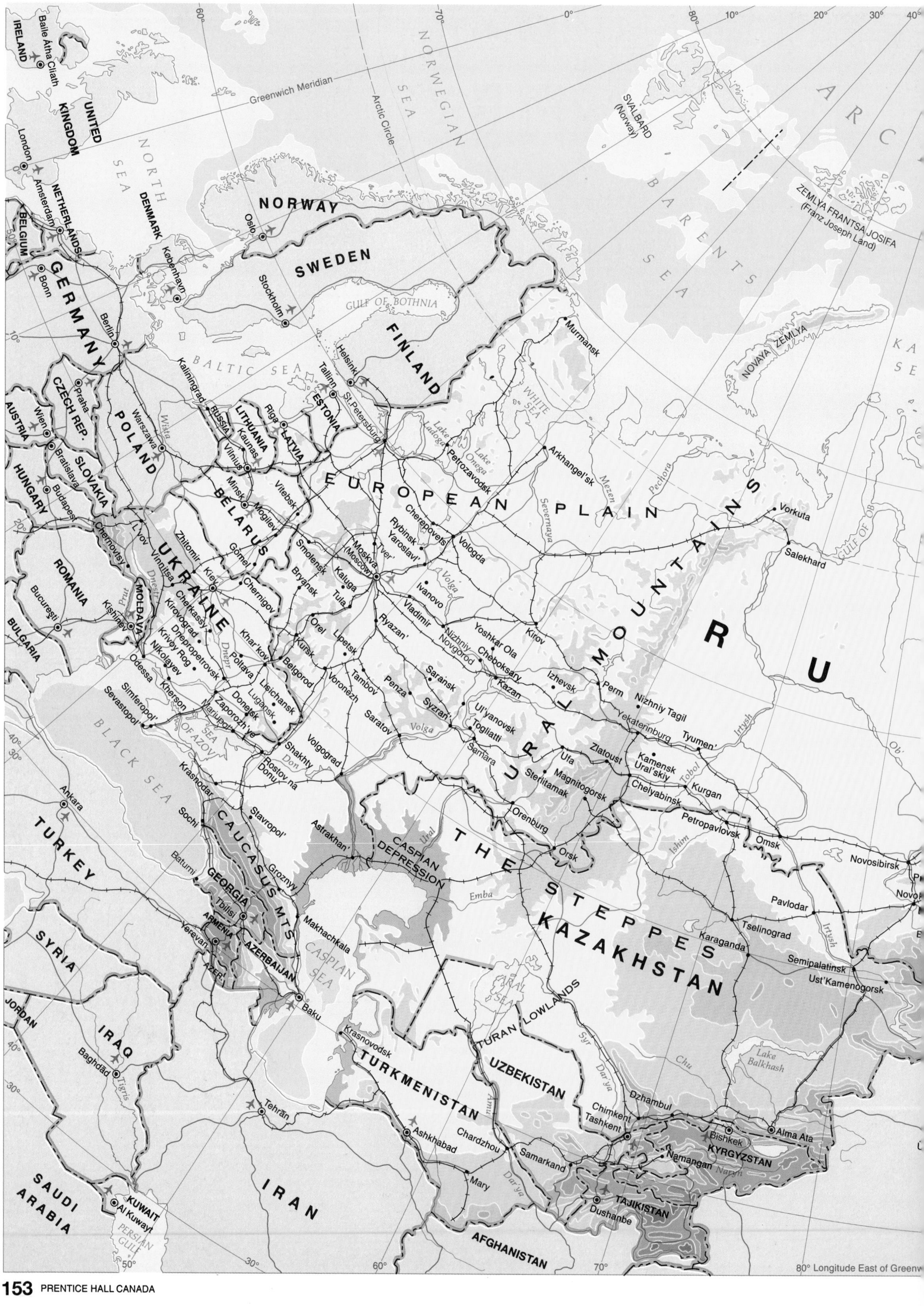
IRELAND
Baile Átha Cliath
UNITED KINGDOM
London
NORTH SEA
NETHERLANDS
Amsterdam
BELGIUM
DENMARK
København
GERMANY
Bonn
Berlin
Greenwich Meridian
NORWEGIAN SEA
Arctic Circle
NORWAY
Oslo
SWEDEN
Stockholm
GULF OF BOTHNIA
FINLAND
Helsinki
BALTIC SEA
SVALBARD (Norway)
BARENTS SEA
ZEMLYA FRANTSA JOSIFA (Franz Joseph Land)
NOVAYA ZEMLYA
Murmansk
WHITE SEA
Tallinn
ESTONIA
St. Petersburg
Riga
LATVIA
LITHUANIA
Kaunas
Vilnius
Kaliningrad
RUSSIA
POLAND
Warszawa
Wisła
CZECH REP.
Praha
AUSTRIA
Wien
SLOVAKIA
Bratislava
HUNGARY
Budapest
ROMANIA
București
BULGARIA
MOLDAVA
Kishinev
Prut
Dnestr
Chernovtsy
L'vov
Vinnitsa
UKRAINE
Zhitomir
Kiev
Cherkassy
Kirovograd
Dnepropetrovsk
Krivoy Rog
Nikolayev
Odessa
Kherson
Simferopol'
Sevastopol'
Zaporozh'ye
Mariupol'
SEA OF AZOV
Donetsk
Lugansk
Lisichansk
Khar'kov
Poltava
Dnepr
Chernigov
Gomel'
BELARUS
Minsk
Mogilev
Vitebsk
Smolensk
Bryansk
Orel
Kursk
Belgorod
Lipetsk
Voronezh
Tula
Kaluga
Moskva (Moscow)
Tver'
Yaroslavl'
Rybinsk
Cherepovets
Vologda
Lake Ladoga
Lake Onega
Petrozavodsk
Arkhangel'sk
Mezen
Pechora
Severnaya
EUROPEAN PLAIN
Vorkuta
Salekhard
GULF OF OB
Ivanovo
Vladimir
Volga
Ryazan'
Nizhniy Novgorod
Yoshkar Ola
Cheboksary
Kazan
Kirov
Izhevsk
Perm
Nizhniy Tagil
Yekaterinburg
Tyumen'
Irtysh
Ob'
Tambov
Penza
Saransk
Syzran'
Ul'yanovsk
Togliatti
Samara
Saratov
Volgograd
Shakhty
Don
Rostov na Donu
BLACK SEA
Krasnodar
Sochi
Stavropol'
CAUCASUS MTS
Astrakhan'
CASPIAN DEPRESSION
Ural
Ufa
Zlatoust
Sterlitamak
Magnitogorsk
Orenburg
Orsk
Kamensk Ural'skiy
Chelyabinsk
Kurgan
Tobol
Petropavlovsk
Ishim
Omsk
Novosibirsk
URAL MOUNTAINS
R U
THE STEPPES
KAZAKHSTAN
Emba
Pavlodar
Tselinograd
Karaganda
Semipalatinsk
Ust'Kamenogorsk
Irtysh
Lake Balkhash
Chu
TURKEY
Ankara
Batumi
GEORGIA
Tbilisi
ARMENIA
Yerevan
AZERBAIJAN
AZER.
Groznyy
Makhachkala
CASPIAN SEA
Baku
ARAL SEA
TURAN LOWLANDS
Syr Dar'ya
Krasnovodsk
TURKMENISTAN
UZBEKISTAN
Amu Dar'ya
Ashkhabad
Chardzhou
Mary
Samarkand
Tashkent
Chimkent
Dzhambul
Bishkek
Alma Ata
KYRGYZSTAN
Namangan
Naryn
TAJIKISTAN
Dushanbe
AFGHANISTAN
SYRIA
JORDAN
IRAQ
Baghdād
Tigris
Tehrān
IRAN
SAUDI ARABIA
KUWAIT
Al Kuwayt
PERSIAN GULF
80° Longitude East of Greenw

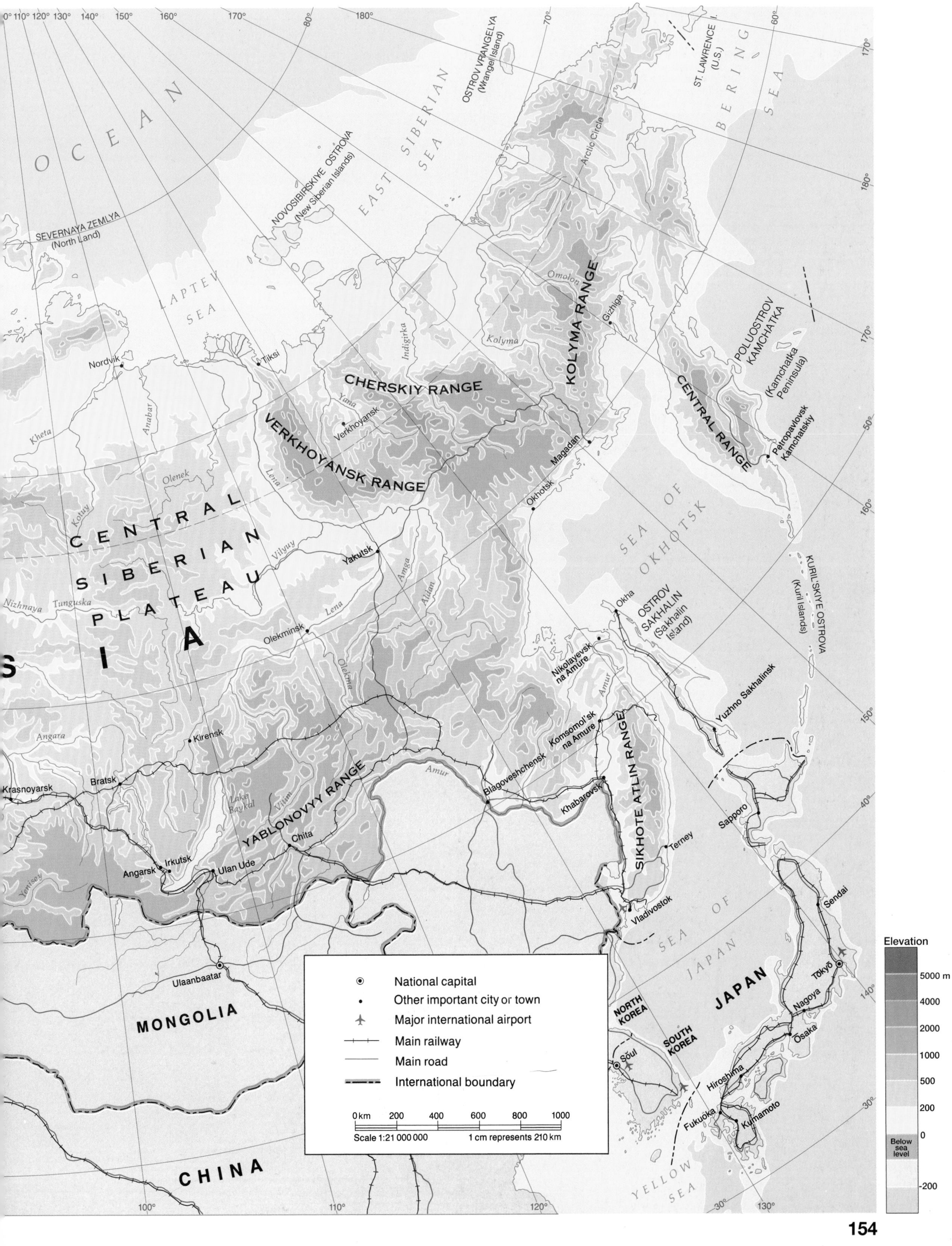
OCEAN
SEVERNAYA ZEMLYA
(North Land)
NOVOSIBIRSKIYE OSTROVA
(New Siberian Islands)
LAPTEV SEA
EAST SIBERIAN SEA
OSTROV VRANGELYA
(Wrangel Island)
ST. LAWRENCE I.
(U.S.)
BERING SEA
Arctic Circle
Nordvik
Tiksi
Anabar
Kheta
Olenek
Kotuy
Indigirka
Kolyma
Omolon
Gizhiga
CHERSKIY RANGE
VERKHOYANSK RANGE
KOLYMA RANGE
Verkhoyansk
Yana
Lena
Magadan
Okhotsk
POLUOSTROV KAMCHATKA
(Kamchatka Peninsula)
CENTRAL RANGE
Petropavlovsk Kamchatskiy
SEA OF OKHOTSK
CENTRAL SIBERIAN PLATEAU
ASIA
Vilyuy
Yakutsk
Amga
Aldan
Nizhnaya Tunguska
Olekminsk
Olekma
OSTROV SAKHALIN
(Sakhalin Island)
Okha
KURIL'SKIYE OSTROVA
(Kuril Islands)
Nikolayevsk na Amure
Amur
Komsomol'sk na Amure
Yuzhno Sakhalinsk
Angara
Kirensk
Bratsk
Krasnoyarsk
Lake Baykal
Vitim
YABLONOVYY RANGE
Blagoveshchensk
Khabarovsk
SIKHOTE ATLIN RANGE
Sapporo
Terney
Chita
Irkutsk
Angarsk
Ulan Ude
Yenisey
Vladivostok
Sendai
SEA OF JAPAN
Ulaanbaatar
MONGOLIA
NORTH KOREA
SOUTH KOREA
JAPAN
Tōkyō
Nagoya
Ōsaka
Hiroshima
Sŏul
Fukuoka
Kumamoto
YELLOW SEA
CHINA
National capital
Other important city or town
Major international airport
Main railway
Main road
International boundary
0 km 200 400 600 800 1000
Scale 1:21 000 000
1 cm represents 210 km
Elevation
5000 m
4000
2000
1000
500
200
0
Below sea level
-200

WORLD-REGIONAL

Northern Eurasia stretches from the polar ice to arid central Asia, and covers almost half the globe in longitude, from 20°E to 170°W. Its political history in the 20th century has had a very large influence on world affairs. In 1917, the Czar of Russia was deposed. Communists quickly gained power and created the Union of Soviet Socialist Republics (U.S.S.R.). From about 1950 to 1988, the Soviet Union, the world's major communist state, engaged in confrontation with the United States, the world's major capitalist state. Both sides, in what came to be called the Cold War, contributed to political struggles in other countries, but refrained from war with one another.

After the Soviet Union collapsed, Russia and ten other former Soviet states formed the Commonwealth of Independent States (C.I.S.) in 1992, designating Minsk in Belarus as their administrative centre. Many problems remain unsolved: the balance between human rights and political stability; the development of a market economy; the protection of ethnic minorities; and the preservation of the environment.

A view of modern Moscow

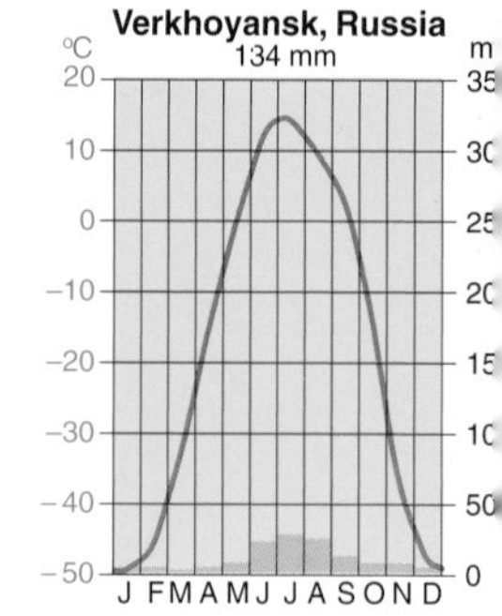

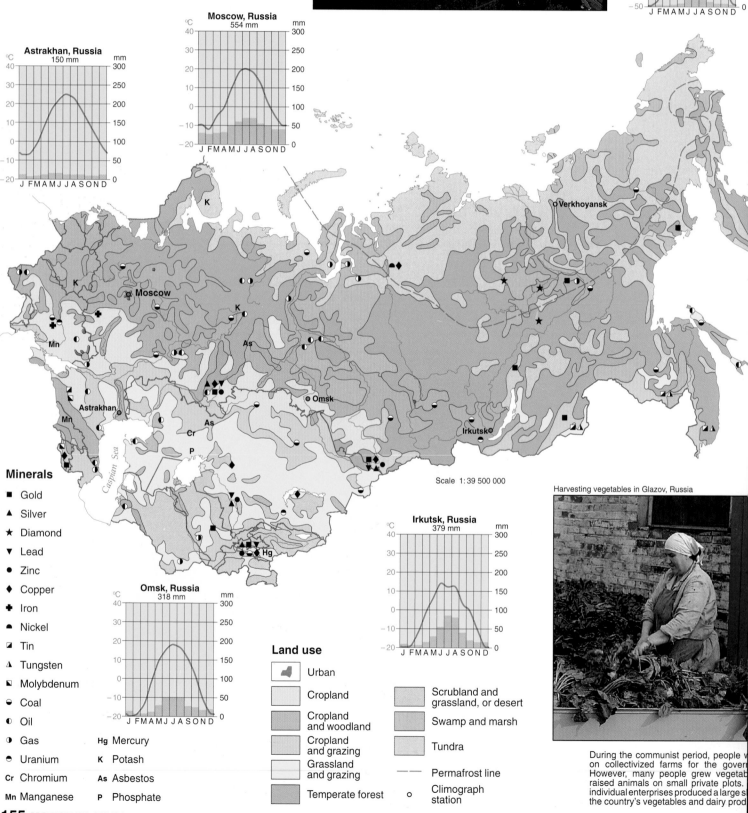

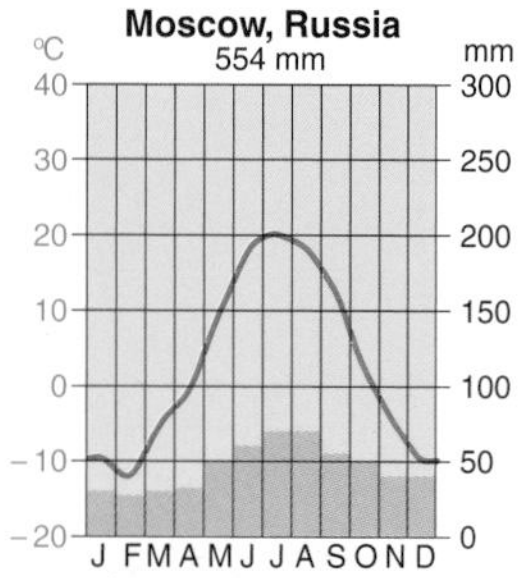

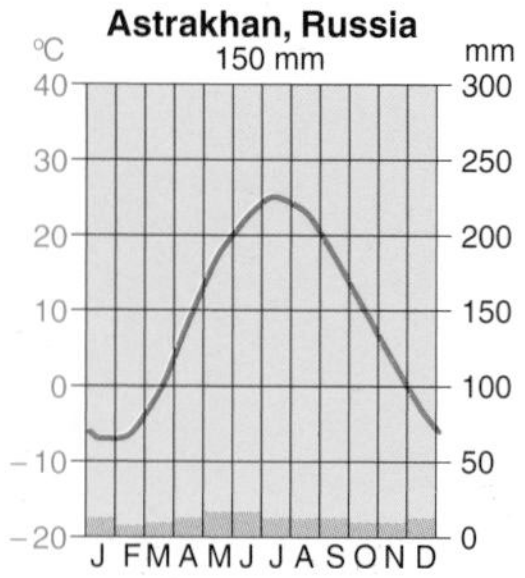

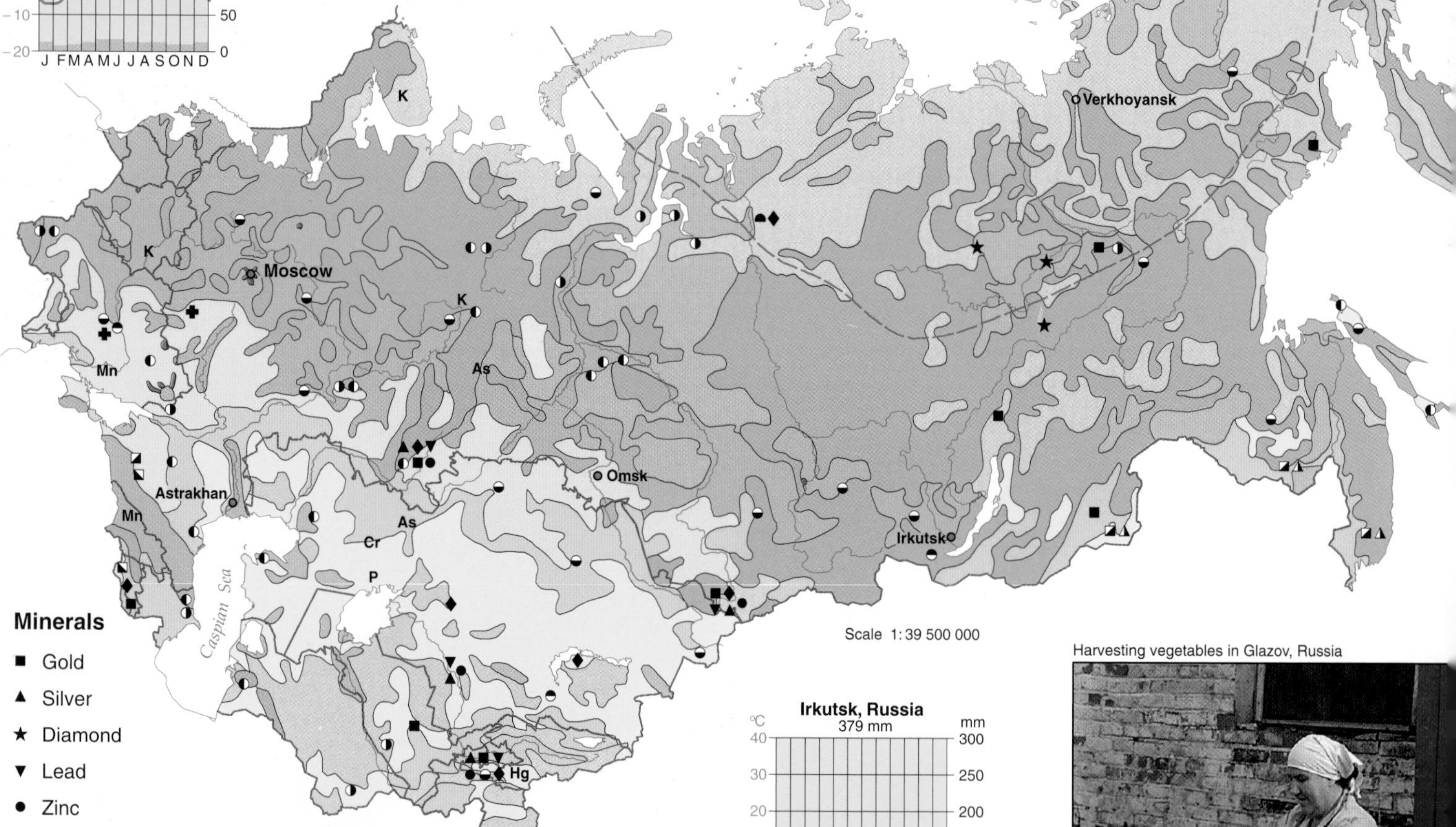

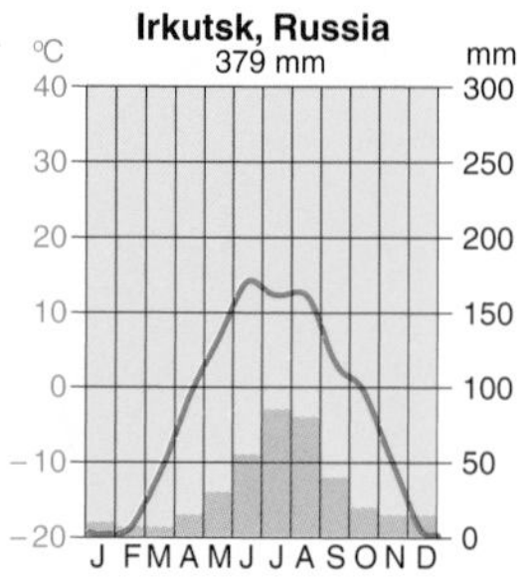

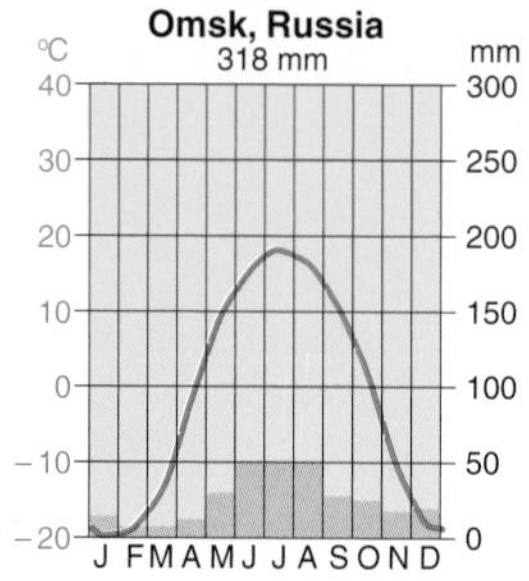

Harvesting vegetables in Glazov, Russia

During the communist period, people w on collectivized farms for the govern However, many people grew vegetab raised animals on small private plots. individual enterprises produced a large s the country's vegetables and dairy prod

Ethnic diversity

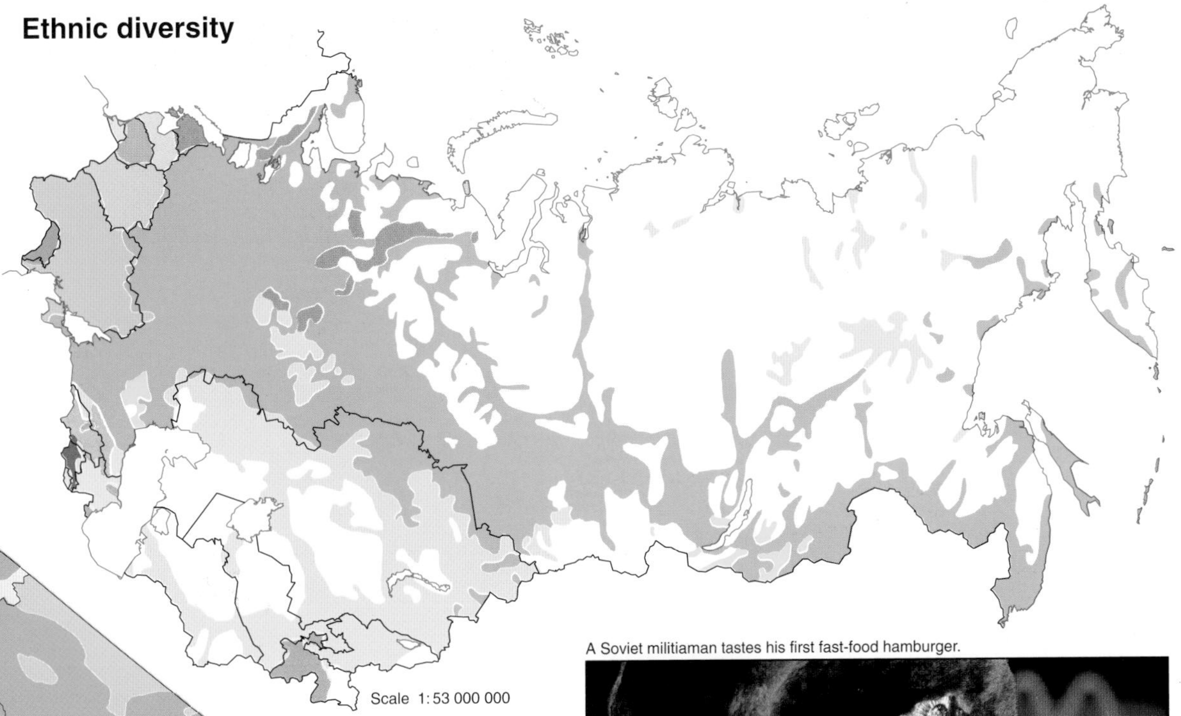

uages

- Russian
- Belorussian
- Ukrainian
- Armenian
- Caucasian
- Moldavian
- Finnish
- Latvian
- Lithuanian
- Iranian
- Turkic
- Mongolian
- Sparsely populated

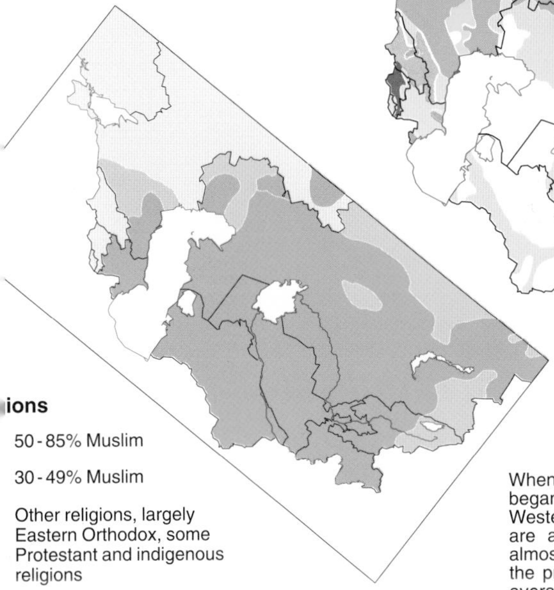

ions

- 50-85% Muslim
- 30-49% Muslim
- Other religions, largely Eastern Orthodox, some Protestant and indigenous religions

A Soviet militiaman tastes his first fast-food hamburger.

When the Cold War ended, private enterprise began to appear in the former Soviet Union. Western franchises such as fast-food outlets are a fairly recent phenomenon. Although almost all ingredients are produced in Russia, the price of a hamburger is still high for the average citizen.

ensed Soviet history

- March revolution. Bolshevik takeover in November
- Lenin founds Soviet Republic. Civil war erupts
- Union treaty is established with Stalin in power
- Collectivization of land begins
- Famine strikes Ukraine
- Germany invades. Soviet Union joins with Allies
- Soviet domination in Eastern Europe established at end of World War II. Estonia, Latvia, Lithuania annexed by Soviet Union
- Kruschev becomes leader of Soviet Union
- Warsaw Pact unites Albania, Bulgaria, Czechoslovakia, East Germany, Hungary, Poland, Romania and Russia in a military command
- Soviet Union invades Hungary to crush uprising
- Soviet Union invades Czechoslovakia to end reforms
- Soviet army invades Afghanistan
- Cold War begins to thaw. Mikhail Gorbachev, Premier of the Soviet Union, introduces reforms
- Soviet hold on Eastern Europe weakens
- Unification of East and West Germany. Multi-party political system introduced in Soviet Union
- Independence of Latvia, Lithuania and Estonia. Soviet Union dissolves
- Eleven former republics form C.I.S. Estonia, Latvia, Lithuania and Georgia decline
- Georgia joins C.I.S. Russia's voters approve new constitution but give overwhelming support to the nationalist Vladimir Zhironovsky and his Liberal Democratic party

GDP and exports

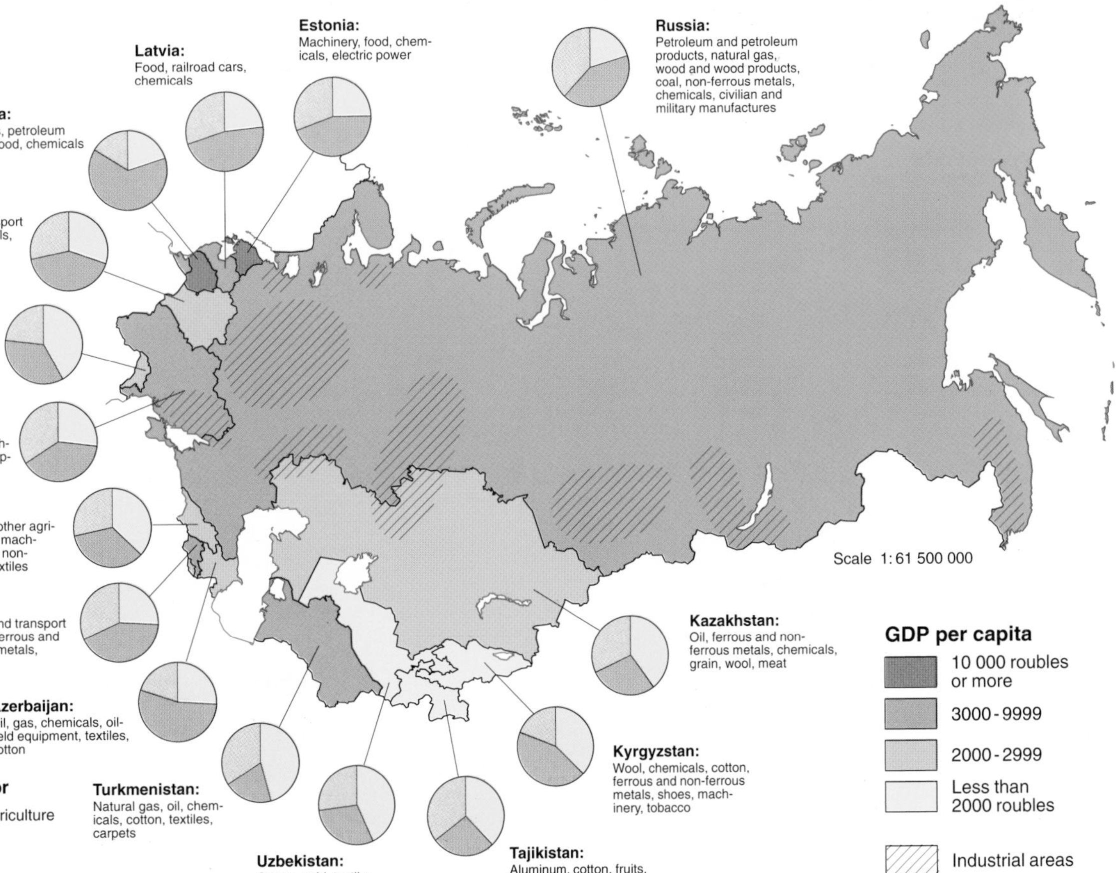

GDP by sector

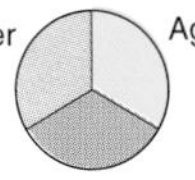

GDP per capita

- 10 000 roubles or more
- 3000-9999
- 2000-2999
- Less than 2000 roubles
- Industrial areas

WORLD-REGIONAL

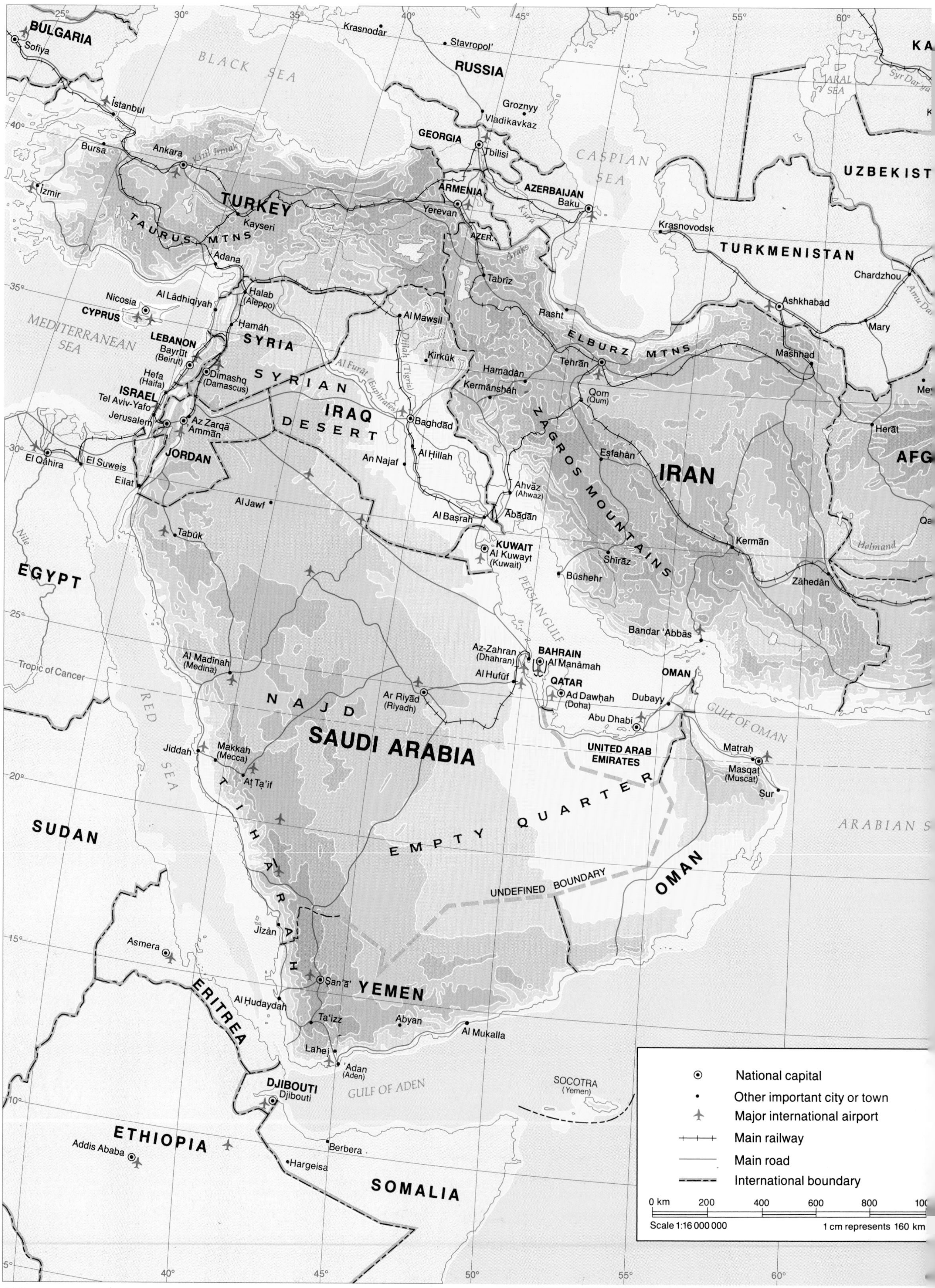

BULGARIA
Sofiya
İstanbul
Bursa
Ankara
İzmir
Kizil Irmak
TURKEY
Kayseri
TAURUS MTNS
Adana
BLACK SEA
Krasnodar
Stavropol'
RUSSIA
Groznyy
Vladikavkaz
GEORGIA
Tbilisi
CASPIAN SEA
ARMENIA
Yerevan
AZERBAIJAN
Baku
AZER.
Kura
Araks
ARAL SEA
Syr Dar'ya
UZBEKIST
KA
Krasnovodsk
TURKMENISTAN
Chardzhou
Amu Dar
Ashkhabad
Mary
Mashhad
Tabrīz
Rasht
ELBURZ MTNS
Tehrān
Qom (Qum)
Hamadān
Kermānshāh
Eşfahān
IRAN
ZAGROS MOUNTAINS
Herāt
AFG
Kermān
Helmand
Zāhedān
Shīrāz
Būshehr
Bandar 'Abbās
Nicosia
CYPRUS
Al Lādhiqīyah
Ḥalab (Aleppo)
Ḥamāh
SYRIA
MEDITERRANEAN SEA
LEBANON
Bayrūt (Beirut)
Hefa (Haifa)
Dimashq (Damascus)
ISRAEL
Tel Aviv-Yafo
Jerusalem
Az Zarqā'
'Ammān
JORDAN
SYRIAN DESERT
Al Mawşil
Kirkūk
Dijlah (Tigris)
Al Furāt (Euphrates)
IRAQ
Baghdād
Al Ḥillah
An Najaf
Ahvāz (Ahwaz)
Al Başrah
Ābādān
KUWAIT
Al Kuwayt (Kuwait)
PERSIAN GULF
El Qâhira
El Suweis
Eilat
Al Jawf
Tabūk
Nile
EGYPT
Tropic of Cancer
Al Madīnah (Medina)
NAJD
RED SEA
Ar Riyāḍ (Riyadh)
SAUDI ARABIA
Az-Zahran (Dhahran)
Al Hufūf
BAHRAIN
Al Manāmah
QATAR
Ad Dawḥah (Doha)
OMAN
Dubayy
Abu Dhabi
UNITED ARAB EMIRATES
GULF OF OMAN
Maṭraḥ
Masqaṭ (Muscat)
Şūr
ARABIAN S
Jiddah
Makkah (Mecca)
Aṭ Ṭā'if
EMPTY QUARTER
TIHĀMAH
SUDAN
UNDEFINED BOUNDARY
OMAN
Jīzān
Asmera
ERITREA
Şan'ā'
YEMEN
Al Ḥudaydah
Ta'izz
Abyan
Al Mukalla
Lahej
'Adan (Aden)
GULF OF ADEN
SOCOTRA (Yemen)
DJIBOUTI
Djibouti
ETHIOPIA
Addis Ababa
Berbera
Hargeisa
SOMALIA
National capital
Other important city or town
Major international airport
Main railway
Main road
International boundary
0 km
200
400
600
800
Scale 1:16 000 000
1 cm represents 160 km

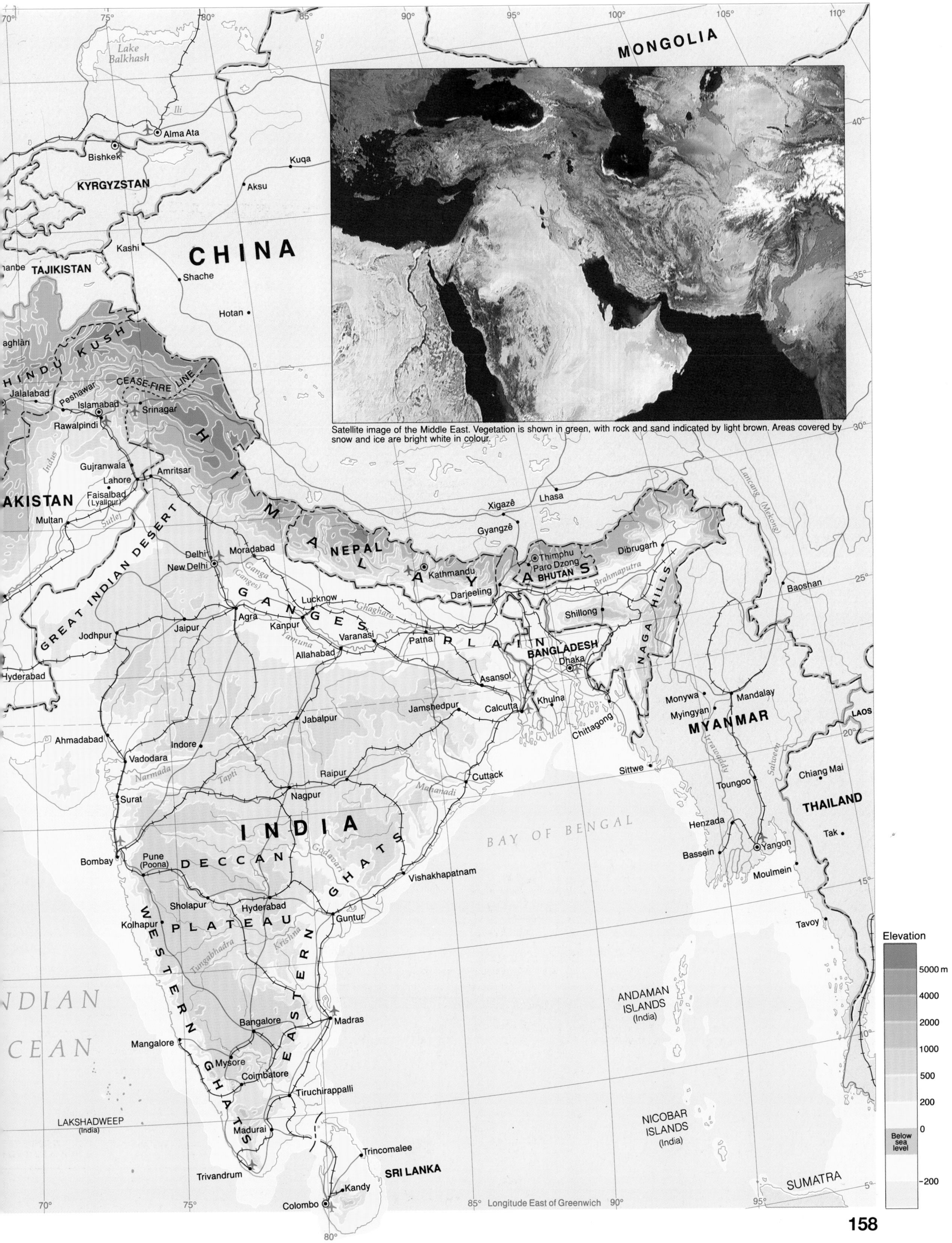

Satellite image of the Middle East. Vegetation is shown in green, with rock and sand indicated by light brown. Areas covered by snow and ice are bright white in colour.

WORLD-REGIONAL

This large area, mainly south of the Himalayas, has two main peninsulas, the Indian subcontinent and the Arabian peninsula. The variations in climate are extreme: the region includes both the Saudi Arabian desert and the rainiest city on earth, Cherrapunji, in the foothills of the Himalayas. Southwest Asia is also culturally diverse. It is home to many religions, languages and ethnic groups.

In recent decades, oil has played a major role in both the economy and the politics of the region. Almost half of the world's proven oil reserves are found along the shores of the Persian Gulf. However, the oil surplus in the 1980s cut back exports and lowered prices.

The Indian subcontinent is one of the most densely populated areas in the world. Although only the size of Québec, India alone has almost one sixth of the world's population.

More modern agricultural methods are helping India feed its large population

Land use

- Urban
- Cropland
- Cropland and woodland
- Cropland and grazing
- Grassland and grazing
- Temperate forest
- Tropical forest
- Scrubland and grassland, or desert
- Swamp and marsh
- Tundra
- ○ Climograph station

Minerals

- ◆ Copper
- ✚ Iron
- Aluminum
- Coal
- Oil
- Gas
- Cr Chromium
- Mn Manganese
- B Borax

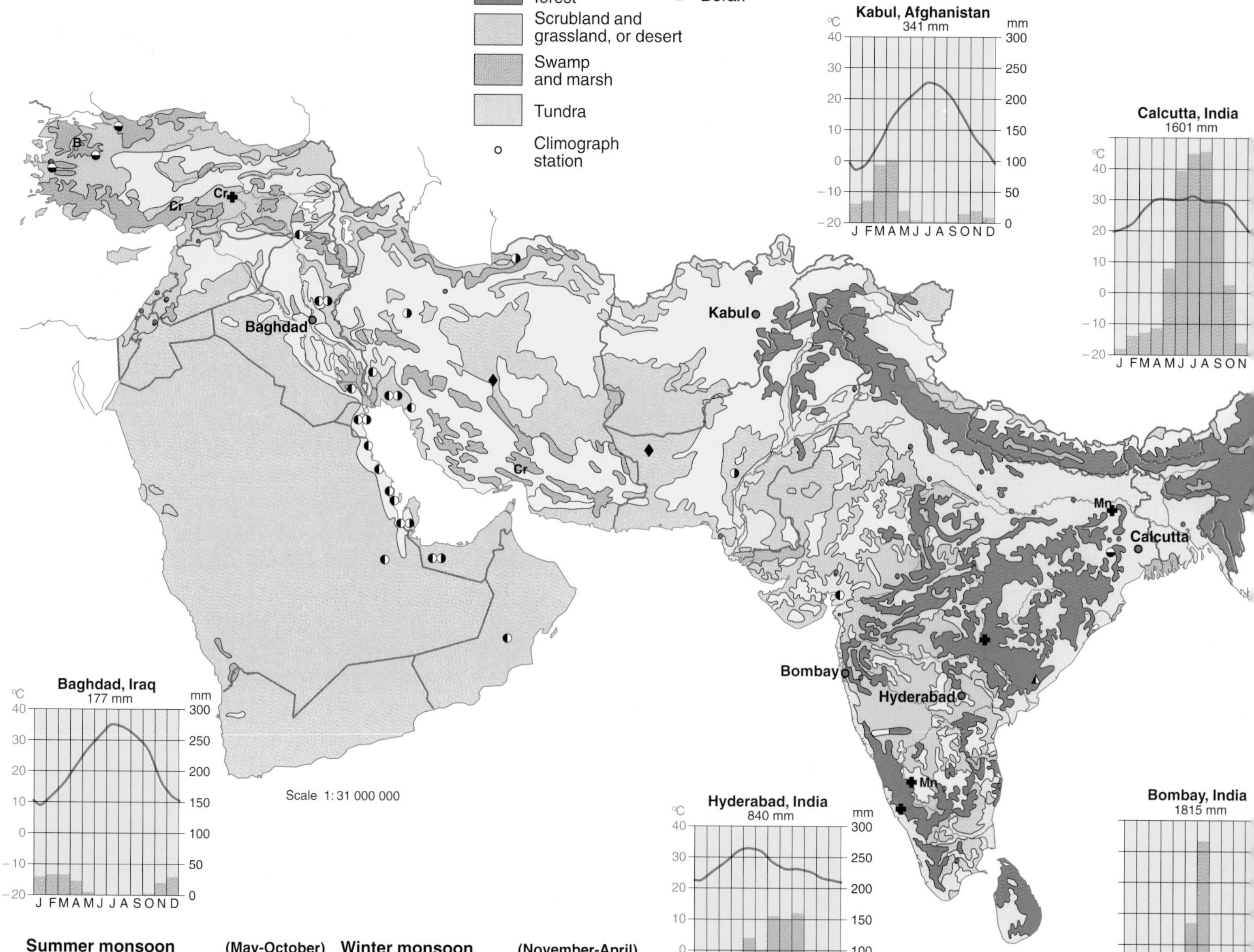

Scale 1: 31 000 000

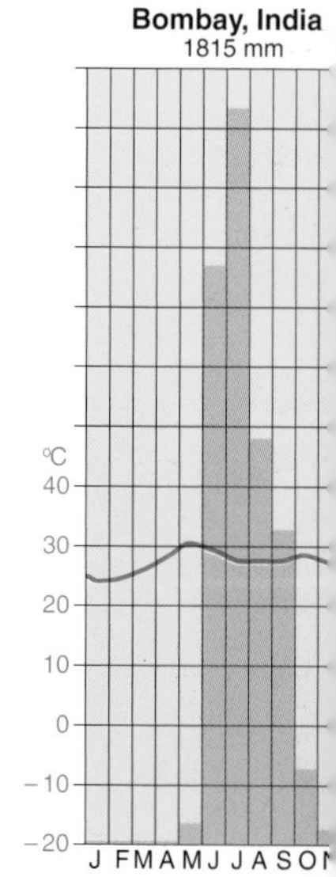

Summer monsoon **(May-October)**

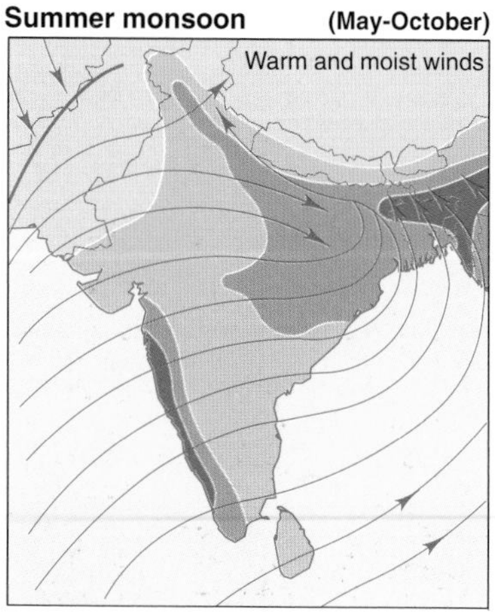

Winter monsoon **(November-April)**

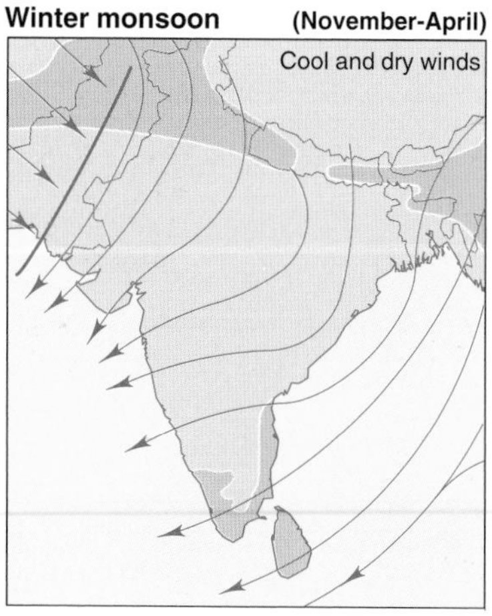

Six months rainfall (mm)

- 2000
- 1000
- 250

The climate of this region is dominated by the monsoon. The word monsoon comes from the Arabic *mawsim*, which means a wind that changes direction with the seasons. Today, a monsoon refers to a wind that brings rain. In summer, the monsoon blows from the sea towards land, and in the winter from the land towards the sea. Its duration and intensity vary from year to year. Ocean currents, the Himalayas and the spin of the earth also affect the wind system.

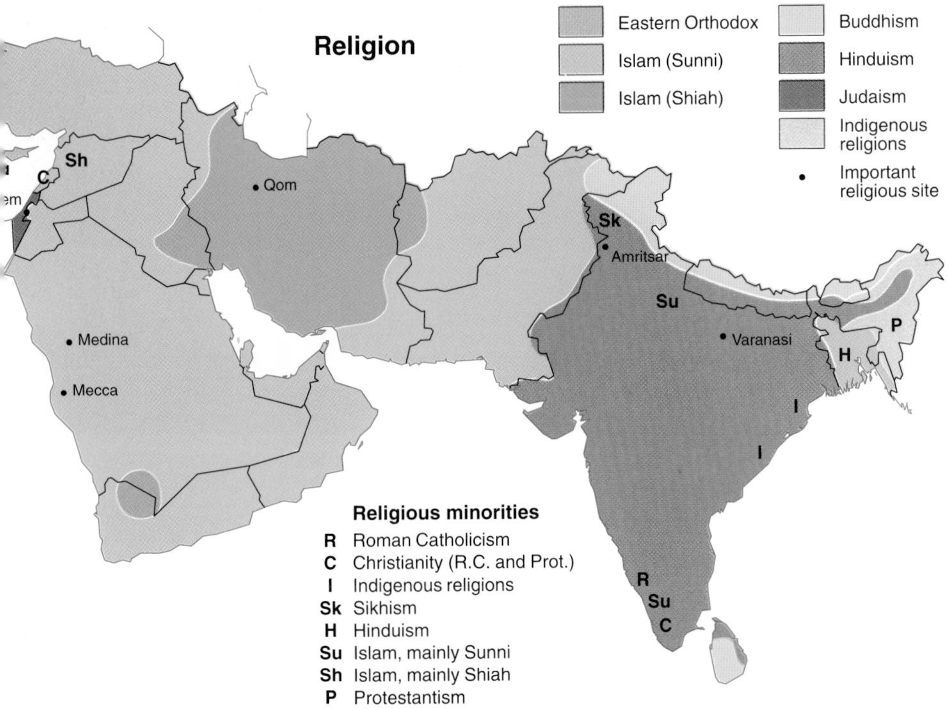

The Godin Tepe archeological dig in Iran. Southwest Asia contains the ruins of some of the oldest known civilizations.

Israel

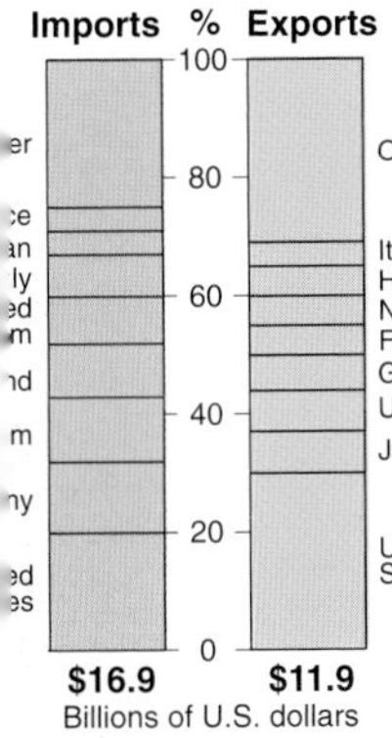

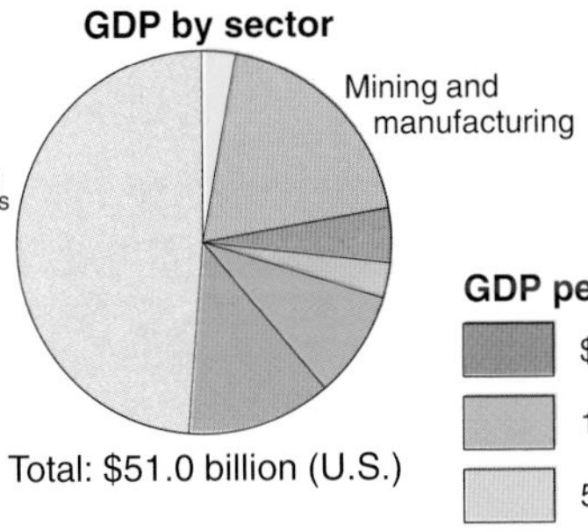

...n imports:
...hinery and transport ...pment 31%
...eral ores 16%
...d 10%
...micals 10%
... 10%

...n exports:
...hinery 30%
...nonds 27%
...micals 13%
...les 8%
...d, beverages 5%
...ber and plastics 4%

India

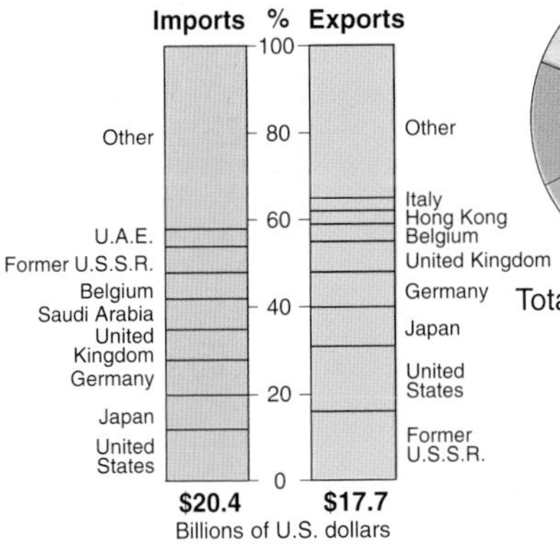

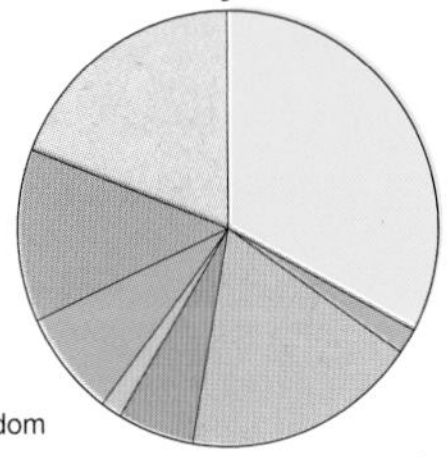

Main imports:
Mineral fuels 25%
Machinery 15%
Diamonds and other precious stones 6%
Iron and steel 6%
Electrical equipment 5%
Transport equipment 4%

Main exports:
Diamonds, other precious stones, jewellery 16%
Clothing 12%
Leather goods 8%
Chemical products 8%
Cotton yarn and fabric 6%

GDP per capita ($U.S.)
- $10 000 or more
- 1000-9999
- 530-999
- Less than $350

Annual change in GDP (1980-90 average)
- ↑ Positive
- ↓ Negative
- — No data

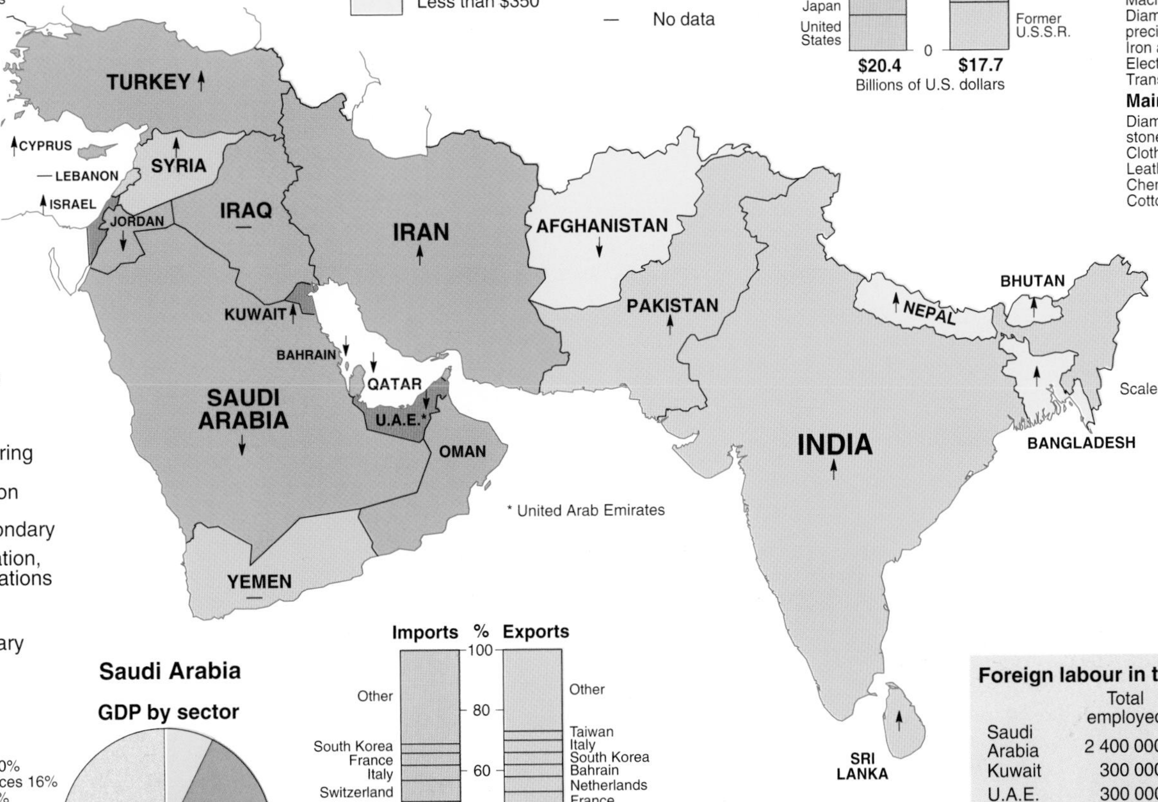

* United Arab Emirates

GDP by sector
- Agriculture
- Mining
- Manufacturing
- Construction
- Other secondary
- Transportation, communications
- Trade
- Other tertiary

Saudi Arabia

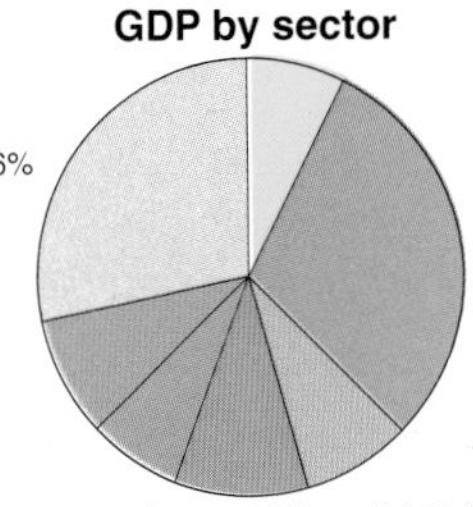

Main imports:
Transport equipment 20%
Machinery and appliances 16%
Textiles and clothing 9%
Metals 9%
Chemicals 8%
Precious stones and jewellery 7%

Main exports:
Crude petroleum 74%
Other fuel-related products 15%
Chemicals 6%

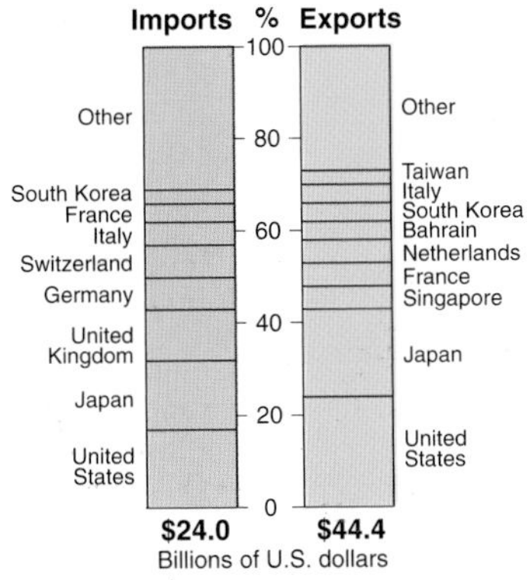

Foreign labour in the Gulf states

	Total employed	Foreign employed
Saudi Arabia	2 400 000	1 300 000
Kuwait	300 000	250 000
U.A.E.	300 000	240 000
Oman	350 000	70 000
Qatar	86 000	54 000
Bahrain	60 000	30 000

Origins of foreign labour

Yemen, Palestine, Egypt, Jordan	75%
Pakistan, India, other nations	25%

For explanation of terms, see glossary.

WORLD-REGIONAL

Although not a geographical designation, "Middle East" is a widely used political term for this region. Many areas in the Middle East are affected by two extremes—a lack of rainfall and an abundance of oil. Population growth, industrialization and irrigation all strain existing water supplies. Economic disparity in the region is another issue. While oil-rich states tend to have a larger GNP, not all countries with oil have a high per capita income because of population size and military expenditures.

Underdevelopment and population growth are problems in this region. Except in Israel, where recent increases have been from immigration, the natural population growth rate is over 2.5 percent per year. Muslim states are also struggling to defend traditional customs and beliefs from the impact of technology and Western-style materialism.

There have been many recent conflicts, including the Gulf War of 1990 and internal conflicts involving Kurds in Iraq and Turkey. However, in 1993 Israel and the Palestinians, long at war, formally recognized each other as states, a first step towards peace.

Night view of oil refinery in Abadan, Iran

Saudi Arabia, its five partners in the Gulf Co-operation Council (Oman, United Arab Emirates, Kuwait, Qatar and Bahrain), and Iran and Iraq are key oil-producing countries. The rapid rise in the price of oil in the 1970s benefited these nations, but overproduction in the late 1980s lowered oil prices and hurt the region's economy. The Gulf War, which damaged the economies of Iraq and Kuwait, also caused extensive environmental damage.

Oilfields, pipelines and exports

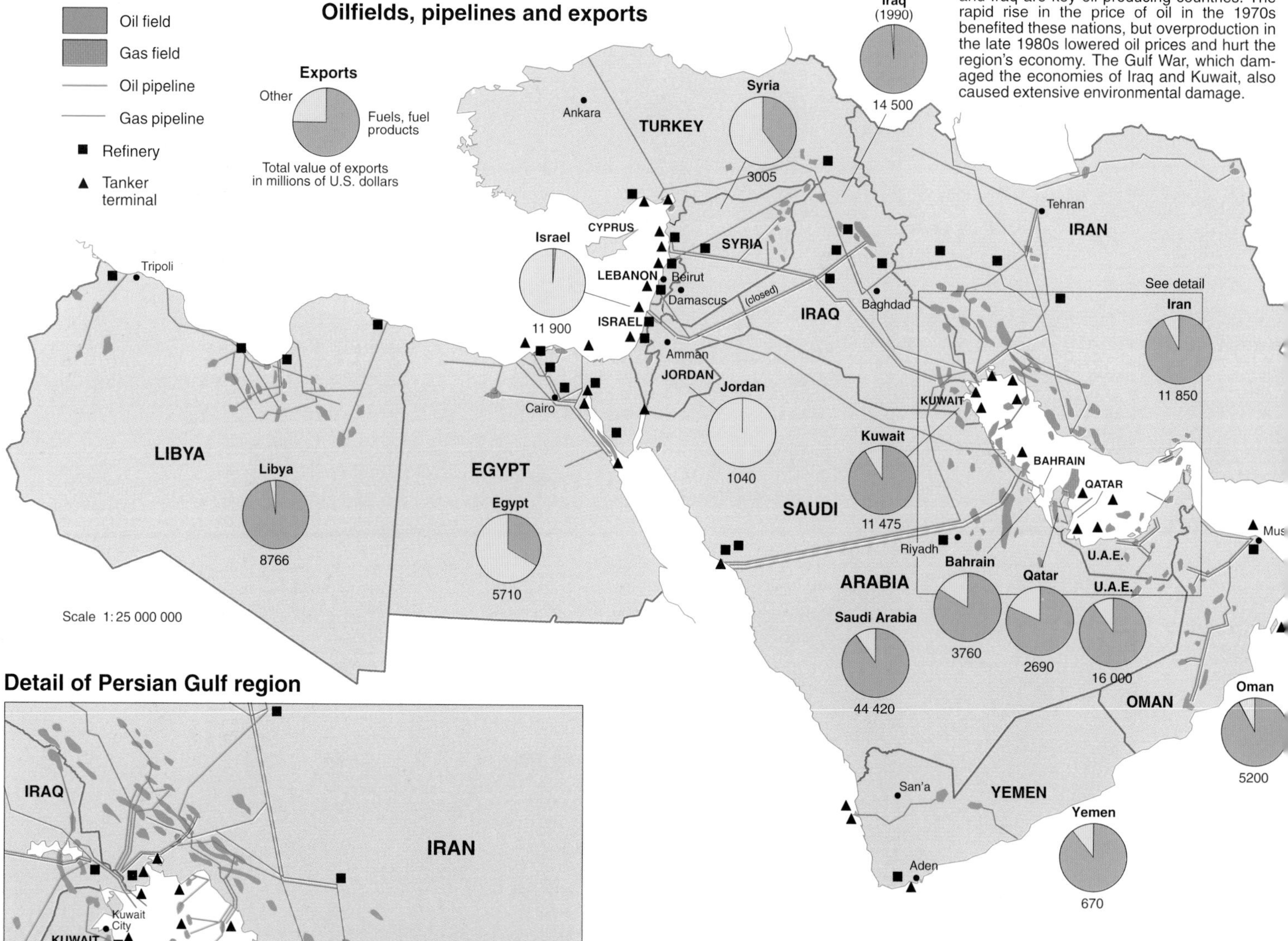

Detail of Persian Gulf region

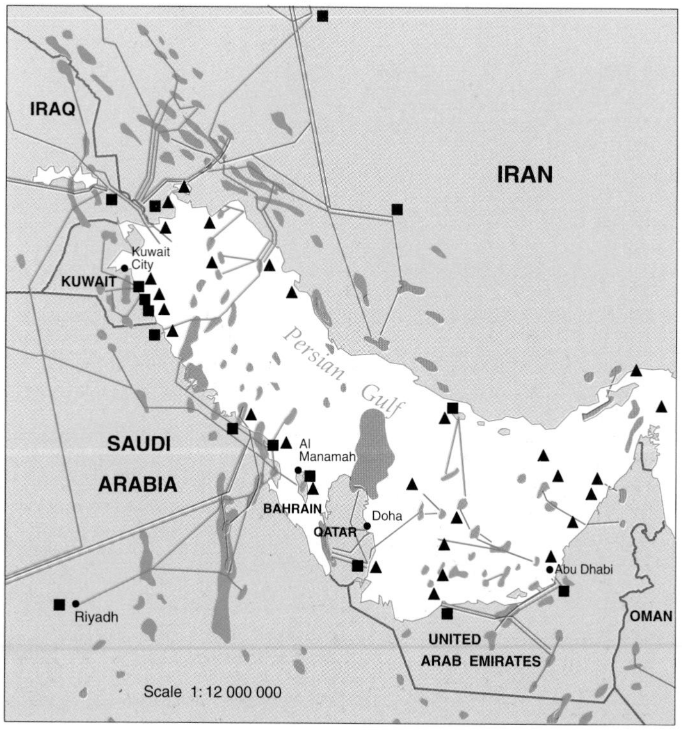

Oil reserves in the Middle East
(68.8% of world reserves)

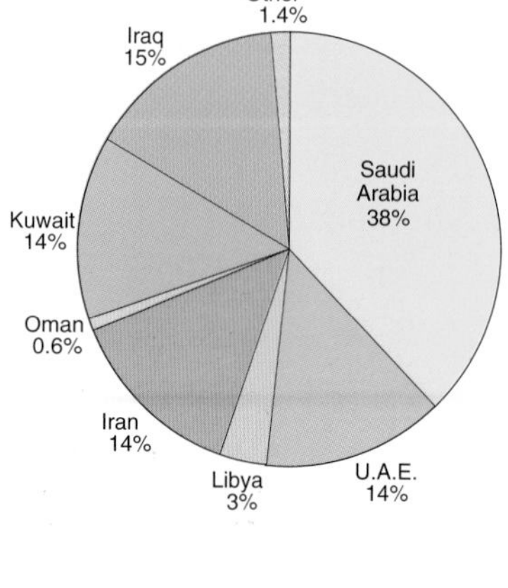

Destinations of Middle East oil exports

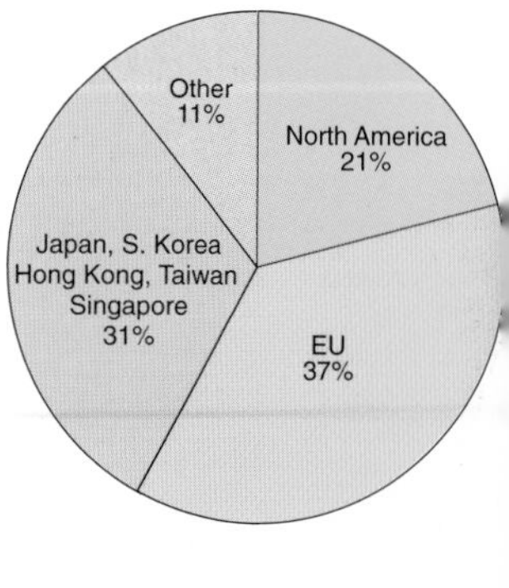

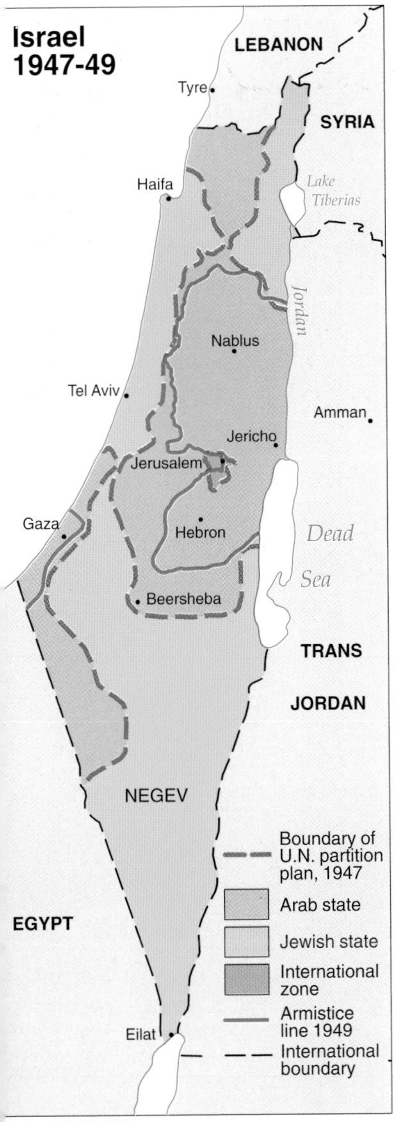

Scale 1: 3 330 000

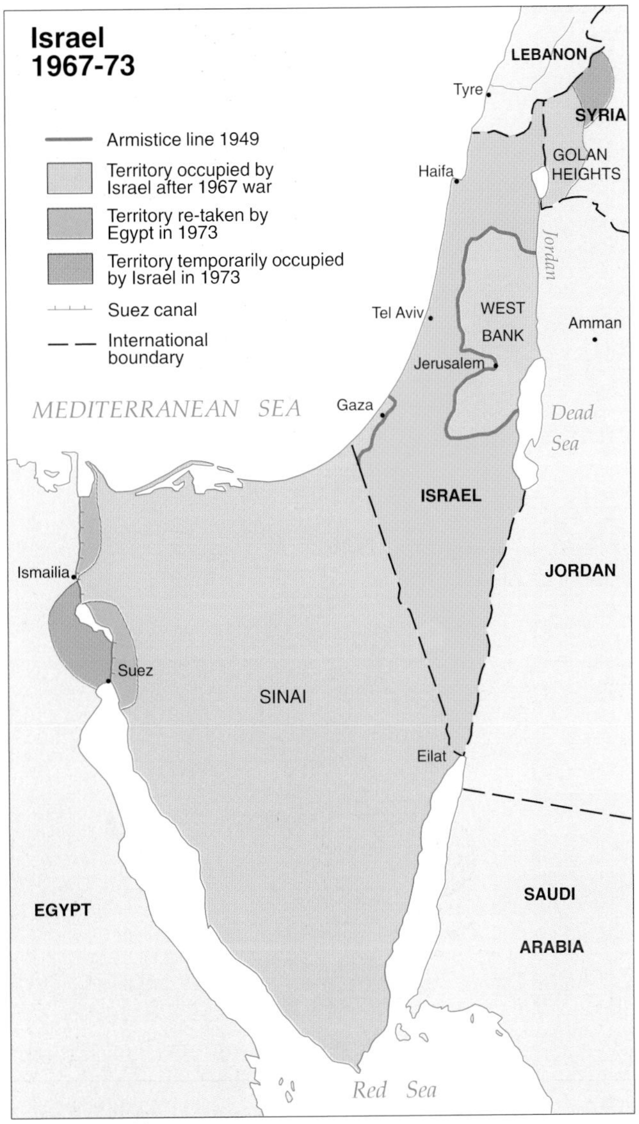

Scale 1: 4 750 000

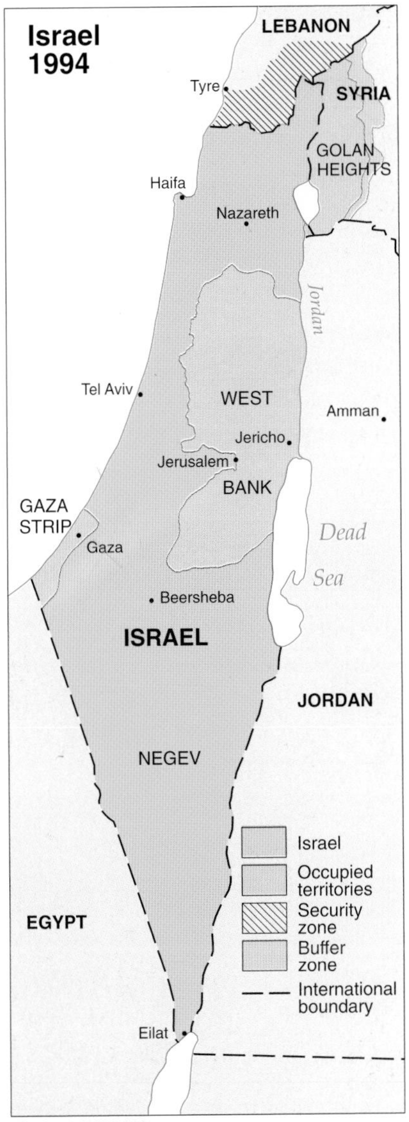

Scale 1: 3 330 000

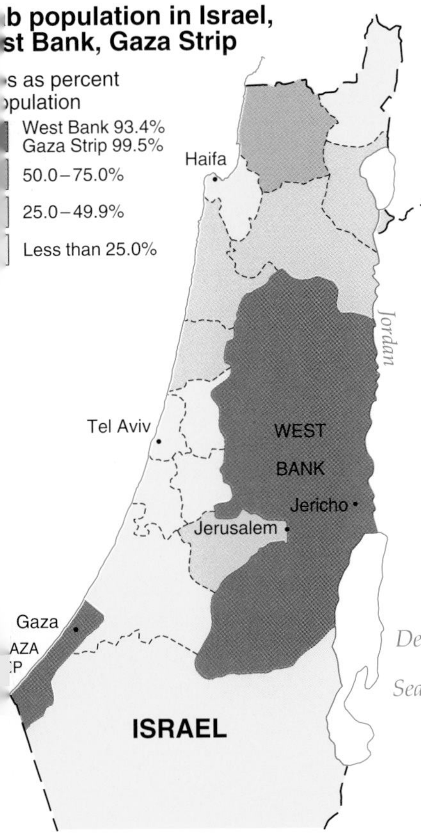

Timeline of Israeli-Arab conflict in the Middle East

1948 United Nations decides to create two states – Israel and Palestine. Palestinians and Arab allies attack Israel and are defeated. Many Palestinians become refugees

1956 Egypt nationalizes Suez Canal. Israel attacks Egypt

1964 PLO founded, vowing to destroy Israel. (PLO was originally an organization of the Arab States.) Terrorist arm called El Fatah, led by Yasser Arafat, begins operations against Israel

1967 Six-day war when Israel captures Sinai, Gaza, Golan Heights, West Bank and east Jerusalem from neighbouring countries. U.N. calls for withdrawal from these territories but Israel refuses

1969 El Fatah takes control of PLO

1970 Jordan attacks Palestinian terrorists operating in Jordan and they retreat to Lebanon

1973 Egypt and Syria attack Israel. Israel resists successfully

1978 Camp David Accord is signed by Egypt, Israel and United States, offering limited self-rule to Palestinians in occupied territories. PLO declines

1979 Egypt-Israel Peace Treaty. Sinai returns to Egypt; Gaza remains with Israel. Egypt denounced by other Arab states

1982 Israel invades Lebanon to eliminate Palestinian guerillas

1987 West Bank and Gaza uprisings against Israeli rule

1988 Arafat recognizes Israel's right to exist

1990 Iraq's invasion of Kuwait is supported by Arafat

1991 U.S.-led coalition defeats Iraq

1993 Agreement signed providing self-rule for Gaza and parts of West Bank

Beginning in the late 18th century, Zionist organizations advocated a homeland for Jewish people. The first Zionist settlement in Palestine was built in 1882. In 1948, after the atrocities of the Holocaust, the United Nations took up the cause of a Jewish homeland and Israel was created out of Palestine, an area occupied by the British since 1918. But the native Arab Palestinians, backed by neighbouring Muslim states, refused to recognize Israel, and there were frequent wars. Today, more than two million Arabs live in Israel, many in the West Bank and Gaza—regions that are moving towards independence. Israel's delicate political situation creates high military costs, which are a considerable economic burden.

Farming in Israel

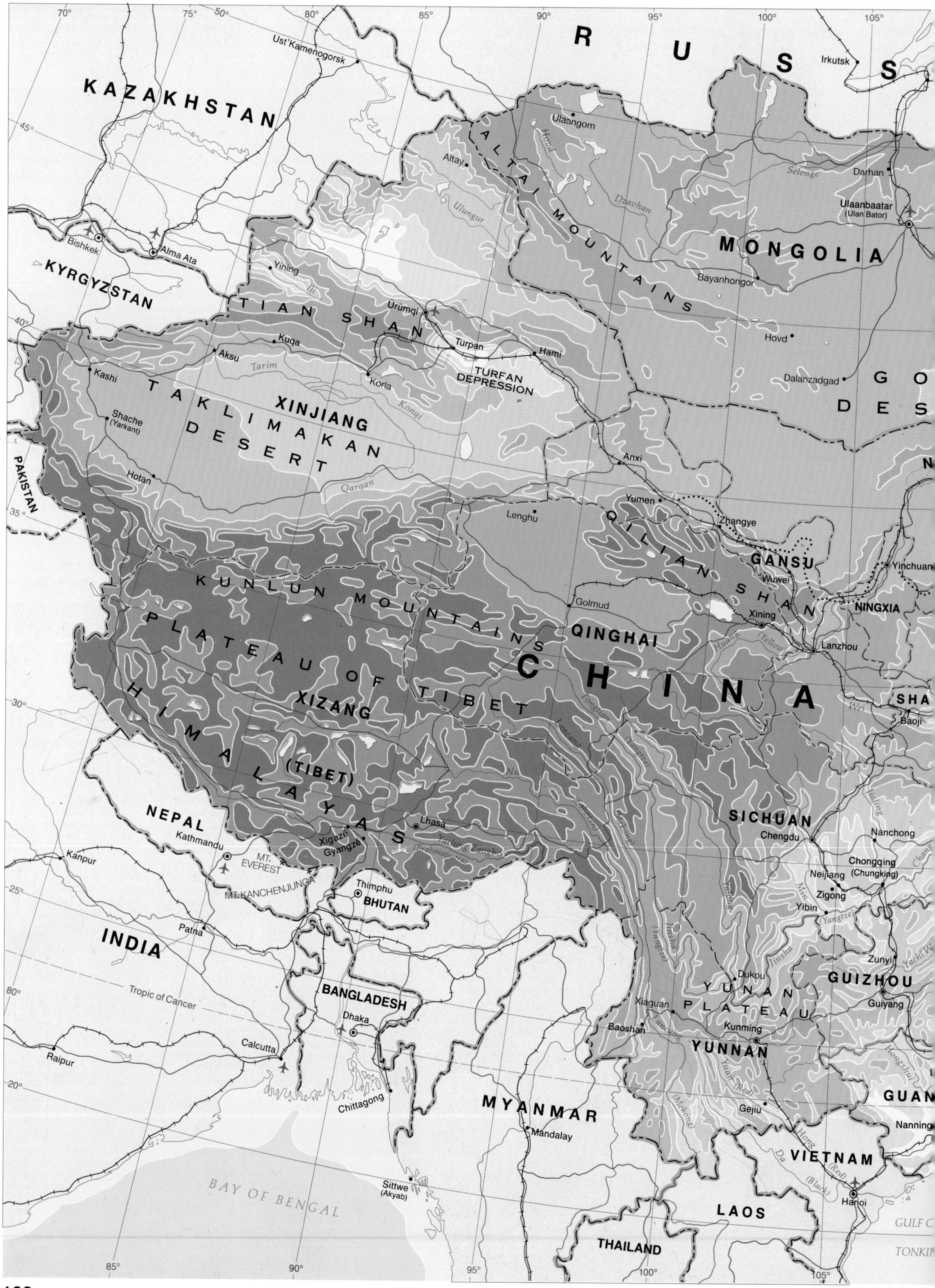
RUSSIA
KAZAKHSTAN
KYRGYZSTAN
PAKISTAN
MONGOLIA
CHINA
XINJIANG
XIZANG
(TIBET)
QINGHAI
GANSU
NINGXIA
SICHUAN
GUIZHOU
YUNNAN
NEPAL
BHUTAN
INDIA
BANGLADESH
MYANMAR
THAILAND
LAOS
VIETNAM
ALTAI MOUNTAINS
TIAN SHAN
TAKLIMAKAN DESERT
TURFAN DEPRESSION
KUNLUN MOUNTAINS
PLATEAU OF TIBET
HIMALAYAS
QILIAN SHAN
YUNAN PLATEAU
MT. EVEREST
MT. KANCHENJUNGA
BAY OF BENGAL
Tropic of Cancer
Ust'Kamenogorsk
Irkutsk
Ulaangom
Altay
Darhan
Ulaanbaatar
(Ulan Bator)
Bayanhongor
Hovd
Dalanzadgad
Bishkek
Alma Ata
Yining
Ürümqi
Turpan
Hami
Kuqa
Aksu
Korla
Kashi
Shache
(Yarkant)
Hotan
Anxi
Yumen
Zhangye
Lenghu
Yinchuan
Wuwei
Golmud
Xining
Lanzhou
Baoji
Lhasa
Xigazê
Gyangzê
Kathmandu
Thimphu
Kanpur
Patna
Dhaka
Calcutta
Raipur
Chittagong
Sittwe
(Akyab)
Mandalay
Chengdu
Nanchong
Chongqing
(Chungking)
Neijiang
Zigong
Yibin
Zunyi
Dukou
Guiyang
Xiaguan
Baoshan
Kunming
Gejiu
Nanning
Hanoi
Selenge
Dzavhan
Ulungur
Ili
Tarim
Konqi
Qarqan
Huang
(Yellow)
Wei
Tongtian
Lancang
Nu
Yarlung Zangbo
(Brahmaputra)
Min
(Yangtze)
Jinsha
Hongshui
Yuan
(Red)
(Mekong)
Hong
(Red)
Da
(Black)
70°
75°
80°
85°
90°
95°
100°
105°
45°
40°
35°
30°
25°
20°

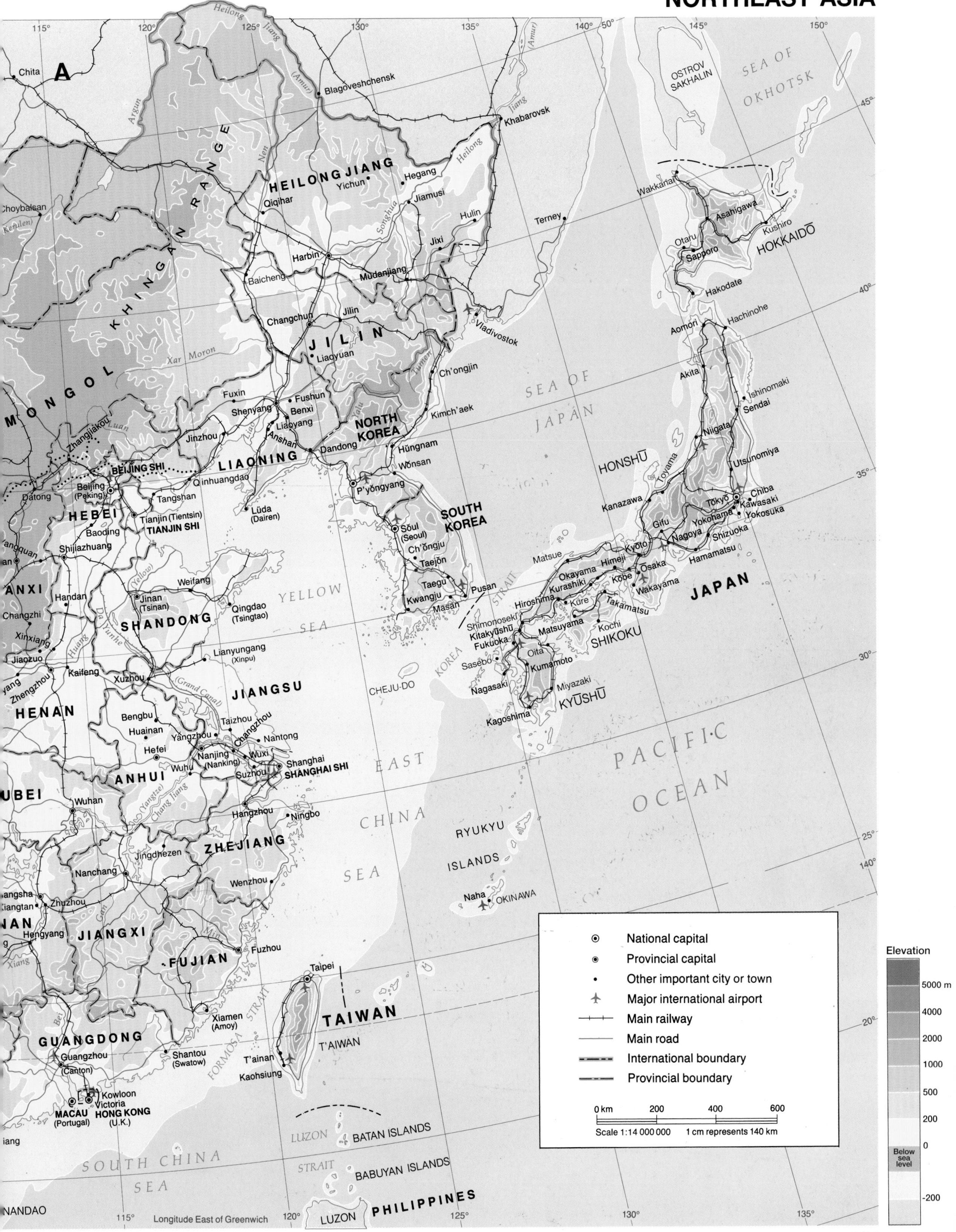
A
Chita
Blagoveshchensk
Khabarovsk
OSTROV SAKHALIN
SEA OF OKHOTSK
HEILONGJIANG
Yichun
Hegang
Jiamusi
Qiqihar
Hulin
Jixi
Harbin
Mudanjiang
Baicheng
Changchun
Jilin
JILIN
Liaoyuan
Vladivostok
Ch'ongjin
Terney
Wakkanai
Asahigawa
Kushiro
Otaru
Sapporo
HOKKAIDŌ
Hakodate
Aomori
Hachinohe
Akita
Ishinomaki
Sendai
Niigata
Utsunomiya
SEA OF JAPAN
Kimch'aek
Fuxin
Shenyang
Fushun
Benxi
Liaoyang
Anshan
Jinzhou
NORTH KOREA
Dandong
Hŭngnam
Wŏnsan
LIAONING
Qinhuangdao
BEIJING SHI
Beijing (Peking)
Tangshan
P'yŏngyang
Datong
HEBEI
Tianjin (Tientsin)
TIANJIN SHI
Lüda (Dairen)
Sŏul (Seoul)
SOUTH KOREA
Ch'ŏngju
Taejŏn
Baoding
Shijiazhuang
HONSHŪ
Toyama
Kanazawa
Tōkyō
Chiba
Kawasaki
Yokohama
Yokosuka
Gifu
Nagoya
Shizuoka
Kyōto
Hamamatsu
Matsue
Himeji
Okayama
Kōbe
Ōsaka
Kurashiki
Wakayama
Hiroshima
Kure
Takamatsu
JAPAN
Weifang
Jinan (Tsinan)
Qingdao (Tsingtao)
YELLOW SEA
SHANDONG
Taegu
Pusan
Kwangju
Masan
KOREA STRAIT
Shimonoseki
Kitakyūshū
Fukuoka
Matsuyama
Kōchi
SHIKOKU
Ōita
Kumamoto
Sasebo
Nagasaki
Miyazaki
KYŪSHŪ
Kagoshima
CHEJU-DO
Handan
Changzhi
Xinxiang
Jiaozuo
Kaifeng
Zhengzhou
HENAN
Lianyungang (Xinpu)
Xuzhou
JIANGSU
Bengbu
Huainan
Taizhou
Changzhou
Yangzhou
Nantong
Hefei
Nanjing (Nanking)
Wuxi
Shanghai
SHANGHAI SHI
Wuhu
Suzhou
ANHUI
PACIFIC OCEAN
EAST CHINA SEA
Wuhan
Hangzhou
Ningbo
RYUKYU ISLANDS
Jingdhezen
ZHEJIANG
Nanchang
Wenzhou
Naha
OKINAWA
Zhuzhou
JIANGXI
Hengyang
FUJIAN
Fuzhou
Taipei
TAIWAN
T'AIWAN
Xiamen (Amoy)
FORMOSA STRAIT
GUANGDONG
Guangzhou (Canton)
Shantou (Swatow)
T'ainan
Kaohsiung
Kowloon
Victoria
MACAU (Portugal)
HONG KONG (U.K.)
LUZON
BATAN ISLANDS
STRAIT
BABUYAN ISLANDS
PHILIPPINES
SOUTH CHINA SEA
NANDAO
Longitude East of Greenwich
National capital
Provincial capital
Other important city or town
Major international airport
Main railway
Main road
International boundary
Provincial boundary
0 km
200
400
600
Scale 1:14 000 000
1 cm represents 140 km
Elevation
5000 m
4000
2000
1000
500
200
0
Below sea level
-200

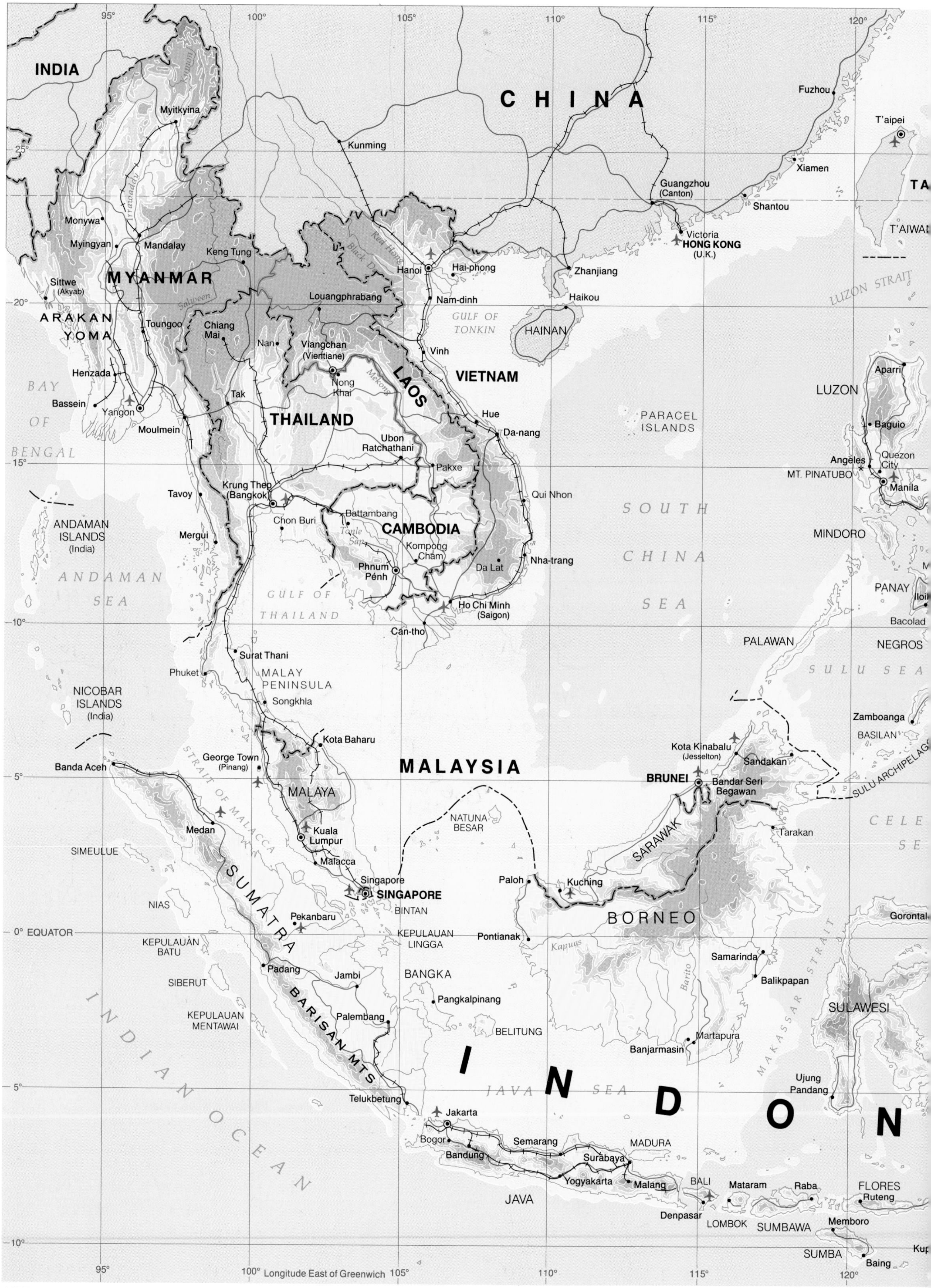
INDIA
C H I N A
MYANMAR
ARAKAN YOMA
THAILAND
LAOS
VIETNAM
CAMBODIA
MALAYSIA
BRUNEI
SINGAPORE
I N D O N
Myitkyina
Kunming
Monywa
Myingyan
Mandalay
Keng Tung
Sittwe
(Akyab)
Toungoo
Chiang Mai
Nan
Louangphrabang
Viangchan
(Vientiane)
Nong Khai
Henzada
Bassein
Yangon
Moulmein
Tak
Ubon Ratchathani
Pakxe
Krung Thep
(Bangkok)
Chon Buri
Battambang
Kompong Cham
Phnum Pénh
Tavoy
Mergui
Hanoi
Hai-phong
Nam-dinh
Vinh
Hue
Da-nang
Qui Nhon
Nha-trang
Da Lat
Ho Chi Minh
(Saigon)
Can-tho
Guangzhou
(Canton)
Victoria
HONG KONG
(U.K.)
Zhanjiang
Haikou
HAINAN
Fuzhou
T'aipei
Xiamen
Shantou
TA
T'AIWA
LUZON STRAIT
GULF OF TONKIN
PARACEL ISLANDS
Red Hong
Black Da
Mekong
Salween
Irrawaddy
Tonle Sap
BAY OF BENGAL
ANDAMAN ISLANDS
(India)
ANDAMAN SEA
GULF OF THAILAND
SOUTH CHINA SEA
NICOBAR ISLANDS
(India)
LUZON
Aparri
Baguio
Angeles
MT. PINATUBO
Quezon City
Manila
MINDORO
PANAY
Iloi
Bacolad
NEGROS
PALAWAN
SULU SEA
Zamboanga
BASILAN
SULU ARCHIPELAGO
Surat Thani
Phuket
MALAY PENINSULA
Songkhla
Kota Baharu
George Town
(Pinang)
Banda Aceh
MALAYA
Kuala Lumpur
Malacca
Singapore
STRAIT OF MALACCA
Medan
SIMEULUE
NIAS
KEPULAUAN BATU
SIBERUT
KEPULAUAN MENTAWAI
SUMATRA
BARISAN MTS
Pekanbaru
Padang
Jambi
Palembang
Telukbetung
BINTAN
KEPULAUAN LINGGA
BANGKA
Pangkalpinang
BELITUNG
NATUNA BESAR
Kota Kinabalu
(Jesselton)
Sandakan
Bandar Seri Begawan
SARAWAK
Tarakan
Kuching
Paloh
Pontianak
BORNEO
Kapuas
Barito
Samarinda
Balikpapan
MAKASSAR STRAIT
Martapura
Banjarmasin
CELE
SE
Gorontal
SULAWESI
Ujung Pandang
JAVA SEA
Jakarta
Bogor
Bandung
Semarang
Surabaya
MADURA
Yogyakarta
Malang
JAVA
BALI
Denpasar
Mataram
LOMBOK
SUMBAWA
Raba
FLORES
Ruteng
Memboro
SUMBA
Baing
Kup
INDIAN OCEAN
95°
100°
105°
110°
115°
120°
25°
20°
15°
10°
5°
0° EQUATOR
Longitude East of Greenwich

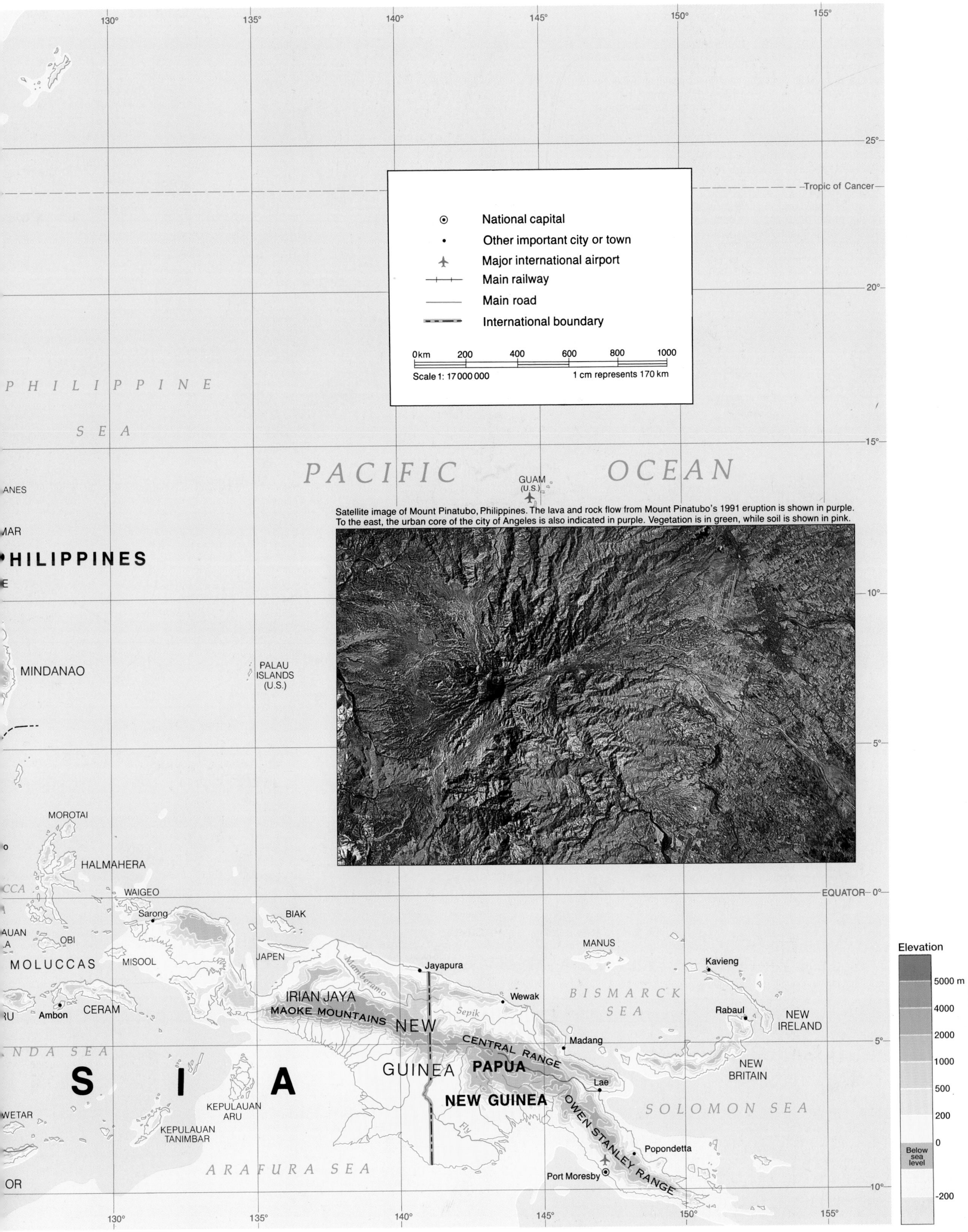

Satellite image of Mount Pinatubo, Philippines. The lava and rock flow from Mount Pinatubo's 1991 eruption is shown in purple. To the east, the urban core of the city of Angeles is also indicated in purple. Vegetation is in green, while soil is shown in pink.

WORLD-REGIONAL

World economic power is shifting toward Asia and the Pacific Rim. Only a few years ago, this region was characterized by illiteracy, malnutrition, a high birth rate and political instability. Although rural areas still lag economically, most Asian countries are experiencing record economic growth, and are expected to grow more rapidly in the 1990s than many more industrialized nations. Rising consumer demand and a booming export trade have raised standards of living; in some Asian nations, living standards exceed European levels.

Most of the economic growth has been based on an essential resource—people. A vast pool of relatively cheap but well-educated labour has transformed this region from an agricultural society to a diverse economy based on manufacturing and services. These economic advances are being made in conjunction with increased food production, limited population growth, expanded educational opportunities and improved access to basic health services.

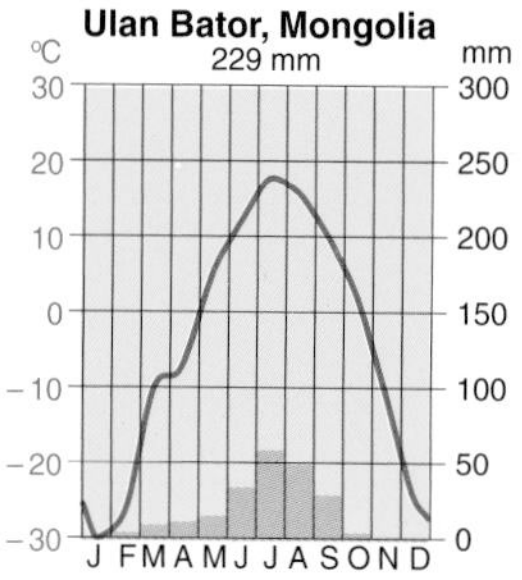

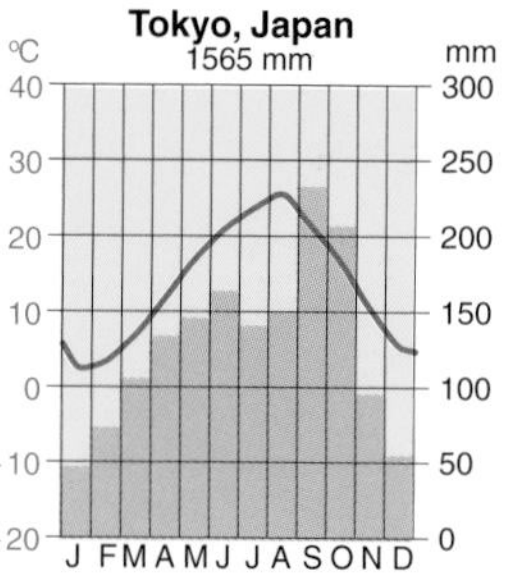

Land use

- Urban
- Cropland
- Cropland and woodland
- Cropland and grazing
- Grassland and grazing
- Tropical forest
- Temperate forest
- Scrubland and grassland, or desert
- Swamp and marsh
- Tundra

Minera

- Golc
- Cop
- Iron
- Tin
- Tung
- Moly
- Coal
- Oil
- Gas
- Cr Chrc
- Hg Merc
- Sb Antir

○ Climograph statio

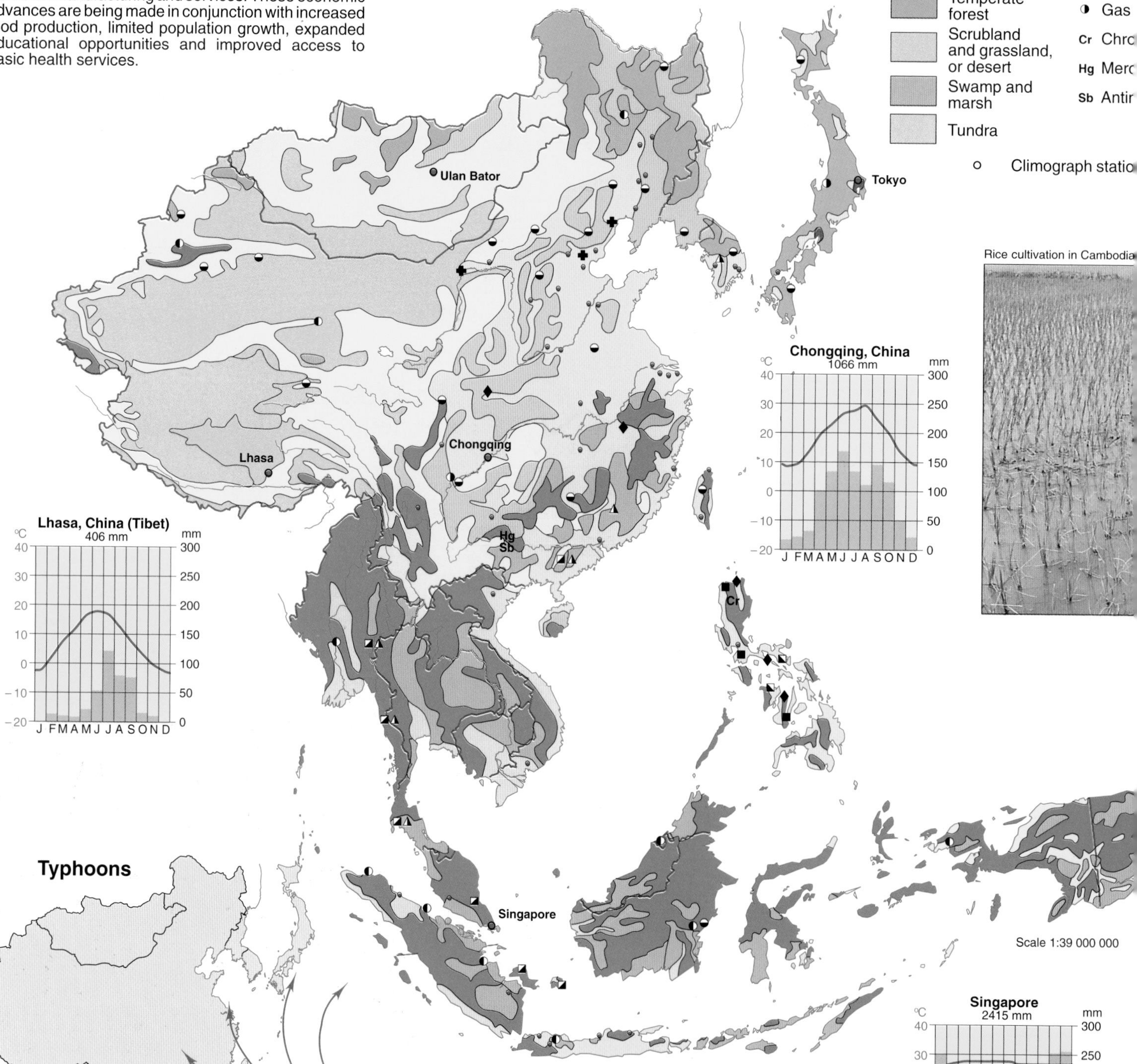

Rice cultivation in Cambodia

Typhoons

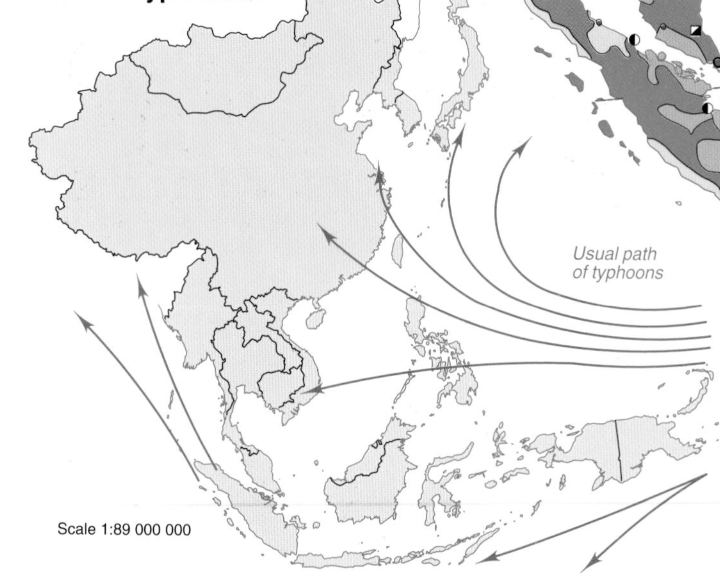

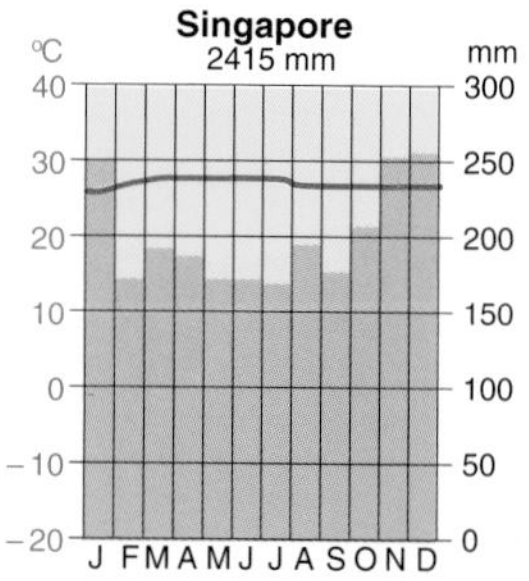

Extending in latitude from 10°S to 50°N, Eastern Asia experiences a wide range of climatic conditions—from the cold deserts of Mongolia to the tropical rainforests of Indonesia. Tropical storms called typhoons affect many coastal areas. They are most severe in autumn but can occur in any month, and as frequently as 20 times a year. Although large in extent, these storms are usually not as violent as hurricanes in North and Central America. However, the torrential rains that accompany typhoons are so heavy that the amount of precipitation can exceed the rainfall during the summer monsoons. Some areas depend on typhoons to alleviate dry conditions.

EASTERN ASIA 1

Eastern Asia includes one of the oldest civilizations in the world, China, the history of which can be traced back 5000 years. Technology and the arts flourished in China. Priests and merchants carried their ideas throughout Eastern Asia. There was a link to Europe by the Silk Route but it was often interrupted by thieves or by government policy. Portuguese, then British colonists settled in the area in the 16th and 17th centuries.

Many of the countries in this region regained independence after World War II, but others remained colonies longer. Brunei, for example, became independent on January 1, 1984. Hong Kong, a British colony for 150 years, and Macao, a Portuguese territory, revert to China in 1997 and 1999. The southern nations in the region have only recently emerged both politically and economically. Brunei's economy is based on the production of oil and gas, and Singapore, which split from the Malaysian federation in 1965, is already one of the richest Asian nations.

Colonization and independence

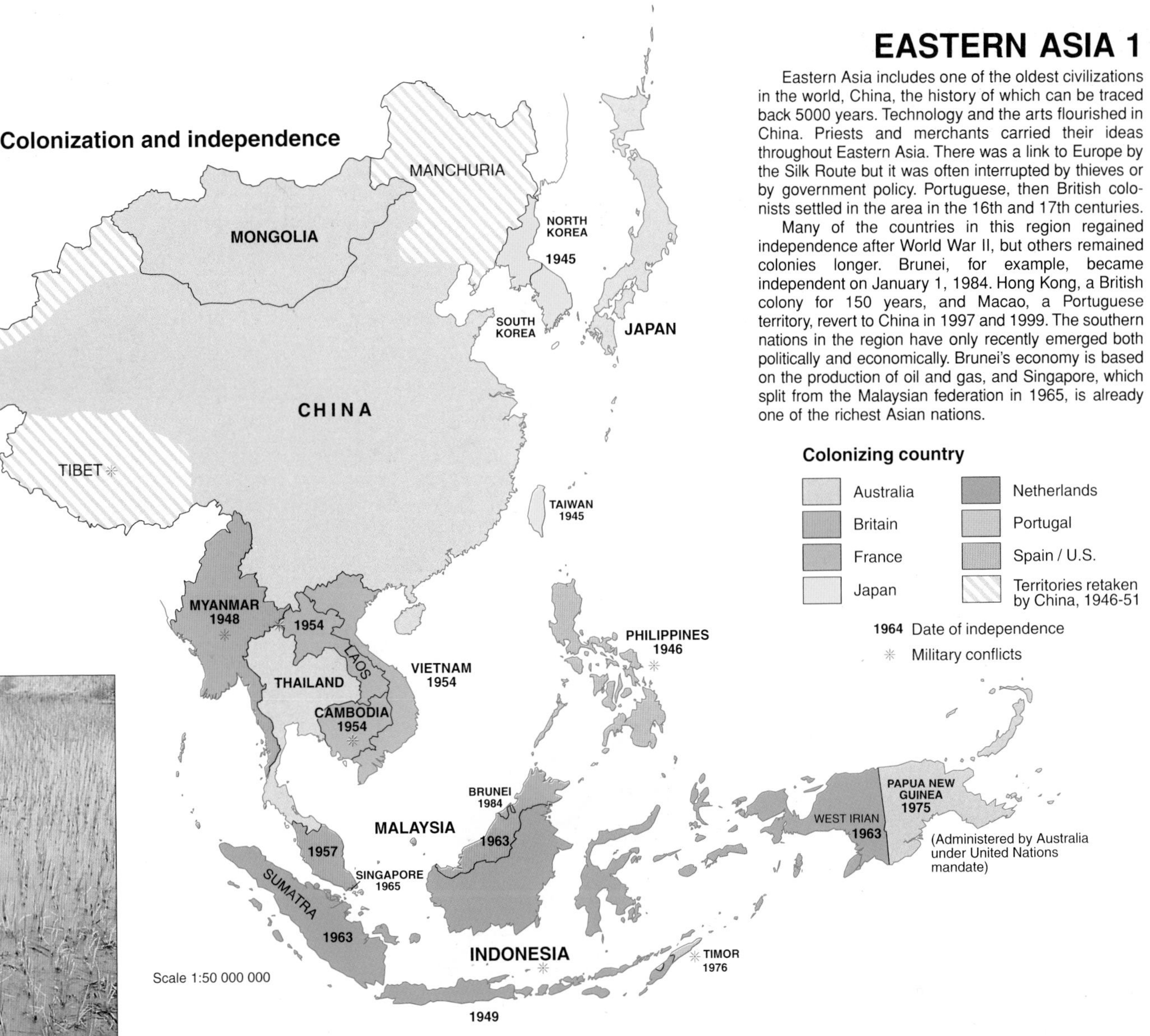

Earthquakes, volcanoes and tidal waves

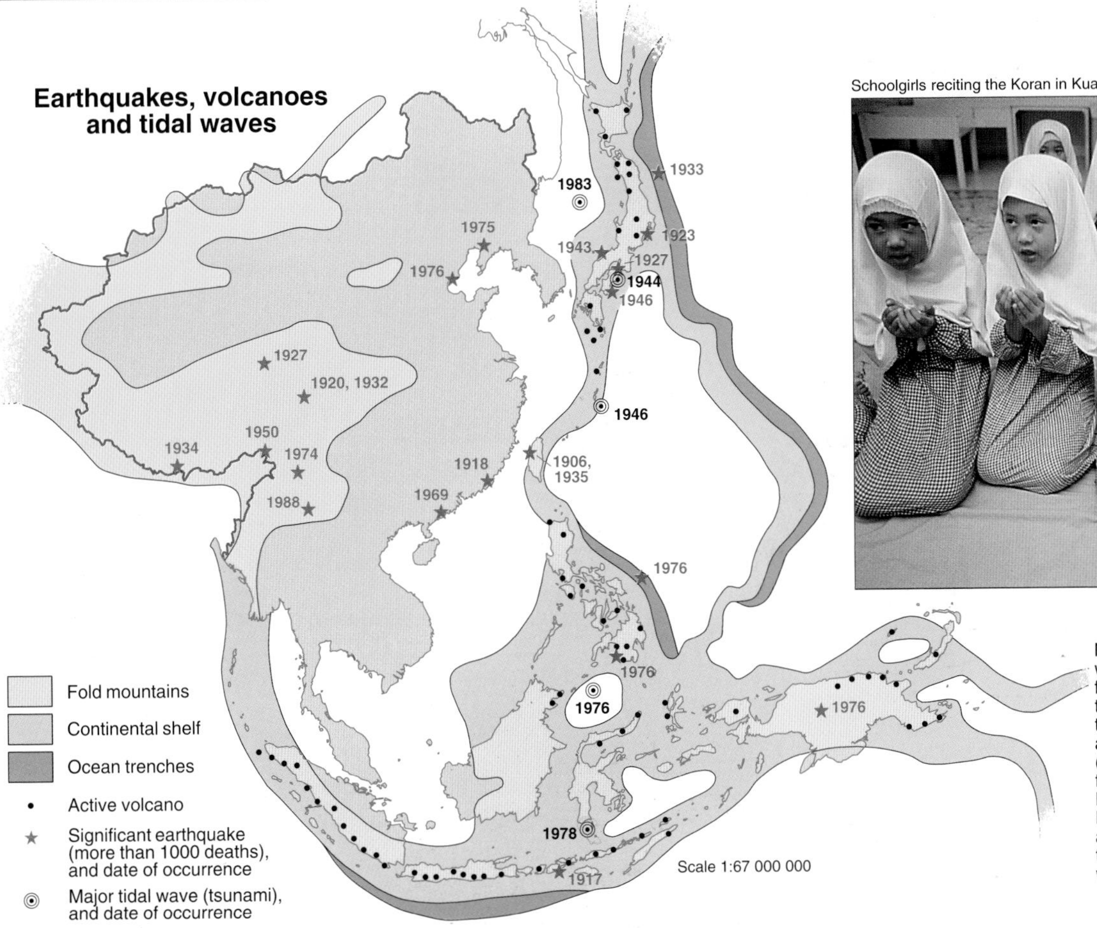

Schoolgirls reciting the Koran in Kuala Lumpur, Malaysia

Most mountain-building activities along the western side of the Pacific Ocean have resulted from the convergence of tectonic plates. Along the boundaries, volcanic activity releases tremendous amounts of energy, often accompanied by earthquakes and tsunamis (tidal waves). This area is part of "the ring of fire"—an active volcanic area surrounding the Pacific Ocean. In 1991, Mount Pinatubo in the Philippines erupted, sending ash, rock, debris and dust as high as 30 km into the atmosphere; the spread of this volcanic cloud affected world weather patterns.

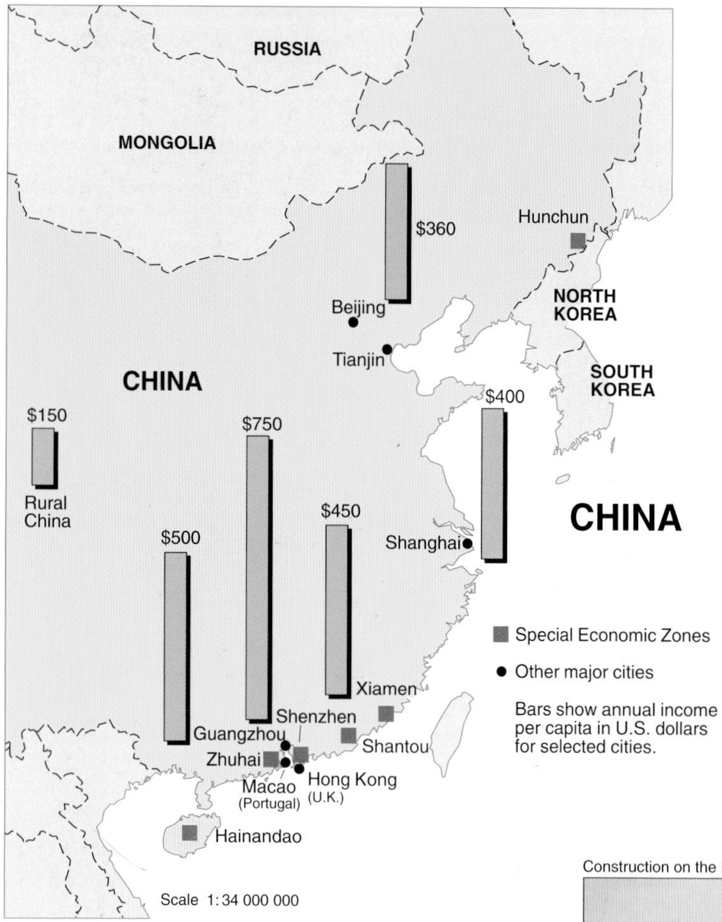

CHINA

Almost one person in every five on earth is a citizen of China, one of the world's oldest civilizations and one of its poorest nations. Only about 20 percent of the people live in urban areas. Since the death of Mao Zedong in 1976, China has expanded its trade and is actively seeking to modernize its economy. But after 200 years of isolation, the Chinese economy still has far to go. Recent growth has been quick but uneven. The establishment of Special Economic Zones, where a free market system is allowed, has attracted foreign investment. In China's current drive for economic improvement, however, pollution controls and human rights seem of less concern.

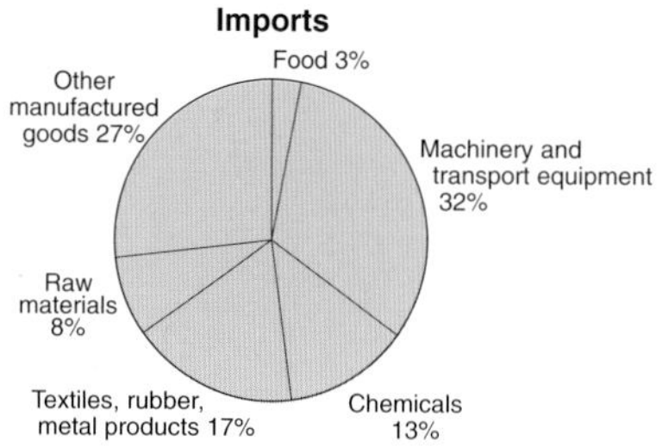

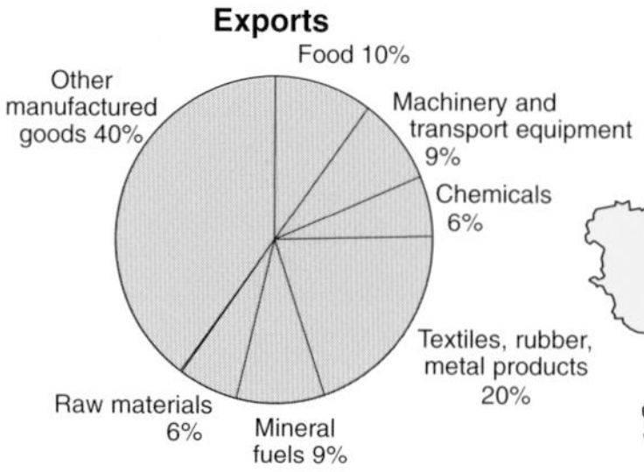

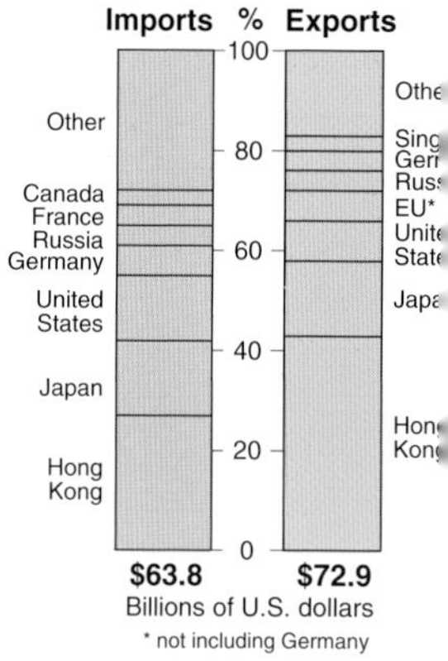

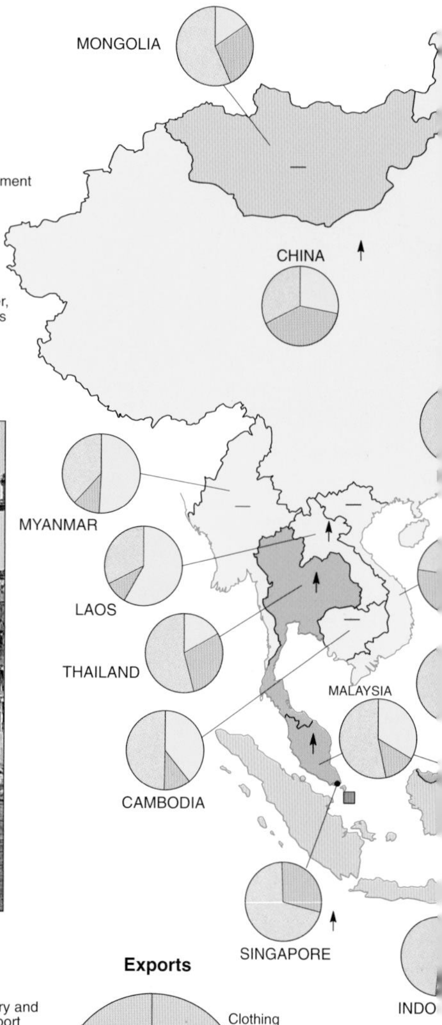

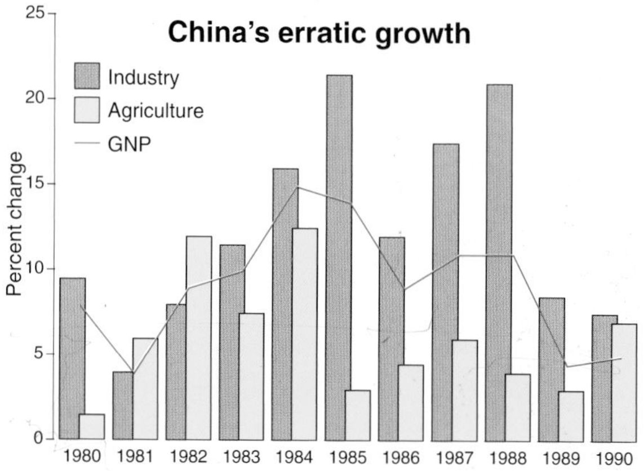

Construction on the Lisang River in China

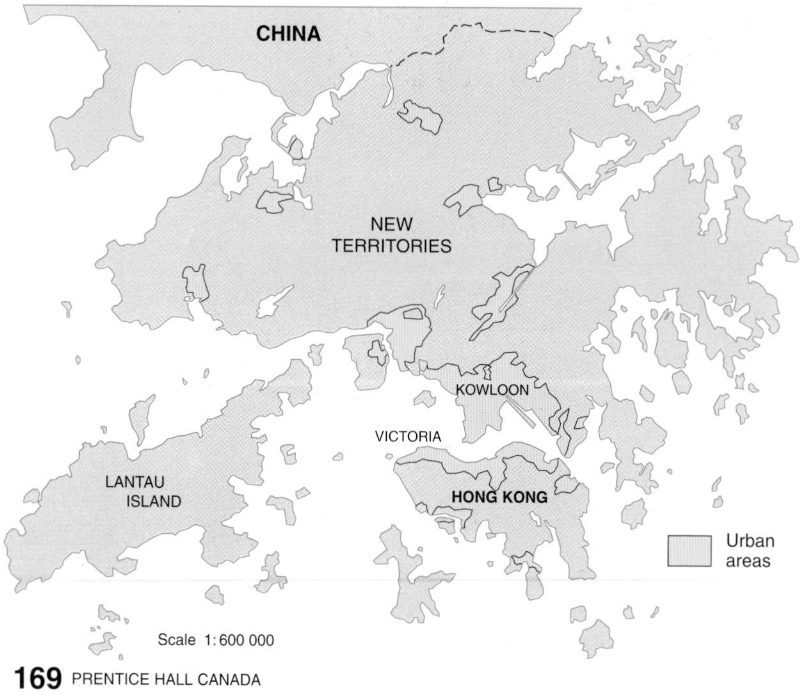

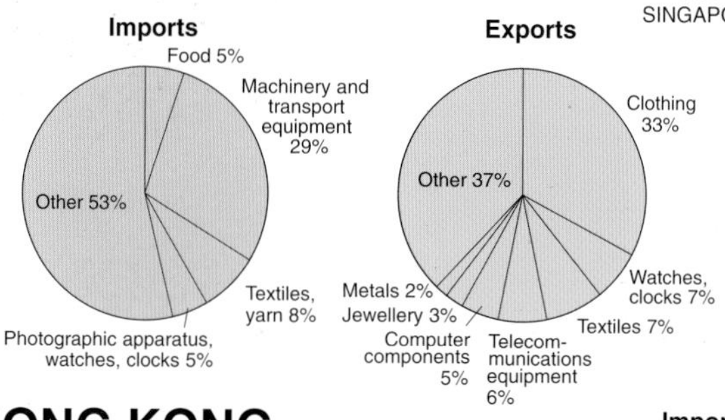

HONG KONG

After 150 years as a British colony, Hong Kong reverts to China in 1997. Since 1949, the colony has become the world's largest container port and largest exporter of clothing. With a population density of 5400 people per square kilometre on a land mass of only 1067 km^2, Hong Kong has built its reputation on trade and manufacturing, and more recently on high technology goods. Hong Kong has also successfully marketed its financial services, construction, printing and publishing. Many Hong Kong natives have emigrated to Canada in recent years.

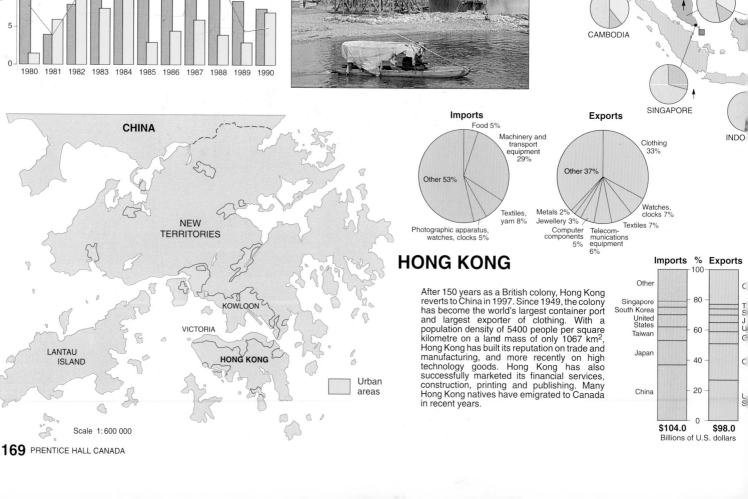

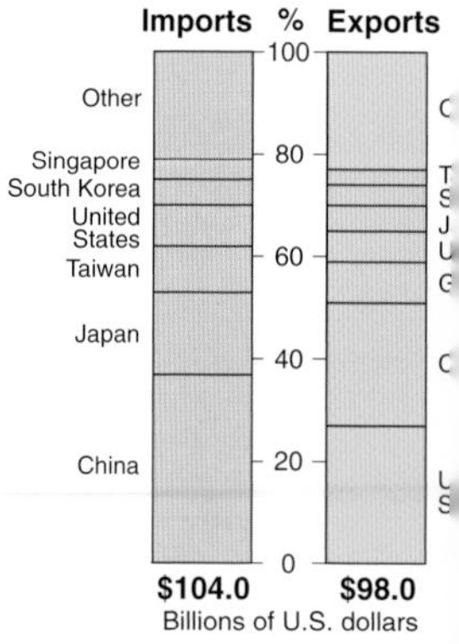

This region is sometimes called the Far East, but the rm Eastern Asia is preferred because it eliminates the estern European point of view and is geographically ccurate. Eastern Asia has experienced major conomic growth in the past 15 years. During the 980s, seven of the top ten fastest-growing economies, alled the Seven Tigers, were in this region. The newly dustrialized nations — Hong Kong, Singapore, South orea and Taiwan — are growing more quickly than apan, and their exports, combined, now exceed apanese exports.

Eastern Asia is still a region of great disparity, owever. At opposite extremes are Cambodia, one of e world's least developed countries, and Japan, hich has the highest Human Development Index ting, according to the United Nations.

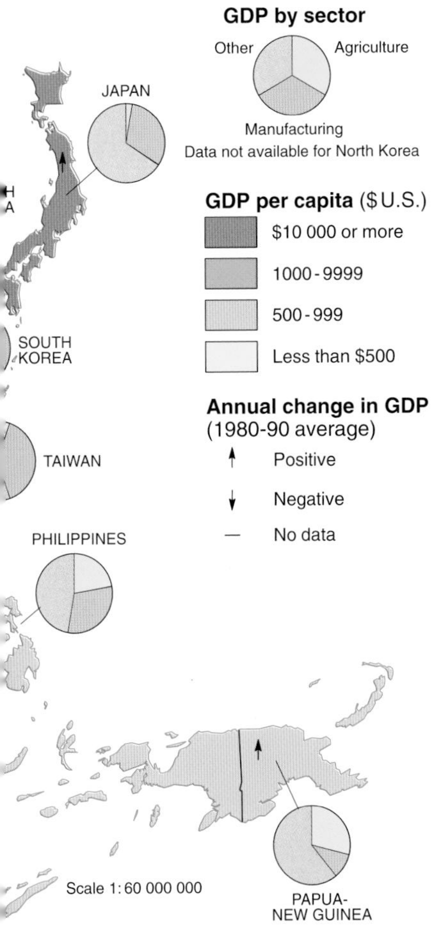

SOUTH KOREA

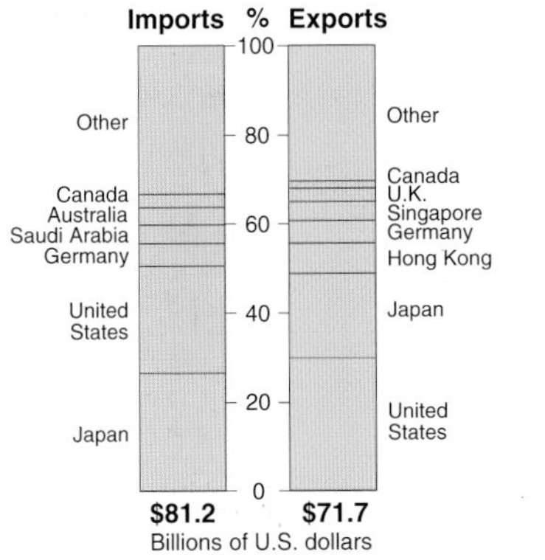

A colony of Japan until 1945, South Korea emerged out of two major world conflicts — World War II and the Korean War — to become one of the world's top ten trading nations. Relying on an inexpensive but well-educated labour force, South Korea produces mainly textiles, clothing and footwear. Recently, however, Korean companies have marketed motor vehicles (the Hyundai), computers, aerospace technology, nuclear power and financial services. The country's economic growth outpaced every other nation's in the 1980s, and the 1990s have ushered in greater democracy and freedom of information.

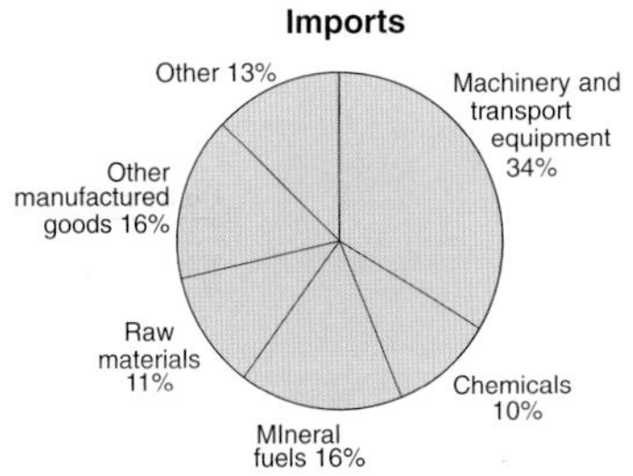

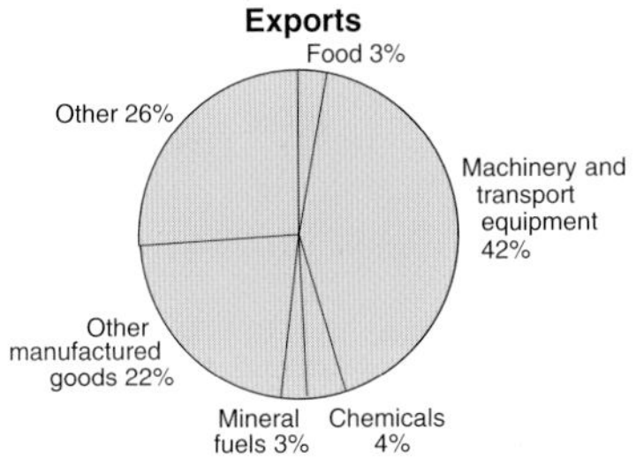

Pulp arrives from Canada at the port of Toma Komai, Japan.

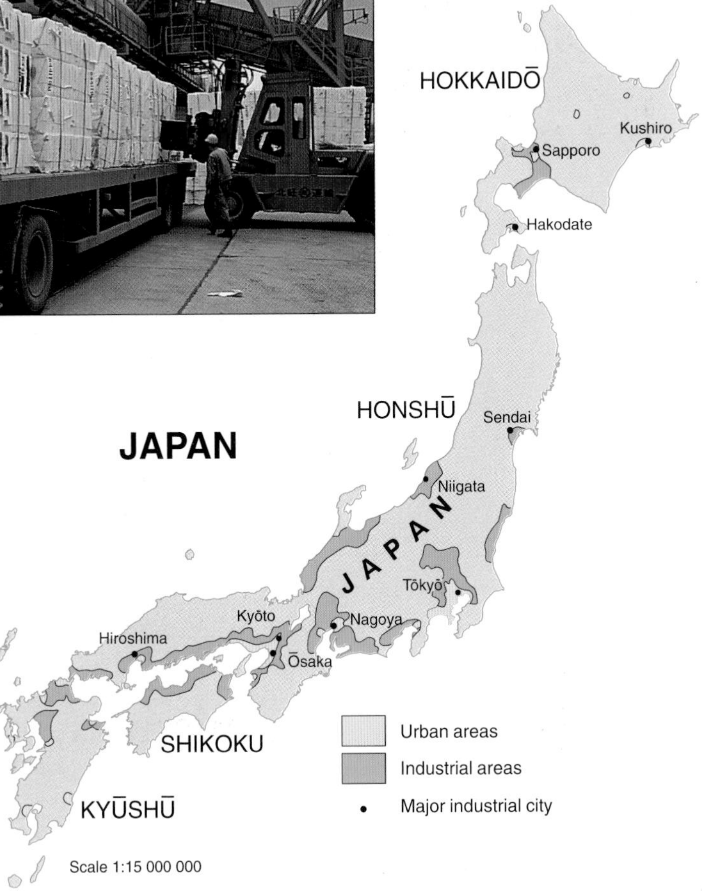

JAPAN

n has been called an economic miracle. Despite a of natural resources, the country has risen from t in World War II to become a major economic r. Japan is the world's most successful trialized nation, producing steel, automobiles, machinery and high technology goods — mostly xport. Japan also has the world's largest fishing With only 6 percent of the land available for ation, agriculture represents only 3 percent of the s Domestic Product. Japan does face some omic challenges, however: it depends on imported s population is aging; and the high value of its ncy drives up the prices of its goods.

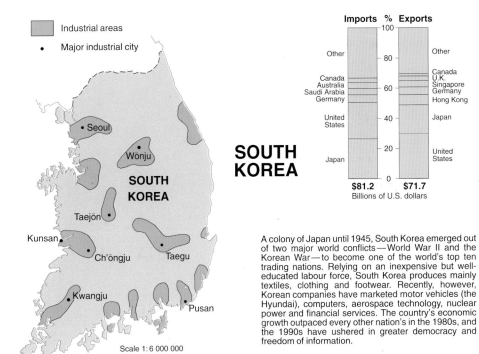

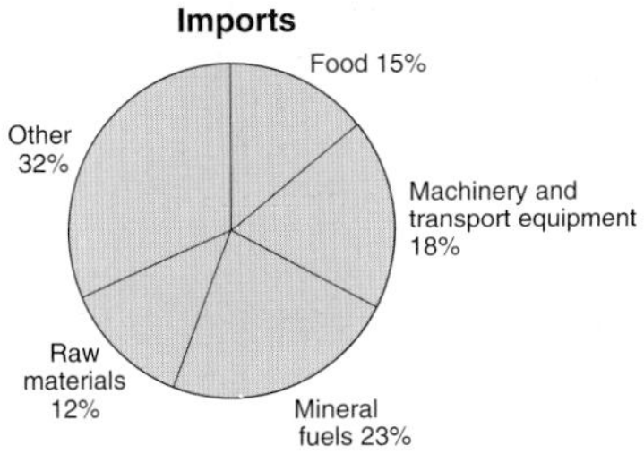

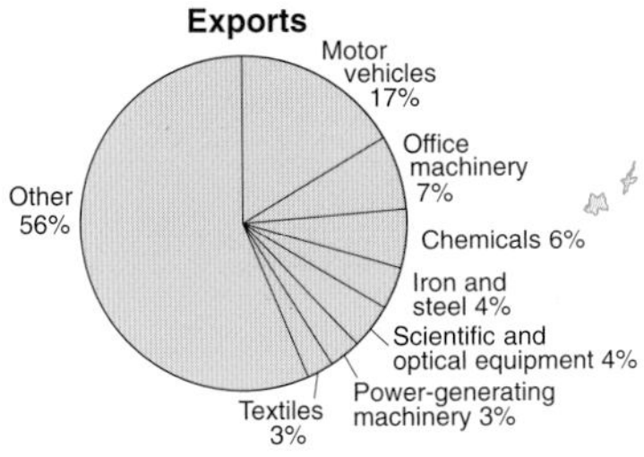

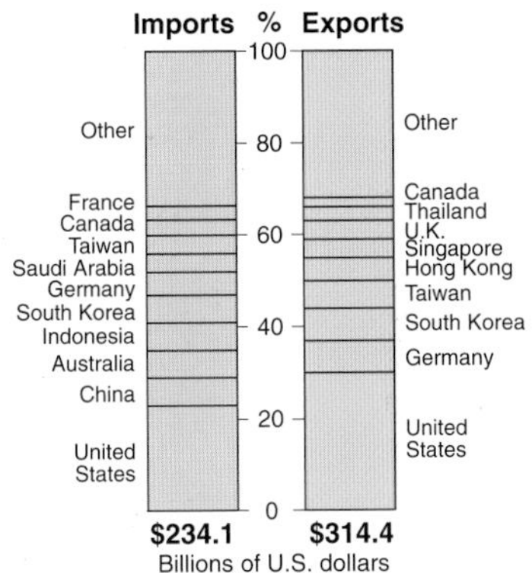

Satellite image of northeastern Australia. Light green indicates vegetation. The Barrier Reef is shown as a thin line of brown off the eastern coast. Clouds are shown as flecks of white.

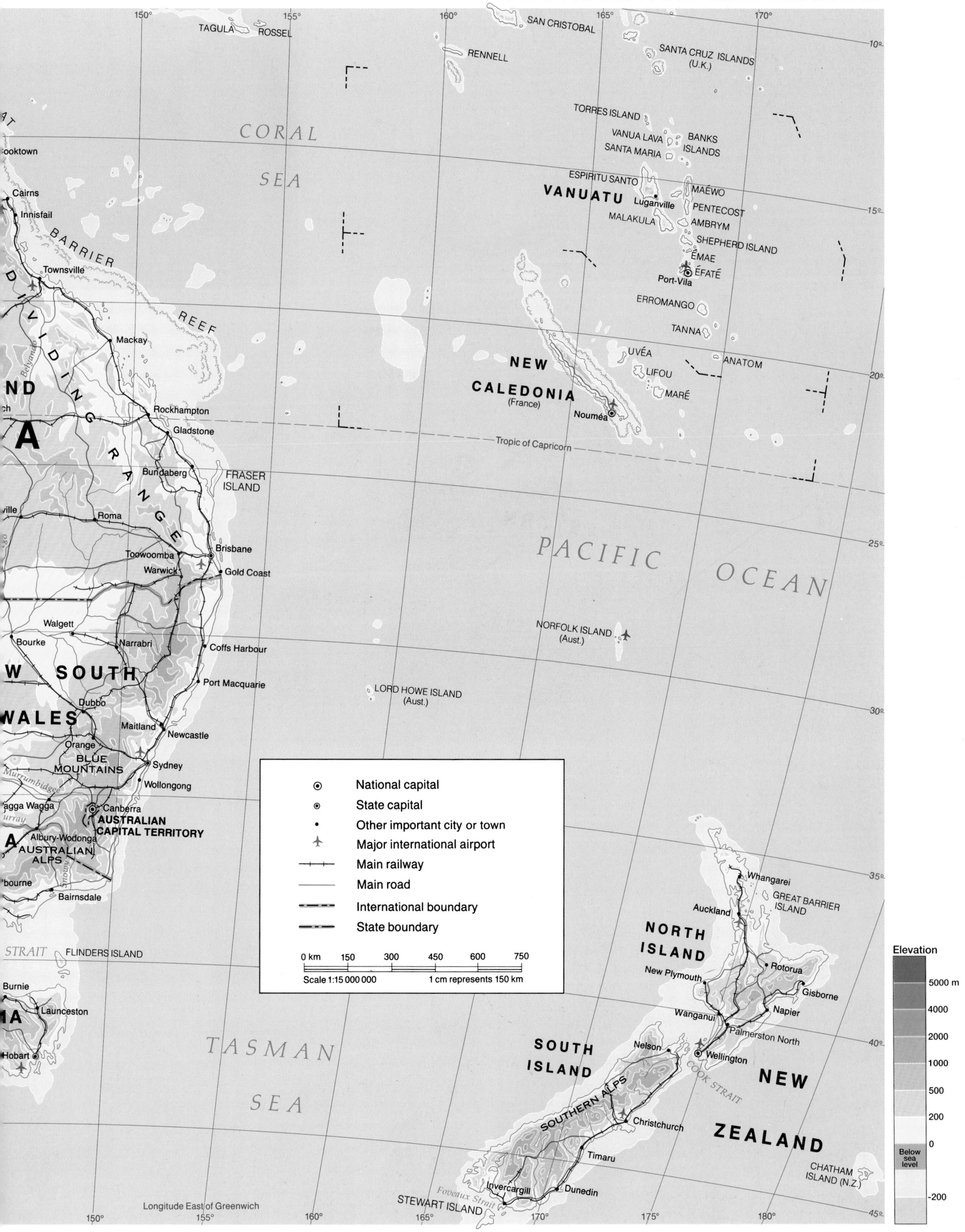
150°
155°
160°
165°
170°
175°
180°
10°
15°
20°
25°
30°
35°
40°
45°
TAGULA
ROSSEL
SAN CRISTOBAL
RENNELL
SANTA CRUZ ISLANDS (U.K.)
TORRES ISLAND
VANUA LAVA
BANKS ISLANDS
SANTA MARIA
ESPIRITU SANTO
VANUATU
Luganville
MAÉWO
PENTECOST
MALAKULA
AMBRYM
SHEPHERD ISLAND
ÉMAE
ÉFATÉ
Port-Vila
ERROMANGO
TANNA
ANATOM
UVÉA
LIFOU
MARÉ
NEW CALEDONIA (France)
Nouméa
Tropic of Capricorn
CORAL SEA
ooktown
Cairns
Innisfail
BARRIER REEF
Townsville
DIVIDING RANGE
Mackay
Belyando
ND
ch
A
Rockhampton
Gladstone
Bundaberg
FRASER ISLAND
ville
Roma
Brisbane
Toowoomba
Warwick
Gold Coast
PACIFIC OCEAN
Walgett
Bourke
Narrabri
Coffs Harbour
NORFOLK ISLAND (Aust.)
W SOUTH WALES
Port Macquarie
LORD HOWE ISLAND (Aust.)
Dubbo
Maitland
Newcastle
Orange
BLUE MOUNTAINS
Sydney
Murrumbidgee
Wollongong
agga Wagga
Canberra
AUSTRALIAN CAPITAL TERRITORY
urray
Albury-Wodonga
AUSTRALIAN ALPS
bourne
Bairnsdale
STRAIT
FLINDERS ISLAND
Burnie
Launceston
IA
Hobart
TASMAN SEA
National capital
State capital
Other important city or town
Major international airport
Main railway
Main road
International boundary
State boundary
0 km 150 300 450 600 750
Scale 1:15 000 000
1 cm represents 150 km
Whangarei
GREAT BARRIER ISLAND
Auckland
NORTH ISLAND
Rotorua
New Plymouth
Gisborne
Napier
Wanganui
Palmerston North
SOUTH ISLAND
Nelson
Wellington
COOK STRAIT
NEW ZEALAND
SOUTHERN ALPS
Christchurch
Timaru
CHATHAM ISLAND (N.Z.)
Invercargill
Dunedin
Foveaux Strait
STEWART ISLAND
Longitude East of Greenwich
Elevation
5000 m
4000
2000
1000
500
200
0
Below sea level
-200

WORLD-REGIONAL

Australia is the sixth largest country in the world. Although the central part is mainly arid, tropical rainforests are found in the north where rainfall is reliable. Mediterranean and subtropical climates prevail in the southwest and southeast. Despite the country's vast expanse, 85 percent of Australians are urban dwellers.

Like Canada, Australia is an exporting nation. It is the world's leading wool producer, largest beef exporter and a major player in the world wheat trade. Minerals such as iron ore, bauxite (the principle ore of aluminum), coal, nickel and zinc are also important to this resource-rich nation. Australia's standard of living and life expectancy are among the highest in the world. Trade with Pacific Rim countries is vital and growing.

New Zealand is a small, mountainous country with some active volcanoes. The economy is based on pasture-land agriculture; farms produce meat, wool, butter and cheese. However, the establishment of the European Union has cut the previously stable United Kingdom market for exports from 70 percent to 10 percent in recent years, forcing New Zealand to diversify its economy.

Land use

- Urban
- Cropland
- Cropland and woodland
- Cropland and grazing
- Grassland and grazing
- Tropical forest
- Temperate forest
- Scrubland and grassland, or desert
- Swamp and marsh
- Tundra
- ○ Climograph station

Minerals

- ■ Gold
- ▲ Silver
- ▼ Lead
- ● Zinc
- ◆ Copper
- ✚ Iron
- Tin
- Aluminum
- Tungsten
- Coal
- Oil
- Gas
- Uranium
- Mn Manganese

The Australian outback

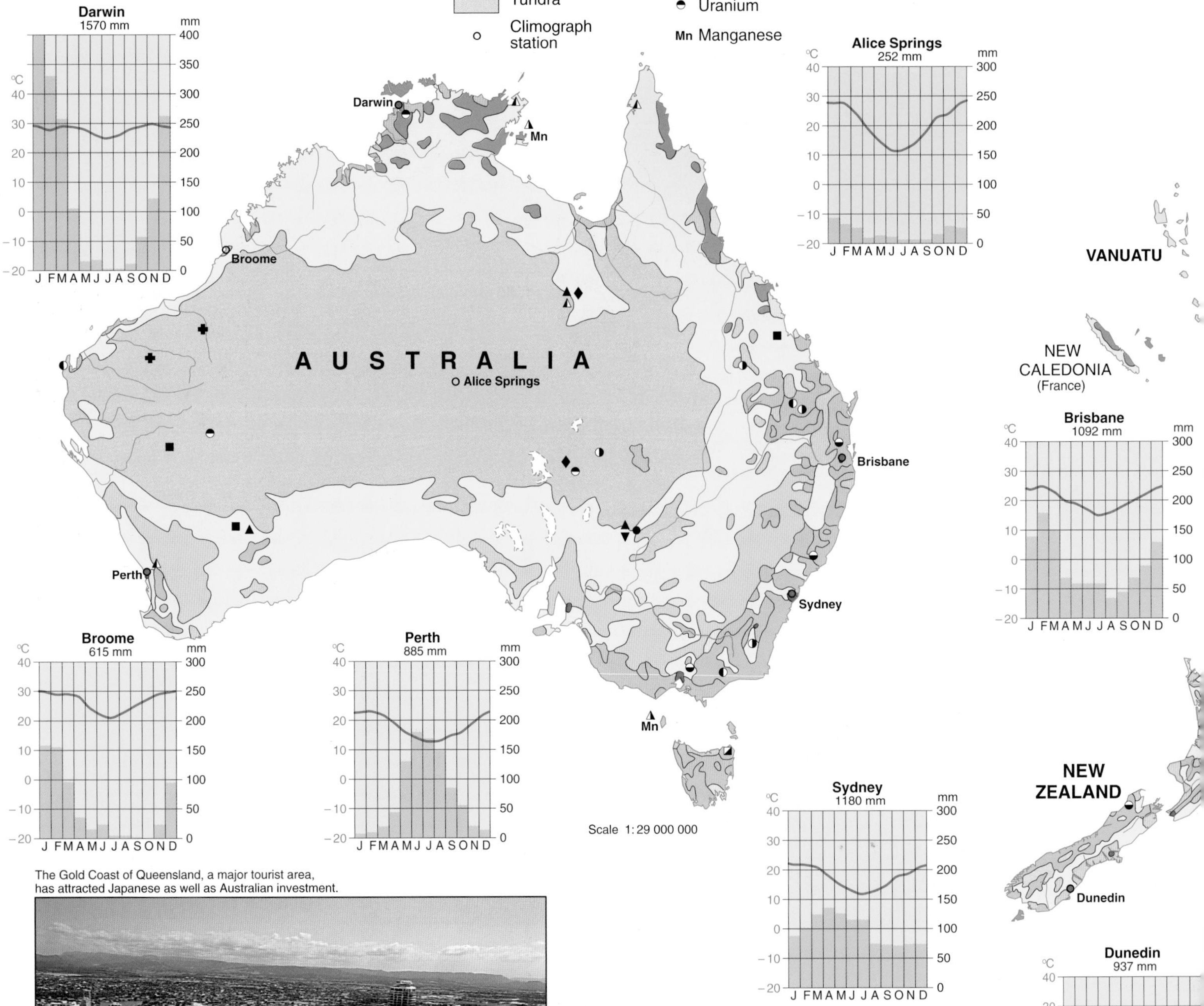

The Gold Coast of Queensland, a major tourist area, has attracted Japanese as well as Australian investment.

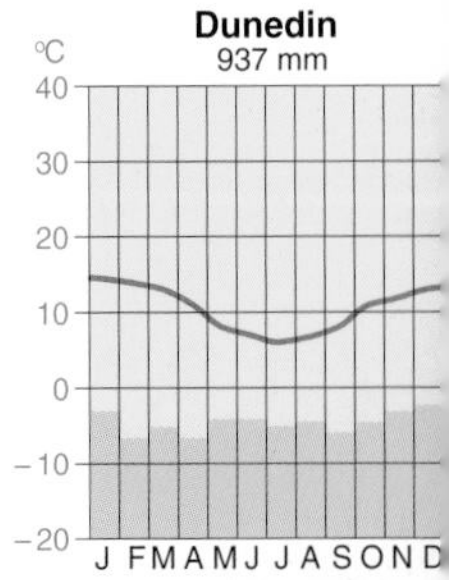

The unique landscapes of Australia and New Zealand are home to plants and animals found only in these countries. The more unusual animals include the kangaroo, the duck-billed platypus and the koala bear. Both countries consider wildlife conservation a priority.

Other countries in this region co-operate with Australia and New Zealand on environmental policies. A South Pacific Forum meets regularly to discuss issues such as toxic waste dumps, global warming and the thinning ozone level over Antarctica. These nations have also established a nuclear-free zone.

AUSTRALIA–NEW ZEALAND

The Aborigines have occupied Australia for at least 40 000 years. By the late 18th century, they numbered over 300 000. But by 1900, conflict with settlers and low resistance to European diseases had reduced the Aborigine population to fewer than 70 000 people. According to the 1986 census, less than 2 percent of the country's population is of aboriginal descent. Today, some Aborigines live in large reserves in north and central Australia. Various groups are successfully seeking compensation and land ownership.

After James Cook's voyage to the east coast of Australia in 1770, England claimed the continent and transported convicts there to ease pressure on British jails. But throughout the 19th and 20th centuries, many people from the British Isles emigrated willingly. Although the Australian government enforced a "White Australia" policy for most of the 20th century, immigration has recently increased from Pacific Rim countries.

Over 80 percent of New Zealanders are of British origin. The Maoris (New Zealand aboriginals) make up about 9 percent of the population, and have enjoyed a revival of their culture in recent years.

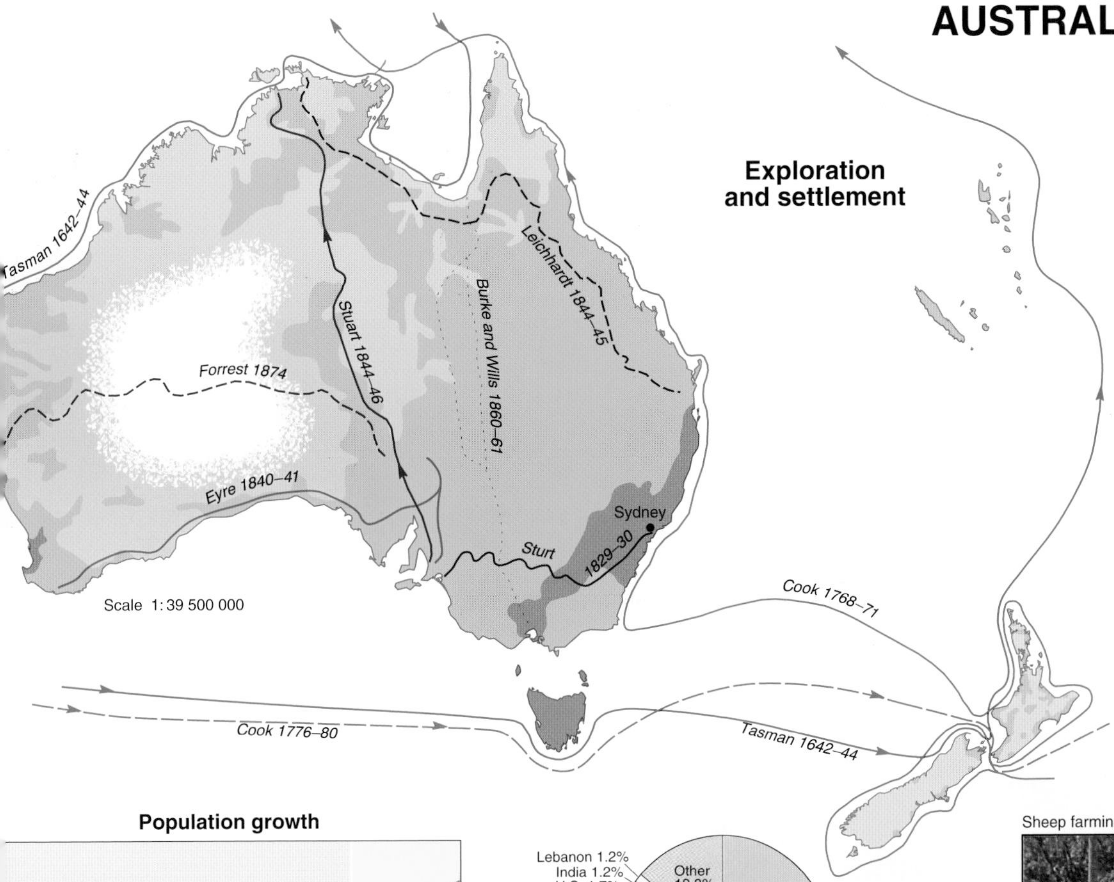

Australian settlement
- 1788 - 1838
- 1838 - 1888
- After 1888

New Zealand settlement
- Before 1840
- After 1840

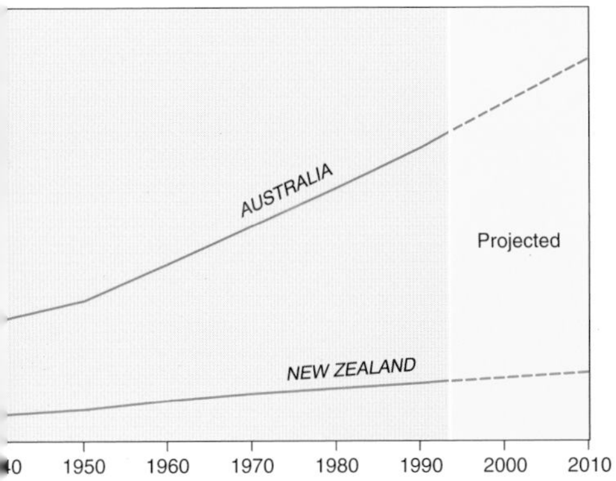

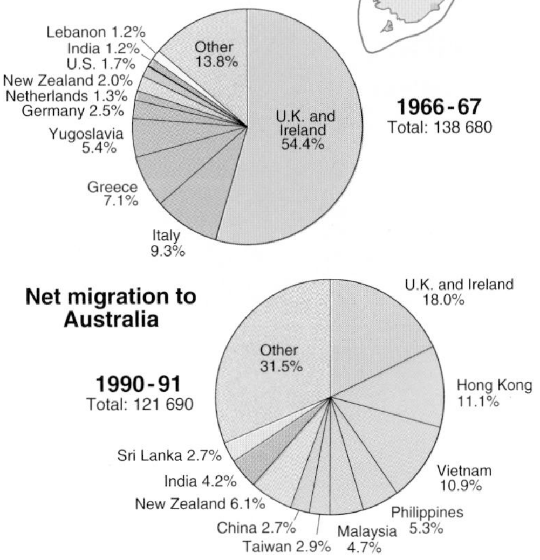

Sheep farming is an important industry in New Zealand.

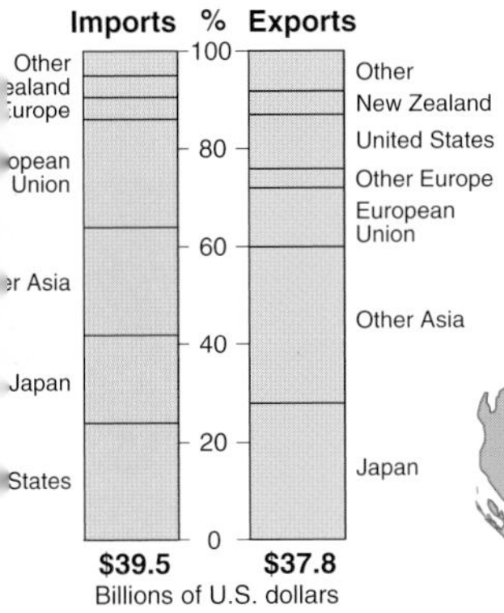

Australia

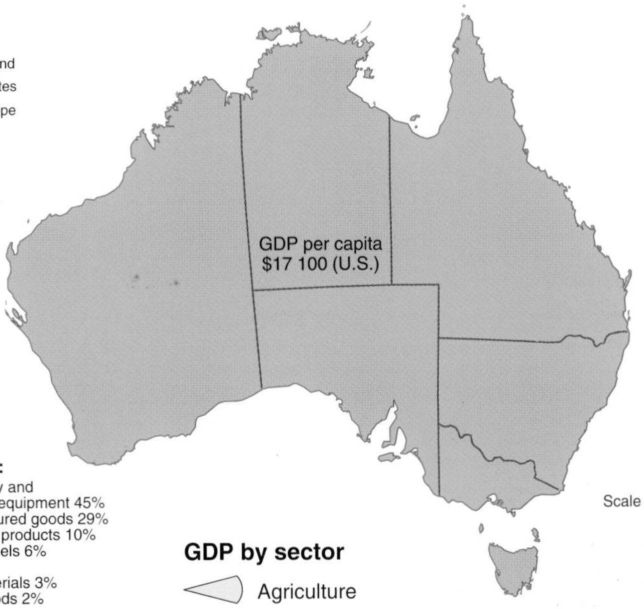

GDP by sector

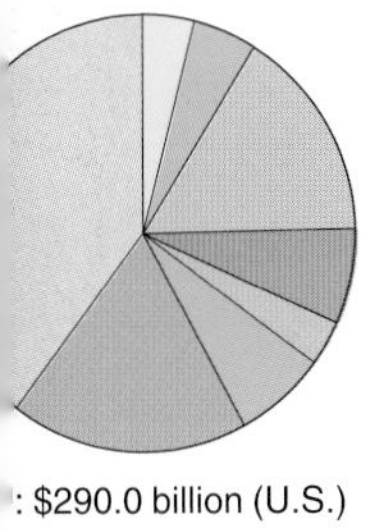

Imports:
Machinery and transport equipment 45%
Manufactured goods 29%
Chemical products 10%
Mineral fuels 6%
Food 5%
Raw materials 3%
Other goods 2%

Exports:
Raw materials 23%
Manufactured goods 21%
Mineral fuels 20%
Food 18%
Machinery and transport equipment 9%
Other goods 9%

GDP by sector
- Agriculture
- Mining
- Manufacturing
- Construction
- Other secondary
- Transportation, communications
- Trade
- Other tertiary

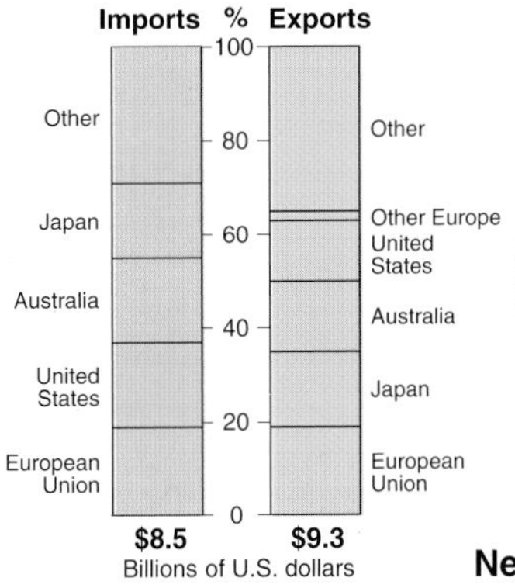

GDP by sector

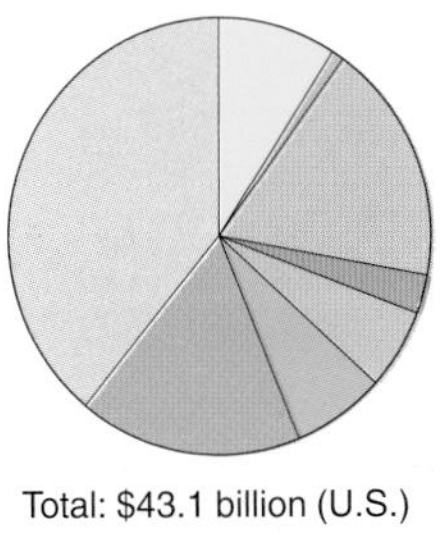

New Zealand

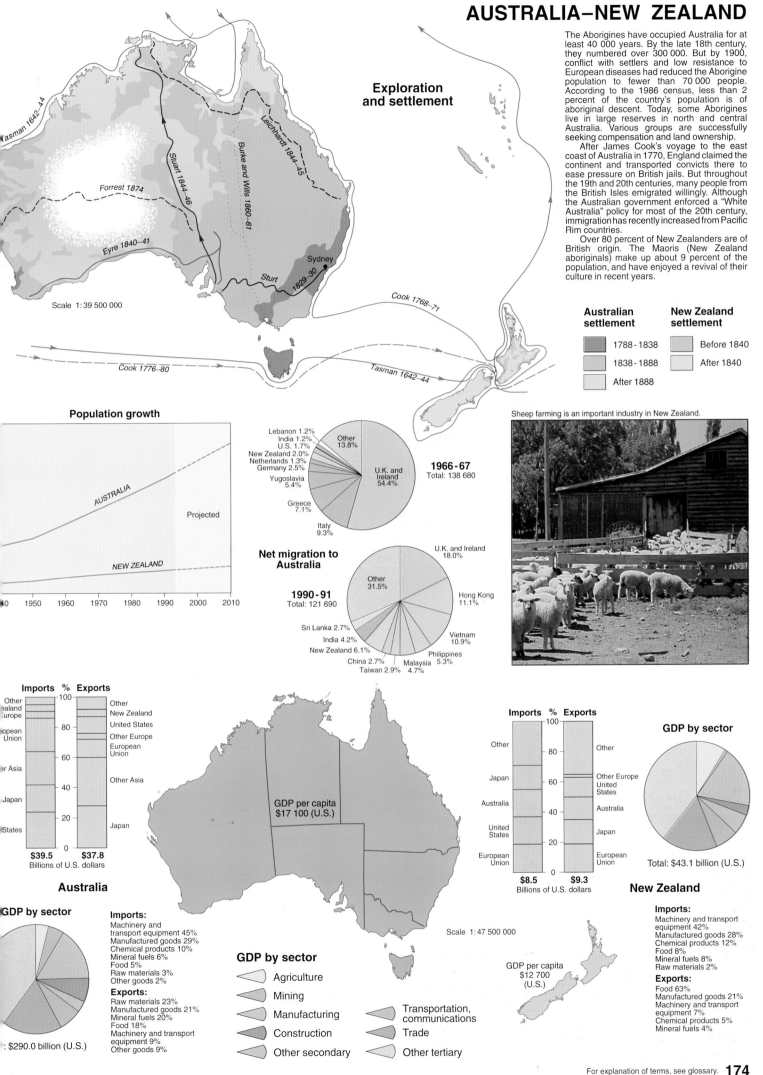

Imports:
Machinery and transport equipment 42%
Manufactured goods 28%
Chemical products 12%
Food 8%
Mineral fuels 8%
Raw materials 2%

Exports:
Food 63%
Manufactured goods 21%
Machinery and transport equipment 7%
Chemical products 5%
Mineral fuels 4%

WORLD-REGIONAL

Although the term Pacific Rim originally referred only to countries with coastlines bordering the Pacific Ocean, it has come to include Southeast and South Asia, in addition to East Asia, North and South America, Australia and New Zealand. The increased awareness of this region in North America reflects its increasing political, cultural and economic significance in both Canada and the United States. Asia promises to be the economic powerhouse of the future and to provide new markets for North American goods. By the year 2000, it is estimated that Asia will have 60 percent of the world's people, 50 percent of its production, and 40 percent of its consumption, and a GNP larger than that of Western Europe. Over the last 20 years, immigration, international investment and the work of organizations such as Asia-Pacific Economic Co-operation (APEC) have strengthened political, cultural and economic ties in this region.

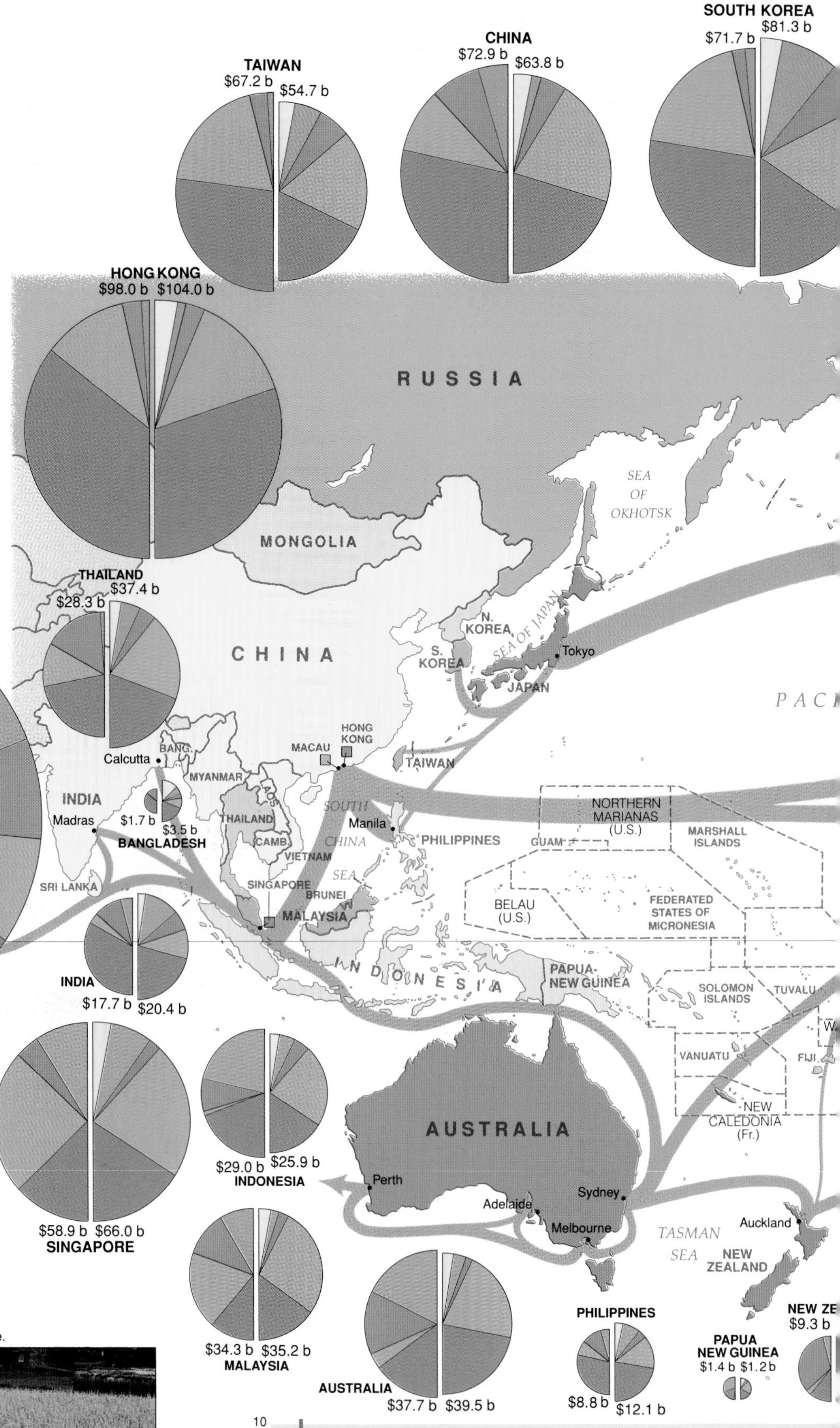

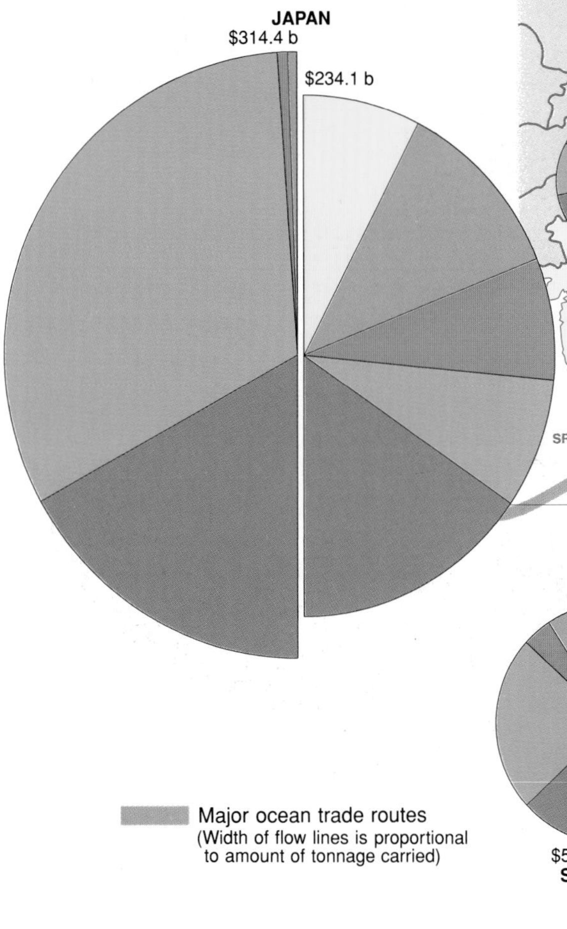

Major ocean trade routes
(Width of flow lines is proportional to amount of tonnage carried)

Although Japan produces enough rice for domestic consumption, under the 1993 GATT agreement it permitted rice imports for the first time.

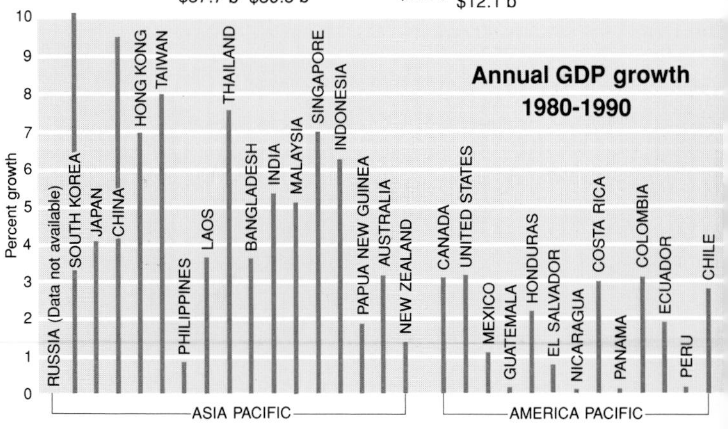

PACIFIC RIM

Honda plant in Alliston, Ontario

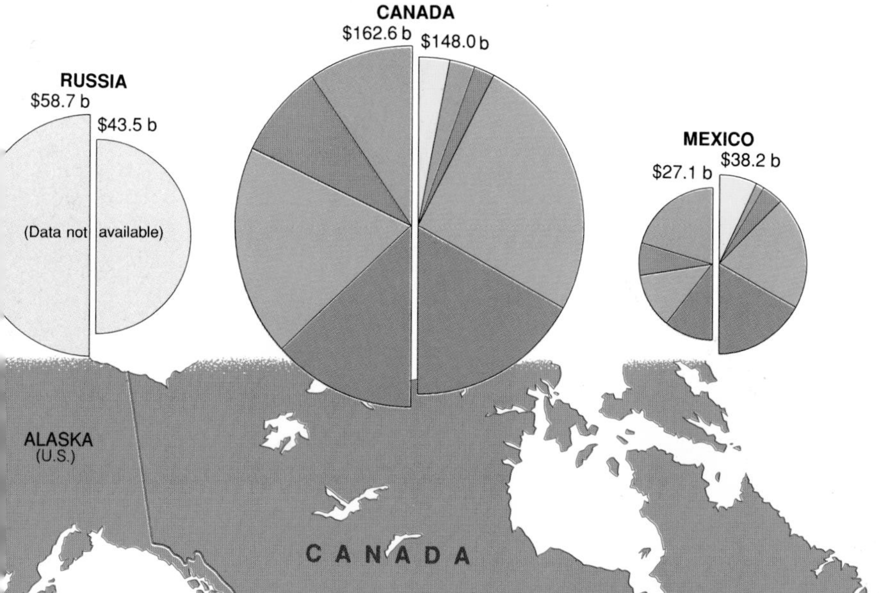

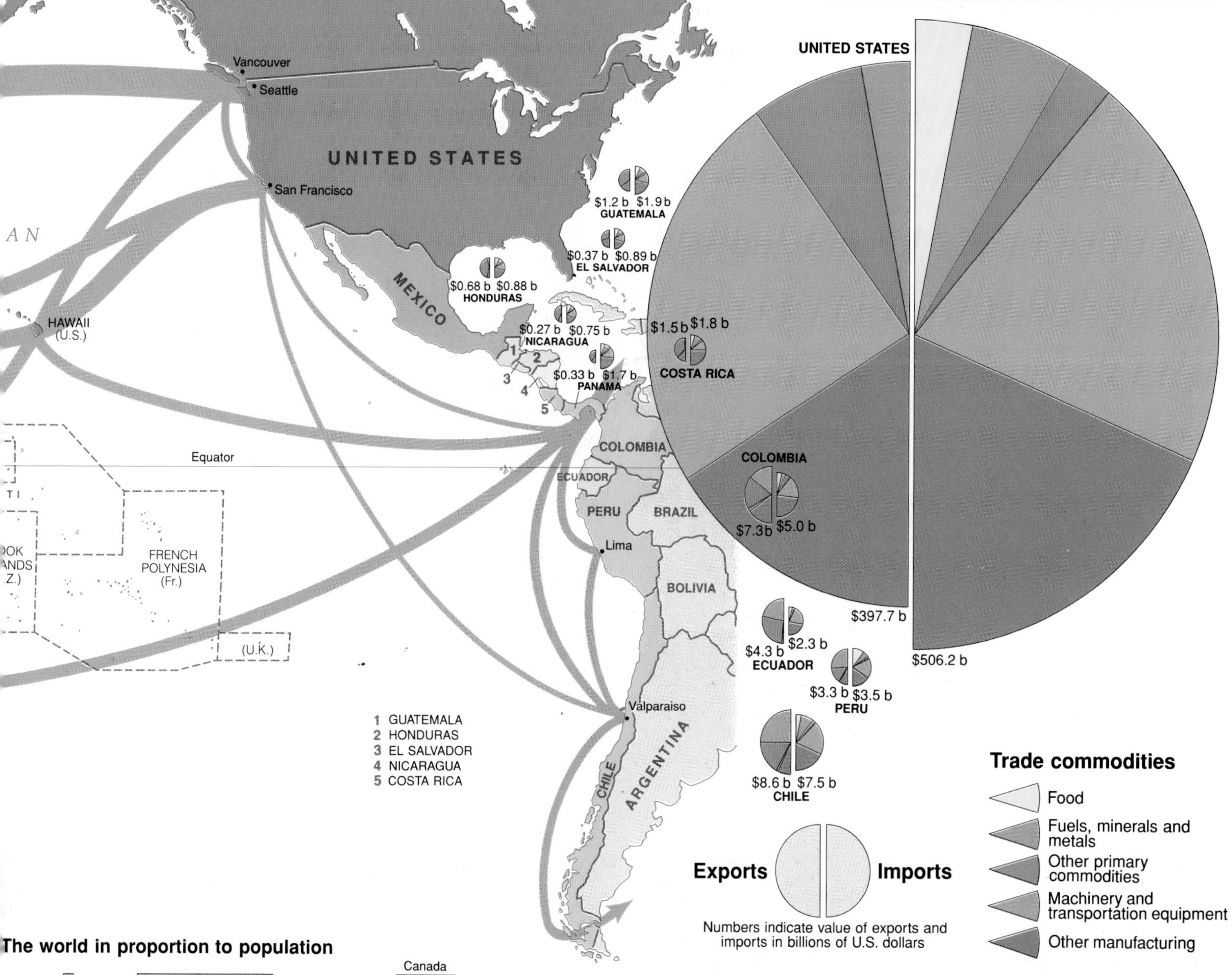

Trade commodities

- Food
- Fuels, minerals and metals
- Other primary commodities
- Machinery and transportation equipment
- Other manufacturing

The world in proportion to population

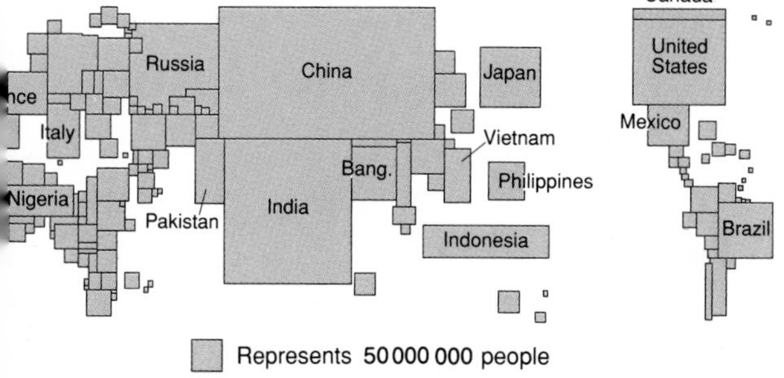

Represents 50 000 000 people

Trade and market access are critical to the economic future of Pacific Rim countries. Trading blocs (such as the EU, ASEAN and NAFTA) can prevent this free flow of goods. Many Pacific Rim companies have become multinational corporations by establishing branches in other countries. This strategy allows these companies to work within the restrictive economic environments created by the blocs.

GDP per capita ($U.S.)

- $10 000 or more
- 2000-9999
- 1000-1999
- 500-999
- Less than 500

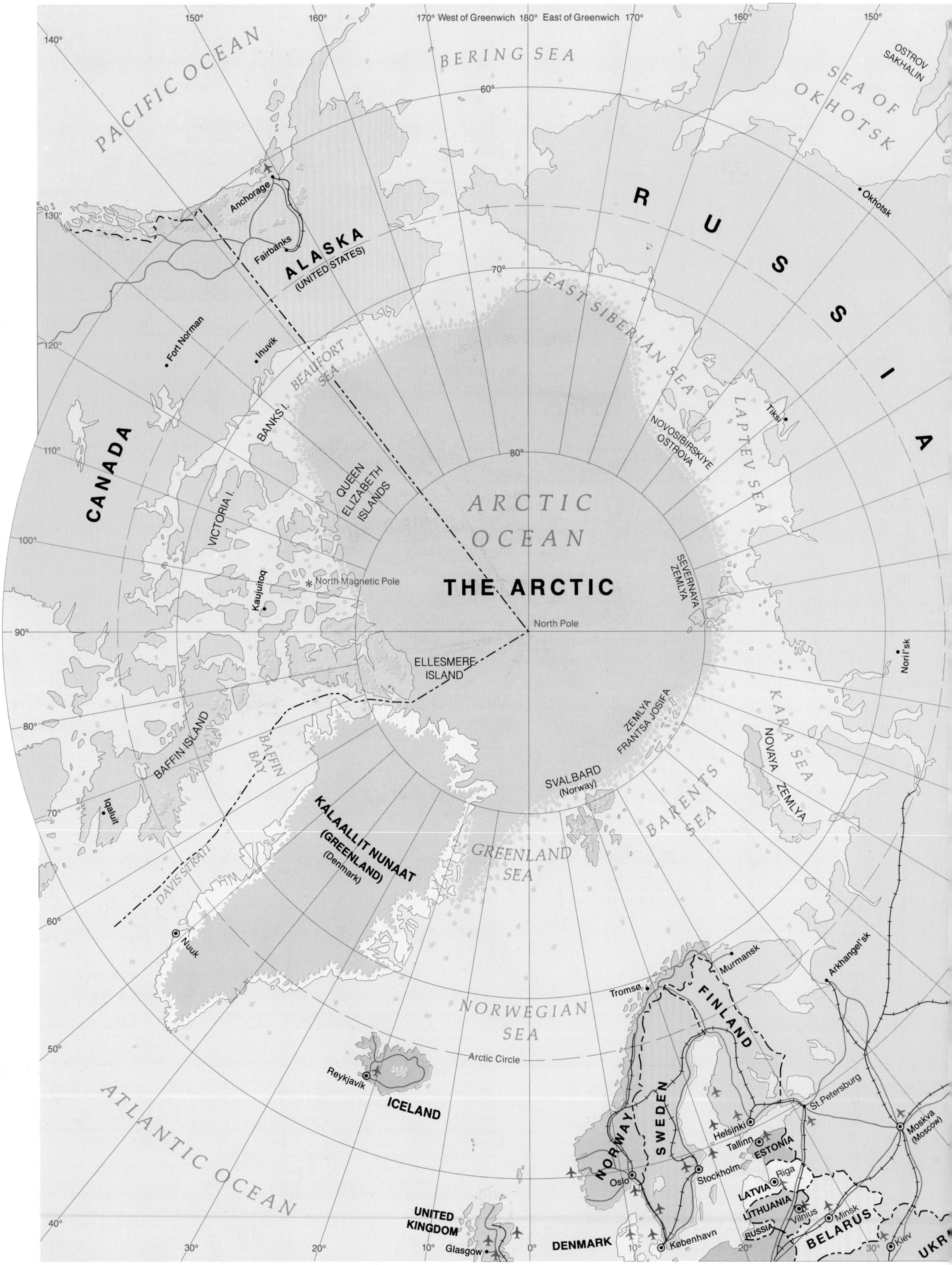
170° West of Greenwich 180° East of Greenwich 170°
PACIFIC OCEAN
BERING SEA
SEA OF OKHOTSK
OSTROV SAKHALIN
Okhotsk
Anchorage
Fairbanks
ALASKA (UNITED STATES)
RUSSIA
EAST SIBERIAN SEA
Fort Norman
Inuvik
BEAUFORT SEA
BANKS I.
CANADA
NOVOSIBIRSKIYE OSTROVA
LAPTEV SEA
Tiksi
QUEEN ELIZABETH ISLANDS
VICTORIA I.
ARCTIC OCEAN
SEVERNAYA ZEMLYA
North Magnetic Pole
Kaujuitoq
THE ARCTIC
North Pole
Noril'sk
ELLESMERE ISLAND
ZEMLYA FRANTSA JOSIFA
KARA SEA
NOVAYA ZEMLYA
BAFFIN ISLAND
BAFFIN BAY
SVALBARD (Norway)
BARENTS SEA
Iqaluit
KALAALLIT NUNAAT (GREENLAND) (Denmark)
GREENLAND SEA
DAVIS STRAIT
Nuuk
Murmansk
Arkhangel'sk
Tromsø
FINLAND
NORWEGIAN SEA
Arctic Circle
Reykjavik
ICELAND
St. Petersburg
ATLANTIC OCEAN
NORWAY
SWEDEN
Helsinki
Tallinn
ESTONIA
Moskva (Moscow)
Oslo
Stockholm
Riga
LATVIA
LITHUANIA
Vilnius
Minsk
RUSSIA
BELARUS
UNITED KINGDOM
Glasgow
DENMARK
København
Kiev
UKR

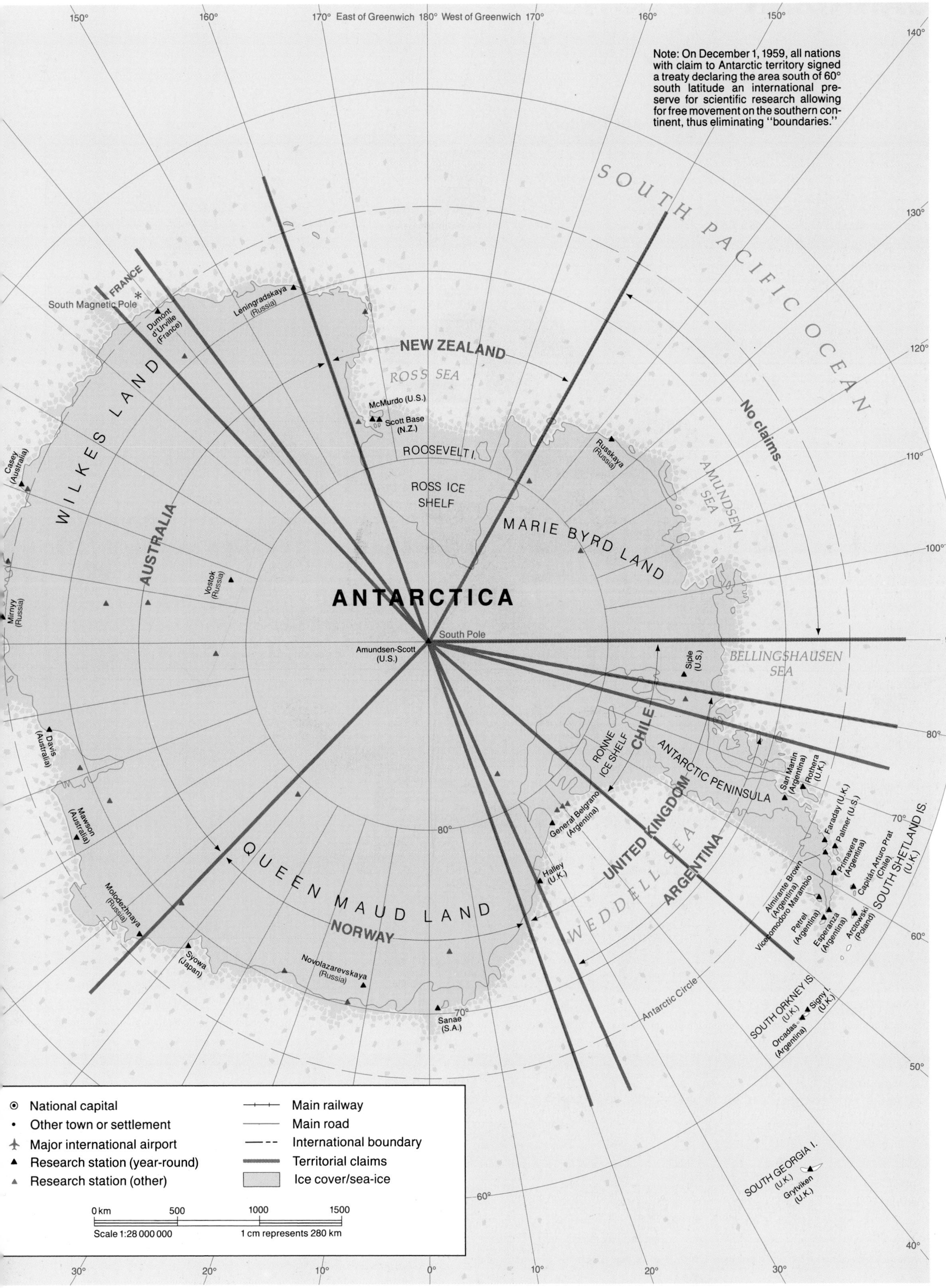

170° East of Greenwich 180° West of Greenwich 170°
150°
160°
140°
130°
120°
110°
100°
90°
80°
70°
60°
50°
40°
30°
20°
10°
0°
Note: On December 1, 1959, all nations with claim to Antarctic territory signed a treaty declaring the area south of 60° south latitude an international preserve for scientific research allowing for free movement on the southern continent, thus eliminating "boundaries."
SOUTH PACIFIC OCEAN
No claims
AMUNDSEN SEA
BELLINGSHAUSEN SEA
FRANCE
South Magnetic Pole
Dumont d'Urville (France)
Leningradskaya (Russia)
NEW ZEALAND
ROSS SEA
McMurdo (U.S.)
Scott Base (N.Z.)
ROOSEVELT I.
ROSS ICE SHELF
Russkaya (Russia)
MARIE BYRD LAND
WILKES LAND
AUSTRALIA
Casey (Australia)
Mirnyy (Russia)
Vostok (Russia)
ANTARCTICA
South Pole
Amundsen-Scott (U.S.)
Siple (U.S.)
CHILE
RONNE ICE SHELF
ANTARCTIC PENINSULA
San Martin (Argentina)
Rothera (U.K.)
Faraday (U.K.)
Palmer (U.S.)
Primavera (Argentina)
Capitán Arturo Prat (Chile)
SOUTH SHETLAND IS. (U.K.)
Almirante Brown (Argentina)
Vicecomodoro Marambio
Petrel (Argentina)
Esperanza (Argentina)
Arctowski (Poland)
UNITED KINGDOM
ARGENTINA
WEDDELL SEA
General Belgrano (Argentina)
Halley (U.K.)
Davis (Australia)
Mawson (Australia)
Molodezhnaya (Russia)
QUEEN MAUD LAND
NORWAY
80°
Syowa (Japan)
Novolazarevskaya (Russia)
Sanae (S.A.)
70°
Antarctic Circle
SOUTH ORKNEY IS.
Orcadas (Argentina)
Signy (U.K.)
SOUTH GEORGIA I. (U.K.)
Grytviken (U.K.)
60°
National capital
Other town or settlement
Major international airport
Research station (year-round)
Research station (other)
Main railway
Main road
International boundary
Territorial claims
Ice cover/sea-ice
0 km 500 1000 1500
Scale 1:28 000 000
1 cm represents 280 km

CANADIAN STATISTICS

Land and Freshwater Area by Province or Territory

Province or Territory	Land and Freshwater Area (km^2)	Percent of Total Area	Fresh Water (Percent of Area)	Province or Territory	Land and Freshwater Area (km^2)	Percent of Total Area	Fresh Water (Percent of Area)
Newfoundland	405 720	4.1	8.4	Saskatchewan	652 330	6.5	12.5
Prince Edward Island	5 660	0.1	0	Alberta	661 190	6.6	2.5
Nova Scotia	55 490	0.6	4.8	British Columbia	947 800	9.5	1.9
New Brunswick	73 440	0.7	1.8	Yukon Territory	483 450	4.8	0.9
Québec	1 540 680	15.5	11.9	Northwest Territories	3 426 320	34.4	3.9
Ontario	1 068 580	10.7	16.6				
Manitoba	649 950	6.5	15.6	**Canada**	**9 970 610**	**100.0**	**7.6**

Source: Energy, Mines and Resources Canada, Geographical Mapping Division, 1983

Climate

Station	Altitude (m)	Temperature (degrees Celsius)						Precipitation			Seasonal Changes		
		January Average	July Average	Mean Annual	Record High	Record Low	Heating Degree-Days	Total (mm)	Snowfall (cm)	Days With Precipitation	Frost-Free Days	Last Frost in Spring	First Frost in Autumn
St. John's	141	−4	15	5	31	−23	4 798	1 511	364	210	130	June 3	Oct. 12
Goose Bay	44	−16	16	0	38	−39	6 494	877	409	176	122	June 4	Oct. 5
Sydney	60	−4	18	6	35	−26	4 433	1 340	288	179	145	May 23	Oct. 16
Yarmouth	38	−3	16	7	30	−21	4 012	1 283	205	157	174	May 2	Oct. 24
Charlottetown	57	−7	18	6	34	−28	4 565	1 128	305	169	150	May 17	Oct. 15
Saint John	30	−7	17	6	34	−30	4 693	1 306	217	149	170	May 3	Oct. 21
Chatham (N.B.)	34	−9	19	5	38	−35	4 827	1 051	309	152	122	May 22	Sept. 21
Schefferville	512	−23	13	−5	32	−51	8 197	723	335	188	73	June 18	Aug. 31
Québec	75	−12	19	4	36	−36	5 024	1 089	326	164	132	May 18	Sept. 28
Montréal (McGill)	57	−9	22	7	36	−34	4 437	999	243	164	183	April 22	Oct. 23
Val d'Or	338	−17	17	2	34	−44	5 955	902	280	180	98	June 4	Sept. 11
Kapuskasing	229	−18	17	1	36	−46	6 330	871	321	186	83	June 13	Sept. 5
Ottawa	126	−11	21	6	38	−36	4 635	851	216	152	142	May 11	Oct. 1
Toronto	116	−4	22	9	41	−33	3 658	790	141	134	192	April 20	Oct. 30
Windsor	194	−4	22	9	38	−26	3 557	836	104	137	173	April 29	Oct. 20
Thunder Bay	196	−15	18	2	36	−41	5 708	738	222	141	101	May 31	Sept. 10
Sioux Lookout	374	−19	18	1	36	−46	6 205	742	237	165	113	May 29	Sept. 20
Winnipeg	240	−18	20	2	41	−45	5 864	535	131	120	118	May 25	Sept. 21
Churchill	35	−28	12	−7	33	−45	9 193	397	192	141	81	June 22	Sept. 12
Regina	574	−17	19	2	43	−50	5 909	398	115	112	107	May 27	Sept. 12
Medicine Hat	721	−12	20	5	42	−46	4 859	348	121	90	125	May 17	Sept. 20
Calgary	1 079	−11	17	3	36	−45	5 948	437	154	113	106	May 28	Sept. 12
Edmonton	676	−15	18	3	34	−48	5 600	447	132	124	127	May 14	Sept. 19
Penticton	342	−3	20	9	41	−27	3 509	296	69	101	143	May 10	Oct. 1
Vancouver	5	2	17	10	33	−18	3 031	1 068	52	158	212	March 31	Oct. 30
Victoria	20	3	16	10	36	−16	2 967	856	45	154	202	April 13	Nov. 2
Estevan Point	6	5	14	9	29	−14	3 192	3 027	34	202	226	April 5	Nov. 18
Dawson	324	−29	16	−5	35	−58	8 232	326	136	120	92	May 26	Aug. 27
Whitehorse	698	−19	14	−1	34	−52	6 855	260	127	118	87	June 5	Sept. 1
Fort Smith	203	−27	16	−4	35	−54	7 803	331	145	127	64	June 15	Aug. 19
Yellowknife	208	−29	16	−6	32	−51	8 601	250	119	114	108	May 30	Sept. 16
Inuvik	61	−29	13	−10	32	−57	10 183	260	174	128	45	June 26	Aug. 11
Iqaluit	21	−26	8	−9	24	−46	9 820	415	247	135	59	June 30	Aug. 29
Alert	63	−32	4	−18	20	−49	13 093	156	145	99	4	July 14	July 19

Source: Environment Canada, Meteorological Branch, 1978

Population Growth in Canada

Census Year	Population (10^3)	Increase From Last Census Year 10^3	Increase From Last Census Year Percent	Annual Rate of Growth (percent)	Census Year	Population (10^3)	Increase From Last Census Year 10^3	Increase From Last Census Year Percent	Annual Rate of Growth (percent)
1851	2 436.3	na	na	na	1941	11 506.7	1 129.9	10.9	1.0
1861	3 229.6	793.3	32.6	2.9	1951[1]	14 009.4	2 502.8	21.8	1.7
1871	3 689.3	459.6	14.2	1.3	1961	18 238.2	2 157.5	13.4	2.5
1881	4 324.8	635.6	17.2	1.6	1966	20 014.9	1 776.6	9.7	1.9
1891	4 833.2	508.4	11.8	1.1	1971	21 568.3	1 553.4	7.8	1.5
1901	5 371.3	538.1	11.1	1.1	1976	22 992.6	1 424.3	6.6	1.3
1911	7 206.6	1 835.3	34.2	3.0	1981	24 343.2	1 350.6	5.9	1.1
1921	8 787.9	1 581.3	21.9	2.0	1986	25 354.1	1 010.9	4.2	0.8
1931	10 376.8	1 588.8	18.1	1.7	1991	27 297.0	1 936.0	7.6	1.5

[1] Newfoundland included for the first time. Excluding Newfoundland, the increase would have been 2 141 358 or 18.6%.
Source: Census of Canada

Population Growth by Province and Territory (10^3)

Year	Nfld.	P.E.I.	N.S.	N.B.	Qué.	Ont.	Man.	Sask.	Alta.	B.C.	Y.T.	N.W.T.	**Canada**
1867	na	88	364	271	1 123	1 525	15	na	na	32	na	45	**3 463**
1871	na	94	388	286	1 191	1 621	25	na	na	36	na	48	**3 689**
1881	na	109	441	321	1 360	1 927	62	na	na	49	na	56	**4 325**
1891	na	109	450	321	1 489	2 114	153	na	na	98	na	99	**4 833**
1901	na	103	460	331	1 649	2 183	255	91	73	179	27	20	**5 371**
1911	na	94	492	352	2 006	2 527	461	492	374	393	9	7	**7 207**
1921	na	88.6	523.8	387.9	2 360.5	2 933.7	610.1	757.5	588.5	524.6	4.1	8.1	**8 787.4**
1931	na	88.0	512.8	408.2	2 874.7	3 431.7	700.1	921.8	731.6	694.3	4.2	9.3	**10 376.7**
1941	na	95.0	578.0	457.4	3 331.9	3 787.7	729.7	896.0	796.2	817.8	5.0	12.0	**11 506.7**
1951	361.4	98.4	642.6	515.7	4 055.7	4 597.6	776.5	831.7	939.5	1 165.2	9.1	16.0	**14 009.4**
1961	457.9	104.6	737.0	597.9	5 259.2	6 236.1	921.7	952.2	1 332.0	1 629.1	14.6	23.0	**18 265.3**
1971	522.1	111.6	789.0	634.6	6 027.8	7 703.1	988.2	926.2	1 627.9	2 184.6	18.4	34.8	**21 568.3**
1981	567.7	122.5	847.4	696.4	6 438.4	8 625.1	1 026.2	968.3	2 237.7	2 744.5	23.2	45.7	**24 343.2**
1986	568.3	126.6	873.2	710.4	6 540.3	9 113.5	1 071.2	1 010.2	2 375.3	2 889.2	23.5	52.2	**25 354.1**
1990	573.4	130.3	894.2	723.2	6 769.0	9 743.3	1 091.6	999.5	2 471.6	3 126.6	26.1	53.8	**26 602.6**
1991	568.0	130.0	900	724	6 896	10 085	1 092.0	989.0	2 546.0	3 282	26.9	54.8	**27 297**
1992	577.5	130.5	906.3	729.3	6 925.2	10 098.6	1 096.8	993.2	2 562.7	3 297.6	27.9	56.5	**27 402.2**

Source: Census of Canada

Demographics of Canada

Period	Total Population Growth (10^3)	Births (10^3)	Deaths (10^3)	Natural Increase (10^3)	Natural Increase (percent)	Immigration (10^3)	Emigration (10^3)	Net Migration (10^3)	Net Migration Rate (percent)	Population at End of Period (10^3)
1851–1861	793	1 281	670	611	77.0	352	170	182	23.0	3 230
1861–1871	460	1 370	760	610	132.6	260	410	–150	–32.6	3 689
1871–1881	636	1 480	790	690	108.5	350	404	–54	–8.5	4 325
1881–1891	508	1 524	870	654	128.7	680	826	–146	–28.7	4 833
1891–1901	538	1 548	880	668	124.2	250	380	–130	–24.2	5 371
1901–1911	1 835	1 925	900	1 025	55.9	1 550	740	810	44.1	7 207
1911–1921	1 581	2 340	1 070	1 270	80.3	1 400	1 089	311	19.7	8 788
1921–1931	1 589	2 420	1 060	1 360	85.6	1 200	970	230	14.5	10 377
1931–1941	1 130	2 294	1 072	1 222	108.1	149	241	–92	–8.1	11 507
1941–1951[1]	2 503	3 212	1 220	1 992	92.3	548	382	166	7.7	14 009
1951–1956[2]	2 071	2 106	633	1 473	71.1	783	185	598	28.9	16 081
1956–1961	2 157	2 362	687	1 675	77.7	760	378	482	22.3	18 238
1961–1966	1 777	2 249	731	1 518	85.4	539	280	259	14.6	20 015
1966–1971	1 553	1 856	766	1 090	70.2	890	427	463	29.8	21 568
1971–1976	1 424	1 758	823	934	65.6	841	352	489	34.4	22 993
1976–1981	1 350	1 820	842	978	75.9	588	278	310	24.1	24 343
1981–1986	1 011	1 873	885	988	97.7	500	235	264	26.1	25 354
1986–1991	1 943	2 328	1 142	1 186	61.0	2 396	1 640	756	39.0	27 297

[1] In the 1941–1951 period, Newfoundland demographic information is included for the first time, but for 1949–50 and 1950–51 only.
[2] Starting in 1951, the census was taken every five years.
Source: Statistics Canada, Ottawa, 1993

Demographics, Provinces and Territories, 1992

			Canada	Nfld.	P.E.I.	N.S.	N.B.	Qué.	Ont.	Man.	Sask.	Alta.	B.C.	Yukon	N.W.T.
Birth Rate/1 000	1992	(P)	15.1	13.5	15.7	14.3	13.6	14.6	15.4	16.0	16.1	16.6	14.2	20.0	27.4
Death Rate/1 000	1992	(P)	7.3	6.9	8.7	8.5	7.7	7.2	7.3	8.4	8.4	5.7	7.4	4.4	4.1
Natural Increase Rate/1 000	1992	(P)	7.9	6.6	7.0	5.9	6.0	7.4	8.1	7.6	7.8	10.9	6.8	15.7	23.3
International Migration															
Immigration Rate/1 000	1992	(P)	8.6	1.2	1.1	2.0	1.0	6.9	12.4	4.4	2.6	6.6	10.3	3.7	2.0
Emigration Rate/1 000	1992	(PP)	1.4	0.3	0.2	0.5	1.0	0.7	1.8	1.8	0.8	2.2	1.6	2.2	1.8
Net Migration Rate/1 000	1992	(PP)	7.1	1.0	0.9	1.5	0.0	6.2	10.6	2.5	1.8	4.4	8.6	1.5	0.2
Interprovincial Migration															
In-migration Rate/1 000	1992	(PP)	13.1	18.0	24.4	23.3	19.6	4.1	8.4	16.7	19.5	26.9	25.5	96.1	70.7
Out-migration Rate/1 000	1992	(PP)	13.1	21.9	35.9	23.2	22.5	6.1	8.7	24.3	29.3	25.5	14.3	71.4	72.3
Net Migration Rate/1 000	1992	(PP)	0.0	–3.9	–11.4	0.0	–2.9	–2.0	-0.3	–7.6	–9.8	1.4	11.2	24.8	–1.6
Total Net Migration Rate/1 000	1992	(PP)	7.1	–2.9	–10.5	1.5	–3.0	4.2	10.3	–5.1	–8.0	5.8	19.8	26.3	–1.4
Total Growth Rate/1 000	1992	(PP)	15.0	3.7	–3.5	7.4	3.0	11.6	18.4	2.5	–0.3	16.8	26.6	42.0	21.9

(P) Preliminary data
(PP) Preliminary postcensal estimates
Source: Statistics Canada—Catalogue No. 91–210, Vol. 10.

Population of Census Metropolitan Areas

City (in order of population, 1991)	Population (10^3)						Percent Change 1986–91
	1951	1961	1971	1981	1986	1991	
Toronto	1 261.9	1 919.4	2 628.0	2 998.9	3 432.0[2]	3 893.0	13.4
Montréal	1 539.3	2 215.6	2 743.2	2 828.4	2 921.4	3 127.2	7.0
Vancouver	586.2	826.8	1 082.4	1 268.2	1 380.7	1 602.5	16.1
Ottawa-Hull	311.6	457.0	602.5	718.0	819.3	920.9	12.4
Edmonton	193.6	359.8	495.7	657.1	774.0[2]	839.9	8.5
Calgary	142.3	279.1	403.3	592.7	671.5[2]	754.0	12.3
Winnipeg	357.2	476.5	540.3	548.8	625.3	652.4	4.3
Québec	289.3	379.1	480.5	576.1	603.3	645.6	7.0
Hamilton	281.9	401.1	498.5	542.1	557.0	599.8	7.7
London	167.7	226.7	286.0	283.7	342.3	381.5	11.5
St. Catharines-Niagara	189.0	257.8	303.4	304.4	343.3	364.6	6.2
Kitchener	107.5	154.9	226.8	287.8	311.2	356.4	14.5
Halifax	138.4	193.4	222.6	277.7	295.9[2]	320.5	8.3
Victoria	114.9	155.8	195.8	233.5	255.2[2]	288.0	12.9
Windsor	182.6	217.2	258.6	246.1	254.0	262.1	3.2
Oshawa	na	na	120.3	154.2	203.5	240.1	18.0
Saskatoon	55.7	95.6	126.4	154.2	200.7	210.0	4.6
Regina	72.7	113.7	140.7	164.3	186.5	191.7	2.8
St. John's, Nfld.	80.9	106.7	131.8	154.8	161.9	171.9	6.2
Chicoutimi-Jonquière	91.2	127.6	133.7	135.2	158.5	160.9	1.6
Sudbury	80.5	127.4	155.4	149.9	148.9	157.6	5.8
Sherbrooke	na	na	na	74.1[1]	130.0	139.2	7.1
Trois-Rivières	46.1	53.5	55.9	111.5	128.9	136.3	5.8
Saint John, N.B.	80.7	98.1	106.7	114.0	121.3	125.0	3.1
Thunder Bay	73.7	102.1	112.1	121.4	122.2	124.4	1.8

[1] Sherbrooke was not a CMA until the 1986 census. Consequently, the 1981 figure represents only the city of Sherbrooke and not the CMA.
[2] indicates that a boundary change in this year may affect the population number.
Source: Statistics Canada, Ottawa, 1991

Population by Age Group (10^3)

Year	Total population	under 5 years	5–9	10–14	15–24	25–34	35–44	45–54	55–64	65 years and over
1911	7 207	890	785	702	1 398	1 219	863	620	393	336
1931	10 377	1 074	1 133	1 074	1 952	1 495	1 334	1 075	662	576
1951	14 010	1 722	1 398	1 131	2 147	2 174	1 869	1 407	1 077	1 086
1971	21 568	1 817	2 254	2 310	4 004	2 890	2 527	2 292	1 731	1 745
1991	27 296	1 907	1 908	1 878	3 831	4 866	4 372	2 966	2 400	3 170

Source: Census of Canada, 1991

Projected Population by Age Group (10^3)

Year	Total population	under 5 years	5–9	10–14	15–24	25–34	35–44	45–54	55–64	65 years and over
2011	30 325	1 622	1 632	1 694	3 799	3 985	4 042	4 684	4 066	4 800
2036	30 998	1 455	1 492	1 558	3 357	3 566	3 810	4 006	3 809	7 945

Source: Statistics Canada, 1991

Refugees to Canada, 1959 to 1992

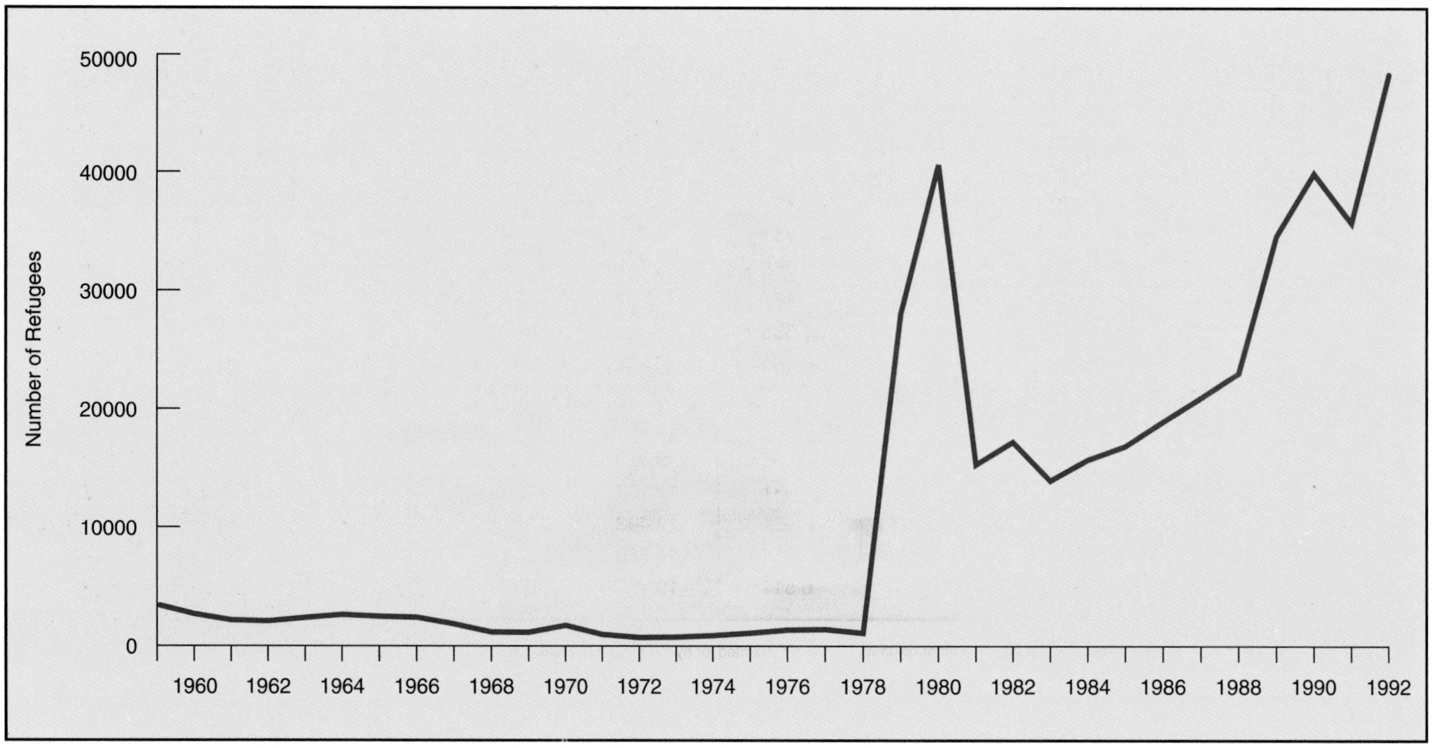

Source: Employment and Immigration Canada

Refugees to Canada by Country

Country	1980	1986	1990
Eastern Europe	**4 062**	**5 388**	**15 381**
Czechoslovakia	1 015	697	1 149
Hungary	296	545	500
Poland	477	3 620	11 874
Romania	307	442	1 015
USSR	1 914	40	1 151
Middle East	**37**	**1 139**	**3 515**
Iran	16	874	2 013
Iraq	5	183	514
Lebanon	10	26	725
Africa	**191**	**1 249**	**4 048**
Ethiopia	72	905	2 173
Ghana	0	38	145
Nigeria	2	5	28
Somalia	6	35	1 069
South Africa	16	53	50
Sudan	9	19	171
Uganda	3	54	71
Indochina	**34 637**	**6 065**	**6 547**
Cambodia	3 261	1 665	712
Laos	6 264	617	565
Vietnam	25 112	3 783	5 270

Country	1980	1986	1990
Central America (including Mexico and the Caribbean)	**308**	**3 934**	**5 393**
Cuba	293	41	8
El Salvador	1	2 459	3 750
Guatemala	2	710	76
Haiti	8	6	41
Honduras	0	39	185
Nicaragua	3	670	548
South America	**396**	**407**	**811**
Argentina	21	9	12
Chile	355	182	667
Guyana	1	154	12
Uruguay	14	22	3
Other	**717**	**965**	**3 483**
Afghanistan	7	539	972
Sri Lanka	1	265	1 234
Total Refugees	**40 348**	**19 147**	**39 628**

Source: Employment and Immigration Canada, Ottawa, 1991

Origin of Immigrants

Year[1]	Total Europe	Total Asia	Hong Kong	United States	Africa	Other North, Central, and South America	Australasia	All other countries	Total
1955	90 771	3 662	10 395	1 700	2 625				109 946
1960	82 706	4 218	11 247	1 657	3 115				104 111
1965	107 816	11 684	15 143	2 711	6 309				146 758
1970	75 328	21 451	24 424	4 388	9 666				147 713
1975	72 898	47 382	20 155	2 174	27 472				187 881
1980	41 168	71 602	9 926	1 555	11 612				143 117
1985	18 859	38 597	7 380	6 669	3 545	15 504	506	622	84 302
1986	22 709	41 600	5 893	7 275	4 770	21 638	503	724	99 219
1987	37 563	67 337	16 170	7 967	8 501	28 901	753	1 076	152 098
1988	40 689	81 136	23 281	6 537	9 380	22 365	745	1 077	161 929
1989	52 105	93 261	19 908	6 931	12 199	25 464	894	1 147	192 001
1990	51 945	111 744	29 261	6 084	13 442	28 368	988	1 659	214 230
1991	46 664	116 547	21 305	6 167	15 668	36 449	943	2 113	224 551
1992	44 944	135 418	35 762	6 511	19 493	38 708	1 101	2 525	248 700

[1] Prior to 1980, any differences between the total and the sum of the other columns would be accounted for by the remaining blank columns.
Source: 1991 Census (Catalogue 93-332)

Destination of Immigrants

Year	Nfld	P.E.I.	N.S.	N.B.	Qué.	Ont.	Man.	Sask.	Alta.	B.C.	Yukon	N.W.T.	**Canada**
1950	204	164	1 167	663	13 575	39 041	3 654	2 897	6 424	6 091	13	19	**73 912**
1960	305	83	1 210	634	23 774	54 491	4 337	2 087	6 949	10 120	39	81	**104 111**
1970	630	185	2 007	1 070	23 261	80 732	5 826	1 709	10 405	21 683	77	128	**147 713**
1975	1 106	235	2 124	2 093	28 042	98 471	7 134	2 837	16 277	29 272	97	193	**187 881**
1980	541	190	1 616	1 207	22 541	62 264	7 684	3 604	18 841	24 440	96	93	**143 117**
1985	325	113	974	609	14 884	40 730	3 415	1 905	9 001	12 239	36	71	**84 302**
1990	546	176	1 563	842	40 842	113 438	6 637	2 361	18 944	28 726	80	75	**214 230**
1991	644	145	1 495	658	48 153	116 960	5 585	2 430	16 860	31 419	78	124	**224 551**
1992	791	152	2 332	748	48 032	136 216	4 979	2 448	17 310	35 458	123	111	**248 707**

Source: 1991 Census (Catalogue 93-332)

Employment by Industry (10^3)

Year	Primary (non-agricultural)	Manufacturing	Construction	Transportation, Communication, and Other Utilities	Trade	Finance, Insurance, and Real Estate	Services	Public Administration	**Total Non-Agricultural**
1970	216	1 768	467	698	1 328	379	2 040	510	**7 406**
1975	220	1 871	603	812	1 637	474	2 520	665	**8 802**
1980	300	2 111	624	906	1 837	611	3 096	744	**10 229**
1985	289	1 960	579	876	1 985	629	3 630	798	**10 746**
1990	283	2 001	778	951	2 247	755	4 299	831	**12 143**
1991	280	1 865	695	916	2 169	760	4 376	832	**11 892**
1992	257	1 788	681	922	2 155	763	4 408	834	**11 808**

Source: Statistics Canada, 1991, 1992

Ethnic Origin

Origin(s)	Nfld.	P.E.I.	N.S.	N.B.	Qué.	Ont.	Man.	Sask.	Alta.	B.C.	Y.T.	N.W.T.	Canada
Single	**465 650**	**72 930**	**532 845**	**503 820**	**6 237 905**	**6 698 995**	**669 405**	**558 675**	**1 451 000**	**1 952 855**	**14 160**	**41 540**	**19 199 790**
British	442 805	56 405	391 810	236 385	286 080	2 536 515	183 485	160 720	493 195	812 470	5 295	5 885	5 611 050
French	9 700	11 845	55 310	235 010	5 077 830	527 580	53 580	30 070	74 615	68 790	875	1 390	6 146 000
Western European	1 830	1 995	34 395	7 790	55 045	502 865	125 530	137 035	248 850	237 355	1 510	1 265	1 355 485
Northern European	520	195	1 645	1 575	3 930	51 885	14 915	24 400	48 350	65 335	480	355	213 600
Eastern European	580	320	5 030	1 630	60 670	391 720	106 405	84 800	168 660	125 370	760	855	946 810
Southern European	955	120	5 415	2 170	290 160	902 725	22 570	5 520	51 250	97 600	235	305	1 379030
Other European	95	50	1 545	625	78 350	134 220	12 400	1 290	8 560	13 900	50	55	251 140
Arab	280	310	2 780	800	62 960	58 835	960	645	12 680	3 760	10	20	144 050
West Asian	85	25	375	265	23 435	43 785	1 030	3 050	9 110	10	25		81 660
South Asian	885	185	2 315	790	29 240	231 385	8 770	2 945	40 030	103 545	60	155	420 295
East and Southeast Asian	1 090	125	3 330	1 800	84 710	448 225	39 870	11 925	114 860	254 420	245	620	961 225
African	20	25	210	35	2 705	18 410	650	545	1 885	1 925	10	10	26 430
Pacific Islands	—	—	25	—	25	440	25	—	1 395	5 275	20	—	7 215
Latin, Central and South American	15	85	345	100	28 290	37 350	2 905	1 135	7 170	8 090	25	15	85 535
Caribbean	60	—	190	105	26 755	59 860	1 745	275	3 615	1 745	10	30	94 395
Black	110	25	10 825	1 045	41 165	150 685	4 530	1 125	9 125	5 875	25	70	224 620
Inuit	2 670	10	95	70	6 850	620	365	75	560	290	60	18 430	30 085
Métis	320	25	250	170	8 670	4 680	18 850	14 145	20 485	5 065	165	2 320	75 150
North American Indian	2 350	370	7 185	4 030	49 880	65 710	55 125	52 050	47 400	69 060	3 550	8 665	365 375
Other origins	1 260	810	9 775	9 405	21 145	531 500	15 690	29 495	95 250	63 860	780	1 065	780 035
Multiple origins	**98 290**	**55 165**	**358 105**	**212 675**	**572 395**	**3 278 055**	**409 985**	**417 360**	**1 068 180**	**1 294 650**	**13 495**	**15 890**	**7 794 250**
Total population	**563 935**	**128 100**	**890 950**	**716 495**	**6 810 300**	**9 977 050**	**1 079 390**	**976 040**	**2 519 180**	**3 247 505**	**27 655**	**57 430**	**26 994 045**

Source: Statistics Canada (93-315), 1991

Labour Force by Province, 1992

Industry	Employment (10^3)									
	Nfld.	P.E.I.	N.S.	N.B.	Qué.	Ont.	Man.	Sask.	Alta.	B.C.
Agriculture	—	3.7	7.4	6.7	64.1	113.9	40.7	76.0	89.4	29.3
Forestry	—	—	3.6	3.0	13.7	9.7	—	—	3.9	22.5
Fishing and trapping	10.2	2.5	8.9	5.0	—	—	—	—	—	6.9
Mining industries	3.2	—	4.1	3.6	19.2	29.6	6.3	10.4	67.8	12.6
Manufacturing	16.1	4.0	41.1	36.5	515.5	840.9	51.4	25.6	92.1	164.9
Construction	10.1	3.2	18.0	16.0	146.9	253.9	20.7	18.6	81.8	111.5

Source: Statistics Canada, 1993

Religion in Canada

Religion	Nfld.	P.E.I.	N.S.	N.B.	Qué.	Ont.	Man.	Sask.	Alta.	B.C.	Y.T.	N.W.T.	Canada
Catholic	**208 900**	**60 635**	**331 340**	**386 580**	**5 861 205**	**3 544 515**	**327 785**	**316 935**	**666 785**	**603 080**	**5 580**	**21 910**	**12 335 255**
Roman Catholic	208 855	60 620	331 015	386 485	5 855 980	3 506 820	293 955	296 725	640 485	595 315	5 535	21 830	12 203 620
Ukrainian Catholic	35	15	320	95	3 990	36 545	33 410	20 155	26 170	7 525	45	8	128 390
Protestant	**343 960**	**62 000**	**482 180**	**287 435**	**398 725**	**4 428 305**	**550 120**	**521 680**	**1 219 240**	**1 446 475**	**11 905**	**28 680**	**9 780 710**
Anglican	147 520	6 690	128 375	61 245	96 065	1 059 910	94 190	69 930	173 160	328 580	4 090	18 355	2 188 115
Baptist	1 360	5 315	98 490	80 935	27 505	264 625	20 170	15 415	63 735	84 090	995	715	663 360
Jehovah's Witnesses	2 415	465	5 630	3 540	33 420	56 535	5 720	8 130	18 360	33 665	285	205	168 375
Latter-day Saints (Mormons)	195	235	2 865	920	3 495	23 240	1 910	2 845	46 830	17 945	160	125	100 770
Lutheran	410	115	11 470	1 585	10 700	227 915	55 130	82 160	137 145	108 190	665	720	636 210
Mennonite and Hutterite	30	20	560	240	1 670	52 645	73 450	29 190	32 305	39 150	110	80	229 460
Pentecostal	40 125	1 320	10 730	22 770	28 955	167 175	21 210	17 710	52 990	70 620	600	2 220	436 435
Presbyterian	2 155	10 990	31 225	10 135	18 865	422 160	16 050	11 575	48 385	63 985	360	400	636 295
Salvation Army	44 490	285	4 920	1 430	1 220	38 820	2 835	2 185	5 780	10 120	90	180	112 345
United Church	97 395	25 995	153 040	75 570	62 030	1 410 535	200 375	222 120	419 600	420 755	2 400	3 300	3 093 120
Protestant, not otherwise specified	1 010	5 185	17 890	12 160	56 545	323 520	22 405	17 185	81 555	89 645	955	890	628 945
Eastern Orthodox	**365**	**115**	**2 290**	**765**	**89 285**	**187 910**	**20 655**	**19 505**	**42 720**	**23 540**	**80**	**160**	**387 395**
Jewish	**125**	**85**	**1 950**	**875**	**97 735**	**175 640**	**13 670**	**1 375**	**9 945**	**16 565**	**45**	**60**	**318 070**
Eastern non-Christian	**1 225**	**335**	**4 720**	**1 660**	**97 550**	**379 625**	**16 590**	**6 515**	**78 460**	**159 965**	**315**	**500**	**747 455**
Buddhist	105	60	1 485	365	31 640	65 325	5 260	1 885	20 745	36 430	45	80	163 415
Hindu	445	20	970	605	14 120	106 705	3 470	1 680	10 765	18 140	10	70	157 015
Islam	305	60	1 435	250	44 930	145 560	3 525	1 185	31 000	24 930	30	50	253 260
Sikh	130	65	330	45	4 525	50 085	3 495	565	13 550	74 550	40	60	147 440
Native Indian or Inuit	**10**	**20**	**45**	**75**	**165**	**2 780**	**1 175**	**1 990**	**2 030**	**2 320**	**175**	**60**	**10 840**
No religious affiliation	**9 275**	**4 880**	**68 010**	**38 740**	**262 800**	**1 247 640**	**148 170**	**107 225**	**496 150**	**987 985**	**9 470**	**6 015**	**3 386 365**
Total	**563 935**	**128 100**	**890 950**	**716 495**	**6 810 300**	**9 977 050**	**1 079 390**	**976 040**	**2 519 180**	**3 247 505**	**27 655**	**57 430**	**26 994 045**

Source: Statistics Canada, 1991

Unemployment Rate by Province[1]

Annual Average	Nfld.	P.E.I.	N.S.	N.B.	Qué.	Ont.	Man.	Sask.	Alta.	B.C.	**Canada**
1966	5.8	0.0	4.7	5.3	4.1	2.6	2.8	1.5	2.5	4.6	**3.4**
1970	7.3	0.0	5.3	6.3	7.0	4.4	5.3	4.2	5.1	7.7	**5.7**
1975	14.0	8.0	7.7	9.8	8.1	6.3	4.5	2.9	4.1	8.5	**6.9**
1980	13.3	10.6	9.7	11.0	9.8	6.8	5.5	4.4	3.7	6.8	**7.5**
1985	20.8	13.3	13.6	15.1	11.8	8.0	8.2	8.1	10.0	14.1	**10.5**
1986	19.2	13.4	13.1	14.3	11.0	7.0	7.7	7.7	9.8	12.5	**9.5**
1987	17.9	13.2	12.3	13.1	10.3	6.1	7.4	7.4	9.6	11.9	**8.8**
1988	16.4	13.0	10.2	12.0	9.4	5.0	7.8	7.5	8.0	10.3	**7.8**
1989	15.8	14.1	9.9	12.5	9.3	5.1	7.5	7.4	7.2	9.1	**7.5**
1990	17.1	14.9	10.5	12.1	10.1	6.3	7.2	7.0	7.0	8.3	**8.1**
1991	18.4	16.8	12.0	12.7	11.9	9.6	8.8	7.4	8.2	9.9	**10.3**
1992	20.2	17.7	13.1	12.8	12.8	10.8	9.6	8.2	9.5	10.4	**11.3**

[1] No data available for Yukon or Northwest Territories
Source: Statistics Canada (11-210), 1991/1992

Participation Rates by Province, 1992

Province	Males over 25	Females over 25	All population over 25
Canada	75	57	66
Nfld.	61	47	54
N.S.	68	52	60
N.B.	67	52	59
P.E.I.	73	59	66
Qué.	72	54	63
Ont.	75	60	67
Man.	74	58	66
Sask.	75	59	67
Alta.	80	64	72
B.C.	74	59	66

Source: Statistics Canada, Ottawa, 1993, *Canadian Economic Observer*

Family Structure

Province	Lone-parent Families	Common-law Families	Married Couples	Total Number of Families	Average Family Size	Families Without Children at Home
Nfld.	11.9	6.6	81.4	150 715	3.3	25.1
P.E.I.	12.9	5.9	81.1	33 895	3.2	30.3
N.S.	13.5	8.2	78.2	244 615	3.1	33.8
N.B.	13.4	8.0	78.6	198 010	3.1	31.9
Qué.	14.3	16.3	69.4	1 883 230	3.0	34.1
Ont.	12.6	6.7	80.7	2 726 740	3.1	35.0
Man.	13.1	7.4	79.4	285 935	3.1	35.8
Sask.	11.7	6.9	81.4	257 555	3.2	36.7
Alta.	12.4	8.9	78.6	667 985	3.1	34.4
B.C.	12.1	9.6	78.3	887 660	3.0	40.3
Canada[1]	**12.9**	**9.8**	**77.2**	**7 356 170**	**3.1**	**35.1**

[1] Includes Yukon and Northwest Territories.
Source: Statistics Canada, 1991 Census

AIDS in Canada (as of July 1992)

Province or Territory	Number of Reported Cases			
	Males	Females	Total	Deaths
Nfld.	22	4	26	16
P.E.I.	4	0	4	4
N.S.	89	8	97	62
N.B.	45	6	51	19
Qué.	1 796	219	2 015	1 090
Ont.	2 504	96	2 600	1 937
Man.	71	3	74	49
Sask.	54	3	57	38
Alta.	401	18	419	144
B.C.	1 189	23	1 212	752
Y.T.	2	0	2	1
N.W.T.	3	0	3	0
Canada	**6 180**	**380**	**6 560**	**4 112**

Source: Federal Centre for AIDS (Health and Welfare Canada)

Educational Attainment

Province or Territory	Year	Percent of Population Over 25			
		Less than Grade 9	Between Grades 9 and 13	Some Post-Secondary[1]	University Degree
Newfoundland	1951	72.6	26.8	—	0.6
	1971	52.4	35.4	10	2.3
	1991	25.4	37.0	29.8	7.9
Prince Edward Island	1951	55.9	43.1	—	1.0
	1971	42.7	39.0	15.0	3.3
	1991	18.5	38.6	33.5	9.4
Nova Scotia	1951	50.9	47.5	—	1.5
	1971	36.1	43.9	15.6	4.4
	1991	15.5	40.0	33.0	11.5
New Brunswick	1951	63.3	35.6	—	1.1
	1971	49.1	32.5	14.6	3.8
	1991	23.7	38.1	28.8	9.4
Québec	1951	62.8	34.9	—	2.3
	1971	50.7	30.5	13.7	5.2
	1991	23.6	38.6	26.2	11.5
Ontario	1951	50.6	46.4	—	3.0
	1971	34.7	42.7	16.7	5.9
	1991	14.1	39.3	31.9	14.6
Manitoba	1951	55.0	43.6	—	1.5
	1971	39.2	40.4	15.6	4.8
	1991	17.6	40.1	30.8	11.5
Saskatchewan	1951	61.0	37.8	—	1.2
	1971	43.9	37.8	14.6	3.8
	1991	18.8	39.6	31.9	9.8
Alberta	1951	50.6	47.8	—	1.6
	1971	30.1	44.2	19.6	6.1
	1991	10.5	38.4	37.3	13.7
British Columbia	1951	42.2	55.4	—	2.4
	1971	27.2	47.3	19.4	5.7
	1991	10.4	38.7	38.2	12.7
Yukon Territory	1951	43.6	54.4	—	1.9
	1971	28.0	45.5	21.6	4.9
	1991	7.9	30.6	46.7	14.8
Northwest Territories	1951	74.0	23.6	—	2.4
	1971	54.4	27.6	13.3	4.6
	1991	26.5	23.4	32.6	12.5
Canada	1951	**54.9**	**42.9**	—	**2.3**
	1971	**39.4**	**39.2**	**16.0**	**5.3**
	1991	**16.6**	**38.9**	**31.6**	**12.8**

[1] This category was not included in the 1951 census.
Source: Census of Canada, 1991

Federal Government Revenue, Selected Sources ($10^6$$)

Expenditure	1965	1975	1980	1985	1990	1992
Personal Income Tax	2 642	12 709	21 296	34 764	63 110	67 098
Corporate Income Tax	1 759	5 748	8 130	9 210	11 400	11 000
Non-Resident Taxes	170	481	867	1 054	1 550	1 470
General Sales Taxes	1 917	3 515	5 429	9 383	16 700	17 648
Alcohol Taxes	688	1 195	1 510	2 639	3 305	5 130
Customs Duties (on imported goods)	686	1 887	3 188	3 975	4 150	4 055
Total Revenue	9 000	32 354	53 796	83 060	128 894	137 654

Source: Finance Canada, Statistics Canada, 1993

Federal Government Expenditures, Selected ($10^6$$)

Revenue	1965	1975	1980	1985	1990	1992
Protection	1 744	3 397	6 372	11 876	14 985	15 729
Health	475	2 231	4 391	7 134	7 444	7 123
Social Services	247	9 858	19 578	34 445	48 976	54 504
Education	296	1 039	2 513	3 973	3 874	4 123
Environment	11	246	343	422	707	647
Foreign Affairs and Aid	159	748	1 076	2 050	3 495	3 662
Transfers to Provinces	433	2 688	4 387	5 799	10 179	10 146
Debt Charges	1 107	3 968	10 784	25 417	42 630	43 114
Total Expenditures	8 832	37 464	67 829	1 169 111	159 285	168 062
Surplus (Deficit)	168	(5 110)	(14 033)	(33 851)	(30 391)	(30 408)

Source: Finance Canada, Statistics Canada, 1993

National Political Party Representation

Region	Election Year														
	1949	1953	1957	1958	1962	1963	1965	1968	1972	1974	1979	1980	1984	1988	1993
Atlantic Canada — Total Seats	**34**	**33**	**33**	**33**	**33**	**33**	**32**	**32**	**32**	**31**	**32**	**32**	**32**	**32**	**32**
Liberal	26	27	12	8	14	20	15	7	10	13	12	19	7	20	31
Progressive Conservative	7	5	21	25	18	13	18	25	22	17	17	13	25	12	1
New Democratic Party	1	1	–	–	1	–	–	–	–	1	2	–	–	–	–
Québec — Total Seats	**70**	**70**	**71**	**75**	**75**	**75**	**73**	**74**	**73**	**74**	**75**	**75**	**75**	**75**	**75**
Liberal	68	66	62	25	35	47	56	56	56	60	67	74	17	12	19
Progressive Conservative	2	4	9	50	14	8	8	4	2	3	2	1	58	63	1
Bloc Québécois	–	–	–	–	–	–	–	–	–	–	–	–	–	–	54
Others[1]	–	–	–	–	–	–	–	–	–	–	–	–	–	–	1
Ontario — Total Seats	**82**	**85**	**85**	**85**	**85**	**85**	**85**	**88**	**87**	**88**	**95**	**95**	**95**	**99**	**99**
Liberal	56	51	21	15	44	52	51	64	36	55	32	52	15	42	98
Progressive Conservative	25	33	61	67	35	27	25	17	40	25	57	38	66	47	–
New Democratic Party	1	1	3	3	6	6	9	6	11	8	6	5	13	10	–
Reform Party	–	–	–	–	–	–	–	–	–	–	–	–	–	–	1
Western Canada — Total Seats (including Territories)	**71**	**72**	**72**	**72**	**72**	**72**	**72**	**70**	**68**	**68**	**80**	**80**	**80**	**89**	**89**
Liberal	43	27	10	1	7	10	9	28	7	13	3	2	2	8	29
Progressive Conservative	7	9	21	66	49	47	46	26	42	49	60	51	61	48	–
New Democratic Party	11	21	22	5	12	11	12	16	19	6	18	27	17	33	9
Social Credit	10	15	19	–	4	4	5	–	–	–	–	–	–	–	–
Reform Party	–	–	–	–	–	–	–	–	–	–	–	–	–	–	51
Canada — Total Seats in House of Commons	**262**	**265**	**265**	**265**	**265**	**265**	**265**	**265**	**265**	**282**	**282**	**282**	**282**	**295**	**295**
Liberal	193	171	105	49	100	129	131	155	109	141	114	147	40	82	177
Progressive Conservative	41	51	112	208	116	95	97	72	107	95	136	103	211	170	2
New Democratic Party	13	23	25	8	19	17	21	22	31	16	26	32	30	43	9
Social Credit	10	15	19	–	30	24	14	14	15	11	6	–	–	–	–
Bloc Québécois	–	–	–	–	–	–	–	–	–	–	–	–	–	–	54
Reform Party	–	–	–	–	–	–	–	–	–	–	–	–	–	–	52
Others[1]	5	5	4	–	–	–	2	2	2	1	–	–	1	–	1

[1] Includes independents and those who may have broken party ranks and run on a specific issue.
Sources: Beck, J.M. *Pendulum of Power*, Scarborough, Prentice-Hall Canada, Ltd., 1968; Public Archives

WORLD TRADE STATISTICS

Exports by Type, Selected Countries (% of total exports)

Country	Agricultural Products		Mineral Products		Food Beverages Tobacco		Textiles		Wood, Paper		Chemicals		Processed Metals		Other Manufacturing	
	1980	1990	1980	1990	1980	1990	1980	1990	1980	1990	1980	1990	1980	1990	1980	1990
Australia	24.0	14.8	19.1	22.3	22.0	21.2	2.4	2.5	1.2	0.8	9.9	5.4	6.9	10.4	15.2	22.8
Brunei	0.0	0.0	98.2	97.0	0.0	0.0	0.0	0.0	0.0	0.0	0.0	0.0	0.0	0.0	1.7	2.9
Canada	9.7	6.4	18.2	10.9	3.5	3.0	1.0	0.9	17.5	15.9	9.1	8.9	29.4	41.6	11.8	13.9
Chile	8.7	15.2	11.1	12.8	8.9	11.3	0.2	0.8	8.6	8.9	5.4	3.7	2.8	6.3	54.2	41.1
China	na	8.9	na	8.1	na	6.3	na	30.9	na	1.0	na	9.9	na	22.2	na	13.0
Colombia	69.1	38.0	2.0	28.1	7.4	4.1	7.4	10.3	2.1	1.9	5.3	10.3	3.7	2.0	3.0	5.3
Ecuador	18.8	42.0	55.5	44.0	14.4	6.0	6.6	0.6	1.3	1.0	7.9	5.3	1.2	0.8	0.3	0.5
Egypt	19.8	17.8	57.9	22.2	1.5	2.5	10.3	27.8	0.0	0.7	7.1	13.2	2.0	2.6	1.3	13.2
France	6.7	6.3	1.1	0.6	10.4	10.3	6.5	6.3	2.8	3.5	19.2	17.9	40.0	43.7	13.3	11.2
Germany[1]	1.0	1.1	2.6	0.7	4.7	4.2	5.6	5.7	2.8	3.5	17.0	16.2	52.9	58.1	13.5	10.2
Hong Kong	3.0	0.4	2.3	0.2	1.8	2.3	36.7	40.5	1.5	2.8	6.0	5.4	34.4	38.6	14.6	10.1
India	25.1	13.2	14.9	25.8	7.8	5.2	29.5	30.3	0.5	0.3	5.2	10.2	12.1	9.5	4.9	5.6
Indonesia	18.4	11.6	67.6	41.2	2.2	3.3	0.7	13.5	1.5	13.9	5.8	8.1	0.7	3.3	3.1	4.9
Israel	10.2	5.6	30.9	31.1	5.2	5.2	8.3	6.3	1.0	0.7	18.0	16.2	21.9	30.8	4.5	4.1
Italy	3.3	2.2	0.7	0.3	4.5	4.5	17.0	18.5	2.4	2.5	16.0	11.4	42.9	47.3	13.6	13.4
Japan	0.5	0.3	0.3	0.1	1.0	0.5	4.7	2.3	0.9	0.9	8.3	8.8	66.6	77.6	17.6	9.5
Malaysia	27.9	11.2	24.8	17.0	12.9	8.6	2.7	6.0	6.6	6.8	2.4	3.7	12.3	41.2	10.3	5.0
Mexico	24.8	9.8	46.8	34.6	4.3	3.3	2.4	2.6	1.5	1.8	6.1	9.8	8.8	27.6	5.3	10.5
Netherlands	7.7	8.3	8.8	3.4	14.9	15.6	4.7	4.4	2.6	3.6	32.9	24.5	21.7	30.2	6.7	10.0
Peru	8.5	8.7	35.4	24.5	9.5	18.9	7.5	9.3	0.6	0.1	9.1	9.1	2.3	2.7	26.0	26.7
Philippines	11.0	11.2	18.1	5.8	27.3	12.2	7.0	8.0	6.6	5.1	2.8	6.2	4.4	12.9	22.8	38.6
Saudi Arabia	0.0	0.0	96.7	70.1	0.0	0.1	0.0	0.1	0.0	0.0	2.6	23.6	0.0	0.2	0.0	0.4
Singapore	11.3	3.6	1.1	0.4	5.6	3.6	4.3	4.9	3.3	2.2	32.6	25.7	30.6	54.8	11.5	4.8
South Korea	5.3	2.6	0.5	0.2	2.7	1.3	35.7	30.0	3.7	1.0	9.0	8.6	27.3	45.2	15.5	10.9
Spain	10.0	8.4	1.0	0.7	8.9	7.1	9.3	7.6	5.2	3.6	14.4	16.6	32.5	44.2	18.7	11.7
Sweden	1.3	0.8	2.1	1.5	1.7	1.8	2.8	2.0	20.0	19.1	11.3	12.0	48.7	52.2	12.1	10.8
Switzerland	0.6	0.5	5.7	4.7	3.2	2.7	7.1	5.6	2.6	2.9	21.2	22.7	48.0	49.6	11.5	11.4
Thailand	33.2	20.0	3.8	4.7	23.7	22.3	9.8	20.3	1.2	1.6	2.5	4.3	7.6	18.8	18.3	8.2
United Kingdom	2.0	1.8	15.6	7.6	5.4	5.5	5.3	5.0	2.1	2.9	17.1	17.4	41.5	48.2	11.0	11.7
United States	15.4	7.8	4.6	2.5	5.5	5.2	2.7	2.4	3.7	4.2	12.9	14.0	47.1	55.1	8.1	8.9
Venezuela	0.3	0.6	66.3	48.1	0.1	0.3	0.0	0.2	0.1	0.3	30.1	37.1	0.5	1.5	2.7	11.9

[1] West Germany only.
Source: United Nations, 1980, 1990 International Trade Statistics Yearbook, Vol. 1

Share of Global Exports of High Technology Products (%)

Ranking	Computers 1980	Computers 1990	Telecommunications Equipment 1980	Telecommunications Equipment 1990
1	United States (38.6)	United States (24.2)	Germany[1] (16.7)	Japan (28.4)
2	Germany[1] (11.5)	Japan (17.3)	Sweden (15.3)	United States (15.9)
3	United Kingdom (10.4)	United Kingdom (8.7)	United States (10.9)	Sweden (6.9)
4	France (8.6)	Germany[1] (6.6)	Japan (10.3)	Germany[1] (6.5)
5	Italy (6.6)	Taiwan (6.3)	Netherlands (9.3)	Canada (4.7)
6	Japan (4.3)	Singapore (6.1)	Belgium (7.4)	Taiwan (3.9)
7	Canada (3.4)	Netherlands (4.2)	France (6.5)	South Korea (3.4)
8	Sweden (2.9)	France (4.0)	Canada (5.1)	Netherlands (3.3)
9	Hong Kong (1.9)	Italy (3.3)	United Kingdom (4.1)	France (3.1)
10	Netherlands (1.6)	South Korea (2.5)	Italy (2.7)	United Kingdom (3.1)

[1] West Germany only.
Source: Handbook of International Economic Statistics

continued on p. 190 (bottom)

Imports by Type, Selected Countries (% of total imports)

Country	Foods & Beverages		Unprocessed Resources		Fuels		Machinery		Transport Equipment		Consumer Goods		Other	
	1980	1990	1980	1990	1980	1990	1980	1990	1980	1990	1980	1990	1980	1990
Australia	4.4	4.2	28.9	22.9	10.5	5.1	23.4	26.5	15.2	17.6	16.1	16.4	1.3	7.2
Brunei	13.0	na	33.0	na	1.4	na	22.2	na	13.3	na	14.4	na	3.8	na
Canada	6.7	5.7	22.8	21.3	12.2	6.1	22.0	25.0	25.3	26.4	9.4	12.7	1.6	2.8
Chile	13.1	na	22.5	na	18.3	na	15.7	na	14.9	na	11.9	na	3.6	na
China	na	7.7	na	44.4	na	2.4	na	27.6	na	11.6	na	6.1	na	0.5
Colombia	9.2	5.9	35.1	46.5	12.1	4.7	23.4	25.4	15.0	11.5	4.4	3.6	6.8	2.4
Ecuador	7.6	8.0	34.6	44.8	1.0	4.1	28.3	23.6	21.2	12.3	6.4	6.6	0.8	0.6
Egypt	26.6	25.3	40.6	43.8	1.0	2.3	17.0	17.0	10.4	6.8	4.4	4.8	0.1	0.0
France	8.8	8.6	32.1	31.3	26.4	9.4	13.7	20.4	8.1	13.9	10.4	16.2	0.5	0.1
Germany[1]	10.6	9.2	31.3	30.3	22.1	8.2	11.6	18.6	7.6	13.6	14.2	18.1	2.5	2.1
Hong Kong	11.2	6.2	45.2	36.4	5.5	2.4	14.2	18.3	4.9	3.1	18.5	33.2	0.5	0.5
India (1988)	8.8	7.8	31.5	47.8	44.6	16.8	9.5	16.7	4.4	2.9	1.1	1.8	0.0	6.1
Indonesia	11.9	3.9	36.1	41.4	15.7	8.6	21.4	32.5	11.6	11.1	3.0	2.2	0.3	0.3
Israel	9.1	6.7	39.3	47.3	26.3	8.8	14.8	18.5	6.5	8.8	3.8	8.6	0.2	1.4
Italy	11.3	10.6	33.5	34.6	27.7	10.5	11.1	17.8	9.8	13.7	6.2	10.2	0.5	2.6
Japan	9.8	12.4	29.2	32.2	49.9	24.4	4.7	10.9	1.9	5.6	3.9	13.2	0.7	1.2
Malaysia (1988)	9.6	9.0	30.1	33.0	15.1	5.3	26.8	39.2	11.7	6.2	6.1	7.0	0.7	0.3
Mexico (1989)	7.5	13.2	35.4	37.1	2.1	3.8	28.8	27.3	18.6	6.3	4.6	11.4	3.2	0.7
Netherlands	11.0	10.3	30.7	31.1	23.5	10.0	12.3	20.4	7.3	10.7	13.8	16.8	1.4	0.7
Peru (1988)	17.5	33.6	34.8	35.9	2.0	4.0	27.2	16.1	13.3	6.3	4.8	3.5	0.5	0.4
Philippines (1988)	5.5	7.3	28.3	35.9	28.3	13.1	16.1	15.0	8.7	5.1	1.9	2.6	11.2	20.9
Saudi Arabia	12.2	14.5	31.2	22.9	0.6	0.3	21.4	16.1	15.4	22.1	19.1	23.1	0.4	1.1
Singapore	5.9	4.8	27.8	23.4	28.6	15.8	19.7	33.4	8.3	7.7	8.2	13.7	1.3	1.1
South Korea	7.4	3.7	38.2	39.9	29.8	15.8	16.4	30.9	6.1	5.4	2.0	3.8	0.1	0.5
Spain	9.3	9.1	28.3	29.3	38.4	11.6	12.7	22.3	5.4	16.4	5.8	11.2	0.1	0.1
Sweden	6.1	5.4	27.3	26.8	24.0	8.9	18.4	25.2	9.1	14.2	14.5	18.4	0.6	1.0
Switzerland	7.1	5.6	37.8	33.1	11.0	4.5	16.0	21.2	9.2	11.4	18.7	23.5	0.2	0.8
Thailand (1988)	3.9	4.9	31.0	40.1	30.3	7.6	16.1	28.2	9.1	11.3	3.2	3.7	6.5	4.3
United Kingdom	11.6	9.1	36.1	29.2	13.4	6.1	13.8	21.9	12.3	15.0	11.4	16.1	1.4	2.6
United States	7.6	5.5	21.0	18.9	32.7	13.2	10.5	20.8	14.1	19.0	12.1	19.3	1.7	3.2
Venezuela	11.2	8.6	33.7	40.0	1.0	2.3	28.0	30.9	14.3	9.8	11.6	7.7	0.2	0.8

[1] West Germany only.
Source: United Nations, 1980, 1990 International Trade Statistics Yearbook, Vol. 1

Share of Global Exports of High Technology Products (%) – continued

Ranking	Aerospace		Machine Tools and Robotics	
	1980	1990	1980	1990
1	United States (47.6)	United States (50.3)	Germany[1] (25.8)	Japan (21.8)
2	United Kingdom (19.7)	France (17.5)	United States (14.1)	Germany[1] (20.2)
3	Germany[1] (9.1)	United Kingdom (7.7)	Japan (11.3)	United States (13.1)
4	France (6.0)	Germany[1] (4.1)	Switzerland (9.1)	Italy (9.7)
5	Canada (4.4)	Canada (4.0)	Italy (8.7)	Switzerland (7.3)
6	Netherlands (2.5)	Netherlands (2.5)	United Kingdom (6.9)	United Kingdom (4.5)
7	Italy (2.1)	Italy (1.3)	France (6.3)	France (3.6)
8	Belgium (1.2)	Japan (0.9)	Sweden (2.4)	Netherlands (2.5)
9	Switzerland (0.8)	Sweden (0.9)	Canada (1.8)	Sweden (2.5)
10	Japan (0.6)	Switzerland (0.5)	Netherlands (1.8)	Taiwan (2.1)

[1] West Germany only.
Source: Handbook of International Economic Statistics

WORLD STATISTICS

Country and Official Name	Capital city	Land area ($10^3 km^2$)	Population (10^6)	Density (persons/km^2)	Annual population growth rate (1986–1991 average)[1]	Urban population (as a percentage of total population)	Population under 15/over 60 (%)	Life expectancy Male/Female (years)
Afghanistan *Republic of Afghanistan*	Kabul (Kabol)	652	18.0	28	3.9	19	45 / 4.2	44.0 / 43.0
Albania *Republic of Albania*	Tiranë	28.7	3.4	117	1.8	36	35 / 7	72.0 / 79.0
Algeria *Democratic and Popular Republic of Algeria*	Algiers	2 381	26.4	12	2.8	52	44 / 6	65.8 / 66.3
Andorra *Principality of Andorra*	Andorra la Vella	0.18	0.05	122	3.2	65	17 / 14	74.0 / 81.0
Angola *Republic of Angola*	Luanda	1 247	10.6	9	2.7	27	42 / 5	44.9 / 48.1
Antigua and Barbuda *Antigua and Barbuda*	Saint John's	0.4	0.06	145	1.4	33	44 / 8	70.0 / 73.0
Argentina *Argentine Republic*	Buenos Aires	2 780	33	12	1.4	86	30 / 12	68.0 / 74.0
Armenia *Republic of Armenia*	Yerevan	30	3.4	115	1.6	66	30 / 10	67.9 / 73.4
Aruba	Oranjestad	0.2	0.07	358	1.0	na	26 / 10	71.6 / 76.8
Australia *Commonwealth of Australia*	Canberra	7 682	17.6	2.3	1.6	86	22 / 15	73.3 / 79.6
Austria *Republic of Austria*	Vienna	84	7.9	94	0.3	58	20 / 19	72.5 / 79.0
Azerbaijan *Azerbaijani Republic*	Baku (Baky)	86	7.2	84	0.8	66	33 / 8	66.9 / 74.8
Bahamas *The Commonwealth of the Bahamas*	Nassau	14	0.26	19	2.0	60	38 / 6	69.0 / 76.0
Bahrain *State of Bahrain*	Manama	0.7	0.53	767	3.2	83	33 / 4	71.0 / 76.0
Bangladesh *People's Republic of Bangladesh*	Dhaka	144	111	771	2.4	16	47 / 6	56.4 / 55.4
Barbados *Barbados*	Bridgetown	0.4	0.30	602	0.3	45	29 / 13	71.9 / 76.9
Belarus *Republic of Belarus*	Minsk	207	10.3	50	0.7	66	23 / 16	63.9 / 74.7
Belgium *Kingdom of Belgium*	Brussels	30	10.0	328	0.2	97	18 / 21	72.4 / 79.1
Belize *Belize*	Belmopan	23	0.2	9	2.5	52	46 / 6	67.0 / 72.0
Benin *Republic of Benin*	Porto-Novo (official) Contonou (de facto)	112	4.9	44	3.1	38	49 / na	49.0 / 52.0
Bermuda *Bermuda*	Hamilton	0.05	0.06	1 117	0.8	100	23 / 12	73.0 / 79.0
Bhutan *Kingdom of Bhutan*	Thimphu	47	1.5	32	2.4	5	39 / 7	49.2 / 47.8
Bolivia *Republic of Bolivia*	La Paz (admin) Sucre (judicial)	1 099	7.7	7	2.8	51	41 / 6	51.0 / 55.0
Bosnia-Hercegovina *Republic of Bosnia and Hercegovina*	Sarajevo	51	4.4	86	0.6	56	28 / 8	68.0 / 73.0
Botswana *Republic of Botswana*	Gaborone	582	1.4	2.3	3.4	22	57 / 7	52.7 / 59.3
Brazil *Federative Republic of Brazil*	Brasília	8 512	151	18	2.1	75	39 / 6	63.5 / 69.1
Brunei *State of Brunei, Abode of Peace*	Bandar Seri Begawan	5.7	0.3	47	3.1	64	39 / 5	72.6 / 76.4
Bulgaria *Republic of Bulgaria*	Sofia	111	8.9	81	0.1	68	21 / 17	68.0 / 74.7
Burkina Faso *Burkina Faso*	Ouagadougou	274	9.5	35	2.8	9	48 / 6	47.6 / 50.9
Burundi *Republic of Burundi*	Bujumbura	28	5.7	203	2.6	6	42 / 6	50.0 / 54.0
Cambodia *State of Cambodia (Kampuchea)*	Phnom Penh	182	9	49	2.4	12	44 / 5	46.5 / 49.4
Cameroon *Republic of Cameroon*	Yaoundé	465	13	27	3.4	43	43 / 6	53.5 / 56.5
Canada *Canada*	Ottawa	9 970	27	2.7	1.2	77	21 / 15	73.3 / 80.0
Cape Verde *Republic of Cape Verde*	Praia	4	0.3	86	1.3	30	23 / 9	63.0 / 67.0
Central African Republic *Central African Republic*	Bangui	622	3	4.7	2.2	47	43 / 5	48.0 / 53.0
Chad *Republic of Chad*	N'Djamena	1 284	6	4.6	2.5	30	41 / 4	45.9 / 49.1
Chile *Republic of Chile*	Santiago	756	14	18	1.7	86	32 / 8	68.1 / 75.1
China *People's Republic of China*	Beijing	9 572	1 166	122	1.8	56	28 / 8	68.4 / 71.4
Colombia *Republic of Colombia*	Santafé de Bogotá	1 142	34	30	2.0	61	36 / 6	66.4 / 72.3
Comoros *Federal Islamic Republic of Comoros*	Moroni	2	0.5	267	3.4	30	47 / 6	54.0 / 59.0
Congo *Republic of the Congo*	Brazzaville	342	2.7	7.9	3.1	41	45 / 5	52.1 / 57.3
Costa Rica *Republic of Costa Rica*	San José	51	3	62	2.6	86	38 / 6	72.9 / 77.6
Côte d'Ivoire *Republic of Côte d'Ivoire*	Abidjan	321	13	40	3.9	40	45 / 4	52.0 / 56.0
Croatia *Republic of Croatia*	Zagreb	57	5	85	0.6	56	21 / 15	67.0 / 74.0
Cuba *Republic of Cuba*	Havana	111	11	98	1.0	73	30 / 11	72.7 / 76.1
Cyprus *Republic of Cyprus / Turkish Republic of Northern Cyprus*	Nicosia/Nicosia	9	0.8	82	1.1	64	25 / 15	73.9 / 78.3
Czech Republic *Czech Republic*	Prague	79	10	127	0.2	78	24 / 16	68.9 / 75.1
Denmark *Kingdom of Denmark*	Copenhagen	43	5	120	0.1	87	17 / 20	72.0 / 77.7
Djibouti *Republic of Djibouti*	Djibouti	23	0.6	24	3.5	81	38 / na	47.4 / 50.7
Dominica *Commonwealth of Dominica*	Roseau	0.8	0.07	95	1.3	na	40 / 10	73.0 / 79.0
Dominican Republic *Dominican Republic*	Santo Domingo	48	7.5	154	2.3	60	41 / 6	65.0 / 69.0
Ecuador *Republic of Ecuador*	Quito	271	11	39	2.8	56	39 / 6	63.4 / 67.6
Egypt *Arab Republic of Egypt*	Cairo	998	56	56	2.7	47	40 / 6	59.0 / 60.0
El Salvador *Republic of El Salvador*	San Salvador	21	5.4	260	1.7	44	46 / 5	63.0 / 68.0
Equatorial Guinea *Republic of Equatorial Guinea*	Malabo	28	0.4	13	2.3	30	42 / 6	48.0 / 52.0
Eritrea *Eritrea*	Asmara	93.7	3.3	35	3.0	na	na / na	—/—
Estonia *Republic of Estonia*	Tallinn	45	1.6	35	0.8	72	22 / 17	65.8 / 75.0

[1] Growth rate is the difference between birth rate and death rate.

Persons per physician[2]	Economically active population (%)	Labour force in agriculture, forestry, fishing (%)	Land area in farms (%)	Irrigated land as a (%) of arable land	GNP per capita ($/person)	Expenditure on education (% of GNP)	% of students enrolled in primary/secondary education	Military expenditure (% of GNP)	External public debt (10^6 $U.S.)	External debt service ratio (%)	Tourism receipts (% of GNP)	Tourist arrivals, 1990 (10^3)	Newspaper circulation/1000 population	Persons per TV	Persons per motor vehicle
5 000	31	60	61.0	34	220	1.8	19/na	7.7	4 964	4.2	0.03	8	10	160	236
575	46	55	40.0	73	1 280	—	na/na	4.1	162	2.0	na	na	42	13	146
1 200	24	14	16.7	5	2 060	9.4	88/53	5.1	24 316	58.1	0.12	1 137	56	16	20
450	55	1	2.0	—	13 550	—	na/na	...	...	...	na	na	0	8	1.5
14 000	42	72	3.4	89	620	5.2	66/10	23.9	7 152	6.9	na	46	14	22	59
1 600	43	9	9.0	—	4 600	3.7	na/na	...	167	8.5	69.0	197	75	3	5
350	38	6	73.1	7	2 370	1.5	na/na	3.4	46 146	25.2	1.2	2 728	503	5	6
250	62	6	43.3	—	na	—	na/na	...	...	...	na	na	—	3	13
920	43	0.2	—	—	11 840	—	na/na	—	...	...	49.0	433	—	4	3
440	50	5	60.8	4	17 080	4.8	97/79	2.3	...	...	0.14	2 215	405	2.7	1.7
350	46	8	87.4	0	19 240	6.5	93/na	1.1	...	...	8.8	19 011	na	2.9	2.5
250	45	na	48.3	—	na	—	na/na	...	...	...	na	na	416	3	13
800	51	4	2.6	10	11 510	4.4	na/na	0.5	563	12.4	46.0	1 562	141	5	3
700	54	2	5.2	—	6 380	5.4	96/70	6.5	1 240	5.7	3.5	195	na	2.7	4
6 200	30	57	63.5	30	200	2.2	63/17	1.6	11 464	15.8	0.01	115	9	325	1 200
1 050	49	5	45.9	—	6 540	6.9	99/89	0.6	569	12.1	30.0	432	161	4	6
250	53	19	44.9	—	—	—	na/na	...	...	...	na	na	530	3	13
310	42	2	44.6	0	15 440	4.9	97/89	2.5	...	...	2.3	3 163	213	2.5	2.5
2 000	30	28	10.0	5	1 970	—	na/na	2.0	140.7	5.3	24.0	222	na	16	33
16 000	48	64	29.3	0	360	5.1	52/13	2.0	1 262	4.8	2.8	50	3	290	120
675	60	2	4.4	—	27 790	3.2	na/na	—	403	10.5	30.0	434	300	2.0	2.6
10 000	45	93	3.4	30	190	3.7	na/na	...	80.4	...	0.7	2	0	6 200	350
2 125	31	42	20.6	5	620	2.3	83/27	4.0	3 683	28.6	2.2	217	na	18	30
450	23	4	49.4	—	na	—	na/na	...	...	...	na	na	31	4	6
7 200	37	43	5.9	0	2 040	8.2	93/24	2.8	509.8	4.4	2.5	844	na	150	27
690	43	23	44.5	4	2 680	3.7	84/15	1.3	82 098	14.6	0.4	1 079	na	4	9
1 500	38	4	2.8	33	13 260	—	na/na	8.1	25	0.1	1.0	625	0	2.6	2.3
325	52	17	55.6	33	5 310	5.5	86/59	11.9	9 564	16.0	0.8	4 500	267	4	6
30 000	51	92	32.6	0	330	2.3	28/5	2.1	750	5.2	0.3	46	1	220	375
19 000	53	93	85.8	6	210	3.2	51/3	2.6	850	41.2	0.3	109	4	1 210	265
27 000	50	74	16.5	3	130	—	na/na	...	1 400	29.7	—	na	0	180	750
12 500	39	74	3.3	0	940	3.3	75/15	1.3	4 784	12.8	0.2	100	5	2 400	90
440	52	3.5	7.4	1.8	20 440	7.0	96/93	2.0	...	...	1.2	15 258	221	1.7	1.6
4 200	23	33	6.2	5	890	2.9	95/12	11.8	144.3	3.8	na	na	0	70	20
17 000	49	66	0.8	—	390	2.9	46/na	1.7	815	6.2	0.7	6	0.1	450	150
47 600	35	77	45.8	0	190	1.8	38/na	4.3	430.4	2.2	1.1	29	0.3	1 100	400
930	36	18	38.0	30	1 940	3.6	89/58	3.4	10 339	14.9	2.1	950	67	4	13
650	55	71	17.4	49	370	2.4	100/na	3.7	45 319	8.7	0.5	10 484	na	9	225
1 080	34	29	27.0	13	1 250	2.9	69/na	2.1	14 680	33.2	0.9	813	265	6	25
12 300	29	53	44.3	—	480	6.5	55/na	...	177.1	1.0	—	8	0	4 100	70
4 350	29	52	0.7	3	1 010	5.1	na/na	5.1	4 380	14.8	0.3	46	7	400	50
1 200	37	25	60.0	41	1 910	4.4	86/36	0.5	3 076	21.1	5.2	435	105	5	12
18 000	39	61	8.6	3	730	6.9	na/na	1.5	10 050	12.7	0.5	196	11	15	50
450	32	5	57.0	—	5 600	—	na/na	...	...	...	2.2	na	130	4	6
310	44	18	78.3	34	2 000	6.6	95/69	3.9	4 980	11.5	1.3	300	160	5	30
484	48	14	35.6	34	8 050	3.6	99/84	0.9	1 542	6.9	22.0	1 561	165	2.4	4
320	51	12	54.1	6	na	5.4	na/na	6.8	5 346	9.1	0.4	na	325	2.7	5
375	57	5	65.3	17	22 100	7.6	97/84	2.2	...	...	2.9	1 275	360	2.0	2.7
5 500	44	77	0.3	—	430	2.7	37/13	8.1	144.6	5.3	3.1	47	8	40	35
2 700	38	25	26.3	—	1 980	5.7	na/na	...	75.9	3.8	15.0	45	0	16	21
950	34	22	49.8	23	820	1.5	73/na	0.8	3 440	6.3	12.9	1 533	37	10	28
850	35	31	29.6	33	965	2.6	na/na	1.7	9 854	26.8	1.9	332	na	14	32
650	28	36	3.0	100	600	6.8	na/na	5.0	34 242	20.9	6.4	2 411	na	15	38
1 600	35	40	69.0	21	960	2.0	70/15	4.0	1 898	20.3	1.2	194	52	12	31
61 000	39	58	—	—	330	—	na/na	...	206.0	2.2	na	na	na	140	41
na	na	85(e)	4(e)	—	—	—	20(e)/na	...	...	...	na	na	na	—	—
250	55	12	31.1	—	na	—	na/na	...	...	...	na	na	1 600	3	13

[2] Statistics for the newly created Soviet Republics and those of the newly created states from the former Yugoslavia are those for the former nation as a whole, where they were given as averages (eg. birth, death). They may not be applicable to each newly formed republic.
(e) = estimate

Country and Official Name	Capital city	Land area ($10^3 km^2$)	Population (10^6)	Density (persons/km^2)	Annual population growth rate (1986–1991 average)[1]	Urban population (as a percentage of total population)	Population under 15/over 60 (%)	Life expectancy Male/Female (years)
Ethiopia *Transitional Government of Ethiopia*	Addis Ababa	1 251	54	43	2.9	13	47 / 6	45.4 / 48.7
Fiji *Republic of Fiji*	Suva	18	0.7	41	0.7	39	38 / 6	68.3 / 72.8
Finland *Republic of Finland*	Helsinki	338	5.0	15	0.3	60	19 / 18	70.8 / 78.9
France *French Republic*	Paris	544	57	105	0.5	74	19 / 20	72.7 / 80.9
French Guiana	Cayenne	87	0.1	1.4	5.1	81	33 / 6	63.4 / 69.7
French Polynesia	Papeete	4	0.3	52	2.4	55	36 / 5	66.1 / 71.3
Gabon *Gabonese Republic*	Libreville	268	1.2	4.7	3.5	44	35 / 9	51.9 / 55.2
Gambia *Republic of the Gambia*	Banjul	11	0.9	86	2.9	21	44 / 6	43.4 / 46.6
Georgia *Republic of Georgia*	Tbilisi	70	5.4	79	0.8	66	25 / 14	69.0 / 76.3
Germany *Federal Republic of Germany*	Bonn (Berlin has been designated)	357	79	222	0.4	84	15 / 21	72.1 / 78.7
Ghana *Republic of Ghana*	Accra	239	15	64	3.2	33	45 / 6	54.2 / 57.8
Greece *Hellenic Republic*	Athens	132	10	78	0.5	63	24 / 17	72.6 / 77.6
Grenada *Grenada*	St. George's	0.3	0.1	261	1.3	na	39 / 10	69.0 / 74.0
Guadeloupe *Department of Guadeloupe*	Agana	1.7	0.4	225	1.9	48	25 / 12	70.0 / 77.0
Guam	Basse-Terre	0.5	0.1	257	2.3	40	30 / 7	69.5 / 75.6
Guatemala *Republic of Guatemala*	Guatemala City	109	9.4	87	2.8	39	45 / 5	59.7 / 64.4
Guinea *Republic of Guinea*	Conakry	246	7.2	30	2.5	24	43 / 5	44.0 / 45.0
Guinea-Bissau *Republic of Guinea-Bissau*	Bissau	36	1.0	28	2.1	29	44 / 7	41.9 / 45.1
Guyana *Co-operative Republic of Guyana*	Georgetown	215	0.7	3.5	0.2	28	41 / 6	61.0 / 68.0
Haiti *Republic of Haiti*	Port-au-Prince	28	6.8	244	2.0	28	40 / 8	52.0 / 55.0
Honduras *Republic of Honduras*	Tegucigalpa	112	5.0	45	2.0	44	48 / 5	63.0 / 67.0
Hong Kong *Hong Kong*	—	1	5.7	5 700	1.2	100	23 / na	75.0 / 81.0
Hungary *Republic of Hungary*	Budapest	93	10	111	–0.3	61	21 / 19	65.1 / 73.7
Iceland *Republic of Iceland*	Reykjavík	103	0.3	2.5	1.2	91	25 / 15	75.7 / 80.3
India *Republic of India*	New Delhi	3 165	890	282	2.1	27	40 / 7	58.1 / 59.1
Indonesia *Republic of Indonesia*	Jakarta	1 919	185	96	1.8	31	41 / 6	55.6 / 58.9
Iran *Islamic Republic of Iran*	Tehran	1 638	60	37	3.2	57	46 / 5	64.0 / 65.0
Iraq *Republic of Iraq*	Baghdad	435	19	43	3.1	71	45 / 5	46.0 / 57.0
Ireland *Éire, Ireland*	Dublin	70	3.6	50	–0.3	57	29 / 15	71.0 / 76.7
Israel *State of Israel*	Jerusalem	21	5.2	253	2.0	92	33 / 13	73.9 / 77.5
Italy *Italian Republic*	Rome	301	57	190	0.1	69	21 / 17	73.2 / 79.7
Jamaica *Jamaica*	Kingston	11	2.4	223	0.7	52	38 / 10	71.4 / 75.8
Japan *Nihon (Japan)*	Tokyo	378	124	329	0.4	77	22 / 15	75.9 / 81.8
Jordan *Hashemite Kingdom of Jordan*	Amman	90	3.6	41	3.7	61	52 / 4	70.0 / 73.0
Kazakhstan *Republic of Kazakhstan*	Alma-Ata	2 717	17	6.3	1.6	66	32 / 9	64.2 / 73.2
Kenya *Republic of Kenya*	Nairobi	583	27	46	4.1	24	51 / 4	59.0 / 63.0
Kiribati *Republic of Kiribati*	Bairiki	0.8	0.07	92	2.2	35	40 / 5	52.0 / 58.0
Korea, North *Democratic People's Republic of Korea*	P'yongyang-si	123	22	181	1.8	59	29 / 7	66.0 / 72.0
Korea, South *Republic of Korea*	Seoul	99	44	440	1.0	63	25 / 8	67.4 / 75.4
Kuwait *State of Kuwait*	Kuwait City	18	1.2	67	na	96	37 / 2	72.0 / 76.0
Kyrgyzstan *Republic of Kyrgyzstan*	Bishkek (Frunze)	199	4.5	23	2.3	66	38 / 8	64.5 / 72.8
Laos *Lao People's Democratic Republic*	Vientiane	237	4.4	19	2.9	19	44 / 6	47.8 / 50.8
Latvia *Republic of Latvia*	Riga	65	2.7	42	0.8	71	21 / 17	64.2 / 74.6
Lebanon *Republic of Lebanon*	Beirut	10	2.8	274	0.6	81	43 / 8	65.1 / 69.0
Lesotho *Kingdom of Lesotho*	Maseru	30	1.9	62	2.6	20	41 / 7	54.0 / 63.0
Liberia *Republic of Liberia*	Monrovia	99	2.8	28	4.1	46	43 / 6	54.0 / 57.0
Libya *Socialist People's Libyan Arab Jamahiriya*	Tripoli	1 757	4.4	2.5	2.8	70	50 / 4	59.1 / 62.5
Liechtenstein *Principality of Liechtenstein*	Vaduz	0.2	0.03	185	1.3	na	23 / 12	66.1 / 72.9
Lithuania *Republic of Lithuania*	Vilnius	65	3.8	58	0.8	68	23 / 16	66.9 / 76.3
Luxembourg *Grand Duchy of Luxembourg*	Luxembourg	3	0.3	150	0.7	84	19 / 18	70.6 / 77.9
Macau *Macau*	Macau	0.02	0.4	20 388	2.7	95	25 / 10	75.1 / 80.3
Macedonia *Republic of Macedonia*	Skopje	26	2	80	0.6	—	29 / 9	68.0 / 72.0
Madagascar *Democratic Republic of Madagascar*	Antananarivo	587	13	22	3.3	25	44 / 6	54.0 / 57.0
Malawi *Republic of Malawi*	Lilongwe	118	9.5	80	3.7	12	45 / 6	48.4 / 49.7
Malaysia *Malaysia*	Kuala Lumpur	330	19	56	2.5	43	40 / 6	68.8 / 73.3
Maldives *Republic of Maldives*	Male	0.3	0.2	772	3.4	25	47 / 5	66.0 / 64.1
Mali *Republic of Mali*	Bamako	1 248	8.5	7	1.8	19	46 / 6	45.0 / 47.0
Malta *Republic of Malta*	Valletta	0.3	0.4	1 140	0.9	85	24 / 14	73.7 / 78.1
Marshall Islands *Republic of the Marshall Islands*	Majuro	0.2	0.05	276	4.3	47	51 / 4	59.6 / 62.6
Martinique *Department of Martinique*	Fort-de-France	1.1	0.4	327	1.1	81	23 / 14	71.0 / 77.0

[1] Growth rate is the difference between birth rate and death rate.

Persons per physician[2]	Economically active population (%)	Labour force in agriculture, forestry, fishing (%)	Land area in farms (%)	Irrigated land as a (%) of arable land	GNP per capita ($/person)	Expenditure on education (% of GNP)	% of students enrolled in primary/secondary education	Military expenditure (% of GNP)	External public debt (10^6 $U.S.)	External debt service ratio (%)	Tourism receipts (% of GNP)	Tourist arrivals, 1990 (10^3)	Newspaper circulation/1000 population	Persons per TV	Persons per motor vehicle
37 000	41	88	5.7	1	120	4.4	28/na	12.8	3 116	30.6	0.4	73	1	510	800
2 700	34	44	15.2	1	1 770	5.0	98/na	2.2	292.4	8.4	17.0	279	56	72	11
525	51	8	40.5	3	16 100	5.8	na/na	1.6	...	...	0.9	866	520	2.0	2.3
400	43	5	61.8	6	19 450	5.3	100/83	3.7	...	...	1.8	51 462	na	2.0	2
374	42	11	0.3	—	1 825	19.2	na/na	—	45	2.6	na	na	45	18	3
1 100	40	10	10.4	19	12 900	9.7	na/na	—	345	6.2	6.0	132	125	7.4	4
2 050	44	70	0.3	—	3 225	5.6	na/na	4.5	2 945	4.9	0.1	108	14	30	35
9 950	47	74	16.5	7	260	4.0	62/na	0.7	304.2	22.1	11.4	101	1	na	130
250	45	26	45.7	—	na	—	na/na	...	...	...	na	na	670	3	13
360	50	3	47.7	4	22 730	—	na/na	2.8	...	...	0.8	17 045	420	2.1	2.2
22 250	45	59	10.8	1	390	3.4	na/na	0.6	2 670	18.2	1.4	146	13	86	140
300	40	22	26.9	41	6 000	2.7	98/85	5.9	21 291	18.3	4.2	8 873	na	4	4
1 900	40	14	40.2	—	2 125	4.6	na/na	...	91.1	2.1	20.0	82	0	3	16
682	44	7	84.1	14	3 200	15.0	na/na	—	41	2.5	—	331	50	2.5	3
823	50	1	9.8	—	7 000	8.5	na/na	—	...	...	—	780	na	1.6	1.9
2 240	34	49	38.1	6	900	1.8	58/13	1.6	2 179	10.3	2.2	508	na	19	30
10 500	40	78	6.5	4	480	3.3	26/7	1.2	2 230	6.0	—	na	0	106	260
7 200	30	72	4.7	—	180	2.8	45/3	2.4	544.3	26.9	—	na	12	na	175
6 800	36	20	26.2	27	370	8.8	na/na	2.7	1 663	55.2	10.0	67	77	20	25
6 000	41	57	57.0	14	370	1.8	44/na	1.9	745	3.3	3.1	120	7	240	125
1 700	39	35	23.5	6	590	4.9	91/21	3.2	3 159	31.8	0.9	202	49	26	40
950	50	1	6.8	33	11 500	2.8	94/66	—	8 066	3.0	7.5	5 933	na	4	17
325	45	17	88.5	3	6 200	6.0	91/73	6.3	18 046	43.2	1.6	20 510	237	2.5	5
375	50	10	—	—	21 150	5.4	na/na	—	...	...	2.2	142	520	3	1.8
2 425	39	66	49.7	26	350	3.2	na/na	3.1	61 097	20.4	0.5	1 707	21	43	225
7 400	43	55	25.3	48	560	0.9	99/41	1.7	44 974	22.6	1.9	2 178	22	90	70
3 000	26	25	63.8	41	2 450	3.1	94/48	7.9	1 797	...	0.4	154	na	25	26
3 325	24	13	13.1	49	1 940	3.8	84/39	30.7	16 146	24.7	0.2	747	30	18	20
680	37	13	82.6	—	9 600	7.0	89/79	1.6	...	...	4.3	3 666	200	3.5	4
345	35	3	28.2	63	10 870	8.6	na/na	12.8	17 959	32.8	2.7	1 063	350	4	5
234	42	8	78.4	34	16 900	5.0	na/na	2.4	...	...	2.0	26 679	105	4	2.2
2 100	44	23	54.8	17	1 510	6.6	99/59	1.1	3 873	23.6	21.0	841	51	5	22
610	53	7	13.9	69	25 400	4.8	100/96	1.0	...	...	0.1	1 879	430	1.8	2.1
630	23	5	4.1	18	1 280	5.9	93/68	12.7	6 486	21.5	13.0	2 633	73	13	13
250	39	19	72.7	—	na	—	na/na	...	...	...	na	na	400	4	13
7 325	40	78	11.9	3	370	6.5	91/na	2.7	4 810	21.1	5.0	801	na	100	85
4 200	45	71	—	—	760	8.7	na/na	...	12	0.2	3.7	3	na	6	147
370	45	43	—	82	990	3.6	na/na	20.0	3 708	240.5	na	na	na	90	na
1 100	44	16	21.2	68	5 400	3.6	100/79	4.3	17 814	6.2	1.5	2 959	na	5	16
675	39	1	0.3	50	16 200	5.0	85/na	6.2	610	8.3	0.2	50	200	2.4	3.6
250	33	33	51.0	—	na	—	na/na	...	...	...	na	na	375	4	13
6 500	49	76	7.1	14	200	1.0	70/na	10.5	1 053	11.5	na	25	8	130	200
250	55	17	38.5	—	na	—	na/na	...	...	...	na	na	1 600	3	13
771	27	19	27.0	41	690	—	na/na	8.2	545	...	na	na	na	3	5
16 000	35	23	12.3	—	470	4.0	70/14	2.3	371.8	2.4	1.1	171	10	35	75
24 600	34	68	3.8	2	440	5.7	na/na	4.8	1 126	—	0.6	na	na	58	210
700	24	19	5.1	13	5 300	10.1	na/na	14.9	2 607	4.7	0.01	96	10	8	6
950	50	2	23.1	—	21 000	—	na/na	—	...	...	—	78	575	3	1.6
250	52	6	52.3	—	na	—	na/na	...	...	...	na	na	725	3	13
530	43	3	53.1	—	28 700	4.7	na/na	0.9	...	...	2.7	820	380	4	1.9
1 850	50	0.2	—	—	8 000	—	na/na	...	251	1.8	3.7	1 138	575	7	17
422	23	8	51.3	—	3 110	—	na/na	...	...	...	2.2	na	23	4	6
7 500	44	78	3.5	35	230	1.8	64/na	1.5	3 677	33.5	2.6	53	65	100	135
27 000	41	82	14.1	1	200	3.3	50/na	2.3	1 366	19.3	0.5	117	3	na	300
2 700	38	31	31.2	33	2 350	5.6	na/na	2.9	16 107	9.6	4.0	7 477	150	7	9
7 750	27	25	63.5	—	440	—	na/na	...	63.9	4.1	—	195	11	45	240
21 000	35	82	1.8	10	270	3.3	19/na	2.0	2 306	7.1	1.1	44	5	825	280
450	37	3	41.2	8	6 650	3.6	97/75	1.1	121.8	0.5	—	872	200	2.7	3
2 225	14	1	—	—	1 500	—	na/na	—	...	...	—	7	0	na	50
600	46	7	86.9	60	4 100	13.5	na/na	—	27	1.2	1.6	282	83	8	2.3

[2] Statistics for the newly created Soviet Republics and those of the newly created states from the former Yugoslavia are those for the former nation as a whole, where they were given as averages (eg. birth, death). They may not be applicable to each newly formed republic.

Country and Official Name	Capital city	Land area ($10^3 km^2$)	Population (10^6)	Density (persons/km^2)	Annual population growth rate (1986–1991 average)[1]	Urban population (as a percentage of total population)	Population under 15/over 60 (%)	Life expectancy Male/Female (years)
Mauritania *Islamic Republic of Mauritania*	Nouakchott	1 030	2.1	2	2.7	47	46 / 5	44.0 / 50.0
Mauritius *Republic of Mauritius*	Port Louis	2	1.1	530	1.1	41	33 / 7	64.7 / 72.2
Mexico *United Mexican States*	Mexico City	1 958	85	43	1.6	73	38 / 6	67.8 / 73.9
Micronesia *Federated States of Micronesia*	Palikir	0.7	0.1	162	3.2	20	46 / 6	68.0 / 73.0
Moldova *Republic of Moldova*	Chisinau (Kishinyov)	34	4.4	131	0.9	66	28 / 12	65.2 / 72.0
Mongolia *Mongolia*	Ulaanbaatar (Ulan Bator)	1 567	2.2	1.4	2.0	60	42 / 6	61.2 / 63.8
Morocco *Kingdom of Morocco*	Rabat	459	26	57	2.7	48	42 / 6	61.6 / 65.0
Mozambique *Republic of Mozambique*	Maputo	812	15	18	2.1	27	44 / 4	46.9 / 50.2
Myanmar (Burma) *Union of Myanmar*	Yangon (Rangoon)	677	43	64	2.1	25	41 / 6	60.0 / 63.5
Namibia *Republic of Namibia*	Windhoek	823	1.4	1.7	2.6	28	44 / 4	57.5 / 60.0
Nepal *Kingdom of Nepal*	Kathmandu	147	20	135	2.5	10	41 / 6	55.4 / 52.6
Netherlands *Kingdom of the Netherlands*	The Hague	42	15	363	0.6	88	18 / 17	73.7 / 80.2
Netherlands Antilles *Netherlands Antilles*	Willemstad	0.8	0.2	239	0.4	92	30 / 9	71.1 / 75.8
New Caledonia *Territory of New Caledonia and Dependencies*	Nouméa	19	0.2	9.4	2.1	60	33 / 7	66.5 / 71.8
New Zealand *New Zealand (Aotearoa)*	Wellington	270	3.5	13	0.7	84	23 / 16	72.0 / 77.9
Nicaragua *Republic of Nicaragua*	Managua	131	4.1	32	3.4	60	48 / 5	60.0 / 65.0
Niger *Republic of Niger*	Niamey	1 187	8.3	7	3.4	20	47 / 5	44.9 / 48.1
Nigeria *Federal Republic of Nigeria*	Abuja	924	123	133	3.3	35	43 / 4	50.8 / 54.3
Norway *Kingdom of Norway*	Oslo	324	4.3	13	0.4	75	24 / 21	73.4 / 79.8
Oman *Sultanate of Oman*	Muscat	306	1.6	5.4	3.8	9	44 / 5	65.0 / 68.0
Pakistan *Islamic Republic of Pakistan*	Islamabad	880	130	148	3.4	32	45 / 6	59.3 / 60.7
Panama *Republic of Panama*	Panama City	76	2.5	33	2.1	53	30 / 9	72.0 / 76.0
Papua New Guinea *Independent State of Papua New Guinea*	Port Moresby	463	3.8	8.3	2.2	15	35 / 7	54.0 / 56.0
Paraguay *Republic of Paraguay*	Asunción	407	4.5	11	2.9	48	41 / 6	65.1 / 69.5
Peru *Republic of Peru*	Lima	1 285	22	18	2.5	70	41 / 6	61.5 / 65.3
Philippines *Republic of Philippines*	Manila	300	64	212	2.4	43	42 / 5	62.7 / 66.5
Poland *Republic of Poland*	Warsaw	313	38	123	0.4	62	25 / 15	66.8 / 75.5
Portugal *Republic of Portugal*	Lisbon	92	10	113	0.4	34	26 / 16	70.6 / 77.6
Puerto Rico *Commonwealth of Puerto Rico*	San Juan	9	3.6	393	1.0	71	32 / 11	71.5 / 78.4
Qatar *State of Qatar*	Doha	11	0.5	46	4.0	89	28 / 2	65.2 / 67.6
Réunion *Department of Réunion*	Saint-Denis	2.5	0.6	250	1.8	63	30 / 8	69.0 / 78.3
Romania *Romania*	Bucharest	238	22	98	0.4	53	26 / 14	66.5 / 72.4
Russia *Russian Federation*	Moscow	17 075	150	8.8	0.8	66	23 / 15	63.9 / 74.3
Rwanda *Republic of Rwanda*	Kigali	26	7.4	280	3.5	8	46 / 5	48.8 / 52.2
St. Kitts and Nevis *Federation of St. Kitts and Nevis*	Basseterre	0.3	0.04	160	-0.3	49	37 / 14	64.0 / 71.0
St. Lucia *St. Lucia*	Castries	0.6	0.1	218	2.0	47	37 / 10	68.6 / 74.4
St. Vincent and the Grenadines *St. Vincent and the Grenadines*	Kingstown	0.4	0.1	280	1.3	25	42 / 6	68.0 / 72.0
San Marino *Most Serene Republic of San Marino*	San Marino	0.06	0.02	387	0.6	90	24 / 15	70.7 / 76.2
São Tomé and Principe *Democratic Republic of São Tomé and Principe*	Sao Tomé	1	0.1	126	2.2	40	46 / 7	64.0 / 67.0
Saudi Arabia *Kingdom of Saudi Arabia*	Riyadh	2 240	15	6.8	4.0	77	47 / 6	65.0 / 68.0
Senegal *Republic of Senegal*	Dakar	196	7.7	39	2.6	38	48 / 5	48.3 / 50.3
Seychelles *Republic of Seychelles*	Victoria	0.5	0.07	157	0.7	47	34 / 10	67.3 / 74.2
Sierra Leone *Republic of Sierra Leone*	Freetown	72	4.4	61	2.6	32	41 / 8	41.4 / 44.6
Singapore *Republic of Singapore*	Singapore	0.6	2.8	4 489	1.1	100	23 / 10	70.3 / 75.8
Slovak Republic *Slovak Republic*	Bratislava	49	5.3	108	0.6	74	24 / 16	— / —
Slovenia *Republic of Slovenia*	Ljubljana	20	2.0	99	0.6	56	23 / 15	67.0 / 75.0
Solomon Islands *Solomon Islands*	Honiara	28	0.3	12	2.9	15	47 / 5	59.9 / 61.4
Somalia *Somali Democratic Republic*	Mogadishu	638	7.8	12	3.1	36	46 / 5	45.4 / 48.6
South Africa *Republic of South Africa*	Pretoria	1 226	39	32	2.0	60	37 / 6	61.0 / 67.0
Spain *Kingdom of Spain*	Madrid	505	39	77	0.7	78	26 / 15	73.3 / 79.7
Sri Lanka *Democratic Socialist Republic of Sri Lanka*	Colombo and Sri Jayewardenepura Kotte	66	18	270	1.3	21	35 / 7	69.1 / 73.4
Sudan *Republic of the Sudan*	Khartoum	2 504	30	12	3.6	30	45 / 5	52.0 / 54.0
Suriname *Republic of Suriname*	Paramaribo	164	0.4	2.5	1.5	65	39 / 7	67.8 / 72.8
Swaziland *Kingdom of Swaziland*	Mbabane	17	0.8	47	3.3	30	47 / 5	56.2 / 59.8
Sweden *Kingdom of Sweden*	Stockholm	450	8.7	20	0.6	84	18 / 23	74.4 / 80.2
Switzerland *Swiss Confederation*	Bern	41	6.9	167	0.7	60	19 / 18	74.0 / 80.8
Syria *Syrian Arab Republic*	Damascus	185	13	70	3.4	50	49 / 5	65.2 / 69.2

[1] Growth rate is the difference between birth rate and death rate.

Persons per physician[2]	Economically active population (%)	Labour force in agriculture, forestry, fishing (%)	Land area in farms (%)	Irrigated land as a (%) of arable land	GNP per capita ($/person)	Expenditure on education (% of GNP)	% of students enrolled in primary/secondary education	Military expenditure (% of GNP)	External public debt (10^6 $U.S.)	External debt service ratio (%)	Tourism receipts (% of GNP)	Tourist arrivals, 1990 (10^3)	Newspaper circulation/1000 population	Persons per TV	Persons per motor vehicle
9 550	31	66	0.2	6	500	4.9	na/na	4.3	1 898	8.3	1.6	na	na	1 825	150
1 190	45	18	91.5	17	2 250	4.1	93/na	0.2	739	4.4	11.0	292	75	8	40
600	30	22	72.7	22	2 500	3.8	100/44	0.5	76 204	18.3	2.5	6 393	150	7	9
2 540	13	2	12.2	0	980	—	na/na	—	...	...	—	na	0	17	na
250	48	33	73.5	—	na	—	na/na	...	...	...	na	na	560	3	13
390	47	59	79.6	6	na	7.0	94/na	...	14 196	115.0	na	147	84	18	na
4 875	29	39	17.6	15	660	7.3	55/29	5.5	22 097	19.4	5.3	4 024	na	21	30
40 000	49	84	17.8	4	80	1.2	45/na	9.7	4 053	5.3	na	na	3	450	130
3 825	40	68	18.6	11	400	1.6	na/na	3.7	4 446	28.6	0.03	21	12	42	600
4 450	31	43	0.8	0.5	1 200	1.9	na/na	—	22	...	na	na	14	48	11
21 000	45	43	16.7	36	170	2.8	64/23	1.2	1 557	14.2	1.9	255	na	550	575
415	47	4	48.1	—	17 300	6.8	100/82	2.9	...	...	1.4	5 795	320	3	2.5
840	38	1	—	—	7 060	5.2	na/na	—	1 013	7.4	—	na	na	6	2.9
1 180	40	12	15.8	—	10 100	13.4	na/na	—	313	5.2	9.0	87	120	5	2.4
373	47	12	64.5	56	12 700	5.9	100/86	2.2	...	...	2.5	976	310	3	1.9
1 675	33	32	47.7	8	460	3.9	73/23	17.2	8 067	2.6	0.7	106	na	18	50
38 500	34	76	2.9	1	310	3.1	24/4	7.6	1 326	3.7	0.7	21	0.5	310	140
6 950	31	43	37.1	3	270	1.5	na/na	0.5	33 709	19.9	0.08	190	na	12	90
860	50	2	63.5	11	23 100	7.5	98/85	3.3	...	...	1.5	1 955	510	2.9	2.2
1 050	38	62	0.3	87	5 220	3.7	83/42	20.3	2 205	12.6	0.9	149	42	1.5	9
1 989	28	45	24.0	80	380	2.6	na/na	6.8	16 532	17.1	0.4	424	9	60	160
850	35	24	29.3	7	1 830	6.2	90/48	3.4	3 987	2.6	3.8	214	60	4	13
11 900	25	77	0.8	—	860	4.7	na/na	1.4	1 509	16.8	0.8	41	60	375	70
1 600	34	43	53.9	3	1 110	1.5	93/24	1.4	1 736	11.6	2.3	280	18	12	104
1 000	34	34	11.6	37	1 160	3.5	95/42	4.9	3 343	5.4	0.8	317	na	11	35
1 060	40	42	30.1	36	730	2.9	99/54	2.2	24 108	16.4	2.9	893	56	10	60
480	49	28	59.9	0.8	4 350	4.6	97/76	8.9	39 282	4.3	0.2	3 400	185	4	6
375	48	17	56.1	22	4 900	4.9	92/37	3.3	14 432	15.9	7.0	8 020	na	6	5
365	31	4	39.3	57	6 475	8.2	na/na	—	...	...	6.5	2 554	142	4	2.2
570	55	3	0.5	100	15 850	5.6	96/71	9.3	172	5.0	na	100	145	2.8	2.8
625	39	10	84.4	13	6 000	15.1	na/na	—	63	1.0	4.0	200	100	2.8	3
550	47	28	61.9	35	3 200	2.1	na/na	6.1	19	—	0.1	6 533	134	7	16
250	53	13	12.5	—	9 200	—	na/na	...	...	...	0.01	na	1 120	3	13
33 200	46	90	51.3	0	310	4.2	65/7	1.6	692.3	10.5	0.3	43	na	na	340
1 600	40	26	45.3	—	3 330	3.0	na/na	...	36.1	...	47.0	76	0	6	10
2 650	37	26	38.0	25	1 900	7.2	na/na	...	60.3	...	56.0	147	0	3	17
2 900	36	26	34.8	—	1 610	5.8	na/na	...	57.3	3.1	—	54	0	7	15
375	53	2	76.5	—	17 000	—	na/na	...	...	...	—	582	na	3	na
2 800	32	54	100.0	6	380	4.3	na/na	1.6	129	28.0	2.0	1	0	na	45
850	36	4	1.0	39	6 020	7.6	56/31	16.0	1 187	3.4	2.1	827	50	4	3.2
17 000	34	na	59.1	3	710	3.7	48/13	2.0	2 953	14.9	2.9	246	6	125	60
1 400	44	10	27.8	—	4 670	9.1	na/na	5.6	151.2	7.3	38.0	104	47	8	12
13 200	37	65	38.1	2	240	2.0	na/na	0.7	606	...	2.0	98	2	175	105
850	56	0.3	9.0	100	12 300	3.4	100/na	5.1	3 853	1.6	13.0	4 842	300	5	7
340	50	13	—	5	na	—	na/na	...	...	...	0.4	na	315	—	—
422	40	3	42.8	—	10 700	—	na/na	...	...	...	0.4	na	100	4	6
10 000	14	46	3.4	—	580	—	na/na	...	105.3	10.4	—	9	0	na	95
13 300	42	71	—	11	150	4.7	11/4	3.2	1 922	25.0	0.9	46	na	2 300	245
1 340	39	14	70.2	9	2 525	0.4	na/na	4.4	...	...	0.9	1 029	37	11	7
280	39	10	87.8	22	10 900	4.3	100/74	2.1	...	...	4.0	34 300	na	2.0	3
7 200	43	41	30.6	60	470	3.0	100/na	3.2	4 911	10.1	1.6	298	44	30	60
9 340	35	64	13.3	15	420	4.0	na/na	2.2	9 156	2.6	0.04	33	45	120	175
1 800	23	9	1.0	100	3 050	9.5	99/40	3.0	54	2.0	0.8	29	43	10	8
8 000	41	72	44.6	39	820	6.2	82/na	1.7	251.2	5.7	4.0	300	25	62	24
320	53	3	21.7	4	23 700	7.3	100/86	2.6	...	...	1.5	731	590	2.3	2.2
329	52	6	29.1	6	32 800	4.9	na/na	2.1	...	...	3.1	13 200	475	2.9	2.1
1 350	26	22	32.8	14	990	4.1	97/52	11.6	14 959	25.2	2.0	562	21	17	46

[2] Statistics for the newly created Soviet Republics and those of the newly created states from the former Yugoslavia are those for the former nation as a whole, where they were given as averages (eg. birth, death). They may not be applicable to each newly formed republic.

Country and Official Name	Capital city	Land area ($10^3 km^2$)	Population (10^6)	Density (persons/km^2)	Annual population growth rate (1986–1991 average)[1]	Urban population (as a percentage of total population)	Population under 15/over 60 (%)	Life expectancy Male/Female (years)
Taiwan *Republic of China*	Taipei	36	21	573	1.1	75	32 / 7	71.3 / 76.8
Tajikistan *Republic of Tajikistan*	Dushanbe	143	5.6	39	3.2	55	43 / 6	67.0 / 72.1
Tanzania *United Republic of Tanzania*	Dar es Salaam and Dodoma	943	26	27	2.8	27	46 / 6	50.0 / 55.0
Thailand *Kingdom of Thailand*	Bangkok	513	57	111	2.0	23	29 / 7	65.1 / 69.2
Togo *Republic of Togo*	Lomé	57	3.7	65	2.9	25	50 / 5	54.0 / 58.0
Tonga *Kingdom of Tonga*	Nuku'alofa	0.8	0.1	125	1.3	30	41 / 6	61.0 / 64.8
Trinidad and Tobago *Republic of Trinidad and Tobago*	Port of Spain	5	1.3	246	1.3	69	34 / 8	69.7 / 74.7
Tunisia *Republic of Tunisia*	Tunis	155	8.4	55	2.1	54	40 / 7	66.4 / 68.7
Turkey *Republic of Turkey*	Ankara	779	59	77	2.1	61	37 / 7	68.0 / 72.0
Turkmenistan *Republic of Turkmenistan*	Ashkhabad	488	3.9	8	2.6	55	41 / 6	62.9 / 69.7
Tuvalu *Tuvalu*	Fongafale	0.02	0.001	395	2.0	34	32 / 8	61.0 / 63.0
Uganda *Republic of Uganda*	Kampala	241	17	72	3.0	10	48 / 4	51.4 / 54.7
Ukraine *Ukraine*	Kiev (Kyyiv)	604	52	87	0.1	66	22 / 18	66.0 / 75.0
United Arab Emirates *United Arab Emirates*	Abu Dhabi	78	2.0	26	3.1	78	32 / 2	69.0 / 74.0
United Kingdom *United Kingdom of Great Britain and Northern Ireland*	London	244	58	237	0.3	89	21 / 20	72.4 / 78.0
United States *United States of America*	Washington	9 529	255	27	0.9	75	23 / 16	72.0 / 78.8
Uruguay *Oriental Republic of Uruguay*	Montevideo	176	3.1	18	0.6	86	27 / 16	68.9 / 75.3
Uzbekistan *Republic of Uzbekistan*	Tashkent	447	21	48	2.6	55	41 / 6	66.2 / 72.6
Vanuatu *Republic of Vanuatu*	Vila	12	0.2	13	2.4	18	45 / 5	61.1 / 59.3
Venezuela *Republic of Venezuela*	Caracas	912	20	22	2.5	84	41 / 5	66.9 / 73.1
Vietnam *Socialist Republic of Vietnam*	Hanoi	330	69	210	2.0	20	39 / 7	63.0 / 67.0
Virgin Islands (U.S.)	Charlotte Amalie	0.4	0.1	292	–0.1	45	36 / 7	66.7 / 70.7
Western Samoa *Independent State of Western Samoa*	Apia	3	0.2	57	2.5	21	41 / 6	64.0 / 69.0
Yemen *Republic of Yemen*	San'a	532	12	23	0.7	29	46 / 6	49.0 / 51.0
Yugoslavia *Federal Republic of Yugoslavia*	Belgrade	102	10.3	101	0.6	56	24 / 12	68.2 / 73.2
Zaire *Republic of Zaire*	Kinshasa	2 345	41	18	2.4	40	45 / 5	52.3 / 55.7
Zambia *Republic of Zambia*	Lusaka	752	8.3	11	3.8	50	49 / 5	54.4 / 56.5
Zimbabwe *Republic of Zimbabwe*	Harare	391	9.9	25	2.7	28	51 / 3	59.0 / 62.6

[1] Growth rate is the difference between birth rate and death rate.

Persons per physician[2]	Economically active population (%)	Labour force in agriculture, forestry, fishing (%)	Land area in farms (%)	Irrigated land as a (%) of arable land	GNP per capita ($/person)	Expenditure on education (% of GNP)	% of students enrolled in primary/secondary education	Military expenditure (% of GNP)	External public debt (10^6 $U.S.)	External debt service ratio (%)	Tourism receipts (% of GNP)	Tourist arrivals, 1990 (10^3)	Newspaper circulation/1000 population	Persons per TV	Persons per motor vehicle
965	50	21	78.5	38	7 940	3.6	na/na	5.4	2 751	1.1	1.1	1 934	200	3	8
250	46	43	30.1	—	na	—	na/na	...	...	...	na	na	310	3	13
20 000	48	84	8.5	4	120	3.7	48/na	4.1	5 294	18.1	2.3	153	8	310	250
5 000	54	56	27.6	22	1 425	3.2	na/na	2.7	12 572	10.5	5.5	5 299	90	17	20
13 000	41	70	7.1	—	410	5.2	72/na	3.3	1 096	9.8	1.5	103	na	150	50
2 150	25	45	44.5	—	1 010	4.2	na/na	—	49.2	3.7	9.0	21	75	na	25
1 215	38	11	25.8	30	3 475	4.9	91/71	1.6	1 808	12.8	2.1	194	140	5	4
1 850	30	22	61.1	9	1 420	6.3	95/23	2.8	6 506	22.2	8.2	3 204	33	12	15
1 190	38	45	39.9	9	1 630	1.8	84/40	4.1	38 595	24.6	3.4	4 799	55	5	25
250	43	42	73.4	—	na	—	na/na	...	...	...	na	na	320	3	13
2 150	55	1	—	—	530	—	100/na	...	...	...	—	1	0	na	na
21 000	45	83	11.3	0	220	3.4	53/na	1.5	2 301	34.8	0.3	50	4	205	360
250	51	25	68.5	—	na	—	na/na	...	...	...	na	na	485	3	13
650	47	5	0.2	17	19 850	2.1	100/na	5.3	1 473	2.1	—	616	195	11	4
611	50	2	75.8	2	16 070	4.7	99/79	4.2	...	...	1.6	18 021	388	2.9	2.6
404	50	3	41.0	10	21 700	6.8	95/88	5.8	...	...	0.7	39 772	251	1.2	1.3
344	40	4	90.7	9	2 560	3.1	88/55	2.2	3 044	29.8	3.3	1 267	na	5	8
250	43	29	59.4	—	na	—	na/na	...	...	...	na	na	322	3	13
8 000	47	77	15.0	—	1 050	5.5	na/na	...	30.6	1.5	—	35	0	135	22
600	37	12	34.3	8	2 560	4.2	87/44	1.0	24 643	17.0	0.7	525	na	6	8
3 050	47	67	27.4	32	210	3.0	88/na	19.4	15 072	30.1	0.6	180	38	30	na
625	39	1	20.9	—	11 740	7.5	na/na	—	...	...	—	523	240	3	2.4
3 600	26	60	24.8	—	730	5.9	na/na	—	91.7	5.1	—	48	0	27	40
5 550	26	71	0.1	23	640	6.1	na/na	9.9	5 040	5.0	0.3	52	11	110	30
450	25	5	na	0	4 200	na	na/na	3.6	13 492	6.6	2.2	7 880	100	4	6
23 200	38	67	2.6	0	230	0.9	60/17	0.8	8 851	6.2	0.1	46	1	1 750	190
8 437	34	70	1.3	0.5	420	5.5	80/15	1.4	4 784	10.6	0.2	141	17	45	50
6 950	42	65	76.6	8	640	8.5	100/na	6.7	2 449	19.1	0.7	606	22	68	36

[2] Statistics for the newly created Soviet Republics and those of the newly created states from the former Yugoslavia are those for the former nation as a whole, where they were given as averages (eg. birth, death). They may not be applicable to each newly formed republic.

GLOSSARY

(The terms that appear in the definitions in italic type are also defined in this glossary.)

Aboriginal population Those persons who reported at least one aboriginal status (Indian, *Métis* or *Inuit* ancestry) and/or reported being *registered* under the *Indian Act* of Canada. Similar definitions exist in other nations such as Australia and New Zealand.

Aboriginal rights rights claimed by *First Nations peoples* as the original inhabitants of a country

Abyssal plain a large, generally flat area of the ocean floor at a depth of 4000 m or more

Acid deposition also called "acid rain" or acid *precipitation.* It is produced by sulphur and nitrogen emissions from the burning of *fossil fuels.* Coal and oil used in energy production, industrial boilers and automobile engines all promote acid rain. When washed from the *atmosphere*, the precipitation increases the level of acidity in lakes, streams and *soils*, and kills vegetation, fish and wildlife.

Agriculture farming; this involves the work of cultivating soil, producing crops and raising animals.

Air mass a large body of air with generally the same temperature and moisture conditions throughout. Warm and moist or cool and moist air masses usually develop over large bodies of water, and hot and dry or cold and dry air masses develop over large land areas.

Alluvial soil an *azonal* soil developed from materials deposited by moving water. Alluvial *soil* is often found in the *deltas* of rivers. It is usually young, rich in *minerals* and very valuable for *agriculture.*

Alpine occurring at high altitudes (for example, alpine forest or alpine climate)

Anticline an arch-like upfolding of *rock* resulting from pressure in the earth's *crust*

Aquifer an underground layer of *permeable rock*, such as sandstone or limestone, that contains water. The water accumulates in this layer because its movement is blocked by *non-porous* rock. Aquifers become underground reservoirs for water.

Arable land land suitable for ploughing and cultivation. The term does not include *pastureland* or forested areas not capable of growing crops.

Arctic the high *latitudes* in the northern or southern *hemispheres*

Arctic climatic region a climatic region in Canada's Far North characterized by low *precipitation*, very cold winters, and cold summers. The average temperature is rarely above freezing.

Asthenosphere a layer of the earth's interior extending from 80 to 250 km beneath the surface. The asthenosphere is partially molten and *convection currents* can move it.

Atmosphere the envelope of gases that surrounds the earth. It consists mostly of nitrogen, oxygen and carbon dioxide, along with some other gases and water vapour.

Azonal soil one of the three major *soil* groups, known as orders. these soils are young and have indistinct *horizons* (layers). *Alluvial soil* is an example.

Badlands an area in a semi-arid environment where deep gullies and ravines have been formed by water *erosion.* In Canada, the largest area of badlands is in southern Alberta.

Barometric pressure a measurement of the air pressure in the atmosphere which, in the metric system, is measured in kilopascals.

Barrel a unit of measurement in the *imperial system* equal to approximately 160 L or 0.16 mL^3

Bathymetric contour. See *Isobath*

Bedrock the solid *rock* that usually lies beneath the *soil*

Billion In North America, this number represents one thousand million; elsewhere, the term means one million million. To avoid confusion, the term "one thousand million" can be used to indicate the North American billion.

Bioaccumulation a process whereby chemical substances increase in concentration as they are passed along to higher levels of the *food chain*

Biochemical oxygen demand (BOD) the amount of oxygen dissolved in water that bacteria require to decompose *organic wastes*

Biosphere that part of the earth which supports life; it consists of two layers, the *atmosphere* and the *lithosphere*

Birth rate the number of live births per thousand people, usually measured over a period of one year

Boreal the *coniferous* forest area of Canada, Russia, the Baltic Republics and Scandinavia; the term means "of the north".

Boreal climatic region a climatic region that stretches across Canada south of the Arctic. Moderate *precipitation* and broad range in temperature are its key characteristics.

Broadleaf trees trees with wide, flat leaves rather than needle-like leaves. In Canada, most broadleaf trees lose their leaves in winter. Examples include oak, maple, birch and poplar.

Calorie (cal) a unit of energy in the *imperial system* of measurement, defined as the amount of heat required to raise the temperature of one kilogram of water one celsius degree.
1 cal = 4.187 *joules.* The term is often used to indicate energy value of foods.

Cambrian period the first geological period of the Palaeozoic Era, extending from about 600 000 000 to 500 000 000 years ago. Much of the world was covered by water and invertebrates (small water animals) flourished.

Canadian International Development Agency (CIDA) a Canadian government organization formed to assist *developing countries* with technical advice and loans of both money and personnel

Canadian Shield an area of *Precambrian rock*, mostly *igneous*, that covers almost half of Canada. It is the dominant physical feature in the Near North region.

Canyon a steep-walled gorge on land, or a deep trough in the ocean

Capital money, property, or goods that can be used to produce income for a person, company or country

Census Metropolitan Area (CMA) in Canada, a city with a population over 100 000

CFCs chlorofluorocarbons. Synthetic gases containing chlorine, fluorine, and carbon. When released, they may reduce the amount of *ozone* in the *atmosphere.* Common sources of CFCs include some foam materials (e.g., foam cups), some refrigerants, aerosol sprays, and cleaning solvents.

Chernozem a fertile black or dark brown *soil*, rich in *humus.* Chernozems are found in the Canadian prairies, the Ukraine, Eastern Europe, and the United States.

CIDA. See *Canadian International Development Agency*

Clear cutting the harvesting of all trees that are large enough for commercial use

Climatic region an area in which the general conditions of temperature and *precipitation* are reasonably similar. For example, the term "desert" can be used to describe a particular area of dry conditions.

Climograph a combination line and bar graph used to illustrate temperature and *precipitation* for a climatic station. Temperature is shown as a line and precipitation as a series of bars, each representing one month.

Confederation the union of independent political units to form one nation. Canada's Confederation took place on July 1, 1867 when the British North America Act established the Dominion of Canada by joining Nova Scotia, New Brunswick, Quebec and Ontario. In 1991, the Confederation of Independent States (CIS) joined many of the former Soviet republics.

Coniferous forest a type of forest containing mainly needle-leaf trees that have cones and usually have *softwood* trunks. Examples of coniferous trees include douglas fir, cedar, spruce and hemlock. These forests thrive in sub-arctic areas with infertile acidic *soil*.

Constitution the system of fundamental laws and principles of government, in written form. Canada repatriated its constitution in April 1982.

Consumer demand purchase of goods by consumers, which creates a need for more production

Continental climate a type of climate where no large body of water moderates the temperature. This results in a wide range of temperature and low *precipitation*, usually in summer. Winnipeg in Canada and Verkhoyansk in Russia have a continental climate.

Continental crust the solid layer above the *lithosphere* that underlies the major continents. It is thicker than the solid layer under the oceans.

Continental drift the movement of the large, rigid *plates* in the earth's *lithosphere*. The theory of continental drift was developed by Alfred Wegener in the early twentieth century. But it was not until the 1960s that the theory became widely accepted, through the work of a Canadian scientist, J. Tuzo Wilson.

Continentality the influence of a large land mass on climate. **See** *Continental climate*.

Convection current The movement of material in liquids and gases caused by differences in heat. Warmer and less dense materials tend to rise and cooler, more dense materials fall. In the tropics, convection currents refer to a situation where the air heats up, rises, cools, condenses and produces precipitation.

Cordilleran climatic region a climatic region along the west coast of Canada, characterized by variations in temperature and *precipitation* over short distances. These variations are caused by differences in altitude, *relief*, and exposure within the region.

Core the centre of the earth. The temperature of the innermost part, the *solid core,* exceeds 4000°C. This core contains high concentrations of iron and nickel. The *liquid core* that surrounds the solid core is not actually liquid, but it is more fluid than the solid core.

Cropland land used to raise crops, such as wheat, rice, corn or sugar cane

Crown land in Canada, land owned by the nation. The government derives income from the sale or lease of this land.

Crude oil oil in its natural, unrefined form, as it comes from the earth

Crust the relatively thin outer layer of the earth containing both the ocean basins and the continents. It is different in chemical and physical properties from the mantle beneath.

Crustacean a class of invertebrates with a hard outer shell and jointed appendages. Examples include shrimp, lobster and crab.

Cultural mosaic The coexistence of different customs, arts, and languages in a society. Toronto and Vancouver are both examples of cities that have a diverse culture. Toronto has the distinction of being the most culturally diverse city in the world.

Cyclonic storm a low pressure area, often accompanied by warm and cold fronts, which brings *precipitation* to the middle *latitudes*

Death rate the number of deaths per 1000 people, usually measured over a period of one year. Canada's death rate is 7.3; in Guinea and Afghanistan, the death rate is at least 20.0.

Deciduous forest a forest consisting mainly of trees that regularly shed all their leaves, usually in the autumn. The trees are dormant during the winter. Examples of deciduous trees include maple, oak, and birch.

Delta a river deposit formed at the point where the river exits into a large body of water. Fine materials such as silt and mud are deposited, producing a land feature of a triangular shape (like the Greek letter "delta", Δ, hence its name).

Deposition the laying down of materials carried by water, ice or wind

Desertification a process by which deserts extend into surrounding areas. Desertification can be caused by climatic change or by human activities. It is most significant in the Sahel region of Africa.

Developed countries one of two basic classifications of countries (**see** *developing countries*) determined by examining factors such as economic development, *Gross National Product* per capita, income per capita, potential for development, energy use, *literacy*, and quantity and quality of food. Based on their available resources, developed countries are thought to be able to provide a reasonable quality of life for all their inhabitants.

Developing countries one of two basic classifications of countries, determined by comparison with *developed countries* (**see** *developed countries*). Developing countries face many challenges, particularly in the area of providing a reasonable quality of life for their inhabitants.

Devonian period the fourth geological period of the Palaeozoic Era, extending from 400 000 000 to 345 000 000 years ago. This was the period in which fish developed.

Differential erosion the wearing away of different types of *rock* in the same location at different rates. This process forms distinct landscapes. Rocks containing resistant *minerals* are less easily worn down by water, wind and ice. In humid climates, certain types of limestone easily resist this process whereas sandstone and shale break down easily.

Diplomatic relations a relationship between countries where they exchange ambassadors to talk over political and economic questions of interest to both. The term is usually used in the negative, when two countries "break off" diplomatic relations because talks have broken down.

Domestic trade the movement of goods and services within a country. For example, the movement of goods among provinces of Canada would be considered domestic trade.

Drainage basin an area drained by a river or series of rivers into a common body of water. In Canada, the Hudson Bay drainage basin is the largest, draining nearly 3.8 million square kilometres, from southern Alberta to Baffin Island. Other examples are the Pacific and St. Lawrence drainage basins.

Drumlin a hill, usually oval or tear-shaped, formed by glaciers. Drumlins are often found in groups or fields. In Ontario, many are found near Peterborough and Guelph.

EC. See *European Union*

Ecoclimatic region a large area in which similar climatic conditions have resulted in similar ecological responses (*soil*, vegetation, animal life)

Economic disparity a large difference in economic strength between two areas. For example, a great economic disparity exists between *developed* and the *developing* countries.

Ecosystem a living community in the water or on land. The concept emphasizes the interactions among the host environment and the plants and animals it contains. Ecosystems can be as small as a pond or as large as a *tropical rainforest.*

Effluent liquid waste from *industry* or domestic activity

Emigration the movement of people **out of** an area or country

Emirate a country ruled by a Middle Eastern Muslim ruler called an **emir**. Seven emirates joined together in 1971 to form an independent country called the United Arab Emirates.

Endangered species an animal or plant species threatened with imminent *extinction* in a specific area

Equator an imaginary circle, indicated by a line, which is equally distant from the north and south poles and divides the earth into northern and southern hemispheres. It is the basis for distances measured on the surface of the earth called *latitude.*

Equatorial scale a scale for the drawing of maps that is accurate at or near the *equator.* The earth cannot be exactly represented on a flat surface because its round shape creates distortions. In some map projections, the scale varies with the distance from the equator.

Equinox a date on which the sun is directly overhead at noon at the *equator.* March 21 and September 21. (Plural: equinoxes)

Erosion the wearing down and carrying away of material from the earth's surface by the agency of water, wind and ice. (Verb: erode)

Escarpment a steep slope or cliff formed by *faulting* or *differential erosion.* Examples include the Niagara Escarpment in Ontario and the escarpments on either side of the Great Rift Valley in Africa. The Great Rift Valley was formed by faulting and the Niagara Escarpment was formed by the *erosion* of several types of *rock* of differing resistance.

Esker a long narrow *ridge* of rounded and *sorted* materials, usually quite coarse, such as sand and gravel. Eskers are formed in or under *glaciers*, as *meltwaters* deposit materials and create streams.

Ethnicity membership in a group of people as distinguished by customs, language, and background.

European Union an alliance of twelve member nations and some associate members. It aims to integrate the members' economies, and to have a common currency, co-ordinated social development and eventual political unity. Full members of the community include Belgium, Denmark, France, Germany, Greece, Ireland, Italy, Luxembourg, the Netherlands, Portugal, Spain and the United Kingdom. Formerly called the European Community or the Common Market.

Exports goods and *services* sold in other countries. Canada's chief exports include automobile parts, grains, *minerals*, and forest products.

External aid assistance that one nation provides another, usually in the form of goods, money, or technical expertise

External debt. See *Foreign debt*

External trade trade with other countries (as compared to *domestic trade* which takes place **within** a country). Trade between Canada and the United States is external trade whereas trade between British Columbia and Ontario is domestic trade.

Extinct species a plant or animal that no longer exists. An example would be the passenger pigeon. (Noun: extinction)

Extirpated species a plant or animal that no longer exists in one location but is still found elsewhere. This official designation has been assigned by the Committee on the Status of Endangered Wildlife in Canada to a species or sub-species of plant or animal which was formerly native to Canada and is no longer known in Canada but may be found elsewhere in the world.

Extractive industry an industry that takes *raw materials* from the natural environment for *refining* or *manufacturing.* Examples include mining, fishing and lumbering.

Extrusive rock igneous rock formed on the surface of the earth from quickly cooling lava or ash. Examples include basalt, pumice, and breccia.

Far North. See *Arctic*

Fast ice ice that covers a body of water and shows little or no movement during the winter season

Fault a fracture in the earth's *crust* along which *rock strata* have moved vertically or horizontally. Faults may trap oil or gas. (Also: faulting)

Feedlot a method of raising beef cattle for market in which cattle are confined to a small area and fed highly efficient grains

First Nations all peoples who are descendants of Canada's original inhabitants, including North American Indians, Inuit and Métis.

Flora the plant life of a region

Folding the bending of *rock* layers, often resulting in the formation of "fold" mountains. The Rocky Mountains and the Andes are examples of fold mountains. **See also** *Anticline* and *Syncline*

Food chain a flow of energy and nutrients through an *ecosystem* in a pattern of feeding relationships. For example, *plankton* are eaten by small fish which are eaten by large fish which are eaten by people.

Foreign debt one nation's debt to another. Also called External debt

Foreign investment the investment of money in a country by individuals or companies from another country. Many *developing* countries have large amounts of foreign investment.

Fossil the imprint or remains of a prehistoric plant or animal, usually found in *sedimentary rock*

Fossil fuel any energy source originating from prehistoric plants and animals, mostly found in or associated with *sedimentary rock.* The main fossil fuels are coal, peat, natural gas and petroleum. The burning of fossil fuels is often associated with air *pollution* and *acid deposition.*

Front the surface or line between masses of air that have differing characteristics. A warm front marks the advance of warm air into cooler air. A cold front marks the advance of cold air into warmer areas. An occluded front occurs when a cold front runs underneath a warm front.

Frost-free period the total number of days between the average dates of the last frost in the spring and the first frost in the autumn

Gauging station a point at which information on water flows is measured. The data collected are used to interpret patterns in stream flow and to predict their consequences.

GDP. See *Gross Domestic Product*

Generating station a plant where electricity is produced. Power sources include falling water, coal, petroleum, natural gas, nuclear fission or any other energy source

Generator a machine that converts mechanical energy (for example, falling water) into electrical energy

Geologic time the division of the earth's history of approximately 4.5 billion years into eras and periods

Geological province a large area whose *rock* structure, rock type, and/or rock age show common characteristics

Glacier a slow-moving mass of ice found at high *latitudes* and high altitudes

Gneiss a coarse-grained *metamorphic rock* that resembles granite and usually consists of alternating layers of different minerals which give the rock a banded appearance

GNP. See *Gross National Product*

Grassland a region where vegetation consists mainly of grasses because moisture is insufficient to support trees. Grassland areas often have regional names, such as Pampas (Argentina), Steppe (Ukraine), and Prairie (North America).

Greenhouse effect a warming of the *atmosphere* caused by retention of the energy from the earth's *ecosystem.* Human activity has affected the gases that make up the atmosphere and may be causing an acceleration of this natural process.

Gross Domestic Product the total dollar value of all goods and *services* produced in a country in a given year. A high Gross Domestic Product (GDP) indicates a high level of development.

Gross National Product the total dollar value of all goods and *services* produced in a country in a given year *plus* transfer payments from other nations. In most situations, the Gross National Product (GNP) and the *Gross Domestic Product* are very close in value.

Ground water the water in the *soil* and in the *bedrock* underlying the soil

Growing degree days a figure compiled by adding the number of degrees by which the average temperature of each day in a year exceeds 6°C. This figure is important to farmers when planting crops.

Growing season the season in which warmth and moisture is sufficient for crops to grow

Habitat an environment that supports plant, animal or human life

Hardwood wood produced by most deciduous trees such as oak and maple

Hazardous waste discarded materials that pose a risk to humans or the environment. Special disposal techniques are generally required.

Heath an open area, often found at high altitudes and in high *latitudes*, usually covered in grasses or short shrubs

Heating degree days a figure compiled by adding the number of degrees by which the temperature of each day in a year goes below 18°C. At 18°C, the furnace is usually turned on in homes and businesses.

Hemisphere any half of a globe or sphere. The earth has traditionally been divided into hemispheres by the *equator* (northern and southern hemispheres) and by the prime meridian and International Date Line (eastern and western hemispheres).

Horizon a distinct layer in a *soil* profile. Most soil profiles have three horizons. Sub-categories of soil within the three basic horizons are indicated by the letters A, B and C.

Humidity the amount of moisture in the air, expressed as a percentage of the total amount of moisture the air could hold. For example, 95 percent humidity indicates that *precipitation* will occur shortly because the air has reached nearly 100 percent of its carrying capacity.

Humus the upper layer of the *soil* consisting of decaying and decayed *organic materials*

Hurricane an intense *cyclonic storm*. It is usually established over *tropical* waters, but it may migrate into *temperate* areas. For reference purposes, hurricanes are given names (e.g., Hurricane Andrew, Hurricane Hugo, Hurricane Hazel).

Hydro-electric power electricity produced by the natural movement of falling water, such as at Niagara Falls in Ontario or Churchill Falls in Labrador

Hydrocarbon a substance containing only compounds of carbon and hydrogen. There are thousands of these compounds; examples include methane and *crude oil*.

Ice cap a large mass of ice, smaller than an ice sheet. The largest ice cap today is found in Greenland.

Ice floe free floating ice detached from the polar ice mass.

Ice sheet a very large mass of ice that may cover whole continents. The last continental ice sheet covered much of North America until about 20 000 years ago.

Igneous rock *rock* formed by the cooling of molten materials from the interior of the earth. **See also** *Intrusive rock* and *Extrusive rock*.

Immigration the movement of people **into** an area or country

Imperial system a standard system of measurement for weights and distances which was established by British law, but is now used mainly in the United States. In Canada it was replaced by the metric system.

Impermeable rock. See *Non-porous rock*

Imports goods and *services* purchased from another area, usually another country. Canada's main imports include automobiles, petroleum, and electrical goods.

Improved land areas of the earth's surface that have been either cleared of trees or ploughed for the growing of crops

Indian Act a federal statute specifying government responsibilities regarding *First Nations*. The Act has been amended many times since it was passed in 1876. In the past 25 years, the federal government and First Nations associations have met frequently to discuss major changes to the Act. In 1985, an amendment to the Act restored full *registered* status to many individuals who had lost their status, especially through marriage outside of the First Nations.

Industrial mineral a *mineral* such as stone, sand or gravel, used in construction

Industry *extractive* processes (mining, forestry, fishing), *manufacturing* and *services* such as commerce and insurance. The term does **not** include *agriculture*.

Infant mortality rate the number of deaths of children under one year of age per 1000 births in a given year. In *developed* countries, this figure is very low (less than 10 per thousand), but it may be as high as 200 per thousand in nations where there are few health services.

Inflation a general increase in the price of goods and *services* over time. In periods of high inflation, the purchasing power of money decreases. High rates of inflation can erode gains made by weak or struggling economies.

Inorganic materials in *soil*, materials such as *rock* fragments, liquids, and gases. Inorganic and *organic materials* combine to produce soil.

Inshore fishery the sector of the fishing *industry* that uses small boats (under 10 tonnes) that return to the home port each day. It operates less than 15 km from shore.

Intensive farming a form of *agriculture* that gains high yields per hectare by investing much labour and capital. In Canada, examples include the use of greenhouses, *feedlots* and irrigated plots of land.

Internal drainage a drainage system with no outlet to the ocean. It usually occurs when land in the interior of the continent is below sea level. Examples include the Dead Sea in Israel, the Caspian Sea in Russia and Lake Eyre in Australia.

International date line an imaginary line that follows 180° *longitude* (approximately). The area of the world just east of the line is one day ahead of the area just west of the line. The line varies slightly to avoid splitting islands or countries into separate days.

Intrazonal soil one of the three major *soil* orders. Soils in this group develop in areas where *bedrock* or drainage has more influence on development than climate or natural vegetation.

Intrusion the penetration of molten material into the earth's *crust*, usually along a fracture or *fault*

Intrusive rock *igneous rock* formed **within** the earth's *crust* from magma (molten rock that has not reached the earth's surface). Examples include granite, gabbro and serpentine.

Inuit the *First Nations peoples* who live in the Far North region of Canada and Greenland

Isobar a line joining points on a map with the same atmospheric pressure. **See also** *Atmosphere*

Isobath a contour that shows *relief* on the floor of a body of water

Isoline a line on a map joining points with the same numerical value. ***(Note:** The prefix "iso" means equal and is used in terms such as isotherm (equal temperature) and isobar (equal pressure).)*

Isostatic rebound the vertical movement of parts of the earth's *crust* in an attempt to maintain vertical equilibrium within the *crust*. When the continental *glaciers* receded, the land mass of Canada rose after the weight of the ice was gone. In some places, the process continues.

Isotherm a line joining points on a map that have the same temperature

Jet stream a tube-like band of high velocity wind found in the upper *atmosphere*. The location of the jet stream has a significant influence on weather patterns. The polar jet stream often marks the division between cold *arctic* air and warm *tropical* air.

Joule (J) a metric unit of energy defined as the work done by the force of one newton when it moves its point of application a distance of one metre

Labour force the total number of people working or looking for paid work

Land claims the case presented by *First Nations* for ownership and control of lands on which they live or have lived. Some claims have been settled, but many more have not yet been settled either through negotiation between the government and the First Nations groups or through the courts.

Land use the type of human activity that a given land area is used for. Common land uses include *agriculture*, and residential, commercial and industrial uses.

Landed value the dollar value of fish caught, before they are marketed and processed

Landform any feature of the earth's surface formed by earth movements or by the wearing down of the surface of the earth

Latitude distance north and south of the *equator*, measured in degrees. The north pole is at 90°N and the south pole is at 90°S. All lines of latitude run parallel to the *equator*.

Leaching the natural removal of *soluble minerals* in *soil* from the A *horizon* to the B horizon by *percolating* water. The process is most significant in humid climates.

Lichen a living partnership between an alga and a fungus. Lichens exhibit a great range of forms and are plentiful in many locations, but they are most commonly associated with *tundra* areas.

Life expectancy the average number of years that an individual is expected to live. The availability of food and health care are major factors in determining life expectancy in a country.

Literacy the ability to read and write. Various scales are used to measure literacy, which makes world-wide comparisons difficult. National and international organizations are trying to raise literacy rates throughout the world.

Lithosphere the solid outer layer of the earth including the top part of the *mantle* and all of the earth's *crust*

Longitude distance east and west of the *prime meridian*, measured in degrees. The lines of longitude, called meridians, join the north and south poles.

Mantle a concentric layer of the earth's interior, nearly 3000 km thick, between the *crust* and the *core*

Manufacturing *industry* that changes *raw materials* into finished products, usually with the assistance of machinery. Automobile manufacturing is one of Canada's biggest industries.

Market economy an economic system that depends on *consumer demand* to stimulate the production of goods for the market. It differs from a *centrally planned* economy, in which the state determines what goods will be produced.

Mass transit the movement of people, usually in an *urban* area by bus, streetcar, subway, LRT (light rail transit) or commuter train.

Meltwater water produced by the thawing of ice or the melting of snow. Meltwaters from the last glacial period covered large areas of Manitoba, southern Ontario and Quebec, and the Clay Belt of the Near North.

Metallic mineral a *mineral* that yields a metal when processed. Examples include iron, gold, silver, copper and uranium ore.

Metamorphic rock a type of *rock* that results from changes in other rock types. For example, heat and/or pressure on existing *sedimentary* or *igneous* rocks produces marble, *gneiss* and quartzite.

Meteorologist a person who studies the phenomena of the *atmosphere;* many meteorologists are employed to forecast weather patterns.

Métis a person of mixed European and First Nations descent

Mid Atlantic Ridge a system of submarine mountains (*seamounts*) that thrust up from the Atlantic's *abyssal plain*. The ridge runs parallel to the ocean margins and marks the edges of tectonic *plates*.

Mid Atlantic Rift a steep-sided trough within the *Mid Atlantic Ridge*

Migration the movement of people, birds, or animals from one location to another

Military regime a type of government controlled by the military. It often results from a coup (overthrow of the previous government).

Millibar (mbar) a unit of atmospheric pressure used on weather maps. 1000 mbar = 1 bar

Mineral a naturally occurring crystalline substance with a specific chemical composition and regular internal structure, made up of two or more elements. Examples include quartz and feldspar. Most *rocks* consist of combinations of minerals.

Mineral fuel a fuel produced from *minerals*. Examples include uranium and *fossil fuels* such as coal and oil.

Mixed farming a type of *agriculture* which combines crops and livestock

Moisture deficiency an amount of moisture needed for plant growth that is **not** provided by *precipitation*. When high temperatures increase evaporation, a moisture deficiency usually occurs. Irrigation is often used in *agriculture* to reduce the deficiency.

Monsoon a wind that changes with the season. It develops from the differences in pressure over land and water in summer and winter and causes distinctive wet and dry seasons. The summer monsoon of the Indian subcontinent is the most notable because it has been associated with rain.

Moraine materials deposited by a *glacier*, often in the form of hills. *End moraines* are found at the farthest point of the advance of the glacier.

Muskeg a bog or marsh, formed by the deposit of thick layers of decaying organic matter, usually found in an arctic or sub-arctic environment

NAFTA the North American Free Trade Agreement. An organization of three countries—Canada, Mexico and the United States—whose objective is to reduce and remove tariffs among themselves in order to stimulate trade and economic development.

Native peoples. See *First Nations*.

Natural increase the difference between the number of births and deaths, often given per 1000 people. Data is usually kept for countries, but more detailed information is sometimes available for cities, provinces, or states as well.

Nautical mile a unit of distance used in navigation, equal to 1.853 km. The distance is one minute of an arc on a Great Circle drawn on a sphere the size of the earth.

Needleleaf trees with needles rather than flat leaves. The needles fall and are replaced throughout the year, not only in the autumn. Most needleleaf trees are evergreens such as pine, spruce and fir.

Net migration the difference between *immigration* and *emigration*

Newly Industrialized Countries (NIE) Hong Kong, Singapore, Taiwan and South Korea. These countries, which are sometimes called "the Four Dragons", have experienced great *industrial* and *export* growth in the past twenty years.

Nomadic without permanent residence. Nomadic peoples may be hunters following the movement of game or herders seeking pasture for domestic animals.

Non-governmental organization (NGO) a private group that raises funds to provide assistance to *developing* countries. Examples include Oxfam and Save the Children.

Non-porous rock also called impermeable *rock.* A dense type of rock that impedes or prevents liquids from passing through its few pore spaces. Examples are shale and some forms of limestone.

North-West Rebellion a conflict between the *Métis* and the federal government in 1885 in what is now Saskatchewan

Northwest Passage a sea route across northern Canada. Early explorers searched for this route to Cathay (China). The route is difficult to traverse due to ice conditions.

Nuclear power electricity produced by using the heat from nuclear fission to cause steam to drive a generator

Nuclear reactor a device designed to produce electricity by using the heat from nuclear fission. In Canada, several active reactors, mostly in Ontario, produce power and a few small reactors are used for research.

OECD the Organization for Economic Co-operation and Development, founded in 1961 to encourage economic and social welfare among members and stimulate aid to developing nations. Its headquarters are in Paris; current members include: Australia, Austria, Belgium, Canada, Denmark, France, Finland, Germany, Greece, Iceland, Ireland, Italy, Japan, Luxembourg, the Netherlands, New Zealand, Norway, Portugal, Spain, Sweden, Switzerland, Turkey, United Kingdom and United States of America. The former Yugoslavia was an "observer" member.

Offshore fishery the sector of the fishing *industry* that usually operates more than 80 km from shore. Its ships typically weigh more than 50 tonnes and stay at sea for several days before returning to the home port.

Oil (or gas) field an area in which oil or gas has been discovered. New oil and gas fields have been found off the east coast of Canada and in the Beaufort Sea.

Oil sands sands saturated with heavy *crude oil.* Large oil sands have been found near Athabasca in Alberta.

Old growth forest forest with many large old trees, including standing dead trees, as well as fallen trees. The treetops are characterized by multiple layers of canopy created by trees of varying age.

Oregon Territory one of the last areas to be designated as belonging to Canada or the United States. The area included southern British Columbia, Oregon and Washington state.

Organization for Economic Co-operation and Development. See *OECD*

Organic materials living materials, such as plants and animals

Organic soil an incompletely developed *soil* containing mostly dead *organic materials.* This soil is often found in areas of poor drainage such as marshes and swamps.

Organic wastes discarded materials that originated from plants and animals

Outcrop *bedrock* that protrudes from the surface of the land, often without vegetation or *soil* covering

Overmature forest Similar to *old growth forest,* but with significant numbers of standing dead trees and numerous fallen logs. Many trees have started to decay.

Ozone layer a layer of the *stratosphere* where oxygen (O_2) converts to ozone (O_3). Ozone absorbs much ultraviolet radiation; when the ozone layer is reduced in thickness, these harmful rays increase skin damage and lead to increases in the incidence of skin cancer.

Pacific Rim the countries around the Pacific Ocean, from Chile to Alaska on the east side and from Australia to Japan on the west side. In recent years, rapid industrialization and economic growth has occurred in some of the Pacific Rim countries.

Pack ice seasonal ice formed by the joining of several *ice floes*

Pangaea according to the theory of *continental drift,* the original land mass that broke apart to create the present continents

Parkland a transitional vegetation zone between *grassland* and *boreal* forest containing differing combinations of both grassland and forest

Participation rate the *labour force* expressed as a percent of the total population 15 years of age and older

Percolation the downward movement of water through the *soil* and through joints in the *bedrock*

Permafrost ground that does not completely thaw in summer. The surface layer thaws but the ground underneath remains frozen and does not permit *meltwater* to drain downward. Boggy conditions often result.

Permeable rock also called porous rock. A type of *rock* with many pores or spaces that allow liquids to pass through.

Pesticide chemical used to kill unwanted plants and animals. Some authorities include herbicides, insecticides, algicides and fungicides in this definition.

Plankton microscopic plant and animal life found in water. An important food source for fish and whales.

Plantation farming the use of land for a single crop. The term is usually associated with the production of tea, rubber, coffee or sugar in the tropics.

Plate a section of the earth's *crust* that "floats" in the earth's *mantle* in much the same way as an iceberg floats in water. Plate movement (*continental drift*) can cause earthquakes and volcanoes.

Plateau an upland area with a fairly flat surface and steep slopes. Rivers often dissect plateau surfaces.

Plate tectonics the study of the movement of *plates* in the earth's *lithosphere*

Podzol a shallow, highly leached and acidic *soil* usually associated with *coniferous* forests. Although generally poor for *agriculture,* podzols can be improved by adding lime. **See also** *Leaching*

Pollution the release of substances into the environment that harm living organisms and damage resources. Pollution of land, water or air can result in a range of problems from an impairment of the quality of life (for example, closed beaches) to hazards to human health (for example, an increased risk of cancer).

Population density determined by dividing the total population of a region by the total area:

$$\text{Population density} = \frac{\text{Population}}{\text{Area (km}^2\text{)}}$$

Population distribution the pattern of habitation in an area

Population profile a diagram showing the structure of a population, usually according to age and sex

Porous rock. See *Permeable rock*

Post-glacial lake a body of water formed by the *meltwaters* of a receding *glacier.* Where ice blocked the normal drainage routes, extensive ponding occurred. Following the last continental glaciation, Lake Agassiz in Manitoba and Lake Iroquois in Ontario were formed.

Prairie township system the laying out of the flat lands of the Canadian west into blocks for settlement

Precambrian that period of the earth's prehistory dating from the formation of the earth to approximately 600 million years ago

Precipitation moisture that accumulates in clouds and then falls to earth as rain, snow, hail, sleet or ice pellets

Prevailing winds the most common winds in a location. In most parts of Canada, the westerlies are the prevailing winds. In some places in the world, wind patterns change by the season.

Primary industry *industry* that works directly with natural resources, including fishing, forestry, mining and *agriculture*

Prime meridian an imaginary line at zero degrees *longitude.* It passes through Greenwich, England, so it is sometimes called the Meridian of Greenwich. All meridians are numbered east and west of this line.

Pulp ground-up, wet cellulose used to make paper. Pulp is usually made from wood, but can also be prepared from linen or rags. It has the consistency of porridge.

Quaternary the most recent period of the Cenozoic Era, also known as the age of humans

Quota a permitted number or proportion. For example, an *import* quota limits the quantity of a product that will be accepted into a country.

Rainforest a thick, luxuriant evergreen forest found in areas where high *precipitation* is evenly distributed throughout the year. *Tropical* rainforests are found the Amazon and Congo basins. *Temperate* rainforests are found in China, Australia, New Zealand, and the southeastern part of the United States.

RAMSAR the Convention on Conservation of Wetlands of International Importance, drafted in Ramsar, Iran, in 1971. The convention acknowledges the value of *wetlands* as areas of critical biological significance.

Rangeland land used for grazing cattle

Raw material material that a *manufacturing industry* processes into a more finished state. For example, iron is a raw material for the steel industry, and steel is a raw material for the ship-building industry.

Reef a *ridge* or strip of coral lying close to the surface of an ocean

Refinery a processing plant for *raw materials*, such as oil, sugar, or metals. (Also: refining)

Refugee an individual who has been compelled to move to another region or country because of political, economic or environmental crises. In recent years, many people have come to Canada as refugees.

Regeneration the process of reproducing or renewing. In the forest *industry*, regeneration refers to the planting of trees in areas cleared of forest.

Registered Indian a person who is registered under and subject to the *Indian Act*

Relief this term is used in two senses: (1) The general physical variations of the land; and (2) the difference between the highest and lowest points in a given area.

Remote sensing the gathering of information by the use of electronic or other sensing devices in satellites

Reserve land belonging to the federal government which *registered Indians* have the right to occupy and use

Resource-based product a product formed mainly from one of the basic resources of the land or sea. Paper, canned salmon and flour are examples.

Retail business that sells products or *services* to consumers

Ridge an upland area, usually long and narrow, with steep sides. Ridges are found both on land and in the ocean.

Rift a depression in the land, also called a *rift valley*. Rifts occur in the ocean when *plates* separate; an example is the *Mid Atlantic Rift.*

Ring of Fire the area around the Pacific Ocean in which *tectonic* activity is greatest. The area is named for its large number of volcanoes.

Rock a consolidated mixture of one or more *minerals*. There are three basic types of rock—*igneous*, *sedimentary*, and *metamorphic*.

Rock strata layers or beds of one kind of *rock*, usually *sedimentary* (Singular: stratum)

Run-off moisture, either from *precipitation* or from melting snow, that flows over the surface and eventually joins streams and rivers. The term is also used to describe the excess water that may accumulate during specific times of the year, for example, the "spring run-off".

Rural concerning the area outside towns and cities

Satellite image an image, similar to a photograph but recorded on bands of the electro-magnetic spectrum, taken from a satellite. Landsat images, taken from 900 km above the earth, are of value to geographers, hydrologists, geologists, and wildlife and *agriculture* specialists.

Sea floor spreading the movement of *plates* away from a central *rift* on the sea floor to create new sea floors

Sea ice a covering of thick ice over a large area of water. Sea ice is common in the Arctic Ocean.

Seamount a mountain that rises from the sea floor but does not break the surface of the water. Seamounts are often underwater volcanoes with steep slopes and pointed peaks or crests.

Sedges coarse, grasslike plants that grow in clumps, usually in areas of poor drainage

Sedimentary rock a type of *rock* formed by the compression of deposits from water, wind, and ice. Examples include shale, sandstone, limestone, and conglomerate.

Seismic zone an area of the earth's *crust* that experiences horizontal or vertical movement, often associated with earthquakes and volcanoes. Areas of high seismic activity are found along *fault* lines and on the edges of *tectonic plates.*

Services economic activities in which no goods are produced. Examples of people in the *labour force* who provide services are sales personnel, bank employees, teachers, doctors, bus drivers and accountants.

"Seven Tigers" a convenient way of referring to seven nations of Eastern Asia that are growing quickly and actively pursuing higher standards of living: Hong Kong, Indonesia, Singapore, South Korea, Japan, Taiwan, and Thailand

Seven Years' War a European war (1756–1763) that involved many countries including Austria, England, France, Prussia, Russia, Sweden and Saxony. One of the more famous battles of the Seven Years' War was that between Wolfe and Montcalm on the Plains of Abraham above Quebec.

Shaded relief map a map that illustrates the general physical variations of land using graphic representations

Sheikdom a region under the rule of a sheik, who is the head of an Arab family, clan, tribe or village

Snow belts areas that have higher than average snowfall due to the *prevailing winds*, their elevation, and the presence of a body of water. For example, there is a snow belt in southern Ontario in the lee of the Great Lakes.

Softwood wood produced by most *coniferous* trees such as pine

Soil a complex mixture of *organic* and *inorganic materials*, differentiated into layers and capable of supporting plant life

Solar energy energy produced directly or indirectly from the sun

Solar radiation radiant heat from the sun, emitted in the form of short waves. It is measured in megajoules per square metre. **See also** *Joule*

Soluble able to be dissolved. For example, certain *minerals* are soluble in water and are carried down through the *soil* from one *horizon* to the next.

Sorted materials materials that have been grouped together by size and weight by the action of water (or wind)

Specialty crop a crop that thrives in a particular area, for example, peaches from the Niagara Fruit Belt, vegetables from the Fraser *delta* area of British Columbia, or bananas from Central America

Stratosphere the layer of the *atmosphere* from 10 to 50 km above the earth's surface. Below this level, temperatures decrease with altitude, whereas in the stratosphere, temperatures increase with altitude.

Subduction the downward movement of an oceanic *plate* into the *asthenosphere*. Eventually, the plate melts into this molten mass. Subduction usually occurs along converging plate boundaries.

Submarine cable an undersea cable used to transmit telephone calls. Submarine cables may soon be replaced by satellites.

Subsistence farming a type of *agriculture* in which livestock is raised and crops are cultivated for consumption rather than for sale

Surface current the movement of water near its surface, caused by differences in water density and salinity and the rotation of the earth

Sustainable development a level of development that ensures potential for future generations. The environment and its resources are not overwhelmed by human activity.

Sustained yield the use of a renewable resource such as fish, trees, air, and water at a rate that allows the resource to renew itself. Production is not interrupted by shortages.

Syncline *rock* layers (often *sedimentary*) which have been folded downwards to form a valley, or rock layers in a valley formed between two *anticlines*

Tectonics all the processes that deform the earth's *crust*

Temperate referring to the region south of the Arctic Circle and north of the Antarctic Circle but outside of the tropics — a "middle" zone of the earth

Tertiary period the first period of the Cenozoic Era, extending from about 65 million years ago to the beginning of the glacial periods about 1 million years ago. This period involved the development of mammals.

Thermal energy electricity produced by burning *fossil fuels*

Till a mixture of *unsorted* materials deposited by *glaciers*. These materials vary in size from clay particles to boulders.

Time zone a geographical area within which clocks are set to a standard time. Every 15° of *longitude* is a different time zone. Canada extends through seven of the world's time zones.

Transform fault a fracture in the earth's surface associated with *plate* movements. It extends at right angles to the *rifts* that occur between plates.

Treeline the end of the area of forest. This is not actually a "line" but an area in which trees begin to appear. It is more defined when associated with higher altitude than with higher *latitude*.

Tropic of Cancer an imaginary line drawn 23 1/2 degrees north of the *equator* indicating the northernmost extent of the apparent movement of the sun

Tropic of Capricorn an imaginary line drawn 23 1/2 degrees south of the *equator* indicating the southernmost extent of the apparent movement of the sun

Tropical the region of the world between the Tropic of Cancer and the Tropic of Capricorn

Troposphere the layer of air directly above the earth's surface. It contains more than 95% of the earth's air and extends an average of 10 km upwards, although the range is from 7 km to 17 km depending upon *latitude*. Temperatures decrease with altitude.

Tsunami a tidal wave caused by an earthquake. Because of the high incidence of earthquakes in the *Ring of Fire*, tsunamis are mainly associated with the Pacific Ocean

Tundra a term used to describe the climate, vegetation, or *soil* of the arctic and sub-arctic regions between the forested areas and those with permanent snow and ice. Tundras have little vegetation except for mosses and *lichens*, and often have *permafrost*.

Typhoon an intense *cyclonic storm* that usually occurs in *tropical* areas, particularly in the Pacific Ocean. A similar phenomenon is known as a *hurricane* in the western *hemisphere*.

Unemployment rate the number of persons in the total *labour force* who are seeking work but have not found it, usually expressed as a percent

Unsorted materials materials that have been carried and then deposited by moving ice.

Upper mantle. See *Lithosphere*

Urban referring to a city or town. Canada's definition of an urban area is an area that contains at least 1000 people with a population density of 400 or more per square kilometre.

Urbanization the process of change in an area from a *rural* to an *urban* landscape

Value added the total change in the monetary value of goods at each stage of the *manufacturing* process. For example, when automobiles are assembled from parts, the total value of the automobile is greater than the sum of the value of the parts.

Water table the level beneath the earth's surface below which the *rock* and *soil* are saturated. The depth of the water table varies greatly, depending upon a number of factors, including type of rock, availability of water, slope and the influence of humans.

Watt the power that produces energy at the rate of one *joule* per second

Weather station a location equipped with instruments to record atmospheric conditions. In Canada, continuous data is relayed to Environment Canada for analysis and forecasting.

Wetland land whose *water table* is at or very near the surface. This includes bogs, swamps, marshes and areas of shallow water. These areas are valuable for migrating birds and also act as "filters" for water before it enters rivers or streams.

Wholesale the business of selling products and *services* in large quantities, not to the final consumer but to *retail* businesses which then sell them to consumers

Wild land an area not developed for human use, including forested areas, *parkland*, open woodland, and areas north of the *treeline*

Wind chill factor a measurement that combines the effect of low temperatures and high winds. This combination can cause frostbite or hypothermia (loss of body heat leading to death).

Wisconsin Ice Sheet the most recent continental *ice sheet* that began to recede about 15 000 years ago. It covered much of North America with ice up to 2 000 m thick.

Zonal soil the most predominant of the three major *soil* orders (zonal, *azonal* and *intrazonal*). Soils in this group are well developed and have distinct *horizons*. They reflect the influence of climate and natural vegetation. Examples include *chernozems* and *podzols*.

GAZETTEER

The names of places and features in this atlas are listed in alphabetical order. The number printed in boldface type immediately to the left of the name is the number of the page where the place or feature is best shown (some names can be found on several pages). The latitude and longitude are given for the location of the name on the map. They are listed in the two columns to the left of the page number.

Slashes are used to indicate two names for a single entry. For example, Cairo is listed twice as Cairo/El Qâhira and as El Qâhira/Cairo. When two or more entries have the same name, the country or Canadian province is given in parentheses. For example, London is listed twice, as London (U.K.) and as London (Ont.). A closed square (■) indicates a country, and an open square (□) indicates an administrative division within a country, for example, a province or state.

Names given with page references but without co-ordinates are located only on thematic maps.

Abbreviations

The following abbreviations have been used on the map plates and in the Gazetteer.

Alb.	Albania
Alta.	Alberta
Aust.	Austria
Austral.	Australia
Assoc.	Association
B.	Bay
B.C.	British Columbia
Bang.	Bangladesh
Bel.	Belgium
Bulg.	Bulgaria
Cam.	Cameroon
C.A.R.	Central African Republic
CMA	Census Metropolitan Area
Co.	Company
Corp.	Corporation
C.P.	Canadian Pacific
C.P.I.	Consumer Price Index
CUSO	Canadian University Services Overseas
Czech.	Czechoslovakia
Dem. Yemen	Democratic Yemen
Den.	Denmark
Dom. Rep.	Dominican Republic
D.P.R.	Democratic People's Republic
(e.)	Estimate based on a compilation from a number of sources
E.E.C.	European Economic Community
Eq. Guinea	Equatorial Guinea
Est.	Estimated
Fr.	France
Fr. Guiana	French Guiana
Germ.	Germany
G.N.P.	Gross National Product
Hond.	Honduras
Hung.	Hungary
I.	Island
Is.	Islands
Inc.	Incorporated
Int.	International
Jam.	Jamaica
L.	Lake
Leb.	Lebanon
Ltd.	Limited
Lux.	Luxembourg
Man.	Manitoba
M.L.A.	Member of the Legislative Assembly
M.P.	Member of Parliament
Mt.	Mount or Mountain
Mts.	Mountains
Mun.	Municipal
na	Not available or not applicable
NATO	North Atlantic Treaty Organization
N.B.	New Brunswick
neg.	Negligible
Neth.	Netherlands
Nfld.	Newfoundland
N.S.	Nova Scotia
N.W.T.	Northwest Territories
N.Z.	New Zealand
Ont.	Ontario
P.	Post Office
P.D.R.	People's Democratic Republic
P.E.I.	Prince Edward Island
Port.	Portugal
P.R.	Puerto Rico
Qué.	Québec
R.	River
Rom.	Romania
S.A.	South Africa
Sask.	Saskatchewan
Switz.	Switzerland
U.A.E.	United Arab Emirates
U.K.	United Kingdom
U.S.	United States
U.S.S.R.	Union of Soviet Socialist Republics
Ven.	Venezuela
Y.T.	Yukon Territory
Yugo.	Yugoslavia

A

26N	77W	**128**	Abaco I.
30N	48E	**157**	Ābādān
13N	20E	**142**	Abéché
49N	122W	**70**	Abbotsford
45N	98W	**123**	Aberdeen
5N	4W	**141**	Abidjan
32N	99W	**123**	Abilene
49N	82W	**66**	Abitibi R.
24N	54E	**157**	Abu Dhabi
9S	68W	**132**	Abunã R.
15N	45E	**157**	Abyan
17N	100W	**127**	Acapulco
5N	0	**141**	Accra
25N	51E	**157**	Ad Dawḩah / Doha
12N	45E	**157**	‘Adan / Aden
37N	35E	**157**	Adana
9N	38E	**142**	Addis Ababa
35S	138E	**171**	Adelaide
12N	45E	**157**	Aden / ‘Adan
13N	50E	**157**	Aden, Gulf of
45N	75W	**124**	Adirondacks, Mts.
43N	16E	**147**	Adriatic Sea
40N	25E	**148**	Aegean Sea
33N	65E	**157**	Afghanistan ■
17N	8E	**141**	Agadez
30N	9W	**141**	Agadir
27N	78E	**158**	Agra
22N	102W	**127**	Aguascalientes
25N	5E	**141**	Ahagger Mts.
23N	72E	**157**	Ahmadabad
31N	49E	**157**	Ahvāz / Ahwaz
20N	10E	**141**	Air Plateau
51N	114W	**75**	Airdrie
44N	79W	**79**	Ajax
30N	20E	**142**	Ajdābiyā
52N	81W	**60**	Akimiski I.
39N	140E	**164**	Akita
68N	135W	**59**	Aklavik
41N	81W	**124**	Akron
41N	80E	**163**	Aksu
61N	79W	**60**	Akulivik
21N	93E	**165**	Akyab / Sittwe
30N	47E	**157**	Al Başrah
34N	42E	**157**	Al Furāt R. / Euphrates R.
23N	46E	**157**	Al Hillah
14N	43E	**157**	Al Ḩudaydah
25N	49E	**157**	Al Hufūf
24N	23E	**142**	Al Jawf (Libya)
30N	40E	**157**	Al Jawf (Saudi Arabia)
29N	48E	**157**	Al Kuwayt / Kuwait
35N	36E	**157**	Al Lādhiqīyah
24N	40E	**157**	Al Madīnah / Medina
26N	50E	**157**	Al Manāmah
36N	43E	**157**	Al Mawşil
14N	49E	**157**	Al Mukalla
32N	87W	**124**	Alabama □
31N	88W	**124**	Alabama R.
65N	150W	**177**	Alaska □
58N	145W	**121**	Alaska, Gulf of
41N	20E	**147**	Albania ■
35S	118E	**171**	Albany (Austral.)
42N	73W	**124**	Albany (U.S.)
52N	84W	**66**	Albany R.
2N	31W	**144**	Albert, L.
54N	115W	**75**	Alberta □
57N	10E	**147**	Ålborg
35N	106W	**123**	Albuquerque
36S	147E	**172**	Albury-Wodonga
59N	132E	**153**	Aldan R.
36N	37E	**157**	Aleppo / Halab
82N	62W	**60**	Alert
31N	30E	**142**	Alexandria / El Iskandarîya (Egypt)
45N	74W	**80**	Alexandria (Ont.)
36N	3E	**141**	Alger / Algiers
27N	3E	**141**	Algeria ■
36N	3E	**141**	Algiers / Alger
45N	78W	**66**	Algonquin Provincial Park
23S	153E	**171**	Alice Springs
25N	82E	**158**	Allahabad
44N	80E	**79**	Alliston
48N	71W	**66**	Alma
43N	77E	**153**	Alma Ata
65S	63W	**178**	Almirante Brown (research station)
45N	76W	**80**	Almonte
43N	85E	**163**	Altai Mountains (Tian Shan)
48N	88E	**163**	Altay
64N	72W	**60**	Amadjuak L.
35N	101W	**123**	Amarillo
2S	54W	**131**	Amazon R. / Amazonas R.
0	50W	**132**	Amazon R., Mouths of
2S	54W	**131**	Amazonas R. / Amazon R.
1S	78W	**131**	Ambato
3S	128E	**166**	Ambon
16S	168E	**172**	Ambrym, I.
61N	131E	**154**	Amga R.
45N	64W	**85**	Amherst
42N	83W	**79**	Amherstburg
32N	36E	**157**	‘Ammān
48N	78W	**66**	Amos
24N	118E	**164**	Amoy / Xiamen
31N	75E	**158**	Amritsar
52N	5E	**147**	Amsterdam
40N	62E	**153**	Amu Dar'ya R.
71N	124W	**59**	Amundsen Gulf
72N	115W	**178**	Amundsen Sea
90S	0	**178**	Amundsen-Scott (research station)
52N	139E	**154**	Amur R.
32N	44E	**157**	An Najaf
72N	113E	**154**	Anabar R.
16S	48W	**132**	Anapolis
		92	Anatolian Plateau
20S	169E	**172**	Anatom, I.
44N	80W	**79**	Ancaster
62N	150W	**121**	Anchorage
12N	92E	**158**	Andaman Is.
13N	96E	**165**	Andaman Sea
10N	75W	**131**	Andes Mountains
42N	1E	**147**	Andorra ■
25N	78W	**128**	Andros I.
58N	97E	**154**	Angara R.
52N	104E	**154**	Angarsk
15N	120E	**165**	Angeles
12S	18E	**143**	Angola ■
18N	63W	**128**	Anguilla
32N	117E	**164**	Anhui □
40N	33E	**157**	Ankara
36N	7E	**141**	Annaba
45N	66W	**85**	Annapolis Royal
38N	76W	**124**	Annapolis
41N	123E	**164**	Anshan
19S	47E	**144**	Antananarivo
67S	60W	**178**	Antarctic Peninsula
90S	0	**178**	Antarctica
49N	63W	**85**	Anticosti National Park
49N	63W	**85**	Anticosti, Île d'
45N	62W	**85**	Antigonish
17N	61W	**128**	Antigua and Barbuda ■
24S	70W	**133**	Antofagasta
12S	49E	**144**	Antsiranana
11N	96E	**163**	Anxi
40N	140E	**164**	Aomori
22N	17E	**142**	Aozou
0	70W	**131**	Apaporis R.
18N	121E	**165**	Aparri
35N	84W	**124**	Appalachian Mts.
43N	12E	**149**	Apennines
44N	88W	**124**	Appleton
12S	74W	**131**	Apurímac R.
20S	55W	**134**	Aquidauana
24N	46E	**157**	Ar Riyād / Riyadh
23N	65E	**157**	Arabian Sea
11S	37W	**132**	Aracaju
21S	30W	**134**	Araçatuba
9S	135E	**166**	Arafura Sea
5S	48W	**132**	Araguaia R.
1N	51W	**132**	Araguari R.
20N	94E	**165**	Arakan Yoma
39N	47E	**157**	Araks R.
44N	60E	**153**	Aral Sea
7N	69W	**131**	Arauca R.
67N		**177**	Arctic
73N	85W	**60**	Arctic Bay
67N		**177**	Arctic Circle
78N	160W	**177**	Arctic Ocean
62S	59W	**178**	Arctowski (research station)
16S	71W	**131**	Arequipa
47N	54W	**86**	Argentia
35S	66W	**133**	Argentina ■
51N	118E	**164**	Argun R.
56N	10E	**147**	Århus
18S	70W	**131**	Arica (Chile)
2S	71W	**131**	Arica (Colombia)
11S	58W	**131**	Arinos R.
8S	60W	**131**	Aripuanã R.
34N	111W	**123**	Arizona □
64N	41E	**153**	Arkangel'sk
35N	92W	**124**	Arkansas □
33N	91W	**124**	Arkansas R.
40N	44E	**153**	Armenia ■
4N	75W	**131**	Armenia (Colombia)
50N	119W	**70**	Armstrong
48N	54W	**86**	Arnold's Cove
45N	76W	**66**	Arnprior
44N	80W	**79**	Arthur
8S	133E	**166**	Aru Is.
13N	70W	**131**	Aruba
3S	36E	**144**	Arusha
48N	71W	**66**	Arvida
43N	142E	**164**	Asahigawa
23N	87E	**158**	Asansol
45N	72W	**80**	Asbestos
51N	103W	**76**	Asessippi Provincial Park
23S	117E	**171**	Ashburton R.
50N	121W	**70**	Ashcroft
38N	58E	**153**	Ashkhabad
39N	107E	**123**	Aspen
15N	39E	**142**	Asmera
49N	106W	**76**	Assiniboia
50N	101W	**76**	Assiniboine R.
50N	115W	**70**	Assiniboine, Mt., Provincial Park
49N	48E	**153**	Astrakhan'
25S	57W	**133**	Asunción
24N	33E	**142**	Aswān
27N	31E	**142**	Asyût
21N	40E	**157**	Aţ Ţā'if
		91	Atacama Desert
21N	13E	**141**	Atar
17N	33E	**142**	Atbara
17N	35E	**142**	Atbara R.
54N	113W	**75**	Athabasca
59N	109W	**65**	Athabasca, L.
58N	112W	**75**	Athabasca R.
38N	23E	**148**	Athens / Athínai
48N	91W	**65**	Atikokan
34N	84W	**124**	Atlanta
39N	74W	**124**	Atlantic City
0	30W	**85**	Atlantic Ocean
30N	10E	**141**	Atlas Mountains
59N	133W	**69**	Atlin
60N	134W	**69**	Atlin L.
59N	134W	**69**	Atlin Provincial Park
7N	77W	**131**	Atrato R.
37S	174E	**172**	Auckland
33N	82W	**124**	Augusta (Georgia)
44N	69W	**124**	Augusta (Maine)
39N	105W	**123**	Aurora (Colorado)
44N	79W	**79**	Aurora (Ont.)
30N	97W	**123**	Austin
23S	135E	**171**	Australia ■
36S	146E	**172**	Australian Alps
35S	149E	**172**	Australian Capital Territory □
47N	14E	**147**	Austria ■
67N	67W	**60**	Auyuittuq National Park
47N	53W	**86**	Avalon Peninsula
80N	90W	**59**	Axel Heiberg I.
13S	74W	**131**	Ayacucho
26S	139E	**171**	Ayers Rock
45N	76W	**80**	Aylmer (E. Ont.)
42N	80W	**79**	Aylmer (S. Ont.)
32N	36E	**157**	Az Zarqā'
26N	50E	**157**	Az-Zahran / Dhahran
40N	48E	**153**	Azerbaijan ■
46N	36E	**153**	Azov, Sea of

B

54N	126W	**69**	Babine L.
19N	122E	**164**	Babuyan Is.
4S	44W	**131**	Bacabal
65N	104W	**59**	Back R.
10N	122E	**165**	Bacolod
74N	70W	**177**	Baffin Bay
68N	75W	**60**	Baffin I.
33N	44E	**157**	Baghdād
36N	69E	**158**	Baghlān
16N	120E	**165**	Baguio
24N	75W	**128**	Bahamas ■
		132	Bahía = Salvador
38S	62W	**133**	Bahía Blanca
7N	20E	**142**	Bahr Aouk R.
7N	31E	**142**	Bahr el Jebel / White Nile R.
12N	20E	**142**	Bahr Salamat R.
26N	50E	**157**	Bahrain ■
46N	123W	**164**	Baicheng
53N	80W	**60**	Baie James (James Bay)
50N	56W	**86**	Baie Verte
49N	68W	**66**	Baie-Comeau
47N	70W	**80**	Baie-St-Paul
10S	120E	**165**	Baing
37S	147E	**172**	Bairnsdale
64N	96W	**59**	Baker L.
35N	119W	**123**	Bakersfield
40N	49W	**153**	Baku
48N	66W	**85**	Baldwin Provincial Park
39N	3E	**147**	Balearic Is. / Islas Baleares
57N	67W	**60**	Baleine, à la, R.
8S	115E	**165**	Bali, I.
1S	117W	**165**	Balikpapan
44N	22E	**150**	Balkan Mts.
46N	75E	**153**	Balkhash, L.
37S	144E	**171**	Ballarat
45N	79W	**79**	Balsam L.
18N	101W	**127**	Balsas R.
56N	20E	**147**	Baltic Sea
39N	76W	**124**	Baltimore
12N	8W	**141**	Bamako
5N	20E	**142**	Bambari
6N	10E	**141**	Bamenda
5N	95E	**165**	Banda Aceh
6S	130E	**166**	Banda Sea
27N	56E	**157**	Bandar ‘Abbās
5N	115E	**165**	Bandar Seri Begawan
3S	17E	**143**	Bandundu
7S	107E	**165**	Bandung
51N	115W	**70**	Banff
51N	116W	**70**	Banff National Park
13N	77E	**158**	Bangalore
5N	23E	**142**	Bangassou
32N	20E	**142**	Banghāzī / Benghazi
1N	125E	**165**	Bangka, I.
13N	100E	**165**	Bangkok / Krung Thep
24N	90E	**158**	Bangladesh ■
45N	69W	**124**	Bangor
4N	18E	**141**	Bangui
14N	6W	**141**	Bani R.
3S	114E	**165**	Banjarmasin
13N	16W	**141**	Banjul
73N	121W	**59**	Banks I. (N.W.T.)
10S	142E	**172**	Banks Is. (Vanuatu)
39N	115E	**164**	Baoding
34N	107E	**163**	Baoji
25N	99E	**163**	Baoshan
40N	110E	**163**	Baotou
44N	67W	**124**	Bar Harbor
48N	58W	**86**	Barachois Pond Provincial Park
13N	59W	**128**	Barbados ■
42N	2E	**147**	Barcelona (Spain)
10N	64W	**131**	Barcelona (Venezuela)
73N	39E	**177**	Barents Sea
41N	17E	**147**	Bari
8N	70W	**131**	Barinas
2S	102E	**165**	Barisan Mts.
2S	114E	**165**	Barito R.
29S	119E	**171**	Barlee, L.
53N	83E	**153**	Barnaul
10N	69W	**131**	Barquisimeto
22S	44W	**134**	Barra Mansa
7N	73W	**131**	Barrancabermeja
11N	74W	**131**	Barranquilla
54N	114W	**75**	Barrhead
44N	79W	**79**	Barrie
51N	120W	**70**	Barrière
20S	115E	**171**	Barrow I.
6N	122E	**165**	Basilan, I.
39S	146E	**172**	Bass Strait
16N	61W	**128**	Basse-Terre
17N	95E	**165**	Bassein
42N	9E	**147**	Bastia
20N	122E	**150**	Batan Is.
47N	65W	**85**	Bathurst
70N	128W	**59**	Bathurst, Cape
11S	130E	**171**	Bathurst I. (Australia)
76N	100W	**59**	Bathurst I. (N.W.T.)
30N	91W	**124**	Baton Rouge
13N	103E	**165**	Battambang
53N	110W	**75**	Battle R.
53N	108W	**75**	Battleford
0	98E	**165**	Batu Is.
41N	41E	**153**	Batumi
12N	5E	**141**	Bauchi Plateau
22S	49W	**134**	Bauru
48N	52W	**86**	Bay de Verte
46N	101E	**163**	Bayanhongor
53N	108E	**154**	Baykal, L.
33N	35E	**157**	Bayrūt / Beirut
43N	82W	**79**	Bear Creek
54N	93W	**65**	Bearskin Lake
46N	70W	**80**	Beauceville
72N	140W	**59**	Beaufort Sea
30N	94W	**124**	Beaumont
55N	107W	**76**	Beauval
55N	119W	**75**	Beaverlodge
44N	79W	**79**	Beaverton
46N	72W	**80**	Bécancour
31N	2W	**141**	Béchar
24N	113E	**164**	Bei R.
39N	116E	**164**	Beijing / Peking
39N	116E	**164**	Beijing Shi □
19S	35E	**144**	Beira

33N	35E	**157**	Beirut / Bayrūt
53N	25E	**153**	Belarus ■
		176	Belau
56N	78W	**60**	Belcher Is.
1S	48W	**132**	Belém
5N	45E	**142**	Belet Weyne
54N	6W	**147**	Belfast
50N	5E	**147**	Belgium ■
51N	37E	**153**	Belgorod
44N	20E	**148**	Belgrade / Beograd
3S	108E	**165**	Belitung, I.
17N	88W	**127**	Belize ■
17N	88W	**127**	Belize (city)
52N	126W	**69**	Bella Coola
52N	55W	**86**	Belle Isle
51N	56W	**86**	Belle Isle, Strait of
44N	77W	**79**	Belleville
49N	123W	**123**	Bellingham
66S	80W	**178**	Bellingshausen Sea
6N	75W	**131**	Bello
17N	88W	**127**	Belmopan
20S	44W	**134**	Belo Horizonte
46N	73W	**80**	Beloeil
21S	146E	**172**	Belyando R.
36S	144E	**172**	Bendigo
15N	90E	**158**	Bengal, Bay of
33N	117E	**164**	Bengbu
32N	20E	**142**	Benghazi / Banghāzī
13S	67W	**131**	Beni R.
10N	2E	**141**	Benin ■
6N	5E	**141**	Benin City
9N	11E	**141**	Benue R.
41N	123E	**164**	Benxi
44N	20E	**148**	Beograd / Belgrade
10N	45E	**142**	Berbera
4N	15E	**141**	Berbérati
52N	97W	**65**	Berens River
60N	5E	**147**	Bergen
58N	167E	**177**	Bering Sea
52N	13E	**147**	Berlin
30S	68W	**133**	Bermejo R.
24S	64W	**133**	Bermejo Teuco R.
32N	65W	**122**	Bermuda ■
47N	7E	**147**	Bern
27N	90E	**158**	Bhutan ■
1S	136E	**166**	Biak, I.
54N	89W	**65**	Big Trout Lake
43N	3W	**147**	Bilbao
45N	108W	**123**	Billings
42N	76W	**124**	Binghampton
1N	104E	**165**	Bintan I.
37S	74W	**133**	Bío Bío R.
10N	23E	**142**	Birao
50N	97W	**76**	Birds Hill Provincial Park
50N	122W	**70**	Birkenhead Lake Provincial Park
52N	2W	**147**	Birmingham (U.K.)
33N	86W	**124**	Birmingham (U.S.)
45N	2W	**147**	Biscay, Bay of
43N	75E	**153**	Bishkek
34N	5E	**141**	Biskra
46N	100W	**123**	Bismarck
4S	146E	**166**	Bismarck Sea
11N	15W	**141**	Bissau
52N	85E	**153**	Biysk
22N	105E	**165**	Black R. / Da R.
44N	104W	**123**	Black Hills
44N	79W	**79**	Black R.
44N	31E	**157**	Black Sea
10N	3W	**141**	Black Volta R.
45N	66W	**85**	Blacks Harbour
50N	127E	**154**	Blagoveshchensk
46N	7E	**149**	Blanc, Mt.
37S	63W	**133**	Blanca, Bahía
15S	35E	**144**	Blantyre
42S	82W	**79**	Blenheim
29S	26E	**143**	Bloemfontein
52N	97W	**65**	Bloodvein River
34S	148E	**172**	Blue Mountains
15N	32E	**142**	Blue Nile R. / El Bahr el Azraq
27S	49W	**134**	Blumenau
7N	11W	**141**	Bo
3N	61W	**131**	Boa Vista
45N	79W	**79**	Bobcaygeon
10N	7W	**141**	Bobo Dioulasso
67N	14E	**147**	Bodø
6S	107E	**165**	Bogor
4N	74W	**131**	Bogotá
43N	116W	**123**	Boise
49N	100W	**76**	Boissevain
17S	64W	**131**	Bolivia ■
44N	80W	**79**	Bolton
19N	72E	**158**	Bombay
12N	68W	**131**	Bonaire
48N	65W	**85**	Bonaventure
48N	53W	**86**	Bonavista
48N	53W	**86**	Bonavista Bay
48N	53W	**86**	Bonavista, Cape
10N	15E	**141**	Bongor
50N	7E	**147**	Bonn
71N	90W	**59**	Boothia, Gulf of
71N	94W	**59**	Boothia Peninsula
6S	42W	**132**	Borborema, Plateau of
44N	0	**147**	Bordeaux
46N	63W	**79**	Borden
1N	115E	**165**	Borneo □
4N	49E	**142**	Bosaso
44N	17E	**147**	Bosnia-Hercegovina ■
42N	71W	**124**	Boston
63N	20W	**148**	Bothnia, Gulf of
22S	24E	**143**	Botswana ■
7N	5W	**141**	Bouaké
30S	146E	**172**	Bourke
51N	112W	**75**	Bow R.
44N	79W	**79**	Bowmanville
53N	121W	**70**	Bowron Lake Provincial Park
45N	79W	**66**	Bracebridge
44N	79W	**79**	Bradford
27N	93E	**158**	Brahmaputra R. / Yarlung Zangbo R.
43N	75W	**79**	Brampton
49N	100W	**76**	Brandon
43N	80W	**79**	Brantford
46N	61W	**85**	Bras d'Or L.
15S	48E	**132**	Brasilia
48N	17E	**147**	Bratislava
56N	101E	**154**	Bratsk

10S	50W	**132**	Brazil ■
13S	48W	**132**	Brazilian Highlands
34N	102W	**123**	Brazos R.
4S	15E	**143**	Brazzaville
48N	4W	**147**	Brest
13N	59W	**128**	Bridgetown (Barbados)
45N	65W	**85**	Bridgetown (N.S.)
44N	64W	**85**	Bridgewater
44N	77W	**79**	Brighton (Ont.)
27S	153E	**172**	Brisbane
55N	125W	**70**	British Columbia
49N	16E	**147**	Brno
57N	101W	**65**	Brochet
44N	75W	**80**	Brockville
32S	141E	**171**	Broken Hill
		92	Broken Ridge
45N	72W	**80**	Brome, Lac
50N	112W	**75**	Brooks
		91	Brooks Range
18S	122E	**171**	Broome
45N	74W	**80**	Brownsburg
26N	98W	**123**	Brownsville
45N	81W	**79**	Bruce Peninsula National Park
4N	115E	**165**	Brunei ■
51N	4E	**147**	Bruxelles / Brussels
53N	34E	**153**	Bryansk
7N	73W	**131**	Bucaramanga
48N	56W	**86**	Buchans
44N	26E	**148**	Bucharest / Bucureşti
45N	75W	**80**	Buckingham
44N	26E	**148**	Bucureşti / Bucharest
47N	19E	**147**	Budapest
4N	77W	**131**	Buenaventura
34S	58W	**133**	Buenos Aires
43N	78W	**124**	Buffalo
56N	108W	**75**	Buffalo Narrows
51N	105W	**76**	Buffalo Pound Provincial Park
51N	23E	**148**	Bug R.
51N	117W	**70**	Bugaboo Provincial Park
3S	29E	**144**	Bujumbura
2S	29E	**144**	Bukavu
1S	32E	**144**	Bukoba
20S	28E	**144**	Bulawayo
42N	25E	**148**	Bulgaria ■
2N	22E	**143**	Bumba
33S	115E	**171**	Bunbury
25S	152E	**172**	Bundaberg
31N	32E	**142**	Bûr Sâ'îd / Port Said
10N	46E	**142**	Burao
47N	57W	**86**	Burgeo
47N	55W	**86**	Burin
12N	1W	**141**	Burkina Faso ■
43N	79W	**79**	Burlington (Ont.)
44N	73W	**124**	Burlington (U.S.)
49N	122W	**70**	Burnaby
41S	146E	**172**	Burnie
54N	126W	**69**	Burns Lake
44N	78W	**79**	Burnt R.
40N	29E	**157**	Bursa
3S	126E	**166**	Buru, I.
3S	30E	**144**	Burundi ■
29N	51E	**157**	Būshehr
46N	112W	**123**	Butte
47N	53W	**86**	Butter Pot Provincial Park
5S	122E	**165**	Butung, I.

C

10N	71W	**131**	Cabimas
5S	12E	**156**	Cabinda □
6S	12E	**143**	Cabinda (City)
47N	59W	**85**	Cabot Strait
36N	6W	**147**	Cadiz
39N	9E	**147**	Cagliari
17S	145E	**172**	Cairns
30N	31E	**142**	Cairo / El Qâhira
37N	89W	**124**	Cairo (U.S.)
7S	78W	**131**	Cajamarca
24S	69W	**133**	Calama
22N	88E	**158**	Calcutta
44N	80W	**79**	Caledon
43N	80W	**79**	Caledonia
51N	114W	**75**	Calgary CMA
3N	76W	**131**	Cali
37N	120W	**123**	California □
27N	115W	**127**	California, Baya
27N	111W	**127**	California, Golfo de
12S	77W	**131**	Callao
21N	78W	**128**	Camagüey
14N	104E	**165**	Cambodia ■
43N	80W	**79**	Cambridge
69N	105W	**59**	Cambridge Bay
6N	12E	**141**	Cameroon ■
6N	6E	**141**	Cameroon Mountain
50N	125W	**70**	Campbell River
44N	78W	**79**	Campbellford
47N	66W	**85**	Campbellton
19N	90W	**127**	Campeche
19N	93W	**127**	Campeche, Bahia de
7S	35W	**132**	Campina Grande
22S	47W	**134**	Campinas
20S	54W	**134**	Campo Grande
45N	67W	**85**	Campobello Island
21S	41W	**134**	Campos
53N	113W	**75**	Camrose
10N	105E	**165**	Can-tho
46N	66W	**85**	Canaan R.
60N	95W	**121**	Canada ■
35N	95W	**123**	Canadian R.
29N	17W	**141**	Canary Is.
35S	149E	**172**	Canberra
20N	88W	**127**	Cancún
55N	69W	**60**	Caniapiscau R.
51N	115W	**70**	Canmore
30S	51W	**134**	Canoas
45N	61W	**85**	Canso
43N	8W	**149**	Cantabrian Mts.
42N	82W	**124**	Canton (Ohio)
24N	114E	**164**	Canton / Guangzhou
49N	66W	**85**	Cap-Chat
20N	72W	**128**	Cap-Haïtien
46N	61E	**85**	Cape Breton Highlands National Park
46N	60W	**85**	Cape Breton I.
27N	80W	**124**	Cape Canaveral
42N	70W	**124**	Cape Cod
64N	76W	**60**	Cape Dorset

35N	75W	**124**	Cape Hatteras
61N	70W	**78**	Cape Hopes Advance
34S	18E	**143**	Cape of Good Hope
50N	128W	**69**	Cape Scott Provincial Park
48N	59W	**86**	Cape St. George
46N	64W	**85**	Cape Tormentine
34S	18E	**145**	Cape Town
63S	61W	**178**	Capitán Arturo Prat (research station)
0	73W	**131**	Caquetá R.
10N	67W	**131**	Caracas
48N	65W	**85**	Caraquet
47N	53W	**86**	Carbonear
51N	3W	**147**	Cardiff
49N	113W	**75**	Cardston
15N	75W	**128**	Caribbean Sea
54N	122W	**70**	Cariboo Mountains
45N	76W	**80**	Carleton Place
32N	104W	**123**	Carlsbad
50N	102W	**75**	Carlyle
62N	136W	**59**	Carmacks
48N	126W	**70**	Carmanah Pacific Provincial Park
24S	113E	**171**	Carnarvon
26S	122E	**171**	Carnegie, L.
55N	123W	**70**	Carp Lake Provincial Park
50N	21E	**150**	Carpathians (Mts.)
14S	139E	**171**	Carpentaria, Gulf of
39N	119W	**123**	Carson City
10N	75W	**131**	Cartagena (Colombia)
37N	1W	**147**	Cartagena (Spain)
53N	57W	**66**	Cartwright
8S	36W	**132**	Caruaru
33N	7W	**141**	Casablanca
41N	123W	**123**	Cascade Range
48N	66W	**85**	Cascapédia R.
25S	54W	**134**	Cascavel
66S	110E	**178**	Casey (research station)
42N	106W	**123**	Casper
47N	47E	**153**	Caspian Depression
43N	53E	**153**	Caspian Sea
45N	75E	**80**	Casselman
59N	129W	**69**	Cassiar
58N	128W	**69**	Cassiar Mountains
49N	117W	**70**	Castlegar
14N	61W	**128**	Castries
24N	75W	**128**	Cat Island
48N	53W	**86**	Catalina
28S	64W	**133**	Catamarca
14N	124E	**166**	Catanduanes, I.
49N	120W	**70**	Cathedral Provincial Park
45N	39E	**150**	Caucasus Mts.
6N	64W	**131**	Caura R.
47N	63W	**85**	Cavendish
29S	51W	**133**	Caxias do Sul
5N	52W	**131**	Cayenne
19N	80W	**128**	Cayman Is. □
43N	80W	**79**	Cayuga
10N	124E	**165**	Cebu
53N	100W	**76**	Cedar Lake
42N	91W	**124**	Cedar Rapids
32S	133E	**171**	Ceduna
3N	123E	**165**	Celebes Sea
7N	20E	**142**	Central African Republic ■
5S	142E	**166**	Central Range (Papua New Guinea)
57N	160E	**154**	Central Range (Russia)
56N	33E	**150**	Central Russian Upland
67N	100W	**154**	Central Siberian Plateau
3S	129E	**166**	Ceram, I.
11S	76W	**131**	Cerro de Pasco
36N	5W	**141**	Ceuta
41N	129E	**164**	Ch'ongjin
39N	125E	**164**	Ch'ongju
15N	17E	**142**	Chad ■
13N	14E	**141**	Chad, L.
36S	67W	**133**	Chadileuvú R.
		92	Chagos-Laccadive Ridge
48N	65W	**85**	Chaleur Bay
48N	64W	**85**	Chandler
31N	117E	**164**	Chang R. / Chang Jiang R. / Yangtze R.
43N	124E	**164**	Changchun
28N	113E	**164**	Changsha
36N	113E	**164**	Changzhi
31N	120E	**164**	Changzhou
47N	59W	**86**	Channel-Port aux Basques
47N	84W	**66**	Chapleau
39N	63E	**153**	Chardzhou
10N	16E	**141**	Chari R.
47N	71W	**80**	Charlesbourg
32N	80W	**124**	Charleston (South Carolina)
38N	81W	**124**	Charleston (West Virginia)
26S	146E	**172**	Charleville
35N	80W	**124**	Charlotte
46N	63W	**85**	Charlottetown
47N	65W	**85**	Chatham (N.B.)
42N	82W	**79**	Chatham (Ont.)
50S	74W	**172**	Chatham I.
35N	85W	**124**	Chattanooga
46N	71W	**80**	Chaudière R.
56N	47E	**153**	Cheboksary
45N	61W	**85**	Chedabucto Bay
33N	126E	**164**	Cheju-do
55N	61E	**153**	Chelyabinsk
34N	104E	**163**	Chengdu
59N	38E	**153**	Cherepovets
49N	32E	**153**	Cherkassy
51N	31E	**153**	Chernigov
48N	26E	**153**	Chernovtsy
70N	140E	**154**	Cherskiy Range
38N	76W	**124**	Chesapeake Bay
44N	81W	**79**	Chesley
63N	90W	**59**	Chesterfield Inlet
46N	61W	**85**	Chéticamp
55N	121W	**75**	Chetwynd
41N	104W	**123**	Cheyenne
18N	99E	**165**	Chiang Mai
35N	140E	**164**	Chiba
50N	74W	**66**	Chibougamau
49N	73W	**66**	Chibougamau Provincial Park
49N	66W	**85**	Chic-Chocs Mts.
48N	66W	**85**	Chic-Chocs Provincial Park

42N 87W **124** Chicago
6S 80W **131** Chiclayo
45S 67W **133** Chico R.
50S 69W **133** Chico R. (Gobernad)
48N 71W **66** Chicoutimi-Jonquière CMA
45N 65W **85** Chignecto Bay
28N 106W **127** Chihuahua
52N 124W **70** Chilcotin R.
35S 72W **133** Chile ■
51N 124W **70** Chilko L.
36S 72W **133** Chillán
49N 122W **70** Chilliwack
43S 74W **133** Chiloé, Isla de
9S 78W **131** Chimbote
42N 69E **153** Chimkent
30N 110E **163** China ■
54N 79W **60** Chisasibi
52N 113E **154** Chita
22N 91E **158** Chittagong
13N 101E **165** Chon Buri
29N 106E **163** Chongqing / Chungking
45S 75W **133** Chonos, Archipiélago de los
48N 114E **164** Choybalsan
43S 172E **172** Christchurch
45N 71E **153** Chu R.
43S 65W **133** Chubut R.
29N 106E **163** Chungking / Chongqing
58N 94W **65** Churchill
53N 64W **66** Churchill Falls
58N 95W **65** Churchill R. (Man.)
53N 62W **66** Churchill R. (Nfld.)
22N 80W **128** Cienfuegos
39N 84W **124** Cincinnati
29N 100W **127** Ciudad Acuna
8N 63W **131** Ciudad Bolivar
19N 92W **127** Ciudad del Carmen
7N 63W **131** Ciudad Guayana
31N 106W **127** Ciudad Juárez
23N 97W **127** Ciudad Madero
27N 110W **127** Ciudad Obregón
23N 99W **127** Ciudad Victoria
58N 112W **75** Claire, L.
46N 61W **85** Clark's Harbour
51N 120W **70** Clearwater
54N 101W **76** Clearwater Provincial Park
57N 109W **75** Clearwater River Provincial Park
41N 81W **124** Cleveland
43N 81W **79** Clinton
20S 140E **171** Cloncurry
46N 23E **148** Cluj-Napoca
60 Clyde River = Kangirtugaapik
58N 132W **69** Coast Mountains
36N 121W **123** Coast Range
31N 86W **124** Coastal Plain
45N 71W **80** Coaticook
62N 83W **60** Coats I.
18N 94W **127** Coatzacoalcos
47N 79W **66** Cobalt
43N 78W **79** Cobourg
17S 66W **131** Cochabamba
92 Cocos Ridge
47N 117W **123** Coeur d'Alene
30S 153E **172** Coffs Harbour
11N 77E **158** Coimbatore
38S 143E **172** Colac
54N 110W **75** Cold Lake
50N 119W **70** Coldstream
50N 115W **75** Coleman
44N 80W **79** Collingwood
3N 73W **131** Colombia ■
7N 80E **158** Colombo
9N 80W **128** Colón
37N 106W **123** Colorado □
37S 68W **133** Colorado R. (Argentina)
34N 114W **123** Colorado R. (California)
32N 100W **123** Colorado R. (Texas)
39N 105W **123** Colorado Springs
34N 81W **124** Columbia
38N 77W **124** Columbia, District of □
52N 118W **70** Columbia, Mt.
51N 118W **70** Columbia R. (B.C.)
46N 121W **123** Columbia R. (U.S.)
32N 85W **124** Columbus
46S 67W **133** Comodoro Rivadavia
12S 44E **144** Comoros ■
49N 125W **70** Comox
9N 13W **141** Conakry
37S 73W **133** Concepción
48N 53W **86** Conception Bay
43N 71W **124** Concord
31S 58W **133** Concordia
1S 16E **143** Congo ■
1S 16E **143** Congo R. / Zaïre R.
41N 72W **124** Connecticut □
44N 28E **148** Constanta
36N 6E **141** Constantine
14S 41W **132** Contos R.
46N 73W **80** Contrecoeur
41S 174E **172** Cook Strait
15S 145E **172** Cooktown
55N 12E **147** Copenhagen/København
46N 80W **66** Copper Cliff
68N 15W **59** Coppermine
66N 115W **59** Coppermine R.
30S 72W **133** Coquimbo
64N 83W **60** Coral Harbour
15S 150E **172** Coral Sea
3N 57W **132** Corantijn R.
44N 76W **133** Corcovado, Golfo
31S 64W **133** Córdoba (Argentina)
38N 5W **147** Córdoba (Spain)
52N 8W **147** Cork
49N 58W **86** Corner Brook
45N 74W **80** Cornwall
75N 95W **59** Cornwallis I.
11N 69W **131** Coro
52N 111W **75** Coronation
67N 115W **59** Coronation Gulf
27S 58W **133** Corrientes
28N 97W **123** Corpus Christi
42N 9E **147** Corsica
19S 57W **132** Corumba
43N 8W **147** Coruña, La
10N 84W **128** Costa Rica ■
6N 2E **141** Cotonou
49N 125W **70** Courtenay
45N 72W **80** Cowansville
49N 58W **86** Cox's Cove
20N 86W **127** Cozumel I.
49N 115W **70** Cranbrook
57N 106W **65** Cree L.
44N 80W **79** Creemore
55N 102W **76** Creighton
49N 116W **70** Creston
35N 25E **147** Crete
52N 115W **75** Crimson Lake Provincial Park
54N 97W **65** Cross Lake
46N 17E **147** Croatia ■
50N 115W **70** Crowsnest Pass
7S 73W **131** Cruzeiro do Sul
22N 79W **128** Cuba ■
16S 18E **143** Cubango R.
8N 72W **131** Cúcuta
3S 79W **131** Cuenca
19N 99W **127** Cuernavaca
15S 56W **131** Cuiabá
24N 107W **127** Culiacán
10N 64W **131** Cumaná
65N 66W **60** Cumberland Sound
1N 56W **132** Cumina R.
13N 68W **131** Curaçao
35S 71W **133** Curico
25S 49W **134** Curitiba
20N 86E **158** Cuttack
13S 72W **131** Cuzco
50N 110W **75** Cypress Hills Provincial Park (Alta.)
50N 109W **75** Cypress Hills Provincial Park (Sask.)
35N 33E **157** Cyprus ■
45N 81W **79** Cyprus Lake Provincial Park
49N 17E **147** Czech Republic ■

D

12N 108W **165** Da Lat
22N 105E **165** Da R. / Black R.
32N 117E **164** Da Yunhe R. / Grand Canal R.
16N 108E **165** Da-nang
39N 121E **164** Dairen / Lüda
14N 17W **141** Dakar
44N 102E **163** Dalanzadgad
48N 66W **85** Dalhousie
33N 97W **124** Dallas
7N 6W **141** Daloa
35N 85W **124** Dalton
14S 132E **171** Daly R.
33N 36E **157** Damascus / Dimashq
20S 116E **171** Dampier
40N 124E **164** Dandong
51N 107W **76** Danielson Provincial Park
48N 15E **147** Danube R. / Donau R. / Dunărea R.
46N 72W **80** Danville
7S 39E **144** Dar es Salaam
27N 88E **158** Darjeeling
31S 144E **171** Darling R.
33N 23E **142** Darnah
44N 63W **85** Dartmouth
12S 131E **171** Darwin
40N 113E **164** Datong
57N 26E **148** Daugava R.
51N 100W **76** Dauphin
51N 100W **76** Dauphin L.
7N 125E **166** Davao
41N 90W **124** Davenport
8N 82W **128** David
68S 18E **178** Davis (research station)
65N 58W **60** Davis Strait
64N 139W **59** Dawson
55N 120W **75** Dawson Creek
39N 84W **124** Dayton
29N 81W **124** Daytona Beach
58N 130W **69** Dease Lake
18N 75E **158** Deccan Plateau
46N 77W **66** Deep River
49N 57W **86** Deer Lake
39N 75W **124** Delaware □
28N 77E **158** Delhi (India)
43N 80W **79** Delhi (Ont.)
49N 101W **76** Deloraine
63N 115W **59** Denendeh □
55N 9E **147** Denmark ■
8S 115E **165** Denpasar
39N 105W **123** Denver
17S 123E **171** Derby
41N 93W **124** Des Moines
11N 39E **142** Desē
47S 68W **133** Deseado R.
22N 10E **141** Desert El Djouf
54N 124W **70** Desolation Sound Provincial Marine Park
42N 83W **124** Detroit
75N 85W **59** Devon I.
26N 50E **157** Dhahran / Az-Zahran
26N 85E **158** Dhaka
25S 141E **171** Diamantina R.
35S 69W **133** Diamente R.
27N 95E **158** Dibrugarh
51N 107W **76** Diefenbaker, L.
44N 66W **85** Digby
35N 44E **157** Dijlah R. / Tigris R.
8S 125E **166** Dili
33N 36E **157** Dimashq / Damascus
43N 19E **150** Dinaric Alps
51N 111W **75** Dinosaur Provincial Park
9N 41E **142** Dirē Dawa
26S 113E **171** Dirk Hartog I.
23S 122E **171** Disappointment L.
46N 71W **80** Disraeli
38N 77W **124** District of Columbia □
20S 45W **134** Divinópolis
21N 12E **141** Djado
6N 25E **142** Djema
12N 43E **142** Djibouti ■
11N 43E **142** Djibouti City
10N 3E **141** Djougou
49N 34E **153** Dnepr R.
48N 35E **153** Dnepropetrovsk
48N 28E **153** Dnestr R.
8N 16E **141** Doba
37N 100W **123** Dodge City
6S 35E **144** Dodoma
25N 51E **157** Doha / Ad Dawhah
49N 72W **66** Dolbeau
4N 41E **142** Dolo
15N 61W **128** Dominica ■
19N 70W **128** Dominican Republic ■
48N 42E **153** Don R.
49N 13E **147** Donau R. / Danube R. / Dunărea R.
48N 37E **153** Donetsk
19N 32E **142** Dongola
46N 71W **80** Donnacona
4N 9E **141** Douala
51N 106W **76** Douglas Provincial Park
39N 75W **124** Dover (U.S.)
31S 30E **144** Drakensberg Mountains
53N 115W **75** Drayton Valley
51N 13E **147** Dresden (Germany)
42N 82W **79** Dresden (Ont.)
51N 112W **75** Drumheller
46N 72W **80** Drummondville
49N 93W **65** Dryden
15S 127E **171** Drysdale R.
63N 101W **59** Dubawnt L.
62N 102W **59** Dubawnt R.
25N 55E **157** Dubayy
32S 148E **172** Dubbo
53N 6W **147** Dublin
42N 18E **147** Dubrovnik
42N 72W **80** Duchesnay Provincial Park
51N 101W **76** Duck Mountain Provincial Park (Man.)
51N 102W **76** Duck Mountain Provincial Park (Sask.)
26N 102E **163** Dukou
46N 92W **124** Duluth
66S 140E **178** Dumont d'Urville (research station)
44N 22E **148** Dunărea R. / Danube R. / Donau R.
48N 123W **70** Duncan
44N 80W **79** Dundalk
43N 80W **79** Dundas
46S 170E **172** Dunedin (New Zealand)
27N 83W **124** Dunedin (U.S.)
48N 67W **85** Dunière Provincial Park
43N 79W **79** Dunnville
24S 43W **134** Duque de Caxias
24N 104W **127** Durango
30S 31E **144** Durban
44N 81W **79** Durham (Ont.)
36N 78W **124** Durham (U.S.)
38N 68E **153** Dushanbe
43N 71E **153** Dzambul
48N 95E **163** Dzavhan R.

E

51N 93W **65** Ear Falls
45N 71W **80** East-Angus
30N 126E **164** East China Sea
33S 28E **144** East London (South Africa)
47N 62W **85** East Point
73N 160E **154** East Siberian Sea
46N 71W **80** East-Broughton
12N 79W **158** Eastern Ghats
52N 78W **60** Eastmain R.
45N 91W **124** Eau Claire
42N 2W **147** Ebro R.
2S 78W **131** Ecuador ■
56N 3W **147** Edinburgh
53N 113W **75** Edmonton CMA
47N 68W **85** Edmundston
53N 116W **75** Edson
1S 30E **144** Edward, L.
65N 117W **65** Edzo
17S 168E **172** Éfaté, I.
28N 31E **142** Egypt ■
30N 34E **157** Eilat
27N 13W **141** El Aaiún
15N 34E **142** El Bahr el Azraq / Blue Nile R.
29N 31E **142** El Faiyûm
14N 25E **142** El Fasher
30N 31E **142** El Giza
30N 3E **141** El Golea
31N 30E **142** El Iskandarîya / Alexandria
32N 31E **142** El Mansûra
28N 31E **142** El Minya
13N 30E **142** El Obeid
32N 106W **123** El Paso
30N 31E **142** El Qâhira / Cairo
14N 89W **127** El Salvador ■
30N 32E **142** El Suweis / Suez
26N 33E **142** El Uqsor / Luxor
53N 11E **147** Elbe R.
43N 41E **150** Elbrus, Mt.
37N 50E **157** Elburz Mountains
25N 76W **128** Eleuthera
53N 113W **75** Elk Island National Park
50N 115W **70** Elk Lakes Provincial Park
85N 70W **60** Ellesmere Island National Park
79N 80W **60** Ellesmere I.
46N 82W **66** Elliot Lake
44N 80W **79** Elmvale
17S 168E **172** Émae, I.
47N 56E **153** Emba R.
49N 97W **76** Emerson
27S 56W **134** Encarnacion
45N 63W **85** Enfield
53N 2W **147** England □
50N 4E **147** English Channel
32N 117W **127** Ensenada
0 33E **144** Entebbe
7N 8E **141** Enugu
2S 8E **141** Equatorial Guinea ■
42N 80W **124** Erie
42N 81W **79** Erie, L.
17N 37E **142** Eritrea ■
26S 143E **172** Erromango, I.
55N 35W **145** Esbjerg
33N 53E **157** Esfahān
61N 94W **59** Eskimo Point
1N 80W **131** Esmeraldas
46N 81W **66** Espanola
33S 122E **171** Esperance
51S 70W **178** Esperanza (research station)
15S 167E **172** Espiritu Santo
51N 7E **147** Essen
4N 59W **131** Essequibo R.
42N 83W **79** Essex
51N 102W **76** Esterhazy
49N 103W **76** Estevan
51N 108W **75** Eston
58N 25E **153** Estonia ■
60S 68W **133** Estrecho de Magalianes
46N 71W **80** Etchemin R.
8N 40E **142** Ethiopia ■
10N 33E **142** Ethiopian Highlands
44N 123W **123** Eugene
34N 42E **157** Euphrates R. / Al Furāt R.
80N 85W **60** Eureka (Can.)
41N 124W **123** Eureka (U.S.)
30E 55N **153** European Plain
38N 87W **124** Evansville
27N 87E **163** Everest, Mt.
26N 82W **124** Everglades
43N 81W **79** Exeter
49N 56W **86** Exploits R.
4S 35E **144** Eyasi, L.
28S 137E **171** Eyre, L.

F

18N 22E **142** Fada
65N 148W **121** Fairbanks
56N 118W **75** Fairview
59N 137W **69** Fairweather, Mt.
32N 73E **158** Faisalbad / Lyallpur
57S 60W **133** Falkland Is.
65S 63W **178** Faraday (research station)
47N 96W **124** Fargo
45N 73W **80** Farnham
62N 7W **147** Faroe Is.
12N 6E **141** Faru
45N 82W **79** Fathom Five National Marine Park
18N 19W **142** Faya-Largeau
12S 39W **131** Feira de Santana
80N 44W **79** Fergus
49N 115W **70** Fernie
34N 5W **141** Fès
58N 72W **60** Feuilles, aux, R.
21S 47E **144** Fianarantsoa
13S 38W **132** Fiera-de-Santana
175 Fiji
62N 30E **148** Finland ■
57N 125W **69** Finlay R.
43N 11E **147** Firenze / Florence
18S 125E **171** Fitzroy Crossing
17S 123E **171** Fitzroy R.
35N 112W **123** Flagstaff
54N 102W **76** Flin Flon
40S 148E **172** Flinders I.
19S 141E **171** Flinders R.
32S 138E **171** Flinders Ranges
43N 83W **124** Flint
43N 11E **147** Florence / Firenze
49N 126W **165** Flores I.
6S 43W **132** Floriano
27N 48W **124** Florianopolis
28N 82W **124** Florida □
25N 80W **124** Florida, Straits of
8S 143E **166** Fly R.
50N 54W **86** Fogo I.
47N 83W **66** Foleyet
59N 107W **65** Fond-du-lac
46N 82W **79** Forest
49N 64W **85** Forillon National Park
26S 58W **133** Formosa
25N 120E **164** Formosa Strait
58N 111W **75** Fort Chipewyan
14N 61W **128** Fort de France
44N 79W **79** Fort Erie
48N 93W **65** Fort Frances
65N 123W **59** Fort Franklin
60 Fort George = Chisasibi
66N 128W **59** Fort Good Hope
26N 80W **124** Fort Lauderdale
60N 123W **65** Fort Liard
57N 111W **75** Fort Mackay
50S 113W **75** Fort Macleod
56N 111W **75** Fort McMurray
67N 135W **59** Fort McPherson
27N 82W **124** Fort Myers
58N 122W **75** Fort Nelson
65N 125W **59** Fort Norman
48N 107W **123** Fort Peck Reservoir
0 30E **144** Fort Portal
61N 177W **65** Fort Providence
51N 104W **76** Fort Qu'Appelle
61N 113W **65** Fort Resolution
60 Fort Rupert = Waskagheganish
56N 87W **60** Fort Severn
61N 121W **65** Fort Simpson
60N 112W **65** Fort Smith (N.W.T.)
35N 94W **124** Fort Smith (U.S.)
54N 124W **70** Fort St. James
56N 121W **75** Fort St. John
58N 116W **75** Fort Vermilion
41N 85W **124** Fort Wayne
32N 97W **123** Fort Worth
46N 76W **82** Fort-Coulonge
3S 38W **132** Fortaleza
22S 117E **171** Fortescue R.
47N 55W **86** Fortune
47N 55W **86** Fortune Bay
46S 165E **172** Foveaux Strait
54N 116W **75** Fox Creek
68N 77W **60** Foxe Basin
21S 48W **134** Franca
47N 3E **147** France ■
21S 27E **143** Francistown
54N 125W **69** François L.
38N 84W **124** Frankfort
52N 14E **147** Frankfurt
6 Franklin, District of □
81N 55E **153** Franz Joseph Land / Zemlya Frantsa Josifa

25S	153E	**172**	Fraser I.
51N	122W	**70**	Fraser R.
46N	66W	**85**	Fredericton
26N	78W	**128**	Freeport
8N	13W	**141**	Freetown
4N	53W	**132**	French Guiana ■
		176	French Polynesia □
36N	120W	**123**	Fresno
		60	Frobisher Bay = Iqaluit
56N	108W	**76**	Frobisher L.
30S	139E	**171**	Frome L.
44N	76W	**80**	Frontenac Provincial Park
26N	118E	**164**	Fujian □
33N	130E	**164**	Fukuoka
45N	66W	**85**	Fundy, Bay of
45N	65W	**85**	Fundy National Park
42	124E	**164**	Fushun
42N	121E	**164**	Fuxin
26N	119E	**164**	Fuzhou

G

0	10E	**143**	Gabon ■
24S	26E	**143**	Gaborone
6N	6W	**141**	Gagnoa
52N	68W	**66**	Gagnon
30N	83W	**124**	Gainesville
31S	136E	**171**	Gairdner, L.
36N	108W	**123**	Gallup
29N	94W	**124**	Galveston
53N	9W	**147**	Galway
13N	16W	**141**	Gambia ■
27N	115E	**164**	Gan R.
44N	76W	**80**	Gananoque
49N	54W	**86**	Gander
49N	54W	**86**	Gander L.
28N	79E	**158**	Ganga R. / Ganges R.
28N	78E	**158**	Ganges Plain
36N	104E	**163**	Gansu □
16N	0	**141**	Gao
9S	36W	**132**	Garanhuns
54N	94W	**65**	Garden Hill
50N	122W	**70**	Garibaldi Provincial Park
44N	1E	**147**	Garonne R.
9N	14E	**141**	Garoua
41N	87W	**124**	Gary
25S	117E	**171**	Gascoyne R.
49N	64W	**85**	Gaspé
49N	64W	**85**	Gaspé, Cap de
49N	65W	**85**	Gaspésie, Péninsule de la
49N	65W	**85**	Gaspésie Provincial Park
45N	76W	**80**	Gatineau
45N	76W	**80**	Gatineau Provincial Park
46N	76W	**80**	Gatineau R.
54N	18E	**147**	Gdańsk
38S	144E	**171**	Geelong
23N	103E	**163**	Gejiu
76S	35W	**178**	General Belgrano (research station)
46N	6E	**147**	Geneva / Genève
44N	9E	**147**	Genoa / Genova
58N	66W	**60**	George, R.
5N	100E	**165**	George Town / Pinang
20N	80W	**128**	Georgetown (Cayman Islands)
7N	58W	**131**	Georgetown (Guyana)
46N	62W	**85**	Georgetown (P.E.I.)
41N	45E	**153**	Georgia ■ (C.I.S.)
32N	82W	**124**	Georgia □ (U.S.)
49N	124W	**70**	Georgia (Strait of)
45N	81W	**79**	Georgian Bay
45N	81W	**79**	Georgian Bay Islands National Park
22S	139E	**171**	Georgina R.
28S	114E	**171**	Geraldton (Australia)
49N	87W	**66**	Geraldton (Ont.)
52N	12E	**147**	Germany ■
27N	84E	**158**	Ghaghara R.
6N	1W	**141**	Ghana ■
32N	3E	**141**	Ghardaïa
25N	10E	**141**	Ghāt
12N	79E	**158**	Ghats, Eastern
15N	75E	**158**	Ghats, Western
36N	5W	**147**	Gibraltar □
36N	5W	**141**	Gibraltar, Strait of
25S	123E	**171**	Gibson Desert
35N	136E	**164**	Gifu
43N	5W	**147**	Gijón
33N	111W	**123**	Gila R.
17S	132E	**171**	Gilbert R.
56N	94W	**65**	Gillam
51N	97W	**76**	Gimli
3N	46E	**142**	Gioher
38S	177E	**172**	Gisborne
4S	30E	**144**	Gitega
63N	160E	**154**	Gizhiga
68N	96W	**59**	Gjoa Haven
46N	60W	**85**	Glace Bay
51N	117W	**70**	Glacier National Park
24S	151E	**172**	Gladstone
56N	4W	**147**	Glasgow
42N	81W	**79**	Glencoe
7N	40E	**142**	Goba
48S	70W	**133**	Gobernad
47N	107E	**163**	Gobi Desert
18N	81E	**158**	Godavari R.
43N	81W	**79**	Goderich
33S	69W	**133**	Godoy Cruz
55N	94W	**65**	Gods River
		177	Godthab = Nuuk
16S	49W	**132**	Goiânia
28S	153E	**172**	Gold Coast
50N	125W	**69**	Gold River
51N	118W	**70**	Golden
49N	122W	**70**	Golden Ears Provincial Park
33N	95E	**163**	Golmud
52N	31E	**153**	Gomel
12N	37E	**142**	Gonder
51N	102W	**76**	Good Spirit Lake Provincial Park
53N	60W	**66**	Goose Bay
0	123E	**165**	Gorontalo
57N	12E	**147**	Göteborg
57N	18E	**147**	Gotland
48N	74W	**66**	Gouin, Réservoir
18S	42W	**132**	Governador Valadares
29S	59W	**133**	Goya
54N	133W	**69**	Graham I.
5S	46W	**132**	Grajaú R.
27S	64W	**133**	Gran Chaco
45N	72W	**80**	Granby
27N	78W	**128**	Grand Bahama I.
47N	55W	**86**	Grand Bank
51N	96W	**76**	Grand Beach Provincial Park
43N	81W	**79**	Grand Bend
35N	117E	**164**	Grand Canal R. / Da Yunhe R.
36N	144W	**123**	Grand Canyon
49N	55W	**85**	Grand Falls (N.B.)
54N	64W	**86**	Grand Falls (Nfld.)
49N	118W	**70**	Grand Forks
46N	66W	**85**	Grand L. (N.B.)
54N	61W	**86**	Grand L. (Nfld.)
44N	67W	**85**	Grand Manan I.
42N	79W	**79**	Grand R.
43N	86W	**124**	Grand Rapids
55N	119W	**75**	Grande Prairie
52S	67W	**133**	Grande, Bahía
12S	44W	**132**	Grande, R.
54N	77W	**60**	Grande, La, R.
49N	65W	**85**	Grande-Vallée
55N	126W	**69**	Granisle
55N	101W	**76**	Grass River Provincial Park
49N	107W	**76**	Grasslands National Park
50N	106W	**76**	Gravelbourg
52N	4E	**147**	's-Gravenhage / The Hague
45N	79W	**79**	Gravenhurst
33S	130E	**171**	Great Australian Bight
24N	79W	**128**	Great Bahama Bank
36S	175E	**172**	Great Barrier I.
18S	147E	**172**	Great Barrier Reef
		91	Great Basin
65N	120W	**59**	Great Bear L.
20S	145E	**171**	Great Dividing Range
33N	5E	**141**	Great Eastern Erg
47N	111W	**123**	Great Falls
21N	73W	**128**	Great Inagua I.
27N	73E	**158**	Great Indian Desert
43N	105W	**123**	Great Plains
41N	112W	**123**	Great Salt L.
22S	125E	**171**	Great Sandy Desert
61N	115W	**65**	Great Slave L.
28S	125E	**171**	Great Victoria Desert
40N	115E	**163**	Great Wall
40N	23E	**148**	Greece ■
41N	105W	**123**	Greeley
44N	88W	**124**	Green Bay
39N	110W	**123**	Green R.
66N	45W	**177**	Greenland / Kalaallit Nunaat □
73N	10W	**177**	Greenland Sea
5N	9W	**141**	Greenville
53N	104W	**76**	Greenwater Lake Provincial Park
12N	61W	**128**	Grenada ■
	0	**147**	Greenwich Meridian
51N	55W	**86**	Grey Is.
48N	57W	**86**	Grey R. (Nfld.)
43N	80W	**79**	Grimsby
75N	82W	**60**	Grise Fiord
14S	136E	**171**	Groote Eylandt
49N	58W	**86**	Gros Morne National Park
43N	45E	**153**	Groznyy
54S	37W	**178**	Grytviken (research station)
20N	103W	**127**	Guadalajara
29N	118W	**127**	Guadalupe
16N	61W	**128**	Guadeloupe ■
2N	70W	**131**	Guainia R.
13N	145E	**166**	Guam □
23N	113E	**164**	Guangdong □
24N	109E	**163**	Guangxi □
24N	114E	**164**	Guangzhou / Canton
20N	75W	**128**	Guantánamo (U.S. base)
13S	63W	**131**	Guaporé R. / Iténez R.
15N	90W	**127**	Guatemala ■
14N	90W	**127**	Guatemala (city)
4N	67W	**131**	Guaviare R.
2S	80W	**131**	Guayaquil
3S	80W	**131**	Guayaquil, Golfo de
28N	111W	**127**	Guaymas
43N	80W	**79**	Guelph
5N	65W	**131**	Guiana Highlands
2N	42E	**142**	Guiba R. / Juba R.
25N	110E	**164**	Guilin
10N	10W	**141**	Guinea ■
4N	2E	**141**	Guinea, Gulf of
		92	Guineau Plateau
12N	15W	**141**	Guinea-Bissau ■
26N	106E	**163**	Guiyang
27N	107E	**163**	Guizhou □
32N	74E	**158**	Gujranwala
31N	89W	**124**	Gulfport
3N	33E	**142**	Gulu
16N	80E	**158**	Guntur
12S	49W	**132**	Gurupi
2N	47W	**132**	Gurupi R.
5N	59W	**131**	Guyana ■
		69	Gwaii Haanas = South Moresby
20S	30E	**144**	Gwelo
55N	123W	**75**	Gwillim Lake Provincial Park
29N	90E	**163**	Gyangzê
52N	99W	**76**	Gypsumville

H

21N	106E	**165**	Ha-noi
23N	82W	**128**	Habana / Havana
41N	142E	**164**	Hachinohe
43N	80W	**79**	Hagersville
52N	4E	**147**	Hague, The / 's-Gravenhage
20N	106E	**165**	Hai-phong
53N	132W	**69**	Haida Gwaii / Queen Charlotte Is.
32N	35E	**157**	Haifa / Hefa
20N	110E	**164**	Haikou
19N	109E	**164**	Hainandao □
19N	72W	**128**	Haiti ■
41N	140E	**164**	Hakodate
36N	37E	**157**	Halab / Aleppo
44N	63W	**85**	Halifax CMA
75S	26W	**178**	Halley (research station)
18S	127E	**171**	Halls Creek
0	128E	**166**	Halmahera I.
43N	80W	**79**	Halton Hills
35N	48E	**157**	Hamadān
35N	37E	**157**	Hamāh
34N	137E	**164**	Hamamatsu
52N	118W	**70**	Hamber Provincial Park
53N	10E	**147**	Hamburg
23S	117E	**171**	Hamersley Range
43N	94E	**163**	Hami
38S	142E	**171**	Hamilton (Austral.)
43N	80W	**79**	Hamilton CMA (Ont.)
36N	114E	**164**	Handan
30N	120E	**164**	Hangzhou
40N	92W	**123**	Hanibal
51N	112W	**75**	Hanna
52N	9E	**147**	Hannover
44N	81W	**79**	Hanover
53N	60W	**66**	Happy Valley
18S	31E	**144**	Harare
45N	126E	**164**	Harbin
47N	56W	**86**	Harbour Breton
9N	42W	**142**	Hārer
9N	44E	**142**	Hargeisa
5N	7W	**141**	Harper
40N	77W	**124**	Harrisburg
41N	72W	**124**	Hartford
46N	68W	**85**	Hartland
49N	68W	**66**	Hauterive
23N	82W	**128**	Havana / Habana
34N	114W	**123**	Havasu L.
44N	78W	**79**	Havelock
49N	0	**147**	Havre, Le
50N	63W	**66**	Havre-St-Pierre
20N	156W	**123**	Hawaii □
20N	156W	**123**	Hawaiian Is.
45N	74W	**80**	Hawkesbury
34S	145E	**172**	Hay
59N	118W	**75**	Hay R.
61N	115W	**65**	Hay River
57N	94W	**65**	Hayes R.
49N	83W	**66**	Hearst
52N	53W	**86**	Heart's Content
39N	116E	**164**	Hebei □
57N	7W	**147**	Hebrides, Is.
57N	63W	**60**	Hebron
53N	130W	**69**	Hecate Strait
51N	96W	**76**	Hecla Provincial Park
32N	35E	**157**	Hefa / Haifa
32N	117E	**164**	Hefei
47N	130E	**164**	Hegang
54N	125E	**164**	Heilong Jiang / Amur R.
48N	126E	**164**	Heilongjiang □
46N	112W	**123**	Helena
31N	63E	**157**	Helmand R.
60N	25E	**148**	Helsinki
34N	114E	**164**	Henan □
26N	112E	**164**	Hengyang
55N	82W	**60**	Henrietta Maria, Cape
17N	95E	**165**	Henzada
34N	62E	**157**	Herāt
48N	113E	**164**	Herlen R. / Kerulen R.
29N	111W	**127**	Hermosillo
27N	106W	**127**	Hidalgo del Parral
58N	117W	**75**	High Level
55N	116W	**75**	High Prairie
50N	114W	**75**	High River
19N	155W	**123**	Hilo
30N	80E	**158**	Himalayas
35N	134E	**164**	Himeji
36N	72E	**158**	Hindu Kush
56N	119W	**75**	Hines Creek
53N	117W	**75**	Hinton
34N	132E	**164**	Hiroshima
11N	107E	**165**	Ho Chi Minh / Saigon
43S	147E	**172**	Hobart
41N	111E	**164**	Hohhot
43N	143E	**164**	Hokkaidō □
21N	76W	**128**	Holguin
70N	117W	**59**	Holman
14N	86W	**128**	Honduras ■
23N	104E	**165**	Hong R. / Red R.
22N	114E	**164**	Hong Kong (U.K.) □
25N	107E	**163**	Hongshui R.
49N	64W	**85**	Honguedo, Détroit d'
21N	158W	**123**	Honolulu
36N	138E	**164**	Honshū
49N	121W	**70**	Hope
55N	60W	**66**	Hopedale
50N	55W	**86**	Horse Is.
37N	80E	**163**	Hotan
54N	126W	**69**	Houston (B.C.)
30N	95W	**124**	Houston (U.S.)
44N	102E	**163**	Hovd
49N	91E	**163**	Hovd R.
40N	114E	**164**	Huainan
12S	16E	**143**	Huambo
12S	75W	**131**	Huancayo
38N	111E	**164**	Huang R. / Yellow R.
31N	112E	**164**	Hubei □
60N	86W	**60**	Hudson B.
53N	102W	**76**	Hudson Bay
43N	74W	**124**	Hudson R.
62N	70W	**60**	Hudson Strait
56N	121W	**75**	Hudson's Hope
16N	107E	**165**	Hue
15N	92W	**127**	Huehuetenango
46N	133E	**164**	Hulin
45N	75W	**80**	Hull
52N	105W	**76**	Humboldt
41N	116W	**123**	Humboldt R.
27N	112E	**164**	Hunan □
51N	121W	**70**	100 Mile House
47N	20E	**147**	Hungary ■
39N	127E	**164**	Hŭngnam
45N	74W	**80**	Huntingdon
45N	79W	**66**	Huntsville (Ont.)
34N	87W	**124**	Huntsville (U.S.)
45N	83W	**79**	Huron, L.
17N	78E	**158**	Hyderabad (India)
25N	68E	**158**	Hyderabad (Pakistan)

I

7N	4E	**141**	Ibadan
4N	75W	**131**	Ibagué
45N	73W	**80**	Iberville
2N	67W	**131**	Içana R.
65N	19W	**147**	Iceland ■
44N	114W	**123**	Idaho □
7N	5E	**141**	Ife
69N	81W	**60**	Igloolik
49N	91W	**65**	Ignace
26S	55W	**134**	Iguaçu Falls
26S	52W	**134**	Iguaçu R.
43N	78E	**158**	Ili R.
40N	89W	**124**	Illinois □
10N	123E	**165**	Iloilo
8N	4E	**141**	Ilorin
5S	47W	**132**	Imperatriz
20N	78E	**158**	India ■
5S	75E	**158**	Indian Ocean
40N	86W	**124**	Indiana □
39N	86W	**124**	Indianapolis
68N	145E	**154**	Indigirka R.
		92	Indo-Gangetic Plain
5S	115E	**165**	Indonesia ■
22N	75E	**158**	Indore
32N	71E	**158**	Indus R.
43N	81W	**79**	Ingersoll
47N	60W	**85**	Ingonish
52N	114W	**75**	Innisfail (Alta.)
17S	146E	**172**	Innisfail (Austral.)
58N	78W	**60**	Inukjuak
68N	133W	**59**	Inuvik
46S	168E	**172**	Invercargill
50N	116W	**70**	Invermere
46N	61W	**85**	Inverness
37N	18E	**147**	Ionian Sea
42N	93W	**124**	Iowa □
20S	43W	**134**	Ipatinga
62N	66W	**60**	Iqaluit (Frobisher Bay)
20S	70W	**133**	Iquique
3S	73W	**131**	Iquitos
33N	53E	**157**	Iran ■
20N	101W	**127**	Irapuato
33N	44E	**157**	Iraq ■
53N	8W	**147**	Ireland ■
4S	137E	**166**	Irian Jaya □
7S	36E	**144**	Iringa
4S	54W	**132**	Iriri R.
52N	106E	**154**	Irkutsk
45N	75W	**80**	Iroquois
48N	80W	**66**	Iroquois Falls
23N	106E	**165**	Irrawaddy R.
59N	68E	**153**	Irtysh R.
78N	104W	**59**	Isachsen
3N	43E	**142**	Iscia Baidoa
53N	66E	**153**	Ishim R.
38N	142E	**164**	Ishinomaki
3N	27E	**142**	Isiro
57N	131W	**69**	Iskut R.
33N	73E	**158**	Islamabad
~~39N~~	~~3E~~	~~**147**~~	~~Islas Baleares / Balearic Is.~~
4N	44E	**144**	Islia Baidoa
32N	35E	**157**	Israel ■
41N	29E	**157**	İstanbul
42N	14E	**147**	Italy ■
6S	46W	**132**	Itapecuru R.
13S	63W	**131**	Iténez R. / Guaporé R.
56N	41E	**153**	Ivanovo
7N	5W	**141**	Ivory Coast ■
62N	77W	**60**	Ivujivik
56N	53E	**153**	Izhevsk
38N	27E	**157**	Izmir

J

28N	80E	**158**	Jabalpur
32N	90W	**124**	Jackson
30N	81W	**124**	Jacksonville
50N	63W	**85**	Jacques-Cartier, Détroit de
6S	38W	**132**	Jaguaribe R.
27N	76E	**158**	Jaipur
6S	106E	**165**	Jakarta
34N	71E	**158**	Jalalabad
18N	77W	**128**	Jamaica ■
1S	103E	**165**	Jambi
51N	80W	**60**	James Bay
22N	86E	**158**	Jamshedpur
36N	136E	**164**	Japan ■
40N	135E	**164**	Japan, Sea of
1S	136E	**166**	Japen, I.
1S	72W	**131**	Japura R.
1N	53W	**131**	Jari R.
52N	118W	**70**	Jasper
52N	118W	**70**	Jasper National Park
1N	60W	**131**	Jauaperi R.
7S	110E	**165**	Java □
4S	115E	**165**	Java Sea
2S	140E	**166**	Jayapura
38N	92W	**124**	Jefferson City
16S	42W	**132**	Jequitinhonha R.
41N	35E	**157**	Jerusalem
6N	116E	**165**	Jesselton / Kota Kinabalu
32N	106E	**163**	Jialing R.
46N	130E	**164**	Jiamusi
33N	120E	**164**	Jiangsu □
27N	116E	**164**	Jiangxi □
35N	113E	**164**	Jiaozuo
21N	39E	**157**	Jiddah
43N	127E	**164**	Jilin □
43N	126E	**164**	Jilin (city)
8N	36E	**142**	Jima
37N	117E	**164**	Jinan / Tsinan
29N	117E	**164**	Jingdezhen
0	34E	**144**	Jinja
27N	100E	**163**	Jinsha R. / Yangtze R.
41N	121E	**164**	Jinzhou
7S	63W	**131**	Jiparaná R.
45N	131E	**164**	Jixi
17N	42E	**157**	Jizān
7S	35W	**132**	João Pessoa
26N	73E	**158**	Jodhpur
26S	28E	**143**	Johannesburg
26S	49W	**134**	Joinvile
46N	73W	**80**	Joliette
31N	36E	**157**	Jordan ■
48N	124W	**70**	Juan de Fuca (Strait of)
34N	82W	**133**	Juan Fernández, Islas
9S	41W	**132**	Juàzeiro
5N	32E	**142**	Juba
2N	42E	**144**	Juba R. / Guiba R.
16N	95W	**127**	Juchitan
21S	43W	**134**	Juiz de Fora
34S	61W	**133**	Junin

7S 69W **131** Juruá R.
11S 58W **131** Juruena R.

K

34N 69E **158** Kābul
15N 28E **144** Kabwe
10N 7E **141** Kaduna
16N 14W **141** Kaédi
31N 130E **164** Kagoshima
34N 114E **164** Kaifeng
66N 45W **177** Kalaallit Nunaat / Greenland □
24S 19W **143** Kalahari Desert
43N 83W **124** Kalamazoo
6S 29E **144** Kalemie
30S 121E **171** Kalgoorlie
54N 20E **153** Kaliningrad
54N 36E **153** Kaluga
57N 160E **154** Kamchatka Peninsula / Poluostrov Kamchatka
56N 62E **153** Kamensk Ural'skiy
50N 120W **70** Kamloops
0 32E **144** Kampala
51N 116W **70** Kananaskis Provincial Park
6S 22E **143** Kananga
36N 136E **164** Kanazawa
27N 88E **163** Kanchenjunga, Mt.
7N 80E **158** Kandy
21N 158W **123** Kaneohe
35S 137E **171** Kangaroo I.
58N 66W **60** Kangiqsualujjuaq
62N 72W **60** Kangiqsujuaq
60N 70W **60** Kangirsuk
70N 68W **60** Kangirtugaapik
10N 9W **141** Kankan
12N 8E **141** Kano
26N 80E **158** Kanpur
38N 98W **123** Kansas □
39N 94W **124** Kansas City
23N 120E **164** Kaohsiung
14N 16W **141** Kaolack
0 110E **165** Kapuas R.
49N 82W **66** Kapuskasing
75N 70E **177** Kara Sea
25N 67E **158** Karachi
50N 73E **153** Karaganda
17S 28E **144** Kariba, L.
6S 21E **143** Kasai R.
52N 82W **60** Kashechewan
39N 76E **163** Kashi
16N 36E **142** Kassala
14S 132E **171** Katherine
27N 85E **158** Kathmandu
13N 7E **141** Katsina
22N 159W **123** Kauai
75N 95W **59** Kaujuitoq
19S 19E **143** Kaukau Veld
55N 24E **153** Kaunas
2S 151E **166** Kavieng
35N 139E **164** Kawasaki
14N 11W **141** Kayes
38N 35E **157** Kayseri
50N 70E **153** Kazakhstan■
55N 49E **153** Kazan
59N 127W **69** Kechika R.
48N 67W **85** Kedgwick
26S 15E **143** Keetmanshoop
6 Keewatin, District of □
43N 65W **85** Kejimkujik National Park
50N 119W **70** Kelowna
55N 86E **153** Kemerovo
45N 75W **80** Kemptville
20N 98E **165** Keng Tung
50N 94W **65** Kenora
37N 85W **124** Kentucky □
45N 64W **85** Kentville
1N 38E **144** Kenya ■
0 37E **144** Kenya, Mt.
8S 133E **166** Kepulauan Aru
0 98E **165** Kepulauan Batu
0 105E **165** Kepulauan Lingga
3S 99E **165** Kepulauan Mentawai
2S 126E **166** Kepulauan Sula
8S 132E **166** Kepulauan Tanimbar
30N 57E **157** Kermān
34N 47E **157** Kermānshāh
48N 113E **164** Kerulen R. / Herlen R.
25N 82W **124** Key West
48N 135E **153** Khabarovsk
50N 36E **153** Khar'khov
15N 32E **142** Khartoum
46N 32E **153** Kherson
72N 98E **154** Kheta R.
47N 120E **164** Khingan Range
22N 89E **158** Khulna
52N 116W **70** Kicking Horse Pass
50N 30E **153** Kiev / Kiyev
2S 30E **144** Kigali
5S 18E **143** Kikwit
3S 37E **144** Kilimanjaro, Mt.
46N 81W **66** Killarney Provincial Park
49N 116W **70** Kimberley (B.C.)
28S 24E **143** Kimberley (S. Africa)
17S 127E **171** Kimberley Plateau
41N 129E **164** Kimch'aek
52N 120W **70** Kinbasket L.
44N 81W **79** Kincardine
51N 109W **75** Kindersley
39S 144E **171** King I.
69N 100W **59** King William I.
18N 77W **128** Kingston (Jamaica)
44N 76W **80** Kingston (Ont.)
13N 61W **128** Kingstown
42N 82W **79** Kingsville
4S 15E **143** Kinshasa
57N 109E **154** Kirensk
153 Kirghiz S.S.R. = Kyrgystan
176 Kiribati
48N 80W **66** Kirkland Lake
35N 44E **157** Kirkūk
58N 49W **153** Kirov
48N 35E **153** Kirovograd
0 25E **143** Kisangani
47N 29E **153** Kishinev
0 42E **144** Kismayu
0 35E **144** Kisumu
34N 131E **164** Kitakyūshū
43N 80W **79** Kitchener CMA
54N 128W **69** Kitimat
13S 28E **144** Kitwe
2S 29E **144** Kiuv, L.
50N 30E **153** Kiyev / Kiev
41N 34E **157** Kizil Irmak R.
64N 12E **149** Kjolen Mts.
51N 126W **69** Klinaklini R.
61N 138W **59** Kluane National Park
36N 84W **124** Knoxville
34N 135E **169** Kōbe
55N 12E **147** København / Copenhagen
34N 133E **164** Kochi
49N 117W **70** Kokanee Glacier Provincial Park
62N 34E **150** Kola Peninsula
16N 74E **158** Kolhapur
10S 25E **143** Kolwezi
66N 152E **154** Kolyma R.
65N 155E **154** Kolyma Range
10N 5W **141** Komoé R.
12N 106E **165** Kompong Cham
50N 137E **154** Komsomol'sk na Amure
41N 88E **163** Konqi R.
50N 117W **70** Kootenay L.
51N 116W **70** Kootenay National Park
50N 116W **70** Kootenay R.
40N 127E **164** Korea, North ■
36N 128E **164** Korea, South ■
35N 128E **164** Korea Strait
9N 5W **141** Korhogo
42N 86E **163** Korla
6N 102E **165** Kota Baharu
6N 116E **165** Kota Kinabalu / Jesselton
68N 102E **154** Kotuy R.
47N 65W **85** Kouchibouguac National Park
22N 114E **164** Kowloon
45N 39E **153** Krasnodar
40N 53E **153** Krasnovodsk
56N 93E **154** Krasnoyarsk
16N 78E **158** Krishna R.
47N 33E **153** Krivoy Rog
13N 100E **165** Krung Thep / Bangkok
3N 101E **165** Kuala Lumpur
1N 110E **165** Kuching
32N 130E **164** Kumamoto
6N 1W **141** Kumasi
5N 9E **141** Kumba
27N 97E **165** Kumon R.
17S 12E **143** Kunene R.
35N 85E **163** Kunlun Mts.
25N 102E **163** Kunming
10S 123E **165** Kupang
42N 83E **163** Kuqa
40N 47E **157** Kura R.
34N 134E **164** Kurashiki
34N 132E **164** Kure
55N 65E **153** Kurgan
45N 150E **154** Kuril Is. / Kuril'skiye Ostrova
51N 36E **153** Kursk
43N 145E **164** Kushiro
58N 67W **60** Kuujjuaq
55N 78W **60** Kuujjuarapik
29N 47E **157** Kuwait ■
29N 47E **157** Kuwait / Al Kuwayt
58N 125W **69** Kwadacha Wilderness Provincial Park
16N 23E **143** Kwando R.
35N 127E **164** Kwangju
7S 16E **143** Kwango R.
10S 15E **143** Kwanza R.
5S 18E **143** Kwilu R.
8S 33E **144** Kyoga, L.
35N 135E **164** Kyōto
42N 75E **153** Kyrgyzstan ■
33N 131E **164** Kyūshū □

L

52N 55W **86** L'Anse aux Meadows
46N 73W **80** L'Épiphanie
50N 24E **153** L'vov
48N 70W **66** La Baie
43N 8W **147** La Coruña
44N 91W **124** La Crosse
53N 75W **60** La Grande Rivière
56N 109W **75** La Loche
47N 73W **66** La Mauricie National Park
16S 68W **131** La Paz (Bolivia)
24N 111W **127** La Paz (Mexico)
35S 58W **133** La Plata
29S 66W **133** La Rioja
55N 105W **94** La Ronge, Lac
49N 79W **66** La Sarre
50N 55W **86** La Scie
30S 71W **133** La Serena
47N 77W **66** La Vérendrye Provincial Park
11N 12W **141** Labé
53N 67W **66** Labrador City
53N 58W **66** Labrador Sea
50N 96W **76** Lac du Bonnet
55N 108W **76** Lac Île-à-la-Crosse Provincial Park
55N 112W **75** Lac la Biche
55N 104W **76** Lac La Ronge
55N 104W **76** Lac La Ronge Provincial Park
45N 72W **80** Lac-Brome
45N 71W **80** Lac-Mégantic
34S 144E **172** Lachlan R.
45N 74W **80** Lachute
52N 113W **75** Lacombe
61N 30E **153** Ladoga, L.
47N 83W **66** Lady Evelyn-Smoothwater Provincial Park
48N 123W **70** Ladysmith
6S 147E **166** Lae
6N 4E **141** Lagos
13N 45E **157** Lahej
31N 74E **158** Lahore
30N 94W **124** Lake Charles
48N 124W **70** Lake Cowichan
92 Lake Eyre Basin
62N 70W **60** Lake Harbour
51N 116W **70** Lake Louise
49N 95W **65** Lake of the Woods
47N 84W **66** Lake Superior Provincial Park
10N 73E **158** Lakshadweep (Is.)
1S 10E **143** Lambaréné
2S 41E **144** Lamu
25N 100E **163** Lancang R. / Mekong R.
74N 84W **60** Lancaster Sound
49N 122W **70** Langley
42N 84W **124** Lansing
36N 104E **163** Lanzhou
17N 105E **165** Laos ■
76N 125E **154** Laptev Sea
8S 123E **165** Larantuka
27N 99W **123** Laredo
49N 58W **86** Lark Harbour
28N 16W **141** Las Palmas
36N 115W **123** Las Vegas
29N 91E **163** Lasa / Lhasa
57N 24E **153** Latvia ■
41S 147E **172** Launceston
47N 71W **66** Laurentides Provincial Park
43N 71W **124** Lawrence
34N 98W **123** Lawton
49N 0 **147** Le Havre
51N 110W **75** Leader
57N 100W **65** Leaf Rapids
42N 83W **79** Leamington
34N 36E **157** Lebanon ■
53N 113W **75** Leduc
19S 140E **171** Leichardt R.
51N 12E **147** Leipzig
67N 124E **154** Lena R.
39N 94E **163** Lenghu
69S 159E **178** Leningradskaya (research station)
22N 102W **127** Léon (Mexico)
13N 86W **127** León (Nicaragua)
29S 28E **144** Lesotho ■
56N 115W **75** Lesser Slave L.
55N 115W **75** Lesser Slave Lake Provincial Park
49N 112W **75** Lethbridge
47N 71W **80** Lévis
38N 84W **124** Lexington
11N 125E **166** Leyte, I.
29N 91E **163** Lhasa / Lasa
34N 119E **164** Lianyungang / Xinpu
41N 123E **164** Liao R.
42N 122E **164** Liaoning □
41N 123E **164** Liaoyang
43N 125E **164** Liaoyuan
61N 130W **59** Liard R.
6N 9W **141** Liberia ■
0 9E **143** Libreville
27N 17E **142** Libya ■
29N 20E **142** Libyan Desert
14S 36E **144** Lichinga
47N 9E **147** Liechtenstein ■
46N 76W **80** Lièvre, R.
21S 168E **172** Lifou, I.
11S 26E **144** Likasi
49N 122W **70** Lillooet
14S 34E **144** Lilongwe
12S 77W **131** Lima
40S 70W **133** Limay R.
52N 8W **147** Limerick
39N 103W **128** Limón
23S 28E **144** Limpopo R.
41N 96W **124** Lincoln
10S 40E **144** Lindi
44N 79W **79** Lindsay
0 105E **165** Lingga Is.
52N 39E **153** Lipetsk
2N 22E **142** Lisala
38N 9W **147** Lisboa / Lisbon
49N 38E **153** Lisichansk
43N 80W **79** Listowel
55N 24E **153** Lithuania ■
34N 92W **124** Little Rock
24N 109E **163** Liuzhou
44N 65W **85** Liverpool
17S 26E **144** Livingstone
46N 14E **147** Ljubljana
4N 72W **131** Llanos
53N 110W **75** Lloydminster
23S 69W **133** Loa R.
12S 13E **143** Lobito
43N 65W **85** Lockeport
4N 36W **144** Lodwar
51N 19E **147** Lódź
50N 121W **70** Logan Lake
60N 140W **59** Logan, Mt.
31S 52W **134** Logoa dos Patos
10N 16E **141** Logone R.
47N 2E **147** Loire R.
8S 116E **165** Lombok, I.
6S 1E **141** Lomé
43N 81W **79** London CMA (Ont.)
51N 0 **147** London (U.K.)
23S 51W **134** Londrina
23N 75W **128** Long Island (Bahamas)
41N 73W **124** Long Island (U.S.)
42N 80W **79** Long Point Provincial Park
50N 57W **86** Long Range Mountains
45N 73W **80** Longeuil
49N 86W **66** Longlac
23S 144E **172** Longreach
31S 159E **172** Lord Howe I.
92 Lord Howe Rise
37S 72W **133** Los Angeles (Chile)
34N 118W **123** Los Angeles (U.S.)
75S 64W **133** Los Estados, Isla de
26N 109W **127** Los Mochis
20N 103E **165** Louangphrabang
46N 60W **85** Louisbourg National Park
31N 92W **124** Louisiana □
38N 85W **124** Louisville
49N 118W **70** Lower Arrow L.
0 25E **144** Lualaba R.
42N 117E **164** Luan R.
9S 13E **143** Luanda
12S 32E **144** Luangwa R.
15N 14E **143** Lubango
33N 102W **123** Lubbock
11S 27E **144** Lubumbashi
27N 81E **158** Lucknow (India)
44N 81W **79** Lucknow (Ont.)
39N 121E **164** Lüda / Dairen
26S 15E **143** Lüderitz
49N 39E **153** Lugansk
15S 167E **172** Luganville
21E 2N **143** Lulonga R.
50N 105W **74** Lumsden
44N 64W **85** Lunenburg
31N 104E **164** Luoyang
15S 28E **144** Lusaka
50N 6E **147** Luxembourg ■
49N 6E **147** Luxembourg (city)
26N 33E **142** Luxor / El Uqsor
16N 121E **165** Luzon, I.
20N 121E **165** Luzon Strait
32N 74E **158** Lyallpur / Faisalbad
57N 101W **65** Lynn Lake
45N 5E **147** Lyon
50N 121E **70** Lytton

M

0 51W **132** Macapa
72N 102W **59** M'Clintock Channel
22N 113E **164** Macau □
23S 133E **171** MacDonnell Ranges
42N 22E **148** Macedonia ■
9S 35W **132** Maceió
21S 149E **172** Mackay
22S 129E **171** Mackay, L.
55N 122W **70** Mackenzie
6 Mackenzie, District of □
77N 110W **59** Mackenzie King I.
67N 130W **59** Mackenzie Mountains
67N 130W **59** Mackenzie R.
33N 83W **124** Macon
20S 47E **144** Madagascar ■
5S 145E **166** Madang
32N 17W **141** Madeira Is.
5S 61W **131** Madeira, R.
47N 61W **85** Madeleine, Îles de la
43N 89W **124** Madison
44N 97W **79** Madoc
13N 80E **158** Madras
11S 66W **131** Madre de Dios R.
40N 3W **147** Madrid
7S 113E **165** Madura, I.
10N 78E **158** Madurai
15S 168E **172** Maéwo, I.
8S 40E **144** Mafia I.
59N 151E **154** Magadan
5N 75W **131** Magdalena R.
60S 68W **133** Magallanes, Estrecho de (Magellan's Strait)
76N 100W **177** Magnetic Pole, North
65S 140E **178** Magnetic Pole, South
53N 59E **153** Magnitogorsk
45N 72W **80** Magog
15S 46E **144** Mahajanga
20N 86E **158** Mahanadi R.
44N 64W **85** Mahone Bay
2S 18E **143** Mai-Ndombe, L.
12N 13E **141** Maiduguri
45N 69W **124** Maine □
34S 152E **172** Maitland
44N 81W **79** Maitland R.
1S 118E **165** Makassar Strait
43N 47E **153** Makhachkala
21N 40E **157** Makkah / Mecca
8N 8E **141** Makurdi
4N 9E **141** Malabo
3N 101E **165** Malacca
3N 101E **165** Malacca, Strait of
36N 4W **147** Málaga
10N 32E **142** Malakal
16S 167E **172** Malakula, I.
8S 112E **165** Malang
10S 16E **143** Malange
13S 34E **144** Malawi ■
12S 34E **144** Malawi, L.
7N 100E **165** Malay Peninsula
4N 102E **165** Malaya □
5N 110E **165** Malaysia ■
5N 74E **145** Maldives ■
15N 2W **141** Mali ■
3S 40E **144** Malindi
39N 3E **147** Mallorca
55N 13E **147** Malmö
36N 14E **147** Malta ■
2S 138E **166** Mamberamo R.
19S 34E **144** Mamica
14S 66W **131** Mamoré R.
10N 12W **141** Mamou
8N 8W **141** Man
1N 125E **166** Manado
12N 86W **127** Managua
3S 60W **131** Manaus
53N 2W **147** Manchester (U.K.)
43N 71W **124** Manchester (U.S.)
92 Manchurian Plain
22N 96E **165** Mandalay
13N 74E **158** Mangalore
22S 45E **144** Mangoky R.
51N 68W **66** Manicouagan, Réservoir
14N 121E **165** Manila
55N 97W **76** Manitoba □
51N 98W **76** Manitoba, L.
46N 76W **66** Maniwaki
5N 75W **131** Manizales
57N 117W **75** Manning
49N 120W **70** Manning Provincial Park
62N 80W **60** Mansel I.
2S 147E **166** Manus I.
19N 115W **127** Manzanillo
4S 137E **166** Maoke Mountains
50N 109W **75** Maple Creek
49N 122W **70** Maple Ridge
26S 32E **144** Maputo
38S 57W **133** Mar del Plata
10N 71W **131** Maracaibo
9N 71W **131** Maracaibo, L.
10N 67W **131** Maracay
13N 8E **141** Maradi
5S 77W **131** Maranón R.
48N 86W **66** Marathon
21S 168E **172** Maré, I.
92 Mariana Trench
80S 120W **178** Marie Byrd Land
22S 50W **134** Marília
23S 52W **134** Maringá

22S 60W **133** Mariscal Estigarribia
47N 38E **153** Mariupol'
44N 79W **79** Markham
44N 77W **79** Marmora
4N 54W **132** Maroni R.
10N 14E **141** Maroua
14N 22E **142** Marra Mountains
31N 8W **141** Marrakech
3N 38E **144** Marsabit
43N 5E **147** Marseille
176 Marshall Islands
3S 115E **165** Martapura
42N 71W **124** Martha's Vineyard
14N 61W **128** Martinique □
38N 62E **153** Mary
52N 56W **86** Mary's Harbour
39N 76W **124** Maryland □
47N 55W **86** Marystown
26N 14E **141** Marzūq
35N 128E **164** Masan
12N 123E **165** Masbate
29S 27E **144** Maseru
36N 59E **157** Mashhad
46N 73W **80** Maskinongé
23N 58E **157** Masqaṭ / Muscat
42N 72W **124** Massachusetts □
54N 132W **69** Masset
46N 0 **149** Massif Centrale
46N 73W **66** Mastigouche Provincial Park
6S 13E **143** Matadi
49N 77W **66** Matagami
25N 98W **127** Matamoros
49N 67W **85** Matane
49N 67W **85** Matane Provincial Park
23N 81W **128** Matanzas
48N 67W **85** Matapédia
48N 67W **85** Matapédia R.
8S 116E **165** Mataram
12S 57W **132** Mato Grosso, Plateau of
23N 58E **157** Maṭraḥ
32N 27W **142** Matrûh
70 Matsqui = Abbotsford
36N 133E **164** Matsue
33N 132E **164** Matsuyama
50N 81W **66** Mattagami R.
9N 63W **131** Maturín
21N 156W **123** Maui I.
21N 10W **141** Mauritania ■
20S 58E **144** Mauritius ■
67S 63E **178** Mawson (research station)
18N 67W **128** Mayaguez
63N 136W **59** Mayo
23N 106W **127** Mazatlán
26S 31E **144** Mbabane
0 18E **143** Mbandaka
9S 34E **144** Mbeya
6S 23E **143** Mbuji Mayi
53N 120W **70** McBride
24S 113E **171** McLeod, L.
77S 170E **178** McMurdo (research station)
70 McNaughton L. = Kinbasket L.
36N 114W **123** Mead, L.
54N 108W **75** Meadow Lake
54N 109W **75** Meadow Lake Provincial Park
44N 80W **79** Meaford
21N 40E **157** Mecca / Makkah
3N 98E **165** Medan
6N 75W **131** Medellín
50N 110W **75** Medicine Hat
24N 40E **157** Medina / Al Madīnah
40N 5E **147** Mediterranean Sea
26S 118E **171** Meekatharra
48N 57W **86** Meelpaeg L.
25N 100E **163** Mekong R. / Lancang R.
38S 145E **172** Melbourne
53N 104W **76** Melfort
49N 101W **76** Melita
51N 103W **76** Melville
53N 60W **66** Melville, L.
11S 131E **171** Melville I. (Australia)
75N 112W **59** Melville I. (Canada)
68N 84W **60** Melville Peninsula
9S 119E **165** Memboro
35N 90W **124** Memphis
45N 72W **80** Memphrémagog, Lac
33S 69W **133** Mendoza
14S 18E **143** Menongue
3S 99E **165** Mentawai Is.
34S 57W **133** Mercedes
65N 64W **60** Mercy, Cape
11N 97E **165** Mergui
20N 89W **127** Mérida (Mexico)
8N 71W **131** Merida (Venezuela)
50N 120W **70** Merritt
38N 15E **147** Messina
5N 72W **131** Meta R.
32N 115W **127** Mexicali
20N 100W **127** Mexico ■
19N 99W **127** México City
25N 90W **127** Mexico, Gulf of
34N 64E **157** Meymaneh
65N 48E **153** Mezen R.
25N 80W **124** Miami
44N 85W **124** Michigan □
44N 87W **124** Michigan, L.
176 Micronesia
44N 80W **79** Midland
45N 9E **147** Milan / Milano
34S 142E **171** Mildura
43N 80W **79** Milton
43N 88W **124** Milwaukee
29N 104E **163** Min R.
45N 64W **85** Minas Basin
8N 125E **166** Mindanao □
13N 121E **165** Mindoro, I.
50N 64W **66** Mingan Archipelago Provincial Park
45N 93W **124** Minneapolis
46N 94W **124** Minnesota □
54N 27E **153** Minsk
46N 66W **85** Minto
45N 74W **80** Mirabel
47N 66W **85** Miramichi R.
66S 93E **178** Mirnyy (research station)
2S 130E **166** Misool, I.
32N 15E **141** Miṣratah
48N 83W **66** Missinaibi Lake Provincial Park
50N 83W **66** Missinaibi R.
43N 79W **79** Mississauga
33N 90W **124** Mississippi □
34N 91W **124** Mississippi R.
38N 92W **124** Missouri □
41N 96W **123** Missouri R.
49N 72W **66** Mistassini
51N 74W **66** Mistassini, Lac
50N 74W **66** Mistassini Provincial Park
43N 81W **79** Mitchell
16S 142E **171** Mitchell R.
16N 39E **142** Mits'iwa
72N 77W **60** Mittimatalik
10S 26E **144** Mitumba, Chaine des
32N 132E **164** Miyazaki
30N 88W **124** Mobile
15S 12E **143** Moçâmedes
2N 45E **142** Mogadishu / Muqdisho
54N 30E **153** Mogilev
47N 28E **150** Moldova ■
67S 45E **178** Molodezhnava (research station)
27S 22E **143** Molopo R.
4S 124E **166** Molucca Sea
1S 127E **166** Moluccas, I.
4S 39E **144** Mombasa
43N 7E **147** Monaco ■
50N 118W **70** Monashee Provincial Park
46N 65W **85** Moncton
47N 103E **163** Mongolia ■
57N 121W **70** Monkman Provincial Park
42N 83W **124** Monroe
6N 10W **141** Monrovia
45N 72W **80** Mont Orford Provincial Park
47N 71W **80** Mont Ste-Anne Provincial Park
46N 74W **66** Mont Tremblant Provincial Park
48N 68W **85** Mont-Joli
47N 75W **66** Mont-Laurier
46N 62W **85** Montague
47N 110W **123** Montana □
18N 78W **128** Montego Bay
36N 122W **123** Monterey
9N 76W **131** Monteria
25N 100W **127** Monterrey
13S 44W **132** Montes Claros
35S 56W **134** Montevideo
32N 86W **124** Montgomery
47N 70W **80** Montmagny
44N 72W **124** Montpelier
45N 73W **80** Montréal CMA
54N 106W **76** Montreal L.
16N 62W **128** Montserrat ■
22N 95E **165** Monywa
51N 80W **60** Moose Factory
50N 105W **76** Moose Jaw
49N 102W **76** Moose Mountain Provincial Park
51N 80W **86** Moose R.
51N 80W **60** Moosonee
15N 4W **141** Mopti
29N 79E **158** Moradabad
19N 101W **127** Morelia
53N 131W **69** Moresby I.
39N 80W **124** Morgantown
32N 6W **141** Morocco ■
6S 38E **144** Morogoro
11S 42S **144** Moroni
2N 128E **166** Morotai, I
49N 97W **76** Morris
45N 75W **80** Morrisburg
55N 37E **153** Moscow / Moskva
5S 37W **132** Mossoró
16N 97E **165** Moulmein
9N 16E **141** Moundou
51N 115W **70** Mount Assiniboine Provincial Park
47N 67W **85** Mount Carleton Provincial Park
57N 130W **69** Mount Edziza Provincial Park
44N 81W **79** Mount Forest
20S 139E **171** Mount Isa
27S 118E **171** Mount Magnet
51N 118W **70** Mount Revelstoke National Park
53N 119W **70** Mount Robson Provincial Park
20S 35E **144** Mozambique ■
15S 41E **144** Mozambique (city)
20S 39E **144** Mozambique Channel
10S 40E **144** Mtwara
44N 129E **164** Mudanjiang
12S 28E **144** Mufulira
46N 61W **85** Mulgrave
30N 71E **158** Multan
13S 30E **144** Munchinga Mountains
59N 125W **69** Muncho Lake Provincial Park
48N 11E **147** Munich / München
2N 45E **142** Muqdisho / Mogadishu
27S 114E **171** Murchison R.
38N 1W **147** Murcia
49N 65W **85** Murdochville
69N 33E **153** Murmansk
46N 62W **85** Murray Harbour
36S 144E **171** Murray R.
34S 147E **172** Murrumbidgee R.
23N 58E **157** Muscat / Masqaṭ
43N 86W **124** Muskegon
2S 33E **144** Mwanza
9S 29W **144** Mweru, L.
21N 97E **165** Myanmar ■
11S 34W **144** Mzuzu
21N 95E **165** Myingyan
25N 97E **165** Myitkyina
12N 76E **158** Mysore

N

12N 15E **141** N'Djamena
13N 123E **165** Naga
25N 93E **158** Naga Hills
32N 130E **164** Nagasaki
35N 137E **164** Nagoya
21N 79E **158** Nagpur
26N 127E **164** Naha
61N 125W **59** Nahanni National Park
54N 132W **69** Naikoon Provincial Park
56N 61W **60** Nain
1S 36E **144** Nairobi
24N 41E **157** Najd
50N 86W **66** Nakina
0 36E **144** Nakuru
50N 117W **70** Nakusp
20N 106E **165** Nam-dinh
41N 71E **153** Namangan
19S 13E **143** Namib Desert
22S 18E **143** Namibia ■
15N 39E **144** Nampula
19N 101E **165** Nan
49N 124W **70** Nanaimo
28N 116E **164** Nanchang
30N 106E **163** Nanchong
32N 118E **164** Nanjing / Nanking
22N 108E **163** Nanning
47N 1W **147** Nantes
42N 80W **79** Nanticoke
50N 114W **75** Nanton
32N 121E **164** Nantong
0 37E **144** Nanyuki
37N 122W **123** Napa
44N 77W **79** Napanee
39S 177E **172** Napier
41N 14E **147** Naples / Napoli
2S 74W **131** Napo R.
41N 14E **147** Napoli / Naples
22N 74E **158** Narmada R.
30S 149E **172** Narrabri
33S 117E **171** Narrogin
68N 17E **147** Narvik
41N 75E **153** Naryn R.
36N 86W **124** Nashville
55N 129W **69** Nass R.
25N 77W **128** Nassau
23N 32E **142** Nasser L.
5S 35W **132** Natal
50N 61W **66** Natashquan
4N 108E **165** Natuna-Besar
9N 39E **142** Nazret
13S 29E **144** Ndola
41N 100W **123** Nebraska □
54N 124W **70** Nechako R.
53N 97W **65** Negginan
39S 65W **133** Negro R. (Argentina)
1S 63W **131** Negro R. (Brazil)
10N 122E **165** Negros, I.
42N 112E **164** Nei Mongol □
29N 105E **163** Neijiang
3N 75W **131** Neiva
49N 117W **70** Nelson (B.C.)
41S 173E **172** Nelson (N.Z.)
57N 95W **65** Nelson R.
49N 125E **164** Nen R.
28N 84E **158** Nepal ■
47N 66W **85** Nepisiguit R.
52N 5E **147** Netherlands ■
12N 69W **128** Netherlands Antilles □
66N 71W **60** Nettilling L.
39S 68W **133** Neuquén
39N 117W **123** Nevada □
7N 57W **131** New Amsterdam
5N 152E **166** New Britain
47N 66W **85** New Brunswick □
21S 165E **172** New Caledonia □
28N 77E **158** New Delhi
45N 62W **85** New Glasgow
4S 136E **166** New Guinea □
43N 71W **124** New Hampshire □
55N 127W **69** New Hazelton
6N 153E **166** New Ireland
40N 74W **124** New Jersey □
47N 79W **66** New Liskeard
34N 106W **123** New Mexico ■
30N 90W **124** New Orleans
52N 91W **65** New Osnaburgh
39S 174E **172** New Plymouth
48N 66W **85** New Richmond
75N 142E **154** New Siberian Is. / Novosibirskiye Ostrova
33S 146E **172** New South Wales □
46N 60W **85** New Waterford
70 New Westminster = Burnaby
42N 76W **124** New York □
40N 74W **124** New York (city)
40S 176E **172** New Zealand ■
40N 74W **124** Newark
33S 151E **172** Newcastle (Austral.)
47N 65W **85** Newcastle (N.B.)
44N 79W **79** Newcastle (Ont.)
54N 6W **147** Newcastle (U.K.)
53N 58W **86** Newfoundland □
23S 119E **171** Newman
44N 79W **79** Newmarket
37N 76W **124** Newport News
7N 13E **141** Ngaoundéré
2N 15E **143** Ngoko R.
12N 109E **165** Nha-trang
43N 79W **79** Niagara
43N 79W **124** Niagara Falls
43N 79W **79** Niagara-on-the-Lake
13N 2E **141** Niamey
1N 97E **165** Nias, I.
11N 85W **128** Nicaragua ■
12N 85W **128** Nicaragua, L.
9N 93E **158** Nicobar Is.
35N 33E **157** Nicosia
13N 10E **141** Niger ■
16N 0 **141** Niger R.
8N 8E **141** Nigeria ■
38N 139E **164** Niigata
47N 32E **153** Nikolayev
53N 140E **154** Nikolayevsk na Amure
30N 31E **142** Nile R.
30N 121E **164** Ningbo
38N 106E **163** Ningxia □
15N 9W **141** Nioro du Sahel
53N 104W **76** Nipawin
54N 104W **76** Nipawin Provincial Park
49N 87W **65** Nipigon
49N 88W **65** Nipigon, L.
64N 95E **154** Nizhnaya Tunguska R.
57N 44W **153** Nizhniy Novgorod
58N 60E **153** Nizhniy Tagil
5N 10E **143** Nkongsamba
32N 112W **127** Nogales
64N 166W **121** Nome
18N 103E **165** Nong Khai
51N 95W **65** Nopiming Provincial Park
46N 74W **80** Nora R.
48N 79W **66** Noranda, Rouyn
75N 111E **154** Nordvik
36N 76W **124** Norfolk
29S 168E **172** Norfolk I.
69N 88E **154** Noril'sk
65N 127W **59** Norman Wells
49N 58W **86** Norris Point
53N 108W **75** North Battleford
46N 79W **66** North Bay
47N 60W **85** North, Cape
35N 80W **124** North Carolina □
70 North Cowichan = Duncan
47N 100W **123** North Dakota □
38S 175E **172** North I.
40N 127E **164** North Korea ■
79N 100E **154** North Land / Severnaya Zemlya
76N 102W **59** North Magnetic Pole
42N 104W **123** North Platte R.
90N **177** North Pole
54N 110W **75** North Saskatchewan R.
56N 4E **147** North Sea
46N 60W **85** North Sydney
50N 123W **70** North Vancouver
53N 60W **66** North West River
32S 117E **171** Northam
53N 12E **149** Northern European Plain
54N 7W **147** Northern Ireland □
176 Northern Marianas
16S 133E **171** Northern Territory □
69N 140W **59** Northern Yukon National Park
46N 64W **85** Northumberland Strait
65N 100W **59** Northwest Territories □
67N 11E **147** Norway ■
54N 98W **65** Norway House
66N 1E **177** Norwegian Sea
44N 78W **79** Norwood
49N 55W **86** Notre Dame Bay
46N 73W **80** Notre-Dame-des-Prairies
46N 68W **85** Notre-Dame-du-Lac
45N 80W **79** Nottawasaga B.
44N 80W **79** Nottawasaga R.
21N 17W **141** Nouadhibou
18N 16W **141** Nouakchott
22S 166E **172** Nouméa
53N 78W **60** Nouveau Comptoir / Wemindiji
23S 44W **134** Nova Iguacu
45N 63W **85** Nova Scotia □
75N 56E **177** Novaya Zemlya
30S 51W **134** Novo Humburgo
53N 87E **153** Novokuznetsk
71S 11E **178** Novolazarevskaya (research station)
55N 83E **153** Novosibirsk
75N 142E **154** Novosibirskiye Ostrova / New Siberian Is.
31N 95E **163** Nu R. / Salween R.
22N 32E **142** Nubian Desert
27N 99W **127** Nuevo Laredo
32S 125E **171** Nullarbor Plain
65N 95W **59** Nunavut □
49N 11E **147** Nürnberg
64N 51W **177** Nuuk
12N 25E **142** Nyala
1S 37E **144** Nyeri
8N 9W **141** Nzérékoré

O

22N 158W **123** Oahu I.
43N 80W **79** Oakville
17N 96W **127** Oaxaca
67N 73E **153** Ob, Gulf of
59N 80E **153** Ob' R.
54N 94W **65** Obasquia Provincial Park
1S 127E **166** Obi I.
55N 10E **147** Odense
32N 103W **123** Odessa (U.S.)
46N 30E **153** Odessa (Ukraine)
53N 14E **147** Odra R.
8N 4E **141** Ogbomosho
41N 112W **123** Ogden
40N 84W **124** Ohio □
39N 82W **124** Ohio R.
33N 131E **164** Oita
57N 43E **148** Oka R.
50N 120W **70** Okanagan L.
34N 134E **164** Okayama
27N 81W **124** Okeechobee, L.
53N 143E **154** Okha
59N 143E **154** Okhotsk
55N 145E **177** Okhotsk, Sea of
26N 128E **164** Okinawa
35N 97W **123** Oklahoma □
35N 97W **123** Oklahoma City
19S 22E **143** Okovango Swamp
37S 60W **133** Olavarría
68N 140W **59** Old Crow
50N 106W **76** Old Wives L.
50N 112W **75** Oldman R.
58N 122E **154** Olekma R.
60N 121E **154** Olekminsk
68N 114E **154** Olenek R.
7S 35W **132** Olinda
47N 123W **123** Olympia
41N 96W **124** Omaha
23N 58E **157** Oman ■
24N 58E **157** Oman, Gulf of
15N 32E **142** Omdurman
56N 126W **69** Omineca R.
65N 161E **154** Omolon R.
55N 73E **153** Omsk
62N 35E **153** Onega, L.
67N 39E **148** Onega R.
6N 6E **141** Onitsha
52N 88W **79** Ontario □

43N 78W 79 Ontario, L.
27S 135E 171 Oodnadatta
54N 126W 69 Ootsa L.
42N 9W 147 Oporto / Porto
35N 0 141 Oran
33S 149E 172 Orange
30S 20E 143 Orange R.
44N 80W 79 Orangeville
60S 47W 178 Orcadas (research station)
18S 128E 171 Ord R.
44N 121W 123 Oregon □
53N 36E 153 Orel
51N 55E 153 Orenburg
44N 79W 79 Orillia
8N 66W 131 Orinoco R.
10N 60W 131 Orinoco R., Mouths of
19N 97W 127 Orizaba
59N 3W 147 Orkney Is.
27N 81W 124 Orlando
47N 71W 80 Orléans, Î.d'
46N 66W 85 Oromocto
51N 58E 153 Orsk
18S 67W 131 Oruro
34N 135E 164 Ōsaka
44N 79W 79 Oshawa CMA
60N 10E 147 Oslo
49N 119W 70 Osogoos
40S 73W 133 Osorno
51N 143E 154 Ostrov Sakhalin / Sakhalin I.
72N 180E 154 Ostrov Vrangelya / Wrangel I.
43N 141E 164 Otaru
10N 0 141 Oti R.
45N 76W 80 Ottawa R.
45N 75W 80 Ottawa-Hull CMA
12N 1W 141 Ouagadougou
32N 5E 141 Ouargla
5N 20E 142 Oubangui R. / Ubangi R.
34S 22E 143 Oudtshoorn
34N 2W 141 Oujda
65N 25E 147 Oulu
51N 107W 76 Outlook
43N 6W 147 Oviedo
44N 81W 79 Owen Sound
7S 146E 166 Owen Stanley Range
55N 95W 65 Oxford House
2N 11E 143 Oyem

P

38N 127E 164 P'yŏnyang
0 140E 146 Pacific Ocean
49N 125W 69 Pacific Rim National Park
1S 100E 165 Padang
2S 5E 143 Pagula Is.
56N 97W 65 Paint Lake Provincial Park
0 101E 165 Pakanbaru
30N 70E 158 Pakistan ■
15N 106E 165 Pakxe
7N 134E 166 Palau Is.
9N 118E 165 Palawan, I.
3S 105E 165 Palembang
38N 13E 147 Palermo
39N 2E 177 Palma
64S 65W 178 Palmer (research station)
44N 81W 79 Palmerston
40S 175E 172 Palmerston North
3N 76W 131 Palmira
2N 109E 165 Paloh
37S 64W 133 Pampas
8N 80W 128 Panama ■
9N 79W 128 Panama Canal
9N 79W 128 Panamá City
8N 79W 128 Panama, Golfo de
11N 122E 165 Panay, I.
2S 106E 165 Pangkalpinang
66N 66W 60 Pangnirtung
46N 75W 66 Papineau-Labelle
8S 145E 166 Papua New Guinea ■
2S 50W 132 Para R.
17N 113E 165 Paracel Is.
14S 62W 131 Paraguá R.
23S 57W 133 Paraguay ■
25S 58W 133 Paraguay R.
22S 43W 134 Paraíbo R.
6N 55W 132 Paramaribo
32S 60W 133 Paraná
30S 59W 133 Paraná R.
19S 51W 132 Paranaíba R.
23S 53W 134 Paranapanema R.
13S 40W 132 Paraquacu R.
49N 2E 147 Paris
43N 81W 79 Parkhill
49N 124W 70 Parksville
4S 42W 132 Parnaíba R.
27N 89E 158 Paro Dzong
45N 64W 85 Parrsboro
70N 124W 59 Parry, Cape
76N 115W 59 Parry Is.
45N 80W 66 Parry Sound
55N 122W 70 Parsnip R.
0 54W 132 Paru R.
27S 52W 134 Passo Fundo
2S 77W 131 Pastaza R.
2N 77W 131 Pasto
91 Patagonia
25N 85E 158 Patna
31S 52W 134 Patos, Logoa dos
69N 124W 59 Paulatuk
52N 77E 153 Pavlodar
16S 68W 131 Paz, La (Bolivia)
24N 111W 127 Paz, La (Mexico)
59N 114W 75 Peace R.
56N 117W 75 Peace River
49N 119W 70 Peachland
66N 52E 153 Pechora R.
32N 106W 123 Pecos R.
67N 135W 59 Peel R.
40N 116E 164 Peking / Beijing
42N 82W 79 Pelee I.
68N 90W 59 Pelly Bay
62N 131W 59 Pelly R.
31S 52W 134 Pelotas
13S 41E 144 Pemba
5S 40E 144 Pemba I.
54N 115W 75 Pembina R.
46N 77W 66 Pembroke
56N 3W 149 Pennines
41N 78W 124 Pennsylvania □
30N 87W 124 Pensacola
15S 168E 172 Pentecost, I.
49N 119W 70 Penticton
53N 45E 153 Penza
40N 89W 124 Peoria
48N 64W 85 Percé
4N 75W 131 Pereira
46S 71W 133 Perito Moreno
58N 57E 153 Perm'
27N 50E 157 Persian Gulf
32S 116E 171 Perth (Australia)
45N 76W 66 Perth (Ont.)
8S 75W 131 Peru ■
34N 71E 158 Peshawar
46N 77W 66 Petawawa
56N 109W 75 Peter Pond L.
44N 78W 79 Peterborough
45N 78W 79 Peterborough Provincial Park
48N 64W 85 Petit Cap
46N 65W 85 Petitcodiac
46N 75W 80 Petite-Nation Provincial Park
65S 55W 178 Petrel (research station)
43N 82W 79 Petrolia
55N 69E 153 Petropavlovsk
53N 158E 154 Petropavlovsk Kamchatskiy
23S 43W 134 Petrópolis
61N 34E 153 Petrozavodsk
40N 75W 124 Philadelphia
16N 130E 166 Philippine Sea
12N 123E 166 Philippines ■
12N 105E 165 Phnum Pénh
33N 112W 123 Phoenix
8N 98E 165 Phuket
44N 77W 79 Picton
45N 62W 85 Pictou
44N 10W 123 Pierre
29S 31E 144 Pietermaritzburg
44N 78W 79 Pigeon L.
24S 61W 133 Pilcomayo R.
5N 100E 165 Pinang / George Town
22N 83W 128 Pinar del Rio
15N 120E 165 Pinatubo, Mt.
49N 114W 75 Pincher Creek
4S 46W 132 Pindaré R.
61N 114W 65 Pine Point
43N 81W 79 Pinery Provincial Park
22N 83W 128 Pinos, Isla de
22S 47W 134 Piracicaba
40N 80W 124 Pittsburgh
5S 80W 131 Piura
47N 54W 86 Placentia
47N 55W 86 Placentia Bay
47N 67W 85 Plaster Rock
35S 58W 133 Plata, La
12S 57W 132 Plateau of Mato Grosso
34N 85E 163 Plateau of Tibet
41N 99W 123 Platte R.
46N 71W 80 Plessisville
42N 24E 148 Plovdiv
43N 112W 123 Pocatello
42N 82W 79 Point Pelee National Park
4S 12E 143 Pointe-Noire
52N 20E 147 Poland ■
55N 85W 60 Polar Bear Provincial Park
49N 34E 153 Poltava
55N 160E 154 Poluostrov Kamchatka / Kamchatka Peninsula
18N 66W 128 Ponce
60 Pond Inlet = Mittimatalik
52N 113W 75 Ponoka
25S 50W 134 Ponta Grossa
22S 56W 134 Ponta Porã
0 109E 165 Pontianak
18N 74E 158 Poona / Pune
18S 67W 131 Poopé, L.
3N 77W 131 Popayan
8S 148E 166 Popondetta
49N 125W 70 Port Alberni
50N 127W 69 Port Alice
30N 94W 124 Port Arthur
50N 57W 86 Port au Choix
32S 138E 171 Port Augusta
46N 64W 85 Port Borden
42N 80W 60 Port Burwell
42N 79W 79 Port Colborne
49N 123W 70 Port Coquitlam
43N 80W 79 Port Dover
44N 81W 79 Port Elgin
34S 25E 144 Port Elizabeth
0 9E 143 Port Gentil
4N 7E 141 Port Harcourt
50N 127W 69 Port Hardy
45N 61W 85 Port Hawkesbury
20S 118E 171 Port Hedland
44N 78W 79 Port Hope
34S 136E 171 Port Lincoln
31S 152E 172 Port Macquarie
51N 128W 69 Port McNeill
50N 64W 85 Port Menier
9S 147E 166 Port Moresby
10N 61W 128 Port of Spain
44N 79W 79 Port Perry
33S 138E 171 Port Pirie
31N 32E 142 Port Said / Bûr Sa'îd
42N 81W 79 Port Stanley
19N 37E 142 Port Sudan
18N 72W 128 Port-au-Prince
50N 67W 66 Port-Cartier
48N 65W 85 Port-Daniel
18S 168E 172 Port-Vila
50N 98W 76 Portage-la-Prairie
45N 122W 123 Portland
47N 72W 66 Portneuf Provincial Park
4S 52W 134 Pôrto Alegre
6N 2E 141 Porto Novo
42N 9W 147 Porto / Oporto
8S 64W 131 Pôrto Velho
40N 7W 147 Portugal ■
27S 27W 134 Posadas
60 Poste de la Baleine = Kunjjuarapik
19S 66W 131 Potosí
60N 67W 60 Povungnituk
37N 110W 123 Powell L.
50N 124W 70 Powell River
52N 17E 147 Poznań
50N 14E 147 Prague / Praha
27S 61W 133 Pres. Roque Saénz Peña
45N 75W 80 Prescott
22S 52W 134 Presidente Prudente
26S 28E 144 Pretoria
64S 61W 178 Primavera (research station)
55N 110W 75 Primrose L.
53N 106W 76 Prince Albert
54N 106W 76 Prince Albert National Park
67N 76W 60 Prince Charles I.
46N 63W 85 Prince Edward Island □
47N 63W 85 Prince Edward Island National Park
54N 123W 70 Prince George
73N 99W 59 Prince of Wales I.
77N 120W 59 Prince Patrick I.
54N 130W 69 Prince Rupert
49N 120W 70 Princeton
1N 7E 141 Príncipe
54N 86E 153 Prokop'yevsk
42N 71W 124 Providence
40N 112W 123 Provo
52N 110W 75 Provost
70N 148W 121 Prudhoe Bay
47N 27E 153 Prut R.
19N 98W 127 Puebla
38N 104W 123 Pueblo
45S 72W 133 Puerto Aisén
15N 88W 127 Puerto Barrios
48S 66W 133 Puerto Deseado
41S 73W 133 Puerto Montt
52S 73W 133 Puerto Natales
20N 71W 128 Puerto Plata
18N 66W 128 Puerto Rico □
50S 69W 133 Puerto Santo Cruz
48N 86W 66 Pukaskwa National Park
18N 74E 158 Pune / Poona
53S 71W 133 Punta Arenas (Chile)
10S 85W 128 Puntarenas (Costa Rica)
8S 65W 131 Purus R.
35N 129E 164 Pusan
3S 70W 131 Putumayo R.
43N 1W 149 Pyrenees

Q

32N 65E 157 Qandahār
38N 85E 163 Qarqan R.
25N 51E 157 Qatar ■
30N 27W 142 Qattara Depression
26N 33W 142 Qena
47N 100E 163 Qilian Shan
36N 120E 164 Qingdao / Tsingtao
36N 98E 163 Qinghai □
40N 119E 164 Qinhuangdao
47N 124E 164 Qiqihar
35N 51E 157 Qom / Qum
51N 103W 76 Qu'Appelle R.
61N 69W 60 Quaqtaq
50N 70W 122 Québec □
46N 71W 80 Québec CMA
53N 132W 69 Queen Charlotte Is. / Haida Gwaii
51N 128W 69 Queen Charlotte Sound
76N 95W 59 Queen Elizabeth Is.
86S 160W 178 Queen Maud Land □
68N 102W 59 Queen Maude Gulf
22S 142E 172 Queensland
18S 37E 144 Quelimane
53N 122W 70 Quesnel
52N 121W 70 Quesnel L.
48N 91W 65 Quetico Provincial Park
30N 67E 157 Quetta
15N 91W 127 Quezaltenango
15N 121E 165 Quezon City
14N 109E 165 Qui Nhon
52N 104W 76 Quill Ls.
26S 144E 172 Quilpie
0 78W 131 Quito
35N 51E 157 Qum / Qom

R

8S 119E 165 Raba
34N 6W 141 Rabat
4S 152E 166 Rabaul
46N 53W 86 Race, Cape
54N 78W 60 Radisson
49N 104W 76 Radville
65N 116W 65 Rae
21N 82E 158 Raipur
35N 78W 124 Raleigh
51N 112W 75 Ralston
34S 71W 133 Rancagua
62N 93W 59 Rankin Inlet
44N 103W 123 Rapid City
37N 48E 157 Rasht
33N 73E 158 Rawalpindi
8S 35W 132 Recife
51N 56W 86 Red Bay
51N 113W 75 Red Deer
51N 111W 75 Red Deer R. (Alta.)
49N 57W 86 Red Indian L.
51N 94W 65 Red Lake
50N 96W 76 Red R. (Man.)
48N 97W 123 Red R. (U.S.)
23N 104E 165 Red R. / Hong R.
25N 36E 157 Red Sea
27N 0 141 Reggane
50N 105W 76 Regina CMA
57N 102W 65 Reindeer L.
45N 76W 66 Renfrew
11S 161E 172 Rennell, I.
39N 120W 123 Reno
45N 73W 80 Repentigny
66N 86W 60 Repulse Bay
48N 75W 66 Réservoir Gouin
53N 67W 66 Réservoir Manicouagan
27S 59W 133 Resistencia
48N 67W 85 Restigouche
48N 67W 85 Restigouche R.
21S 56E 144 Réunion
59 Resolute = Kaujuitoq
51N 118W 70 Revelstoke
51N 119W 70 Revelstoke L.
51N 118W 70 Revelstoke Mountain National Park
19N 113W 127 Revillagigedo, Islas de
64N 22E 147 Reykjavik
26N 97W 127 Reynosa
51N 6E 147 Rhein R. / Rhine R.
41N 71W 124 Rhode I. □
36N 28E 148 Rhodes
44N 4E 147 Rhône R.
21S 48W 134 Ribeirão Prêto
44N 78W 79 Rice L.
46N 73W 80 Richelieu R.
47N 65W 85 Richibucto
37N 77W 124 Richmond
45N 76W 80 Rideau L.
45N 76W 80 Rideau R.
51N 100W 76 Riding Mountain National Park
56N 24E 153 Rīga
52N 114W 75 Rimbey
48N 68W 85 Rimouski
48N 68W 85 Rimouski Provincial Park
10S 67W 131 Rio Branco
30N 105W 123 Rio Bravo del Norte
33S 64W 133 Río Cuarto
23S 43W 134 Rio de Janeiro
35S 58W 133 Río de la Plata
32S 52W 134 Rio Grande (city)
30N 105W 123 Rio Grande R. / Rio Bravo del Norte
52S 69W 133 Río Gallegos
51N 97W 76 Riverton
48N 69W 85 Rivière-du-Loup
49N 64W 85 Rivière-du-Renard
24N 46E 157 Riyadh / Ar Riyād
37N 80W 124 Roanoke
37N 76W 124 Roanoke R.
48N 72W 66 Roberval
53N 119W 70 Robson, Mt.
35S 54W 134 Rocha
44N 92W 124 Rochester (Mn.)
43N 77W 124 Rochester (N.Y.)
42N 91W 124 Rock Island
42N 89W 124 Rockford
23S 150E 172 Rockhampton
45N 75W 80 Rockland
52N 115W 75 Rocky Mountain House
54N 126W 70 Rocky Mountains (B.C.)
45N 115W 123 Rocky Mountains (U.S.)
42N 24E 150 Rodope Mts.
51N 56W 86 Roddickton
51N 117W 70 Rogers Pass
26S 148E 172 Roma (Austral.)
42N 12E 147 Roma / Rome (Italy)
47N 26E 148 Romania ■
42N 12E 147 Rome / Roma (Italy)
42N 82W 79 Rondeau Provincial Park
55N 105W 76 Ronge, Lac la
78S 60W 178 Ronne Ice Shelf
79S 162W 178 Roosevelt I.
14S 135E 171 Roper R.
33S 61W 133 Rosario
15N 61W 128 Roseau
51N 108W 76 Rosetown
82S 180 178 Ross Ice Shelf
74S 178E 178 Ross Sea
11S 155E 172 Rossel, I.
44N 65W 85 Rossignol, L.
49N 117W 70 Rossland
47N 39E 153 Rostov na Donu
67S 70W 178 Rothera (research station)
10S 123E 171 Roti, I.
38S 176E 172 Rotorua
52N 4E 147 Rotterdam
46N 75W 80 Rouge R.
48N 56W 86 Round Pond
48N 79W 66 Rouyn, Noranda
51N 105W 76 Rowans Ravine Provincial Park
7S 37E 144 Rufiji R.
8S 33E 144 Rukwa, L.
51N 78W 60 Rupert, de, R.
51N 102W 76 Russell
62N 95E 154 Russia ■
75S 136W 178 Russkaya (research station)
8S 120E 165 Ruteng
11S 36E 144 Ruvuma R.
2S 30E 144 Rwanda ■
54N 39E 153 Ryazan'
58N 39E 153 Rybinsk
26N 128E 164 Ryukyu Is.

S

52N 4E 147 's-Gravenhage/The Hague
27N 14E 141 Sabhā
43N 65W 85 Sable, Cape
44N 60W 85 Sable I.
72N 126W 59 Sachs Harbour
46N 64W 85 Sackville
38N 121W 123 Sacramento
32N 9W 141 Safi
20N 5E 141 Sahara Desert
11N 107E 165 Saigon / Ho Chi Minh
45N 66W 85 Saint John CMA
17N 62W 128 St. John's (Antigua and Barbuda)
48N 103W 123 Sakakawea L.
51N 143E 154 Sakhalin I. / Ostrov Sakhalin
45N 74W 80 Salaberry-de-Valleyfield
28S 64W 133 Salado R.
35S 60W 133 Salado R. (Buenos Aires)
35S 67W 133 Salado R. (San Justo)
66N 66E 153 Salekhard
45N 123W 123 Salem
37N 127W 123 Salinas
62N 74W 60 Salluit
50N 119W 70 Salmon Arm
54N 123W 70 Salmon R.
40N 112W 123 Salt Lake City
25S 65W 133 Salta
25N 102W 128 Saltillo
13S 38W 132 Salvador (Bahia)
16N 97E 165 Salween R.

47N 13E 147 Salzburg
12N 125E 166 Samar, I.
54N 50E 153 Samara
0 117E 165 Samarinda
39N 67E 153 Samarkand
29N 98W 123 San Antonio
41S 65W 133 San Antonio Oeste
34N 117W 123 San Bernardino
34S 71W 133 San Bernardo
42S 71W 133 San Carlos de Bariloche
8N 73W 131 San Cristóbal
10S 161E 172 San Cristobal, I.
32N 117W 123 San Diego
37N 122W 123 San Francisco
19N 70W 128 San Francisco de Macoris
34S 59W 133 San Isidro
41S 62W 133 San Jorge, Golfo
10N 84W 128 San José (Costa Rica)
37N 122W 123 San José (U.S.)
23S 47W 134 San José dos Campos
32S 68W 133 San Juan (Argentina)
18N 66W 128 San Juan (Puerto Rico)
30S 60W 133 San Justo
33S 66W 133 San Luis
35N 121W 123 San Luis Obispo
22N 101W 127 San Luis Potosi
44N 12E 147 San Marino ■
68S 68W 178 San Martin (research station)
42S 64W 133 San Matias, Golfo
14N 87W 127 San Miguel
27S 65W 133 San Miguel de Tucumán
18N 69W 127 San Pedro de Macoris
15N 88W 127 San Pedro Sula
35S 68W 133 San Rafael
13N 89W 127 San Salvador
24S 65W 133 San Salvador de Jujuy
15N 44E 157 Şa'ña'
70S 1W 178 Sanae (research station)
5N 12E 141 Sanaga R.
6N 118E 165 Sandakan
53N 131W 69 Sandspit
53N 93W 65 Sandy Lake
2N 17E 143 Sangha R.
6N 22E 143 Sankuru R.
14N 89W 127 Santa Ana
34N 119W 123 Santa Barbara
22N 80W 128 Santa Clara
17S 63W 131 Santa Cruz (Bolivia)
29N 18W 141 Santa Cruz de Tenerife
10S 166E 172 Santa Cruz Is.
50S 71W 133 Santa Cruz R.
31S 60W 133 Santa Fe (Argentina)
35N 106W 123 Santa Fe (U.S.)
29S 53W 134 Santa Maria
14S 165E 172 Santa Maria, I.
11N 74W 131 Santa Marta
36S 64W 133 Santa Rosa
3S 55W 132 Santarém
33S 70W 133 Santiago (Chile)
19N 70W 128 Santiago (Dominican Republic)
20N 75W 128 Santiago de Cuba
28S 64W 133 Santiago del Estero
24S 47W 134 Santo André
28S 54W 134 Santo Ângelo
18N 70W 128 Santo Domingo
24S 46W 134 Santos
22S 48W 134 Sao Carlos
6S 52W 132 São Felix do Xingu
15S 44W 132 São Francisco R.
21S 50W 134 São José do Rio Prêto
23S 46W 134 São José dos Campos
2S 44W 132 São Luis
8S 57W 132 São Manuel R. / Teles Pires R.
23S 46W 134 São Paulo
0 6E 143 São Tomé
0 6E 143 Sao Tome and Principe ■
44N 5E 147 Saône R.
43N 141E 164 Sapporo
44N 18E 147 Sarajevo
54N 45E 153 Saransk
51N 46E 153 Saratov
2N 113E 165 Sarawak □
40N 9E 147 Sardinia
9N 18E 142 Sarh
43N 82W 79 Sarnia
1S 131E 166 Sarong
49N 79W 84 Sarre, La
33N 130E 164 Sasebo
54N 106W 76 Saskatchewan □
50N 108W 76 Saskatchewan Landing Provincial Park
54N 103W 76 Saskatchewan R.
52N 106W 76 Saskatoon CMA
6N 7W 141 Sassandra R.
41N 8E 147 Sassari
26N 44E 157 Saudi Arabia ■
12S 58W 131 Sauêruina R.
44N 81W 79 Saugeen R.
46N 84W 66 Sault Ste. Marie
10S 20E 143 Saurimo
32N 81W 124 Savannah
34N 82W 124 Savannah R.
21S 34E 144 Save R.
10S 122E 171 Sawu, I.
92 Sayan Mts.
2N 44E 144 Scebeli R.
54N 67W 66 Schefferville
50N 55W 86 Scie, La
91 Scotia Ridge
57N 4W 147 Scotland □
45N 71W 80 Scotstown
66S 100E 178 Scott Base (research station)
42N 76W 124 Scranton
44N 79W 79 Scugog L.
43N 81W 79 Seaforth
47N 122W 123 Seattle
49N 123W 70 Sechelt
13N 7W 141 Ségou
49N 2E 147 Seine R.
22S 28E 144 Selebi Phikwe
49N 104E 163 Selenge R.
42N 80W 76 Selkirk
52N 118W 70 Selkirk Mountains
6S 67W 131 Selvas
7S 110E 165 Semarang
50N 80E 153 Semipalatinsk
38N 141E 164 Sendai
14N 14W 141 Senegal ■
16N 14W 141 Sénégal R.
37N 127E 164 Seoul / Sŏul
4S 142E 166 Sepik R.
50N 66W 66 Sept-Îles
2S 35E 144 Serengeti
22S 27E 144 Serowe
29S 51W 134 Serra do Mar
51N 92W 65 Seul, Lac
44N 33E 153 Sevastopol'
56N 87W 65 Severn R.
63N 44E 153 Severnaya Dvina R.
79N 100E 154 Severnaya Zemlya / North Land
37N 6W 147 Sevilla
10S 46E 144 Seychelles ■
92 Seychelles Ridge
34N 10E 141 Sfax
35N 109E 163 Shaanxi □
38N 77E 163 Shache / Yarkant
47N 40E 153 Shakhty
36S 118E 164 Shandong □
31N 121E 164 Shanghai
31N 121E 164 Shanghai Shi □
23N 116E 164 Shantou / Swatow
37N 112E 164 Shanxi □
27N 111E 164 Shaoyang
45N 77W 80 Sharbot Lake
49N 108W 75 Shaunavon
46N 73W 80 Shawinigan
6N 44E 142 Shebelē R.
46N 65W 85 Shediac
45N 62W 85 Sheet Harbour
44N 80W 79 Shelburne
53N 106W 76 Shellbrook
41N 123E 164 Shenyang
17S 168E 172 Shepherd I.
45N 72W 80 Sherbrooke
60N 1W 147 Shetland Is.
38N 114E 164 Shijiazhuang
33N 133E 164 Shikoku □
25N 92E 158 Shillong
34N 130E 164 Shimonoseki
48N 65W 85 Shippegan
29N 52E 157 Shīrāz
17S 35E 144 Shire R.
35N 138E 164 Shizuoka
18N 76E 158 Sholapur
32N 94W 124 Shreveport
51N 119W 70 Shuswap, L.
34N 109E 163 Sian / Xi'an
1S 99E 165 Siberut, I.
31N 104E 163 Sichuan □
37N 16E 147 Sicily
49N 124W 70 Sidney
9N 12W 141 Sierra Leone ■
19N 104W 127 Sierra Madre del Sur
30N 108W 127 Sierra Madre Occidental
29N 102W 127 Sierra Madre Oriental
60S 46W 178 Signy I. (research station)
12N 9W 141 Siguiri
11N 6W 141 Sikasso
45N 135E 154 Sikhote Atlin Range
50N 119W 70 Silver Star Provincial Park
43N 80W 79 Simcoe
44N 79W 79 Simcoe, L.
2N 95E 165 Simeulue, I.
45N 34E 153 Simferopol'
25S 136E 171 Simpson Desert
1N 104E 165 Singapore ■
1N 104E 165 Singapore (city)
42N 96W 124 Sioux City
43N 96W 124 Sioux Falls
50N 92W 65 Sioux Lookout
76S 82W 178 Siple (research station)
49N 57W 86 Sir Richard Squires Provincial Park
21N 93E 165 Sittwe / Akyab
57N 130W 69 Skeena Mountains
54N 129W 69 Skeena R.
42N 21E 148 Skopje
55N 115W 75 Slave Lake
59N 113W 65 Slave R.
49N 20E 147 Slovakia ■
47N 15E 147 Slovenia ■
54N 63W 66 Smallwood Reservoir
54N 127W 69 Smithers
45N 76W 80 Smiths Falls
55N 118W 75 Smoky R.
54N 32E 153 Smolensk
43N 116W 123 Snake R.
55N 100W 76 Snow Lake
37S 148E 172 Snowy R.
44N 40E 153 Sochi
12N 54E 157 Socotra
42N 23E 148 Sofia / Sofiya
9N 1E 141 Sokodé
13N 5E 141 Sokoto
175 Solomon Islands
7S 150E 166 Solomon Sea
7N 47E 142 Somalia ■
73N 93W 59 Somerset I.
46N 129E 164 Songhua R.
13N 100E 165 Songkhla
48N 123W 70 Sooke
46N 73W 80 Sorel
23S 47W 134 Sorocaba
37N 127E 164 Sŏul / Seoul
46N 62W 85 Souris
49N 101W 76 Souris R.
30S 25E 143 South Africa ■
32S 139E 171 South Australia □
41N 86W 124 South Bend
33N 81W 124 South Carolina
10N 113E 165 South China Sea
45N 100W 123 South Dakota □
55S 38W 178 South Georgia, I. □
44S 170E 172 South I.
36N 128E 164 South Korea ■
65S 140E 178 South Magnetic Pole
53N 131W 69 South Moresby National Park
45N 75W 80 South Nation R.
63S 45W 178 South Orkney Is. □
62S 130W 178 South Pacific Ocean
90S 178 South Pole
51N 110W 75 South Saskatchewan R.
62S 59W 178 South Shetland Is. □
44N 81W 79 Southampton (Ont.)
51N 1W 147 Southampton (U.K.)
64N 84W 60 Southampton I.
56N 103W 76 Southend
44S 170E 172 Southern Alps
57N 98W 65 Southern Indian L.
62S 30E 171 Southern Ocean
7N 5W 147 Spain ■
50N 115W 70 Sparwood
57N 129W 69 Spatsizi Plateau Wilderness Provincial Park
69N 93W 59 Spence Bay
34S 137E 171 Spencer Gulf
56N 119W 75 Spirit River
52N 96W 65 Split Lake
47N 117W 123 Spokane
42N 72W 124 Springfield
45N 64W 85 Springhill
29N 29W 144 Springs
53N 114W 75 Spruce Grove
49N 99W 76 Spruce Woods Provincial Park
49N 123W 70 Squamish
49N 57W 86 Squires, Sir Richard, Provincial Park
7N 81E 158 Sri Lanka ■
34N 75E 158 Srinagar
46N 72W 80 St-Casimir
48N 72W 66 St-Félicien
46N 71W 80 St-François, Lac
46N 72W 80 St-François R.
46N 72W 80 St-Georges
46N 73W 80 St-Hyacinthe
45N 73W 80 St-Jean
46N 74W 80 St-Jérôme
46N 71W 80 St-Joseph-de-Beauce
47N 73W 66 St-Maurice Provincial Park
46N 73W 80 St-Maurice, Riv.
47N 56W 86 St-Pierre et Miquelon □
46N 73W 80 St-Pierre, Lac
48N 67W 85 St-Quentin
48N 56W 86 St. Alban's
54N 114W 75 St. Albert (Alta.)
45N 75W 80 St. Albert (Qué.)
45N 67W 85 St. Andrews
18N 77W 128 St. Ann's Bay
51N 55W 86 St. Anthony
43N 79W 79 St. Catharines-Niagara C.M.A.
42N 82W 79 St. Clair, L.
60N 140W 59 St. Elias, Mt.
46N 4E 145 St. Étienne
45N 67W 85 St. George
48N 59W 86 St. George, Cape
12N 62W 128 St. George's (Grenada)
48N 58W 86 St. George's (Nfld.)
48N 59W 86 St. George's Bay
16S 11W 143 St. Helena I.
47N 68W 85 St. Jacques
45N 66W 85 St. John R.
47N 52W 86 St. John's CMA (Nfld.)
18N 63W 128 St. Kitts and Nevis
48N 62W 85 St. Lawrence, Gulf of
63N 170W 154 St. Lawrence I.
56 St. Lawrence Islands National Park
49N 66W 85 St. Lawrence R.
47N 68W 85 St. Leonard
16N 16W 141 St. Louis (Senegal)
38N 90W 124 St. Louis (U.S.)
14N 61W 128 St. Lucia ■
43N 81W 79 St. Mary's
50N 116W 70 St. Mary's Alpine Provincial Park
47N 54W 86 St. Mary's Bay (Nfld.)
44N 66W 85 St. Mary's Bay (N.S.)
54N 112W 75 St. Paul (Alta.)
45N 93W 124 St. Paul (U.S.)
60N 30E 153 St. Petersburg (Russia)
41N 80W 124 St. Petersburg (U.S.)
45N 67W 85 St. Stephen
54N 95W 65 St. Theresa Point
42N 81W 79 St. Thomas
13N 61W 128 St. Vincent and the Grenadines ■
51S 59W 133 Stanley
59N 5E 147 Stavanger
45N 42E 153 Stavropol'
44N 80W 79 Stayner
46N 74W 80 Ste-Adèle
46N 74W 80 Ste-Agathe-des-Monts
47N 72W 80 Ste-Anne R.
47N 45W 80 Ste-Foy
46N 71W 80 Ste-Marie
50N 97W 76 Ste. Anne
49N 66W 85 Ste. Anne-des-Monts
40N 107W 123 Steamboat Springs
49N 96W 76 Steinbach
45N 62W 85 Stellarton
48N 58W 86 Stephenville
50N 60E 153 Steppes, The
53N 56E 153 Sterlitamak
52N 112W 75 Stettler
56N 130W 69 Stewart
47S 168E 172 Stewart I.
63N 135W 59 Stewart R.
45N 63W 85 Stewiacke
58N 132W 69 Stikine R.
44N 77W 79 Stirling
59N 18E 147 Stockholm
58N 125W 70 Stone Mountain Provincial Park
44N 78W 79 Stony L.
44N 79W 79 Stouffville
43N 81W 79 Stratford
50N 125W 69 Strathcona Provincial Park
43N 81W 79 Strathroy
46N 80W 66 Sturgeon Falls
44N 78W 79 Sturgeon L.
19S 65W 131 Sucre
15N 30E 142 Sudan ■
46N 81W 66 Sudbury CMA
30N 33E 142 Suez / El Suweis
60N 44E 148 Sukhona R.
27N 69E 158 Sukkur
2S 126E 166 Sula Is.
2S 120E 165 Sulawesi
6N 121E 165 Sulu Archipelago
8N 120E 165 Sulu Sea
1N 107E 165 Sumatra □
9S 119E 165 Sumba, I.
8S 117E 165 Sumbawa, I.
53N 89W 65 Summer Beaver
49N 119W 70 Summerland
46N 63W 85 Summerside
47N 87W 65 Superior, L.
22N 59E 157 Şur
7S 112E 165 Surabaya
21N 73E 158 Surat
9N 99E 165 Surat Thani
63N 73E 148 Surgut R.
4N 56W 132 Suriname ■
49N 123W 70 Surrey
45N 65W 85 Sussex
30N 73E 158 Sutlej R.
44N 79W 79 Sutton
31N 120E 164 Suzhou
78N 17E 153 Svalbard
79N 97W 59 Sverdrup Is.
54N 115W 75 Swan Hills Provincial Park
52N 101W 76 Swan River
23N 116E 164 Swatow / Shantou
26S 31E 144 Swaziland ■
67N 15E 147 Sweden ■
50N 107W 76 Swift Current
53N 118W 75 Switzer, William A., Provincial Park
46N 8E 147 Switzerland ■
43N 82W 79 Sydenham R.
34S 151E 172 Sydney (Australia)
46N 60W 85 Sydney (N.S.)
46N 60W 85 Sydney Mines
69S 39E 178 Syowa (research station)
45N 65E 153 Syr Dar'ya (river)
43N 76W 124 Syracuse
35N 38E 157 Syria ■
33N 40E 157 Syrian Desert
53N 48E 153 Syzran'

T

23N 120E 164 T'ainan
25N 121E 164 Taipei
24N 121E 164 T'aiwan □
13N 44E 157 Ta'izz
49N 112W 75 Taber
5S 33E 144 Tabora
38N 46E 157 Tabrīz
28N 36E 157 Tabūk
47N 122W 123 Tacoma
58N 98W 59 Tadoule L.
36N 128E 164 Taegu
36N 127E 164 Taejŏn
11S 153E 172 Tagula, I.
15N 5E 141 Tahoua
50N 126W 69 Tahsis
24N 121E 164 Taiwan ■
38N 112E 164 Taiyuan
32N 120E 164 Taizhou
39N 70E 153 Tajikistan ■
40N 6W 147 Tajo R.
17N 99E 165 Tak
34N 134E 164 Takamatsu
56N 126W 69 Takla L.
40N 83E 163 Taklimakan Desert
5N 2W 141 Takoradi
35S 72W 133 Talca
36S 73W 133 Talcahuano
30N 84W 124 Tallahassee
59N 24E 153 Tallinn
9N 0 141 Tamale
23N 5E 141 Tamanrasset
52N 41E 153 Tambov
28N 82W 124 Tampa
61N 24E 148 Tampere
22N 98W 127 Tampico
12N 37E 142 Tana, L.
1S 40E 144 Tana R.
5S 39E 144 Tanga
6S 30E 144 Tanganyika, L.
36N 5W 141 Tanger
39N 118E 164 Tangshan
8S 132E 166 Tanimbar Is.
19S 168E 172 Tanna, I.
15N 14E 141 Tanout
31N 31E 142 Tanta
6S 34E 144 Tanzania ■
15N 92W 127 Tapachula
4S 55W 132 Tapajós R.
21N 76E 158 Tapti R.
32N 13E 141 Tarābulus / Tripoli
4N 117E 165 Tarakan
22S 65W 133 Tarija
41N 82E 163 Tarim R.
46N 3E 147 Tarn R.
58N 27E 148 Tartu
41N 69E 153 Tashkent
36S 160E 172 Tasman Sea
42S 146E 172 Tasmania □
57N 127W 69 Tatlatui Provincial Park
49N 18W 150 Tatra, Mt.
60N 140W 69 Tatshenshini Provincial Park
37N 37E 157 Taurus Mountains
14N 98E 165 Tavoy
67N 82E 153 Taz R.
41N 45E 153 Tbilisi
2N 65W 131 Tefé
14N 87W 127 Tegucigalpa
35N 51E 157 Tehrān
16N 95W 127 Tehuantepec, Golfo de
39N 9W 147 Tejo R.
32N 34E 157 Tel Aviv-Yafo
8S 57W 132 Teles Pires R. (Sao Manuel)
5S 105E 165 Telukbetung
39S 73W 133 Temuco
19S 134E 171 Tennant Creek
36N 86W 124 Tennessee □
34N 87W 124 Tennessee R.
21N 105W 127 Tepic
5S 42W 132 Teresina
45N 137E 154 Terney
48N 54W 86 Terra Nova National Park
54N 128W 69 Terrace

SOURCES

Information for the map plates and world statistics was adapted from the following sources. A full reference is given the first time a source is listed. An abbreviated version is given thereafter.

1/2 Cappon, Lester J. (ed-in-chief), *Atlas of Early American History: The Revolutionary Era,* 1760-1790 (Princeton: Princeton University Press, 1976). Department of Energy, Mines, and Resources, *The National Atlas of Canada,* 4th ed. (Toronto: MacMillan of Canada, 1974). Kerr, D.G.G., *A Historical Atlas of Canada* (Toronto: Thomas Nelson and Sons Canada Ltd., 1966).

3/4 Kerr, D.G.G., *A Historical Atlas of Canada* (Toronto: Thomas Nelson and Sons Canada Ltd., 1966)
First Nations map updated using current information from Canadian First Nations; *The Geography of Indian Cultures,* National Geographic Society map, Washington, D.C. (1982); *Handbook of North American Indians,* Vols 5 & 6. Washington: Smithsonian Institute (1981).

5/6 *The National Atlas of Canada,* 4th ed., 1974; Statistics Canada.

7/8 Department of the Environment, Canadian Hydrographic Service, Map #400, 1974. Statistics Canada, Canadian Census, 1991.

9/10 Statistics Canada (Canadian Census 1871, 1911, 1991); Census Metropolitan Areas and Census Agglomerations (1991), Cat. No. 93-303; *Canada Yearbook 1993.* Warkentin, John (ed), *Canada: A Geographical Interpretation* (Toronto: Methuen Publications, 1968).

11/12 Department of Energy, Mines and Resources. "Indian Treaties" map (1977). Indian and Northern Affairs, Canada: Information Sheet Nos. 1 and 55; "Population Projections of Registered Indians", 1986 - 2011 (1990); "Comprehensive Land Claims in Canada", map. Department of the Secretary of State of Canada, "Canada's Off-Reserve Aboriginal Population: A Statistical Overview", 1991. Statistics Canada, Aboriginal Data: Age and Sex, Cat. No. 94-327.

13/14 Employment and Immigration Canada. Ministry of Citizenship and Culture. Statistics Canada: Census of Canada 1991; Internal migration, 1992: 93-316; Immigration by Country of Origin: 93-316.

15/16 *The National Atlas of Canada.* The Reader's Digest Association (Canada) Ltd., 1981.

17/18 *The National Atlas of Canada* (4th ed., 1974).

19/20 *British Palaeozoic Fossils* (London: British Museum, 1974). *The Canadian Oxford School Atlas,* 4th ed. (Toronto: Oxford University Press (Canada), 1977). *Life Nature Library* (New York: Time Inc., 1963): *The Fishes; The Mammals. The National Atlas of Canada* (4th ed., 1974).

21/22 *The Atlas of Canada,* 1981. Energy, Mines and Resources Canada, Mineral Policy Sector; Uranium Division, Electricity Branch; "Principal Mineral Areas of Canada" map (900A). 1991. Mining Association of Canada. Statistics Canada: Exports by Country (Jan.-Dec. 1992), Cat. No. 65-003; Gas Utilities (1991), Cat. No. 57-205.

23/24 Alberta Energy, Electricity Power Branch. B.C. Hydro. Hydro-Quebec. Manitoba Hydro. New Brunswick Electric Power Commission. Newfoundland and Labrador Hydro. Nova Scotia Power. Ontario Hydro, Power System Operations. Saskatchewan Power Corporation. Saskatchewan Power Corporation. Lloyd Payne, Dundas Hydro. Energy Resources Conservation Board, Pub. No. 91-50, 1991. Atomic Energy Control Board. open files.

25/26 Hare, F. Kenneth and Morley K. Thomas, *Climate Canada* (Toronto: John Wiley and Sons Canada, 1975). *The National Atlas of Canada* (4th ed., 1974); (5th ed., "Energy" map. 1978).

27/28 *The Atlas of Canada,* 1981. *Atlas of the World* (Washington: The National Geographic Society, 1981). Environment Canada, "Mapping Weather", 1978. Wright, Peter, *The Prentice-Hall Concise Book of the Weather* (Toronto: Prentice Hall Canada, 1981).

29/30 Environment Canada, "Groundwater - Nature's Hidden Treasure", 1990; Inland Waters Directorate; Waste Management Directorate, 1990. *Great International Atlas,* 1981. International Joint Commission on the Great Lakes, 1985. Ministry of Supply and Services, "Still Waters: The Chilling Effects of Acid Rain", 1981. *The National Atlas of Canada* (4th ed., 1974).

31/32 Bridges, E.M. *World Soils* (London: Cambridge University Press, 1970). *The National Atlas of Canada* (4th ed., 1974).

33/34 Agriculture Canada, *Canadian Farm Economics,* Vol. 23, No. 1, 1991. *The National Atlas of Canada* (4th ed., 1974). Statistics Canada, National Accounts and Environment Division and Agriculture Division: Agriculture Economic Statistics, Cat. No. 21-603E; Agricultural Profile of Canada, Cat. No. 93-351; Census of Agriculture, Cat. No. 10-545E, 1991; Census Overview of Canadian Agriculture: 1971-1991, Cat. No. 93-348; Trends and Highlights of Canadian Agriculture and Its People, Cat. No. 96-303E.

35/36 *The National Atlas of Canada* (4th ed., 1974). Statistics Canada, 1991.

37/38 *The National Atlas of Canada* (4th ed., 1974).

39/40 Canadian Pulp and Paper Association, 1990. Environment Canada, Forestry Service, "Canada's Eight Forest Regions" map, 1974; "Canada's Forests" map, 1979. Energy, Mines and Resources Canada, "Canada Pulp and Paper Mills" map, 1986. Forestry Canada, 1990. "Trees and Forests of Canada", published by Maclean's Magazine for The Canadian Pulp and Paper Association, 1981. Minister of Supply and Services Canada, The State of Canada's Environment-1991, 1991. Statistics Canada.

41/42 Canadian Pulp and Paper Association, 1990; "Farming Canada's Forests," 1991. Forestry Canada, 1990. MacMillan-Bloedel Ltd., "The Forest Industry of British Columbia," 1976. Sierra Club of Western Canada and The Wilderness Society, "Ancient Rainforests at Risk" map, 1991.

43/44 Farley, Albert L., *Atlas of British Columbia* (Vancouver: University of British Columbia Press, 1979). Fisheries and Oceans Canada, Annual Report; Biological Sciences Directorate; "Rebuilding the Atlantic Fish Stocks", 1990. *Life Nature Library:* Ecology (New York: Time Inc., 1963). *The National Atlas of Canada* (4th ed., 1974). North Atlantic Fisheries Organization; Northwest Atlantic Fisheries Organization (NAFO).

45/46 Canadian Manufacturers' Association. Economic Council of Canada. Statistics Canada, 1992; Monthly Survey of Manufacturing (Dec. 1992), Cat. No. 13-001; Census of Manufactures (1984), Cat. No. 31-209; Summary of Canadian International Trade (Dec. 1992), Cat. No. 65-001.

47/48 Statistics Canada, Earnings of Men and Women in 1991, Cat. No. 13-217; Canadian Economic Observer (1991), Cat. No. 11-210; Provincial GDP by Industry 1984-1991, Cat. No. 15-203; Industry and Class of Worker (1991), Cat. No. 93-326; Perspectives (Spring 1992, Spring 1993), Cat. No. 75-001E; *The Daily* (March 1993), Cat. No. 11-001E.

49/50 Environics Research Group Ltd., 1991. National Council of Welfare, Poverty Profile 1980-1990, and Update for 1991. National Literacy Secretariat. Statistics Canada, The Nation, Cat. No. 93-325; Age, Sex, Marital Status and Common-law Status, Cat. No. 92-325E; (Canadian Census, 1991); Canadian Crime Statistics (1991), Cat. No. 85-205; International Travel (1990), Cat. No. 66-201; Canadian Economic Observer, Cat. No. 11-210.

51/52 Statistics Canada, Trucking in Canada (1991), Cat. No. 53-222; Passenger Bus and Urban Transit Statistics (1992), Cat. No. 53-003; Shipping in Canada (1991), Cat. No. 54-205; Environment and Wealth Accounts Division, The State of Canada's Environment-1991; Air Travel, Cat. No. 51-204; Railway Operating Statistics (1992). Cat. No. 52-003. Transport Canada; Transport Canada, Dangerous Goods Directorate.

53/54 Environment Canada, Regulatory Affairs and Program Integration Branch, Environmental Protection; MUD/MUNDAT databases. Statistics Canada, 1981; Canadian Census, 1991; State of Environment Report for Canada, Cat. No. SOE 91-1.

55/56 State of the Environment Report of Canada, 1986. Committee on the Status of Endangered Wildlife in Canada. Canadian Parks Service, "Still Waters: The Chilling Effects of Acid Rain (Ottawa: Ministry of Supply and Services, 1981). Environment Canada, Canadian Wildlife Service, Migratory Birds and Conservation Branch; Lands Directorate, 1982.

57/58 Canadian International Development Agency, 1991-1992 Estimates (Part III, Expenditure Plan), 1991; 1992. Statistics Canada, Imports, Merchandise Trade, 1992, Cat. No.65-203; Exports, Merchandise Trade, 1992, Cat. No. 65-202.

59/60, 65/66, 69/70, 75/76, 79/80, 85/86 *The Gazetteer Atlas of Canada,* 1980. Statistics Canada, Canadian Census, 1991.

61/62 *Climate Canada,* 1979. Draper, A., *Global Atlas* (Toronto: Gage Educational Publishing Company, 1993). *The National Atlas of Canada* (4th ed., 1974). World Wildlife Fund. Statistics Canada, 1991.

63, 64; 73, 74; 83, 84 These maps are based on information taken from the National Topographic System map sheets 33-L©1983, 31-F-7©1987, 92-N©1986, 72-I-10©1988, 31-H©1992, 21–I-13©1992. Her Majesty the Queen in Right of Canada, with permission of Energy, Mines and Resources Canada. Environment Canada, Atmospheric Environment Service, 1989, "The Greenhouse Effect: Impacts of the Arctic". The State of Canada's Environment-1991.

67/68 Bickerstaff, A., W.L. Wallace, and F. Evert, Growth of Forests in Canada - Part 2, Environment Canada, Canadian Forestry Service, 1981. Climate Canada, 1979. Energy, Mines and Resources Canada, "Canada Pulp and Paper Mills" map, 1986; "Principal Mineral Areas of Canada" map (900A), 1991. The National Atlas of Canada (4th ed., 1974). The State of Canada's Environment-1991.

71/72 Alberta Department of Tourism and Economic Development. Atlas of British Columbia, 1979. Climate Canada, 1979. Ministry of Tourism of British Columbia, "Visitor '89... A Travel Survey of Visitors to British Columbia", 1990. Statistics Canada, The Nation, Cat. No. 93-315; Shipping in Canada 1991, Cat. No. 54-205. Canada Yearbook 1992. Quarterly Demographic Statistics, Cat. No. 91-002.

77/78 *Canada: A Geographical Interpretation,* 1968. Climate Canada, 1979. The State of Canada's Environment-1991. Statistics Canada, 1991; The Nation, Cat. No. 93-315; Grain Trade of Canada 1991-1992, Cat. No. 22-201; Agricultural Profile of British Columbia - Parts 1 and 2, Cat. No. 95-393; Agricultural Profile of Alberta - Parts 1 and 1. Cat. No. 95-382; Agricultural Profile of Manitoba - Parts 1 and 2, Cat. No. 95-363; Agricultural Profile of Saskatchewan - Parts 1 and 2, Cat. No. 95-370; Quarterly Demographic Statistics, Cat. No. 91-002.

81/82 *Canada Yearbook 1992. Climate Canada,* 1979. 1994 *Corpus Almanac & Canadian Sourcebook* (Don Mills: Southam Group, 1994). Environment Canada, Canada Land Data Systems Division, Land Directorate. Statistics Canada, 1991; The Nation, Cat. No. 93-315; Census of Manufactures (1984), Cat. No. 31-209; Census Divisions and Census Subdivisions Reference Maps, Cat. No. 92-319.

87/88 *Climate Canada,* 1979. Fisheries and Oceans Canada. *The National Atlas of Canada* (4th ed., 1974). *The Oceanographic Atlas of the North Atlantic Ocean* (Washington: U.S. Naval and Oceanographic Office, 1965). Provincial Tourist Offices, 1993. Statistics Canada, Agricultural Profile of Canada - Part 1. Cat. No. 93-350; Quarterly Demographic Statistics. Cat. No. 91-002.

89/90 *The Atlas of Canada.* Environment Canada. *National Geographic,* "Atlantic Ocean Floor:, map published in Vol. 147 No. 6 (June 1968). The Reader's Digest Association (Canada) Ltd., 1981.

93/94 *The New International Atlas,* (Chicago: Rand McNally and Co., 1980). *The Times Atlas of the World,* (Boston: Houghton-Mifflin, 1967).

95/96 Lean, Geoffrey and Don Hinrichsen, Adam Markham, *World Wildlife Fund Atlas of the Environment* (New York: Prentice Hall, 1990). Strahler, Arthur N., *Introduction to Physical Geography* (New York: John Wiley and Sons, 1965). *World Soils,* 1970.

97/98 Banks, A.S. and W. Overstreet (eds), *The Political Handbook of the World* (New York: McGraw Hill Book Co., 1980). *World Almanac and Book of Facts 1993. The World Factbook* (Washington: Central Intelligence Agency, 1993).

99/100 *Great International Atlas, 1981. Europa Yearbook* (London: Europa Publications, 1993); *World Factbook, 1993.*

101/102 United Nations Population Division, *Demographic Yearbook, 1992. The World Factbook. Yearbook of the Encyclopedia Britannica.*

103/104 *The New International Atlas, 1980. World Development Report 1993* (New York: Oxford University Press for The World Bank, 1993). Food dependency data: United Nations Development Programme. Fishing: United Nations Food and Agricultural Organization, Yearbook on Fishery Statistics.

105/106 *Europa Yearbook.*

107/108 Moynihan, M., *World Atlas* (New York: Business International Corporation, 1992). *1990 Yearbook of Tourism* (Madrid: World Tourism Organization, 1992). Imports, exports and manufacturing data: *Europa Yearbook; United Nations International Trade Statistics Yearbook.*

109/110 *BP Statistical Review of World Energy* (London: British Petroleum Co., 1992). Davis, C.A., Geothermal Resources Council, 1990. Gipe, Paul, American Wind Energy Association, 1992. Golob, Richard and Eric Brus, *The Almanac of Renewable Energy* (New York: H. Holt, 1993). Huttrer, G.W., "Geothermal Electric Power - A 1990 World Status Update," *Geothermal Resources Council Bulletin* 19, No. 7. 1989 *Energy Statistics Yearbook* (New York: United Nations, 1991). World Resources Institute, World Bank, The United Nations Development Program, 1987.

111/112 *United Nations: Demographic Yearbook; FAO Yearbook; Unesco Yearbook; World Health Organization Yearbook.*

113/114 *International Labour Organization Statistical Yearbook. The New International Atlas* (Chicago: Rand McNally, 1989). Literacy: *United Nations Statistical Yearbook:* Women in work force.

115/116 Human Development Index, United Nations Development Programme (New York: United Nations, 1990). U.S. Committee for Refugees, World Refugee Survey (Dec. 1992). *World Development Report 1993.* Military spending: *The Military Balance* (Washington: International Institute for Strategic Studies, 1993); *Statesman's Yearbook* (London: MacMillan, 1992).

117/118 *World Wildlife Fund Atlas of the Environment,* 1990. *The World Environment 1972-1992: Two Decades of Challenge* (London: Chapman & Hall, 1992). Percentage of land protected: U.N. Development Programme.

119/120 Atmospheric Environment Service, 1992. Personal communication with David Broadhurst, 1993. National Aeronautics and Space Administration, 1993. United Nations Environment Programme 1987. United Nations Environment Programme, *World Atlas of Desertification,* (London, Baltimore: Edward Arnold, 1992). The Ozone Layer. United Nations Environment Programme/Global Environmental Monitoring System, Environmental Library, No. 2. Nairobi. "The Ozone Layer" fact sheet. *World Resources 1990-91* and *1992-93* (New York: Oxford University Press, 1991 and 1993). "Scientific Assessment of Ozone Depletion: 1991," World Meterological Organization, Global Ozone Research and Monitoring Project-Report No. 25.

125/126 *The New Cosmopolitan World Atlas* (Chicago: Rand McNally, 1993). *The Times Atlas of the World,* 9th ed. (London: Times Books, 1992). Bureau of the Census (U.S.). 1990 Census Profile, No. 1 - March 1991. U.S. Department of Commerce, Bureau of Economic Analysis: Local Area Personal Income, 1984-89, Volume 1 Summary, July 1991; Office of Trade and Investment Analysis, 1991, U.S. Foreign Trade Highlights. U.S. Department of Labor, Monthly Labor Review, Sept. 1990. Imports and exports data: UN International Trade Statistics Yearbook.

129/130 *The Times Atlas of the World,* 1992. *The New Cosmopolitan World Atlas, 1993.*

135/136 *The Times Atlas of the World,,* 1992. *The New Cosmopolitan World Atlas, 1993.*

139/140 *Atlas of Desertification, 1992. Encyclopedia Britannica Yearbook, 1993. The New Cosmopolitan World Atlas, 1993. The Times Atlas of the World, 1992. World Wildlife Fund Atlas of the Environment, 1990. Oxford Atlas of Modern World History* (Oxford: Oxford University Press, 1989). GDP: *Europa Yearbook.* Imports and Exports: *UN International Trade Statistics Yearbook.*

151/152 *Europe, The Magazine of the European Community,* Dec. 1991. *The New Cosmopolitan World Atlas, 1993.* Library of Congress Congressional Research Service, CRS Review, March/April 1990, p.10. *The Times Atlas of the World,* 1992. GDP: *Europa Yearbook.* Imports and Exports: *UN International Trade Statistics Yearbook.*

155/156 Chaliand, Gerard and Jean-Pierre Rageau, *Strategic Atlas,* rev. ed. (New York: Harper & Row, Publishers, 1990). *Goode's World Atlas,* 18th ed. (Chicago: Rand McNally, 1990). *The New Cosmopolitan World Atlas,* 1993. Rogers, Alisdair, *Atlas of Social Issues* (Oxford: Ilex Publishers, 1990). *The Times Atlas of the World,* 1992. *The World Factbook* 1993-94. GDP: *Europa Yearbook.* Imports and Exports: *UN International Trade Statistics Yearbook.*

159/160 *The New Cosmopolitan World Atlas, 1993. The Times Atlas of the World, 1992.* GDP: *Europa Yearbook.* Imports and Exports: *UN International Trade Statistics Yearbook.*

161/162 Boustani, Rafic and Philippe Fargues, *The Atlas of the Arab World: Geopolitics and Society* (New York: Facts on File, 1990). "Energy Map of the Middle East", Petroleum Economist, Series No. 5, 2nd. ed., 1993. *Atlas of the Middle East* (New York: Macmillan Publishing Company, 1988). *The Middle East Today* (Wellesley, Mass.: World Eagle, 1989). *The New Cosmopolitan World Atlas,* 1993. *The Times Atlas of the World,* 1992. Maps of the Israel and area: *New York Times,* Sept. 8, 1993.

167/168, 169/70 *Defense Monitor,* Vol. XXI, No. 6 (October 1992). *The Economist. Historical Atlas of the World* (Chicago: Rand McNally, 1981). *The New Cosmopolitan World Atlas,* 1993. *State of the World Atlas. The Times Atlas of the World, 1992;* comprehensive ed. (Boston: Houghton Mifflin, 1967).

169/170 *Atlas of Japan* (Tokyo: Teikoku-Shoin, 1968). Prybla, Jan, "China's Economic Dynamos" *Current History,* Vol. 1, No. 566 (Sept. '92). People's Republic of China, State Statistical Bureau, *Statistical Yearbook of China 1990,* (Chinese version, Beijing 1990). *World Development Report 1993. World Market Atlas, 1992.* GDP: *Europa Yearbook.*

173/174 *Australians: A Historical Atlas* (Victoria, Melbourne and Sydney: Fairfax, Syme & Weldon Associates, 1987). Immigration Update March Quarter 1992, Australian government. *The New Cosmopolitan World Atlas,* 1993. *The Times Atlas of the World,* 1992. Imports and Exports: *UN International Trade Statistics Yearbook.*

175/176 *World Development Report 1993.* Import/Export totals: *UN International Trade Statistics Yearbook.* GDP change: *Encyclopedia Britannica Yearbook, 1993.*

WORLD STATISTICS (pages 191–198)

Demographic Yearbook (New York: United Nations, 1992).

Energy Statistical Yearbook (Washington: European Community Information Services, 1992).

The Europa Yearbook (London: Europa Publications, 1993).

1993 Information Please Environmental Almanac (Boston: Houghton Mifflin, 1993).

The Military Balance (Washington: International Institute for Strategic Studies, 1993).

Statesman's Yearbook (London: MacMillan, 1992).

Statistical Record of the Environment (Detroit: Gale Research Inc., 1992).

Statistical Yearbook (New York: UNESCO, 1992).

Third World Guide (Montevideo (Uruguay): Garamond Press, 1990).

World Bank Atlas (Washington: World Bank, 1992).

World Census of Agriculture (New York: United Nations, 1992).

World Development Report (Oxford University Press, 1992).

The World Factbook (Washington: Central Intelligence Agency, 1992).

The World in Figures (Boston: G.K. Hall, 1992).

Yearbook of the Encyclopedia Britannica (New York: Encyclopedia Britannica, 1993).

Yearbook of Trade Statistics (Washington: International Monetary Fund, 1993).

CANADA INDEX

(c) = chart (g) = graph (i) = illustration (p) = photo
(See World index to compare Canada with other nations.)

WORLD INDEX

(c) = chart (g) = graph (i) = illustration (p) = photo
(See separate Canada index for most Canadian entries. Canada dealt with here only in relation to world patterns.)

11:00 am
Noon
1:00 pm
2:00 pm
3:00 pm
4:00 pm
5:00 pm
6:00 pm
7:00 pm
8:00 pm
9:00 pm
10:00 pm
11:00 pm
Midnig
Greenwich meridian
10:00 pm
8:30 pm
Greenwich Observatory
8:30 pm
Monday
Sunday
TIME ZONES
Standard time with even hours difference from Greenwich
Standard time with odd hours difference from Greenwich
Areas where time varies from standard time by half an hour or more
Many countries seasonally adjust their time to take advantage of the varying amounts of daylight during the year.
13:00
14:00
15:00
16:00
17:00
18:00
19:00
20:00
21:00
22:00
23:00
24:00